203 NYA

Mechanics

Custom

College Edition

NELSON

NELSON

ISBN-13: 978-0-17-658214-2
ISBN-10: 0-17-658214-2

Consists of Selections from:

PHYSICS SCI ENGINEERS
MODERN PHYSICS
SERWAY/JEWETT
ISBN 10: 1-133-95405-7, © 2014

Contents

Appendices

Pedagogical Color Chart

Mechanics and Thermodynamics

Displacement and position vectors

Displacement and position component vectors

Linear ($\vec{v}$) and angular ($\vec{\omega}$) velocity vectors

Velocity component vectors

Force vectors ($\vec{F}$)
Force component vectors

Acceleration vectors ($\vec{a}$)
Acceleration component vectors

Energy transfer arrows

W_{eng}

Q_c

Q_h

Process arrow

Linear ($\vec{p}$) and angular ($\vec{L}$) momentum vectors

Linear and angular momentum component vectors

Torque vectors ($\vec{\tau}$)

Torque component vectors

Schematic linear or rotational motion directions

Dimensional rotational arrow

Enlargement arrow

Springs

Pulleys

Electricity and Magnetism

Electric fields
Electric field vectors
Electric field component vectors

Magnetic fields
Magnetic field vectors
Magnetic field component vectors

Positive charges

Negative charges

Resistors

Batteries and other DC power supplies

Switches

Capacitors

Inductors (coils)

Voltmeters

Ammeters

AC Sources

Lightbulbs

Ground symbol

Current

Light and Optics

Light ray

Focal light ray

Central light ray

Converging lens

Diverging lens

Mirror

Curved mirror

Objects

Images

v

Some Physical Constants

Quantity	Symbol	Value[a]
Atomic mass unit	u	$1.660\ 538\ 782\ (83) \times 10^{-27}$ kg
		$931.494\ 028\ (23)$ MeV/c^2
Avogadro's number	N_A	$6.022\ 141\ 79\ (30) \times 10^{23}$ particles/mol
Bohr magneton	$\mu_B = \dfrac{e\hbar}{2m_e}$	$9.274\ 009\ 15\ (23) \times 10^{-24}$ J/T
Bohr radius	$a_0 = \dfrac{\hbar^2}{m_e e^2 k_e}$	$5.291\ 772\ 085\ 9\ (36) \times 10^{-11}$ m
Boltzmann's constant	$k_B = \dfrac{R}{N_A}$	$1.380\ 650\ 4\ (24) \times 10^{-23}$ J/K
Compton wavelength	$\lambda_C = \dfrac{h}{m_e c}$	$2.426\ 310\ 217\ 5\ (33) \times 10^{-12}$ m
Coulomb constant	$k_e = \dfrac{1}{4\pi\epsilon_0}$	$8.987\ 551\ 788\ldots \times 10^9$ N·m^2/C^2 (exact)
Deuteron mass	m_d	$3.343\ 583\ 20\ (17) \times 10^{-27}$ kg
		$2.013\ 553\ 212\ 724\ (78)$ u
Electron mass	m_e	$9.109\ 382\ 15\ (45) \times 10^{-31}$ kg
		$5.485\ 799\ 094\ 3\ (23) \times 10^{-4}$ u
		$0.510\ 998\ 910\ (13)$ MeV/c^2
Electron volt	eV	$1.602\ 176\ 487\ (40) \times 10^{-19}$ J
Elementary charge	e	$1.602\ 176\ 487\ (40) \times 10^{-19}$ C
Gas constant	R	$8.314\ 472\ (15)$ J/mol·K
Gravitational constant	G	$6.674\ 28\ (67) \times 10^{-11}$ N·m^2/kg^2
Neutron mass	m_n	$1.674\ 927\ 211\ (84) \times 10^{-27}$ kg
		$1.008\ 664\ 915\ 97\ (43)$ u
		$939.565\ 346\ (23)$ MeV/c^2
Nuclear magneton	$\mu_n = \dfrac{e\hbar}{2m_p}$	$5.050\ 783\ 24\ (13) \times 10^{-27}$ J/T
Permeability of free space	μ_0	$4\pi \times 10^{-7}$ T·m/A (exact)
Permittivity of free space	$\epsilon_0 = \dfrac{1}{\mu_0 c^2}$	$8.854\ 187\ 817\ldots \times 10^{-12}$ C^2/N·m^2 (exact)
Planck's constant	h	$6.626\ 068\ 96\ (33) \times 10^{-34}$ J·s
	$\hbar = \dfrac{h}{2\pi}$	$1.054\ 571\ 628\ (53) \times 10^{-34}$ J·s
Proton mass	m_p	$1.672\ 621\ 637\ (83) \times 10^{-27}$ kg
		$1.007\ 276\ 466\ 77\ (10)$ u
		$938.272\ 013\ (23)$ MeV/c^2
Rydberg constant	R_H	$1.097\ 373\ 156\ 852\ 7\ (73) \times 10^7$ m^{-1}
Speed of light in vacuum	c	$2.997\ 924\ 58 \times 10^8$ m/s (exact)

Note: These constants are the values recommended in 2006 by CODATA, based on a least-squares adjustment of data from different measurements. For a more complete list, see P. J. Mohr, B. N. Taylor, and D. B. Newell, "CODATA Recommended Values of the Fundamental Physical Constants: 2006." *Rev. Mod. Phys.* **80:2**, 633–730, 2008.

[a]The numbers in parentheses for the values represent the uncertainties of the last two digits.

Solar System Data

Body	Mass (kg)	Mean Radius (m)	Period (s)	Mean Distance from the Sun (m)
Mercury	3.30×10^{23}	2.44×10^6	7.60×10^6	5.79×10^{10}
Venus	4.87×10^{24}	6.05×10^6	1.94×10^7	1.08×10^{11}
Earth	5.97×10^{24}	6.37×10^6	3.156×10^7	1.496×10^{11}
Mars	6.42×10^{23}	3.39×10^6	5.94×10^7	2.28×10^{11}
Jupiter	1.90×10^{27}	6.99×10^7	3.74×10^8	7.78×10^{11}
Saturn	5.68×10^{26}	5.82×10^7	9.29×10^8	1.43×10^{12}
Uranus	8.68×10^{25}	2.54×10^7	2.65×10^9	2.87×10^{12}
Neptune	1.02×10^{26}	2.46×10^7	5.18×10^9	4.50×10^{12}
Pluto[a]	1.25×10^{22}	1.20×10^6	7.82×10^9	5.91×10^{12}
Moon	7.35×10^{22}	1.74×10^6	—	—
Sun	1.989×10^{30}	6.96×10^8	—	—

[a]In August 2006, the International Astronomical Union adopted a definition of a planet that separates Pluto from the other eight planets. Pluto is now defined as a "dwarf planet" (like the asteroid Ceres).

Physical Data Often Used

Average Earth–Moon distance	3.84×10^8 m
Average Earth–Sun distance	1.496×10^{11} m
Average radius of the Earth	6.37×10^6 m
Density of air (20°C and 1 atm)	1.20 kg/m^3
Density of air (0°C and 1 atm)	1.29 kg/m^3
Density of water (20°C and 1 atm)	1.00×10^3 kg/m^3
Free-fall acceleration	9.80 m/s^2
Mass of the Earth	5.97×10^{24} kg
Mass of the Moon	7.35×10^{22} kg
Mass of the Sun	1.99×10^{30} kg
Standard atmospheric pressure	1.013×10^5 Pa

Note: These values are the ones used in the text.

Some Prefixes for Powers of Ten

Power	Prefix	Abbreviation	Power	Prefix	Abbreviation
10^{-24}	yocto	y	10^1	deka	da
10^{-21}	zepto	z	10^2	hecto	h
10^{-18}	atto	a	10^3	kilo	k
10^{-15}	femto	f	10^6	mega	M
10^{-12}	pico	p	10^9	giga	G
10^{-9}	nano	n	10^{12}	tera	T
10^{-6}	micro	μ	10^{15}	peta	P
10^{-3}	milli	m	10^{18}	exa	E
10^{-2}	centi	c	10^{21}	zetta	Z
10^{-1}	deci	d	10^{24}	yotta	Y

Standard Abbreviations and Symbols for Units

Symbol	Unit	Symbol	Unit
A	ampere	K	kelvin
u	atomic mass unit	kg	kilogram
atm	atmosphere	kmol	kilomole
Btu	British thermal unit	L	liter
C	coulomb	lb	pound
°C	degree Celsius	ly	light-year
cal	calorie	m	meter
d	day	min	minute
eV	electron volt	mol	mole
°F	degree Fahrenheit	N	newton
F	farad	Pa	pascal
ft	foot	rad	radian
G	gauss	rev	revolution
g	gram	s	second
H	henry	T	tesla
h	hour	V	volt
hp	horsepower	W	watt
Hz	hertz	Wb	weber
in.	inch	yr	year
J	joule	Ω	ohm

Mathematical Symbols Used in the Text and Their Meaning

Symbol	Meaning
$=$	is equal to
$\equiv$	is defined as
$\neq$	is not equal to
$\propto$	is proportional to
$\sim$	is on the order of
$>$	is greater than
$<$	is less than
$\gg (\ll)$	is much greater (less) than
$\approx$	is approximately equal to
Δx	the change in x
$\displaystyle\sum_{i=1}^{N} x_i$	the sum of all quantities x_i from $i = 1$ to $i = N$
$|x|$	the absolute value of x (always a nonnegative quantity)
$\Delta x \rightarrow 0$	Δx approaches zero
$\dfrac{dx}{dt}$	the derivative of x with respect to t
$\dfrac{\partial x}{\partial t}$	the partial derivative of x with respect to t
$\displaystyle\int$	integral

Conversions

Length

1 in. = 2.54 cm (exact)
1 m = 39.37 in. = 3.281 ft
1 ft = 0.304 8 m
12 in. = 1 ft
3 ft = 1 yd
1 yd = 0.914 4 m
1 km = 0.621 mi
1 mi = 1.609 km
1 mi = 5 280 ft
$1\ \mu m = 10^{-6}\ m = 10^3\ nm$
1 light-year = 9.461×10^{15} m

Area

$1\ m^2 = 10^4\ cm^2 = 10.76\ ft^2$
$1\ ft^2 = 0.092\ 9\ m^2 = 144\ in.^2$
$1\ in.^2 = 6.452\ cm^2$

Volume

$1\ m^3 = 10^6\ cm^3 = 6.102 \times 10^4\ in.^3$
$1\ ft^3 = 1\ 728\ in.^3 = 2.83 \times 10^{-2}\ m^3$
$1\ L = 1\ 000\ cm^3 = 1.057\ 6\ qt = 0.035\ 3\ ft^3$
$1\ ft^3 = 7.481\ gal = 28.32\ L = 2.832 \times 10^{-2}\ m^3$
$1\ gal = 3.786\ L = 231\ in.^3$

Mass

1 000 kg = 1 t (metric ton)
1 slug = 14.59 kg
$1\ u = 1.66 \times 10^{-27}\ kg = 931.5\ MeV/c^2$

Force

1 N = 0.224 8 lb
1 lb = 4.448 N

Velocity

1 mi/h = 1.47 ft/s = 0.447 m/s = 1.61 km/h
1 m/s = 100 cm/s = 3.281 ft/s
1 mi/min = 60 mi/h = 88 ft/s

Acceleration

$1\ m/s^2 = 3.28\ ft/s^2 = 100\ cm/s^2$
$1\ ft/s^2 = 0.304\ 8\ m/s^2 = 30.48\ cm/s^2$

Pressure

$1\ bar = 10^5\ N/m^2 = 14.50\ lb/in.^2$
1 atm = 760 mm Hg = 76.0 cm Hg
$1\ atm = 14.7\ lb/in.^2 = 1.013 \times 10^5\ N/m^2$
$1\ Pa = 1\ N/m^2 = 1.45 \times 10^{-4}\ lb/in.^2$

Time

$1\ yr = 365\ days = 3.16 \times 10^7\ s$
$1\ day = 24\ h = 1.44 \times 10^3\ min = 8.64 \times 10^4\ s$

Energy

$1\ J = 0.738\ ft \cdot lb$
1 cal = 4.186 J
$1\ Btu = 252\ cal = 1.054 \times 10^3\ J$
$1\ eV = 1.602 \times 10^{-19}\ J$
$1\ kWh = 3.60 \times 10^6\ J$

Power

1 hp = 550 ft·lb/s = 0.746 kW
1 W = 1 J/s = 0.738 ft·lb/s
1 Btu/h = 0.293 W

Some Approximations Useful for Estimation Problems

1 m ≈ 1 yd
1 kg ≈ 2 lb
$1\ N \approx \frac{1}{4}\ lb$
$1\ L \approx \frac{1}{4}\ gal$

1 m/s ≈ 2 mi/h
$1\ yr \approx \pi \times 10^7\ s$
60 mi/h ≈ 100 ft/s
$1\ km \approx \frac{1}{2}\ mi$

Note: See Table A.1 of Appendix A for a more complete list.

The Greek Alphabet

Alpha	A	α	Iota	I	ι	Rho	P	ρ
Beta	B	β	Kappa	K	κ	Sigma	Σ	σ
Gamma	Γ	γ	Lambda	Λ	λ	Tau	T	τ
Delta	Δ	δ	Mu	M	μ	Upsilon	Y	υ
Epsilon	E	ϵ	Nu	N	ν	Phi	Φ	ϕ
Zeta	Z	ζ	Xi	Ξ	ξ	Chi	X	χ
Eta	H	η	Omicron	O	o	Psi	Ψ	ψ
Theta	Θ	θ	Pi	Π	π	Omega	Ω	ω

To the Student

It is appropriate to offer some words of advice that should be of benefit to you, the student. Before doing so, we assume you have read the Preface, which describes the various features of the text and support materials that will help you through the course.

 ## How to Study

Instructors are often asked, "How should I study physics and prepare for examinations?" There is no simple answer to this question, but we can offer some suggestions based on our own experiences in learning and teaching over the years.

First and foremost, maintain a positive attitude toward the subject matter, keeping in mind that physics is the most fundamental of all natural sciences. Other science courses that follow will use the same physical principles, so it is important that you understand and are able to apply the various concepts and theories discussed in the text.

 ## Concepts and Principles

It is essential that you understand the basic concepts and principles before attempting to solve assigned problems. You can best accomplish this goal by carefully reading the textbook before you attend your lecture on the covered material. When reading the text, you should jot down those points that are not clear to you. Also be sure to make a diligent attempt at answering the questions in the Quick Quizzes as you come to them in your reading. We have worked hard to prepare questions that help you judge for yourself how well you understand the material. Study the **What If?** features that appear in many of the worked examples carefully. They will help you extend your understanding beyond the simple act of arriving at a numerical result. The Pitfall Preventions will also help guide you away from common misunderstandings about physics. During class, take careful notes and ask questions about those ideas that are unclear to you. Keep in mind that few people are able to absorb the full meaning of scientific material after only one reading; several readings of the text and your notes may be necessary. Your lectures and laboratory work supplement the textbook and should clarify some of the more difficult material. You should minimize your memorization of material. Successful memorization of passages from the text, equations, and derivations does not necessarily indicate that you understand the material. Your understanding of the material will be enhanced through a combination of efficient study habits, discussions with other students and with instructors, and your ability to solve the problems presented in the textbook. Ask questions whenever you believe that clarification of a concept is necessary.

 ## Study Schedule

It is important that you set up a regular study schedule, preferably a daily one. Make sure that you read the syllabus for the course and adhere to the schedule set by your instructor. The lectures will make much more sense if you read the corresponding text material *before* attending them. As a general rule, you should devote about two hours of study time for each hour you are in class. If you are having trouble with the course, seek the advice of the instructor or other students who have taken the course. You may find it necessary to seek further instruction from experienced students. Very often, instructors offer review sessions in addition to regular class periods. Avoid the practice of delaying study until a day or two before an exam. More often than not, this approach has disastrous results. Rather than undertake an all-night study session before a test, briefly review the basic concepts and equations, and then get a good night's rest. If you believe that you need additional help in understanding the concepts, in preparing for exams, or in problem solving, we suggest that you acquire a copy of the *Student Solutions Manual/Study Guide* that accompanies this textbook.

Visit the *Physics for Scientists and Engineers* Web site at **www.cengagebrain.com/ shop/ISBN/9781133954156** to see samples of select student supplements. You can purchase any Cengage Learning product at your local college store or at our preferred online store **CengageBrain.com.**

Use the Features

You should make full use of the various features of the text discussed in the Preface. For example, marginal notes are useful for locating and describing important equations and concepts, and **boldface** indicates important definitions. Many useful tables are contained in the appendices, but most are incorporated in the text where they are most often referenced. Appendix B is a convenient review of mathematical tools used in the text.

Answers to Quick Quizzes and odd-numbered problems are given at the end of the textbook, and solutions to selected end-of-chapter questions and problems are provided in the *Student Solutions Manual/Study Guide*. The table of contents provides an overview of the entire text, and the index enables you to locate specific material quickly. Footnotes are sometimes used to supplement the text or to cite other references on the subject discussed.

After reading a chapter, you should be able to define any new quantities introduced in that chapter and discuss the principles and assumptions that were used to arrive at certain key relations. The chapter summaries and the review sections of the *Student Solutions Manual/Study Guide* should help you in this regard. In some cases, you may find it necessary to refer to the textbook's index to locate certain topics. You should be able to associate with each physical quantity the correct symbol used to represent that quantity and the unit in which the quantity is specified. Furthermore, you should be able to express each important equation in concise and accurate prose.

Problem Solving

R. P. Feynman, Nobel laureate in physics, once said, "You do not know anything until you have practiced." In keeping with this statement, we strongly advise you to develop the skills necessary to solve a wide range of problems. Your ability to solve problems will be one of the main tests of your knowledge of physics; therefore, you should try to solve as many problems as possible. It is essential that you understand basic concepts and principles before attempting to solve problems. It is good practice to try to find alternate solutions to the same problem. For example, you can solve problems in mechanics using Newton's laws, but very often an alternative method that draws on energy considerations is more direct. You should not deceive yourself into thinking that you understand a problem merely because you have seen it solved in class. You must be able to solve the problem and similar problems on your own.

The approach to solving problems should be carefully planned. A systematic plan is especially important when a problem involves several concepts. First, read the problem several times until you are confident you understand what is being asked. Look for any key words that will help you interpret the problem and perhaps allow you to make certain assumptions. Your ability to interpret a question properly is an integral part of problem solving. Second, you should acquire the habit of writing down the information given in a problem and those quantities that need to be found; for example, you might construct a table listing both the quantities given and the quantities to be found. This procedure is sometimes used in the worked examples of the textbook. Finally, after you have decided on the method you believe is appropriate for a given problem, proceed with your solution. The General Problem-Solving Strategy will guide you through complex problems. If you follow the steps of this procedure *(Conceptualize, Categorize, Analyze, Finalize)*, you will find it easier to come up with a solution and gain more from your efforts. This strategy, located at the end of Chapter 2 (pages 45–47), is used in all worked examples in the remaining chapters so that you can learn how to apply it. Specific problem-solving strategies for certain types of situations are included in the text and appear with a special heading. These specific strategies follow the outline of the General Problem-Solving Strategy.

Often, students fail to recognize the limitations of certain equations or physical laws in a particular situation. It is very important that you understand and remember the assumptions that underlie a particular theory or formalism. For example, certain equations in kinematics apply only to a particle moving with constant acceleration. These equations are not valid for describing motion whose acceleration is not constant, such as the motion of an object connected to a spring or the motion of an object through a fluid. Study the Analysis Models for Problem Solving in the chapter summaries carefully so that you know how each model can be applied to a specific situation. The analysis models provide you with a logical structure for solving problems and help you develop your thinking skills to become more like those of a physicist. Use the analysis model approach to save you hours of looking for the correct equation and to make you a faster and more efficient problem solver.

Experiments

Physics is a science based on experimental observations. Therefore, we recommend that you try to supplement the text by performing various types of "hands-on" experiments either at home or in the laboratory. These experiments can be used to test ideas and models discussed in class or in the textbook. For example, the common Slinky toy is excellent for studying traveling waves, a ball swinging on the end of a long string can be used to investigate pendulum motion, various masses attached to the end of a vertical spring or rubber band can be used to determine its elastic nature, an old pair of polarized sunglasses and some discarded lenses and a magnifying glass are the components of various experiments in optics, and an approximate measure of the free-fall acceleration can be determined simply by measuring with a stopwatch the time interval required for a ball to drop from a known height. The list of such experiments is endless. When physical models are not available, be imaginative and try to develop models of your own.

New Media

If available, we strongly encourage you to use the **Enhanced WebAssign** product that is available with this textbook. It is far easier to understand physics if you see it in action, and the materials available in Enhanced WebAssign will enable you to become a part of that action.

It is our sincere hope that you will find physics an exciting and enjoyable experience and that you will benefit from this experience, regardless of your chosen profession. Welcome to the exciting world of physics!

The scientist does not study nature because it is useful; he studies it because he delights in it, and he delights in it because it is beautiful. If nature were not beautiful, it would not be worth knowing, and if nature were not worth knowing, life would not be worth living.

—**Henri Poincaré**

Mechanics

The Honda FCX Clarity, a fuel-cell-powered automobile available to the public, albeit in limited quantities. A fuel cell converts hydrogen fuel into electricity to drive the motor attached to the wheels of the car. Automobiles, whether powered by fuel cells, gasoline engines, or batteries, use many of the concepts and principles of mechanics that we will study in this first part of the book. Quantities that we can use to describe the operation of vehicles include position, velocity, acceleration, force, energy, and momentum. *(PRNewsFoto/American Honda)*

Physics, the most fundamental physical science, is concerned with the fundamental principles of the Universe. It is the foundation upon which the other sciences—astronomy, biology, chemistry, and geology—are based. It is also the basis of a large number of engineering applications. The beauty of physics lies in the simplicity of its fundamental principles and in the manner in which just a small number of concepts and models can alter and expand our view of the world around us.

The study of physics can be divided into six main areas:

1. *classical mechanics,* concerning the motion of objects that are large relative to atoms and move at speeds much slower than the speed of light
2. *relativity,* a theory describing objects moving at any speed, even speeds approaching the speed of light
3. *thermodynamics,* dealing with heat, work, temperature, and the statistical behavior of systems with large numbers of particles
4. *electromagnetism,* concerning electricity, magnetism, and electromagnetic fields
5. *optics,* the study of the behavior of light and its interaction with materials
6. *quantum mechanics,* a collection of theories connecting the behavior of matter at the submicroscopic level to macroscopic observations

The disciplines of mechanics and electromagnetism are basic to all other branches of classical physics (developed before 1900) and modern physics (c. 1900–present). The first part of this textbook deals with classical mechanics, sometimes referred to as *Newtonian mechanics* or simply *mechanics.* Many principles and models used to understand mechanical systems retain their importance in the theories of other areas of physics and can later be used to describe many natural phenomena. Therefore, classical mechanics is of vital importance to students from all disciplines. ■

Physics and Measurement

Stonehenge, in southern England, was built thousands of years ago. Various theories have been proposed about its function, including a burial ground, a healing site, and a place for ancestor worship. One of the more intriguing theories suggests that Stonehenge was an observatory, allowing measurements of some of the quantities discussed in this chapter, such as position of objects in space and time intervals between repeating celestial events. *(Stephen Inglis/Shutterstock.com)*

WebAssign Interactive content from this and other chapters may be assigned online in Enhanced WebAssign.

Like all other sciences, physics is based on experimental observations and quantitative measurements. The main objectives of physics are to identify a limited number of fundamental laws that govern natural phenomena and use them to develop theories that can predict the results of future experiments. The fundamental laws used in developing theories are expressed in the language of mathematics, the tool that provides a bridge between theory and experiment.

When there is a discrepancy between the prediction of a theory and experimental results, new or modified theories must be formulated to remove the discrepancy. Many times a theory is satisfactory only under limited conditions; a more general theory might be satisfactory without such limitations. For example, the laws of motion discovered by Isaac Newton (1642–1727) accurately describe the motion of objects moving at normal speeds but do not apply to objects moving at speeds comparable to the speed of light. In contrast, the special theory of relativity developed later by Albert Einstein (1879–1955) gives the same results as Newton's laws at low speeds but also correctly describes the motion of objects at speeds approaching the speed of light. Hence, Einstein's special theory of relativity is a more general theory of motion than that formed from Newton's laws.

Classical physics includes the principles of classical mechanics, thermodynamics, optics, and electromagnetism developed before 1900. Important contributions to classical physics

2

were provided by Newton, who was also one of the originators of calculus as a mathematical tool. Major developments in mechanics continued in the 18th century, but the fields of thermodynamics and electromagnetism were not developed until the latter part of the 19th century, principally because before that time the apparatus for controlled experiments in these disciplines was either too crude or unavailable.

A major revolution in physics, usually referred to as *modern physics,* began near the end of the 19th century. Modern physics developed mainly because many physical phenomena could not be explained by classical physics. The two most important developments in this modern era were the theories of relativity and quantum mechanics. Einstein's special theory of relativity not only correctly describes the motion of objects moving at speeds comparable to the speed of light; it also completely modifies the traditional concepts of space, time, and energy. The theory also shows that the speed of light is the upper limit of the speed of an object and that mass and energy are related. Quantum mechanics was formulated by a number of distinguished scientists to provide descriptions of physical phenomena at the atomic level. Many practical devices have been developed using the principles of quantum mechanics.

Scientists continually work at improving our understanding of fundamental laws. Numerous technological advances in recent times are the result of the efforts of many scientists, engineers, and technicians, such as unmanned planetary explorations, a variety of developments and potential applications in nanotechnology, microcircuitry and high-speed computers, sophisticated imaging techniques used in scientific research and medicine, and several remarkable results in genetic engineering. The effects of such developments and discoveries on our society have indeed been great, and it is very likely that future discoveries and developments will be exciting, challenging, and of great benefit to humanity.

1.1 Standards of Length, Mass, and Time

To describe natural phenomena, we must make measurements of various aspects of nature. Each measurement is associated with a physical quantity, such as the length of an object. The laws of physics are expressed as mathematical relationships among physical quantities that we will introduce and discuss throughout the book. In mechanics, the three fundamental quantities are length, mass, and time. All other quantities in mechanics can be expressed in terms of these three.

If we are to report the results of a measurement to someone who wishes to reproduce this measurement, a *standard* must be defined. It would be meaningless if a visitor from another planet were to talk to us about a length of 8 "glitches" if we do not know the meaning of the unit glitch. On the other hand, if someone familiar with our system of measurement reports that a wall is 2 meters high and our unit of length is defined to be 1 meter, we know that the height of the wall is twice our basic length unit. Whatever is chosen as a standard must be readily accessible and must possess some property that can be measured reliably. Measurement standards used by different people in different places—throughout the Universe—must yield the same result. In addition, standards used for measurements must not change with time.

In 1960, an international committee established a set of standards for the fundamental quantities of science. It is called the **SI** (Système International), and its fundamental units of length, mass, and time are the *meter, kilogram,* and *second,* respectively. Other standards for SI fundamental units established by the committee are those for temperature (the *kelvin*), electric current (the *ampere*), luminous intensity (the *candela*), and the amount of substance (the *mole*).

Length

We can identify **length** as the distance between two points in space. In 1120, the king of England decreed that the standard of length in his country would be named the *yard* and would be precisely equal to the distance from the tip of his nose to the end of his outstretched arm. Similarly, the original standard for the foot adopted by the French was the length of the royal foot of King Louis XIV. Neither of these standards is constant in time; when a new king took the throne, length measurements changed! The French standard prevailed until 1799, when the legal standard of length in France became the **meter** (m), defined as one ten-millionth of the distance from the equator to the North Pole along one particular longitudinal line that passes through Paris. Notice that this value is an Earth-based standard that does not satisfy the requirement that it can be used throughout the Universe.

As recently as 1960, the length of the meter was defined as the distance between two lines on a specific platinum–iridium bar stored under controlled conditions in France. Current requirements of science and technology, however, necessitate more accuracy than that with which the separation between the lines on the bar can be determined. In the 1960s and 1970s, the meter was defined as 1 650 763.73 wavelengths[1] of orange-red light emitted from a krypton-86 lamp. In October 1983, however, the meter was redefined as **the distance traveled by light in vacuum during a time of 1/299 792 458 second.** In effect, this latest definition establishes that the speed of light in vacuum is precisely 299 792 458 meters per second. This definition of the meter is valid throughout the Universe based on our assumption that light is the same everywhere.

Table 1.1 lists approximate values of some measured lengths. You should study this table as well as the next two tables and begin to generate an intuition for what is meant by, for example, a length of 20 centimeters, a mass of 100 kilograms, or a time interval of 3.2×10^7 seconds.

Mass

The SI fundamental unit of **mass,** the **kilogram** (kg), is defined as **the mass of a specific platinum–iridium alloy cylinder kept at the International Bureau of Weights and Measures at Sèvres, France.** This mass standard was established in 1887 and

> **Pitfall Prevention 1.1**
>
> **Reasonable Values** Generating intuition about typical values of quantities when solving problems is important because you must think about your end result and determine if it seems reasonable. For example, if you are calculating the mass of a housefly and arrive at a value of 100 kg, this answer is *unreasonable* and there is an error somewhere.

Table 1.1 Approximate Values of Some Measured Lengths

	Length (m)
Distance from the Earth to the most remote known quasar	1.4×10^{26}
Distance from the Earth to the most remote normal galaxies	9×10^{25}
Distance from the Earth to the nearest large galaxy (Andromeda)	2×10^{22}
Distance from the Sun to the nearest star (Proxima Centauri)	4×10^{16}
One light-year	9.46×10^{15}
Mean orbit radius of the Earth about the Sun	1.50×10^{11}
Mean distance from the Earth to the Moon	3.84×10^{8}
Distance from the equator to the North Pole	1.00×10^{7}
Mean radius of the Earth	6.37×10^{6}
Typical altitude (above the surface) of a satellite orbiting the Earth	2×10^{5}
Length of a football field	9.1×10^{1}
Length of a housefly	5×10^{-3}
Size of smallest dust particles	$\sim 10^{-4}$
Size of cells of most living organisms	$\sim 10^{-5}$
Diameter of a hydrogen atom	$\sim 10^{-10}$
Diameter of an atomic nucleus	$\sim 10^{-14}$
Diameter of a proton	$\sim 10^{-15}$

---

[1]We will use the standard international notation for numbers with more than three digits, in which groups of three digits are separated by spaces rather than commas. Therefore, 10 000 is the same as the common American notation of 10,000. Similarly, $\pi = 3.14159265$ is written as 3.141 592 65.

Table 1.2 Approximate Masses of Various Objects	
	Mass (kg)
Observable Universe	$\sim 10^{52}$
Milky Way galaxy	$\sim 10^{42}$
Sun	1.99×10^{30}
Earth	5.98×10^{24}
Moon	7.36×10^{22}
Shark	$\sim 10^{3}$
Human	$\sim 10^{2}$
Frog	$\sim 10^{-1}$
Mosquito	$\sim 10^{-5}$
Bacterium	$\sim 1 \times 10^{-15}$
Hydrogen atom	1.67×10^{-27}
Electron	9.11×10^{-31}

Table 1.3 Approximate Values of Some Time Intervals	
	Time Interval (s)
Age of the Universe	4×10^{17}
Age of the Earth	1.3×10^{17}
Average age of a college student	6.3×10^{8}
One year	3.2×10^{7}
One day	8.6×10^{4}
One class period	3.0×10^{3}
Time interval between normal heartbeats	8×10^{-1}
Period of audible sound waves	$\sim 10^{-3}$
Period of typical radio waves	$\sim 10^{-6}$
Period of vibration of an atom in a solid	$\sim 10^{-13}$
Period of visible light waves	$\sim 10^{-15}$
Duration of a nuclear collision	$\sim 10^{-22}$
Time interval for light to cross a proton	$\sim 10^{-24}$

Figure 1.1 (a) The National Standard Kilogram No. 20, an accurate copy of the International Standard Kilogram kept at Sèvres, France, is housed under a double bell jar in a vault at the National Institute of Standards and Technology. (b) A cesium fountain atomic clock. The clock will neither gain nor lose a second in 20 million years.

has not been changed since that time because platinum–iridium is an unusually stable alloy. A duplicate of the Sèvres cylinder is kept at the National Institute of Standards and Technology (NIST) in Gaithersburg, Maryland (Fig. 1.1a). Table 1.2 lists approximate values of the masses of various objects.

Time

Before 1967, the standard of **time** was defined in terms of the *mean solar day*. (A solar day is the time interval between successive appearances of the Sun at the highest point it reaches in the sky each day.) The fundamental unit of a **second** (s) was defined as $\left(\frac{1}{60}\right)\left(\frac{1}{60}\right)\left(\frac{1}{24}\right)$ of a mean solar day. This definition is based on the rotation of one planet, the Earth. Therefore, this motion does not provide a time standard that is universal.

In 1967, the second was redefined to take advantage of the high precision attainable in a device known as an *atomic clock* (Fig. 1.1b), which measures vibrations of cesium atoms. One second is now defined as **9 192 631 770 times the period of vibration of radiation from the cesium-133 atom.**[2] Approximate values of time intervals are presented in Table 1.3.

In addition to SI, another system of units, the *U.S. customary system*, is still used in the United States despite acceptance of SI by the rest of the world. In this system, the units of length, mass, and time are the foot (ft), slug, and second, respectively. In this book, we shall use SI units because they are almost universally accepted in science and industry. We shall make some limited use of U.S. customary units in the study of classical mechanics.

In addition to the fundamental SI units of meter, kilogram, and second, we can also use other units, such as millimeters and nanoseconds, where the prefixes *milli-* and *nano-* denote multipliers of the basic units based on various powers of ten. Prefixes for the various powers of ten and their abbreviations are listed in Table 1.4 (page 6). For example, 10^{-3} m is equivalent to 1 millimeter (mm), and 10^{3} m corresponds to 1 kilometer (km). Likewise, 1 kilogram (kg) is 10^{3} grams (g), and 1 mega volt (MV) is 10^{6} volts (V).

The variables length, time, and mass are examples of *fundamental quantities*. Most other variables are *derived quantities*, those that can be expressed as a mathematical combination of fundamental quantities. Common examples are *area* (a product of two lengths) and *speed* (a ratio of a length to a time interval).

---

[2]Period is defined as the time interval needed for one complete vibration.

Table 1.4	Prefixes for Powers of Ten				
Power	Prefix	Abbreviation	Power	Prefix	Abbreviation
10^{-24}	yocto	y	10^{3}	kilo	k
10^{-21}	zepto	z	10^{6}	mega	M
10^{-18}	atto	a	10^{9}	giga	G
10^{-15}	femto	f	10^{12}	tera	T
10^{-12}	pico	p	10^{15}	peta	P
10^{-9}	nano	n	10^{18}	exa	E
10^{-6}	micro	μ	10^{21}	zetta	Z
10^{-3}	milli	m	10^{24}	yotta	Y
10^{-2}	centi	c			
10^{-1}	deci	d			

▶ A table of the letters in the Greek alphabet is provided on the back endpaper of this book.

Another example of a derived quantity is **density.** The density ρ (Greek letter rho) of any substance is defined as its *mass per unit volume:*

$$\rho \equiv \frac{m}{V} \qquad (1.1)$$

In terms of fundamental quantities, density is a ratio of a mass to a product of three lengths. Aluminum, for example, has a density of 2.70×10^3 kg/m³, and iron has a density of 7.86×10^3 kg/m³. An extreme difference in density can be imagined by thinking about holding a 10-centimeter (cm) cube of Styrofoam in one hand and a 10-cm cube of lead in the other. See Table 14.1 in Chapter 14 for densities of several materials.

Quick Quiz 1.1 In a machine shop, two cams are produced, one of aluminum and one of iron. Both cams have the same mass. Which cam is larger? **(a)** The aluminum cam is larger. **(b)** The iron cam is larger. **(c)** Both cams have the same size.

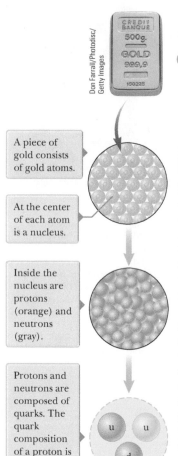

A piece of gold consists of gold atoms.

At the center of each atom is a nucleus.

Inside the nucleus are protons (orange) and neutrons (gray).

Protons and neutrons are composed of quarks. The quark composition of a proton is shown here.

Don Farrall/Photodisc/ Getty Images

Figure 1.2 Levels of organization in matter.

1.2 Matter and Model Building

If physicists cannot interact with some phenomenon directly, they often imagine a **model** for a physical system that is related to the phenomenon. For example, we cannot interact directly with atoms because they are too small. Therefore, we build a mental model of an atom based on a system of a nucleus and one or more electrons outside the nucleus. Once we have identified the physical components of the model, we make predictions about its behavior based on the interactions among the components of the system or the interaction between the system and the environment outside the system.

As an example, consider the behavior of *matter*. A sample of solid gold is shown at the top of Figure 1.2. Is this sample nothing but wall-to-wall gold, with no empty space? If the sample is cut in half, the two pieces still retain their chemical identity as solid gold. What if the pieces are cut again and again, indefinitely? Will the smaller and smaller pieces always be gold? Such questions can be traced to early Greek philosophers. Two of them—Leucippus and his student Democritus—could not accept the idea that such cuttings could go on forever. They developed a model for matter by speculating that the process ultimately must end when it produces a particle that can no longer be cut. In Greek, *atomos* means "not sliceable." From this Greek term comes our English word *atom.*

The Greek model of the structure of matter was that all ordinary matter consists of atoms, as suggested in the middle of Figure 1.2. Beyond that, no additional structure was specified in the model; atoms acted as small particles that interacted with one another, but internal structure of the atom was not a part of the model.

In 1897, J. J. Thomson identified the electron as a charged particle and as a constituent of the atom. This led to the first atomic model that contained internal structure. We shall discuss this model in Chapter 42.

Following the discovery of the nucleus in 1911, an atomic model was developed in which each atom is made up of electrons surrounding a central nucleus. A nucleus of gold is shown in Figure 1.2. This model leads, however, to a new question: Does the nucleus have structure? That is, is the nucleus a single particle or a collection of particles? By the early 1930s, a model evolved that described two basic entities in the nucleus: protons and neutrons. The proton carries a positive electric charge, and a specific chemical element is identified by the number of protons in its nucleus. This number is called the **atomic number** of the element. For instance, the nucleus of a hydrogen atom contains one proton (so the atomic number of hydrogen is 1), the nucleus of a helium atom contains two protons (atomic number 2), and the nucleus of a uranium atom contains 92 protons (atomic number 92). In addition to atomic number, a second number—**mass number,** defined as the number of protons plus neutrons in a nucleus—characterizes atoms. The atomic number of a specific element never varies (i.e., the number of protons does not vary), but the mass number can vary (i.e., the number of neutrons varies).

Is that, however, where the process of breaking down stops? Protons, neutrons, and a host of other exotic particles are now known to be composed of six different varieties of particles called **quarks,** which have been given the names of *up, down, strange, charmed, bottom,* and *top.* The up, charmed, and top quarks have electric charges of $+\frac{2}{3}$ that of the proton, whereas the down, strange, and bottom quarks have charges of $-\frac{1}{3}$ that of the proton. The proton consists of two up quarks and one down quark as shown at the bottom of Figure 1.2 and labeled u and d. This structure predicts the correct charge for the proton. Likewise, the neutron consists of two down quarks and one up quark, giving a net charge of zero.

You should develop a process of building models as you study physics. In this study, you will be challenged with many mathematical problems to solve. One of the most important problem-solving techniques is to build a model for the problem: identify a system of physical components for the problem and make predictions of the behavior of the system based on the interactions among its components or the interaction between the system and its surrounding environment.

1.3 Dimensional Analysis

In physics, the word *dimension* denotes the physical nature of a quantity. The distance between two points, for example, can be measured in feet, meters, or furlongs, which are all different ways of expressing the dimension of length.

The symbols we use in this book to specify the dimensions of length, mass, and time are L, M, and T, respectively.[3] We shall often use brackets [ ] to denote the dimensions of a physical quantity. For example, the symbol we use for speed in this book is v, and in our notation, the dimensions of speed are written $[v] = L/T$. As another example, the dimensions of area A are $[A] = L^2$. The dimensions and units of area, volume, speed, and acceleration are listed in Table 1.5. The dimensions of other quantities, such as force and energy, will be described as they are introduced in the text.

Table 1.5   **Dimensions and Units of Four Derived Quantities**

Quantity	Area (A)	Volume (V)	Speed (v)	Acceleration (a)
Dimensions	L^2	L^3	L/T	L/T^2
SI units	m^2	m^3	m/s	m/s^2
U.S. customary units	ft^2	ft^3	ft/s	ft/s^2

---

[3]The *dimensions* of a quantity will be symbolized by a capitalized, nonitalic letter such as L or T. The *algebraic symbol* for the quantity itself will be an italicized letter such as *L* for the length of an object or *t* for time.

In many situations, you may have to check a specific equation to see if it matches your expectations. A useful procedure for doing that, called **dimensional analysis,** can be used because dimensions can be treated as algebraic quantities. For example, quantities can be added or subtracted only if they have the same dimensions. Furthermore, the terms on both sides of an equation must have the same dimensions. By following these simple rules, you can use dimensional analysis to determine whether an expression has the correct form. Any relationship can be correct only if the dimensions on both sides of the equation are the same.

To illustrate this procedure, suppose you are interested in an equation for the position x of a car at a time t if the car starts from rest at $x = 0$ and moves with constant acceleration a. The correct expression for this situation is $x = \frac{1}{2}at^2$ as we show in Chapter 2. The quantity x on the left side has the dimension of length. For the equation to be dimensionally correct, the quantity on the right side must also have the dimension of length. We can perform a dimensional check by substituting the dimensions for acceleration, L/T^2 (Table 1.5), and time, T, into the equation. That is, the dimensional form of the equation $x = \frac{1}{2}at^2$ is

$$L = \frac{L}{T^2} \cdot T^2 = L$$

The dimensions of time cancel as shown, leaving the dimension of length on the right-hand side to match that on the left.

A more general procedure using dimensional analysis is to set up an expression of the form

$$x \propto a^n t^m$$

where n and m are exponents that must be determined and the symbol $\propto$ indicates a proportionality. This relationship is correct only if the dimensions of both sides are the same. Because the dimension of the left side is length, the dimension of the right side must also be length. That is,

$$[a^n t^m] = L = L^1 T^0$$

Because the dimensions of acceleration are L/T^2 and the dimension of time is T, we have

$$(L/T^2)^n T^m = L^1 T^0 \quad \rightarrow \quad (L^n T^{m-2n}) = L^1 T^0$$

The exponents of L and T must be the same on both sides of the equation. From the exponents of L, we see immediately that $n = 1$. From the exponents of T, we see that $m - 2n = 0$, which, once we substitute for n, gives us $m = 2$. Returning to our original expression $x \propto a^n t^m$, we conclude that $x \propto at^2$.

Quick Quiz 1.2 True or False: Dimensional analysis can give you the numerical value of constants of proportionality that may appear in an algebraic expression.

Pitfall Prevention 1.2
Symbols for Quantities Some quantities have a small number of symbols that represent them. For example, the symbol for time is almost always t. Other quantities might have various symbols depending on the usage. Length may be described with symbols such as x, y, and z (for position); r (for radius); a, b, and c (for the legs of a right triangle); ℓ (for the length of an object); d (for a distance); h (for a height); and so forth.

Example 1.1 **Analysis of an Equation**

Show that the expression $v = at$, where v represents speed, a acceleration, and t an instant of time, is dimensionally correct.

SOLUTION

Identify the dimensions of v from Table 1.5:
$$[v] = \frac{L}{T}$$

▶ **1.1** continued

Identify the dimensions of a from Table 1.5 and multiply by the dimensions of t: $\qquad [at] = \dfrac{L}{T^2}\, T = \dfrac{L}{T}$

Therefore, $v = at$ is dimensionally correct because we have the same dimensions on both sides. (If the expression were given as $v = at^2$, it would be dimensionally *incorrect*. Try it and see!)

Example 1.2   **Analysis of a Power Law**

Suppose we are told that the acceleration a of a particle moving with uniform speed v in a circle of radius r is proportional to some power of r, say r^n, and some power of v, say v^m. Determine the values of n and m and write the simplest form of an equation for the acceleration.

SOLUTION

Write an expression for a with a dimensionless constant of proportionality k: $\qquad\qquad a = kr^n v^m$

Substitute the dimensions of a, r, and v:

$$\frac{L}{T^2} = L^n \left(\frac{L}{T}\right)^m = \frac{L^{n+m}}{T^m}$$

Equate the exponents of L and T so that the dimensional equation is balanced: $\qquad n + m = 1 \text{ and } m = 2$

Solve the two equations for n: $\qquad\qquad\qquad\qquad n = -1$

Write the acceleration expression: $\qquad\qquad\qquad a = kr^{-1}v^2 = k\dfrac{v^2}{r}$

In Section 4.4 on uniform circular motion, we show that $k = 1$ if a consistent set of units is used. The constant k would not equal 1 if, for example, v were in km/h and you wanted a in m/s^2 .

1.4 Conversion of Units

Sometimes it is necessary to convert units from one measurement system to another or convert within a system (for example, from kilometers to meters). Conversion factors between SI and U.S. customary units of length are as follows:

1 mile	= 1 609 m	= 1.609 km	1 ft = 0.304 8 m = 30.48 cm
1 m	= 39.37 in.	= 3.281 ft	1 in. = 0.025 4 m = 2.54 cm (exactly)

A more complete list of conversion factors can be found in Appendix A.

Like dimensions, units can be treated as algebraic quantities that can cancel each other. For example, suppose we wish to convert 15.0 in. to centimeters. Because 1 in. is defined as exactly 2.54 cm, we find that

$$15.0 \text{ in.} = (15.0 \text{ in.})\left(\frac{2.54 \text{ cm}}{1 \text{ in.}}\right) = 38.1 \text{ cm}$$

where the ratio in parentheses is equal to 1. We express 1 as 2.54 cm/1 in. (rather than 1 in./2.54 cm) so that the unit "inch" in the denominator cancels with the unit in the original quantity. The remaining unit is the centimeter, our desired result.

Pitfall Prevention 1.3
Always Include Units When performing calculations with numerical values, include the units for every quantity and carry the units through the entire calculation. Avoid the temptation to drop the units early and then attach the expected units once you have an answer. By including the units in every step, you can detect errors if the units for the answer turn out to be incorrect.

Q uick Quiz 1.3 The distance between two cities is 100 mi. What is the number of kilometers between the two cities? **(a)** smaller than 100 **(b)** larger than 100 **(c)** equal to 100

Example 1.3   Is He Speeding?

On an interstate highway in a rural region of Wyoming, a car is traveling at a speed of 38.0 m/s. Is the driver exceeding the speed limit of 75.0 mi/h?

SOLUTION

Convert meters in the speed to miles:

$$(38.0 \text{ m/s}) \left(\frac{1 \text{ mi}}{1\,609 \text{ m}} \right) = 2.36 \times 10^{-2} \text{ mi/s}$$

Convert seconds to hours:

$$(2.36 \times 10^{-2} \text{ mi/s}) \left(\frac{60 \text{ s}}{1 \text{ min}} \right) \left(\frac{60 \text{ min}}{1 \text{ h}} \right) = 85.0 \text{ mi/h}$$

The driver is indeed exceeding the speed limit and should slow down.

WHAT IF? What if the driver were from outside the United States and is familiar with speeds measured in kilometers per hour? What is the speed of the car in km/h?

Answer We can convert our final answer to the appropriate units:

$$(85.0 \text{ mi/h}) \left(\frac{1.609 \text{ km}}{1 \text{ mi}} \right) = 137 \text{ km/h}$$

Figure 1.3 shows an automobile speedometer displaying speeds in both mi/h and km/h. Can you check the conversion we just performed using this photograph?

Figure 1.3 The speedometer of a vehicle that shows speeds in both miles per hour and kilometers per hour.

1.5 Estimates and Order-of-Magnitude Calculations

Suppose someone asks you the number of bits of data on a typical musical compact disc. In response, it is not generally expected that you would provide the exact number but rather an estimate, which may be expressed in scientific notation. The estimate may be made even more approximate by expressing it as an *order of magnitude*, which is a power of ten determined as follows:

1. Express the number in scientific notation, with the multiplier of the power of ten between 1 and 10 and a unit.
2. If the multiplier is less than 3.162 (the square root of 10), the order of magnitude of the number is the power of 10 in the scientific notation. If the multiplier is greater than 3.162, the order of magnitude is one larger than the power of 10 in the scientific notation.

We use the symbol $\sim$ for "is on the order of." Use the procedure above to verify the orders of magnitude for the following lengths:

$$0.008\,6 \text{ m} \sim 10^{-2} \text{ m} \qquad 0.002\,1 \text{ m} \sim 10^{-3} \text{ m} \qquad 720 \text{ m} \sim 10^{3} \text{ m}$$

Usually, when an order-of-magnitude estimate is made, the results are reliable to within about a factor of 10. If a quantity increases in value by three orders of magnitude, its value increases by a factor of about $10^3 = 1\,000$.

Inaccuracies caused by guessing too low for one number are often canceled by other guesses that are too high. You will find that with practice your guesstimates become better and better. Estimation problems can be fun to work because you freely drop digits, venture reasonable approximations for unknown numbers, make simplifying assumptions, and turn the question around into something you can answer in your head or with minimal mathematical manipulation on paper. Because of the simplicity of these types of calculations, they can be performed on a *small* scrap of paper and are often called "back-of-the-envelope calculations."

Example 1.4   **Breaths in a Lifetime**

Estimate the number of breaths taken during an average human lifetime.

SOLUTION

We start by guessing that the typical human lifetime is about 70 years. Think about the average number of breaths that a person takes in 1 min. This number varies depending on whether the person is exercising, sleeping, angry, serene, and so forth. To the nearest order of magnitude, we shall choose 10 breaths per minute as our estimate. (This estimate is certainly closer to the true average value than an estimate of 1 breath per minute or 100 breaths per minute.)

Find the approximate number of minutes in a year:

$$1 \text{ yr} \left(\frac{400 \text{ days}}{1 \text{ yr}} \right) \left(\frac{25 \text{ h}}{1 \text{ day}} \right) \left(\frac{60 \text{ min}}{1 \text{ h}} \right) = 6 \times 10^5 \text{ min}$$

Find the approximate number of minutes in a 70-year lifetime:

$$\text{number of minutes} = (70 \text{ yr})(6 \times 10^5 \text{ min/yr})$$
$$= 4 \times 10^7 \text{ min}$$

Find the approximate number of breaths in a lifetime:

$$\text{number of breaths} = (10 \text{ breaths/min})(4 \times 10^7 \text{ min})$$
$$= 4 \times 10^8 \text{ breaths}$$

Therefore, a person takes on the order of 10^9 breaths in a lifetime. Notice how much simpler it is in the first calculation above to multiply 400×25 than it is to work with the more accurate 365×24.

WHAT IF? What if the average lifetime were estimated as 80 years instead of 70? Would that change our final estimate?

Answer We could claim that $(80 \text{ yr})(6 \times 10^5 \text{ min/yr}) = 5 \times 10^7 \text{ min}$, so our final estimate should be 5×10^8 breaths. This answer is still on the order of 10^9 breaths, so an order-of-magnitude estimate would be unchanged.

1.6 Significant Figures

When certain quantities are measured, the measured values are known only to within the limits of the experimental uncertainty. The value of this uncertainty can depend on various factors, such as the quality of the apparatus, the skill of the experimenter, and the number of measurements performed. The number of **significant figures** in a measurement can be used to express something about the uncertainty. The number of significant figures is related to the number of numerical digits used to express the measurement, as we discuss below.

As an example of significant figures, suppose we are asked to measure the radius of a compact disc using a meterstick as a measuring instrument. Let us assume the accuracy to which we can measure the radius of the disc is ± 0.1 cm. Because of the uncertainty of ± 0.1 cm, if the radius is measured to be 6.0 cm, we can claim only that its radius lies somewhere between 5.9 cm and 6.1 cm. In this case, we say that the measured value of 6.0 cm has two significant figures. Note that *the significant figures include the first estimated digit.* Therefore, we could write the radius as (6.0 ± 0.1) cm.

Zeros may or may not be significant figures. Those used to position the decimal point in such numbers as 0.03 and 0.007 5 are not significant. Therefore, there are one and two significant figures, respectively, in these two values. When the zeros come after other digits, however, there is the possibility of misinterpretation. For example, suppose the mass of an object is given as 1 500 g. This value is ambiguous because we do not know whether the last two zeros are being used to locate the decimal point or whether they represent significant figures in the measurement. To remove this ambiguity, it is common to use scientific notation to indicate the number of significant figures. In this case, we would express the mass as 1.5×10^3 g if there are two significant figures in the measured value, 1.50×10^3 g if there are three significant figures, and 1.500×10^3 g if there are four. The same rule holds for numbers less than 1, so 2.3×10^{-4} has two significant figures (and therefore could be written 0.000 23) and 2.30×10^{-4} has three significant figures (also written as 0.000 230).

In problem solving, we often combine quantities mathematically through multiplication, division, addition, subtraction, and so forth. When doing so, you must make sure that the result has the appropriate number of significant figures. A good rule of thumb to use in determining the number of significant figures that can be claimed in a multiplication or a division is as follows:

> When multiplying several quantities, the number of significant figures in the final answer is the same as the number of significant figures in the quantity having the smallest number of significant figures. The same rule applies to division.

Let's apply this rule to find the area of the compact disc whose radius we measured above. Using the equation for the area of a circle,

$$A = \pi r^2 = \pi (6.0 \text{ cm})^2 = 1.1 \times 10^2 \text{ cm}^2$$

If you perform this calculation on your calculator, you will likely see 113.097 335 5. It should be clear that you don't want to keep all of these digits, but you might be tempted to report the result as 113 cm². This result is not justified because it has three significant figures, whereas the radius only has two. Therefore, we must report the result with only two significant figures as shown above.

For addition and subtraction, you must consider the number of decimal places when you are determining how many significant figures to report:

Pitfall Prevention 1.4
Read Carefully Notice that the rule for addition and subtraction is different from that for multiplication and division. For addition and subtraction, the important consideration is the number of *decimal places,* not the number of *significant figures.*

> When numbers are added or subtracted, the number of decimal places in the result should equal the smallest number of decimal places of any term in the sum or difference.

As an example of this rule, consider the sum

$$23.2 + 5.174 = 28.4$$

Notice that we do not report the answer as 28.374 because the lowest number of decimal places is one, for 23.2. Therefore, our answer must have only one decimal place.

The rule for addition and subtraction can often result in answers that have a different number of significant figures than the quantities with which you start. For example, consider these operations that satisfy the rule:

$$1.000\ 1 + 0.000\ 3 = 1.000\ 4$$
$$1.002 - 0.998 = 0.004$$

In the first example, the result has five significant figures even though one of the terms, 0.000 3, has only one significant figure. Similarly, in the second calculation, the result has only one significant figure even though the numbers being subtracted have four and three, respectively.

In this book, most of the numerical examples and end-of-chapter problems will yield answers having three significant figures. When carrying out estimation calculations, we shall typically work with a single significant figure.

◄ **Significant figure guidelines used in this book**

If the number of significant figures in the result of a calculation must be reduced, there is a general rule for rounding numbers: the last digit retained is increased by 1 if the last digit dropped is greater than 5. (For example, 1.346 becomes 1.35.) If the last digit dropped is less than 5, the last digit retained remains as it is. (For example, 1.343 becomes 1.34.) If the last digit dropped is equal to 5, the remaining digit should be rounded to the nearest even number. (This rule helps avoid accumulation of errors in long arithmetic processes.)

A technique for avoiding error accumulation is to delay the rounding of numbers in a long calculation until you have the final result. Wait until you are ready to copy the final answer from your calculator before rounding to the correct number of significant figures. In this book, we display numerical values rounded off to two or three significant figures. This occasionally makes some mathematical manipulations look odd or incorrect. For instance, looking ahead to Example 3.5 on page 69, you will see the operation -17.7 km $+$ 34.6 km $=$ 17.0 km. This looks like an incorrect subtraction, but that is only because we have rounded the numbers 17.7 km and 34.6 km for display. If all digits in these two intermediate numbers are retained and the rounding is only performed on the final number, the correct three-digit result of 17.0 km is obtained.

Pitfall Prevention 1.5

Symbolic Solutions When solving problems, it is very useful to perform the solution completely in algebraic form and wait until the very end to enter numerical values into the final symbolic expression. This method will save many calculator keystrokes, especially if some quantities cancel so that you never have to enter their values into your calculator! In addition, you will only need to round once, on the final result.

Example 1.5   Installing a Carpet

A carpet is to be installed in a rectangular room whose length is measured to be 12.71 m and whose width is measured to be 3.46 m. Find the area of the room.

SOLUTION

If you multiply 12.71 m by 3.46 m on your calculator, you will see an answer of 43.976 6 m². How many of these numbers should you claim? Our rule of thumb for multiplication tells us that you can claim only the number of significant figures in your answer as are present in the measured quantity having the lowest number of significant figures. In this example, the lowest number of significant figures is three in 3.46 m, so we should express our final answer as 44.0 m².

Summary

Definitions

The three fundamental physical quantities of mechanics are **length, mass,** and **time,** which in the SI system have the units **meter** (m), **kilogram** (kg), and **second** (s), respectively. These fundamental quantities cannot be defined in terms of more basic quantities.

The **density** of a substance is defined as its *mass per unit volume:*

$$\rho \equiv \frac{m}{V} \tag{1.1}$$

continued

Concepts and Principles

The method of **dimensional analysis** is very powerful in solving physics problems. Dimensions can be treated as algebraic quantities. By making estimates and performing order-of-magnitude calculations, you should be able to approximate the answer to a problem when there is not enough information available to specify an exact solution completely.

When you compute a result from several measured numbers, each of which has a certain accuracy, you should give the result with the correct number of **significant figures.**

When **multiplying** several quantities, the number of significant figures in the final answer is the same as the number of significant figures in the quantity having the smallest number of significant figures. The same rule applies to **division.**

When numbers are **added** or **subtracted,** the number of decimal places in the result should equal the smallest number of decimal places of any term in the sum or difference.

Objective Questions   [1.] denotes answer available in *Student Solutions Manual/Study Guide*

1. One student uses a meterstick to measure the thickness of a textbook and obtains 4.3 cm ± 0.1 cm. Other students measure the thickness with vernier calipers and obtain four different measurements: (a) 4.32 cm ± 0.01 cm, (b) 4.31 cm ± 0.01 cm, (c) 4.24 cm ± 0.01 cm, and (d) 4.43 cm ± 0.01 cm. Which of these four measurements, if any, agree with that obtained by the first student?

2. A house is advertised as having 1 420 square feet under its roof. What is its area in square meters? (a) 4 660 m^2 (b) 432 m^2 (c) 158 m^2 (d) 132 m^2 (e) 40.2 m^2

[3.] Answer each question yes or no. Must two quantities have the same dimensions (a) if you are adding them? (b) If you are multiplying them? (c) If you are subtracting them? (d) If you are dividing them? (e) If you are equating them?

4. The price of gasoline at a particular station is 1.5 euros per liter. An American student can use 33 euros to buy gasoline. Knowing that 4 quarts make a gallon and that 1 liter is close to 1 quart, she quickly reasons that she can buy how many gallons of gasoline? (a) less than 1 gallon (b) about 5 gallons (c) about 8 gallons (d) more than 10 gallons

5. Rank the following five quantities in order from the largest to the smallest. If two of the quantities are equal, give them equal rank in your list. (a) 0.032 kg (b) 15 g (c) 2.7 × 10^5 mg (d) 4.1 × 10^{-8} Gg (e) 2.7 × 10^8 μg

6. What is the sum of the measured values 21.4 s + 15 s + 17.17 s + 4.00 3 s? (a) 57.573 s (b) 57.57 s (c) 57.6 s (d) 58 s (e) 60 s

7. Which of the following is the best estimate for the mass of all the people living on the Earth? (a) 2 × 10^8 kg (b) 1 × 10^9 kg (c) 2 × 10^{10} kg (d) 3 × 10^{11} kg (e) 4 × 10^{12} kg

8. (a) If an equation is dimensionally correct, does that mean that the equation must be true? (b) If an equation is not dimensionally correct, does that mean that the equation cannot be true?

9. Newton's second law of motion (Chapter 5) says that the mass of an object times its acceleration is equal to the net force on the object. Which of the following gives the correct units for force? (a) kg · m/s^2 (b) kg · m^2/s^2 (c) kg/m · s^2 (d) kg · m^2/s (e) none of those answers

10. A calculator displays a result as 1.365 248 0 × 10^7 kg. The estimated uncertainty in the result is ±2%. How many digits should be included as significant when the result is written down? (a) zero (b) one (c) two (d) three (e) four

Conceptual Questions   [1.] denotes answer available in *Student Solutions Manual/Study Guide*

[1.] Suppose the three fundamental standards of the metric system were length, *density,* and time rather than length, *mass,* and time. The standard of density in this system is to be defined as that of water. What considerations about water would you need to address to make sure that the standard of density is as accurate as possible?

2. Why is the metric system of units considered superior to most other systems of units?

3. What natural phenomena could serve as alternative time standards?

4. Express the following quantities using the prefixes given in Table 1.4. (a) 3 × 10^{-4} m (b) 5 × 10^{-5} s (c) 72 × 10^2 g

Problems

Section 1.1 Standards of Length, Mass, and Time

Note: Consult the endpapers, appendices, and tables in the text whenever necessary in solving problems. For this chapter, Table 14.1 and Appendix B.3 may be particularly useful. Answers to odd-numbered problems appear in the back of the book.

1. (a) Use information on the endpapers of this book to calculate the average density of the Earth. (b) Where does the value fit among those listed in Table 14.1 in Chapter 14? Look up the density of a typical surface rock like granite in another source and compare it with the density of the Earth.

2. The standard kilogram (Fig. 1.1a) is a platinum–iridium
W cylinder 39.0 mm in height and 39.0 mm in diameter. What is the density of the material?

3. An automobile company displays a die-cast model of its first car, made from 9.35 kg of iron. To celebrate its hundredth year in business, a worker will recast the model in solid gold from the original dies. What mass of gold is needed to make the new model?

4. A proton, which is the nucleus of a hydrogen atom, can be modeled as a sphere with a diameter of 2.4 fm and a mass of 1.67×10^{-27} kg. (a) Determine the density of the proton. (b) State how your answer to part (a) compares with the density of osmium, given in Table 14.1 in Chapter 14.

5. Two spheres are cut from a certain uniform rock. One
W has radius 4.50 cm. The mass of the other is five times greater. Find its radius.

6. What mass of a material with density ρ is required to make a hollow spherical shell having inner radius r_1 and outer radius r_2?

Section 1.2 Matter and Model Building

7. A crystalline solid consists of atoms stacked up in a repeating lattice structure. Consider a crystal as shown in Figure P1.7a. The atoms reside at the corners of cubes of side $L = 0.200$ nm. One piece of evidence for the regular arrangement of atoms comes from the flat surfaces along which a crystal separates, or cleaves, when it is broken. Suppose this crystal cleaves along a face diagonal as shown in Figure P1.7b. Calculate the spacing d between two adjacent atomic planes that separate when the crystal cleaves.

Figure P1.7

8. The mass of a copper atom is 1.06×10^{-25} kg, and the density of copper is 8 920 kg/m³ . (a) Determine the number of atoms in 1 cm³ of copper. (b) Visualize the one cubic centimeter as formed by stacking up identical cubes, with one copper atom at the center of each. Determine the volume of each cube. (c) Find the edge dimension of each cube, which represents an estimate for the spacing between atoms.

Section 1.3 Dimensional Analysis

9. Which of the following equations are dimensionally correct? (a) $v_f = v_i + ax$ (b) $y = (2 \text{ m}) \cos (kx)$, where $k = 2 \text{ m}^{-1}$

10. Figure P1.10 shows a *frustum*
W *of a cone.* Match each of the expressions

(a) $\pi(r_1 + r_2)[h^2 + (r_2 - r_1)^2]^{1/2}$,
(b) $2\pi(r_1 + r_2)$, and
(c) $\pi h(r_1^2 + r_1 r_2 + r_2^2)/3$

with the quantity it describes: (d) the total circumference of the flat circular faces, (e) the volume, or (f) the area of the curved surface.

Figure P1.10

11. Kinetic energy K (Chapter 7) has dimensions kg · m²/s². It can be written in terms of the momentum p (Chapter 9) and mass m as

$$K = \frac{p^2}{2m}$$

(a) Determine the proper units for momentum using dimensional analysis. (b) The unit of force is the newton N, where $1\ \text{N} = 1\ \text{kg} \cdot \text{m/s}^2$. What are the units of momentum p in terms of a newton and another fundamental SI unit?

12. Newton's law of universal gravitation is represented by

$$F = \frac{GMm}{r^2}$$

where F is the magnitude of the gravitational force exerted by one small object on another, M and m are the masses of the objects, and r is a distance. Force has the SI units $\text{kg} \cdot \text{m/s}^2$. What are the SI units of the proportionality constant G?

13. The position of a particle moving under uniform acceleration is some function of time and the acceleration. Suppose we write this position as $x = ka^m t^n$, where k is a dimensionless constant. Show by dimensional analysis that this expression is satisfied if $m = 1$ and $n = 2$. Can this analysis give the value of k?

14. (a) Assume the equation $x = At^3 + Bt$ describes the motion of a particular object, with x having the dimension of length and t having the dimension of time. Determine the dimensions of the constants A and B. (b) Determine the dimensions of the derivative $dx/dt = 3At^2 + B$.

Section 1.4 Conversion of Units

15. A solid piece of lead has a mass of 23.94 g and a volume of 2.10 cm³. From these data, calculate the density of lead in SI units (kilograms per cubic meter).

16. An ore loader moves 1 200 tons/h from a mine to the surface. Convert this rate to pounds per second, using 1 ton = 2 000 lb.

17. A rectangular building lot has a width of 75.0 ft and a length of 125 ft. Determine the area of this lot in square meters.

18. Suppose your hair grows at the rate 1/32 in. per day. Find the rate at which it grows in nanometers per second. Because the distance between atoms in a molecule is on the order of 0.1 nm, your answer suggests how rapidly layers of atoms are assembled in this protein synthesis.

19. *Why is the following situation impossible?* A student's dormitory room measures 3.8 m by 3.6 m, and its ceiling is 2.5 m high. After the student completes his physics course, he displays his dedication by completely wallpapering the walls of the room with the pages from his copy of volume 1 (Chapters 1–22) of this textbook. He even covers the door and window.

20. A pyramid has a height of 481 ft, and its base covers an area of 13.0 acres (Fig. P1.20). The volume of a pyramid is given by the expression $V = \frac{1}{3}Bh$, where B is the area of the base and h is the height. Find the volume of this pyramid in cubic meters. (1 acre = 43 560 ft²)

21. The pyramid described in Problem 20 contains approximately 2 million stone blocks that average 2.50 tons each. Find the weight of this pyramid in pounds.

Figure P1.20 Problems 20 and 21.

22. Assume it takes 7.00 min to fill a 30.0-gal gasoline tank. (a) Calculate the rate at which the tank is filled in gallons per second. (b) Calculate the rate at which the tank is filled in cubic meters per second. (c) Determine the time interval, in hours, required to fill a 1.00-m³ volume at the same rate. (1 U.S. gal = 231 in.³)

23. A *section* of land has an area of 1 square mile and contains 640 acres. Determine the number of square meters in 1 acre.

24. A house is 50.0 ft long and 26 ft wide and has 8.0-ft-high ceilings. What is the volume of the interior of the house in cubic meters and in cubic centimeters?

25. One cubic meter (1.00 m³) of aluminum has a mass of 2.70×10^3 kg, and the same volume of iron has a mass of 7.86×10^3 kg. Find the radius of a solid aluminum sphere that will balance a solid iron sphere of radius 2.00 cm on an equal-arm balance.

26. Let ρ_{Al} represent the density of aluminum and ρ_{Fe} that of iron. Find the radius of a solid aluminum sphere that balances a solid iron sphere of radius r_{Fe} on an equal-arm balance.

27. One gallon of paint (volume = 3.78×10^{-3} m³) covers an area of 25.0 m². What is the thickness of the fresh paint on the wall?

28. An auditorium measures 40.0 m × 20.0 m × 12.0 m. The density of air is 1.20 kg/m³. What are (a) the volume of the room in cubic feet and (b) the weight of air in the room in pounds?

29. (a) At the time of this book's printing, the U.S. national debt is about \$16 trillion. If payments were made at the rate of \$1 000 per second, how many years would it take to pay off the debt, assuming no interest were charged? (b) A dollar bill is about 15.5 cm long. How many dollar bills attached end to end would it take to reach the Moon? The front endpapers give the Earth–Moon distance. *Note:* Before doing these calculations, try to guess at the answers. You may be very surprised.

30. A hydrogen atom has a diameter of 1.06×10^{-10} m. The nucleus of the hydrogen atom has a diameter of approximately 2.40×10^{-15} m. (a) For a scale model, represent the diameter of the hydrogen atom by the playing length of an American football field (100 yards = 300 ft) and determine the diameter of the nucleus in millimeters. (b) Find the ratio of the volume of the hydrogen atom to the volume of its nucleus.

Section 1.5 **Estimates and Order-of-Magnitude Calculations**

Note: In your solutions to Problems 31 through 34, state the quantities you measure or estimate and the values you take for them.

31. Find the order of magnitude of the number of table-tennis balls that would fit into a typical-size room (without being crushed).

32. (a) Compute the order of magnitude of the mass of a bathtub half full of water. (b) Compute the order of magnitude of the mass of a bathtub half full of copper coins.

33. To an order of magnitude, how many piano tuners reside in New York City? The physicist Enrico Fermi was famous for asking questions like this one on oral Ph.D. qualifying examinations.

34. An automobile tire is rated to last for 50 000 miles. To an order of magnitude, through how many revolutions will it turn over its lifetime?

Section 1.6 **Significant Figures**

Note: Appendix B.8 on propagation of uncertainty may be useful in solving some problems in this section.

35. A rectangular plate has a length of (21.3 ± 0.2) cm and a width of (9.8 ± 0.1) cm. Calculate the area of the plate, including its uncertainty.

36. How many significant figures are in the following numbers? (a) 78.9 ± 0.2 (b) 3.788×10^9 (c) 2.46×10^{-6} (d) 0.005 3

37. The *tropical year*, the time interval from one vernal equinox to the next vernal equinox, is the basis for our calendar. It contains 365.242 199 days. Find the number of seconds in a tropical year.

38. Carry out the arithmetic operations (a) the sum of the measured values 756, 37.2, 0.83, and 2; (b) the product $0.003\ 2 \times 356.3$; and (c) the product $5.620 \times \pi$.

Note: The next 13 problems call on mathematical skills from your prior education that will be useful throughout this course.

39. Review. In a community college parking lot, the number of ordinary cars is larger than the number of sport utility vehicles by 94.7%. The difference between the number of cars and the number of SUVs is 18. Find the number of SUVs in the lot.

40. Review. While you are on a trip to Europe, you must purchase hazelnut chocolate bars for your grandmother. Eating just one square each day, she makes each large bar last for one and one-third months. How many bars will constitute a year's supply for her?

41. Review. A child is surprised that because of sales tax she must pay $1.36 for a toy marked $1.25. What is the effective tax rate on this purchase, expressed as a percentage?

42. Review. The average density of the planet Uranus is 1.27×10^3 kg/m³. The ratio of the mass of Neptune to

that of Uranus is 1.19. The ratio of the radius of Neptune to that of Uranus is 0.969. Find the average density of Neptune.

43. Review. The ratio of the number of sparrows visiting a bird feeder to the number of more interesting birds is 2.25. On a morning when altogether 91 birds visit the feeder, what is the number of sparrows?

44. Review. Find every angle θ between 0 and 360° for which the ratio of $\sin \theta$ to $\cos \theta$ is -3.00.

45. Review. For the right triangle shown in Figure P1.45, what are (a) the length of the unknown side, (b) the tangent of θ, and (c) the sine of ϕ?

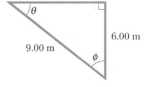

Figure P1.45

46. Review. Prove that one solution of the equation

$$2.00x^4 - 3.00x^3 + 5.00x = 70.0$$

is $x = -2.22$.

47. Review. A pet lamb grows rapidly, with its mass proportional to the cube of its length. When the lamb's length changes by 15.8%, its mass increases by 17.3 kg. Find the lamb's mass at the end of this process.

48. Review. A highway curve forms a section of a circle. A car goes around the curve as shown in the helicopter view of Figure P1.48. Its dashboard compass shows that the car is initially heading due east. After it travels $d = 840$ m, it is heading $\theta = 35.0°$ south of east. Find the radius of curvature of its path. *Suggestion:* You may find it useful to learn a geometric theorem stated in Appendix B.3.

Figure P1.48

49. Review. From the set of equations

$$p = 3q$$
$$pr = qs$$
$$\tfrac{1}{2}pr^2 + \tfrac{1}{2}qs^2 = \tfrac{1}{2}qt^2$$

involving the unknowns p, q, r, s, and t, find the value of the ratio of t to r.

50. Review. Figure P1.50 on page 18 shows students studying the thermal conduction of energy into cylindrical blocks of ice. As we will see in Chapter 20, this process is described by the equation

$$\frac{Q}{\Delta t} = \frac{k\pi d^2(T_h - T_c)}{4L}$$

For experimental control, in one set of trials all quantities except d and Δt are constant. (a) If d is made three times larger, does the equation predict that Δt will get larger or get smaller? By what factor? (b) What pattern of proportionality of Δt to d does the equation predict? (c) To display this proportionality as a straight line on a graph, what quantities should you plot on the horizontal and vertical axes? (d) What expression represents the theoretical slope of this graph?

Figure P1.50

51. **Review.** A student is supplied with a stack of copy paper, ruler, compass, scissors, and a sensitive balance. He cuts out various shapes in various sizes, calculates their areas, measures their masses, and prepares the graph of Figure P1.51. (a) Consider the fourth experimental point from the top. How far is it from the best-fit straight line? Express your answer as a difference in vertical-axis coordinate. (b) Express your answer as a percentage. (c) Calculate the slope of the line. (d) State what the graph demonstrates, referring to the shape of the graph and the results of parts (b) and (c). (e) Describe whether this result should be expected theoretically. (f) Describe the physical meaning of the slope.

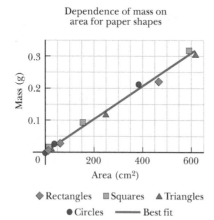

Figure P1.51

52. The radius of a uniform solid sphere is measured to be (6.50 ± 0.20) cm, and its mass is measured to be (1.85 ± 0.02) kg. Determine the density of the sphere in kilograms per cubic meter and the uncertainty in the density.

53. A sidewalk is to be constructed around a swimming pool that measures (10.0 ± 0.1) m by (17.0 ± 0.1) m. If the sidewalk is to measure (1.00 ± 0.01) m wide by (9.0 ± 0.1) cm thick, what volume of concrete is needed and what is the approximate uncertainty of this volume?

Additional Problems

54. Collectible coins are sometimes plated with gold to enhance their beauty and value. Consider a commemorative quarter-dollar advertised for sale at \$4.98. It has a diameter of 24.1 mm and a thickness of 1.78 mm, and it is completely covered with a layer of pure gold 0.180 μm thick. The volume of the plating is equal to the thickness of the layer multiplied by the area to which it is applied. The patterns on the faces of the coin and the grooves on its edge have a negligible effect on its area. Assume the price of gold is \$25.0 per gram. (a) Find the cost of the gold added to the coin. (b) Does the cost of the gold significantly enhance the value of the coin? Explain your answer.

55. In a situation in which data are known to three significant digits, we write 6.379 m = 6.38 m and 6.374 m = 6.37 m. When a number ends in 5, we arbitrarily choose to write 6.375 m = 6.38 m. We could equally well write 6.375 m = 6.37 m, "rounding down" instead of "rounding up," because we would change the number 6.375 by equal increments in both cases. Now consider an order-of-magnitude estimate, in which factors of change rather than increments are important. We write 500 m $\sim 10^3$ m because 500 differs from 100 by a factor of 5 while it differs from 1 000 by only a factor of 2. We write 437 m $\sim 10^3$ m and 305 m $\sim 10^2$ m. What distance differs from 100 m and from 1 000 m by equal factors so that we could equally well choose to represent its order of magnitude as $\sim 10^2$ m or as $\sim 10^3$ m?

56. (a) What is the order of magnitude of the number of microorganisms in the human intestinal tract? A typical bacterial length scale is 10^{-6} m. Estimate the intestinal volume and assume 1% of it is occupied by bacteria. (b) Does the number of bacteria suggest whether the bacteria are beneficial, dangerous, or neutral for the human body? What functions could they serve?

57. The diameter of our disk-shaped galaxy, the Milky Way, is about 1.0×10^5 light-years (ly). The distance to the Andromeda galaxy (Fig. P1.57), which is the spiral galaxy nearest to the Milky Way, is about 2.0 million ly. If a scale model represents the Milky Way and Andromeda

Figure P1.57 The Andromeda galaxy.

galaxies as dinner plates 25 cm in diameter, determine the distance between the centers of the two plates.

58. *Why is the following situation impossible?* In an effort to boost interest in a television game show, each weekly winner is offered an additional $1 million bonus prize if he or she can personally count out that exact amount from a supply of one-dollar bills. The winner must do this task under supervision by television show executives and within one 40-hour work week. To the dismay of the show's producers, most contestants succeed at the challenge.

59. A high fountain of water is located at the center of a circular pool as shown in Figure P1.59. A student walks around the pool and measures its circumference to be 15.0 m. Next, the student stands at the edge of the pool and uses a protractor to gauge the angle of elevation of the top of the fountain to be $\phi = 55.0°$. How high is the fountain?

Figure P1.59
Problems 59 and 60.

60. A water fountain is at the center of a circular pool as shown in Figure P1.59. A student walks around the pool and measures its circumference C. Next, he stands at the edge of the pool and uses a protractor to measure the angle of elevation ϕ of his sightline to the top of the water jet. How high is the fountain?

61. The data in the following table represent measurements of the masses and dimensions of solid cylinders of aluminum, copper, brass, tin, and iron. (a) Use these data to calculate the densities of these substances. (b) State how your results compare with those given in Table 14.1.

Substance	Mass (g)	Diameter (cm)	Length (cm)
Aluminum	51.5	2.52	3.75
Copper	56.3	1.23	5.06
Brass	94.4	1.54	5.69
Tin	69.1	1.75	3.74
Iron	216.1	1.89	9.77

62. The distance from the Sun to the nearest star is about 4×10^{16} m. The Milky Way galaxy (Fig. P1.62) is roughly

Richard Payne/NASA

Figure P1.62 The Milky Way galaxy.

a disk of diameter $\sim 10^{21}$ m and thickness $\sim 10^{19}$ m. Find the order of magnitude of the number of stars in the Milky Way. Assume the distance between the Sun and our nearest neighbor is typical.

63. Assume there are 100 million passenger cars in the United States and the average fuel efficiency is 20 mi/gal of gasoline. If the average distance traveled by each car is 10 000 mi/yr, how much gasoline would be saved per year if the average fuel efficiency could be increased to 25 mi/gal?

64. A spherical shell has an outside radius of 2.60 cm and an inside radius of a. The shell wall has uniform thickness and is made of a material with density 4.70 g/cm³. The space inside the shell is filled with a liquid having a density of 1.23 g/cm³. (a) Find the mass m of the sphere, including its contents, as a function of a. (b) For what value of the variable a does m have its maximum possible value? (c) What is this maximum mass? (d) Explain whether the value from part (c) agrees with the result of a direct calculation of the mass of a solid sphere of uniform density made of the same material as the shell. (e) **What If?** Would the answer to part (a) change if the inner wall were not concentric with the outer wall?

65. Bacteria and other prokaryotes are found deep underground, in water, and in the air. One micron (10^{-6} m) is a typical length scale associated with these microbes. (a) Estimate the total number of bacteria and other prokaryotes on the Earth. (b) Estimate the total mass of all such microbes.

66. Air is blown into a spherical balloon so that, when its radius is 6.50 cm, its radius is increasing at the rate 0.900 cm/s. (a) Find the rate at which the volume of the balloon is increasing. (b) If this volume flow rate of air entering the balloon is constant, at what rate will the radius be increasing when the radius is 13.0 cm? (c) Explain physically why the answer to part (b) is larger or smaller than 0.9 cm/s, if it is different.

67. A rod extending between $x = 0$ and $x = 14.0$ cm has uniform cross-sectional area $A = 9.00$ cm². Its density increases steadily between its ends from 2.70 g/cm³ to 19.3 g/cm³. (a) Identify the constants B and C required in the expression $\rho = B + Cx$ to describe the variable density. (b) The mass of the rod is given by

$$m = \int_{\text{all material}} \rho \, dV = \int_{\text{all } x} \rho A \, dx = \int_{0}^{14.0 \text{ cm}} (B + Cx)(9.00 \text{ cm}^2)\,dx$$

Carry out the integration to find the mass of the rod.

68. In physics, it is important to use mathematical approximations. (a) Demonstrate that for small angles ($< 20°$)

$$\tan \alpha \approx \sin \alpha \approx \alpha = \frac{\pi \alpha'}{180°}$$

where α is in radians and α' is in degrees. (b) Use a calculator to find the largest angle for which $\tan \alpha$ may be approximated by α with an error less than 10.0%.

69. The consumption of natural gas by a company satisfies the empirical equation $V = 1.50t + 0.008\,00t^2$, where V

is the volume of gas in millions of cubic feet and t is the time in months. Express this equation in units of cubic feet and seconds. Assume a month is 30.0 days.

70. A woman wishing to know the height of a mountain **GP** measures the angle of elevation of the mountaintop as 12.0°. After walking 1.00 km closer to the mountain on level ground, she finds the angle to be 14.0°. (a) Draw a picture of the problem, neglecting the height of the woman's eyes above the ground. *Hint:* Use two triangles. (b) Using the symbol y to represent the mountain height and the symbol x to represent the woman's original distance from the mountain, label the picture. (c) Using the labeled picture, write two trigonometric equations relating the two selected variables. (d) Find the height y.

71. A child loves to watch as you fill a transparent plastic **AMT** bottle with shampoo (Fig P1.71). Every horizontal cross section of the bottle is circular, but the diameters of the circles have different values. You pour the brightly colored shampoo into the bottle at a constant rate of 16.5 cm³/s. At what rate is its level in the bottle rising (a) at a point where the diameter of the bottle is 6.30 cm and (b) at a point where the diameter is 1.35 cm?

6.30 cm
1.35 cm

Figure P1.71

Challenge Problems

72. A woman stands at a horizontal distance x from a mountain and measures the angle of elevation of the mountaintop above the horizontal as θ. After walking a distance d closer to the mountain on level ground, she finds the angle to be ϕ. Find a general equation for the height y of the mountain in terms of d, ϕ, and θ, neglecting the height of her eyes above the ground.

73. You stand in a flat meadow and observe two cows (Fig. P1.73). Cow A is due north of you and 15.0 m from your position. Cow B is 25.0 m from your position. From your point of view, the angle between cow A and cow B is 20.0°, with cow B appearing to the right of cow A. (a) How far apart are cow A and cow B? (b) Consider the view seen by cow A. According to this cow, what is the angle between you and cow B? (c) Consider the view seen by cow B. According to this cow, what is the angle between you and cow A? *Hint:* What does the situation look like to a hummingbird hovering above the meadow? (d) Two stars in the sky appear to be 20.0° apart. Star A is 15.0 ly from the Earth, and star B, appearing to the right of star A, is 25.0 ly from the Earth. To an inhabitant of a planet orbiting star A, what is the angle in the sky between star B and our Sun?

Cow A   Cow B

Figure P1.73 Your view of two cows in a meadow. Cow A is due north of you. You must rotate your eyes through an angle of 20.0° to look from cow A to cow B.

Motion in One Dimension

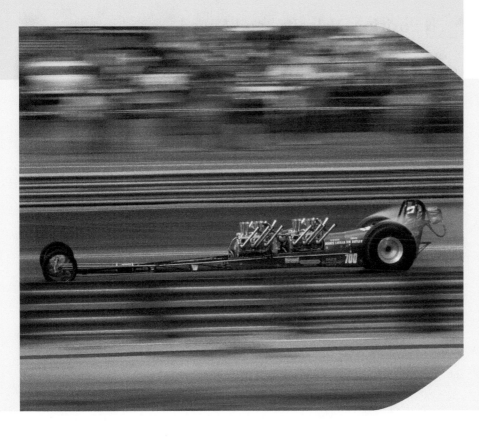

As a first step in studying classical mechanics, we describe the motion of an object while ignoring the interactions with external agents that might be affecting or modifying that motion. This portion of classical mechanics is called *kinematics*. (The word *kinematics* has the same root as *cinema*.) In this chapter, we consider only motion in one dimension, that is, motion of an object along a straight line.

From everyday experience, we recognize that motion of an object represents a continuous change in the object's position. In physics, we can categorize motion into three types: translational, rotational, and vibrational. A car traveling on a highway is an example of translational motion, the Earth's spin on its axis is an example of rotational motion, and the back-and-forth movement of a pendulum is an example of vibrational motion. In this and the next few chapters, we are concerned only with translational motion. (Later in the book we shall discuss rotational and vibrational motions.)

In our study of translational motion, we use what is called the **particle model** and describe the moving object as a *particle* regardless of its size. Remember our discussion of making models for physical situations in Section 1.2. In general, **a particle is a point-like object, that is, an object that has mass but is of infinitesimal size.** For example, if we wish to describe the motion of the Earth around the Sun, we can treat the Earth as a particle and

In drag racing, a driver wants as large an acceleration as possible. In a distance of one-quarter mile, a vehicle reaches speeds of more than 320 mi/h, covering the entire distance in under 5 s. *(George Lepp/ Stone/Getty Images)*

21

obtain reasonably accurate data about its orbit. This approximation is justified because the radius of the Earth's orbit is large compared with the dimensions of the Earth and the Sun.

As an example on a much smaller scale, it is possible to explain the pressure exerted by a gas on the walls of a container by treating the gas molecules as particles, without regard for the internal structure of the molecules.

2.1 Position, Velocity, and Speed

Position ▶

A particle's **position** x is the location of the particle with respect to a chosen reference point that we can consider to be the origin of a coordinate system. The motion of a particle is completely known if the particle's position in space is known at all times.

Consider a car moving back and forth along the x axis as in Figure 2.1a. When we begin collecting position data, the car is 30 m to the right of the reference position $x = 0$. We will use the particle model by identifying some point on the car, perhaps the front door handle, as a particle representing the entire car.

We start our clock, and once every 10 s we note the car's position. As you can see from Table 2.1, the car moves to the right (which we have defined as the positive direction) during the first 10 s of motion, from position Ⓐ to position Ⓑ. After Ⓑ, the position values begin to decrease, suggesting the car is backing up from position Ⓑ through position Ⓕ. In fact, at Ⓓ, 30 s after we start measuring, the car is at the origin of coordinates (see Fig. 2.1a). It continues moving to the left and is more than 50 m to the left of $x = 0$ when we stop recording information after our sixth data point. A graphical representation of this information is presented in Figure 2.1b. Such a plot is called a *position–time graph*.

Notice the *alternative representations* of information that we have used for the motion of the car. Figure 2.1a is a *pictorial representation*, whereas Figure 2.1b is a *graphical representation*. Table 2.1 is a *tabular representation* of the same information. Using an alternative representation is often an excellent strategy for understanding the situation in a given problem. The ultimate goal in many problems is a *math-*

Table 2.1 **Position of the Car at Various Times**

Position	t (s)	x (m)
Ⓐ	0	30
Ⓑ	10	52
Ⓒ	20	38
Ⓓ	30	0
Ⓔ	40	−37
Ⓕ	50	−53

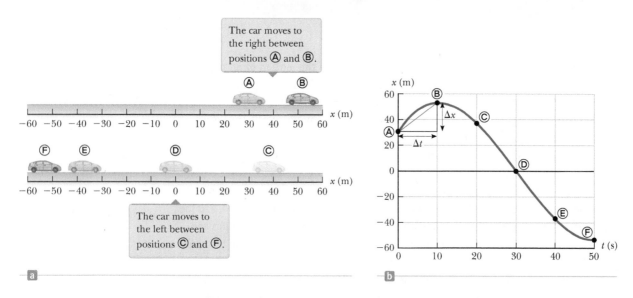

Figure 2.1 A car moves back and forth along a straight line. Because we are interested only in the car's translational motion, we can model it as a particle. Several representations of the information about the motion of the car can be used. Table 2.1 is a tabular representation of the information. (a) A pictorial representation of the motion of the car. (b) A graphical representation (position–time graph) of the motion of the car.

ematical representation, which can be analyzed to solve for some requested piece of information.

Given the data in Table 2.1, we can easily determine the change in position of the car for various time intervals. The **displacement** Δx of a particle is defined as its change in position in some time interval. As the particle moves from an initial position x_i to a final position x_f, its displacement is given by

$$\Delta x \equiv x_f - x_i \qquad (2.1)$$

◀ **Displacement**

We use the capital Greek letter delta (Δ) to denote the *change* in a quantity. From this definition, we see that Δx is positive if x_f is greater than x_i and negative if x_f is less than x_i.

It is very important to recognize the difference between displacement and distance traveled. **Distance** is the length of a path followed by a particle. Consider, for example, the basketball players in Figure 2.2. If a player runs from his own team's basket down the court to the other team's basket and then returns to his own basket, the *displacement* of the player during this time interval is zero because he ended up at the same point as he started: $x_f = x_i$, so $\Delta x = 0$. During this time interval, however, he moved through a *distance* of twice the length of the basketball court. Distance is always represented as a positive number, whereas displacement can be either positive or negative.

Displacement is an example of a vector quantity. Many other physical quantities, including position, velocity, and acceleration, also are vectors. In general, a **vector quantity** requires the specification of both direction and magnitude. By contrast, a **scalar quantity** has a numerical value and no direction. In this chapter, we use positive (+) and negative (−) signs to indicate vector direction. For example, for horizontal motion let us arbitrarily specify to the right as being the positive direction. It follows that any object always moving to the right undergoes a positive displacement $\Delta x > 0$, and any object moving to the left undergoes a negative displacement so that $\Delta x < 0$. We shall treat vector quantities in greater detail in Chapter 3.

One very important point has not yet been mentioned. Notice that the data in Table 2.1 result only in the six data points in the graph in Figure 2.1b. Therefore, the motion of the particle is not completely known because we don't know its position at *all* times. The smooth curve drawn through the six points in the graph is only a *possibility* of the actual motion of the car. We only have information about six instants of time; we have no idea what happened between the data points. The smooth curve is a *guess* as to what happened, but keep in mind that it is *only* a guess. If the smooth curve does represent the actual motion of the car, the graph contains complete information about the entire 50-s interval during which we watch the car move.

It is much easier to see changes in position from the graph than from a verbal description or even a table of numbers. For example, it is clear that the car covers more ground during the middle of the 50-s interval than at the end. Between positions Ⓒ and Ⓓ, the car travels almost 40 m, but during the last 10 s, between positions Ⓔ and Ⓕ, it moves less than half that far. A common way of comparing these different motions is to divide the displacement Δx that occurs between two clock readings by the value of that particular time interval Δt. The result turns out to be a very useful ratio, one that we shall use many times. This ratio has been given a special name: the *average velocity*. The **average velocity** $v_{x,\text{avg}}$ of a particle is defined as the particle's displacement Δx divided by the time interval Δt during which that displacement occurs:

$$v_{x,\text{avg}} \equiv \frac{\Delta x}{\Delta t} \qquad (2.2)$$

◀ **Average velocity**

where the subscript x indicates motion along the x axis. From this definition we see that average velocity has dimensions of length divided by time (L/T), or meters per second in SI units.

Figure 2.2 On this basketball court, players run back and forth for the entire game. The distance that the players run over the duration of the game is nonzero. The displacement of the players over the duration of the game is approximately zero because they keep returning to the same point over and over again.

The average velocity of a particle moving in one dimension can be positive or negative, depending on the sign of the displacement. (The time interval Δt is always positive.) If the coordinate of the particle increases in time (that is, if $x_f > x_i$), Δx is positive and $v_{x,avg} = \Delta x / \Delta t$ is positive. This case corresponds to a particle moving in the positive x direction, that is, toward larger values of x. If the coordinate decreases in time (that is, if $x_f < x_i$), Δx is negative and hence $v_{x,avg}$ is negative. This case corresponds to a particle moving in the negative x direction.

We can interpret average velocity geometrically by drawing a straight line between any two points on the position–time graph in Figure 2.1b. This line forms the hypotenuse of a right triangle of height Δx and base Δt. The slope of this line is the ratio $\Delta x / \Delta t$, which is what we have defined as average velocity in Equation 2.2. For example, the line between positions Ⓐ and Ⓑ in Figure 2.1b has a slope equal to the average velocity of the car between those two times, $(52\ \text{m} - 30\ \text{m})/(10\ \text{s} - 0) = 2.2\ \text{m/s}$.

In everyday usage, the terms *speed* and *velocity* are interchangeable. In physics, however, there is a clear distinction between these two quantities. Consider a marathon runner who runs a distance d of more than 40 km and yet ends up at her starting point. Her total displacement is zero, so her average velocity is zero! Nonetheless, we need to be able to quantify how fast she was running. A slightly different ratio accomplishes that for us. The **average speed** v_{avg} of a particle, a scalar quantity, is defined as the total distance d traveled divided by the total time interval required to travel that distance:

◀ Average speed

$$v_{avg} \equiv \frac{d}{\Delta t} \tag{2.3}$$

The SI unit of average speed is the same as the unit of average velocity: meters per second. Unlike average velocity, however, average speed has no direction and is always expressed as a positive number. Notice the clear distinction between the definitions of average velocity and average speed: average velocity (Eq. 2.2) is the *displacement* divided by the time interval, whereas average speed (Eq. 2.3) is the *distance* divided by the time interval.

Knowledge of the average velocity or average speed of a particle does not provide information about the details of the trip. For example, suppose it takes you 45.0 s to travel 100 m down a long, straight hallway toward your departure gate at an airport. At the 100-m mark, you realize you missed the restroom, and you return back 25.0 m along the same hallway, taking 10.0 s to make the return trip. The magnitude of your average *velocity* is $+75.0\ \text{m}/55.0\ \text{s} = +1.36\ \text{m/s}$. The average *speed* for your trip is $125\ \text{m}/55.0\ \text{s} = 2.27\ \text{m/s}$. You may have traveled at various speeds during the walk and, of course, you changed direction. Neither average velocity nor average speed provides information about these details.

> **Pitfall Prevention 2.1**
> **Average Speed and Average Velocity** The magnitude of the average velocity is *not* the average speed. For example, consider the marathon runner discussed before Equation 2.3. The magnitude of her average velocity is zero, but her average speed is clearly not zero.

Ⓠuick Quiz 2.1 Under which of the following conditions is the magnitude of the average velocity of a particle moving in one dimension smaller than the average speed over some time interval? **(a)** A particle moves in the $+x$ direction without reversing. **(b)** A particle moves in the $-x$ direction without reversing. **(c)** A particle moves in the $+x$ direction and then reverses the direction of its motion. **(d)** There are no conditions for which this is true.

Example 2.1   Calculating the Average Velocity and Speed

Find the displacement, average velocity, and average speed of the car in Figure 2.1a between positions Ⓐ and Ⓕ.

▶ **2.1** continued

SOLUTION

Consult Figure 2.1 to form a mental image of the car and its motion. We model the car as a particle. From the position–time graph given in Figure 2.1b, notice that $x_{\text{Ⓐ}} = 30$ m at $t_{\text{Ⓐ}} = 0$ s and that $x_{\text{Ⓕ}} = -53$ m at $t_{\text{Ⓕ}} = 50$ s.

Use Equation 2.1 to find the displacement of the car: $\Delta x = x_{\text{Ⓕ}} - x_{\text{Ⓐ}} = -53$ m $- 30$ m $= \boxed{-83 \text{ m}}$

This result means that the car ends up 83 m in the negative direction (to the left, in this case) from where it started. This number has the correct units and is of the same order of magnitude as the supplied data. A quick look at Figure 2.1a indicates that it is the correct answer.

Use Equation 2.2 to find the car's average velocity:

$$v_{x,\text{avg}} = \frac{x_{\text{Ⓕ}} - x_{\text{Ⓐ}}}{t_{\text{Ⓕ}} - t_{\text{Ⓐ}}}$$

$$= \frac{-53 \text{ m} - 30 \text{ m}}{50 \text{ s} - 0 \text{ s}} = \frac{-83 \text{ m}}{50 \text{ s}} = \boxed{-1.7 \text{ m/s}}$$

We cannot unambiguously find the average speed of the car from the data in Table 2.1 because we do not have information about the positions of the car between the data points. If we adopt the assumption that the details of the car's position are described by the curve in Figure 2.1b, the distance traveled is 22 m (from Ⓐ to Ⓑ) plus 105 m (from Ⓑ to Ⓕ), for a total of 127 m.

Use Equation 2.3 to find the car's average speed: $v_{\text{avg}} = \dfrac{127 \text{ m}}{50 \text{ s}} = \boxed{2.5 \text{ m/s}}$

Notice that the average speed is positive, as it must be. Suppose the red-brown curve in Figure 2.1b were different so that between 0 s and 10 s it went from Ⓐ up to 100 m and then came back down to Ⓑ. The average speed of the car would change because the distance is different, but the average velocity would not change.

2.2 Instantaneous Velocity and Speed

Often we need to know the velocity of a particle at a particular instant in time t rather than the average velocity over a finite time interval Δt. In other words, you would like to be able to specify your velocity just as precisely as you can specify your position by noting what is happening at a specific clock reading, that is, at some specific instant. What does it mean to talk about how quickly something is moving if we "freeze time" and talk only about an individual instant? In the late 1600s, with the invention of calculus, scientists began to understand how to describe an object's motion at any moment in time.

To see how that is done, consider Figure 2.3a (page 26), which is a reproduction of the graph in Figure 2.1b. What is the particle's velocity at $t = 0$? We have already discussed the average velocity for the interval during which the car moved from position Ⓐ to position Ⓑ (given by the slope of the blue line) and for the interval during which it moved from Ⓐ to Ⓕ (represented by the slope of the longer blue line and calculated in Example 2.1). The car starts out by moving to the right, which we defined to be the positive direction. Therefore, being positive, the value of the average velocity during the interval from Ⓐ to Ⓑ is more representative of the initial velocity than is the value of the average velocity during the interval from Ⓐ to Ⓕ, which we determined to be negative in Example 2.1. Now let us focus on the short blue line and slide point Ⓑ to the left along the curve, toward point Ⓐ, as in Figure 2.3b. The line between the points becomes steeper and steeper, and as the two points become extremely close together, the line becomes a tangent line to the curve, indicated by the green line in Figure 2.3b. The slope of this tangent line

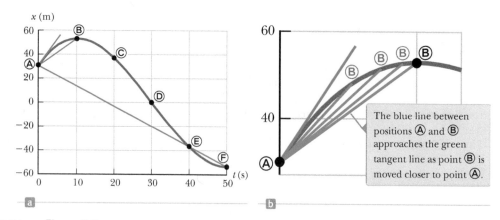

Figure 2.3 (a) Graph representing the motion of the car in Figure 2.1. (b) An enlargement of the upper-left-hand corner of the graph.

The blue line between positions Ⓐ and Ⓑ approaches the green tangent line as point Ⓑ is moved closer to point Ⓐ.

Pitfall Prevention 2.2

Slopes of Graphs In any graph of physical data, the *slope* represents the ratio of the change in the quantity represented on the vertical axis to the change in the quantity represented on the horizontal axis. Remember that *a slope has units* (unless both axes have the same units). The units of slope in Figures 2.1b and 2.3 are meters per second, the units of velocity.

Instantaneous velocity ▶

Pitfall Prevention 2.3

Instantaneous Speed and Instantaneous Velocity In Pitfall Prevention 2.1, we argued that the magnitude of the average velocity is not the average speed. The magnitude of the instantaneous velocity, however, *is* the instantaneous speed. In an infinitesimal time interval, the magnitude of the displacement is equal to the distance traveled by the particle.

represents the velocity of the car at point Ⓐ. What we have done is determine the *instantaneous velocity* at that moment. In other words, the **instantaneous velocity** v_x equals the limiting value of the ratio $\Delta x/\Delta t$ as Δt approaches zero:[1]

$$v_x \equiv \lim_{\Delta t \to 0} \frac{\Delta x}{\Delta t} \tag{2.4}$$

In calculus notation, this limit is called the *derivative* of x with respect to t, written dx/dt:

$$v_x \equiv \lim_{\Delta t \to 0} \frac{\Delta x}{\Delta t} = \frac{dx}{dt} \tag{2.5}$$

The instantaneous velocity can be positive, negative, or zero. When the slope of the position–time graph is positive, such as at any time during the first 10 s in Figure 2.3, v_x is positive and the car is moving toward larger values of x. After point Ⓑ, v_x is negative because the slope is negative and the car is moving toward smaller values of x. At point Ⓑ, the slope and the instantaneous velocity are zero and the car is momentarily at rest.

From here on, we use the word *velocity* to designate instantaneous velocity. When we are interested in *average velocity*, we shall always use the adjective *average*.

The **instantaneous speed** of a particle is defined as the magnitude of its instantaneous velocity. As with average speed, instantaneous speed has no direction associated with it. For example, if one particle has an instantaneous velocity of +25 m/s along a given line and another particle has an instantaneous velocity of −25 m/s along the same line, both have a speed[2] of 25 m/s.

Ⓠuick Quiz 2.2 Are members of the highway patrol more interested in **(a)** your average speed or **(b)** your instantaneous speed as you drive?

Conceptual Example 2.2   **The Velocity of Different Objects**

Consider the following one-dimensional motions: **(A)** a ball thrown directly upward rises to a highest point and falls back into the thrower's hand; **(B)** a race car starts from rest and speeds up to 100 m/s; and **(C)** a spacecraft drifts through space at constant velocity. Are there any points in the motion of these objects at which the instantaneous velocity has the same value as the average velocity over the entire motion? If so, identify the point(s).

---

[1]Notice that the displacement Δx also approaches zero as Δt approaches zero, so the ratio looks like 0/0. While this ratio may appear to be difficult to evaluate, the ratio does have a specific value. As Δx and Δt become smaller and smaller, the ratio $\Delta x/\Delta t$ approaches a value equal to the slope of the line tangent to the x-versus-t curve.

[2]As with velocity, we drop the adjective for instantaneous speed. *Speed* means "instantaneous speed."

▶ **2.2** continued

(A) The average velocity for the thrown ball is zero because the ball returns to the starting point; therefore, its displacement is zero. There is one point at which the instantaneous velocity is zero: at the top of the motion.

(B) The car's average velocity cannot be evaluated unambiguously with the information given, but it must have some value between 0 and 100 m/s. Because the car will have every instantaneous velocity between 0 and 100 m/s at some time during the interval, there must be some instant at which the instantaneous velocity is equal to the average velocity over the entire motion.

(C) Because the spacecraft's instantaneous velocity is constant, its instantaneous velocity at *any* time and its average velocity over *any* time interval are the same.

Example 2.3   **Average and Instantaneous Velocity**

A particle moves along the *x* axis. Its position varies with time according to the expression $x = -4t + 2t^2$, where *x* is in meters and *t* is in seconds.[3] The position–time graph for this motion is shown in Figure 2.4a. Because the position of the particle is given by a mathematical function, the motion of the particle is completely known, unlike that of the car in Figure 2.1. Notice that the particle moves in the negative *x* direction for the first second of motion, is momentarily at rest at the moment $t = 1$ s, and moves in the positive *x* direction at times $t > 1$ s.

(A) Determine the displacement of the particle in the time intervals $t = 0$ to $t = 1$ s and $t = 1$ s to $t = 3$ s.

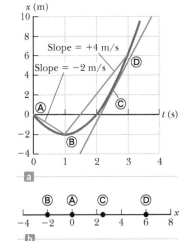

SOLUTION

From the graph in Figure 2.4a, form a mental representation of the particle's motion. Keep in mind that the particle does not move in a curved path in space such as that shown by the red-brown curve in the graphical representation. The particle moves only along the *x* axis in one dimension as shown in Figure 2.4b. At $t = 0$, is it moving to the right or to the left?

During the first time interval, the slope is negative and hence the average velocity is negative. Therefore, we know that the displacement between Ⓐ and Ⓑ must be a negative number having units of meters. Similarly, we expect the displacement between Ⓑ and Ⓓ to be positive.

Figure 2.4 (Example 2.3) (a) Position–time graph for a particle having an *x* coordinate that varies in time according to the expression $x = -4t + 2t^2$. (b) The particle moves in one dimension along the *x* axis.

In the first time interval, set $t_i = t_Ⓐ = 0$ and $t_f = t_Ⓑ = 1$ s and use Equation 2.1 to find the displacement:

$$\Delta x_{Ⓐ \to Ⓑ} = x_f - x_i = x_Ⓑ - x_Ⓐ$$
$$= [-4(1) + 2(1)^2] - [-4(0) + 2(0)^2] = \boxed{-2 \text{ m}}$$

For the second time interval ($t = 1$ s to $t = 3$ s), set $t_i = t_Ⓑ = 1$ s and $t_f = t_Ⓓ = 3$ s:

$$\Delta x_{Ⓑ \to Ⓓ} = x_f - x_i = x_Ⓓ - x_Ⓑ$$
$$= [-4(3) + 2(3)^2] - [-4(1) + 2(1)^2] = \boxed{+8 \text{ m}}$$

These displacements can also be read directly from the position–time graph.

(B) Calculate the average velocity during these two time intervals.

continued

---

[3]Simply to make it easier to read, we write the expression as $x = -4t + 2t^2$ rather than as $x = (-4.00 \text{ m/s})t + (2.00 \text{ m/s}^2)t^{2.00}$. When an equation summarizes measurements, consider its coefficients and exponents to have as many significant figures as other data quoted in a problem. Consider its coefficients to have the units required for dimensional consistency. When we start our clocks at $t = 0$, we usually do not mean to limit the precision to a single digit. Consider any zero value in this book to have as many significant figures as you need.

▶ **2.3** continued

`SOLUTION`

In the first time interval, use Equation 2.2 with $\Delta t = t_f - t_i = t_® - t_@ = 1$ s:

$$v_{x,avg\,(@ \to ®)} = \frac{\Delta x_{@ \to ®}}{\Delta t} = \frac{-2 \text{ m}}{1 \text{ s}} = \boxed{-2 \text{ m/s}}$$

In the second time interval, $\Delta t = 2$ s:

$$v_{x,avg\,(® \to ®)} = \frac{\Delta x_{® \to ®}}{\Delta t} = \frac{8 \text{ m}}{2 \text{ s}} = \boxed{+4 \text{ m/s}}$$

These values are the same as the slopes of the blue lines joining these points in Figure 2.4a.

(C) Find the instantaneous velocity of the particle at $t = 2.5$ s.

`SOLUTION`

Measure the slope of the green line at $t = 2.5$ s (point Ⓒ) in Figure 2.4a:

$$v_x = \frac{10 \text{ m} - (-4 \text{ m})}{3.8 \text{ s} - 1.5 \text{ s}} = \boxed{+6 \text{ m/s}}$$

Notice that this instantaneous velocity is on the same order of magnitude as our previous results, that is, a few meters per second. Is that what you would have expected?

2.3 Analysis Model: Particle Under Constant Velocity

In Section 1.2 we discussed the importance of making models. A particularly important model used in the solution to physics problems is an *analysis model*. An

◀ Analysis model

analysis model is a common situation that occurs time and again when solving physics problems. Because it represents a common situation, it also represents a common type of problem that we have solved before. When you identify an analysis model in a new problem, the solution to the new problem can be modeled after that of the previously-solved problem. Analysis models help us to recognize those common situations and guide us toward a solution to the problem. The form that an analysis model takes is a description of either (1) the behavior of some physical entity or (2) the interaction between that entity and the environment. When you encounter a new problem, you should identify the fundamental details of the problem and attempt to recognize which of the situations you have already seen that might be used as a model for the new problem. For example, suppose an automobile is moving along a straight freeway at a constant speed. Is it important that it is an automobile? Is it important that it is a freeway? If the answers to both questions are no, but the car moves in a straight line at constant speed, we model the automobile as a *particle under constant velocity*, which we will discuss in this section. Once the problem has been modeled, it is no longer about an automobile. It is about a particle undergoing a certain type of motion, a motion that we have studied before.

This method is somewhat similar to the common practice in the legal profession of finding "legal precedents." If a previously resolved case can be found that is very similar legally to the current one, it is used as a model and an argument is made in court to link them logically. The finding in the previous case can then be used to sway the finding in the current case. We will do something similar in physics. For a given problem, we search for a "physics precedent," a model with which we are already familiar and that can be applied to the current problem.

All of the analysis models that we will develop are based on four fundamental simplification models. The first of the four is the particle model discussed in the introduction to this chapter. We will look at a particle under various behaviors and environmental interactions. Further analysis models are introduced in later chapters based on simplification models of a *system*, a *rigid object*, and a *wave*. Once we have introduced these analysis models, we shall see that they appear again and again in different problem situations.

When solving a problem, you should avoid browsing through the chapter looking for an equation that contains the unknown variable that is requested in the problem. In many cases, the equation you find may have nothing to do with the problem you are attempting to solve. It is *much* better to take this first step: **Identify the analysis model that is appropriate for the problem.** To do so, think carefully about what is going on in the problem and match it to a situation you have seen before. Once the analysis model is identified, there are a small number of equations from which to choose that are appropriate for that model, sometimes only one equation. Therefore, **the model tells you which equation(s) to use for the mathematical representation.**

Let us use Equation 2.2 to build our first analysis model for solving problems. We imagine a particle moving with a constant velocity. The model of a **particle under constant velocity** can be applied in *any* situation in which an entity that can be modeled as a particle is moving with constant velocity. This situation occurs frequently, so this model is important.

If the velocity of a particle is constant, its instantaneous velocity at any instant during a time interval is the same as the average velocity over the interval. That is, $v_x = v_{x,avg}$. Therefore, Equation 2.2 gives us an equation to be used in the mathematical representation of this situation:

$$v_x = \frac{\Delta x}{\Delta t} \qquad (2.6)$$

Remembering that $\Delta x = x_f - x_i$, we see that $v_x = (x_f - x_i)/\Delta t$, or

$$x_f = x_i + v_x \Delta t$$

This equation tells us that the position of the particle is given by the sum of its original position x_i at time $t = 0$ plus the displacement $v_x \Delta t$ that occurs during the time interval Δt. In practice, we usually choose the time at the beginning of the interval to be $t_i = 0$ and the time at the end of the interval to be $t_f = t$, so our equation becomes

$$x_f = x_i + v_x t \quad \text{(for constant } v_x) \qquad (2.7)$$

Equations 2.6 and 2.7 are the primary equations used in the model of a particle under constant velocity. Whenever you have identified the analysis model in a problem to be the particle under constant velocity, you can immediately turn to these equations.

Figure 2.5 is a graphical representation of the particle under constant velocity. On this position–time graph, the slope of the line representing the motion is constant and equal to the magnitude of the velocity. Equation 2.7, which is the equation of a straight line, is the mathematical representation of the particle under constant velocity model. The slope of the straight line is v_x and the y intercept is x_i in both representations.

Example 2.4 below shows an application of the particle under constant velocity model. Notice the analysis model icon AM, which will be used to identify examples in which analysis models are employed in the solution. Because of the widespread benefits of using the analysis model approach, you will notice that a large number of the examples in the book will carry such an icon.

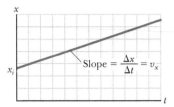

Figure 2.5 Position–time graph for a particle under constant velocity. The value of the constant velocity is the slope of the line.

◀ **Position as a function of time for the particle under constant velocity model**

| Example 2.4 | **Modeling a Runner as a Particle** | AM |

A kinesiologist is studying the biomechanics of the human body. (*Kinesiology* is the study of the movement of the human body. Notice the connection to the word *kinematics*.) She determines the velocity of an experimental subject while he runs along a straight line at a constant rate. The kinesiologist starts the stopwatch at the moment the runner passes a given point and stops it after the runner has passed another point 20 m away. The time interval indicated on the stopwatch is 4.0 s.

(A) What is the runner's velocity?

continued

▶ **2.4** continued

SOLUTION

We model the moving runner as a particle because the size of the runner and the movement of arms and legs are unnecessary details. Because the problem states that the subject runs at a constant rate, we can model him as a *particle under constant velocity.*

Having identified the model, we can use Equation 2.6 to find the constant velocity of the runner:

$$v_x = \frac{\Delta x}{\Delta t} = \frac{x_f - x_i}{\Delta t} = \frac{20\ \text{m} - 0}{4.0\ \text{s}} = \boxed{5.0\ \text{m/s}}$$

(B) If the runner continues his motion after the stopwatch is stopped, what is his position after 10 s have passed?

SOLUTION

Use Equation 2.7 and the velocity found in part (A) to find the position of the particle at time $t = 10$ s:

$$x_f = x_i + v_x t = 0 + (5.0\ \text{m/s})(10\ \text{s}) = \boxed{50\ \text{m}}$$

Is the result for part (A) a reasonable speed for a human? How does it compare to world-record speeds in 100-m and 200-m sprints? Notice the value in part (B) is more than twice that of the 20-m position at which the stopwatch was stopped. Is this value consistent with the time of 10 s being more than twice the time of 4.0 s?

The mathematical manipulations for the particle under constant velocity stem from Equation 2.6 and its descendent, Equation 2.7. These equations can be used to solve for any variable in the equations that happens to be unknown if the other variables are known. For example, in part (B) of Example 2.4, we find the position when the velocity and the time are known. Similarly, if we know the velocity and the final position, we could use Equation 2.7 to find the time at which the runner is at this position.

A particle under constant velocity moves with a constant speed along a straight line. Now consider a particle moving with a constant speed through a distance d along a curved path. This situation can be represented with the model of a **particle under constant speed.** The primary equation for this model is Equation 2.3, with the average speed v_{avg} replaced by the constant speed v:

$$v = \frac{d}{\Delta t} \tag{2.8}$$

As an example, imagine a particle moving at a constant speed in a circular path. If the speed is 5.00 m/s and the radius of the path is 10.0 m, we can calculate the time interval required to complete one trip around the circle:

$$v = \frac{d}{\Delta t} \quad \rightarrow \quad \Delta t = \frac{d}{v} = \frac{2\pi r}{v} = \frac{2\pi(10.0\ \text{m})}{5.00\ \text{m/s}} = 12.6\ \text{s}$$

Analysis Model   **Particle Under Constant Velocity**

Imagine a moving object that can be modeled as a particle. If it moves at a constant speed through a displacement Δx in a straight line in a time interval Δt, its constant velocity is

$$v_x = \frac{\Delta x}{\Delta t} \tag{2.6}$$

The position of the particle as a function of time is given by

$$x_f = x_i + v_x t \tag{2.7}$$

Examples:

- a meteoroid traveling through gravity-free space
- a car traveling at a constant speed on a straight highway
- a runner traveling at constant speed on a perfectly straight path
- an object moving at terminal speed through a viscous medium (Chapter 6)

Analysis Model Particle Under Constant Speed

Imagine a moving object that can be modeled as a particle. If it moves at a constant speed through a distance d along a straight line or a curved path in a time interval Δt, its constant speed is

$$v = \frac{d}{\Delta t} \qquad (2.8)$$

Examples:

- a planet traveling around a perfectly circular orbit
- a car traveling at a constant speed on a curved racetrack
- a runner traveling at constant speed on a curved path
- a charged particle moving through a uniform magnetic field (Chapter 29)

2.4 Acceleration

In Example 2.3, we worked with a common situation in which the velocity of a particle changes while the particle is moving. When the velocity of a particle changes with time, the particle is said to be *accelerating*. For example, the magnitude of a car's velocity increases when you step on the gas and decreases when you apply the brakes. Let us see how to quantify acceleration.

Suppose an object that can be modeled as a particle moving along the x axis has an initial velocity v_{xi} at time t_i at position Ⓐ and a final velocity v_{xf} at time t_f at position Ⓑ as in Figure 2.6a. The red-brown curve in Figure 2.6b shows how the velocity varies with time. The **average acceleration** $a_{x,avg}$ of the particle is defined as the *change* in velocity Δv_x divided by the time interval Δt during which that change occurs:

$$a_{x,avg} \equiv \frac{\Delta v_x}{\Delta t} = \frac{v_{xf} - v_{xi}}{t_f - t_i} \qquad (2.9)$$

◀ **Average acceleration**

As with velocity, when the motion being analyzed is one dimensional, we can use positive and negative signs to indicate the direction of the acceleration. Because the dimensions of velocity are L/T and the dimension of time is T, acceleration has dimensions of length divided by time squared, or L/T². The SI unit of acceleration is meters per second squared (m/s²). It might be easier to interpret these units if you think of them as meters per second per second. For example, suppose an object has an acceleration of +2 m/s². You can interpret this value by forming a mental image of the object having a velocity that is along a straight line and is increasing by 2 m/s during every time interval of 1 s. If the object starts from rest,

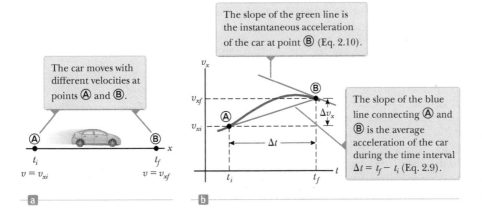

Figure 2.6 (a) A car, modeled as a particle, moving along the x axis from Ⓐ to Ⓑ, has velocity v_{xi} at $t = t_i$ and velocity v_{xf} at $t = t_f$. (b) Velocity–time graph (red-brown) for the particle moving in a straight line.

you should be able to picture it moving at a velocity of +2 m/s after 1 s, at +4 m/s after 2 s, and so on.

In some situations, the value of the average acceleration may be different over different time intervals. It is therefore useful to define the **instantaneous acceleration** as the limit of the average acceleration as Δt approaches zero. This concept is analogous to the definition of instantaneous velocity discussed in Section 2.2. If we imagine that point Ⓐ is brought closer and closer to point Ⓑ in Figure 2.6a and we take the limit of $\Delta v_x / \Delta t$ as Δt approaches zero, we obtain the instantaneous acceleration at point Ⓑ:

◀ Instantaneous acceleration

$$a_x \equiv \lim_{\Delta t \to 0} \frac{\Delta v_x}{\Delta t} = \frac{dv_x}{dt} \qquad (2.10)$$

That is, the instantaneous acceleration equals the derivative of the velocity with respect to time, which by definition is the slope of the velocity–time graph. The slope of the green line in Figure 2.6b is equal to the instantaneous acceleration at point Ⓑ. Notice that Figure 2.6b is a *velocity–time* graph, not a *position–time* graph like Figures 2.1b, 2.3, 2.4, and 2.5. Therefore, we see that just as the velocity of a moving particle is the slope at a point on the particle's x–t graph, the acceleration of a particle is the slope at a point on the particle's v_x–t graph. One can interpret the derivative of the velocity with respect to time as the time rate of change of velocity. If a_x is positive, the acceleration is in the positive x direction; if a_x is negative, the acceleration is in the negative x direction.

Figure 2.7 illustrates how an acceleration–time graph is related to a velocity–time graph. The acceleration at any time is the slope of the velocity–time graph at that time. Positive values of acceleration correspond to those points in Figure 2.7a where the velocity is increasing in the positive x direction. The acceleration reaches a maximum at time $t_Ⓐ$, when the slope of the velocity–time graph is a maximum. The acceleration then goes to zero at time $t_Ⓑ$, when the velocity is a maximum (that is, when the slope of the v_x–t graph is zero). The acceleration is negative when the velocity is decreasing in the positive x direction, and it reaches its most negative value at time $t_Ⓒ$.

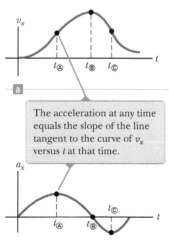

The acceleration at any time equals the slope of the line tangent to the curve of v_x versus t at that time.

Figure 2.7 (a) The velocity–time graph for a particle moving along the x axis. (b) The instantaneous acceleration can be obtained from the velocity–time graph.

Ⓠuick Quiz 2.3 Make a velocity–time graph for the car in Figure 2.1a. Suppose the speed limit for the road on which the car is driving is 30 km/h. True or False? The car exceeds the speed limit at some time within the time interval 0 − 50 s.

For the case of motion in a straight line, the direction of the velocity of an object and the direction of its acceleration are related as follows. When the object's velocity and acceleration are in the same direction, the object is speeding up. On the other hand, when the object's velocity and acceleration are in opposite directions, the object is slowing down.

To help with this discussion of the signs of velocity and acceleration, we can relate the acceleration of an object to the total *force* exerted on the object. In Chapter 5, we formally establish that **the force on an object is proportional to the acceleration of the object:**

$$F_x \propto a_x \qquad (2.11)$$

This proportionality indicates that acceleration is caused by force. Furthermore, force and acceleration are both vectors, and the vectors are in the same direction. Therefore, let us think about the signs of velocity and acceleration by imagining a force applied to an object and causing it to accelerate. Let us assume the velocity and acceleration are in the same direction. This situation corresponds to an object that experiences a force acting in the same direction as its velocity. In this case, the object speeds up! Now suppose the velocity and acceleration are in opposite directions. In this situation, the object moves in some direction and experiences a force acting in the opposite direction. Therefore, the object slows down! It is very useful to equate the direction of the acceleration to the direction of a force because it is easier from our everyday experience to think about what effect a force will have on an object than to think only in terms of the direction of the acceleration.

Quick Quiz 2.4 If a car is traveling eastward and slowing down, what is the direction of the force on the car that causes it to slow down? **(a)** eastward **(b)** westward **(c)** neither eastward nor westward

From now on, we shall use the term *acceleration* to mean instantaneous acceleration. When we mean average acceleration, we shall always use the adjective *average*. Because $v_x = dx/dt$, the acceleration can also be written as

$$a_x = \frac{dv_x}{dt} = \frac{d}{dt}\left(\frac{dx}{dt}\right) = \frac{d^2x}{dt^2} \tag{2.12}$$

That is, in one-dimensional motion, the acceleration equals the *second derivative* of x with respect to time.

> **Pitfall Prevention 2.4**
>
> **Negative Acceleration** Keep in mind that *negative acceleration does not necessarily mean that an object is slowing down*. If the acceleration is negative and the velocity is negative, the object is speeding up!

> **Pitfall Prevention 2.5**
>
> **Deceleration** The word *deceleration* has the common popular connotation of *slowing down*. We will not use this word in this book because it confuses the definition we have given for negative acceleration.

Conceptual Example 2.5   **Graphical Relationships Between *x*, *v*$_x$, and *a*$_x$**

The position of an object moving along the *x* axis varies with time as in Figure 2.8a. Graph the velocity versus time and the acceleration versus time for the object.

SOLUTION

The velocity at any instant is the slope of the tangent to the *x*–*t* graph at that instant. Between $t = 0$ and $t = t_Ⓐ$, the slope of the *x*–*t* graph increases uniformly, so the velocity increases linearly as shown in Figure 2.8b. Between $t_Ⓐ$ and $t_Ⓑ$, the slope of the *x*–*t* graph is constant, so the velocity remains constant. Between $t_Ⓑ$ and $t_Ⓓ$, the slope of the *x*–*t* graph decreases, so the value of the velocity in the v_x–*t* graph decreases. At $t_Ⓓ$, the slope of the *x*–*t* graph is zero, so the velocity is zero at that instant. Between $t_Ⓓ$ and $t_Ⓔ$, the slope of the *x*–*t* graph and therefore the velocity are negative and decrease uniformly in this interval. In the interval $t_Ⓔ$ to $t_Ⓕ$, the slope of the *x*–*t* graph is still negative, and at $t_Ⓕ$ it goes to zero. Finally, after $t_Ⓕ$, the slope of the *x*–*t* graph is zero, meaning that the object is at rest for $t > t_Ⓕ$.

The acceleration at any instant is the slope of the tangent to the v_x–*t* graph at that instant. The graph of acceleration versus time for this object is shown in Figure 2.8c. The acceleration is constant and positive between 0 and $t_Ⓐ$, where the slope of the v_x–*t* graph is positive. It is zero between $t_Ⓐ$ and $t_Ⓑ$ and for $t > t_Ⓕ$ because the slope of the v_x–*t* graph is zero at these times. It is negative between $t_Ⓑ$ and $t_Ⓔ$ because the slope of the v_x–*t* graph is negative during this interval. Between $t_Ⓔ$ and $t_Ⓕ$, the acceleration is positive like it is between 0 and $t_Ⓐ$, but higher in value because the slope of the v_x–*t* graph is steeper.

Notice that the sudden changes in acceleration shown in Figure 2.8c are unphysical. Such instantaneous changes cannot occur in reality.

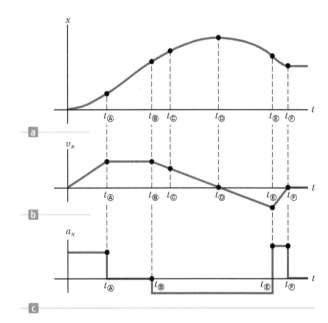

Figure 2.8 (Conceptual Example 2.5) (a) Position–time graph for an object moving along the *x* axis. (b) The velocity–time graph for the object is obtained by measuring the slope of the position–time graph at each instant. (c) The acceleration–time graph for the object is obtained by measuring the slope of the velocity–time graph at each instant.

Example 2.6   Average and Instantaneous Acceleration

The velocity of a particle moving along the x axis varies according to the expression $v_x = 40 - 5t^2$, where v_x is in meters per second and t is in seconds.

(A) Find the average acceleration in the time interval $t = 0$ to $t = 2.0$ s.

The acceleration at Ⓑ is equal to the slope of the green tangent line at $t = 2$ s, which is -20 m/s^2.

SOLUTION

Think about what the particle is doing from the mathematical representation. Is it moving at $t = 0$? In which direction? Does it speed up or slow down? Figure 2.9 is a v_x–t graph that was created from the velocity versus time expression given in the problem statement. Because the slope of the entire v_x–t curve is negative, we expect the acceleration to be negative.

Figure 2.9 (Example 2.6) The velocity–time graph for a particle moving along the x axis according to the expression $v_x = 40 - 5t^2$.

Find the velocities at $t_i = t_Ⓐ = 0$ and $t_f = t_Ⓑ = 2.0$ s by substituting these values of t into the expression for the velocity:

$$v_{xⒶ} = 40 - 5t_Ⓐ{}^2 = 40 - 5(0)^2 = +40 \text{ m/s}$$

$$v_{xⒷ} = 40 - 5t_Ⓑ{}^2 = 40 - 5(2.0)^2 = +20 \text{ m/s}$$

Find the average acceleration in the specified time interval $\Delta t = t_Ⓑ - t_Ⓐ = 2.0$ s:

$$a_{x,\text{avg}} = \frac{v_{xf} - v_{xi}}{t_f - t_i} = \frac{v_{xⒷ} - v_{xⒶ}}{t_Ⓑ - t_Ⓐ} = \frac{20 \text{ m/s} - 40 \text{ m/s}}{2.0 \text{ s} - 0 \text{ s}}$$

$$= \boxed{-10 \text{ m/s}^2}$$

The negative sign is consistent with our expectations: the average acceleration, represented by the slope of the blue line joining the initial and final points on the velocity–time graph, is negative.

(B) Determine the acceleration at $t = 2.0$ s.

SOLUTION

Knowing that the initial velocity at any time t is $v_{xi} = 40 - 5t^2$, find the velocity at any later time $t + \Delta t$:

$$v_{xf} = 40 - 5(t + \Delta t)^2 = 40 - 5t^2 - 10t\,\Delta t - 5(\Delta t)^2$$

Find the change in velocity over the time interval Δt:

$$\Delta v_x = v_{xf} - v_{xi} = -10t\,\Delta t - 5(\Delta t)^2$$

To find the acceleration at any time t, divide this expression by Δt and take the limit of the result as Δt approaches zero:

$$a_x = \lim_{\Delta t \to 0} \frac{\Delta v_x}{\Delta t} = \lim_{\Delta t \to 0}(-10t - 5\,\Delta t) = -10t$$

Substitute $t = 2.0$ s:

$$a_x = (-10)(2.0) \text{ m/s}^2 = \boxed{-20 \text{ m/s}^2}$$

Because the velocity of the particle is positive and the acceleration is negative at this instant, the particle is slowing down.

Notice that the answers to parts (A) and (B) are different. The average acceleration in part (A) is the slope of the blue line in Figure 2.9 connecting points Ⓐ and Ⓑ. The instantaneous acceleration in part (B) is the slope of the green line tangent to the curve at point Ⓑ. Notice also that the acceleration is *not* constant in this example. Situations involving constant acceleration are treated in Section 2.6.

So far, we have evaluated the derivatives of a function by starting with the definition of the function and then taking the limit of a specific ratio. If you are familiar with calculus, you should recognize that there are specific rules for taking derivatives. These rules, which are listed in Appendix B.6, enable us to evaluate derivatives quickly. For instance, one rule tells us that the derivative of any constant is zero. As another example, suppose x is proportional to some power of t such as in the expression

$$x = At^n$$

where A and n are constants. (This expression is a very common functional form.) The derivative of x with respect to t is

$$\frac{dx}{dt} = nAt^{n-1}$$

Applying this rule to Example 2.6, in which $v_x = 40 - 5t^2$, we quickly find that the acceleration is $a_x = dv_x/dt = -10t$, as we found in part (B) of the example.

2.5 Motion Diagrams

The concepts of velocity and acceleration are often confused with each other, but in fact they are quite different quantities. In forming a mental representation of a moving object, a pictorial representation called a *motion diagram* is sometimes useful to describe the velocity and acceleration while an object is in motion.

A motion diagram can be formed by imagining a *stroboscopic* photograph of a moving object, which shows several images of the object taken as the strobe light flashes at a constant rate. Figure 2.1a is a motion diagram for the car studied in Section 2.1. Figure 2.10 represents three sets of strobe photographs of cars moving along a straight roadway in a single direction, from left to right. The time intervals between flashes of the stroboscope are equal in each part of the diagram. So as to not confuse the two vector quantities, we use red arrows for velocity and purple arrows for acceleration in Figure 2.10. The arrows are shown at several instants during the motion of the object. Let us describe the motion of the car in each diagram.

In Figure 2.10a, the images of the car are equally spaced, showing us that the car moves through the same displacement in each time interval. This equal spacing is consistent with the car moving with *constant positive velocity* and *zero acceleration*. We could model the car as a particle and describe it with the particle under constant velocity model.

In Figure 2.10b, the images become farther apart as time progresses. In this case, the velocity arrow increases in length with time because the car's displacement between adjacent positions increases in time. These features suggest the car is moving with a *positive velocity* and a *positive acceleration*. The velocity and acceleration are in the same direction. In terms of our earlier force discussion, imagine a force pulling on the car in the same direction it is moving: it speeds up.

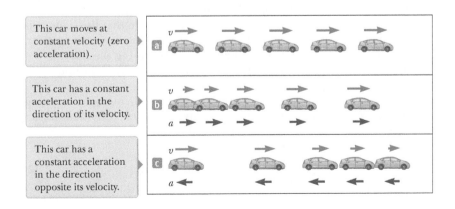

This car moves at constant velocity (zero acceleration).

This car has a constant acceleration in the direction of its velocity.

This car has a constant acceleration in the direction opposite its velocity.

Figure 2.10 Motion diagrams of a car moving along a straight roadway in a single direction. The velocity at each instant is indicated by a red arrow, and the constant acceleration is indicated by a purple arrow.

In Figure 2.10c, we can tell that the car slows as it moves to the right because its displacement between adjacent images decreases with time. This case suggests the car moves to the right with a negative acceleration. The length of the velocity arrow decreases in time and eventually reaches zero. From this diagram, we see that the acceleration and velocity arrows are *not* in the same direction. The car is moving with a *positive velocity*, but with a *negative acceleration*. (This type of motion is exhibited by a car that skids to a stop after its brakes are applied.) The velocity and acceleration are in opposite directions. In terms of our earlier force discussion, imagine a force pulling on the car opposite to the direction it is moving: it slows down.

Each purple acceleration arrow in parts (b) and (c) of Figure 2.10 is the same length. Therefore, these diagrams represent motion of a *particle under constant acceleration.* This important analysis model will be discussed in the next section.

Quick Quiz 2.5 Which one of the following statements is true? **(a)** If a car is traveling eastward, its acceleration must be eastward. **(b)** If a car is slowing down, its acceleration must be negative. **(c)** A particle with constant acceleration can never stop and stay stopped.

2.6 Analysis Model: Particle Under Constant Acceleration

If the acceleration of a particle varies in time, its motion can be complex and difficult to analyze. A very common and simple type of one-dimensional motion, however, is that in which the acceleration is constant. In such a case, the average acceleration $a_{x,avg}$ over any time interval is numerically equal to the instantaneous acceleration a_x at any instant within the interval, and the velocity changes at the same rate throughout the motion. This situation occurs often enough that we identify it as an analysis model: the **particle under constant acceleration.** In the discussion that follows, we generate several equations that describe the motion of a particle for this model.

If we replace $a_{x,avg}$ by a_x in Equation 2.9 and take $t_i = 0$ and t_f to be any later time t, we find that

$$a_x = \frac{v_{xf} - v_{xi}}{t - 0}$$

or

$$v_{xf} = v_{xi} + a_x t \quad \text{(for constant } a_x) \tag{2.13}$$

This powerful expression enables us to determine an object's velocity at *any* time t if we know the object's initial velocity v_{xi} and its (constant) acceleration a_x. A velocity–time graph for this constant-acceleration motion is shown in Figure 2.11b. The graph is a straight line, the slope of which is the acceleration a_x; the (constant) slope is consistent with $a_x = dv_x/dt$ being a constant. Notice that the slope is positive, which indicates a positive acceleration. If the acceleration were negative, the slope of the line in Figure 2.11b would be negative. When the acceleration is constant, the graph of acceleration versus time (Fig. 2.11c) is a straight line having a slope of zero.

Because velocity at constant acceleration varies linearly in time according to Equation 2.13, we can express the average velocity in any time interval as the arithmetic mean of the initial velocity v_{xi} and the final velocity v_{xf}:

$$v_{x,avg} = \frac{v_{xi} + v_{xf}}{2} \quad \text{(for constant } a_x) \tag{2.14}$$

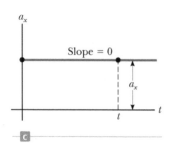

Figure 2.11 A particle under constant acceleration a_x moving along the x axis: (a) the position–time graph, (b) the velocity–time graph, and (c) the acceleration–time graph.

Notice that this expression for average velocity applies *only* in situations in which the acceleration is constant.

We can now use Equations 2.1, 2.2, and 2.14 to obtain the position of an object as a function of time. Recalling that Δx in Equation 2.2 represents $x_f - x_i$ and recognizing that $\Delta t = t_f - t_i = t - 0 = t$, we find that

$$x_f - x_i = v_{x,\text{avg}}\, t = \tfrac{1}{2}(v_{xi} + v_{xf})t$$

$$x_f = x_i + \tfrac{1}{2}(v_{xi} + v_{xf})t \quad (\text{for constant } a_x) \tag{2.15}$$

◀ Position as a function of velocity and time for the particle under constant acceleration model

This equation provides the final position of the particle at time t in terms of the initial and final velocities.

We can obtain another useful expression for the position of a particle under constant acceleration by substituting Equation 2.13 into Equation 2.15:

$$x_f = x_i + \tfrac{1}{2}[v_{xi} + (v_{xi} + a_x t)]t$$

$$x_f = x_i + v_{xi}t + \tfrac{1}{2}a_x t^2 \quad (\text{for constant } a_x) \tag{2.16}$$

◀ Position as a function of time for the particle under constant acceleration model

This equation provides the final position of the particle at time t in terms of the initial position, the initial velocity, and the constant acceleration.

The position–time graph for motion at constant (positive) acceleration shown in Figure 2.11a is obtained from Equation 2.16. Notice that the curve is a parabola. The slope of the tangent line to this curve at $t = 0$ equals the initial velocity v_{xi}, and the slope of the tangent line at any later time t equals the velocity v_{xf} at that time.

Finally, we can obtain an expression for the final velocity that does not contain time as a variable by substituting the value of t from Equation 2.13 into Equation 2.15:

$$x_f = x_i + \tfrac{1}{2}(v_{xi} + v_{xf})\left(\frac{v_{xf} - v_{xi}}{a_x}\right) = x_i + \frac{v_{xf}^2 - v_{xi}^2}{2a_x}$$

$$v_{xf}^2 = v_{xi}^2 + 2a_x(x_f - x_i) \quad (\text{for constant } a_x) \tag{2.17}$$

◀ Velocity as a function of position for the particle under constant acceleration model

This equation provides the final velocity in terms of the initial velocity, the constant acceleration, and the position of the particle.

For motion at *zero* acceleration, we see from Equations 2.13 and 2.16 that

$$\left.\begin{array}{l} v_{xf} = v_{xi} = v_x \\ x_f = x_i + v_x t \end{array}\right\} \quad \text{when } a_x = 0$$

That is, when the acceleration of a particle is zero, its velocity is constant and its position changes linearly with time. In terms of models, when the acceleration of a particle is zero, the particle under constant acceleration model reduces to the particle under constant velocity model (Section 2.3).

Equations 2.13 through 2.17 are **kinematic equations** that may be used to solve any problem involving a particle under constant acceleration in one dimension. These equations are listed together for convenience on page 38. The choice of which equation you use in a given situation depends on what you know beforehand. Sometimes it is necessary to use two of these equations to solve for two unknowns. You should recognize that the quantities that vary during the motion are position x_f, velocity v_{xf}, and time t.

You will gain a great deal of experience in the use of these equations by solving a number of exercises and problems. Many times you will discover that more than one method can be used to obtain a solution. Remember that these equations of kinematics *cannot* be used in a situation in which the acceleration varies with time. They can be used only when the acceleration is constant.

Quick Quiz 2.6 In Figure 2.12, match each v_x–t graph on the top with the a_x–t graph on the bottom that best describes the motion.

Figure 2.12 (Quick Quiz 2.6) Parts (a), (b), and (c) are v_x–t graphs of objects in one-dimensional motion. The possible accelerations of each object as a function of time are shown in scrambled order in (d), (e), and (f).

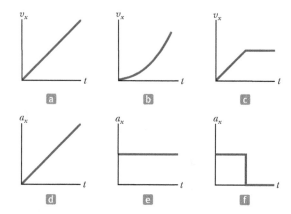

Analysis Model   **Particle Under Constant Acceleration**

Imagine a moving object that can be modeled as a particle. If it begins from position x_i and initial velocity v_{xi} and moves in a straight line with a constant acceleration a_x, its subsequent position and velocity are described by the following kinematic equations:

$$v_{xf} = v_{xi} + a_x t \tag{2.13}$$

$$v_{x,avg} = \frac{v_{xi} + v_{xf}}{2} \tag{2.14}$$

$$x_f = x_i + \tfrac{1}{2}(v_{xi} + v_{xf})t \tag{2.15}$$

$$x_f = x_i + v_{xi}t + \tfrac{1}{2}a_x t^2 \tag{2.16}$$

$$v_{xf}^2 = v_{xi}^2 + 2a_x(x_f - x_i) \tag{2.17}$$

Examples

- a car accelerating at a constant rate along a straight freeway
- a dropped object in the absence of air resistance (Section 2.7)
- an object on which a constant net force acts (Chapter 5)
- a charged particle in a uniform electric field (Chapter 23)

Example 2.7   **Carrier Landing**   AM

A jet lands on an aircraft carrier at a speed of 140 mi/h ($\approx$ 63 m/s).

(A) What is its acceleration (assumed constant) if it stops in 2.0 s due to an arresting cable that snags the jet and brings it to a stop?

SOLUTION

You might have seen movies or television shows in which a jet lands on an aircraft carrier and is brought to rest surprisingly fast by an arresting cable. A careful reading of the problem reveals that in addition to being given the initial speed of 63 m/s, we also know that the final speed is zero. Because the acceleration of the jet is assumed constant, we model it as a *particle under constant acceleration*. We define our x axis as the direction of motion of the jet. Notice that we have no information about the change in position of the jet while it is slowing down.

▶ **2.7** continued

Equation 2.13 is the only equation in the particle under constant acceleration model that does not involve position, so we use it to find the acceleration of the jet, modeled as a particle:

$$a_x = \frac{v_{xf} - v_{xi}}{t} \approx \frac{0 - 63 \text{ m/s}}{2.0 \text{ s}}$$
$$= -32 \text{ m/s}^2$$

(B) If the jet touches down at position $x_i = 0$, what is its final position?

SOLUTION

Use Equation 2.15 to solve for the final position: $x_f = x_i + \frac{1}{2}(v_{xi} + v_{xf})t = 0 + \frac{1}{2}(63 \text{ m/s} + 0)(2.0 \text{ s}) = \boxed{63 \text{ m}}$

Given the size of aircraft carriers, a length of 63 m seems reasonable for stopping the jet. The idea of using arresting cables to slow down landing aircraft and enable them to land safely on ships originated at about the time of World War I. The cables are still a vital part of the operation of modern aircraft carriers.

WHAT IF?   Suppose the jet lands on the deck of the aircraft carrier with a speed faster than 63 m/s but has the same acceleration due to the cable as that calculated in part (A). How will that change the answer to part (B)?

Answer If the jet is traveling faster at the beginning, it will stop farther away from its starting point, so the answer to part (B) should be larger. Mathematically, we see in Equation 2.15 that if v_{xi} is larger, then x_f will be larger.

Example 2.8   **Watch Out for the Speed Limit!**   AM

A car traveling at a constant speed of 45.0 m/s passes a trooper on a motorcycle hidden behind a billboard. One second after the speeding car passes the billboard, the trooper sets out from the billboard to catch the car, accelerating at a constant rate of 3.00 m/s². How long does it take the trooper to overtake the car?

SOLUTION

A pictorial representation (Fig. 2.13) helps clarify the sequence of events. The car is modeled as a *particle under constant velocity*, and the trooper is modeled as a *particle under constant acceleration*.

Figure 2.13   (Example 2.8) A speeding car passes a hidden trooper.

First, we write expressions for the position of each vehicle as a function of time. It is convenient to choose the position of the billboard as the origin and to set $t_{\text{Ⓑ}} = 0$ as the time the trooper begins moving. At that instant, the car has already traveled a distance of 45.0 m from the billboard because it has traveled at a constant speed of $v_x = 45.0$ m/s for 1 s. Therefore, the initial position of the speeding car is $x_{\text{Ⓑ}} = 45.0$ m.

Using the particle under constant velocity model, apply Equation 2.7 to give the car's position at any time t: $x_{\text{car}} = x_{\text{Ⓑ}} + v_{x\,\text{car}}t$

A quick check shows that at $t = 0$, this expression gives the car's correct initial position when the trooper begins to move: $x_{\text{car}} = x_{\text{Ⓑ}} = 45.0$ m.

The trooper starts from rest at $t_{\text{Ⓑ}} = 0$ and accelerates at $a_x = 3.00$ m/s² away from the origin. Use Equation 2.16 to give her position at any time t:

$$x_f = x_i + v_{xi}t + \tfrac{1}{2}a_xt^2$$
$$x_{\text{trooper}} = 0 + (0)t + \tfrac{1}{2}a_xt^2 = \tfrac{1}{2}a_xt^2$$

Set the positions of the car and trooper equal to represent the trooper overtaking the car at position Ⓒ:

$$x_{\text{trooper}} = x_{\text{car}}$$
$$\tfrac{1}{2}a_xt^2 = x_{\text{Ⓑ}} + v_{x\,\text{car}}t$$

continued

▶ **2.8** c o n t i n u e d

Rearrange to give a quadratic equation:

$$\tfrac{1}{2}a_x t^2 - v_{x\,car}t - x_\circledB = 0$$

Solve the quadratic equation for the time at which the trooper catches the car (for help in solving quadratic equations, see Appendix B.2.):

$$t = \frac{v_{x\,car} \pm \sqrt{v_{x\,car}^2 + 2a_x x_\circledB}}{a_x}$$

$$(1)\quad t = \frac{v_{x\,car}}{a_x} \pm \sqrt{\frac{v_{x\,car}^2}{a_x^2} + \frac{2x_\circledB}{a_x}}$$

Evaluate the solution, choosing the positive root because that is the only choice consistent with a time $t > 0$:

$$t = \frac{45.0\ \text{m/s}}{3.00\ \text{m/s}^2} + \sqrt{\frac{(45.0\ \text{m/s})^2}{(3.00\ \text{m/s}^2)^2} + \frac{2(45.0\ \text{m})}{3.00\ \text{m/s}^2}} = \boxed{31.0\ \text{s}}$$

Why didn't we choose $t = 0$ as the time at which the car passes the trooper? If we did so, we would not be able to use the particle under constant acceleration model for the trooper. Her acceleration would be zero for the first second and then $3.00\ \text{m/s}^2$ for the remaining time. By defining the time $t = 0$ as when the trooper begins moving, we can use the particle under constant acceleration model for her movement for all positive times.

WHAT IF?   What if the trooper had a more powerful motorcycle with a larger acceleration? How would that change the time at which the trooper catches the car?

Answer If the motorcycle has a larger acceleration, the trooper should catch up to the car sooner, so the answer for the time should be less than 31 s. Because all terms on the right side of Equation (1) have the acceleration a_x in the denominator, we see symbolically that increasing the acceleration will decrease the time at which the trooper catches the car.

2.7 Freely Falling Objects

Galileo Galilei
Italian physicist and astronomer
(1564–1642)
Galileo formulated the laws that govern the motion of objects in free fall and made many other significant discoveries in physics and astronomy. Galileo publicly defended Nicolaus Copernicus's assertion that the Sun is at the center of the Universe (the heliocentric system). He published *Dialogue Concerning Two New World Systems* to support the Copernican model, a view that the Catholic Church declared to be heretical.

Georgios Kollidas/Shutterstock.com

It is well known that, in the absence of air resistance, all objects dropped near the Earth's surface fall toward the Earth with the same constant acceleration under the influence of the Earth's gravity. It was not until about 1600 that this conclusion was accepted. Before that time, the teachings of the Greek philosopher Aristotle (384–322 BC) had held that heavier objects fall faster than lighter ones.

The Italian Galileo Galilei (1564–1642) originated our present-day ideas concerning falling objects. There is a legend that he demonstrated the behavior of falling objects by observing that two different weights dropped simultaneously from the Leaning Tower of Pisa hit the ground at approximately the same time. Although there is some doubt that he carried out this particular experiment, it is well established that Galileo performed many experiments on objects moving on inclined planes. In his experiments, he rolled balls down a slight incline and measured the distances they covered in successive time intervals. The purpose of the incline was to reduce the acceleration, which made it possible for him to make accurate measurements of the time intervals. By gradually increasing the slope of the incline, he was finally able to draw conclusions about freely falling objects because a freely falling ball is equivalent to a ball moving down a vertical incline.

You might want to try the following experiment. Simultaneously drop a coin and a crumpled-up piece of paper from the same height. If the effects of air resistance are negligible, both will have the same motion and will hit the floor at the same time. In the idealized case, in which air resistance is absent, such motion is referred to as *free-fall* motion. If this same experiment could be conducted in a vacuum, in which air resistance is truly negligible, the paper and the coin would fall with the same acceleration even when the paper is not crumpled. On August 2, 1971, astronaut David Scott conducted such a demonstration on the Moon. He simultaneously released a hammer and a feather, and the two objects fell together to the lunar surface. This simple demonstration surely would have pleased Galileo!

When we use the expression *freely falling object*, we do not necessarily refer to an object dropped from rest. A freely falling object is any object moving freely under the influence of gravity alone, regardless of its initial motion. Objects thrown upward or downward and those released from rest are all falling freely once they are released. Any freely falling object experiences an acceleration directed *downward*, regardless of its initial motion.

We shall denote the magnitude of the *free-fall acceleration*, also called the *acceleration due to gravity*, by the symbol g. The value of g decreases with increasing altitude above the Earth's surface. Furthermore, slight variations in g occur with changes in latitude. At the Earth's surface, the value of g is approximately 9.80 m/s^2. Unless stated otherwise, we shall use this value for g when performing calculations. For making quick estimates, use $g = 10 \text{ m/s}^2$.

If we neglect air resistance and assume the free-fall acceleration does not vary with altitude over short vertical distances, the motion of a freely falling object moving vertically is equivalent to the motion of a particle under constant acceleration in one dimension. Therefore, the equations developed in Section 2.6 for the particle under constant acceleration model can be applied. The only modification for freely falling objects that we need to make in these equations is to note that the motion is in the vertical direction (the y direction) rather than in the horizontal direction (x) and that the acceleration is downward and has a magnitude of 9.80 m/s^2. Therefore, we choose $a_y = -g = -9.80 \text{ m/s}^2$, where the negative sign means that the acceleration of a freely falling object is downward. In Chapter 13, we shall study how to deal with variations in g with altitude.

Quick Quiz 2.7 Consider the following choices: (a) increases, (b) decreases, (c) increases and then decreases, (d) decreases and then increases, (e) remains the same. From these choices, select what happens to **(i)** the acceleration and **(ii)** the speed of a ball after it is thrown upward into the air.

> **Pitfall Prevention 2.6**
>
> **g and g** Be sure not to confuse the italic symbol g for free-fall acceleration with the nonitalic symbol g used as the abbreviation for the unit gram.

> **Pitfall Prevention 2.7**
>
> **The Sign of g** Keep in mind that g is a *positive number*. It is tempting to substitute -9.80 m/s^2 for g, but resist the temptation. Downward gravitational acceleration is indicated explicitly by stating the acceleration as $a_y = -g$.

> **Pitfall Prevention 2.8**
>
> **Acceleration at the Top of the Motion** A common misconception is that the acceleration of a projectile at the top of its trajectory is zero. Although the velocity at the top of the motion of an object thrown upward momentarily goes to zero, *the acceleration is still that due to gravity* at this point. If the velocity and acceleration were both zero, the projectile would stay at the top.

Conceptual Example 2.9   **The Daring Skydivers**

A skydiver jumps out of a hovering helicopter. A few seconds later, another skydiver jumps out, and they both fall along the same vertical line. Ignore air resistance so that both skydivers fall with the same acceleration. Does the difference in their speeds stay the same throughout the fall? Does the vertical distance between them stay the same throughout the fall?

SOLUTION

At any given instant, the speeds of the skydivers are different because one had a head start. In any time interval Δt after this instant, however, the two skydivers increase their speeds by the same amount because they have the same acceleration. Therefore, the difference in their speeds remains the same throughout the fall.

The first jumper always has a greater speed than the second. Therefore, in a given time interval, the first skydiver covers a greater distance than the second. Consequently, the separation distance between them increases.

Example 2.10   Not a Bad Throw for a Rookie!   AM

A stone thrown from the top of a building is given an initial velocity of 20.0 m/s straight upward. The stone is launched 50.0 m above the ground, and the stone just misses the edge of the roof on its way down as shown in Figure 2.14.

(A) Using $t_{Ⓐ} = 0$ as the time the stone leaves the thrower's hand at position Ⓐ, determine the time at which the stone reaches its maximum height.

SOLUTION

You most likely have experience with dropping objects or throwing them upward and watching them fall, so this problem should describe a familiar experience. To simulate this situation, toss a small object upward and notice the time interval required for it to fall to the floor. Now imagine throwing that object upward from the roof of a building. Because the stone is in free fall, it is modeled as a *particle under constant acceleration* due to gravity.

Figure 2.14 (Example 2.10) Position, velocity, and acceleration values at various times for a freely falling stone thrown initially upward with a velocity $v_{yi} = 20.0$ m/s. Many of the quantities in the labels for points in the motion of the stone are calculated in the example. Can you verify the other values that are not?

Ⓑ $t_{Ⓑ} = 2.04$ s
$y_{Ⓑ} = 20.4$ m
$v_{y Ⓑ} = 0$
$a_{y Ⓑ} = -9.80$ m/s²

$t_{Ⓐ} = 0$
$y_{Ⓐ} = 0$
$v_{y Ⓐ} = 20.0$ m/s
$a_{y Ⓐ} = -9.80$ m/s²

Ⓒ $t_{Ⓒ} = 4.08$ s
$y_{Ⓒ} = 0$
$v_{y Ⓒ} = -20.0$ m/s
$a_{y Ⓒ} = -9.80$ m/s²

Ⓐ

Ⓓ $t_{Ⓓ} = 5.00$ s
$y_{Ⓓ} = -22.5$ m
$v_{y Ⓓ} = -29.0$ m/s
$a_{y Ⓓ} = -9.80$ m/s²

50.0 m

$t_{Ⓔ} = 5.83$ s
$y_{Ⓔ} = -50.0$ m
Ⓔ $v_{y Ⓔ} = -37.1$ m/s
$a_{y Ⓔ} = -9.80$ m/s²

Recognize that the initial velocity is positive because the stone is launched upward. The velocity will change sign after the stone reaches its highest point, but the acceleration of the stone will *always* be downward so that it will always have a negative value. Choose an initial point just after the stone leaves the person's hand and a final point at the top of its flight.

Use Equation 2.13 to calculate the time at which the stone reaches its maximum height:

$$v_{yf} = v_{yi} + a_y t \quad \rightarrow \quad t = \frac{v_{yf} - v_{yi}}{a_y}$$

Substitute numerical values:

$$t = t_{Ⓑ} = \frac{0 - 20.0 \text{ m/s}}{-9.80 \text{ m/s}^2} = \boxed{2.04 \text{ s}}$$

(B) Find the maximum height of the stone.

SOLUTION

As in part (A), choose the initial and final points at the beginning and the end of the upward flight.

Set $y_{Ⓐ} = 0$ and substitute the time from part (A) into Equation 2.16 to find the maximum height:

$$y_{max} = y_{Ⓑ} = y_{Ⓐ} + v_{x Ⓐ} t + \tfrac{1}{2} a_y t^2$$

$$y_{Ⓑ} = 0 + (20.0 \text{ m/s})(2.04 \text{ s}) + \tfrac{1}{2}(-9.80 \text{ m/s}^2)(2.04 \text{ s})^2 = \boxed{20.4 \text{ m}}$$

(C) Determine the velocity of the stone when it returns to the height from which it was thrown.

SOLUTION

Choose the initial point where the stone is launched and the final point when it passes this position coming down.

Substitute known values into Equation 2.17:

$$v_{y Ⓒ}^2 = v_{y Ⓐ}^2 + 2a_y(y_{Ⓒ} - y_{Ⓐ})$$

$$v_{y Ⓒ}^2 = (20.0 \text{ m/s})^2 + 2(-9.80 \text{ m/s}^2)(0 - 0) = 400 \text{ m}^2/\text{s}^2$$

$$v_{y Ⓒ} = \boxed{-20.0 \text{ m/s}}$$

▶ **2.10** continued

When taking the square root, we could choose either a positive or a negative root. We choose the negative root because we know that the stone is moving downward at point ©. The velocity of the stone when it arrives back at its original height is equal in magnitude to its initial velocity but is opposite in direction.

(D) Find the velocity and position of the stone at $t = 5.00$ s.

SOLUTION

Choose the initial point just after the throw and the final point 5.00 s later.

Calculate the velocity at Ⓓ from Equation 2.13:

$$v_{y\text{Ⓓ}} = v_{y\text{Ⓐ}} + a_y t = 20.0 \text{ m/s} + (-9.80 \text{ m/s}^2)(5.00 \text{ s}) = \boxed{-29.0 \text{ m/s}}$$

Use Equation 2.16 to find the position of the stone at $t_\text{Ⓓ} = 5.00$ s:

$$y_\text{Ⓓ} = y_\text{Ⓐ} + v_{y\text{Ⓐ}} t + \tfrac{1}{2} a_y t^2$$
$$= 0 + (20.0 \text{ m/s})(5.00 \text{ s}) + \tfrac{1}{2}(-9.80 \text{ m/s}^2)(5.00 \text{ s})^2$$
$$= \boxed{-22.5 \text{ m}}$$

The choice of the time defined as $t = 0$ is arbitrary and up to you to select as the problem solver. As an example of this arbitrariness, choose $t = 0$ as the time at which the stone is at the highest point in its motion. Then solve parts (C) and (D) again using this new initial instant and notice that your answers are the same as those above.

WHAT IF? What if the throw were from 30.0 m above the ground instead of 50.0 m? Which answers in parts (A) to (D) would change?

Answer None of the answers would change. All the motion takes place in the air during the first 5.00 s. (Notice that even for a throw from 30.0 m, the stone is above the ground at $t = 5.00$ s.) Therefore, the height of the throw is not an issue. Mathematically, if we look back over our calculations, we see that we never entered the height of the throw into any equation.

2.8 Kinematic Equations Derived from Calculus

This section assumes the reader is familiar with the techniques of integral calculus. If you have not yet studied integration in your calculus course, you should skip this section or cover it after you become familiar with integration.

The velocity of a particle moving in a straight line can be obtained if its position as a function of time is known. Mathematically, the velocity equals the derivative of the position with respect to time. It is also possible to find the position of a particle if its velocity is known as a function of time. In calculus, the procedure used to perform this task is referred to either as *integration* or as finding the *antiderivative*. Graphically, it is equivalent to finding the area under a curve.

Suppose the v_x–t graph for a particle moving along the x axis is as shown in Figure 2.15 on page 44. Let us divide the time interval $t_f - t_i$ into many small intervals, each of duration Δt_n. From the definition of average velocity, we see that the displacement of the particle during any small interval, such as the one shaded in Figure 2.15, is given by $\Delta x_n = v_{xn,\text{avg}} \Delta t_n$, where $v_{xn,\text{avg}}$ is the average velocity in that interval. Therefore, the displacement during this small interval is simply the area of the shaded rectangle in Figure 2.15. The total displacement for the interval $t_f - t_i$ is the sum of the areas of all the rectangles from t_i to t_f:

$$\Delta x = \sum_n v_{xn,\text{avg}} \Delta t_n$$

where the symbol Σ (uppercase Greek sigma) signifies a sum over all terms, that is, over all values of n. Now, as the intervals are made smaller and smaller, the number of terms in the sum increases and the sum approaches a value equal to the area

Figure 2.15 Velocity versus time for a particle moving along the x axis. The total area under the curve is the total displacement of the particle.

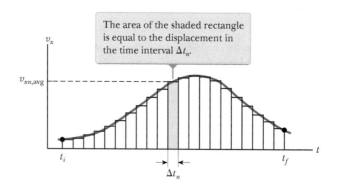

The area of the shaded rectangle is equal to the displacement in the time interval Δt_n.

under the curve in the velocity–time graph. Therefore, in the limit $n \to \infty$, or $\Delta t_n \to 0$, the displacement is

$$\Delta x = \lim_{\Delta t_n \to 0} \sum_n v_{xn,\text{avg}} \, \Delta t_n \tag{2.18}$$

If we know the v_x–t graph for motion along a straight line, we can obtain the displacement during any time interval by measuring the area under the curve corresponding to that time interval.

The limit of the sum shown in Equation 2.18 is called a **definite integral** and is written

Definite integral ▶

$$\lim_{\Delta t_n \to 0} \sum_n v_{xn,\text{avg}} \, \Delta t_n = \int_{t_i}^{t_f} v_x(t) \, dt \tag{2.19}$$

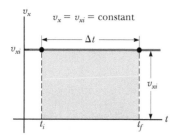

where $v_x(t)$ denotes the velocity at any time t. If the explicit functional form of $v_x(t)$ is known and the limits are given, the integral can be evaluated. Sometimes the v_x–t graph for a moving particle has a shape much simpler than that shown in Figure 2.15. For example, suppose an object is described with the particle under constant velocity model. In this case, the v_x–t graph is a horizontal line as in Figure 2.16 and the displacement of the particle during the time interval Δt is simply the area of the shaded rectangle:

$$\Delta x = v_{xi} \, \Delta t \quad (\text{when } v_x = v_{xi} = \text{constant})$$

Figure 2.16 The velocity–time curve for a particle moving with constant velocity v_{xi}. The displacement of the particle during the time interval $t_f - t_i$ is equal to the area of the shaded rectangle.

Kinematic Equations

We now use the defining equations for acceleration and velocity to derive two of our kinematic equations, Equations 2.13 and 2.16.

The defining equation for acceleration (Eq. 2.10),

$$a_x = \frac{dv_x}{dt}$$

may be written as $dv_x = a_x \, dt$ or, in terms of an integral (or antiderivative), as

$$v_{xf} - v_{xi} = \int_0^t a_x \, dt$$

For the special case in which the acceleration is constant, a_x can be removed from the integral to give

$$v_{xf} - v_{xi} = a_x \int_0^t dt = a_x(t - 0) = a_x t \tag{2.20}$$

which is Equation 2.13 in the particle under constant acceleration model.

Now let us consider the defining equation for velocity (Eq. 2.5):

$$v_x = \frac{dx}{dt}$$

We can write this equation as $dx = v_x \, dt$ or in integral form as

$$x_f - x_i = \int_0^t v_x \, dt$$

Because $v_x = v_{xf} = v_{xi} + a_x t$, this expression becomes

$$x_f - x_i = \int_0^t (v_{xi} + a_x t) \, dt = \int_0^t v_{xi} \, dt + a_x \int_0^t t \, dt = v_{xi}(t - 0) + a_x \left(\frac{t^2}{2} - 0 \right)$$

$$x_f - x_i = v_{xi} t + \tfrac{1}{2} a_x t^2$$

which is Equation 2.16 in the particle under constant acceleration model.

Besides what you might expect to learn about physics concepts, a very valuable skill you should hope to take away from your physics course is the ability to solve complicated problems. The way physicists approach complex situations and break them into manageable pieces is extremely useful. The following is a general problem-solving strategy to guide you through the steps. To help you remember the steps of the strategy, they are *Conceptualize, Categorize, Analyze,* and *Finalize*.

GENERAL PROBLEM-SOLVING STRATEGY

Conceptualize

- The first things to do when approaching a problem are to *think about* and *understand* the situation. Study carefully any representations of the information (for example, diagrams, graphs, tables, or photographs) that accompany the problem. Imagine a movie, running in your mind, of what happens in the problem.

- If a pictorial representation is not provided, you should almost always make a quick drawing of the situation. Indicate any known values, perhaps in a table or directly on your sketch.

- Now focus on what algebraic or numerical information is given in the problem. Carefully read the problem statement, looking for key phrases such as "starts from rest" ($v_i = 0$), "stops" ($v_f = 0$), or "falls freely" ($a_y = -g = -9.80 \text{ m/s}^2$).

- Now focus on the expected result of solving the problem. Exactly what is the question asking? Will the final result be numerical or algebraic? Do you know what units to expect?

- Don't forget to incorporate information from your own experiences and common sense. What should a reasonable answer look like? For example, you wouldn't expect to calculate the speed of an automobile to be 5×10^6 m/s.

Categorize

- Once you have a good idea of what the problem is about, you need to *simplify* the problem. Remove the details that are not important to the solution. For example, model a moving object as a particle. If appropriate, ignore air resistance or friction between a sliding object and a surface.

- Once the problem is simplified, it is important to *categorize* the problem. Is it a simple *substitution problem* such that numbers can be substituted into a simple equation or a definition? If so, the problem is likely to be finished when this substitution is done. If not, you face what we call an *analysis problem:* the situation must be analyzed more deeply to generate an appropriate equation and reach a solution.

- If it is an analysis problem, it needs to be categorized further. Have you seen this type of problem before? Does it fall into the growing list of types of problems that you have solved previously? If so, identify any analysis model(s) appropriate for the problem to prepare for the Analyze step below. We saw the first three analysis models in this chapter: the particle under constant velocity, the particle under constant speed, and the particle under constant acceleration. Being able to classify a problem with an analysis model can make it much easier to lay out a plan to solve it. For example, if your simplification shows that the problem can be treated as a particle under constant acceleration and you have already solved such a problem (such as the examples in Section 2.6), the solution to the present problem follows a similar pattern.

continued

Analyze

- Now you must analyze the problem and strive for a mathematical solution. Because you have already categorized the problem and identified an analysis model, it should not be too difficult to select relevant equations that apply to the type of situation in the problem. For example, if the problem involves a particle under constant acceleration, Equations 2.13 to 2.17 are relevant.

- Use algebra (and calculus, if necessary) to solve symbolically for the unknown variable in terms of what is given. Finally, substitute in the appropriate numbers, calculate the result, and round it to the proper number of significant figures.

Finalize

- Examine your numerical answer. Does it have the correct units? Does it meet your expectations from your conceptualization of the problem? What about the algebraic form of the result? Does it make sense? Examine the variables in the problem to see whether the answer would change in a physically meaningful way if the variables were drastically increased or decreased or even became zero. Looking at limiting cases to see whether they yield expected values is a very useful way to make sure that you are obtaining reasonable results.

- Think about how this problem compared with others you have solved. How was it similar? In what critical ways did it differ? Why was this problem assigned? Can you figure out what you have learned by doing it? If it is a new category of problem, be sure you understand it so that you can use it as a model for solving similar problems in the future.

When solving complex problems, you may need to identify a series of subproblems and apply the problem-solving strategy to each. For simple problems, you probably don't need this strategy. When you are trying to solve a problem and you don't know what to do next, however, remember the steps in the strategy and use them as a guide.

For practice, it would be useful for you to revisit the worked examples in this chapter and identify the *Conceptualize, Categorize, Analyze,* and *Finalize* steps. In the rest of this book, we will label these steps explicitly in the worked examples. Many chapters in this book include a section labeled Problem-Solving Strategy that should help you through the rough spots. These sections are organized according to the General Problem-Solving Strategy outlined above and are tailored to the specific types of problems addressed in that chapter.

To clarify how this Strategy works, we repeat Example 2.7 below with the particular steps of the Strategy identified.

When you **Conceptualize** a problem, try to understand the situation that is presented in the problem statement. Study carefully any representations of the information (for example, diagrams, graphs, tables, or photographs) that accompany the problem. Imagine a movie, running in your mind, of what happens in the problem.

Simplify the problem. Remove the details that are not important to the solution. Then **Categorize** the problem. Is it a simple substitution problem such that numbers can be substituted into a simple equation or a definition? If not, you face an analysis problem. In this case, identify the appropriate analysis model.

Example 2.7 Carrier Landing AM

A jet lands on an aircraft carrier at a speed of 140 mi/h ($\approx$ 63 m/s).

(A) What is its acceleration (assumed constant) if it stops in 2.0 s due to an arresting cable that snags the jet and brings it to a stop?

SOLUTION

Conceptualize
You might have seen movies or television shows in which a jet lands on an aircraft carrier and is brought to rest surprisingly fast by an arresting cable. A careful reading of the problem reveals that in addition to being given the initial speed of 63 m/s, we also know that the final speed is zero.

Categorize
Because the acceleration of the jet is assumed constant, we model it as a *particle under constant acceleration*.

▶ **2.7** continued

....

Analyze

We define our x axis as the direction of motion of the jet. Notice that we have no information about the change in position of the jet while it is slowing down.

Equation 2.13 is the only equation in the particle under constant acceleration model that does not involve position, so we use it to find the acceleration of the jet, modeled as a particle:

$$a_x = \frac{v_{xf} - v_{xi}}{t} = \frac{0 - 63 \text{ m/s}}{2.0 \text{ s}}$$
$$= -32 \text{ m/s}^2$$

(B) If the jet touches down at position $x_i = 0$, what is its final position?

SOLUTION

Use Equation 2.15 to solve for the final position:

$$x_f = x_i + \tfrac{1}{2}(v_{xi} + v_{xf})t = 0 + \tfrac{1}{2}(63 \text{ m/s} + 0)(2.0 \text{ s}) = 63 \text{ m}$$

Finalize

Given the size of aircraft carriers, a length of 63 m seems reasonable for stopping the jet. The idea of using arresting cables to slow down landing aircraft and enable them to land safely on ships originated at about the time of World War I. The cables are still a vital part of the operation of modern aircraft carriers.

WHAT IF?   Suppose the jet lands on the deck of the aircraft carrier with a speed higher than 63 m/s but has the same acceleration due to the cable as that calculated in part (A). How will that change the answer to part (B)?

Answer If the jet is traveling faster at the beginning, it will stop farther away from its starting point, so the answer to part (B) should be larger. Mathematically, we see in Equation 2.15 that if v_{xi} is larger, x_f will be larger.

Now **Analyze** the problem. Select relevant equations from the analysis model. Solve symbolically for the unknown variable in terms of what is given. Substitute in the appropriate numbers, calculate the result, and round it to the proper number of significant figures.

Finalize the problem. Examine the numerical answer. Does it have the correct units? Does it meet your expectations from your conceptualization of the problem? Does the answer make sense? What about the algebraic form of the result? Examine the variables in the problem to see whether the answer would change in a physically meaningful way if the variables were drastically increased or decreased or even became zero.

What If? questions will appear in many examples in the text, and offer a variation on the situation just explored. This feature encourages you to think about the results of the example and assists in conceptual understanding of the principles.

Summary

Definitions

When a particle moves along the x axis from some initial position x_i to some final position x_f, its **displacement** is

$$\Delta x \equiv x_f - x_i \quad (2.1)$$

The **average velocity** of a particle during some time interval is the displacement Δx divided by the time interval Δt during which that displacement occurs:

$$v_{x,\text{avg}} \equiv \frac{\Delta x}{\Delta t} \quad (2.2)$$

The **average speed** of a particle is equal to the ratio of the total distance it travels to the total time interval during which it travels that distance:

$$v_{\text{avg}} \equiv \frac{d}{\Delta t} \quad (2.3)$$

continued

The **instantaneous velocity** of a particle is defined as the limit of the ratio $\Delta x/\Delta t$ as Δt approaches zero. By definition, this limit equals the derivative of x with respect to t, or the time rate of change of the position:

$$v_x \equiv \lim_{\Delta t \to 0} \frac{\Delta x}{\Delta t} = \frac{dx}{dt} \qquad (2.5)$$

The **instantaneous speed** of a particle is equal to the magnitude of its instantaneous velocity.

The **average acceleration** of a particle is defined as the ratio of the change in its velocity Δv_x divided by the time interval Δt during which that change occurs:

$$a_{x,\text{avg}} \equiv \frac{\Delta v_x}{\Delta t} = \frac{v_{xf} - v_{xi}}{t_f - t_i} \qquad (2.9)$$

The **instantaneous acceleration** is equal to the limit of the ratio $\Delta v_x/\Delta t$ as Δt approaches 0. By definition, this limit equals the derivative of v_x with respect to t, or the time rate of change of the velocity:

$$a_x \equiv \lim_{\Delta t \to 0} \frac{\Delta v_x}{\Delta t} = \frac{dv_x}{dt} \qquad (2.10)$$

Concepts and Principles

When an object's velocity and acceleration are in the same direction, the object is speeding up. On the other hand, when the object's velocity and acceleration are in opposite directions, the object is slowing down. Remembering that $F_x \propto a_x$ is a useful way to identify the direction of the acceleration by associating it with a force.

An object falling freely in the presence of the Earth's gravity experiences free-fall acceleration directed toward the center of the Earth. If air resistance is neglected, if the motion occurs near the surface of the Earth, and if the range of the motion is small compared with the Earth's radius, the free-fall acceleration $a_y = -g$ is constant over the range of motion, where g is equal to 9.80 m/s^2.

Complicated problems are best approached in an organized manner. Recall and apply the *Conceptualize, Categorize, Analyze,* and *Finalize* steps of the **General Problem-Solving Strategy** when you need them.

An important aid to problem solving is the use of **analysis models.** Analysis models are situations that we have seen in previous problems. Each analysis model has one or more equations associated with it. When solving a new problem, identify the analysis model that corresponds to the problem. The model will tell you which equations to use. The first three analysis models introduced in this chapter are summarized below.

Analysis Models for Problem–Solving

Particle Under Constant Velocity. If a particle moves in a straight line with a constant speed v_x, its constant velocity is given by

$$v_x = \frac{\Delta x}{\Delta t} \qquad (2.6)$$

and its position is given by

$$x_f = x_i + v_x t \qquad (2.7)$$

Particle Under Constant Speed. If a particle moves a distance d along a curved or straight path with a constant speed, its constant speed is given by

$$v = \frac{d}{\Delta t} \qquad (2.8)$$

Particle Under Constant Acceleration. If a particle moves in a straight line with a constant acceleration a_x, its motion is described by the kinematic equations:

$$v_{xf} = v_{xi} + a_x t \qquad (2.13)$$

$$v_{x,\text{avg}} = \frac{v_{xi} + v_{xf}}{2} \qquad (2.14)$$

$$x_f = x_i + \tfrac{1}{2}(v_{xi} + v_{xf})t \qquad (2.15)$$

$$x_f = x_i + v_{xi}t + \tfrac{1}{2}a_x t^2 \qquad (2.16)$$

$$v_{xf}^2 = v_{xi}^2 + 2a_x(x_f - x_i) \qquad (2.17)$$

1. One drop of oil falls straight down onto the road from the engine of a moving car every 5 s. Figure OQ2.1 shows the pattern of the drops left behind on the pavement. What is the average speed of the car over this section of its motion? (a) 20 m/s (b) 24 m/s (c) 30 m/s (d) 100 m/s (e) 120 m/s

<— 600 m —>

Figure OQ2.1

2. A racing car starts from rest at $t = 0$ and reaches a final speed v at time t. If the acceleration of the car is constant during this time, which of the following statements are true? (a) The car travels a distance vt. (b) The average speed of the car is $v/2$. (c) The magnitude of the acceleration of the car is v/t. (d) The velocity of the car remains constant. (e) None of statements (a) through (d) is true.

3. A juggler throws a bowling pin straight up in the air. After the pin leaves his hand and while it is in the air, which statement is true? (a) The velocity of the pin is always in the same direction as its acceleration. (b) The velocity of the pin is never in the same direction as its acceleration. (c) The acceleration of the pin is zero. (d) The velocity of the pin is opposite its acceleration on the way up. (e) The velocity of the pin is in the same direction as its acceleration on the way up.

4. When applying the equations of kinematics for an object moving in one dimension, which of the following statements *must* be true? (a) The velocity of the object must remain constant. (b) The acceleration of the object must remain constant. (c) The velocity of the object must increase with time. (d) The position of the object must increase with time. (e) The velocity of the object must always be in the same direction as its acceleration.

5. A cannon shell is fired straight up from the ground at an initial speed of 225 m/s. After how much time is the shell at a height of 6.20×10^2 m above the ground and moving downward? (a) 2.96 s (b) 17.3 s (c) 25.4 s (d) 33.6 s (e) 43.0 s

6. An arrow is shot straight up in the air at an initial speed of 15.0 m/s. After how much time is the arrow moving downward at a speed of 8.00 m/s? (a) 0.714 s (b) 1.24 s (c) 1.87 s (d) 2.35 s (e) 3.22 s

7. When the pilot reverses the propeller in a boat moving north, the boat moves with an acceleration directed south. Assume the acceleration of the boat remains constant in magnitude and direction. What happens to the boat? (a) It eventually stops and remains stopped. (b) It eventually stops and then speeds up in the forward direction. (c) It eventually stops and then speeds up in the reverse direction. (d) It never stops but loses speed more and more slowly forever. (e) It never stops but continues to speed up in the forward direction.

8. A rock is thrown downward from the top of a 40.0-m-tall tower with an initial speed of 12 m/s. Assuming negligible air resistance, what is the speed of the rock just before hitting the ground? (a) 28 m/s (b) 30 m/s (c) 56 m/s (d) 784 m/s (e) More information is needed.

9. A skateboarder starts from rest and moves down a hill with constant acceleration in a straight line, traveling for 6 s. In a second trial, he starts from rest and moves along the same straight line with the same acceleration for only 2 s. How does his displacement from his starting point in this second trial compare with that from the first trial? (a) one-third as large (b) three times larger (c) one-ninth as large (d) nine times larger (e) $1/\sqrt{3}$ times as large

10. On another planet, a marble is released from rest at the top of a high cliff. It falls 4.00 m in the first 1 s of its motion. Through what additional distance does it fall in the next 1 s? (a) 4.00 m (b) 8.00 m (c) 12.0 m (d) 16.0 m (e) 20.0 m

11. As an object moves along the x axis, many measurements are made of its position, enough to generate a smooth, accurate graph of x versus t. Which of the following quantities for the object *cannot* be obtained from this graph *alone*? (a) the velocity at any instant (b) the acceleration at any instant (c) the displacement during some time interval (d) the average velocity during some time interval (e) the speed at any instant

12. A pebble is dropped from rest from the top of a tall cliff and falls 4.9 m after 1.0 s has elapsed. How much farther does it drop in the next 2.0 s? (a) 9.8 m (b) 19.6 m (c) 39 m (d) 44 m (e) none of the above

13. A student at the top of a building of height h throws one ball upward with a speed of v_i and then throws a second ball downward with the same initial speed v_i. Just before it reaches the ground, is the final speed of the ball thrown upward (a) larger, (b) smaller, or (c) the same in magnitude, compared with the final speed of the ball thrown downward?

14. You drop a ball from a window located on an upper floor of a building. It strikes the ground with speed v. You now repeat the drop, but your friend down on the ground throws another ball upward at the same speed v, releasing her ball at the same moment that you drop yours from the window. At some location, the balls pass each other. Is this location (a) *at* the halfway point between window and ground, (b) *above* this point, or (c) *below* this point?

15. A pebble is released from rest at a certain height and falls freely, reaching an impact speed of 4 m/s at the floor. Next, the pebble is thrown down with an initial speed of 3 m/s from the same height. What is its speed at the floor? (a) 4 m/s (b) 5 m/s (c) 6 m/s (d) 7 m/s (e) 8 m/s

16. A ball is thrown straight up in the air. For which situation are both the instantaneous velocity and the acceleration zero? (a) on the way up (b) at the top of its flight path (c) on the way down (d) halfway up and halfway down (e) none of the above

17. A hard rubber ball, not affected by air resistance in its motion, is tossed upward from shoulder height, falls to the sidewalk, rebounds to a smaller maximum height, and is caught on its way down again. This motion is represented in Figure OQ2.17, where

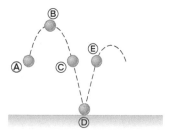

Figure OQ2.17

the successive positions of the ball Ⓐ through Ⓔ are not equally spaced in time. At point Ⓓ the center of the ball is at its lowest point in the motion. The motion of the ball is along a straight, vertical line, but the diagram shows successive positions offset to the right to avoid overlapping. Choose the positive y direction to be upward. (a) Rank the situations Ⓐ through Ⓔ according to the speed of the ball $|v_y|$ at each point, with the largest speed first. (b) Rank the same situations according to the acceleration a_y of the ball at each point. (In both rankings, remember that zero is greater than a negative value. If two values are equal, show that they are equal in your ranking.)

18. Each of the strobe photographs (a), (b), and (c) in Figure OQ2.18 was taken of a single disk moving toward the right, which we take as the positive direction. Within each photograph, the time interval between images is constant. **(i)** Which photograph shows motion with zero acceleration? **(ii)** Which photograph shows motion with positive acceleration? **(iii)** Which photograph shows motion with negative acceleration?

© Cengage Learning/Charles D. Winters

Figure OQ2.18 Objective Question 18 and Problem 23.

Conceptual Questions

1. denotes answer available in *Student Solutions Manual/Study Guide*

1. If the average velocity of an object is zero in some time interval, what can you say about the displacement of the object for that interval?

2. Try the following experiment away from traffic where you can do it safely. With the car you are driving moving slowly on a straight, level road, shift the transmission into neutral and let the car coast. At the moment the car comes to a complete stop, step hard on the brake and notice what you feel. Now repeat the same experiment on a fairly gentle, uphill slope. Explain the difference in what a person riding in the car feels in the two cases. (Brian Popp suggested the idea for this question.)

3. If a car is traveling eastward, can its acceleration be westward? Explain.

4. If the velocity of a particle is zero, can the particle's acceleration be zero? Explain.

5. If the velocity of a particle is nonzero, can the particle's acceleration be zero? Explain.

6. You throw a ball vertically upward so that it leaves the ground with velocity +5.00 m/s. (a) What is its velocity when it reaches its maximum altitude? (b) What is its acceleration at this point? (c) What is the velocity with which it returns to ground level? (d) What is its acceleration at this point?

7. (a) Can the equations of kinematics (Eqs. 2.13–2.17) be used in a situation in which the acceleration varies in time? (b) Can they be used when the acceleration is zero?

8. (a) Can the velocity of an object at an instant of time be greater in magnitude than the average velocity over a time interval containing the instant? (b) Can it be less?

9. Two cars are moving in the same direction in parallel lanes along a highway. At some instant, the velocity of car A exceeds the velocity of car B. Does that mean that the acceleration of car A is greater than that of car B? Explain.

Problems

ENHANCED **WebAssign** The problems found in this chapter may be assigned online in Enhanced WebAssign

1. straightforward; **2.** intermediate; **3.** challenging

1. full solution available in the *Student Solutions Manual/Study Guide*

AMT Analysis Model tutorial available in Enhanced WebAssign

GP Guided Problem

M Master It tutorial available in Enhanced WebAssign

W Watch It video solution available in Enhanced WebAssign

Section 2.1 **Position, Velocity, and Speed**

1. The position versus time for a certain particle moving along the x axis is shown in Figure P2.1. Find the average velocity in the time intervals (a) 0 to 2 s, (b) 0 to 4 s, (c) 2 s to 4 s, (d) 4 s to 7 s, and (e) 0 to 8 s.

Figure P2.1 Problems 1 and 9.

2. The speed of a nerve impulse in the human body is about 100 m/s. If you accidentally stub your toe in the dark, estimate the time it takes the nerve impulse to travel to your brain.

3. A person walks first at a constant speed of 5.00 m/s along a straight line from point Ⓐ to point Ⓑ and then back along the line from Ⓑ to Ⓐ at a constant speed of 3.00 m/s. (a) What is her average speed over the entire trip? (b) What is her average velocity over the entire trip?

4. A particle moves according to the equation $x = 10t^2$, where x is in meters and t is in seconds. (a) Find the average velocity for the time interval from 2.00 s to 3.00 s. (b) Find the average velocity for the time interval from 2.00 to 2.10 s.

5. The position of a pinewood derby car was observed at various times; the results are summarized in the following table. Find the average velocity of the car for (a) the first second, (b) the last 3 s, and (c) the entire period of observation.

t (s)	0	1.0	2.0	3.0	4.0	5.0
x (m)	0	2.3	9.2	20.7	36.8	57.5

Section 2.2 **Instantaneous Velocity and Speed**

6. The position of a particle moving along the x axis varies in time according to the expression $x = 3t^2$, where x is in meters and t is in seconds. Evaluate its position (a) at $t = 3.00$ s and (b) at 3.00 s $+ \Delta t$. (c) Evaluate the limit of $\Delta x/\Delta t$ as Δt approaches zero to find the velocity at $t = 3.00$ s.

7. A position–time graph for a particle moving along the x axis is shown in Figure P2.7. (a) Find the average velocity in the time interval $t = 1.50$ s to $t = 4.00$ s. (b) Determine the instantaneous velocity at $t = 2.00$ s by measuring the slope of the tangent line shown in the graph. (c) At what value of t is the velocity zero?

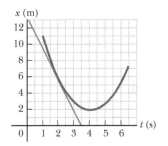

Figure P2.7

8. An athlete leaves one end of a pool of length L at $t = 0$ and arrives at the other end at time t_1. She swims back and arrives at the starting position at time t_2. If she is swimming initially in the positive x direction, determine her average velocities symbolically in (a) the first half of the swim, (b) the second half of the swim, and (c) the round trip. (d) What is her average speed for the round trip?

9. Find the instantaneous velocity of the particle described in Figure P2.1 at the following times: (a) $t = 1.0$ s, (b) $t = 3.0$ s, (c) $t = 4.5$ s, and (d) $t = 7.5$ s.

Section 2.3 **Analysis Model: Particle Under Constant Velocity**

10. Review. The North American and European plates of the Earth's crust are drifting apart with a relative speed of about 25 mm/yr. Take the speed as constant and find when the rift between them started to open, to reach a current width of 2.9×10^3 mi.

11. A hare and a tortoise compete in a race over a straight course 1.00 km long. The tortoise crawls at a speed of 0.200 m/s toward the finish line. The hare runs at a speed of 8.00 m/s toward the finish line for 0.800 km and then stops to tease the slow-moving tortoise as the tortoise eventually passes by. The hare waits for a while after the tortoise passes and then runs toward the finish line again at 8.00 m/s. Both the hare and the tortoise cross the finish line at the exact same instant. Assume both animals, when moving, move steadily at their respective speeds. (a) How far is the tortoise from the finish line when the hare resumes the race? (b) For how long in time was the hare stationary?

12. A car travels along a straight line at a constant speed of 60.0 mi/h for a distance d and then another distance d in the same direction at another constant speed. The average velocity for the entire trip is 30.0 mi/h. (a) What is the constant speed with which the car moved during the second distance d? (b) **What If?** Suppose the second distance d were traveled in the opposite direction; you forgot something and had to return home at the same constant speed as found in part (a). What is the average velocity for this trip? (c) What is the average speed for this new trip?

13. A person takes a trip, driving with a constant speed of 89.5 km/h, except for a 22.0-min rest stop. If the person's average speed is 77.8 km/h, (a) how much time is spent on the trip and (b) how far does the person travel?

Section 2.4 Acceleration

14. **Review.** A 50.0-g Super Ball traveling at 25.0 m/s bounces off a brick wall and rebounds at 22.0 m/s. A high-speed camera records this event. If the ball is in contact with the wall for 3.50 ms, what is the magnitude of the average acceleration of the ball during this time interval?

15. A velocity–time graph for an object moving along the x axis is shown in Figure P2.15. (a) Plot a graph of the acceleration versus time. Determine the average acceleration of the object (b) in the time interval $t = 5.00$ s to $t = 15.0$ s and (c) in the time interval $t = 0$ to $t = 20.0$ s.

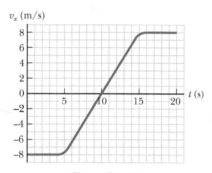

Figure P2.15

16. A child rolls a marble on a bent track that is 100 cm long as shown in Figure P2.16. We use x to represent the position of the marble along the track. On the horizontal sections from $x = 0$ to $x = 20$ cm and from $x = 40$ cm to $x = 60$ cm, the marble rolls with constant speed. On the sloping sections, the marble's speed changes steadily. At the places where the slope changes, the marble stays on the track and does not undergo any sudden changes in speed. The child gives the marble some initial speed at $x = 0$ and $t = 0$ and then watches it roll to $x = 90$ cm, where it turns around, eventually returning to $x = 0$ with the same speed with which the child released it. Prepare graphs of x versus t, v_x versus t, and a_x versus t, vertically aligned with their time axes identical, to show the motion of the marble. You will not be able to place numbers other than zero on the

horizontal axis or on the velocity or acceleration axes, but show the correct graph shapes.

Figure P2.16

17. Figure P2.17 shows a graph of v_x versus t for the motion of a motorcyclist as he starts from rest and moves along the road in a straight line. (a) Find the average acceleration for the time interval $t = 0$ to $t = 6.00$ s. (b) Estimate the time at which the acceleration has its greatest positive value and the value of the acceleration at that instant. (c) When is the acceleration zero? (d) Estimate the maximum negative value of the acceleration and the time at which it occurs.

Figure P2.17

18. (a) Use the data in Problem 5 to construct a smooth graph of position versus time. (b) By constructing tangents to the $x(t)$ curve, find the instantaneous velocity of the car at several instants. (c) Plot the instantaneous velocity versus time and, from this information, determine the average acceleration of the car. (d) What was the initial velocity of the car?

19. A particle starts from rest and accelerates as shown in Figure P2.19. Determine (a) the particle's speed at $t = 10.0$ s and at $t = 20.0$ s, and (b) the distance traveled in the first 20.0 s.

Figure P2.19

20. An object moves along the x axis according to the equation $x = 3.00t^2 - 2.00t + 3.00$, where x is in meters and t is in seconds. Determine (a) the average speed between $t = 2.00$ s and $t = 3.00$ s, (b) the instantaneous speed at $t = 2.00$ s and at $t = 3.00$ s, (c) the average acceleration between $t = 2.00$ s and $t = 3.00$ s, and (d) the instantaneous acceleration at $t = 2.00$ s and $t = 3.00$ s. (e) At what time is the object at rest?

21. A particle moves along the x axis according to the equation $x = 2.00 + 3.00t - 1.00t^2$, where x is in meters and t is in seconds. At $t = 3.00$ s, find (a) the position of the particle, (b) its velocity, and (c) its acceleration.

Section 2.5 Motion Diagrams

22. Draw motion diagrams for (a) an object moving to the right at constant speed, (b) an object moving to the right and speeding up at a constant rate, (c) an object moving to the right and slowing down at a constant rate, (d) an object moving to the left and speeding up at a constant rate, and (e) an object moving to the left and slowing down at a constant rate. (f) How would your drawings change if the changes in speed were not uniform, that is, if the speed were not changing at a constant rate?

23. Each of the strobe photographs (a), (b), and (c) in Figure OQ2.18 was taken of a single disk moving toward the right, which we take as the positive direction. Within each photograph the time interval between images is constant. For each photograph, prepare graphs of x versus t, v_x versus t, and a_x versus t, vertically aligned with their time axes identical, to show the motion of the disk. You will not be able to place numbers other than zero on the axes, but show the correct shapes for the graph lines.

Section 2.6 Analysis Model: Particle Under Constant Acceleration

24. The minimum distance required to stop a car moving at 35.0 mi/h is 40.0 ft. What is the minimum stopping distance for the same car moving at 70.0 mi/h, assuming the same rate of acceleration?

25. An electron in a cathode-ray tube accelerates uniformly from 2.00×10^4 m/s to 6.00×10^6 m/s over 1.50 cm. (a) In what time interval does the electron travel this 1.50 cm? (b) What is its acceleration?

26. A speedboat moving at 30.0 m/s approaches a no-wake buoy marker 100 m ahead. The pilot slows the boat with a constant acceleration of -3.50 m/s² by reducing the throttle. (a) How long does it take the boat to reach the buoy? (b) What is the velocity of the boat when it reaches the buoy?

27. A parcel of air moving in a straight tube with a constant acceleration of -4.00 m/s² has a velocity of 13.0 m/s at 10:05:00 a.m. (a) What is its velocity at 10:05:01 a.m.? (b) At 10:05:04 a.m.? (c) At 10:04:59 a.m.? (d) Describe the shape of a graph of velocity versus time for this parcel of air. (e) Argue for or against the following statement: "Knowing the single value of an object's constant acceleration is like knowing a whole list of values for its velocity."

28. A truck covers 40.0 m in 8.50 s while smoothly slowing down to a final speed of 2.80 m/s. (a) Find its original speed. (b) Find its acceleration.

29. An object moving with uniform acceleration has a velocity of 12.0 cm/s in the positive x direction when its x coordinate is 3.00 cm. If its x coordinate 2.00 s later is -5.00 cm, what is its acceleration?

30. In Example 2.7, we investigated a jet landing on an aircraft carrier. In a later maneuver, the jet comes in for a landing on solid ground with a speed of 100 m/s, and its acceleration can have a maximum magnitude of 5.00 m/s² as it comes to rest. (a) From the instant the jet touches the runway, what is the minimum time interval needed before it can come to rest? (b) Can this jet land at a small tropical island airport where the runway is 0.800 km long? (c) Explain your answer.

31. **Review.** Colonel John P. Stapp, USAF, participated in studying whether a jet pilot could survive emergency ejection. On March 19, 1954, he rode a rocket-propelled sled that moved down a track at a speed of 632 mi/h. He and the sled were safely brought to rest in 1.40 s (Fig. P2.31). Determine (a) the negative acceleration he experienced and (b) the distance he traveled during this negative acceleration.

left, Courtesy U.S. Air Force; right, NASA/Photo Researchers, Inc.

Figure P2.31 (*left*) Col. John Stapp and his rocket sled are brought to rest in a very short time interval. (*right*) Stapp's face is contorted by the stress of rapid negative acceleration.

32. Solve Example 2.8 by a graphical method. On the same graph, plot position versus time for the car and the trooper. From the intersection of the two curves, read the time at which the trooper overtakes the car.

33. A truck on a straight road starts from rest, accelerating at 2.00 m/s² until it reaches a speed of 20.0 m/s. Then the truck travels for 20.0 s at constant speed until the brakes are applied, stopping the truck in a uniform manner in an additional 5.00 s. (a) How long is the truck in motion? (b) What is the average velocity of the truck for the motion described?

34. *Why is the following situation impossible?* Starting from rest, a charging rhinoceros moves 50.0 m in a straight line in 10.0 s. Her acceleration is constant during the entire motion, and her final speed is 8.00 m/s.

35. The driver of a car slams on the brakes when he sees a tree blocking the road. The car slows uniformly with an acceleration of -5.60 m/s² for 4.20 s, making straight skid marks 62.4 m long, all the way to the tree. With what speed does the car then strike the tree?

36. In the particle under constant acceleration model, we identify the variables and parameters v_{xi}, v_{xf}, a_x, t, and $x_f - x_i$. Of the equations in the model, Equations 2.13–2.17, the first does not involve $x_f - x_i$, the second and third do not contain a_x, the fourth omits v_{xf}, and the last leaves out t. So, to complete the set, there should be an equation *not* involving v_{xi}. (a) Derive it from the others. (b) Use the equation in part (a) to solve Problem 35 in one step.

37. A speedboat travels in a straight line and increases in speed uniformly from $v_i = 20.0$ m/s to $v_f = 30.0$ m/s in a displacement Δx of 200 m. We wish to find the time interval required for the boat to move through this

displacement. (a) Draw a coordinate system for this situation. (b) What analysis model is most appropriate for describing this situation? (c) From the analysis model, what equation is most appropriate for finding the acceleration of the speedboat? (d) Solve the equation selected in part (c) symbolically for the boat's acceleration in terms of v_i, v_f, and Δx. (e) Substitute numerical values to obtain the acceleration numerically. (f) Find the time interval mentioned above.

38. A particle moves along the x axis. Its position is given [W] by the equation $x = 2 + 3t - 4t^2$, with x in meters and t in seconds. Determine (a) its position when it changes direction and (b) its velocity when it returns to the position it had at $t = 0$.

39. A glider of length ℓ moves through a stationary photogate on an air track. A photogate (Fig. P2.39) is a device that measures the time interval Δt_d during which the glider blocks a beam of infrared light passing across the photogate. The ratio $v_d = \ell/\Delta t_d$ is the average velocity of the glider over this part of its motion. Suppose the glider moves with constant acceleration. (a) Argue for or against the idea that v_d is equal to the instantaneous velocity of the glider when it is halfway through the photogate in space. (b) Argue for or against the idea that v_d is equal to the instantaneous velocity of the glider when it is halfway through the photogate in time.

Figure P2.39 Problems 39 and 40.

40. A glider of length 12.4 cm moves on an air track with constant acceleration (Fig P2.39). A time interval of 0.628 s elapses between the moment when its front end passes a fixed point Ⓐ along the track and the moment when its back end passes this point. Next, a time interval of 1.39 s elapses between the moment when the back end of the glider passes the point Ⓐ and the moment when the front end of the glider passes a second point Ⓑ farther down the track. After that, an additional 0.431 s elapses until the back end of the glider passes point Ⓑ. (a) Find the average speed of the glider as it passes point Ⓐ. (b) Find the acceleration of the glider. (c) Explain how you can compute the acceleration without knowing the distance between points Ⓐ and Ⓑ.

41. An object moves with constant acceleration 4.00 m/s² and over a time interval reaches a final velocity of 12.0 m/s. (a) If its initial velocity is 6.00 m/s, what is its displacement during the time interval? (b) What is the distance it travels during this interval? (c) If its initial velocity is −6.00 m/s, what is its displacement during

the time interval? (d) What is the total distance it travels during the interval in part (c)?

42. At $t = 0$, one toy car is set rolling on a straight track with initial position 15.0 cm, initial velocity −3.50 cm/s, and constant acceleration 2.40 cm/s². At the same moment, another toy car is set rolling on an adjacent track with initial position 10.0 cm, initial velocity +5.50 cm/s, and constant acceleration zero. (a) At what time, if any, do the two cars have equal speeds? (b) What are their speeds at that time? (c) At what time(s), if any, do the cars pass each other? (d) What are their locations at that time? (e) Explain the difference between question (a) and question (c) as clearly as possible.

43. Figure P2.43 represents part of the performance data of a car owned by a proud physics student. (a) Calculate the total distance traveled by computing the area under the red-brown graph line. (b) What distance does the car travel between the times $t = 10$ s and $t = 40$ s? (c) Draw a graph of its acceleration versus time between $t = 0$ and $t = 50$ s. (d) Write an equation for x as a function of time for each phase of the motion, represented by the segments $0a$, ab, and bc. (e) What is the average velocity of the car between $t = 0$ and $t = 50$ s?

Figure P2.43

44. A hockey player is standing on his skates on a frozen [M] pond when an opposing player, moving with a uniform speed of 12.0 m/s, skates by with the puck. After 3.00 s, the first player makes up his mind to chase his opponent. If he accelerates uniformly at 4.00 m/s², (a) how long does it take him to catch his opponent and (b) how far has he traveled in that time? (Assume the player with the puck remains in motion at constant speed.)

Section 2.7 Freely Falling Objects

Note: In all problems in this section, ignore the effects of air resistance.

45. In Chapter 9, we will define the center of mass of an object and prove that its motion is described by the particle under constant acceleration model when constant forces act on the object. A gymnast jumps straight up, with her center of mass moving at 2.80 m/s as she leaves the ground. How high above this point is her center of mass (a) 0.100 s, (b) 0.200 s, (c) 0.300 s, and (d) 0.500 s thereafter?

46. An attacker at the base of a castle wall 3.65 m high throws a rock straight up with speed 7.40 m/s from a height of 1.55 m above the ground. (a) Will the rock reach the top of the wall? (b) If so, what is its speed at the top? If not, what initial speed must it have to reach the top? (c) Find the change in speed of a rock thrown straight down from the top of the wall at an initial speed of 7.40 m/s and moving between the same two

points. (d) Does the change in speed of the downward-moving rock agree with the magnitude of the speed change of the rock moving upward between the same elevations? (e) Explain physically why it does or does not agree.

47. *Why is the following situation impossible?* Emily challenges David to catch a $1 bill as follows. She holds the bill vertically as shown in Figure P2.47, with the center of the bill between but not touching David's index finger and thumb. Without warning, Emily releases the bill. David catches the bill without moving his hand downward. David's reaction time is equal to the average human reaction time.

Figure P2.47

48. A baseball is hit so that it travels straight upward after being struck by the bat. A fan observes that it takes 3.00 s for the ball to reach its maximum height. Find (a) the ball's initial velocity and (b) the height it reaches.

49. It is possible to shoot an arrow at a speed as high as 100 m/s. (a) If friction can be ignored, how high would an arrow launched at this speed rise if shot straight up? (b) How long would the arrow be in the air?

50. The height of a helicopter above the ground is given by $h = 3.00t^3$, where h is in meters and t is in seconds. At $t = 2.00$ s, the helicopter releases a small mailbag. How long after its release does the mailbag reach the ground?

51. A ball is thrown directly downward with an initial speed of 8.00 m/s from a height of 30.0 m. After what time interval does it strike the ground?

52. A ball is thrown upward from the ground with an initial speed of 25 m/s; at the same instant, another ball is dropped from a building 15 m high. After how long will the balls be at the same height above the ground?

53. A student throws a set of keys vertically upward to her sorority sister, who is in a window 4.00 m above. The second student catches the keys 1.50 s later. (a) With what initial velocity were the keys thrown? (b) What was the velocity of the keys just before they were caught?

54. At time $t = 0$, a student throws a set of keys vertically upward to her sorority sister, who is in a window at distance h above. The second student catches the keys at time t. (a) With what initial velocity were the keys thrown? (b) What was the velocity of the keys just before they were caught?

55. A daring ranch hand sitting on a tree limb wishes to drop vertically onto a horse galloping under the tree. The constant speed of the horse is 10.0 m/s, and the distance from the limb to the level of the saddle is 3.00 m. (a) What must be the horizontal distance between the saddle and limb when the ranch hand makes his move? (b) For what time interval is he in the air?

56. A package is dropped at time $t = 0$ from a helicopter that is descending steadily at a speed v_i. (a) What is the speed of the package in terms of v_i, g, and t? (b) What vertical distance d is it from the helicopter in terms of g and t? (c) What are the answers to parts (a) and (b) if the helicopter is rising steadily at the same speed?

Section 2.8 Kinematic Equations Derived from Calculus

57. Automotive engineers refer to the time rate of change of acceleration as the "jerk." Assume an object moves in one dimension such that its jerk J is constant. (a) Determine expressions for its acceleration $a_x(t)$, velocity $v_x(t)$, and position $x(t)$, given that its initial acceleration, velocity, and position are a_{xi}, v_{xi}, and x_i, respectively. (b) Show that $a_x^2 = a_{xi}^2 + 2J(v_x - v_{xi})$.

58. A student drives a moped along a straight road as described by the velocity–time graph in Figure P2.58. Sketch this graph in the middle of a sheet of graph paper. (a) Directly above your graph, sketch a graph of the position versus time, aligning the time coordinates of the two graphs. (b) Sketch a graph of the acceleration versus time directly below the velocity–time graph, again aligning the time coordinates. On each graph, show the numerical values of x and a_x for all points of inflection. (c) What is the acceleration at $t = 6.00$ s? (d) Find the position (relative to the starting point) at $t = 6.00$ s. (e) What is the moped's final position at $t = 9.00$ s?

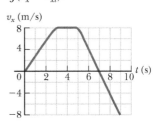

Figure P2.58

59. The speed of a bullet as it travels down the barrel of a rifle toward the opening is given by

$$v = (-5.00 \times 10^7)t^2 + (3.00 \times 10^5)t$$

where v is in meters per second and t is in seconds. The acceleration of the bullet just as it leaves the barrel is zero. (a) Determine the acceleration and position of the bullet as functions of time when the bullet is in the barrel. (b) Determine the time interval over which the bullet is accelerated. (c) Find the speed at which the bullet leaves the barrel. (d) What is the length of the barrel?

Additional Problems

60. A certain automobile manufacturer claims that its deluxe sports car will accelerate from rest to a speed of 42.0 m/s in 8.00 s. (a) Determine the average acceleration of the car. (b) Assume that the car moves with constant acceleration. Find the distance the car travels in the first 8.00 s. (c) What is the speed of the car 10.0 s after it begins its motion if it can continue to move with the same acceleration?

61. The froghopper *Philaenus spumarius* is supposedly the best jumper in the animal kingdom. To start a jump, this insect can accelerate at 4.00 km/s² over a distance of 2.00 mm as it straightens its specially adapted

"jumping legs." Assume the acceleration is constant. (a) Find the upward velocity with which the insect takes off. (b) In what time interval does it reach this velocity? (c) How high would the insect jump if air resistance were negligible? The actual height it reaches is about 70 cm, so air resistance must be a noticeable force on the leaping froghopper.

62. An object is at $x = 0$ at $t = 0$ and moves along the x axis according to the velocity–time graph in Figure P2.62. (a) What is the object's acceleration between 0 and 4.0 s? (b) What is the object's acceleration between 4.0 s and 9.0 s? (c) What is the object's acceleration between 13.0 s and 18.0 s? (d) At what time(s) is the object moving with the lowest speed? (e) At what time is the object farthest from $x = 0$? (f) What is the final position x of the object at $t = 18.0$ s? (g) Through what total distance has the object moved between $t = 0$ and $t = 18.0$ s?

Figure P2.62

63. An inquisitive physics student and mountain climber climbs a 50.0-m-high cliff that overhangs a calm pool of water. He throws two stones vertically downward, 1.00 s apart, and observes that they cause a single splash. The first stone has an initial speed of 2.00 m/s. (a) How long after release of the first stone do the two stones hit the water? (b) What initial velocity must the second stone have if the two stones are to hit the water simultaneously? (c) What is the speed of each stone at the instant the two stones hit the water?

64. In Figure 2.11b, the area under the velocity–time graph and between the vertical axis and time t (vertical dashed line) represents the displacement. As shown, this area consists of a rectangle and a triangle. (a) Compute their areas. (b) Explain how the sum of the two areas compares with the expression on the right-hand side of Equation 2.16.

65. A ball starts from rest and accelerates at 0.500 m/s² while moving down an inclined plane 9.00 m long. When it reaches the bottom, the ball rolls up another plane, where it comes to rest after moving 15.0 m on that plane. (a) What is the speed of the ball at the bottom of the first plane? (b) During what time interval does the ball roll down the first plane? (c) What is the acceleration along the second plane? (d) What is the ball's speed 8.00 m along the second plane?

66. A woman is reported to have fallen 144 ft from the 17th floor of a building, landing on a metal ventilator box that she crushed to a depth of 18.0 in. She suffered only minor injuries. Ignoring air resistance, calculate

(a) the speed of the woman just before she collided with the ventilator and (b) her average acceleration while in contact with the box. (c) Modeling her acceleration as constant, calculate the time interval it took to crush the box.

67. An elevator moves downward in a tall building at a constant speed of 5.00 m/s. Exactly 5.00 s after the top of the elevator car passes a bolt loosely attached to the wall of the elevator shaft, the bolt falls from rest. (a) At what time does the bolt hit the top of the still-descending elevator? (b) In what way is this problem similar to Example 2.8? (c) Estimate the highest floor from which the bolt can fall if the elevator reaches the ground floor before the bolt hits the top of the elevator.

68. *Why is the following situation impossible?* A freight train is lumbering along at a constant speed of 16.0 m/s. Behind the freight train on the same track is a passenger train traveling in the same direction at 40.0 m/s. When the front of the passenger train is 58.5 m from the back of the freight train, the engineer on the passenger train recognizes the danger and hits the brakes of his train, causing the train to move with acceleration -3.00 m/s². Because of the engineer's action, the trains do not collide.

69. The Acela is an electric train on the Washington–New York–Boston run, carrying passengers at 170 mi/h. A velocity–time graph for the Acela is shown in Figure P2.69. (a) Describe the train's motion in each successive time interval. (b) Find the train's peak positive acceleration in the motion graphed. (c) Find the train's displacement in miles between $t = 0$ and $t = 200$ s.

Figure P2.69 Velocity–time graph for the Acela.

70. Two objects move with initial velocity -8.00 m/s, final velocity 16.0 m/s, and constant accelerations. (a) The first object has displacement 20.0 m. Find its acceleration. (b) The second object travels a total distance of 22.0 m. Find its acceleration.

71. At $t = 0$, one athlete in a race running on a long, straight track with a constant speed v_1 is a distance d_1 behind a second athlete running with a constant speed v_2. (a) Under what circumstances is the first athlete able to overtake the second athlete? (b) Find the time t at which the first athlete overtakes the second athlete, in terms of d_1, v_1, and v_2. (c) At what minimum distance d_2 from the leading athlete must the finish line

be located so that the trailing athlete can at least tie for first place? Express d_2 in terms of d_1, v_1, and v_2 by using the result of part (b).

72. A catapult launches a test rocket vertically upward from a well, giving the rocket an initial speed of 80.0 m/s at ground level. The engines then fire, and the rocket accelerates upward at 4.00 m/s^2 until it reaches an altitude of 1 000 m. At that point, its engines fail and the rocket goes into free fall, with an acceleration of -9.80 m/s^2. (a) For what time interval is the rocket in motion above the ground? (b) What is its maximum altitude? (c) What is its velocity just before it hits the ground? (You will need to consider the motion while the engine is operating and the free-fall motion separately.)

73. Kathy tests her new sports car by racing with Stan, an experienced racer. Both start from rest, but Kathy leaves the starting line 1.00 s after Stan does. Stan moves with a constant acceleration of 3.50 m/s^2, while Kathy maintains an acceleration of 4.90 m/s^2. Find (a) the time at which Kathy overtakes Stan, (b) the distance she travels before she catches him, and (c) the speeds of both cars at the instant Kathy overtakes Stan.

74. Two students are on a balcony a distance h above the street. One student throws a ball vertically downward at a speed v_i; at the same time, the other student throws a ball vertically upward at the same speed. Answer the following symbolically in terms of v_i, g, h, and t. (a) What is the time interval between when the first ball strikes the ground and the second ball strikes the ground? (b) Find the velocity of each ball as it strikes the ground. (c) How far apart are the balls at a time t after they are thrown and before they strike the ground?

75. Two objects, A and B, are connected by hinges to a rigid rod that has a length L. The objects slide along perpendicular guide rails as shown in Figure P2.75. Assume object A slides to the left with a constant speed v. (a) Find the velocity v_B of object B as a function of the angle θ. (b) Describe v_B relative to v. Is v_B always smaller than v, larger than v, or the same as v, or does it have some other relationship?

Figure P2.75

76. Astronauts on a distant planet toss a rock into the air. With the aid of a camera that takes pictures at a steady rate, they record the rock's height as a function of time as given in the following table. (a) Find the rock's average velocity in the time interval between each measurement and the next. (b) Using these average velocities to approximate instantaneous velocities at the midpoints of the time intervals, make a graph of velocity as a function of time. (c) Does the rock move with constant acceleration? If so, plot a straight line of best fit on the graph and calculate its slope to find the acceleration.

Time (s)	Height (m)	Time (s)	Height (m)
0.00	5.00	2.75	7.62
0.25	5.75	3.00	7.25
0.50	6.40	3.25	6.77
0.75	6.94	3.50	6.20
1.00	7.38	3.75	5.52
1.25	7.72	4.00	4.73
1.50	7.96	4.25	3.85
1.75	8.10	4.50	2.86
2.00	8.13	4.75	1.77
2.25	8.07	5.00	0.58
2.50	7.90		

77. A motorist drives along a straight road at a constant speed of 15.0 m/s. Just as she passes a parked motorcycle police officer, the officer starts to accelerate at 2.00 m/s^2 to overtake her. Assuming that the officer maintains this acceleration, (a) determine the time interval required for the police officer to reach the motorist. Find (b) the speed and (c) the total displacement of the officer as he overtakes the motorist.

78. A commuter train travels between two downtown stations. Because the stations are only 1.00 km apart, the train never reaches its maximum possible cruising speed. During rush hour the engineer minimizes the time interval Δt between two stations by accelerating at a rate $a_1 = 0.100$ m/s^2 for a time interval Δt_1 and then immediately braking with acceleration $a_2 = -0.500$ m/s^2 for a time interval Δt_2. Find the minimum time interval of travel Δt and the time interval Δt_1.

79. Liz rushes down onto a subway platform to find her train already departing. She stops and watches the cars go by. Each car is 8.60 m long. The first moves past her in 1.50 s and the second in 1.10 s. Find the constant acceleration of the train.

80. A hard rubber ball, released at chest height, falls to the pavement and bounces back to nearly the same height. When it is in contact with the pavement, the lower side of the ball is temporarily flattened. Suppose the maximum depth of the dent is on the order of 1 cm. Find the order of magnitude of the maximum acceleration of the ball while it is in contact with the pavement. State your assumptions, the quantities you estimate, and the values you estimate for them.

Challenge Problems

81. A blue car of length 4.52 m is moving north on a roadway that intersects another perpendicular roadway (Fig. P2.81, page 58). The width of the intersection from near edge to far edge is 28.0 m. The blue car has a constant acceleration of magnitude 2.10 m/s^2 directed south. The time interval required for the nose of the blue car to move from the near (south) edge of the intersection to the north edge of the intersection is 3.10 s. (a) How far is the nose of the blue car from the south edge of the intersection when it stops? (b) For what time interval is *any* part of the blue car within the boundaries of the intersection? (c) A red car is at rest on the perpendicular intersecting roadway. As the nose of the blue car

enters the intersection, the red car starts from rest and accelerates east at 5.60 m/s². What is the minimum distance from the near (west) edge of the intersection at which the nose of the red car can begin its motion if it is to enter the intersection after the blue car has entirely left the intersection? (d) If the red car begins its motion at the position given by the answer to part (c), with what speed does it enter the intersection?

Figure P2.81

82. **Review.** As soon as a traffic light turns green, a car speeds up from rest to 50.0 mi/h with constant acceleration 9.00 mi/h/s. In the adjoining bicycle lane, a cyclist speeds up from rest to 20.0 mi/h with constant acceleration 13.0 mi/h/s. Each vehicle maintains constant velocity after reaching its cruising speed. (a) For what time interval is the bicycle ahead of the car? (b) By what maximum distance does the bicycle lead the car?

83. In a women's 100-m race, accelerating uniformly, Laura takes 2.00 s and Healan 3.00 s to attain their maximum speeds, which they each maintain for the rest of the race. They cross the finish line simultaneously, both setting a world record of 10.4 s. (a) What is the acceleration of each sprinter? (b) What are their respective maximum speeds? (c) Which sprinter is ahead at the 6.00-s mark, and by how much? (d) What is the maximum distance by which Healan is behind Laura, and at what time does that occur?

84. Two thin rods are fastened to the inside of a circular ring as shown in Figure P2.84. One rod of length D is vertical, and the other of length L makes an angle θ with the horizontal. The two rods and the ring lie in a vertical plane. Two small beads are free to slide without friction along the rods. (a) If the two beads are released from rest simultaneously from the positions shown, use your intuition and guess which bead reaches the bottom first. (b) Find an expression for the time interval required for the red bead to fall from point Ⓐ to point Ⓒ in terms of g and D. (c) Find an expression for the time interval required for the blue bead to slide from point Ⓑ to point Ⓒ in terms of g, L, and θ. (d) Show that the two time intervals found in parts (b) and (c) are equal. *Hint:* What is the angle between the chords of the circle Ⓐ Ⓑ and Ⓑ Ⓒ? (e) Do these results surprise you? Was your intuitive guess in part (a) correct? This problem was inspired by an article by Thomas B. Greenslade, Jr., "Galileo's Paradox," *Phys. Teach.* **46**, 294 (May 2008).

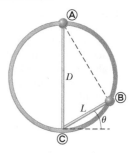

Figure P2.84

85. A man drops a rock into a well. (a) The man hears the sound of the splash 2.40 s after he releases the rock from rest. The speed of sound in air (at the ambient temperature) is 336 m/s. How far below the top of the well is the surface of the water? (b) **What If?** If the travel time for the sound is ignored, what percentage error is introduced when the depth of the well is calculated?

Vectors

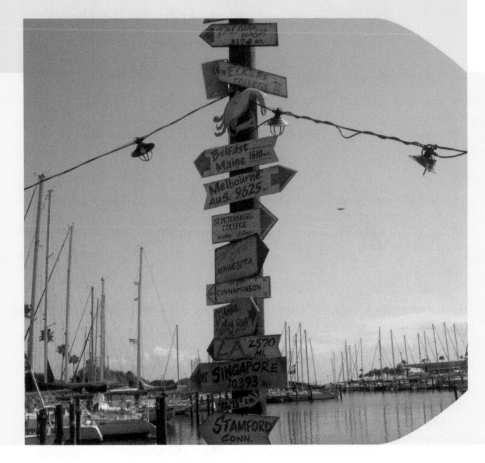

In our study of physics, we often need to work with physical quantities that have both numerical and directional properties. As noted in Section 2.1, quantities of this nature are vector quantities. This chapter is primarily concerned with general properties of vector quantities. We discuss the addition and subtraction of vector quantities, together with some common applications to physical situations.

Vector quantities are used throughout this text. Therefore, it is imperative that you master the techniques discussed in this chapter.

A signpost in Saint Petersburg, Florida, shows the distance and direction to several cities. Quantities that are defined by both a magnitude and a direction are called vector quantities. *(Raymond A. Serway)*

3.1 Coordinate Systems

Many aspects of physics involve a description of a location in space. In Chapter 2, for example, we saw that the mathematical description of an object's motion requires a method for describing the object's position at various times. In two dimensions, this description is accomplished with the use of the Cartesian coordinate system, in which perpendicular axes intersect at a point defined as the origin O (Fig. 3.1). Cartesian coordinates are also called *rectangular coordinates*.

Sometimes it is more convenient to represent a point in a plane by its *plane polar coordinates* (r, θ) as shown in Figure 3.2a (page 60). In this *polar coordinate system*, r is the distance from the origin to the point having Cartesian coordinates (x, y) and θ is the angle between a fixed axis and a line drawn from the origin to the point. The fixed axis is often the positive x axis, and θ is usually measured counterclockwise

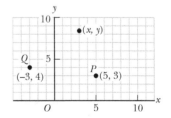

Figure 3.1 Designation of points in a Cartesian coordinate system. Every point is labeled with coordinates (x, y).

59

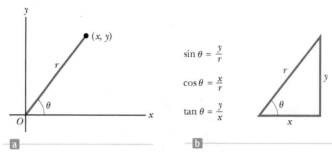

Figure 3.2 (a) The plane polar coordinates of a point are represented by the distance r and the angle θ, where θ is measured counterclockwise from the positive x axis. (b) The right triangle used to relate (x, y) to (r, θ).

from it. From the right triangle in Figure 3.2b, we find that $\sin \theta = y/r$ and that $\cos \theta = x/r$. (A review of trigonometric functions is given in Appendix B.4.) Therefore, starting with the plane polar coordinates of any point, we can obtain the Cartesian coordinates by using the equations

Cartesian coordinates ▶
in terms of polar
coordinates

$$x = r\cos \theta \qquad (3.1)$$
$$y = r\sin \theta \qquad (3.2)$$

Furthermore, if we know the Cartesian coordinates, the definitions of trigonometry tell us that

Polar coordinates in terms ▶
of Cartesian coordinates

$$\tan \theta = \frac{y}{x} \qquad (3.3)$$

$$r = \sqrt{x^2 + y^2} \qquad (3.4)$$

Equation 3.4 is the familiar Pythagorean theorem.

These four expressions relating the coordinates (x, y) to the coordinates (r, θ) apply only when θ is defined as shown in Figure 3.2a—in other words, when positive θ is an angle measured counterclockwise from the positive x axis. (Some scientific calculators perform conversions between Cartesian and polar coordinates based on these standard conventions.) If the reference axis for the polar angle θ is chosen to be one other than the positive x axis or if the sense of increasing θ is chosen differently, the expressions relating the two sets of coordinates will change.

Example 3.1   Polar Coordinates

The Cartesian coordinates of a point in the xy plane are $(x, y) = (-3.50, -2.50)$ m as shown in Figure 3.3. Find the polar coordinates of this point.

SOLUTION

Conceptualize The drawing in Figure 3.3 helps us conceptualize the problem. We wish to find r and θ. We expect r to be a few meters and θ to be larger than 180°.

Categorize Based on the statement of the problem and the Conceptualize step, we recognize that we are simply converting from Cartesian coordinates to polar coordinates. We therefore categorize this example as a substitution problem. Substitution problems generally do not have an extensive Analyze step other than the substitution of numbers into a given equation. Similarly, the Finalize step

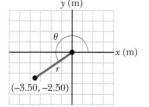

Figure 3.3 (Example 3.1) Finding polar coordinates when Cartesian coordinates are given.

▶ **3.1** continued

consists primarily of checking the units and making sure that the answer is reasonable and consistent with our expectations. Therefore, for substitution problems, we will not label Analyze or Finalize steps.

Use Equation 3.4 to find r:

$$r = \sqrt{x^2 + y^2} = \sqrt{(-3.50 \text{ m})^2 + (-2.50 \text{ m})^2} = \boxed{4.30 \text{ m}}$$

Use Equation 3.3 to find θ:

$$\tan \theta = \frac{y}{x} = \frac{-2.50 \text{ m}}{-3.50 \text{ m}} = 0.714$$

$$\theta = \boxed{216°}$$

Notice that you must use the signs of x and y to find that the point lies in the third quadrant of the coordinate system. That is, $\theta = 216°$, not $35.5°$, whose tangent is also 0.714. Both answers agree with our expectations in the Conceptualize step.

3.2 Vector and Scalar Quantities

We now formally describe the difference between scalar quantities and vector quantities. When you want to know the temperature outside so that you will know how to dress, the only information you need is a number and the unit "degrees C" or "degrees F." Temperature is therefore an example of a *scalar quantity:*

> A **scalar quantity** is completely specified by a single value with an appropriate unit and has no direction.

Other examples of scalar quantities are volume, mass, speed, time, and time intervals. Some scalars are always positive, such as mass and speed. Others, such as temperature, can have either positive or negative values. The rules of ordinary arithmetic are used to manipulate scalar quantities.

If you are preparing to pilot a small plane and need to know the wind velocity, you must know both the speed of the wind and its direction. Because direction is important for its complete specification, velocity is a *vector quantity:*

> A **vector quantity** is completely specified by a number with an appropriate unit (the *magnitude* of the vector) plus a direction.

Another example of a vector quantity is displacement, as you know from Chapter 2. Suppose a particle moves from some point Ⓐ to some point Ⓑ along a straight path as shown in Figure 3.4. We represent this displacement by drawing an arrow from Ⓐ to Ⓑ, with the tip of the arrow pointing away from the starting point. The direction of the arrowhead represents the direction of the displacement, and the length of the arrow represents the magnitude of the displacement. If the particle travels along some other path from Ⓐ to Ⓑ such as shown by the broken line in Figure 3.4, its displacement is still the arrow drawn from Ⓐ to Ⓑ. Displacement depends only on the initial and final positions, so the displacement vector is independent of the path taken by the particle between these two points.

In this text, we use a boldface letter with an arrow over the letter, such as $\vec{\mathbf{A}}$, to represent a vector. Another common notation for vectors with which you should be familiar is a simple boldface character: **A**. The magnitude of the vector $\vec{\mathbf{A}}$ is written either A or $|\vec{\mathbf{A}}|$. The magnitude of a vector has physical units, such as meters for displacement or meters per second for velocity. The magnitude of a vector is *always* a positive number.

Figure 3.4 As a particle moves from Ⓐ to Ⓑ along an arbitrary path represented by the broken line, its displacement is a vector quantity shown by the arrow drawn from Ⓐ to Ⓑ.

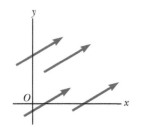

Figure 3.5 These four vectors are equal because they have equal lengths and point in the same direction.

Ⓠuick Quiz 3.1 Which of the following are vector quantities and which are scalar quantities? **(a)** your age **(b)** acceleration **(c)** velocity **(d)** speed **(e)** mass

3.3 Some Properties of Vectors

In this section, we shall investigate general properties of vectors representing physical quantities. We also discuss how to add and subtract vectors using both algebraic and geometric methods.

Equality of Two Vectors

For many purposes, two vectors $\vec{A}$ and $\vec{B}$ may be defined to be equal if they have the same magnitude and if they point in the same direction. That is, $\vec{A} = \vec{B}$ only if $A = B$ and if $\vec{A}$ and $\vec{B}$ point in the same direction along parallel lines. For example, all the vectors in Figure 3.5 are equal even though they have different starting points. This property allows us to move a vector to a position parallel to itself in a diagram without affecting the vector.

Adding Vectors

The rules for adding vectors are conveniently described by a graphical method. To add vector $\vec{B}$ to vector $\vec{A}$, first draw vector $\vec{A}$ on graph paper, with its magnitude represented by a convenient length scale, and then draw vector $\vec{B}$ to the same scale, with its tail starting from the tip of $\vec{A}$, as shown in Figure 3.6. The **resultant vector** $\vec{R} = \vec{A} + \vec{B}$ is the vector drawn from the tail of $\vec{A}$ to the tip of $\vec{B}$.

A geometric construction can also be used to add more than two vectors as shown in Figure 3.7 for the case of four vectors. The resultant vector $\vec{R} = \vec{A} + \vec{B} + \vec{C} + \vec{D}$ is the vector that completes the polygon. In other words, $\vec{R}$ is the vector drawn from the tail of the first vector to the tip of the last vector. This technique for adding vectors is often called the "head to tail method."

When two vectors are added, the sum is independent of the order of the addition. (This fact may seem trivial, but as you will see in Chapter 11, the order is important when vectors are multiplied. Procedures for multiplying vectors are discussed in Chapters 7 and 11.) This property, which can be seen from the geometric construction in Figure 3.8, is known as the **commutative law of addition:**

◀ Commutative law of addition

$$\vec{A} + \vec{B} = \vec{B} + \vec{A} \tag{3.5}$$

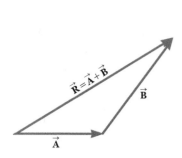

Figure 3.6 When vector $\vec{B}$ is added to vector $\vec{A}$, the resultant $\vec{R}$ is the vector that runs from the tail of $\vec{A}$ to the tip of $\vec{B}$.

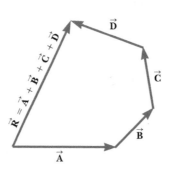

Figure 3.7 Geometric construction for summing four vectors. The resultant vector $\vec{R}$ is by definition the one that completes the polygon.

Figure 3.8 This construction shows that $\vec{A} + \vec{B} = \vec{B} + \vec{A}$ or, in other words, that vector addition is commutative.

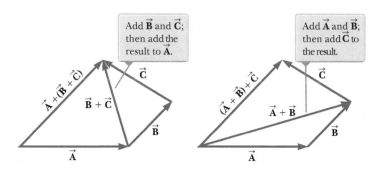

Figure 3.9 Geometric constructions for verifying the associative law of addition.

When three or more vectors are added, their sum is independent of the way in which the individual vectors are grouped together. A geometric proof of this rule for three vectors is given in Figure 3.9. This property is called the **associative law of addition**:

$$\vec{A} + (\vec{B} + \vec{C}) = (\vec{A} + \vec{B}) + \vec{C} \tag{3.6}$$

◄ **Associative law of addition**

In summary, **a vector quantity has both magnitude and direction and also obeys the laws of vector addition** as described in Figures 3.6 to 3.9. When two or more vectors are added together, they must all have the same units and they must all be the same type of quantity. It would be meaningless to add a velocity vector (for example, 60 km/h to the east) to a displacement vector (for example, 200 km to the north) because these vectors represent different physical quantities. The same rule also applies to scalars. For example, it would be meaningless to add time intervals to temperatures.

Negative of a Vector

The negative of the vector $\vec{A}$ is defined as the vector that when added to $\vec{A}$ gives zero for the vector sum. That is, $\vec{A} + (-\vec{A}) = 0$. The vectors $\vec{A}$ and $-\vec{A}$ have the same magnitude but point in opposite directions.

Subtracting Vectors

The operation of vector subtraction makes use of the definition of the negative of a vector. We define the operation $\vec{A} - \vec{B}$ as vector $-\vec{B}$ added to vector $\vec{A}$:

$$\vec{A} - \vec{B} = \vec{A} + (-\vec{B}) \tag{3.7}$$

The geometric construction for subtracting two vectors in this way is illustrated in Figure 3.10a.

Another way of looking at vector subtraction is to notice that the difference $\vec{A} - \vec{B}$ between two vectors $\vec{A}$ and $\vec{B}$ is what you have to add to the second vector

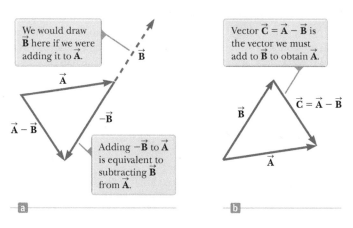

Figure 3.10 (a) Subtracting vector $\vec{B}$ from vector $\vec{A}$. The vector $-\vec{B}$ is equal in magnitude to vector $\vec{B}$ and points in the opposite direction. (b) A second way of looking at vector subtraction.

to obtain the first. In this case, as Figure 3.10b shows, the vector $\vec{A} - \vec{B}$ points from the tip of the second vector to the tip of the first.

Multiplying a Vector by a Scalar

If vector $\vec{A}$ is multiplied by a positive scalar quantity m, the product $m\vec{A}$ is a vector that has the same direction as $\vec{A}$ and magnitude mA. If vector $\vec{A}$ is multiplied by a negative scalar quantity $-m$, the product $-m\vec{A}$ is directed opposite $\vec{A}$. For example, the vector $5\vec{A}$ is five times as long as $\vec{A}$ and points in the same direction as $\vec{A}$; the vector $-\frac{1}{3}\vec{A}$ is one-third the length of $\vec{A}$ and points in the direction opposite $\vec{A}$.

Quick Quiz 3.2 The magnitudes of two vectors $\vec{A}$ and $\vec{B}$ are $A = 12$ units and $B = 8$ units. Which pair of numbers represents the *largest* and *smallest* possible values for the magnitude of the resultant vector $\vec{R} = \vec{A} + \vec{B}$? **(a)** 14.4 units, 4 units **(b)** 12 units, 8 units **(c)** 20 units, 4 units **(d)** none of these answers

Quick Quiz 3.3 If vector $\vec{B}$ is added to vector $\vec{A}$, which *two* of the following choices must be true for the resultant vector to be equal to zero? **(a)** $\vec{A}$ and $\vec{B}$ are parallel and in the same direction. **(b)** $\vec{A}$ and $\vec{B}$ are parallel and in opposite directions. **(c)** $\vec{A}$ and $\vec{B}$ have the same magnitude. **(d)** $\vec{A}$ and $\vec{B}$ are perpendicular.

Example 3.2   A Vacation Trip

A car travels 20.0 km due north and then 35.0 km in a direction 60.0° west of north as shown in Figure 3.11a. Find the magnitude and direction of the car's resultant displacement.

SOLUTION

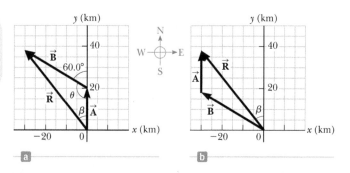

Figure 3.11 (Example 3.2) (a) Graphical method for finding the resultant displacement vector $\vec{R} = \vec{A} + \vec{B}$. (b) Adding the vectors in reverse order $(\vec{B} + \vec{A})$ gives the same result for $\vec{R}$.

Conceptualize The vectors $\vec{A}$ and $\vec{B}$ drawn in Figure 3.11a help us conceptualize the problem. The resultant vector $\vec{R}$ has also been drawn. We expect its magnitude to be a few tens of kilometers. The angle β that the resultant vector makes with the y axis is expected to be less than 60°, the angle that vector $\vec{B}$ makes with the y axis.

Categorize We can categorize this example as a simple analysis problem in vector addition. The displacement $\vec{R}$ is the resultant when the two individual displacements $\vec{A}$ and $\vec{B}$ are added. We can further categorize it as a problem about the analysis of triangles, so we appeal to our expertise in geometry and trigonometry.

Analyze In this example, we show two ways to analyze the problem of finding the resultant of two vectors. The first way is to solve the problem geometrically, using graph paper and a protractor to measure the magnitude of $\vec{R}$ and its direction in Figure 3.11a. (In fact, even when you know you are going to be carrying out a calculation, you should sketch the vectors to check your results.) With an ordinary ruler and protractor, a large diagram typically gives answers to two-digit but not to three-digit precision. Try using these tools on $\vec{R}$ in Figure 3.11a and compare to the trigonometric analysis below!

The second way to solve the problem is to analyze it using algebra and trigonometry. The magnitude of $\vec{R}$ can be obtained from the law of cosines as applied to the triangle in Figure 3.11a (see Appendix B.4).

Use $R^2 = A^2 + B^2 - 2AB \cos \theta$ from the law of cosines to find R:

$$R = \sqrt{A^2 + B^2 - 2AB \cos \theta}$$

Substitute numerical values, noting that $\theta = 180° - 60° = 120°$:

$$R = \sqrt{(20.0 \text{ km})^2 + (35.0 \text{ km})^2 - 2(20.0 \text{ km})(35.0 \text{ km}) \cos 120°}$$

$$= \boxed{48.2 \text{ km}}$$

▶ **3.2** continued

Use the law of sines (Appendix B.4) to find the direction of $\vec{R}$ measured from the northerly direction:

$$\frac{\sin \beta}{B} = \frac{\sin \theta}{R}$$

$$\sin \beta = \frac{B}{R} \sin \theta = \frac{35.0 \text{ km}}{48.2 \text{ km}} \sin 120° = 0.629$$

$$\beta = \boxed{38.9°}$$

The resultant displacement of the car is 48.2 km in a direction 38.9° west of north.

....

Finalize Does the angle β that we calculated agree with an estimate made by looking at Figure 3.11a or with an actual angle measured from the diagram using the graphical method? Is it reasonable that the magnitude of $\vec{R}$ is larger than that of both $\vec{A}$ and $\vec{B}$? Are the units of $\vec{R}$ correct?

Although the head to tail method of adding vectors works well, it suffers from two disadvantages. First, some people find using the laws of cosines and sines to be awkward. Second, a triangle only results if you are adding two vectors. If you are adding three or more vectors, the resulting geometric shape is usually not a triangle. In Section 3.4, we explore a new method of adding vectors that will address both of these disadvantages.

WHAT IF?   Suppose the trip were taken with the two vectors in reverse order: 35.0 km at 60.0° west of north first and then 20.0 km due north. How would the magnitude and the direction of the resultant vector change?

Answer They would not change. The commutative law for vector addition tells us that the order of vectors in an addition is irrelevant. Graphically, Figure 3.11b shows that the vectors added in the reverse order give us the same resultant vector.

3.4 Components of a Vector and Unit Vectors

The graphical method of adding vectors is not recommended whenever high accuracy is required or in three-dimensional problems. In this section, we describe a method of adding vectors that makes use of the projections of vectors along coordinate axes. These projections are called the **components** of the vector or its **rectangular components.** Any vector can be completely described by its components.

Consider a vector $\vec{A}$ lying in the xy plane and making an arbitrary angle θ with the positive x axis as shown in Figure 3.12a. This vector can be expressed as the sum of two other *component vectors* $\vec{A}_x$, which is parallel to the x axis, and $\vec{A}_y$, which is parallel to the y axis. From Figure 3.12b, we see that the three vectors form a right triangle and that $\vec{A} = \vec{A}_x + \vec{A}_y$. We shall often refer to the "components of a vector $\vec{A}$," written A_x and A_y (without the boldface notation). The component A_x represents the projection of $\vec{A}$ along the x axis, and the component A_y represents the projection of $\vec{A}$ along the y axis. These components can be positive or negative. The component A_x is positive if the component vector $\vec{A}_x$ points in the positive x direction and is negative if $\vec{A}_x$ points in the negative x direction. A similar statement is made for the component A_y.

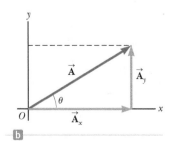

Figure 3.12 (a) A vector $\vec{A}$ lying in the xy plane can be represented by its component vectors $\vec{A}_x$ and $\vec{A}_y$. (b) The y component vector $\vec{A}_y$ can be moved to the right so that it adds to $\vec{A}_x$. The vector sum of the component vectors is $\vec{A}$. These three vectors form a right triangle.

Pitfall Prevention 3.2

x and y Components Equations 3.8 and 3.9 associate the cosine of the angle with the x component and the sine of the angle with the y component. This association is true *only* because we measured the angle θ with respect to the x axis, so do not memorize these equations. If θ is measured with respect to the y axis (as in some problems), these equations will be incorrect. Think about which side of the triangle containing the components is adjacent to the angle and which side is opposite and then assign the cosine and sine accordingly.

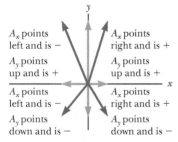

A_x points left and is $-$
A_y points up and is $+$
A_x points left and is $-$
A_y points down and is $-$

A_x points right and is $+$
A_y points up and is $+$
A_x points right and is $+$
A_y points down and is $-$

Figure 3.13 The signs of the components of a vector $\vec{A}$ depend on the quadrant in which the vector is located.

From Figure 3.12 and the definition of sine and cosine, we see that $\cos\theta = A_x/A$ and that $\sin\theta = A_y/A$. Hence, the components of $\vec{A}$ are

$$A_x = A\cos\theta \tag{3.8}$$

$$A_y = A\sin\theta \tag{3.9}$$

The magnitudes of these components are the lengths of the two sides of a right triangle with a hypotenuse of length A. Therefore, the magnitude and direction of $\vec{A}$ are related to its components through the expressions

$$A = \sqrt{A_x^2 + A_y^2} \tag{3.10}$$

$$\theta = \tan^{-1}\left(\frac{A_y}{A_x}\right) \tag{3.11}$$

Notice that the signs of the components A_x and A_y depend on the angle θ. For example, if $\theta = 120°$, A_x is negative and A_y is positive. If $\theta = 225°$, both A_x and A_y are negative. Figure 3.13 summarizes the signs of the components when $\vec{A}$ lies in the various quadrants.

When solving problems, you can specify a vector $\vec{A}$ either with its components A_x and A_y or with its magnitude and direction A and θ.

Suppose you are working a physics problem that requires resolving a vector into its components. In many applications, it is convenient to express the components in a coordinate system having axes that are not horizontal and vertical but that are still perpendicular to each other. For example, we will consider the motion of objects sliding down inclined planes. For these examples, it is often convenient to orient the x axis parallel to the plane and the y axis perpendicular to the plane.

Quick Quiz 3.4 Choose the correct response to make the sentence true: A component of a vector is **(a)** always, **(b)** never, or **(c)** sometimes larger than the magnitude of the vector.

Unit Vectors

Vector quantities often are expressed in terms of unit vectors. A **unit vector** is a dimensionless vector having a magnitude of exactly 1. Unit vectors are used to specify a given direction and have no other physical significance. They are used solely as a bookkeeping convenience in describing a direction in space. We shall use the symbols $\hat{\mathbf{i}}$, $\hat{\mathbf{j}}$, and $\hat{\mathbf{k}}$ to represent unit vectors pointing in the positive x, y, and z directions, respectively. (The "hats," or circumflexes, on the symbols are a standard notation for unit vectors.) The unit vectors $\hat{\mathbf{i}}$, $\hat{\mathbf{j}}$, and $\hat{\mathbf{k}}$ form a set of mutually perpendicular vectors in a right-handed coordinate system as shown in Figure 3.14a. The magnitude of each unit vector equals 1; that is, $|\hat{\mathbf{i}}| = |\hat{\mathbf{j}}| = |\hat{\mathbf{k}}| = 1$.

Consider a vector $\vec{A}$ lying in the xy plane as shown in Figure 3.14b. The product of the component A_x and the unit vector $\hat{\mathbf{i}}$ is the component vector $\vec{A}_x = A_x\hat{\mathbf{i}}$,

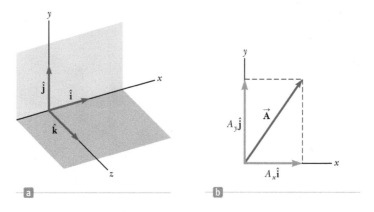

Figure 3.14 (a) The unit vectors $\hat{\mathbf{i}}$, $\hat{\mathbf{j}}$, and $\hat{\mathbf{k}}$ are directed along the x, y, and z axes, respectively. (b) Vector $\vec{A} = A_x\hat{\mathbf{i}} + A_y\hat{\mathbf{j}}$ lying in the xy plane has components A_x and A_y.

which lies on the x axis and has magnitude $|A_x|$. Likewise, $\vec{\mathbf{A}}_y = A_y \hat{\mathbf{j}}$ is the component vector of magnitude $|A_y|$ lying on the y axis. Therefore, the unit-vector notation for the vector $\vec{\mathbf{A}}$ is

$$\vec{\mathbf{A}} = A_x\hat{\mathbf{i}} + A_y\hat{\mathbf{j}} \tag{3.12}$$

For example, consider a point lying in the xy plane and having Cartesian coordinates (x, y) as in Figure 3.15. The point can be specified by the **position vector** $\vec{\mathbf{r}}$, which in unit-vector form is given by

$$\vec{\mathbf{r}} = x\hat{\mathbf{i}} + y\hat{\mathbf{j}} \tag{3.13}$$

This notation tells us that the components of $\vec{\mathbf{r}}$ are the coordinates x and y.

Now let us see how to use components to add vectors when the graphical method is not sufficiently accurate. Suppose we wish to add vector $\vec{\mathbf{B}}$ to vector $\vec{\mathbf{A}}$ in Equation 3.12, where vector $\vec{\mathbf{B}}$ has components B_x and B_y. Because of the bookkeeping convenience of the unit vectors, all we do is add the x and y components separately. The resultant vector $\vec{\mathbf{R}} = \vec{\mathbf{A}} + \vec{\mathbf{B}}$ is

$$\vec{\mathbf{R}} = (A_x\hat{\mathbf{i}} + A_y\hat{\mathbf{j}}) + (B_x\hat{\mathbf{i}} + B_y\hat{\mathbf{j}})$$

or

$$\vec{\mathbf{R}} = (A_x + B_x)\hat{\mathbf{i}} + (A_y + B_y)\hat{\mathbf{j}} \tag{3.14}$$

Because $\vec{\mathbf{R}} = R_x\hat{\mathbf{i}} + R_y\hat{\mathbf{j}}$, we see that the components of the resultant vector are

$$R_x = A_x + B_x$$
$$R_y = A_y + B_y \tag{3.15}$$

Therefore, we see that in the component method of adding vectors, we add all the x components together to find the x component of the resultant vector and use the same process for the y components. We can check this addition by components with a geometric construction as shown in Figure 3.16.

The magnitude of $\vec{\mathbf{R}}$ and the angle it makes with the x axis are obtained from its components using the relationships

$$R = \sqrt{R_x^2 + R_y^2} = \sqrt{(A_x + B_x)^2 + (A_y + B_y)^2} \tag{3.16}$$

$$\tan\theta = \frac{R_y}{R_x} = \frac{A_y + B_y}{A_x + B_x} \tag{3.17}$$

At times, we need to consider situations involving motion in three component directions. The extension of our methods to three-dimensional vectors is straightforward. If $\vec{\mathbf{A}}$ and $\vec{\mathbf{B}}$ both have x, y, and z components, they can be expressed in the form

$$\vec{\mathbf{A}} = A_x\hat{\mathbf{i}} + A_y\hat{\mathbf{j}} + A_z\hat{\mathbf{k}} \tag{3.18}$$

$$\vec{\mathbf{B}} = B_x\hat{\mathbf{i}} + B_y\hat{\mathbf{j}} + B_z\hat{\mathbf{k}} \tag{3.19}$$

The sum of $\vec{\mathbf{A}}$ and $\vec{\mathbf{B}}$ is

$$\vec{\mathbf{R}} = (A_x + B_x)\hat{\mathbf{i}} + (A_y + B_y)\hat{\mathbf{j}} + (A_z + B_z)\hat{\mathbf{k}} \tag{3.20}$$

Notice that Equation 3.20 differs from Equation 3.14: in Equation 3.20, the resultant vector also has a z component $R_z = A_z + B_z$. If a vector $\vec{\mathbf{R}}$ has x, y, and z components, the magnitude of the vector is $R = \sqrt{R_x^2 + R_y^2 + R_z^2}$. The angle θ_x that $\vec{\mathbf{R}}$ makes with the x axis is found from the expression $\cos\theta_x = R_x/R$, with similar expressions for the angles with respect to the y and z axes.

The extension of our method to adding more than two vectors is also straightforward. For example, $\vec{\mathbf{A}} + \vec{\mathbf{B}} + \vec{\mathbf{C}} = (A_x + B_x + C_x)\hat{\mathbf{i}} + (A_y + B_y + C_y)\hat{\mathbf{j}} + (A_z + B_z + C_z)\hat{\mathbf{k}}$. We have described adding displacement vectors in this section because these types of vectors are easy to visualize. We can also add other types of

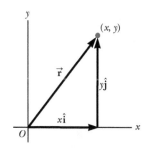

Figure 3.15 The point whose Cartesian coordinates are (x, y) can be represented by the position vector $\vec{\mathbf{r}} = x\hat{\mathbf{i}} + y\hat{\mathbf{j}}$.

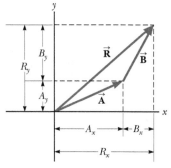

Figure 3.16 This geometric construction for the sum of two vectors shows the relationship between the components of the resultant $\vec{\mathbf{R}}$ and the components of the individual vectors.

Pitfall Prevention 3.3

Tangents on Calculators Equation 3.17 involves the calculation of an angle by means of a tangent function. Generally, the inverse tangent function on calculators provides an angle between $-90°$ and $+90°$. As a consequence, if the vector you are studying lies in the second or third quadrant, the angle measured from the positive x axis will be the angle your calculator returns plus 180°.

vectors, such as velocity, force, and electric field vectors, which we will do in later chapters.

Quick Quiz 3.5 For which of the following vectors is the magnitude of the vector equal to one of the components of the vector? **(a)** $\vec{A} = 2\hat{i} + 5\hat{j}$ **(b)** $\vec{B} = -3\hat{j}$ **(c)** $\vec{C} = +5\hat{k}$

Example 3.3   The Sum of Two Vectors

Find the sum of two displacement vectors $\vec{A}$ and $\vec{B}$ lying in the xy plane and given by

$$\vec{A} = (2.0\hat{i} + 2.0\hat{j}) \text{ m} \quad \text{and} \quad \vec{B} = (2.0\hat{i} - 4.0\hat{j}) \text{ m}$$

SOLUTION

Conceptualize You can conceptualize the situation by drawing the vectors on graph paper. Draw an approximation of the expected resultant vector.

Categorize We categorize this example as a simple substitution problem. Comparing this expression for $\vec{A}$ with the general expression $\vec{A} = A_x\hat{i} + A_y\hat{j} + A_z\hat{k}$, we see that $A_x = 2.0$ m, $A_y = 2.0$ m, and $A_z = 0$. Likewise, $B_x = 2.0$ m, $B_y = -4.0$ m, and $B_z = 0$. We can use a two-dimensional approach because there are no z components.

Use Equation 3.14 to obtain the resultant vector $\vec{R}$: $\vec{R} = \vec{A} + \vec{B} = (2.0 + 2.0)\hat{i} \text{ m} + (2.0 - 4.0)\hat{j} \text{ m}$

Evaluate the components of $\vec{R}$: $R_x = 4.0 \text{ m} \quad R_y = -2.0 \text{ m}$

Use Equation 3.16 to find the magnitude of $\vec{R}$: $R = \sqrt{R_x^2 + R_y^2} = \sqrt{(4.0 \text{ m})^2 + (-2.0 \text{ m})^2} = \sqrt{20} \text{ m} = \boxed{4.5 \text{ m}}$

Find the direction of $\vec{R}$ from Equation 3.17: $\tan\theta = \dfrac{R_y}{R_x} = \dfrac{-2.0 \text{ m}}{4.0 \text{ m}} = -0.50$

Your calculator likely gives the answer $-27°$ for $\theta = \tan^{-1}(-0.50)$. This answer is correct if we interpret it to mean $27°$ clockwise from the x axis. Our standard form has been to quote the angles measured counterclockwise from the $+x$ axis, and that angle for this vector is $\theta = \boxed{333°}$.

Example 3.4   The Resultant Displacement

A particle undergoes three consecutive displacements: $\Delta\vec{r}_1 = (15\hat{i} + 30\hat{j} + 12\hat{k})$ cm, $\Delta\vec{r}_2 = (23\hat{i} - 14\hat{j} - 5.0\hat{k})$ cm, and $\Delta\vec{r}_3 = (-13\hat{i} + 15\hat{j})$ cm. Find unit-vector notation for the resultant displacement and its magnitude.

SOLUTION

Conceptualize Although x is sufficient to locate a point in one dimension, we need a vector $\vec{r}$ to locate a point in two or three dimensions. The notation $\Delta\vec{r}$ is a generalization of the one-dimensional displacement Δx in Equation 2.1. Three-dimensional displacements are more difficult to conceptualize than those in two dimensions because they cannot be drawn on paper like the latter.

For this problem, let us imagine that you start with your pencil at the origin of a piece of graph paper on which you have drawn x and y axes. Move your pencil 15 cm to the right along the x axis, then 30 cm upward along the y axis, and then 12 cm *perpendicularly toward you away*

from the graph paper. This procedure provides the displacement described by $\Delta\vec{r}_1$. From this point, move your pencil 23 cm to the right parallel to the x axis, then 14 cm parallel to the graph paper in the $-y$ direction, and then 5.0 cm perpendicularly away from you toward the graph paper. You are now at the displacement from the origin described by $\Delta\vec{r}_1 + \Delta\vec{r}_2$. From this point, move your pencil 13 cm to the left in the $-x$ direction, and (finally!) 15 cm parallel to the graph paper along the y axis. Your final position is at a displacement $\Delta\vec{r}_1 + \Delta\vec{r}_2 + \Delta\vec{r}_3$ from the origin.

▶ **3.4** c o n t i n u e d

Categorize Despite the difficulty in conceptualizing in three dimensions, we can categorize this problem as a substitution problem because of the careful bookkeeping methods that we have developed for vectors. The mathematical manipulation keeps track of this motion along the three perpendicular axes in an organized, compact way, as we see below.

To find the resultant displacement, add the three vectors:

$$\Delta \vec{r} = \Delta \vec{r}_1 + \Delta \vec{r}_2 + \Delta \vec{r}_3$$
$$= (15 + 23 - 13)\hat{i}\, cm + (30 - 14 + 15)\hat{j}\, cm + (12 - 5.0 + 0)\hat{k}\, cm$$
$$= (25\hat{i} + 31\hat{j} + 7.0\hat{k})\, cm$$

Find the magnitude of the resultant vector:

$$R = \sqrt{R_x^2 + R_y^2 + R_z^2}$$
$$= \sqrt{(25\, cm)^2 + (31\, cm)^2 + (7.0\, cm)^2} = \boxed{40\, cm}$$

Example 3.5   **Taking a Hike**

A hiker begins a trip by first walking 25.0 km southeast from her car. She stops and sets up her tent for the night. On the second day, she walks 40.0 km in a direction 60.0° north of east, at which point she discovers a forest ranger's tower.

(A) Determine the components of the hiker's displacement for each day.

SOLUTION

Conceptualize We conceptualize the problem by drawing a sketch as in Figure 3.17. If we denote the displacement vectors on the first and second days by $\vec{A}$ and $\vec{B}$, respectively, and use the car as the origin of coordinates, we obtain the vectors shown in Figure 3.17. The sketch allows us to estimate the resultant vector as shown.

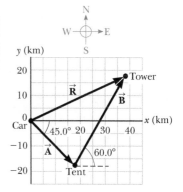

Figure 3.17 (Example 3.5) The total displacement of the hiker is the vector $\vec{R} = \vec{A} + \vec{B}$.

Categorize Having drawn the resultant $\vec{R}$, we can now categorize this problem as one we've solved before: an addition of two vectors. You should now have a hint of the power of categorization in that many new problems are very similar to problems we have already solved if we are careful to conceptualize them. Once we have drawn the displacement vectors and categorized the problem, this problem is no longer about a hiker, a walk, a car, a tent, or a tower. It is a problem about vector addition, one that we have already solved.

. . . .

Analyze Displacement $\vec{A}$ has a magnitude of 25.0 km and is directed 45.0° below the positive x axis.

Find the components of $\vec{A}$ using Equations 3.8 and 3.9:

$$A_x = A \cos(-45.0°) = (25.0\, km)(0.707) = \boxed{17.7\, km}$$
$$A_y = A \sin(-45.0°) = (25.0\, km)(-0.707) = \boxed{-17.7\, km}$$

The negative value of A_y indicates that the hiker walks in the negative y direction on the first day. The signs of A_x and A_y also are evident from Figure 3.17.

Find the components of $\vec{B}$ using Equations 3.8 and 3.9:

$$B_x = B \cos 60.0° = (40.0\, km)(0.500) = \boxed{20.0\, km}$$
$$B_y = B \sin 60.0° = (40.0\, km)(0.866) = \boxed{34.6\, km}$$

(B) Determine the components of the hiker's resultant displacement $\vec{R}$ for the trip. Find an expression for $\vec{R}$ in terms of unit vectors.

SOLUTION

Use Equation 3.15 to find the components of the resultant displacement $\vec{R} = \vec{A} + \vec{B}$:

$$R_x = A_x + B_x = 17.7\, km + 20.0\, km = \boxed{37.7\, km}$$
$$R_y = A_y + B_y = -17.7\, km + 34.6\, km = \boxed{17.0\, km}$$

continued

▶ 3.5 continued

Write the total displacement in unit-vector form: $\vec{R} = (37.7\hat{i} + 17.0\hat{j})$ km

Finalize Looking at the graphical representation in Figure 3.17, we estimate the position of the tower to be about (38 km, 17 km), which is consistent with the components of $\vec{R}$ in our result for the final position of the hiker. Also, both components of $\vec{R}$ are positive, putting the final position in the first quadrant of the coordinate system, which is also consistent with Figure 3.17.

WHAT IF?   After reaching the tower, the hiker wishes to return to her car along a single straight line. What are the components of the vector representing this hike? What should the direction of the hike be?

Answer The desired vector $\vec{R}_{car}$ is the negative of vector $\vec{R}$:

$$\vec{R}_{car} = -\vec{R} = (-37.7\hat{i} - 17.0\hat{j}) \text{ km}$$

The direction is found by calculating the angle that the vector makes with the *x* axis:

$$\tan \theta = \frac{R_{car,y}}{R_{car,x}} = \frac{-17.0 \text{ km}}{-37.7 \text{ km}} = 0.450$$

which gives an angle of $\theta = 204.2°$, or 24.2° south of west.

Summary

Definitions

Scalar quantities are those that have only a numerical value and no associated direction.

Vector quantities have both magnitude and direction and obey the laws of vector addition. The magnitude of a vector is *always* a positive number.

Concepts and Principles

When two or more vectors are added together, they must all have the same units and they all must be the same type of quantity. We can add two vectors $\vec{A}$ and $\vec{B}$ graphically. In this method (Fig. 3.6), the resultant vector $\vec{R} = \vec{A} + \vec{B}$ runs from the tail of $\vec{A}$ to the tip of $\vec{B}$.

A second method of adding vectors involves **components** of the vectors. The *x* component A_x of the vector $\vec{A}$ is equal to the projection of $\vec{A}$ along the *x* axis of a coordinate system, where $A_x = A \cos \theta$. The *y* component A_y of $\vec{A}$ is the projection of $\vec{A}$ along the *y* axis, where $A_y = A \sin \theta$.

If a vector $\vec{A}$ has an *x* component A_x and a *y* component A_y, the vector can be expressed in unit-vector form as $\vec{A} = A_x\hat{i} + A_y\hat{j}$. In this notation, $\hat{i}$ is a unit vector pointing in the positive *x* direction and $\hat{j}$ is a unit vector pointing in the positive *y* direction. Because $\hat{i}$ and $\hat{j}$ are unit vectors, $|\hat{i}| = |\hat{j}| = 1$.

We can find the resultant of two or more vectors by resolving all vectors into their *x* and *y* components, adding their resultant *x* and *y* components, and then using the Pythagorean theorem to find the magnitude of the resultant vector. We can find the angle that the resultant vector makes with respect to the *x* axis by using a suitable trigonometric function.

1. What is the magnitude of the vector $(10\hat{\mathbf{i}} - 10\hat{\mathbf{k}})$ m/s? (a) 0 (b) 10 m/s (c) −10 m/s (d) 10 (e) 14.1 m/s

2. A vector lying in the *xy* plane has components of opposite sign. The vector must lie in which quadrant? (a) the first quadrant (b) the second quadrant (c) the third quadrant (d) the fourth quadrant (e) either the second or the fourth quadrant

3. Figure OQ3.3 shows two vectors $\vec{\mathbf{D}}_1$ and $\vec{\mathbf{D}}_2$. Which of the possibilities (a) through (d) is the vector $\vec{\mathbf{D}}_2 - 2\vec{\mathbf{D}}_1$, or (e) is it none of them?

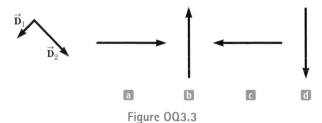

Figure OQ3.3

4. The cutting tool on a lathe is given two displacements, one of magnitude 4 cm and one of magnitude 3 cm, in each one of five situations (a) through (e) diagrammed in Figure OQ3.4. Rank these situations according to the magnitude of the total displacement of the tool, putting the situation with the greatest resultant magnitude first. If the total displacement is the same size in two situations, give those letters equal ranks.

Figure OQ3.4

5. The magnitude of vector $\vec{\mathbf{A}}$ is 8 km, and the magnitude of $\vec{\mathbf{B}}$ is 6 km. Which of the following are possible values for the magnitude of $\vec{\mathbf{A}} + \vec{\mathbf{B}}$? Choose all possible answers. (a) 10 km (b) 8 km (c) 2 km (d) 0 (e) −2 km

6. Let vector $\vec{\mathbf{A}}$ point from the origin into the second quadrant of the *xy* plane and vector $\vec{\mathbf{B}}$ point from the origin into the fourth quadrant. The vector $\vec{\mathbf{B}} - \vec{\mathbf{A}}$ must be in which quadrant, (a) the first, (b) the second, (c) the third, or (d) the fourth, or (e) is more than one answer possible?

7. Yes or no: Is each of the following quantities a vector? (a) force (b) temperature (c) the volume of water in a can (d) the ratings of a TV show (e) the height of a building (f) the velocity of a sports car (g) the age of the Universe

8. What is the *y* component of the vector $(3\hat{\mathbf{i}} - 8\hat{\mathbf{k}})$ m/s? (a) 3 m/s (b) −8 m/s (c) 0 (d) 8 m/s (e) none of those answers

9. What is the *x* component of the vector shown in Figure OQ3.9? (a) 3 cm (b) 6 cm (c) −4 cm (d) −6 cm (e) none of those answers

Figure OQ3.9 Objective Questions 9 and 10.

10. What is the *y* component of the vector shown in Figure OQ3.9? (a) 3 cm (b) 6 cm (c) −4 cm (d) −6 cm (e) none of those answers

11. Vector $\vec{\mathbf{A}}$ lies in the *xy* plane. Both of its components will be negative if it points from the origin into which quadrant? (a) the first quadrant (b) the second quadrant (c) the third quadrant (d) the fourth quadrant (e) the second or fourth quadrants

12. A submarine dives from the water surface at an angle of 30° below the horizontal, following a straight path 50 m long. How far is the submarine then below the water surface? (a) 50 m (b) (50 m)/sin 30° (c) (50 m) sin 30° (d) (50 m) cos 30° (e) none of those answers

13. A vector points from the origin into the second quadrant of the *xy* plane. What can you conclude about its components? (a) Both components are positive. (b) The *x* component is positive, and the *y* component is negative. (c) The *x* component is negative, and the *y* component is positive. (d) Both components are negative. (e) More than one answer is possible.

1. Is it possible to add a vector quantity to a scalar quantity? Explain.

2. Can the magnitude of a vector have a negative value? Explain.

3. A book is moved once around the perimeter of a tabletop with the dimensions 1.0 m by 2.0 m. The book ends up at its initial position. (a) What is its displacement? (b) What is the distance traveled?

4. If the component of vector $\vec{\mathbf{A}}$ along the direction of vector $\vec{\mathbf{B}}$ is zero, what can you conclude about the two vectors?

5. On a certain calculator, the inverse tangent function returns a value between −90° and +90°. In what cases will this value correctly state the direction of a vector in the *xy* plane, by giving its angle measured counterclockwise from the positive *x* axis? In what cases will it be incorrect?

Problems

WebAssign The problems found in this chapter may be assigned online in Enhanced WebAssign

1. straightforward; **2.** intermediate; **3.** challenging

1. full solution available in the *Student Solutions Manual/Study Guide*

AMT Analysis Model tutorial available in Enhanced WebAssign

GP Guided Problem

M Master It tutorial available in Enhanced WebAssign

W Watch It video solution available in Enhanced WebAssign

Section 3.1 Coordinate Systems

1. The polar coordinates of a point are $r = 5.50$ m and $\theta = 240°$. What are the Cartesian coordinates of this point?

2. The rectangular coordinates of a point are given by $(2, y)$, and its polar coordinates are $(r, 30°)$. Determine (a) the value of y and (b) the value of r.

3. Two points in the xy plane have Cartesian coordinates $(2.00, -4.00)$ m and $(-3.00, 3.00)$ m. Determine (a) the distance between these points and (b) their polar coordinates.

4. Two points in a plane have polar coordinates (2.50 m, 30.0°) and (3.80 m, 120.0°). Determine (a) the Cartesian coordinates of these points and (b) the distance between them.

5. The polar coordinates of a certain point are $(r = 4.30$ cm, $\theta = 214°)$. (a) Find its Cartesian coordinates x and y. Find the polar coordinates of the points with Cartesian coordinates (b) $(-x, y)$, (c) $(-2x, -2y)$, and (d) $(3x, -3y)$.

6. Let the polar coordinates of the point (x, y) be (r, θ). Determine the polar coordinates for the points (a) $(-x, y)$, (b) $(-2x, -2y)$, and (c) $(3x, -3y)$.

Section 3.2 Vector and Scalar Quantities

Section 3.3 Some Properties of Vectors

7. A surveyor measures the distance across a straight river by the following method (Fig. P3.7). Starting directly across from a tree on the opposite bank, she walks $d = 100$ m along the riverbank to establish a baseline. Then she sights across to the tree. The angle from her baseline to the tree is $\theta = 35.0°$. How wide is the river?

Figure P3.7

8. Vector $\vec{\mathbf{A}}$ has a magnitude of 29 units and points in the positive y direction. When vector $\vec{\mathbf{B}}$ is added to $\vec{\mathbf{A}}$, the resultant vector $\vec{\mathbf{A}} + \vec{\mathbf{B}}$ points in the negative y direction with a magnitude of 14 units. Find the magnitude and direction of $\vec{\mathbf{B}}$.

9. *Why is the following situation impossible?* A skater glides along a circular path. She defines a certain point on the circle as her origin. Later on, she passes through a point at which the distance she has traveled along the path from the origin is smaller than the magnitude of her displacement vector from the origin.

10. A force $\vec{\mathbf{F}}_1$ of magnitude 6.00 units acts on an object at the origin in a direction $\theta = 30.0°$ above the positive x axis (Fig. P3.10). A second force $\vec{\mathbf{F}}_2$ of magnitude 5.00 units acts on the object in the direction of the positive y axis. Find graphically the magnitude and direction of the resultant force $\vec{\mathbf{F}}_1 + \vec{\mathbf{F}}_2$.

Figure P3.10

11. The displacement vectors $\vec{\mathbf{A}}$ and $\vec{\mathbf{B}}$ shown in Figure P3.11 both have magnitudes of 3.00 m. The direction of vector $\vec{\mathbf{A}}$ is $\theta = 30.0°$. Find graphically (a) $\vec{\mathbf{A}} + \vec{\mathbf{B}}$, (b) $\vec{\mathbf{A}} - \vec{\mathbf{B}}$, (c) $\vec{\mathbf{B}} - \vec{\mathbf{A}}$, and (d) $\vec{\mathbf{A}} - 2\vec{\mathbf{B}}$. (Report all angles counterclockwise from the positive x axis.)

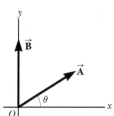

Figure P3.11
Problems 11 and 22.

12. Three displacements are $\vec{\mathbf{A}} = 200$ m due south, $\vec{\mathbf{B}} = 250$ m due west, and $\vec{\mathbf{C}} = 150$ m at 30.0° east of north. (a) Construct a separate diagram for each of the following possible ways of adding these vectors; $\vec{\mathbf{R}}_1 = \vec{\mathbf{A}} + \vec{\mathbf{B}} + \vec{\mathbf{C}}$; $\vec{\mathbf{R}}_2 = \vec{\mathbf{B}} + \vec{\mathbf{C}} + \vec{\mathbf{A}}$; $\vec{\mathbf{R}}_3 = \vec{\mathbf{C}} + \vec{\mathbf{B}} + \vec{\mathbf{A}}$. (b) Explain what you can conclude from comparing the diagrams.

13. A roller-coaster car moves 200 ft horizontally and then rises 135 ft at an angle of 30.0° above the horizontal. It next travels 135 ft at an angle of 40.0° downward. What is its displacement from its starting point? Use graphical techniques.

14. A plane flies from base camp to Lake A, 280 km away in the direction 20.0° north of east. After dropping off supplies, it flies to Lake B, which is 190 km at 30.0° west of north from Lake A. Graphically determine the distance and direction from Lake B to the base camp.

Section 3.4 **Components of a Vector and Unit Vectors**

15. A vector has an x component of -25.0 units and a y component of 40.0 units. Find the magnitude and direction of this vector.

16. Vector $\vec{\mathbf{A}}$ has a magnitude of 35.0 units and points in the direction 325° counterclockwise from the positive x axis. Calculate the x and y components of this vector.

17. A minivan travels straight north in the right lane of a divided highway at 28.0 m/s. A camper passes the minivan and then changes from the left lane into the right lane. As it does so, the camper's path on the road is a straight displacement at 8.50° east of north. To avoid cutting off the minivan, the north–south distance between the camper's back bumper and the minivan's front bumper should not decrease. (a) Can the camper be driven to satisfy this requirement? (b) Explain your answer.

18. A person walks 25.0° north of east for 3.10 km. How far would she have to walk due north and due east to arrive at the same location?

19. Obtain expressions in component form for the position vectors having the polar coordinates (a) 12.8 m, 150°; (b) 3.30 cm, 60.0°; and (c) 22.0 in., 215°.

20. A girl delivering newspapers covers her route by traveling 3.00 blocks west, 4.00 blocks north, and then 6.00 blocks east. (a) What is her resultant displacement? (b) What is the total distance she travels?

21. While exploring a cave, a spelunker starts at the entrance and moves the following distances in a horizontal plane. She goes 75.0 m north, 250 m east, 125 m at an angle $\theta = 30.0°$ north of east, and 150 m south. Find her resultant displacement from the cave entrance. Figure P3.21 suggests the situation but is not drawn to scale.

Figure P3.21

22. Use the component method to add the vectors $\vec{\mathbf{A}}$ and $\vec{\mathbf{B}}$ shown in Figure P3.11. Both vectors have magnitudes of 3.00 m and vector $\vec{\mathbf{A}}$ makes an angle of $\theta = 30.0°$ with the x axis. Express the resultant $\vec{\mathbf{A}} + \vec{\mathbf{B}}$ in unit-vector notation.

23. Consider the two vectors $\vec{\mathbf{A}} = 3\hat{\mathbf{i}} - 2\hat{\mathbf{j}}$ and $\vec{\mathbf{B}} = -\hat{\mathbf{i}} - 4\hat{\mathbf{j}}$. Calculate (a) $\vec{\mathbf{A}} + \vec{\mathbf{B}}$, (b) $\vec{\mathbf{A}} - \vec{\mathbf{B}}$, (c) $|\vec{\mathbf{A}} + \vec{\mathbf{B}}|$, (d) $|\vec{\mathbf{A}} - \vec{\mathbf{B}}|$, and (e) the directions of $\vec{\mathbf{A}} + \vec{\mathbf{B}}$ and $\vec{\mathbf{A}} - \vec{\mathbf{B}}$.

24. A map suggests that Atlanta is 730 miles in a direction of 5.00° north of east from Dallas. The same map shows that Chicago is 560 miles in a direction of 21.0° west of north from Atlanta. Figure P3.24 shows the locations of these three cities. Modeling the Earth as flat, use this information to find the displacement from Dallas to Chicago.

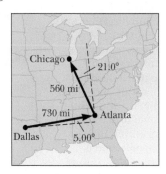

Figure P3.24

25. Your dog is running around the grass in your back yard. He undergoes successive displacements 3.50 m south, 8.20 m northeast, and 15.0 m west. What is the resultant displacement?

26. Given the vectors $\vec{\mathbf{A}} = 2.00\hat{\mathbf{i}} + 6.00\hat{\mathbf{j}}$ and $\vec{\mathbf{B}} = 3.00\hat{\mathbf{i}} - 2.00\hat{\mathbf{j}}$, (a) draw the vector sum $\vec{\mathbf{C}} = \vec{\mathbf{A}} + \vec{\mathbf{B}}$ and the vector difference $\vec{\mathbf{D}} = \vec{\mathbf{A}} - \vec{\mathbf{B}}$. (b) Calculate $\vec{\mathbf{C}}$ and $\vec{\mathbf{D}}$, in terms of unit vectors. (c) Calculate $\vec{\mathbf{C}}$ and $\vec{\mathbf{D}}$ in terms of polar coordinates, with angles measured with respect to the positive x axis.

27. A novice golfer on the green takes three strokes to sink the ball. The successive displacements of the ball are 4.00 m to the north, 2.00 m northeast, and 1.00 m at 30.0° west of south (Fig. P3.27). Starting at the same initial point, an expert golfer could make the hole in what single displacement?

Figure P3.27

28. A snow-covered ski slope makes an angle of 35.0° with the horizontal. When a ski jumper plummets onto the hill, a parcel of splashed snow is thrown up to a maximum displacement of 1.50 m at 16.0° from the vertical in the uphill direction as shown in Figure P3.28. Find the components of its maximum displacement (a) parallel to the surface and (b) perpendicular to the surface.

Figure P3.28

29. The helicopter view in Fig. P3.29 (page 74) shows two people pulling on a stubborn mule. The person on the right pulls with a force $\vec{\mathbf{F}}_1$ of magnitude 120 N and direction of $\theta_1 = 60.0°$. The person on the left pulls with a force $\vec{F}_2$ of magnitude 80.0 N and direction of $\theta_2 = 75.0°$. Find (a) the single force that is equivalent to the two forces shown and (b) the force that a third person would have to exert on the mule to make the resultant force equal to zero. The forces are measured in units of newtons (symbolized N).

Figure P3.29

30. In a game of American football, a quarterback takes the ball from the line of scrimmage, runs backward a distance of 10.0 yards, and then runs sideways parallel to the line of scrimmage for 15.0 yards. At this point, he throws a forward pass downfield 50.0 yards perpendicular to the line of scrimmage. What is the magnitude of the football's resultant displacement?

31. W Consider the three displacement vectors $\vec{A} = (3\hat{i} - 3\hat{j})$ m, $\vec{B} = (\hat{i} - 4\hat{j})$ m, and $\vec{C} = (-2\hat{i} + 5\hat{j})$ m. Use the component method to determine (a) the magnitude and direction of $\vec{D} = \vec{A} + \vec{B} + \vec{C}$ and (b) the magnitude and direction of $\vec{E} = -\vec{A} - \vec{B} + \vec{C}$.

32. W Vector $\vec{A}$ has x and y components of -8.70 cm and 15.0 cm, respectively; vector $\vec{B}$ has x and y components of 13.2 cm and -6.60 cm, respectively. If $\vec{A} - \vec{B} + 3\vec{C} = 0$, what are the components of $\vec{C}$?

33. M The vector $\vec{A}$ has x, y, and z components of 8.00, 12.0, and -4.00 units, respectively. (a) Write a vector expression for $\vec{A}$ in unit-vector notation. (b) Obtain a unit-vector expression for a vector $\vec{B}$ one-fourth the length of $\vec{A}$ pointing in the same direction as $\vec{A}$. (c) Obtain a unit-vector expression for a vector $\vec{C}$ three times the length of $\vec{A}$ pointing in the direction opposite the direction of $\vec{A}$.

34. Vector $\vec{B}$ has x, y, and z components of 4.00, 6.00, and 3.00 units, respectively. Calculate (a) the magnitude of $\vec{B}$ and (b) the angle that $\vec{B}$ makes with each coordinate axis.

35. M Vector $\vec{A}$ has a negative x component 3.00 units in length and a positive y component 2.00 units in length. (a) Determine an expression for $\vec{A}$ in unit-vector notation. (b) Determine the magnitude and direction of $\vec{A}$. (c) What vector $\vec{B}$ when added to $\vec{A}$ gives a resultant vector with no x component and a negative y component 4.00 units in length?

36. W Given the displacement vectors $\vec{A} = (3\hat{i} - 4\hat{j} + 4\hat{k})$ m and $\vec{B} = (2\hat{i} + 3\hat{j} - 7\hat{k})$ m, find the magnitudes of the following vectors and express each in terms of its rectangular components. (a) $\vec{C} = \vec{A} + \vec{B}$ (b) $\vec{D} = 2\vec{A} - \vec{B}$

37. (a) Taking $\vec{A} = (6.00\hat{i} - 8.00\hat{j})$ units, $\vec{B} = (-8.00\hat{i} + 3.00\hat{j})$ units, and $\vec{C} = (26.0\hat{i} + 19.0\hat{j})$ units, determine a and b such that $a\vec{A} + b\vec{B} + \vec{C} = 0$. (b) A

student has learned that a single equation cannot be solved to determine values for more than one unknown in it. How would you explain to him that both a and b can be determined from the single equation used in part (a)?

38. Three displacement vectors of a croquet ball are shown in Figure P3.38, where $|\vec{A}| = 20.0$ units, $|\vec{B}| = 40.0$ units, and $|\vec{C}| = 30.0$ units. Find (a) the resultant in unit-vector notation and (b) the magnitude and direction of the resultant displacement.

Figure P3.38

39. M A man pushing a mop across a floor causes it to undergo two displacements. The first has a magnitude of 150 cm and makes an angle of 120° with the positive x axis. The resultant displacement has a magnitude of 140 cm and is directed at an angle of 35.0° to the positive x axis. Find the magnitude and direction of the second displacement.

40. Figure P3.40 illustrates typical proportions of male (m) and female (f) anatomies. The displacements $\vec{d}_{1m}$ and $\vec{d}_{1f}$ from the soles of the feet to the navel have magnitudes of 104 cm and 84.0 cm, respectively. The displacements $\vec{d}_{2m}$ and $\vec{d}_{2f}$ from the navel to outstretched fingertips have magnitudes of 100 cm and 86.0 cm, respectively. Find the vector sum of these displacements $\vec{d}_3 = \vec{d}_1 + \vec{d}_2$ for both people.

Figure P3.40

41. Express in unit-vector notation the following vectors, each of which has magnitude 17.0 cm. (a) Vector $\vec{E}$ is directed 27.0° counterclockwise from the positive x axis. (b) Vector $\vec{F}$ is directed 27.0° counterclockwise from the positive y axis. (c) Vector $\vec{G}$ is directed 27.0° clockwise from the negative y axis.

42. A radar station locates a sinking ship at range 17.3 km and bearing 136° clockwise from north. From the same station, a rescue plane is at horizontal range 19.6 km, 153° clockwise from north, with elevation 2.20 km. (a) Write the position vector for the ship relative to the plane, letting $\hat{i}$ represent east, $\hat{j}$ north, and $\hat{k}$ up. (b) How far apart are the plane and ship?

43. **Review.** As it passes over Grand Bahama Island, the AMT eye of a hurricane is moving in a direction 60.0° north GP of west with a speed of 41.0 km/h. (a) What is the unit-vector expression for the velocity of the hurricane?

It maintains this velocity for 3.00 h, at which time the course of the hurricane suddenly shifts due north, and its speed slows to a constant 25.0 km/h. This new velocity is maintained for 1.50 h. (b) What is the unit-vector expression for the new velocity of the hurricane? (c) What is the unit-vector expression for the displacement of the hurricane during the first 3.00 h? (d) What is the unit-vector expression for the displacement of the hurricane during the latter 1.50 h? (e) How far from Grand Bahama is the eye 4.50 h after it passes over the island?

44. *Why is the following situation impossible?* A shopper pushing a cart through a market follows directions to the canned goods and moves through a displacement $8.00\hat{i}$ m down one aisle. He then makes a 90.0° turn and moves 3.00 m along the y axis. He then makes another 90.0° turn and moves 4.00 m along the x axis. *Every* shopper who follows these directions correctly ends up 5.00 m from the starting point.

45. **Review.** You are standing on the ground at the origin of a coordinate system. An airplane flies over you with constant velocity parallel to the x axis and at a fixed height of 7.60×10^3 m. At time $t = 0$, the airplane is directly above you so that the vector leading from you to it is $\vec{P}_0 = 7.60 \times 10^3\hat{j}$ m. At $t = 30.0$ s, the position vector leading from you to the airplane is $\vec{P}_{30} = (8.04 \times 10^3\hat{i} + 7.60 \times 10^3\hat{j})$ m as suggested in Figure P3.45. Determine the magnitude and orientation of the airplane's position vector at $t = 45.0$ s.

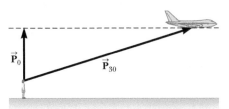

Figure P3.45

46. In Figure P3.46, the line segment represents a path from the point with position vector $(5\hat{i} + 3\hat{j})$ m to the point with location $(16\hat{i} + 12\hat{j})$ m. Point Ⓐ is along this path, a fraction f of the way to the destination. (a) Find the position vector of point Ⓐ in terms of f. (b) Evaluate the expression from part (a) for $f = 0$. (c) Explain whether the result in part (b) is reasonable. (d) Evaluate the expression for $f = 1$. (e) Explain whether the result in part (d) is reasonable.

Figure P3.46 Point Ⓐ is a fraction f of the distance from the initial point (5, 3) to the final point (16, 12).

47. In an assembly operation illustrated in Figure P3.47, a robot moves an object first straight upward and then also to the east, around an arc forming one-quarter of a circle of radius 4.80 cm that lies in an east–west vertical plane. The robot then moves the object upward and to the north, through one-quarter of a circle of radius 3.70 cm that lies in a north–south vertical plane. Find (a) the magnitude of the total displacement of the object and (b) the angle the total displacement makes with the vertical.

Figure P3.47

Additional Problems

48. A fly lands on one wall of a room. The lower-left corner of the wall is selected as the origin of a two-dimensional Cartesian coordinate system. If the fly is located at the point having coordinates (2.00, 1.00) m, (a) how far is it from the origin? (b) What is its location in polar coordinates?

49. As she picks up her riders, a bus driver traverses four successive displacements represented by the expression

$$(-6.30\ b)\hat{i} - (4.00\ b\cos 40°)\hat{i} - (4.00\ b\sin 40°)\hat{j}$$
$$+ (3.00\ b\cos 50°)\hat{i} - (3.00\ b\sin 50°)\hat{j} - (5.00\ b)\hat{j}$$

Here b represents one city block, a convenient unit of distance of uniform size; $\hat{i}$ is east; and $\hat{j}$ is north. The displacements at 40° and 50° represent travel on roadways in the city that are at these angles to the main east–west and north–south streets. (a) Draw a map of the successive displacements. (b) What total distance did she travel? (c) Compute the magnitude and direction of her total displacement. The logical structure of this problem and of several problems in later chapters was suggested by Alan Van Heuvelen and David Maloney, *American Journal of Physics* **67**(3) 252–256, March 1999.

50. A jet airliner, moving initially at 300 mi/h to the east, suddenly enters a region where the wind is blowing at 100 mi/h toward the direction 30.0° north of east. What are the new speed and direction of the aircraft relative to the ground?

51. A person going for a walk follows the path shown in Figure P3.51. The total trip consists of four straight-line paths. At the end of the walk, what is the person's resultant displacement measured from the starting point?

Figure P3.51

52. Find the horizontal and vertical components of the 100-m displacement of a superhero who flies from the

top of a tall building following the path shown in Figure P3.52.

53. **Review.** The biggest
stuffed animal in the world is a snake 420 m long, constructed by Norwegian children. Suppose the snake is laid out in a park as shown in Figure P3.53, forming two straight sides of a 105° angle, with one side 240 m long. Olaf and Inge run a race they invent. Inge runs directly from the tail of the snake to its head, and Olaf starts from the same place at the same moment but runs along the snake. (a) If both children run steadily at 12.0 km/h, Inge reaches the head of the snake how much earlier than Olaf? (b) If Inge runs the race again at a constant speed of 12.0 km/h, at what constant speed must Olaf run to reach the end of the snake at the same time as Inge?

Figure P3.52

Figure P3.53

54. An air-traffic controller observes two aircraft on his radar screen. The first is at altitude 800 m, horizontal distance 19.2 km, and 25.0° south of west. The second aircraft is at altitude 1 100 m, horizontal distance 17.6 km, and 20.0° south of west. What is the distance between the two aircraft? (Place the x axis west, the y axis south, and the z axis vertical.)

55. In Figure P3.55, a spider is resting after starting to spin its web. The gravitational force on the spider makes it exert a downward force of 0.150 N on the junction of the three strands of silk. The junction is supported by different tension forces in the two strands above it so that the resultant force on the junction is zero. The two sloping strands are perpendicular, and we have chosen the x and y directions to be along them. The tension T_x is 0.127 N. Find (a) the tension T_y, (b) the angle the x axis makes with the horizontal, and (c) the angle the y axis makes with the horizontal.

Figure P3.55

56. The rectangle shown in Figure P3.56 has sides parallel to the x and y axes. The position vectors of two corners are $\vec{A}$ = 10.0 m at 50.0° and $\vec{B}$ = 12.0 m at 30.0°. (a) Find the perimeter of the rectangle. (b) Find the magnitude and direction of the vector from the origin to the upper-right corner of the rectangle.

Figure P3.56

57. A vector is given by $\vec{R} = 2\hat{i} + \hat{j} + 3\hat{k}$. Find (a) the magnitudes of the x, y, and z components; (b) the magnitude of $\vec{R}$; and (c) the angles between $\vec{R}$ and the x, y, and z axes.

58. A ferry transports tourists between three islands. It sails from the first island to the second island, 4.76 km away, in a direction 37.0° north of east. It then sails from the second island to the third island in a direction 69.0° west of north. Finally it returns to the first island, sailing in a direction 28.0° east of south. Calculate the distance between (a) the second and third islands and (b) the first and third islands.

59. Two vectors $\vec{A}$ and $\vec{B}$ have precisely equal magnitudes. For the magnitude of $\vec{A} + \vec{B}$ to be 100 times larger than the magnitude of $\vec{A} - \vec{B}$, what must be the angle between them?

60. Two vectors $\vec{A}$ and $\vec{B}$ have precisely equal magnitudes. For the magnitude of $\vec{A} + \vec{B}$ to be larger than the magnitude of $\vec{A} - \vec{B}$ by the factor n, what must be the angle between them?

61. Let $\vec{A}$ = 60.0 cm at 270° measured from the horizontal. Let $\vec{B}$ = 80.0 cm at some angle θ. (a) Find the magnitude of $\vec{A} + \vec{B}$ as a function of θ. (b) From the answer to part (a), for what value of θ does $|\vec{A} + \vec{B}|$ take on its maximum value? What is this maximum value? (c) From the answer to part (a), for what value of θ does $|\vec{A} + \vec{B}|$ take on its minimum value? What is this minimum value? (d) Without reference to the answer to part (a), argue that the answers to each of parts (b) and (c) do or do not make sense.

62. After a ball rolls off the edge of a horizontal table at time $t = 0$, its velocity as a function of time is given by

$$\vec{v} = 1.2\hat{i} - 9.8t\hat{j}$$

where $\vec{v}$ is in meters per second and t is in seconds. The ball's displacement away from the edge of the table, during the time interval of 0.380 s for which the ball is in flight, is given by

$$\Delta\vec{r} = \int_0^{0.380\ \text{s}} \vec{v}\ dt$$

To perform the integral, you can use the calculus theorem

$$\int [A + Bf(x)]dx = \int A\ dx + B\int f(x)\ dx$$

You can think of the units and unit vectors as constants, represented by A and B. Perform the integration to calculate the displacement of the ball from the edge of the table at 0.380 s.

63. **Review.** The instantaneous position of an object is
specified by its position vector leading from a fixed origin to the location of the object, modeled as a particle. Suppose for a certain object the position vector is a function of time given by $\vec{r} = 4\hat{i} + 3\hat{j} - 2t\hat{k}$, where $\vec{r}$ is in meters and t is in seconds. (a) Evaluate $d\vec{r}/dt$. (b) What physical quantity does $d\vec{r}/dt$ represent about the object?

64. Ecotourists use their global positioning system indicator to determine their location inside a botanical garden as latitude 0.002 43 degree south of the equator, longitude 75.642 38 degrees west. They wish to visit a tree at latitude 0.001 62 degree north, longitude 75.644 26 degrees west. (a) Determine the straight-line distance and the direction in which they can walk to reach the tree as follows. First model the Earth as a sphere of radius 6.37×10^6 m to determine the westward and northward displacement components required, in meters. Then model the Earth as a flat surface to complete the calculation. (b) Explain why it is possible to use these two geometrical models together to solve the problem.

65. A rectangular parallelepiped has dimensions a, b, and c as shown in Figure P3.65. (a) Obtain a vector expression for the face diagonal vector $\vec{R}_1$. (b) What is the magnitude of this vector? (c) Notice that $\vec{R}_1$, $c\hat{k}$, and $\vec{R}_2$ make a right triangle. Obtain a vector expression for the body diagonal vector $\vec{R}_2$.

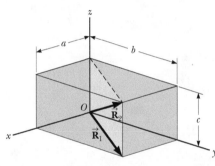

Figure P3.65

66. Vectors $\vec{A}$ and $\vec{B}$ have equal magnitudes of 5.00. The sum of $\vec{A}$ and $\vec{B}$ is the vector $6.00\hat{j}$. Determine the angle between $\vec{A}$ and $\vec{B}$.

Challenge Problem

67. A pirate has buried his treasure on an island with five trees located at the points (30.0 m, −20.0 m), (60.0 m, 80.0 m), (−10.0 m, −10.0 m), (40.0 m, −30.0 m), and (−70.0 m, 60.0 m), all measured relative to some origin, as shown in Figure P3.67. His ship's log instructs you to start at tree A and move toward tree B, but to cover only one-half the distance between A and B. Then move toward tree C, covering one-third the distance between your current location and C. Next move toward tree D, covering one-fourth the distance between where you are and D. Finally move toward tree E, covering one-fifth the distance between you and E, stop, and dig. (a) Assume you have correctly determined the order in which the pirate labeled the trees as A, B, C, D, and E as shown in the figure. What are the coordinates of the point where his treasure is buried? (b) **What If?** What if you do not really know the way the pirate labeled the trees? What would happen to the answer if you rearranged the order of the trees, for instance, to B (30 m, −20 m), A (60 m, 80 m), E (−10 m, −10 m), C (40 m, −30 m), and D (−70 m, 60 m)? State reasoning to show that the answer does not depend on the order in which the trees are labeled.

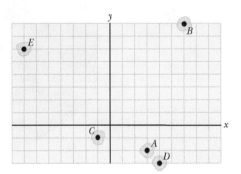

Figure P3.67

Motion in Two Dimensions

Fireworks erupt from the Sydney Harbour Bridge in New South Wales, Australia. Notice the parabolic paths of embers projected into the air. All projectiles follow a parabolic path in the absence of air resistance. *(Graham Monro/ Photolibrary/Jupiter Images)*

In this chapter, we explore the kinematics of a particle moving in two dimensions. Knowing the basics of two-dimensional motion will allow us—in future chapters—to examine a variety of situations, ranging from the motion of satellites in orbit to the motion of electrons in a uniform electric field. We begin by studying in greater detail the vector nature of position, velocity, and acceleration. We then treat projectile motion and uniform circular motion as special cases of motion in two dimensions. We also discuss the concept of relative motion, which shows why observers in different frames of reference may measure different positions and velocities for a given particle.

4.1 The Position, Velocity, and Acceleration Vectors

In Chapter 2, we found that the motion of a particle along a straight line such as the x axis is completely known if its position is known as a function of time. Let us now extend this idea to two-dimensional motion of a particle in the xy plane. We begin by describing the position of the particle. In one dimension, a single numerical value describes a particle's position, but in two dimensions, we indicate its position by its **position vector $\vec{r}$**, drawn from the origin of some coordinate system to the location of the particle in the xy plane as in Figure 4.1. At time t_i, the particle is at point Ⓐ, described by position vector $\vec{r}_i$. At some later time t_f, it is at point Ⓑ, described by position vector $\vec{r}_f$. The path followed by the particle from

Ⓐ to Ⓑ is not necessarily a straight line. As the particle moves from Ⓐ to Ⓑ in the time interval $\Delta t = t_f - t_i$, its position vector changes from $\vec{\mathbf{r}}_i$ to $\vec{\mathbf{r}}_f$. As we learned in Chapter 2, displacement is a vector, and the displacement of the particle is the difference between its final position and its initial position. We now define the **displacement vector** $\Delta\vec{\mathbf{r}}$ for a particle such as the one in Figure 4.1 as being the difference between its final position vector and its initial position vector:

$$\Delta\vec{\mathbf{r}} \equiv \vec{\mathbf{r}}_f - \vec{\mathbf{r}}_i \tag{4.1}$$

◀ **Displacement vector**

The direction of $\Delta\vec{\mathbf{r}}$ is indicated in Figure 4.1. As we see from the figure, the magnitude of $\Delta\vec{\mathbf{r}}$ is *less* than the distance traveled along the curved path followed by the particle.

As we saw in Chapter 2, it is often useful to quantify motion by looking at the displacement divided by the time interval during which that displacement occurs, which gives the rate of change of position. Two-dimensional (or three-dimensional) kinematics is similar to one-dimensional kinematics, but we must now use full vector notation rather than positive and negative signs to indicate the direction of motion.

We define the **average velocity** $\vec{\mathbf{v}}_{\text{avg}}$ of a particle during the time interval Δt as the displacement of the particle divided by the time interval:

$$\vec{\mathbf{v}}_{\text{avg}} \equiv \frac{\Delta\vec{\mathbf{r}}}{\Delta t} \tag{4.2}$$

◀ **Average velocity**

Multiplying or dividing a vector quantity by a positive scalar quantity such as Δt changes only the magnitude of the vector, not its direction. Because displacement is a vector quantity and the time interval is a positive scalar quantity, we conclude that the average velocity is a vector quantity directed along $\Delta\vec{\mathbf{r}}$. Compare Equation 4.2 with its one-dimensional counterpart, Equation 2.2.

The average velocity between points is *independent of the path* taken. That is because average velocity is proportional to displacement, which depends only on the initial and final position vectors and not on the path taken. As with one-dimensional motion, we conclude that if a particle starts its motion at some point and returns to this point via any path, its average velocity is zero for this trip because its displacement is zero. Consider again our basketball players on the court in Figure 2.2 (page 23). We previously considered only their one-dimensional motion back and forth between the baskets. In reality, however, they move over a two-dimensional surface, running back and forth between the baskets as well as left and right across the width of the court. Starting from one basket, a given player may follow a very complicated two-dimensional path. Upon returning to the original basket, however, a player's average velocity is zero because the player's displacement for the whole trip is zero.

Consider again the motion of a particle between two points in the *xy* plane as shown in Figure 4.2 (page 80). The dashed curve shows the path of the particle. As the time interval over which we observe the motion becomes smaller and smaller—that is, as Ⓑ is moved to Ⓑ′ and then to Ⓑ″ and so on—the direction of the displacement approaches that of the line tangent to the path at Ⓐ. The **instantaneous velocity** $\vec{\mathbf{v}}$ is defined as the limit of the average velocity $\Delta\vec{\mathbf{r}}/\Delta t$ as Δt approaches zero:

$$\vec{\mathbf{v}} \equiv \lim_{\Delta t \to 0} \frac{\Delta\vec{\mathbf{r}}}{\Delta t} = \frac{d\vec{\mathbf{r}}}{dt} \tag{4.3}$$

◀ **Instantaneous velocity**

That is, the instantaneous velocity equals the derivative of the position vector with respect to time. The direction of the instantaneous velocity vector at any point in a particle's path is along a line tangent to the path at that point and in the direction of motion. Compare Equation 4.3 with the corresponding one-dimensional version, Equation 2.5.

The magnitude of the instantaneous velocity vector $v = |\vec{\mathbf{v}}|$ of a particle is called the *speed* of the particle, which is a scalar quantity.

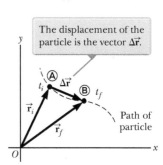

Figure 4.1 A particle moving in the *xy* plane is located with the position vector $\vec{\mathbf{r}}$ drawn from the origin to the particle. The displacement of the particle as it moves from Ⓐ to Ⓑ in the time interval $\Delta t = t_f - t_i$ is equal to the vector $\Delta\vec{\mathbf{r}} = \vec{\mathbf{r}}_f - \vec{\mathbf{r}}_i$.

Figure 4.2 As a particle moves between two points, its average velocity is in the direction of the displacement vector $\Delta \vec{\mathbf{r}}$. By definition, the instantaneous velocity at Ⓐ is directed along the line tangent to the curve at Ⓐ.

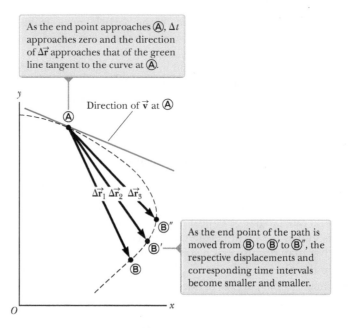

As the end point approaches Ⓐ, Δt approaches zero and the direction of $\Delta \vec{\mathbf{r}}$ approaches that of the green line tangent to the curve at Ⓐ.

Direction of $\vec{\mathbf{v}}$ at Ⓐ

$\Delta \vec{\mathbf{r}}_1$ $\Delta \vec{\mathbf{r}}_2$ $\Delta \vec{\mathbf{r}}_3$

As the end point of the path is moved from Ⓑ to Ⓑ′ to Ⓑ″, the respective displacements and corresponding time intervals become smaller and smaller.

As a particle moves from one point to another along some path, its instantaneous velocity vector changes from $\vec{\mathbf{v}}_i$ at time t_i to $\vec{\mathbf{v}}_f$ at time t_f. Knowing the velocity at these points allows us to determine the average acceleration of the particle. The **average acceleration $\vec{\mathbf{a}}_{avg}$** of a particle is defined as the change in its instantaneous velocity vector $\Delta \vec{\mathbf{v}}$ divided by the time interval Δt during which that change occurs:

◀ **Average acceleration**

$$\vec{\mathbf{a}}_{avg} \equiv \frac{\Delta \vec{\mathbf{v}}}{\Delta t} = \frac{\vec{\mathbf{v}}_f - \vec{\mathbf{v}}_i}{t_f - t_i} \qquad (4.4)$$

Because $\vec{\mathbf{a}}_{avg}$ is the ratio of a vector quantity $\Delta \vec{\mathbf{v}}$ and a positive scalar quantity Δt, we conclude that average acceleration is a vector quantity directed along $\Delta \vec{\mathbf{v}}$. As indicated in Figure 4.3, the direction of $\Delta \vec{\mathbf{v}}$ is found by adding the vector $-\vec{\mathbf{v}}_i$ (the negative of $\vec{\mathbf{v}}_i$) to the vector $\vec{\mathbf{v}}_f$ because, by definition, $\Delta \vec{\mathbf{v}} = \vec{\mathbf{v}}_f - \vec{\mathbf{v}}_i$. Compare Equation 4.4 with Equation 2.9.

When the average acceleration of a particle changes during different time intervals, it is useful to define its instantaneous acceleration. The **instantaneous acceleration $\vec{\mathbf{a}}$** is defined as the limiting value of the ratio $\Delta \vec{\mathbf{v}} / \Delta t$ as Δt approaches zero:

◀ **Instantaneous acceleration**

$$\vec{\mathbf{a}} \equiv \lim_{\Delta t \to 0} \frac{\Delta \vec{\mathbf{v}}}{\Delta t} = \frac{d\vec{\mathbf{v}}}{dt} \qquad (4.5)$$

In other words, the instantaneous acceleration equals the derivative of the velocity vector with respect to time. Compare Equation 4.5 with Equation 2.10.

Various changes can occur when a particle accelerates. First, the magnitude of the velocity vector (the speed) may change with time as in straight-line (one-

Pitfall Prevention 4.1

Vector Addition Although the vector addition discussed in Chapter 3 involves *displacement* vectors, vector addition can be applied to *any* type of vector quantity. Figure 4.3, for example, shows the addition of *velocity* vectors using the graphical approach.

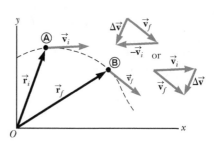

Figure 4.3 A particle moves from position Ⓐ to position Ⓑ. Its velocity vector changes from $\vec{\mathbf{v}}_i$ to $\vec{\mathbf{v}}_f$. The vector diagrams at the upper right show two ways of determining the vector $\Delta \vec{\mathbf{v}}$ from the initial and final velocities.

dimensional) motion. Second, the direction of the velocity vector may change with time even if its magnitude (speed) remains constant as in two-dimensional motion along a curved path. Finally, both the magnitude and the direction of the velocity vector may change simultaneously.

> **Q** uick Quiz 4.1 Consider the following controls in an automobile in motion: gas pedal, brake, steering wheel. What are the controls in this list that cause an acceleration of the car? **(a)** all three controls **(b)** the gas pedal and the brake **(c)** only the brake **(d)** only the gas pedal **(e)** only the steering wheel

4.2 Two-Dimensional Motion with Constant Acceleration

In Section 2.5, we investigated one-dimensional motion of a particle under constant acceleration and developed the particle under constant acceleration model. Let us now consider two-dimensional motion during which the acceleration of a particle remains constant in both magnitude and direction. As we shall see, this approach is useful for analyzing some common types of motion.

Before embarking on this investigation, we need to emphasize an important point regarding two-dimensional motion. Imagine an air hockey puck moving in a straight line along a perfectly level, friction-free surface of an air hockey table. Figure 4.4a shows a motion diagram from an overhead point of view of this puck. Recall that in Section 2.4 we related the acceleration of an object to a force on the object. Because there are no forces on the puck in the horizontal plane, it moves with constant velocity in the x direction. Now suppose you blow a puff of air on the puck as it passes your position, with the force from your puff of air *exactly* in the y direction. Because the force from this puff of air has no component in the x direction, it causes no acceleration in the x direction. It only causes a momentary acceleration in the y direction, causing the puck to have a constant y component of velocity once the force from the puff of air is removed. After your puff of air on the puck, its velocity component in the x direction is unchanged as shown in Figure 4.4b. The generalization of this simple experiment is that **motion in two dimensions can be modeled as two *independent* motions in each of the two perpendicular directions associated with the x and y axes. That is, any influence in the y direction does not affect the motion in the x direction and vice versa.**

The position vector for a particle moving in the xy plane can be written

$$\vec{\mathbf{r}} = x\hat{\mathbf{i}} + y\hat{\mathbf{j}} \tag{4.6}$$

where x, y, and $\vec{\mathbf{r}}$ change with time as the particle moves while the unit vectors $\hat{\mathbf{i}}$ and $\hat{\mathbf{j}}$ remain constant. If the position vector is known, the velocity of the particle can be obtained from Equations 4.3 and 4.6, which give

$$\vec{\mathbf{v}} = \frac{d\vec{\mathbf{r}}}{dt} = \frac{dx}{dt}\hat{\mathbf{i}} + \frac{dy}{dt}\hat{\mathbf{j}} = v_x\hat{\mathbf{i}} + v_y\hat{\mathbf{j}} \tag{4.7}$$

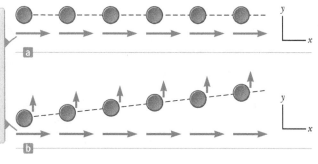

The horizontal red vectors, representing the x component of the velocity, are the same length in both parts of the figure, which demonstrates that motion in two dimensions can be modeled as two independent motions in perpendicular directions.

Figure 4.4 (a) A puck moves across a horizontal air hockey table at constant velocity in the x direction. (b) After a puff of air in the y direction is applied to the puck, the puck has gained a y component of velocity, but the x component is unaffected by the force in the perpendicular direction.

Because the acceleration $\vec{\mathbf{a}}$ of the particle is assumed constant in this discussion, its components a_x and a_y also are constants. Therefore, we can model the particle as a particle under constant acceleration independently in each of the two directions and apply the equations of kinematics separately to the x and y components of the velocity vector. Substituting, from Equation 2.13, $v_{xf} = v_{xi} + a_x t$ and $v_{yf} = v_{yi} + a_y t$ into Equation 4.7 to determine the final velocity at any time t, we obtain

$$\vec{\mathbf{v}}_f = (v_{xi} + a_x t)\hat{\mathbf{i}} + (v_{yi} + a_y t)\hat{\mathbf{j}} = (v_{xi}\hat{\mathbf{i}} + v_{yi}\hat{\mathbf{j}}) + (a_x\hat{\mathbf{i}} + a_y\hat{\mathbf{j}})t$$

Velocity vector as ▶
a function of time for a
particle under constant
acceleration in two
dimensions

$$\vec{\mathbf{v}}_f = \vec{\mathbf{v}}_i + \vec{\mathbf{a}}t \tag{4.8}$$

This result states that the velocity of a particle at some time t equals the vector sum of its initial velocity $\vec{\mathbf{v}}_i$ at time $t = 0$ and the additional velocity $\vec{\mathbf{a}}t$ acquired at time t as a result of constant acceleration. Equation 4.8 is the vector version of Equation 2.13.

Similarly, from Equation 2.16 we know that the x and y coordinates of a particle under constant acceleration are

$$x_f = x_i + v_{xi}t + \tfrac{1}{2}a_x t^2 \qquad y_f = y_i + v_{yi}t + \tfrac{1}{2}a_y t^2$$

Substituting these expressions into Equation 4.6 (and labeling the final position vector $\vec{\mathbf{r}}_f$) gives

$$\vec{\mathbf{r}}_f = (x_i + v_{xi}t + \tfrac{1}{2}a_x t^2)\hat{\mathbf{i}} + (y_i + v_{yi}t + \tfrac{1}{2}a_y t^2)\hat{\mathbf{j}}$$
$$= (x_i\hat{\mathbf{i}} + y_i\hat{\mathbf{j}}) + (v_{xi}\hat{\mathbf{i}} + v_{yi}\hat{\mathbf{j}})t + \tfrac{1}{2}(a_x\hat{\mathbf{i}} + a_y\hat{\mathbf{j}})t^2$$

Position vector as ▶
a function of time for a
particle under constant
acceleration in two
dimensions

$$\vec{\mathbf{r}}_f = \vec{\mathbf{r}}_i + \vec{\mathbf{v}}_i t + \tfrac{1}{2}\vec{\mathbf{a}}t^2 \tag{4.9}$$

which is the vector version of Equation 2.16. Equation 4.9 tells us that the position vector $\vec{\mathbf{r}}_f$ of a particle is the vector sum of the original position $\vec{\mathbf{r}}_i$, a displacement $\vec{\mathbf{v}}_i t$ arising from the initial velocity of the particle, and a displacement $\tfrac{1}{2}\vec{\mathbf{a}}t^2$ resulting from the constant acceleration of the particle.

We can consider Equations 4.8 and 4.9 to be the mathematical representation of a two-dimensional version of the particle under constant acceleration model. Graphical representations of Equations 4.8 and 4.9 are shown in Figure 4.5. The components of the position and velocity vectors are also illustrated in the figure. Notice from Figure 4.5a that $\vec{\mathbf{v}}_f$ is generally not along the direction of either $\vec{\mathbf{v}}_i$ or $\vec{\mathbf{a}}$ because the relationship between these quantities is a vector expression. For the same reason, from Figure 4.5b we see that $\vec{\mathbf{r}}_f$ is generally not along the direction of $\vec{\mathbf{r}}_i$, $\vec{\mathbf{v}}_i$, or $\vec{\mathbf{a}}$. Finally, notice that $\vec{\mathbf{v}}_f$ and $\vec{\mathbf{r}}_f$ are generally not in the same direction.

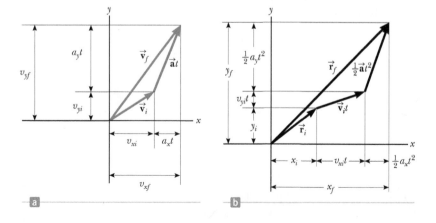

Figure 4.5 Vector representations and components of (a) the velocity and (b) the position of a particle under constant acceleration in two dimensions.

Example 4.1 Motion in a Plane

A particle moves in the xy plane, starting from the origin at $t = 0$ with an initial velocity having an x component of 20 m/s and a y component of -15 m/s. The particle experiences an acceleration in the x direction, given by $a_x = 4.0$ m/s^2.

(A) Determine the total velocity vector at any time.

SOLUTION

Conceptualize The components of the initial velocity tell us that the particle starts by moving toward the right and downward. The x component of velocity starts at 20 m/s and increases by 4.0 m/s every second. The y component of velocity never changes from its initial value of -15 m/s. We sketch a motion diagram of the situation in Figure 4.6. Because the particle is accelerating in the $+x$ direction, its velocity component in this direction increases and the path curves as shown in the diagram. Notice that the spacing between successive images increases as time goes on because the speed is increasing. The placement of the acceleration and velocity vectors in Figure 4.6 helps us further conceptualize the situation.

Figure 4.6 (Example 4.1) Motion diagram for the particle.

Categorize Because the initial velocity has components in both the x and y directions, we categorize this problem as one involving a particle moving in two dimensions. Because the particle only has an x component of acceleration, we model it as a *particle under constant acceleration* in the x direction and a *particle under constant velocity* in the y direction.

Analyze To begin the mathematical analysis, we set $v_{xi} = 20$ m/s, $v_{yi} = -15$ m/s, $a_x = 4.0$ m/s^2, and $a_y = 0$.

Use Equation 4.8 for the velocity vector:

$$\vec{v}_f = \vec{v}_i + \vec{a}t = (v_{xi} + a_x t)\hat{i} + (v_{yi} + a_y t)\hat{j}$$

Substitute numerical values with the velocity in meters per second and the time in seconds:

$$\vec{v}_f = [20 + (4.0)t]\hat{i} + [-15 + (0)t]\hat{j}$$

$$(1) \quad \vec{v}_f = [(20 + 4.0t)\hat{i} - 15\hat{j}]$$

Finalize Notice that the x component of velocity increases in time while the y component remains constant; this result is consistent with our prediction.

(B) Calculate the velocity and speed of the particle at $t = 5.0$ s and the angle the velocity vector makes with the x axis.

SOLUTION

Analyze

Evaluate the result from Equation (1) at $t = 5.0$ s:

$$\vec{v}_f = [(20 + 4.0(5.0))\hat{i} - 15\hat{j}] = (40\hat{i} - 15\hat{j}) \text{ m/s}$$

Determine the angle θ that $\vec{v}_f$ makes with the x axis at $t = 5.0$ s:

$$\theta = \tan^{-1}\left(\frac{v_{yf}}{v_{xf}}\right) = \tan^{-1}\left(\frac{-15 \text{ m/s}}{40 \text{ m/s}}\right) = -21°$$

Evaluate the speed of the particle as the magnitude of $\vec{v}_f$:

$$v_f = |\vec{v}_f| = \sqrt{v_{xf}^2 + v_{yf}^2} = \sqrt{(40)^2 + (-15)^2} \text{ m/s} = 43 \text{ m/s}$$

Finalize The negative sign for the angle θ indicates that the velocity vector is directed at an angle of 21° below the positive x axis. Notice that if we calculate v_i from the x and y components of $\vec{v}_i$, we find that $v_f > v_i$. Is that consistent with our prediction?

(C) Determine the x and y coordinates of the particle at any time t and its position vector at this time.

continued

▶ **4.1** continued

SOLUTION

Analyze

Use the components of Equation 4.9 with $x_i = y_i = 0$ at $t = 0$ and with x and y in meters and t in seconds:

$$x_f = v_{xi}t + \tfrac{1}{2}a_x t^2 = 20t + 2.0t^2$$

$$y_f = v_{yi}t = -15t$$

Express the position vector of the particle at any time t:

$$\vec{\mathbf{r}}_f = x_f \hat{\mathbf{i}} + y_f \hat{\mathbf{j}} = (20t + 2.0t^2)\hat{\mathbf{i}} - 15t\hat{\mathbf{j}}$$

Finalize Let us now consider a limiting case for very large values of t.

WHAT IF? What if we wait a very long time and then observe the motion of the particle? How would we describe the motion of the particle for large values of the time?

Answer Looking at Figure 4.6, we see the path of the particle curving toward the x axis. There is no reason to assume this tendency will change, which suggests that the path will become more and more parallel to the x axis as time grows large. Mathematically, Equation (1) shows that the y component of the velocity remains constant while the x component grows linearly with t. Therefore, when t is very large, the x component of the velocity will be much larger than the y component, suggesting that the velocity vector becomes more and more parallel to the x axis. The magnitudes of both x_f and y_f continue to grow with time, although x_f grows much faster.

Pitfall Prevention 4.2

Acceleration at the Highest Point
As discussed in Pitfall Prevention 2.8, many people claim that the acceleration of a projectile at the topmost point of its trajectory is zero. This mistake arises from confusion between zero vertical velocity and zero acceleration. If the projectile were to experience zero acceleration at the highest point, its velocity at that point would not change; rather, the projectile would move horizontally at constant speed from then on! That does not happen, however, because the acceleration is *not* zero anywhere along the trajectory.

A welder cuts holes through a heavy metal construction beam with a hot torch. The sparks generated in the process follow parabolic paths.

4.3 Projectile Motion

Anyone who has observed a baseball in motion has observed projectile motion. The ball moves in a curved path and returns to the ground. **Projectile motion** of an object is simple to analyze if we make two assumptions: (1) the free-fall acceleration is constant over the range of motion and is directed downward,[1] and (2) the effect of air resistance is negligible.[2] With these assumptions, we find that the path of a projectile, which we call its *trajectory*, is *always* a parabola as shown in Figure 4.7. **We use these assumptions throughout this chapter.**

The expression for the position vector of the projectile as a function of time follows directly from Equation 4.9, with its acceleration being that due to gravity, $\vec{\mathbf{a}} = \vec{\mathbf{g}}$:

$$\vec{\mathbf{r}}_f = \vec{\mathbf{r}}_i + \vec{\mathbf{v}}_i t + \tfrac{1}{2}\vec{\mathbf{g}}t^2 \tag{4.10}$$

where the initial x and y components of the velocity of the projectile are

$$v_{xi} = v_i \cos\theta_i \qquad v_{yi} = v_i \sin\theta_i \tag{4.11}$$

The expression in Equation 4.10 is plotted in Figure 4.8 for a projectile launched from the origin, so that $\vec{\mathbf{r}}_i = 0$. The final position of a particle can be considered to be the superposition of its initial position $\vec{\mathbf{r}}_i$; the term $\vec{\mathbf{v}}_i t$, which is its displacement if no acceleration were present; and the term $\tfrac{1}{2}\vec{\mathbf{g}}t^2$ that arises from its acceleration due to gravity. In other words, if there were no gravitational acceleration, the particle would continue to move along a straight path in the direction of $\vec{\mathbf{v}}_i$. Therefore, the vertical distance $\tfrac{1}{2}\vec{\mathbf{g}}t^2$ through which the particle "falls" off the straight-line path is the same distance that an object dropped from rest would fall during the same time interval.

---

[1]This assumption is reasonable as long as the range of motion is small compared with the radius of the Earth (6.4×10^6 m). In effect, this assumption is equivalent to assuming the Earth is flat over the range of motion considered.

[2]This assumption is often *not* justified, especially at high velocities. In addition, any spin imparted to a projectile, such as that applied when a pitcher throws a curve ball, can give rise to some very interesting effects associated with aerodynamic forces, which will be discussed in Chapter 14.

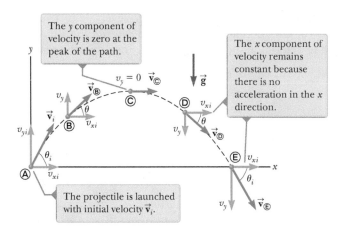

Figure 4.7 The parabolic path of a projectile that leaves the origin with a velocity $\vec{v}_i$. The velocity vector $\vec{v}$ changes with time in both magnitude and direction. This change is the result of acceleration $\vec{a} = \vec{g}$ in the negative y direction.

In Section 4.2, we stated that two-dimensional motion with constant acceleration can be analyzed as a combination of two independent motions in the x and y directions, with accelerations a_x and a_y. Projectile motion can also be handled in this way, with acceleration $a_x = 0$ in the x direction and a constant acceleration $a_y = -g$ in the y direction. Therefore, when solving projectile motion problems, use two analysis models: (1) the particle under constant velocity in the horizontal direction (Eq. 2.7):

$$x_f = x_i + v_{xi}t$$

and (2) the particle under constant acceleration in the vertical direction (Eqs. 2.13–2.17 with x changed to y and $a_y = -g$):

$$v_{yf} = v_{yi} - gt$$

$$v_{y,\text{avg}} = \frac{v_{yi} + v_{yf}}{2}$$

$$y_f = y_i + \tfrac{1}{2}(v_{yi} + v_{yf})t$$

$$y_f = y_i + v_{yi}t - \tfrac{1}{2}gt^2$$

$$v_{yf}^2 = v_{yi}^2 - 2g(y_f - y_i)$$

The horizontal and vertical components of a projectile's motion are completely independent of each other and can be handled separately, with time t as the common variable for both components.

Quick Quiz 4.2 **(i)** As a projectile thrown upward moves in its parabolic path (such as in Fig. 4.8), at what point along its path are the velocity and acceleration vectors for the projectile perpendicular to each other? (a) nowhere (b) the highest point (c) the launch point **(ii)** From the same choices, at what point are the velocity and acceleration vectors for the projectile parallel to each other?

Horizontal Range and Maximum Height of a Projectile

Before embarking on some examples, let us consider a special case of projectile motion that occurs often. Assume a projectile is launched from the origin at $t_i = 0$ with a positive v_{yi} component as shown in Figure 4.9 and returns to *the same horizontal level*. This situation is common in sports, where baseballs, footballs, and golf balls often land at the same level from which they were launched.

Two points in this motion are especially interesting to analyze: the peak point Ⓐ, which has Cartesian coordinates $(R/2, h)$, and the point Ⓑ, which has coordinates $(R, 0)$. The distance R is called the *horizontal range* of the projectile, and the distance h is its *maximum height*. Let us find h and R mathematically in terms of v_i, θ_i, and g.

Figure 4.8 The position vector $\vec{r}_f$ of a projectile launched from the origin whose initial velocity at the origin is $\vec{v}_i$. The vector $\vec{v}_i t$ would be the displacement of the projectile if gravity were absent, and the vector $\tfrac{1}{2}\vec{g}t^2$ is its vertical displacement from a straight-line path due to its downward gravitational acceleration.

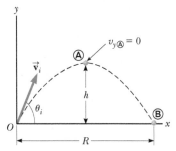

Figure 4.9 A projectile launched over a flat surface from the origin at $t_i = 0$ with an initial velocity $\vec{v}_i$. The maximum height of the projectile is h, and the horizontal range is R. At Ⓐ, the peak of the trajectory, the particle has coordinates $(R/2, h)$.

We can determine h by noting that at the peak $v_{y\text{Ⓐ}} = 0$. Therefore, from the particle under constant acceleration model, we can use the y direction version of Equation 2.13 to determine the time $t_\text{Ⓐ}$ at which the projectile reaches the peak:

$$v_{yf} = v_{yi} - gt \quad \rightarrow \quad 0 = v_i \sin \theta_i - gt_\text{Ⓐ}$$

$$t_\text{Ⓐ} = \frac{v_i \sin \theta_i}{g}$$

Substituting this expression for $t_\text{Ⓐ}$ into the y direction version of Equation 2.16 and replacing $y_f = y_\text{Ⓐ}$ with h, we obtain an expression for h in terms of the magnitude and direction of the initial velocity vector:

$$y_f = y_i + v_{yi}t - \tfrac{1}{2}gt^2 \quad \rightarrow \quad h = (v_i \sin \theta_i)\frac{v_i \sin \theta_i}{g} - \tfrac{1}{2}g\left(\frac{v_i \sin \theta_i}{g}\right)^2$$

$$h = \frac{v_i^2 \sin^2 \theta_i}{2g} \tag{4.12}$$

The range R is the horizontal position of the projectile at a time that is twice the time at which it reaches its peak, that is, at time $t_\text{Ⓑ} = 2t_\text{Ⓐ}$. Using the particle under constant velocity model, noting that $v_{xi} = v_{x\text{Ⓑ}} = v_i \cos \theta_i$, and setting $x_\text{Ⓑ} = R$ at $t = 2t_\text{Ⓐ}$, we find that

$$x_f = x_i + v_{xi}t \quad \rightarrow \quad R = v_{xi}t_\text{Ⓑ} = (v_i \cos \theta_i)2t_\text{Ⓐ}$$

$$= (v_i \cos \theta_i)\frac{2v_i \sin \theta_i}{g} = \frac{2v_i^2 \sin \theta_i \cos \theta_i}{g}$$

Using the identity $\sin 2\theta = 2 \sin \theta \cos \theta$ (see Appendix B.4), we can write R in the more compact form

$$R = \frac{v_i^2 \sin 2\theta_i}{g} \tag{4.13}$$

The maximum value of R from Equation 4.13 is $R_{max} = v_i^2/g$. This result makes sense because the maximum value of $\sin 2\theta_i$ is 1, which occurs when $2\theta_i = 90°$. Therefore, R is a maximum when $\theta_i = 45°$.

Figure 4.10 illustrates various trajectories for a projectile having a given initial speed but launched at different angles. As you can see, the range is a maximum for $\theta_i = 45°$. In addition, for any θ_i other than 45°, a point having Cartesian coordinates $(R, 0)$ can be reached by using either one of two complementary values of θ_i, such as 75° and 15°. Of course, the maximum height and time of flight for one of these values of θ_i are different from the maximum height and time of flight for the complementary value.

Ⓠuick Quiz 4.3 Rank the launch angles for the five paths in Figure 4.10 with respect to time of flight from the shortest time of flight to the longest.

> **Pitfall Prevention 4.3**
> **The Range Equation** Equation 4.13 is useful for calculating R only for a symmetric path as shown in Figure 4.10. If the path is not symmetric, *do not use this equation*. The particle under constant velocity and particle under constant acceleration models are the important starting points because they give the position and velocity components of *any* projectile moving with constant acceleration in two dimensions at *any* time t.

Figure 4.10 A projectile launched over a flat surface from the origin with an initial speed of 50 m/s at various angles of projection.

Problem-Solving Strategy | Projectile Motion

We suggest you use the following approach when solving projectile motion problems.

1. Conceptualize. Think about what is going on physically in the problem. Establish the mental representation by imagining the projectile moving along its trajectory.

2. Categorize. Confirm that the problem involves a particle in free fall and that air resistance is neglected. Select a coordinate system with x in the horizontal direction and y in the vertical direction. Use the particle under constant velocity model for the x component of the motion. Use the particle under constant acceleration model for the y direction. In the special case of the projectile returning to the same level from which it was launched, use Equations 4.12 and 4.13.

3. Analyze. If the initial velocity vector is given, resolve it into x and y components. Select the appropriate equation(s) from the particle under constant acceleration model for the vertical motion and use these along with Equation 2.7 for the horizontal motion to solve for the unknown(s).

4. Finalize. Once you have determined your result, check to see if your answers are consistent with the mental and pictorial representations and your results are realistic.

Example 4.2 | **The Long Jump**

A long jumper (Fig. 4.11) leaves the ground at an angle of 20.0° above the horizontal and at a speed of 11.0 m/s.

(A) How far does he jump in the horizontal direction?

SOLUTION

Conceptualize The arms and legs of a long jumper move in a complicated way, but we will ignore this motion. We conceptualize the motion of the long jumper as equivalent to that of a simple projectile.

Categorize We categorize this example as a projectile motion problem. Because the initial speed and launch angle are given and because the final height is the same as the initial height, we further categorize this problem as satisfying the conditions for which Equations 4.12 and 4.13 can be used. This approach is the most direct way to analyze this problem, although the general methods that have been described will always give the correct answer.

Figure 4.11 (Example 4.2) Romain Barras of France competes in the men's decathlon long jump at the 2008 Beijing Olympic Games.

Analyze

Use Equation 4.13 to find the range of the jumper:

$$R = \frac{v_i^2 \sin 2\theta_i}{g} = \frac{(11.0 \text{ m/s})^2 \sin 2(20.0°)}{9.80 \text{ m/s}^2} = 7.94 \text{ m}$$

(B) What is the maximum height reached?

SOLUTION

Analyze

Find the maximum height reached by using Equation 4.12:

$$h = \frac{v_i^2 \sin^2\theta_i}{2g} = \frac{(11.0 \text{ m/s})^2 (\sin 20.0°)^2}{2(9.80 \text{ m/s}^2)} = 0.722 \text{ m}$$

Finalize Find the answers to parts (A) and (B) using the general method. The results should agree. Treating the long jumper as a particle is an oversimplification. Nevertheless, the values obtained are consistent with experience in sports. We can model a complicated system such as a long jumper as a particle and still obtain reasonable results.

Example 4.3   A Bull's-Eye Every Time   AM

In a popular lecture demonstration, a projectile is fired at a target in such a way that the projectile leaves the gun at the same time the target is dropped from rest. Show that if the gun is initially aimed at the stationary target, the projectile hits the falling target as shown in Figure 4.12a.

SOLUTION

Conceptualize We conceptualize the problem by studying Figure 4.12a. Notice that the problem does not ask for numerical values. The expected result must involve an algebraic argument.

The velocity of the projectile (red arrows) changes in direction and magnitude, but its acceleration (purple arrows) remains constant.

© Cengage Learning/Charles D. Winters

Figure 4.12 (Example 4.3) (a) Multiflash photograph of the projectile–target demonstration. If the gun is aimed directly at the target and is fired at the same instant the target begins to fall, the projectile will hit the target. (b) Schematic diagram of the projectile–target demonstration.

Categorize Because both objects are subject only to gravity, we categorize this problem as one involving two objects in free fall, the target moving in one dimension and the projectile moving in two. The target T is modeled as a *particle under constant acceleration* in one dimension. The projectile P is modeled as a *particle under constant acceleration* in the y direction and a *particle under constant velocity* in the x direction.

Analyze Figure 4.12b shows that the initial y coordinate y_{iT} of the target is $x_T \tan \theta_i$ and its initial velocity is zero. It falls with acceleration $a_y = -g$.

Write an expression for the y coordinate of the target at any moment after release, noting that its initial velocity is zero:

$$(1) \quad y_T = y_{iT} + (0)t - \tfrac{1}{2}gt^2 = x_T \tan \theta_i - \tfrac{1}{2}gt^2$$

Write an expression for the y coordinate of the projectile at any moment:

$$(2) \quad y_P = y_{iP} + v_{yiP}t - \tfrac{1}{2}gt^2 = 0 + (v_{iP} \sin \theta_i)t - \tfrac{1}{2}gt^2 = (v_{iP} \sin \theta_i)t - \tfrac{1}{2}gt^2$$

Write an expression for the x coordinate of the projectile at any moment:

$$x_P = x_{iP} + v_{xiP}t = 0 + (v_{iP} \cos \theta_i)t = (v_{iP} \cos \theta_i)t$$

Solve this expression for time as a function of the horizontal position of the projectile:

$$t = \frac{x_P}{v_{iP} \cos \theta_i}$$

Substitute this expression into Equation (2):

$$(3) \quad y_P = (v_{iP} \sin \theta_i)\left(\frac{x_P}{v_{iP} \cos \theta_i}\right) - \tfrac{1}{2}gt^2 = x_P \tan \theta_i - \tfrac{1}{2}gt^2$$

Finalize Compare Equations (1) and (3). We see that when the x coordinates of the projectile and target are the same—that is, when $x_T = x_P$—their y coordinates given by Equations (1) and (3) are the same and a collision results.

Example 4.4   That's Quite an Arm!

A stone is thrown from the top of a building upward at an angle of 30.0° to the horizontal with an initial speed of 20.0 m/s as shown in Figure 4.13. The height from which the stone is thrown is 45.0 m above the ground.

(A) How long does it take the stone to reach the ground?

SOLUTION

Conceptualize Study Figure 4.13, in which we have indicated the trajectory and various parameters of the motion of the stone.

Categorize We categorize this problem as a projectile motion problem. The stone is modeled as a *particle under constant acceleration* in the y direction and a *particle under constant velocity* in the x direction.

Figure 4.13
(Example 4.4) A stone is thrown from the top of a building.

Analyze We have the information $x_i = y_i = 0$, $y_f = -45.0$ m, $a_y = -g$, and $v_i = 20.0$ m/s (the numerical value of y_f is negative because we have chosen the point of the throw as the origin).

Find the initial x and y components of the stone's velocity:

$$v_{xi} = v_i \cos \theta_i = (20.0 \text{ m/s}) \cos 30.0° = 17.3 \text{ m/s}$$
$$v_{yi} = v_i \sin \theta_i = (20.0 \text{ m/s}) \sin 30.0° = 10.0 \text{ m/s}$$

Express the vertical position of the stone from the particle under constant acceleration model:

$$y_f = y_i + v_{yi} t - \tfrac{1}{2} g t^2$$

Substitute numerical values:

$$-45.0 \text{ m} = 0 + (10.0 \text{ m/s})t + \tfrac{1}{2}(-9.80 \text{ m/s}^2)t^2$$

Solve the quadratic equation for t:

$$t = 4.22 \text{ s}$$

(B) What is the speed of the stone just before it strikes the ground?

SOLUTION

Analyze Use the velocity equation in the particle under constant acceleration model to obtain the y component of the velocity of the stone just before it strikes the ground:

$$v_{yf} = v_{yi} - gt$$

Substitute numerical values, using $t = 4.22$ s:

$$v_{yf} = 10.0 \text{ m/s} + (-9.80 \text{ m/s}^2)(4.22 \text{ s}) = -31.3 \text{ m/s}$$

Use this component with the horizontal component $v_{xf} = v_{xi} = 17.3$ m/s to find the speed of the stone at $t = 4.22$ s:

$$v_f = \sqrt{v_{xf}^2 + v_{yf}^2} = \sqrt{(17.3 \text{ m/s})^2 + (-31.3 \text{ m/s})^2} = 35.8 \text{ m/s}$$

Finalize Is it reasonable that the y component of the final velocity is negative? Is it reasonable that the final speed is larger than the initial speed of 20.0 m/s?

WHAT IF? What if a horizontal wind is blowing in the same direction as the stone is thrown and it causes the stone to have a horizontal acceleration component $a_x = 0.500$ m/s²? Which part of this example, (A) or (B), will have a different answer?

Answer Recall that the motions in the x and y directions are independent. Therefore, the horizontal wind cannot affect the vertical motion. The vertical motion determines the time of the projectile in the air, so the answer to part (A) does not change. The wind causes the horizontal velocity component to increase with time, so the final speed will be larger in part (B). Taking $a_x = 0.500$ m/s², we find $v_{xf} = 19.4$ m/s and $v_f = 36.9$ m/s.

Example 4.5   The End of the Ski Jump

A ski jumper leaves the ski track moving in the horizontal direction with a speed of 25.0 m/s as shown in Figure 4.14. The landing incline below her falls off with a slope of 35.0°. Where does she land on the incline?

SOLUTION

Conceptualize We can conceptualize this problem based on memories of observing winter Olympic ski competitions. We estimate the skier to be airborne for perhaps 4 s and to travel a distance of about 100 m horizontally. We should expect the value of d, the distance traveled along the incline, to be of the same order of magnitude.

Categorize We categorize the problem as one of a particle in projectile motion. As with other projectile motion problems, we use the *particle under constant velocity* model for the horizontal motion and the *particle under constant acceleration* model for the vertical motion.

Figure 4.14 (Example 4.5) A ski jumper leaves the track moving in a horizontal direction.

Analyze It is convenient to select the beginning of the jump as the origin. The initial velocity components are $v_{xi} = 25.0$ m/s and $v_{yi} = 0$. From the right triangle in Figure 4.14, we see that the jumper's x and y coordinates at the landing point are given by $x_f = d \cos \phi$ and $y_f = -d \sin \phi$.

Express the coordinates of the jumper as a function of time, using the particle under constant velocity model for x and the position equation from the particle under constant acceleration model for y:

$$(1) \quad x_f = v_{xi}t$$

$$(2) \quad y_f = v_{yi}t - \tfrac{1}{2}gt^2$$

$$(3) \quad d \cos \phi = v_{xi}t$$

$$(4) \quad -d \sin \phi = -\tfrac{1}{2}gt^2$$

Solve Equation (3) for t and substitute the result into Equation (4):

$$-d \sin \phi = -\tfrac{1}{2}g \left(\frac{d \cos \phi}{v_{xi}} \right)^2$$

Solve for d and substitute numerical values:

$$d = \frac{2v_{xi}^2 \sin \phi}{g \cos^2 \phi} = \frac{2(25.0 \text{ m/s})^2 \sin 35.0°}{(9.80 \text{ m/s}^2) \cos^2 35.0°} = 109 \text{ m}$$

Evaluate the x and y coordinates of the point at which the skier lands:

$$x_f = d \cos \phi = (109 \text{ m}) \cos 35.0° = \boxed{89.3 \text{ m}}$$

$$y_f = -d \sin \phi = -(109 \text{ m}) \sin 35.0° = \boxed{-62.5 \text{ m}}$$

Finalize Let us compare these results with our expectations. We expected the horizontal distance to be on the order of 100 m, and our result of 89.3 m is indeed on this order of magnitude. It might be useful to calculate the time interval that the jumper is in the air and compare it with our estimate of about 4 s.

WHAT IF?   Suppose everything in this example is the same except the ski jump is curved so that the jumper is projected upward at an angle from the end of the track. Is this design better in terms of maximizing the length of the jump?

Answer If the initial velocity has an upward component, the skier will be in the air longer and should therefore travel farther. Tilting the initial velocity vector upward, however, will reduce the horizontal component of the initial velocity. Therefore, angling the end of the ski track upward at a *large* angle may actually *reduce* the distance. Consider the extreme case: the skier is projected at 90° to the horizontal and simply goes up and comes back down at the end of the ski track! This argument suggests that there must be an optimal angle between 0° and 90° that represents a balance between making the flight time longer and the horizontal velocity component smaller.

Let us find this optimal angle mathematically. We modify Equations (1) through (4) in the following way, assuming the skier is projected at an angle θ with respect to the horizontal over a landing incline sloped with an arbitrary angle ϕ:

$$(1) \text{ and } (3) \quad \rightarrow \quad x_f = (v_i \cos \theta)t = d \cos \phi$$

$$(2) \text{ and } (4) \quad \rightarrow \quad y_f = (v_i \sin \theta)t - \tfrac{1}{2}gt^2 = -d \sin \phi$$

▶ **4.5** continued

By eliminating the time t between these equations and using differentiation to maximize d in terms of θ, we arrive (after several steps; see Problem 88) at the following equation for the angle θ that gives the maximum value of d:

$$\theta = 45° - \frac{\phi}{2}$$

For the slope angle in Figure 4.14, $\phi = 35.0°$; this equation results in an optimal launch angle of $\theta = 27.5°$. For a slope angle of $\phi = 0°$, which represents a horizontal plane, this equation gives an optimal launch angle of $\theta = 45°$, as we would expect (see Figure 4.10).

4.4 Analysis Model: Particle in Uniform Circular Motion

Figure 4.15a shows a car moving in a circular path; we describe this motion by calling it **circular motion.** If the car is moving on this path with *constant speed v*, we call it **uniform circular motion.** Because it occurs so often, this type of motion is recognized as an analysis model called the **particle in uniform circular motion.** We discuss this model in this section.

It is often surprising to students to find that even though an object moves at a constant speed in a circular path, *it still has an acceleration.* To see why, consider the defining equation for acceleration, $\vec{a} = d\vec{v}/dt$ (Eq. 4.5). Notice that the acceleration depends on the change in the *velocity.* Because velocity is a vector quantity, an acceleration can occur in two ways as mentioned in Section 4.1: by a change in the *magnitude* of the velocity and by a change in the *direction* of the velocity. The latter situation occurs for an object moving with constant speed in a circular path. The constant-magnitude velocity vector is always tangent to the path of the object and perpendicular to the radius of the circular path. Therefore, the direction of the velocity vector is always changing.

Let us first argue that the acceleration vector in uniform circular motion is always perpendicular to the path and always points toward the center of the circle. If that were not true, there would be a component of the acceleration parallel to the path and therefore parallel to the velocity vector. Such an acceleration component would lead to a change in the speed of the particle along the path. This situation, however, is inconsistent with our setup of the situation: the particle moves with constant speed along the path. Therefore, for *uniform* circular motion, the acceleration vector can only have a component perpendicular to the path, which is toward the center of the circle.

Let us now find the magnitude of the acceleration of the particle. Consider the diagram of the position and velocity vectors in Figure 4.15b. The figure also shows the vector representing the change in position $\Delta\vec{r}$ for an arbitrary time interval. The particle follows a circular path of radius r, part of which is shown by the dashed

Pitfall Prevention 4.4

Acceleration of a Particle in Uniform Circular Motion Remember that acceleration in physics is defined as a change in the *velocity,* not a change in the *speed* (contrary to the everyday interpretation). In circular motion, the velocity vector is always changing in direction, so there is indeed an acceleration.

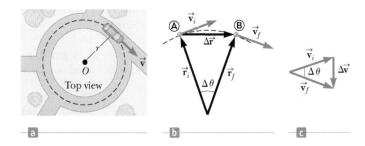

Figure 4.15 (a) A car moving along a circular path at constant speed experiences uniform circular motion. (b) As a particle moves along a portion of a circular path from Ⓐ to Ⓑ, its velocity vector changes from $\vec{v}_i$ to $\vec{v}_f$. (c) The construction for determining the direction of the change in velocity $\Delta\vec{v}$, which is toward the center of the circle for small $\Delta\vec{r}$.

curve. The particle is at Ⓐ at time t_i, and its velocity at that time is $\vec{\mathbf{v}}_i$; it is at Ⓑ at some later time t_f, and its velocity at that time is $\vec{\mathbf{v}}_f$. Let us also assume $\vec{\mathbf{v}}_i$ and $\vec{\mathbf{v}}_f$ differ only in direction; their magnitudes are the same (that is, $v_i = v_f = v$ because it is *uniform* circular motion).

In Figure 4.15c, the velocity vectors in Figure 4.15b have been redrawn tail to tail. The vector $\Delta\vec{\mathbf{v}}$ connects the tips of the vectors, representing the vector addition $\vec{\mathbf{v}}_f = \vec{\mathbf{v}}_i + \Delta\vec{\mathbf{v}}$. In both Figures 4.15b and 4.15c, we can identify triangles that help us analyze the motion. The angle $\Delta\theta$ between the two position vectors in Figure 4.15b is the same as the angle between the velocity vectors in Figure 4.15c because the velocity vector $\vec{\mathbf{v}}$ is always perpendicular to the position vector $\vec{\mathbf{r}}$. Therefore, the two triangles are *similar*. (Two triangles are similar if the angle between any two sides is the same for both triangles and if the ratio of the lengths of these sides is the same.) We can now write a relationship between the lengths of the sides for the two triangles in Figures 4.15b and 4.15c:

$$\frac{|\Delta\vec{\mathbf{v}}|}{v} = \frac{|\Delta\vec{\mathbf{r}}|}{r}$$

where $v = v_i = v_f$ and $r = r_i = r_f$. This equation can be solved for $|\Delta\vec{\mathbf{v}}|$, and the expression obtained can be substituted into Equation 4.4, $\vec{\mathbf{a}}_{avg} = \Delta\vec{\mathbf{v}}/\Delta t$, to give the magnitude of the average acceleration over the time interval for the particle to move from Ⓐ to Ⓑ:

$$|\vec{\mathbf{a}}_{avg}| = \frac{|\Delta\vec{\mathbf{v}}|}{|\Delta t|} = \frac{v|\Delta\vec{\mathbf{r}}|}{r\Delta t}$$

Now imagine that points Ⓐ and Ⓑ in Figure 4.15b become extremely close together. As Ⓐ and Ⓑ approach each other, Δt approaches zero, $|\Delta\vec{\mathbf{r}}|$ approaches the distance traveled by the particle along the circular path, and the ratio $|\Delta\vec{\mathbf{r}}|/\Delta t$ approaches the speed v. In addition, the average acceleration becomes the instantaneous acceleration at point Ⓐ. Hence, in the limit $\Delta t \to 0$, the magnitude of the acceleration is

Centripetal acceleration ▶
for a particle in uniform
circular motion

$$a_c = \frac{v^2}{r} \tag{4.14}$$

An acceleration of this nature is called a **centripetal acceleration** (*centripetal* means *center-seeking*). The subscript on the acceleration symbol reminds us that the acceleration is centripetal.

In many situations, it is convenient to describe the motion of a particle moving with constant speed in a circle of radius r in terms of the **period** T, which is defined as the time interval required for one complete revolution of the particle. In the time interval T, the particle moves a distance of $2\pi r$, which is equal to the circumference of the particle's circular path. Therefore, because its speed is equal to the circumference of the circular path divided by the period, or $v = 2\pi r/T$, it follows that

Period of circular motion ▶
for a particle in uniform
circular motion

$$T = \frac{2\pi r}{v} \tag{4.15}$$

The period of a particle in uniform circular motion is a measure of the number of seconds for one revolution of the particle around the circle. The inverse of the period is the *rotation rate* and is measured in revolutions per second. Because one full revolution of the particle around the circle corresponds to an angle of 2π radians, the product of 2π and the rotation rate gives the **angular speed** ω of the particle, measured in radians/s or s^{-1}:

$$\omega = \frac{2\pi}{T} \tag{4.16}$$

Combining this equation with Equation 4.15, we find a relationship between angular speed and the translational speed with which the particle travels in the circular path:

$$\omega = 2\pi\left(\frac{v}{2\pi r}\right) = \frac{v}{r} \quad \rightarrow \quad v = r\omega \qquad \textbf{(4.17)}$$

Equation 4.17 demonstrates that, for a fixed angular speed, the translational speed becomes larger as the radial position becomes larger. Therefore, for example, if a merry-go-round rotates at a fixed angular speed ω, a rider at an outer position at large r will be traveling through space faster than a rider at an inner position at smaller r. We will investigate Equations 4.16 and 4.17 more deeply in Chapter 10.

We can express the centripetal acceleration of a particle in uniform circular motion in terms of angular speed by combining Equations 4.14 and 4.17:

$$a_c = \frac{(r\omega)^2}{r}$$

$$a_c = r\omega^2 \qquad \textbf{(4.18)}$$

Equations 4.14–4.18 are to be used when the particle in uniform circular motion model is identified as appropriate for a given situation.

Quick Quiz 4.4 A particle moves in a circular path of radius r with speed v. It then increases its speed to $2v$ while traveling along the same circular path. **(i)** The centripetal acceleration of the particle has changed by what factor? Choose one: (a) 0.25 (b) 0.5 (c) 2 (d) 4 (e) impossible to determine **(ii)** From the same choices, by what factor has the period of the particle changed?

> **Pitfall Prevention 4.5**
> **Centripetal Acceleration Is Not Constant** We derived the magnitude of the centripetal acceleration vector and found it to be constant for uniform circular motion, but *the centripetal acceleration vector is not constant*. It always points toward the center of the circle, but it continuously changes direction as the object moves around the circular path.

Analysis Model **Particle in Uniform Circular Motion**

Imagine a moving object that can be modeled as a particle. If it moves in a circular path of radius r at a constant speed v, the magnitude of its centripetal acceleration is

$$a_c = \frac{v^2}{r} \qquad \textbf{(4.14)}$$

and the **period** of the particle's motion is given by

$$T = \frac{2\pi r}{v} \qquad \textbf{(4.15)}$$

The **angular speed** of the particle is

$$\omega = \frac{2\pi}{T} \qquad \textbf{(4.16)}$$

Examples:

- a rock twirled in a circle on a string of constant length
- a planet traveling around a perfectly circular orbit (Chapter 13)
- a charged particle moving in a uniform magnetic field (Chapter 29)
- an electron in orbit around a nucleus in the Bohr model of the hydrogen atom (Chapter 42)

Example 4.6 **The Centripetal Acceleration of the Earth** AM

(A) What is the centripetal acceleration of the Earth as it moves in its orbit around the Sun?

SOLUTION

Conceptualize Think about a mental image of the Earth in a circular orbit around the Sun. We will model the Earth as a particle and approximate the Earth's orbit as circular (it's actually slightly elliptical, as we discuss in Chapter 13).

Categorize The Conceptualize step allows us to categorize this problem as one of a *particle in uniform circular motion*.

Analyze We do not know the orbital speed of the Earth to substitute into Equation 4.14. With the help of Equation 4.15, however, we can recast Equation 4.14 in terms of the period of the Earth's orbit, which we know is one year, and the radius of the Earth's orbit around the Sun, which is 1.496×10^{11} m.

continued

▶ 4.6 continued

Combine Equations 4.14 and 4.15:

$$a_c = \frac{v^2}{r} = \frac{\left(\frac{2\pi r}{T}\right)^2}{r} = \frac{4\pi^2 r}{T^2}$$

Substitute numerical values:

$$a_c = \frac{4\pi^2(1.496 \times 10^{11}\text{ m})}{(1\text{ yr})^2}\left(\frac{1\text{ yr}}{3.156 \times 10^7\text{ s}}\right)^2 = 5.93 \times 10^{-3}\text{ m/s}^2$$

(B) What is the angular speed of the Earth in its orbit around the Sun?

SOLUTION

Analyze

Substitute numerical values into Equation 4.16:

$$\omega = \frac{2\pi}{1\text{ yr}}\left(\frac{1\text{ yr}}{3.156 \times 10^7\text{ s}}\right) = 1.99 \times 10^{-7}\text{ s}^{-1}$$

Finalize The acceleration in part (A) is much smaller than the free-fall acceleration on the surface of the Earth. An important technique we learned here is replacing the speed v in Equation 4.14 in terms of the period T of the motion. In many problems, it is more likely that T is known rather than v. In part (B), we see that the angular speed of the Earth is very small, which is to be expected because the Earth takes an entire year to go around the circular path once.

4.5 Tangential and Radial Acceleration

Let us consider a more general motion than that presented in Section 4.4. A particle moves to the right along a curved path, and its velocity changes *both* in direction and in magnitude as described in Figure 4.16. In this situation, the velocity vector is always tangent to the path; the acceleration vector $\vec{a}$, however, is at some angle to the path. At each of three points Ⓐ, Ⓑ, and Ⓒ in Figure 4.16, the dashed blue circles represent the curvature of the actual path at each point. The radius of each circle is equal to the path's radius of curvature at each point.

As the particle moves along the curved path in Figure 4.16, the direction of the total acceleration vector $\vec{a}$ changes from point to point. At any instant, this vector can be resolved into two components based on an origin at the center of the dashed circle corresponding to that instant: a radial component a_r along the radius of the circle and a tangential component a_t perpendicular to this radius. The *total* acceleration vector $\vec{a}$ can be written as the vector sum of the component vectors:

◀ Total acceleration

$$\vec{a} = \vec{a}_r + \vec{a}_t \tag{4.19}$$

The tangential acceleration component causes a change in the speed v of the particle. This component is parallel to the instantaneous velocity, and its magnitude is given by

◀ Tangential acceleration

$$a_t = \left|\frac{dv}{dt}\right| \tag{4.20}$$

Figure 4.16 The motion of a particle along an arbitrary curved path lying in the *xy* plane. If the velocity vector $\vec{v}$ (always tangent to the path) changes in direction and magnitude, the components of the acceleration $\vec{a}$ are a tangential component a_t and a radial component a_r.

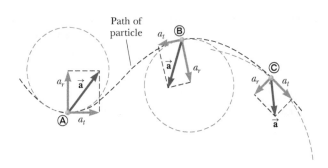

The radial acceleration component arises from a change in direction of the velocity vector and is given by

$$a_r = -a_c = -\frac{v^2}{r}$$

(4.21)   ◀ **Radial acceleration**

where r is the radius of curvature of the path at the point in question. We recognize the magnitude of the radial component of the acceleration as the centripetal acceleration discussed in Section 4.4 with regard to the particle in uniform circular motion model. Even in situations in which a particle moves along a curved path with a varying speed, however, Equation 4.14 can be used for the centripetal acceleration. In this situation, the equation gives the *instantaneous* centripetal acceleration at any time. The negative sign in Equation 4.21 indicates that the direction of the centripetal acceleration is toward the center of the circle representing the radius of curvature. The direction is opposite that of the radial unit vector $\hat{\mathbf{r}}$, which always points away from the origin at the center of the circle.

Because $\vec{\mathbf{a}}_r$ and $\vec{\mathbf{a}}_t$ are perpendicular component vectors of $\vec{\mathbf{a}}$, it follows that the magnitude of $\vec{\mathbf{a}}$ is $a = \sqrt{a_r^2 + a_t^2}$. At a given speed, a_r is large when the radius of curvature is small (as at points Ⓐ and Ⓑ in Fig. 4.16) and small when r is large (as at point Ⓒ). The direction of $\vec{\mathbf{a}}_t$ is either in the same direction as $\vec{\mathbf{v}}$ (if v is increasing) or opposite $\vec{\mathbf{v}}$ (if v is decreasing, as at point Ⓑ).

In uniform circular motion, where v is constant, $a_t = 0$ and the acceleration is always completely radial as described in Section 4.4. In other words, uniform circular motion is a special case of motion along a general curved path. Furthermore, if the direction of $\vec{\mathbf{v}}$ does not change, there is no radial acceleration and the motion is one dimensional (in this case, $a_r = 0$, but a_t may not be zero).

Ⓠuick Quiz 4.5 A particle moves along a path, and its speed increases with time. **(i)** In which of the following cases are its acceleration and velocity vectors parallel? (a) when the path is circular (b) when the path is straight (c) when the path is a parabola (d) never **(ii)** From the same choices, in which case are its acceleration and velocity vectors perpendicular everywhere along the path?

Example 4.7   Over the Rise

A car leaves a stop sign and exhibits a constant acceleration of 0.300 m/s² parallel to the roadway. The car passes over a rise in the roadway such that the top of the rise is shaped like an arc of a circle of radius 500 m. At the moment the car is at the top of the rise, its velocity vector is horizontal and has a magnitude of 6.00 m/s. What are the magnitude and direction of the total acceleration vector for the car at this instant?

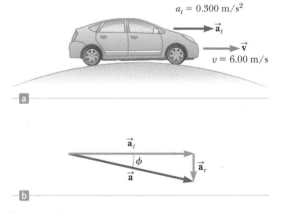

SOLUTION

Conceptualize Conceptualize the situation using Figure 4.17a and any experiences you have had in driving over rises on a roadway.

Categorize Because the accelerating car is moving along a curved path, we categorize this problem as one involving a particle experiencing both tangential and radial acceleration. We recognize that it is a relatively simple substitution problem.

Figure 4.17 (Example 4.7) (a) A car passes over a rise that is shaped like an arc of a circle. (b) The total acceleration vector $\vec{\mathbf{a}}$ is the sum of the tangential and radial acceleration vectors $\vec{\mathbf{a}}_t$ and $\vec{\mathbf{a}}_r$.

The tangential acceleration vector has magnitude 0.300 m/s² and is horizontal. The radial acceleration is given by Equation 4.21, with $v = 6.00$ m/s and $r = 500$ m. The radial acceleration vector is directed straight downward.

continued

▶ **4.7** continued

Evaluate the radial acceleration:

$$a_r = -\frac{v^2}{r} = -\frac{(6.00 \text{ m/s})^2}{500 \text{ m}} = -0.072\ 0 \text{ m/s}^2$$

Find the magnitude of $\vec{a}$:

$$\sqrt{a_r^2 + a_t^2} = \sqrt{(-0.072\ 0 \text{ m/s}^2)^2 + (0.300 \text{ m/s}^2)^2}$$
$$= \boxed{0.309 \text{ m/s}^2}$$

Find the angle ϕ (see Fig. 4.17b) between $\vec{a}$ and the horizontal:

$$\phi = \tan^{-1}\frac{a_r}{a_t} = \tan^{-1}\left(\frac{-0.072\ 0 \text{ m/s}^2}{0.300 \text{ m/s}^2}\right) = \boxed{-13.5°}$$

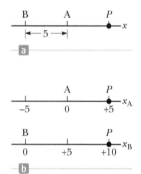

Figure 4.18 Different observers make different measurements. (a) Observer A is located 5 units to the right of Observer B. Both observers measure the position of a particle at P. (b) If both observers see themselves at the origin of their own coordinate system, they disagree on the value of the position of the particle at P.

The woman standing on the beltway sees the man moving with a slower speed than does the woman observing the man from the stationary floor.

Figure 4.19 Two observers measure the speed of a man walking on a moving beltway.

4.6 Relative Velocity and Relative Acceleration

In this section, we describe how observations made by different observers in different frames of reference are related to one another. A frame of reference can be described by a Cartesian coordinate system for which an observer is at rest with respect to the origin.

Let us conceptualize a sample situation in which there will be different observations for different observers. Consider the two observers A and B along the number line in Figure 4.18a. Observer A is located 5 units to the right of observer B. Both observers measure the position of point P, which is located 5 units to the right of observer A. Suppose each observer decides that he is located at the origin of an x axis as in Figure 4.18b. Notice that the two observers disagree on the value of the position of point P. Observer A claims point P is located at a position with a value of $x_A = +5$, whereas observer B claims it is located at a position with a value of $x_B = +10$. Both observers are correct, even though they make different measurements. Their measurements differ because they are making the measurement from different frames of reference.

Imagine now that observer B in Figure 4.18b is moving to the right along the x_B axis. Now the two measurements are even more different. Observer A claims point P remains at rest at a position with a value of +5, whereas observer B claims the position of P continuously changes with time, even passing him and moving behind him! Again, both observers are correct, with the difference in their measurements arising from their different frames of reference.

We explore this phenomenon further by considering two observers watching a man walking on a moving beltway at an airport in Figure 4.19. The woman standing on the moving beltway sees the man moving at a normal walking speed. The woman observing from the stationary floor sees the man moving with a higher speed because the beltway speed combines with his walking speed. Both observers look at the same man and arrive at different values for his speed. Both are correct; the difference in their measurements results from the relative velocity of their frames of reference.

In a more general situation, consider a particle located at point P in Figure 4.20. Imagine that the motion of this particle is being described by two observers, observer A in a reference frame S_A fixed relative to the Earth and a second observer B in a reference frame S_B moving to the right relative to S_A (and therefore relative to the Earth) with a constant velocity $\vec{v}_{BA}$. In this discussion of relative velocity, we use a double-subscript notation; the first subscript represents what is being observed, and the second represents who is doing the observing. Therefore, the notation $\vec{v}_{BA}$ means the velocity of observer B (and the attached frame S_B) as measured by observer A. With this notation, observer B measures A to be moving to the left with a velocity $\vec{v}_{AB} = -\vec{v}_{BA}$. For purposes of this discussion, let us place each observer at her or his respective origin.

We define the time $t = 0$ as the instant at which the origins of the two reference frames coincide in space. Therefore, at time t, the origins of the reference frames will be separated by a distance $v_{BA}t$. We label the position P of the particle relative to observer A with the position vector $\vec{\mathbf{r}}_{PA}$ and that relative to observer B with the position vector $\vec{\mathbf{r}}_{PB}$, both at time t. From Figure 4.20, we see that the vectors $\vec{\mathbf{r}}_{PA}$ and $\vec{\mathbf{r}}_{PB}$ are related to each other through the expression

$$\vec{\mathbf{r}}_{PA} = \vec{\mathbf{r}}_{PB} + \vec{\mathbf{v}}_{BA}t \qquad \textbf{(4.22)}$$

By differentiating Equation 4.22 with respect to time, noting that $\vec{\mathbf{v}}_{BA}$ is constant, we obtain

$$\frac{d\vec{\mathbf{r}}_{PA}}{dt} = \frac{d\vec{\mathbf{r}}_{PB}}{dt} + \vec{\mathbf{v}}_{BA}$$

$$\vec{\mathbf{u}}_{PA} = \vec{\mathbf{u}}_{PB} + \vec{\mathbf{v}}_{BA} \qquad \textbf{(4.23)}$$

where $\vec{\mathbf{u}}_{PA}$ is the velocity of the particle at P measured by observer A and $\vec{\mathbf{u}}_{PB}$ is its velocity measured by B. (We use the symbol $\vec{\mathbf{u}}$ for particle velocity rather than $\vec{\mathbf{v}}$, which we have already used for the relative velocity of two reference frames.) Equations 4.22 and 4.23 are known as **Galilean transformation equations.** They relate the position and velocity of a particle as measured by observers in relative motion. Notice the pattern of the subscripts in Equation 4.23. When relative velocities are added, the inner subscripts (B) are the same and the outer ones (P, A) match the subscripts on the velocity on the left of the equation.

Although observers in two frames measure different velocities for the particle, they measure the *same acceleration* when $\vec{\mathbf{v}}_{BA}$ is constant. We can verify that by taking the time derivative of Equation 4.23:

$$\frac{d\vec{\mathbf{u}}_{PA}}{dt} = \frac{d\vec{\mathbf{u}}_{PB}}{dt} + \frac{d\vec{\mathbf{v}}_{BA}}{dt}$$

Because $\vec{\mathbf{v}}_{BA}$ is constant, $d\vec{\mathbf{v}}_{BA}/dt = 0$. Therefore, we conclude that $\vec{\mathbf{a}}_{PA} = \vec{\mathbf{a}}_{PB}$ because $\vec{\mathbf{a}}_{PA} = d\vec{\mathbf{u}}_{PA}/dt$ and $\vec{\mathbf{a}}_{PB} = d\vec{\mathbf{u}}_{PB}/dt$. That is, the acceleration of the particle measured by an observer in one frame of reference is the same as that measured by any other observer moving with constant velocity relative to the first frame.

◀ **Galilean velocity transformation**

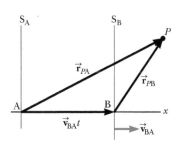

Figure 4.20 A particle located at P is described by two observers, one in the fixed frame of reference S_A and the other in the frame S_B, which moves to the right with a constant velocity $\vec{\mathbf{v}}_{BA}$. The vector $\vec{\mathbf{r}}_{PA}$ is the particle's position vector relative to S_A, and $\vec{\mathbf{r}}_{PB}$ is its position vector relative to S_B.

Example 4.8   A Boat Crossing a River

A boat crossing a wide river moves with a speed of 10.0 km/h relative to the water. The water in the river has a uniform speed of 5.00 km/h due east relative to the Earth.

(A) If the boat heads due north, determine the velocity of the boat relative to an observer standing on either bank.

SOLUTION

Conceptualize Imagine moving in a boat across a river while the current pushes you down the river. You will not be able to move directly across the river, but will end up downstream as suggested in Figure 4.21a.

Categorize Because of the combined velocities of you relative to the river and the river relative to the Earth, we can categorize this problem as one involving relative velocities.

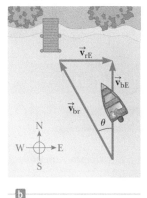

Figure 4.21 (Example 4.8) (a) A boat aims directly across a river and ends up downstream. (b) To move directly across the river, the boat must aim upstream.

Analyze We know $\vec{\mathbf{v}}_{br}$, the velocity of the *boat* relative to the *river*, and $\vec{\mathbf{v}}_{rE}$, the velocity of the *river* relative to the *Earth*. What we must find is $\vec{\mathbf{v}}_{bE}$, the velocity of the *boat* relative to the *Earth*. The relationship between these three quantities is $\vec{\mathbf{v}}_{bE} = \vec{\mathbf{v}}_{br} + \vec{\mathbf{v}}_{rE}$. The terms in the equation must be manipulated as vector quantities; the vectors are shown in Figure 4.21a. The quantity $\vec{\mathbf{v}}_{br}$ is due north; $\vec{\mathbf{v}}_{rE}$ is due east; and the vector sum of the two, $\vec{\mathbf{v}}_{bE}$, is at an angle θ as defined in Figure 4.21a.

continued

▶ **4.8** continued

Find the speed v_{bE} of the boat relative to the Earth using the Pythagorean theorem:

$$v_{bE} = \sqrt{v_{br}^2 + v_{rE}^2} = \sqrt{(10.0 \text{ km/h})^2 + (5.00 \text{ km/h})^2}$$
$$= 11.2 \text{ km/h}$$

Find the direction of $\vec{\mathbf{v}}_{bE}$:

$$\theta = \tan^{-1}\left(\frac{v_{rE}}{v_{br}}\right) = \tan^{-1}\left(\frac{5.00}{10.0}\right) = 26.6°$$

Finalize The boat is moving at a speed of 11.2 km/h in the direction 26.6° east of north relative to the Earth. Notice that the speed of 11.2 km/h is faster than your boat speed of 10.0 km/h. The current velocity adds to yours to give you a higher speed. Notice in Figure 4.21a that your resultant velocity is at an angle to the direction straight across the river, so you will end up downstream, as we predicted.

(B) If the boat travels with the same speed of 10.0 km/h relative to the river and is to travel due north as shown in Figure 4.21b, what should its heading be?

SOLUTION

Conceptualize/Categorize This question is an extension of part (A), so we have already conceptualized and categorized the problem. In this case, however, we must aim the boat upstream so as to go straight across the river.

Analyze The analysis now involves the new triangle shown in Figure 4.21b. As in part (A), we know $\vec{\mathbf{v}}_{rE}$ and the magnitude of the vector $\vec{\mathbf{v}}_{br}$, and we want $\vec{\mathbf{v}}_{bE}$ to be directed across the river. Notice the difference between the triangle in Figure 4.21a and the one in Figure 4.21b: the hypotenuse in Figure 4.21b is no longer $\vec{\mathbf{v}}_{bE}$.

Use the Pythagorean theorem to find v_{bE}:

$$v_{bE} = \sqrt{v_{br}^2 - v_{rE}^2} = \sqrt{(10.0 \text{ km/h})^2 - (5.00 \text{ km/h})^2} = 8.66 \text{ km/h}$$

Find the direction in which the boat is heading:

$$\theta = \tan^{-1}\left(\frac{v_{rE}}{v_{bE}}\right) = \tan^{-1}\left(\frac{5.00}{8.66}\right) = 30.0°$$

Finalize The boat must head upstream so as to travel directly northward across the river. For the given situation, the boat must steer a course 30.0° west of north. For faster currents, the boat must be aimed upstream at larger angles.

WHAT IF?   Imagine that the two boats in parts (A) and (B) are racing across the river. Which boat arrives at the opposite bank first?

Answer In part (A), the velocity of 10 km/h is aimed directly across the river. In part (B), the velocity that is directed across the river has a magnitude of only 8.66 km/h. Therefore, the boat in part (A) has a larger velocity component directly across the river and arrives first.

Summary

Definitions

The **displacement vector** $\Delta\vec{\mathbf{r}}$ for a particle is the difference between its final position vector and its initial position vector:

$$\Delta\vec{\mathbf{r}} \equiv \vec{\mathbf{r}}_f - \vec{\mathbf{r}}_i \tag{4.1}$$

The **average velocity** of a particle during the time interval Δt is defined as the displacement of the particle divided by the time interval:

$$\vec{\mathbf{v}}_{avg} \equiv \frac{\Delta\vec{\mathbf{r}}}{\Delta t} \tag{4.2}$$

The **instantaneous velocity** of a particle is defined as the limit of the average velocity as Δt approaches zero:

$$\vec{\mathbf{v}} \equiv \lim_{\Delta t \to 0} \frac{\Delta\vec{\mathbf{r}}}{\Delta t} = \frac{d\vec{\mathbf{r}}}{dt} \tag{4.3}$$

The **average acceleration** of a particle is defined as the change in its instantaneous velocity vector divided by the time interval Δt during which that change occurs:

$$\vec{a}_{avg} \equiv \frac{\Delta \vec{v}}{\Delta t} = \frac{\vec{v}_f - \vec{v}_i}{t_f - t_i} \tag{4.4}$$

The **instantaneous acceleration** of a particle is defined as the limiting value of the average acceleration as Δt approaches zero:

$$\vec{a} \equiv \lim_{\Delta t \to 0} \frac{\Delta \vec{v}}{\Delta t} = \frac{d\vec{v}}{dt} \tag{4.5}$$

Projectile motion is one type of two-dimensional motion, exhibited by an object launched into the air near the Earth's surface and experiencing free fall. This common motion can be analyzed by applying the particle under constant velocity model to the motion of the projectile in the x direction and the particle under constant acceleration model ($a_y = -g$) in the y direction.

A particle moving in a circular path with constant speed is exhibiting **uniform circular motion.**

Concepts and Principles

If a particle moves with *constant* acceleration $\vec{a}$ and has velocity $\vec{v}_i$ and position $\vec{r}_i$ at $t = 0$, its velocity and position vectors at some later time t are

$$\vec{v}_f = \vec{v}_i + \vec{a}t \tag{4.8}$$

$$\vec{r}_f = \vec{r}_i + \vec{v}_i t + \tfrac{1}{2}\vec{a}t^2 \tag{4.9}$$

For two-dimensional motion in the xy plane under constant acceleration, ea–ch of these vector expressions is equivalent to two component expressions: one for the motion in the x direction and one for the motion in the y direction.

It is useful to think of projectile motion in terms of a combination of two analysis models: (1) the particle under constant velocity model in the x direction and (2) the particle under constant acceleration model in the vertical direction with a constant downward acceleration of magnitude $g = 9.80$ m/s^2.

A particle in uniform circular motion undergoes a radial acceleration $\vec{a}_r$ because the direction of $\vec{v}$ changes in time. This acceleration is called **centripetal acceleration,** and its direction is always toward the center of the circle.

If a particle moves along a curved path in such a way that both the magnitude and the direction of $\vec{v}$ change in time, the particle has an acceleration vector that can be described by two component vectors: (1) a radial component vector $\vec{a}_r$ that causes the change in direction of $\vec{v}$ and (2) a tangential component vector $\vec{a}_t$ that causes the change in magnitude of $\vec{v}$. The magnitude of $\vec{a}_r$ is v^2/r, and the magnitude of $\vec{a}_t$ is $|dv/dt|$.

The velocity $\vec{u}_{PA}$ of a particle measured in a fixed frame of reference S_A can be related to the velocity $\vec{u}_{PB}$ of the same particle measured in a moving frame of reference S_B by

$$\vec{u}_{PA} = \vec{u}_{PB} + \vec{v}_{BA} \tag{4.23}$$

where $\vec{v}_{BA}$ is the velocity of S_B relative to S_A.

Analysis Model for Problem Solving

Particle in Uniform Circular Motion If a particle moves in a circular path of radius r with a constant speed v, the magnitude of its centripetal acceleration is given by

$$a_c = \frac{v^2}{r} \tag{4.14}$$

and the **period** of the particle's motion is given by

$$T = \frac{2\pi r}{v} \tag{4.15}$$

The **angular speed** of the particle is

$$\omega = \frac{2\pi}{T} \tag{4.16}$$

1. Figure OQ4.1 shows a bird's-eye view of a car going around a highway curve. As the car moves from point 1 to point 2, its speed doubles. Which of the vectors (a) through (e) shows the direction of the car's average acceleration between these two points?

Figure OQ4.1

2. Entering his dorm room, a student tosses his book bag to the right and upward at an angle of 45° with the horizontal (Fig. OQ4.2). Air resistance does not affect the bag. The bag moves through point Ⓐ immediately after it leaves the student's hand, through point Ⓑ at the top of its flight, and through point Ⓒ immediately before it lands on the top bunk bed. **(i)** Rank the following horizontal and vertical velocity components from the largest to the smallest. (a) $v_{Ⓐx}$ (b) $v_{Ⓐy}$ (c) $v_{Ⓑx}$ (d) $v_{Ⓑy}$ (e) $v_{Ⓒy}$. Note that zero is larger than a negative number. If two quantities are equal, show them as equal in your list. If any quantity is equal to zero, show that fact in your list. **(ii)** Similarly, rank the following acceleration components. (a) $a_{Ⓐx}$ (b) $a_{Ⓐy}$ (c) $a_{Ⓑx}$ (d) $a_{Ⓑy}$ (e) $a_{Ⓒy}$.

Figure OQ4.2

3. A student throws a heavy red ball horizontally from a balcony of a tall building with an initial speed v_i. At the same time, a second student drops a lighter blue ball from the balcony. Neglecting air resistance, which statement is true? (a) The blue ball reaches the ground first. (b) The balls reach the ground at the same instant. (c) The red ball reaches the ground first. (d) Both balls hit the ground with the same speed. (e) None of statements (a) through (d) is true.

4. A projectile is launched on the Earth with a certain initial velocity and moves without air resistance. Another projectile is launched with the same initial velocity on the Moon, where the acceleration due to gravity is one-sixth as large. How does the maximum altitude of the projectile on the Moon compare with that of the projectile on the Earth? (a) It is one-sixth as large. (b) It is the same. (c) It is $\sqrt{6}$ times larger. (d) It is 6 times larger. (e) It is 36 times larger.

5. Does a car moving around a circular track with constant speed have (a) zero acceleration, (b) an acceleration in the direction of its velocity, (c) an acceleration directed away from the center of its path, (d) an acceleration directed toward the center of its path, or (e) an acceleration with a direction that cannot be determined from the given information?

6. An astronaut hits a golf ball on the Moon. Which of the following quantities, if any, remain constant as a ball travels through the vacuum there? (a) speed (b) acceleration (c) horizontal component of velocity (d) vertical component of velocity (e) velocity

7. A projectile is launched on the Earth with a certain initial velocity and moves without air resistance. Another projectile is launched with the same initial velocity on the Moon, where the acceleration due to gravity is one-sixth as large. How does the range of the projectile on the Moon compare with that of the projectile on the Earth? (a) It is one-sixth as large. (b) It is the same. (c) It is $\sqrt{6}$ times larger. (d) It is 6 times larger. (e) It is 36 times larger.

8. A girl, moving at 8 m/s on in-line skates, is overtaking a boy moving at 5 m/s as they both skate on a straight path. The boy tosses a ball backward toward the girl, giving it speed 12 m/s relative to him. What is the speed of the ball relative to the girl, who catches it? (a) $(8 + 5 + 12)$ m/s (b) $(8 - 5 - 12)$ m/s (c) $(8 + 5 - 12)$ m/s (d) $(8 - 5 + 12)$ m/s (e) $(-8 + 5 + 12)$ m/s

9. A sailor drops a wrench from the top of a sailboat's vertical mast while the boat is moving rapidly and steadily straight forward. Where will the wrench hit the deck? (a) ahead of the base of the mast (b) at the base of the mast (c) behind the base of the mast (d) on the windward side of the base of the mast (e) None of the choices (a) through (d) is true.

10. A baseball is thrown from the outfield toward the catcher. When the ball reaches its highest point, which statement is true? (a) Its velocity and its acceleration are both zero. (b) Its velocity is not zero, but its acceleration is zero. (c) Its velocity is perpendicular to its acceleration. (d) Its acceleration depends on the angle at which the ball was thrown. (e) None of statements (a) through (d) is true.

11. A set of keys on the end of a string is swung steadily in a horizontal circle. In one trial, it moves at speed v in a circle of radius r. In a second trial, it moves at a higher speed $4v$ in a circle of radius $4r$. In the second trial, how does the period of its motion compare with its period in the first trial? (a) It is the same as in the first trial. (b) It is 4 times larger. (c) It is one-fourth as large. (d) It is 16 times larger. (e) It is one-sixteenth as large.

12. A rubber stopper on the end of a string is swung steadily in a horizontal circle. In one trial, it moves at speed v in a circle of radius r. In a second trial, it moves at a higher speed $3v$ in a circle of radius $3r$. In this second trial, is its acceleration (a) the same as in the first trial, (b) three times larger, (c) one-third as large, (d) nine times larger, or (e) one-ninth as large?

13. In which of the following situations is the moving object appropriately modeled as a projectile? Choose all correct answers. (a) A shoe is tossed in an arbitrary direction. (b) A jet airplane crosses the sky with its engines thrusting the plane forward. (c) A rocket leaves the launch pad. (d) A rocket moves through the sky, at much less than the speed of sound, after its fuel has been used up. (e) A diver throws a stone under water.

14. A certain light truck can go around a curve having a radius of 150 m with a maximum speed of 32.0 m/s. To have the same acceleration, at what maximum speed can it go around a curve having a radius of 75.0 m? (a) 64 m/s (b) 45 m/s (c) 32 m/s (d) 23 m/s (e) 16 m/s

Conceptual Questions

1. denotes answer available in *Student Solutions Manual/Study Guide*

1. A spacecraft drifts through space at a constant velocity. Suddenly, a gas leak in the side of the spacecraft gives it a constant acceleration in a direction perpendicular to the initial velocity. The orientation of the spacecraft does not change, so the acceleration remains perpendicular to the original direction of the velocity. What is the shape of the path followed by the spacecraft in this situation?

2. An ice skater is executing a figure eight, consisting of two identically shaped, tangent circular paths. Throughout the first loop she increases her speed uniformly, and during the second loop she moves at a constant speed. Draw a motion diagram showing her velocity and acceleration vectors at several points along the path of motion.

3. If you know the position vectors of a particle at two points along its path and also know the time interval during which it moved from one point to the other, can you determine the particle's instantaneous velocity? Its average velocity? Explain.

4. Describe how a driver can steer a car traveling at constant speed so that (a) the acceleration is zero or (b) the magnitude of the acceleration remains constant.

5. A projectile is launched at some angle to the horizontal with some initial speed v_i, and air resistance is negligible. (a) Is the projectile a freely falling body? (b) What is its acceleration in the vertical direction? (c) What is its acceleration in the horizontal direction?

6. Construct motion diagrams showing the velocity and acceleration of a projectile at several points along its path, assuming (a) the projectile is launched horizontally and (b) the projectile is launched at an angle θ with the horizontal.

7. Explain whether or not the following particles have an acceleration: (a) a particle moving in a straight line with constant speed and (b) a particle moving around a curve with constant speed.

Problems

WebAssign The problems found in this chapter may be assigned online in Enhanced WebAssign

AMT Analysis Model tutorial available in Enhanced WebAssign

GP Guided Problem

M Master It tutorial available in Enhanced WebAssign

W Watch It video solution available in Enhanced WebAssign

1. straightforward; 2. intermediate; 3. challenging

1. full solution available in the *Student Solutions Manual/Study Guide*

Section 4.1 The Position, Velocity, and Acceleration Vectors

1. A motorist drives south at 20.0 m/s for 3.00 min, then turns west and travels at 25.0 m/s for 2.00 min, and finally travels northwest at 30.0 m/s for 1.00 min. For this 6.00-min trip, find (a) the total vector displacement, (b) the average speed, and (c) the average velocity. Let the positive x axis point east.

2. When the Sun is directly overhead, a hawk dives toward the ground with a constant velocity of 5.00 m/s at 60.0° below the horizontal. Calculate the speed of its shadow on the level ground.

3. Suppose the position vector for a particle is given as a function of time by $\vec{r}(t) = x(t)\hat{i} + y(t)\hat{j}$, with $x(t) = at + b$ and $y(t) = ct^2 + d$, where $a = 1.00$ m/s, $b = 1.00$ m, $c = 0.125$ m/s², and $d = 1.00$ m. (a) Calculate the average velocity during the time interval from $t = 2.00$ s to $t = 4.00$ s. (b) Determine the velocity and the speed at $t = 2.00$ s.

4. The coordinates of an object moving in the xy plane vary with time according to the equations $x = -5.00 \sin \omega t$ and $y = 4.00 - 5.00 \cos \omega t$, where ω is a constant, x and y are in meters, and t is in seconds. (a) Determine the components of velocity of the object at $t = 0$. (b) Determine the components of acceleration of the object at $t = 0$. (c) Write expressions for the position vector, the velocity vector, and the acceleration vector of the object at any time $t > 0$. (d) Describe the path of the object in an xy plot.

5. A golf ball is hit off a tee at the edge of a cliff. Its x and y coordinates as functions of time are given by $x = 18.0t$ and $y = 4.00t - 4.90t^2$, where x and y are in meters and t is in seconds. (a) Write a vector expression for the ball's position as a function of time, using the unit vectors $\hat{\mathbf{i}}$ and $\hat{\mathbf{j}}$. By taking derivatives, obtain expressions for (b) the velocity vector $\vec{\mathbf{v}}$ as a function of time and (c) the acceleration vector $\vec{\mathbf{a}}$ as a function of time. (d) Next use unit-vector notation to write expressions for the position, the velocity, and the acceleration of the golf ball at $t = 3.00$ s.

Section 4.2 Two-Dimensional Motion with Constant Acceleration

6. A particle initially located at the origin has an acceleration of $\vec{\mathbf{a}} = 3.00\hat{\mathbf{j}}$ m/s^2 and an initial velocity of $\vec{\mathbf{v}}_i = 5.00\hat{\mathbf{i}}$ m/s. Find (a) the vector position of the particle at any time t, (b) the velocity of the particle at any time t, (c) the coordinates of the particle at $t = 2.00$ s, and (d) the speed of the particle at $t = 2.00$ s.

7. The vector position of a particle varies in time according to the expression $\vec{\mathbf{r}} = 3.00\hat{\mathbf{i}} - 6.00t^2\hat{\mathbf{j}}$, where $\vec{\mathbf{r}}$ is in meters and t is in seconds. (a) Find an expression for the velocity of the particle as a function of time. (b) Determine the acceleration of the particle as a function of time. (c) Calculate the particle's position and velocity at $t = 1.00$ s.

8. It is not possible to see very small objects, such as viruses, using an ordinary light microscope. An electron microscope, however, can view such objects using an electron beam instead of a light beam. Electron microscopy has proved invaluable for investigations of viruses, cell membranes and subcellular structures, bacterial surfaces, visual receptors, chloroplasts, and the contractile properties of muscles. The "lenses" of an electron microscope consist of electric and magnetic fields that control the electron beam. As an example of the manipulation of an electron beam, consider an electron traveling away from the origin along the x axis in the xy plane with initial velocity $\vec{\mathbf{v}}_i = v_i\hat{\mathbf{i}}$. As it passes through the region $x = 0$ to $x = d$, the electron experiences acceleration $\vec{\mathbf{a}} = a_x\hat{\mathbf{i}} + a_y\hat{\mathbf{j}}$, where a_x and a_y are constants. For the case $v_i = 1.80 \times 10^7$ m/s, $a_x = 8.00 \times 10^{14}$ m/s^2, and $a_y = 1.60 \times 10^{15}$ m/s^2, determine at $x = d = 0.010\ 0$ m (a) the position of the electron, (b) the velocity of the electron, (c) the speed of the electron, and (d) the direction of travel of the electron (i.e., the angle between its velocity and the x axis).

9. A fish swimming in a horizontal plane has velocity $\vec{\mathbf{v}}_i = (4.00\hat{\mathbf{i}} + 1.00\hat{\mathbf{j}})$ m/s at a point in the ocean where the position relative to a certain rock is $\vec{\mathbf{r}}_i = (10.0\hat{\mathbf{i}} - 4.00\hat{\mathbf{j}})$ m. After the fish swims with constant acceleration for 20.0 s, its velocity is $\vec{\mathbf{v}} = (20.0\hat{\mathbf{i}} - 5.00\hat{\mathbf{j}})$ m/s. (a) What are the components of the acceleration of the fish? (b) What is the direction of its acceleration with respect to unit vector $\hat{\mathbf{i}}$? (c) If the fish maintains constant acceleration, where is it at $t = 25.0$ s and in what direction is it moving?

10. **Review.** A snowmobile is originally at the point with position vector 29.0 m at 95.0° counterclockwise from the x axis, moving with velocity 4.50 m/s at 40.0°. It moves with constant acceleration 1.90 m/s^2 at 200°. After 5.00 s have elapsed, find (a) its velocity and (b) its position vector.

Section 4.3 Projectile Motion

Note: Ignore air resistance in all problems and take $g = 9.80$ m/s^2 at the Earth's surface.

11. Mayan kings and many school sports teams are named for the puma, cougar, or mountain lion—*Felis concolor*—the best jumper among animals. It can jump to a height of 12.0 ft when leaving the ground at an angle of 45.0°. With what speed, in SI units, does it leave the ground to make this leap?

12. An astronaut on a strange planet finds that she can jump a maximum horizontal distance of 15.0 m if her initial speed is 3.00 m/s. What is the free-fall acceleration on the planet?

13. In a local bar, a customer slides an empty beer mug down the counter for a refill. The height of the counter is 1.22 m. The mug slides off the counter and strikes the floor 1.40 m from the base of the counter. (a) With what velocity did the mug leave the counter? (b) What was the direction of the mug's velocity just before it hit the floor?

14. In a local bar, a customer slides an empty beer mug down the counter for a refill. The height of the counter is h. The mug slides off the counter and strikes the floor at distance d from the base of the counter. (a) With what velocity did the mug leave the counter? (b) What was the direction of the mug's velocity just before it hit the floor?

15. A projectile is fired in such a way that its horizontal range is equal to three times its maximum height. What is the angle of projection?

16. To start an avalanche on a mountain slope, an artillery shell is fired with an initial velocity of 300 m/s at 55.0° above the horizontal. It explodes on the mountainside 42.0 s after firing. What are the x and y coordinates of the shell where it explodes, relative to its firing point?

17. Chinook salmon are able to move through water especially fast by jumping out of the water periodically. This behavior is called *porpoising*. Suppose a salmon swimming in still water jumps out of the water with velocity 6.26 m/s at 45.0° above the horizontal, sails through the air a distance L before returning to the water, and then swims the same distance L underwater in a straight, horizontal line with velocity 3.58 m/s before jumping out again. (a) Determine the average velocity of the fish for the entire process of jumping and swimming underwater. (b) Consider the time interval required to travel the entire distance of $2L$. By what percentage is this time interval reduced by the jumping/swimming process compared with simply swimming underwater at 3.58 m/s?

18. A rock is thrown upward from level ground in such a way that the maximum height of its flight is equal to its horizontal range R. (a) At what angle θ is the rock thrown? (b) In terms of its original range R, what is the range R_{max} the rock can attain if it is launched at

the same speed but at the optimal angle for maximum range? (c) **What If?** Would your answer to part (a) be different if the rock is thrown with the same speed on a different planet? Explain.

19. The speed of a projectile when it reaches its maximum height is one-half its speed when it is at half its maximum height. What is the initial projection angle of the projectile?

20. A ball is tossed from an upper-story window of a build-
W ing. The ball is given an initial velocity of 8.00 m/s at an angle of 20.0° below the horizontal. It strikes the ground 3.00 s later. (a) How far horizontally from the base of the building does the ball strike the ground? (b) Find the height from which the ball was thrown. (c) How long does it take the ball to reach a point 10.0 m below the level of launching?

21. A firefighter, a distance d from a burning building, directs a stream of water from a fire hose at angle θ_i above the horizontal as shown in Figure P4.21. If the initial speed of the stream is v_i, at what height h does the water strike the building?

Figure P4.21

22. A landscape architect is planning an artificial water-fall in a city park. Water flowing at 1.70 m/s will leave the end of a horizontal channel at the top of a vertical wall $h = 2.35$ m high, and from there it will fall into a pool (Fig. P4.22). (a) Will the space behind the waterfall be wide enough for a pedestrian walkway? (b) To sell her plan to the city council, the architect wants to build a model to standard scale, which is one-twelfth actual size. How fast should the water flow in the channel in the model?

Figure P4.22

23. A placekicker must kick a football from a point 36.0 m
AMT (about 40 yards) from the goal. Half the crowd hopes
M the ball will clear the crossbar, which is 3.05 m high. When kicked, the ball leaves the ground with a speed of 20.0 m/s at an angle of 53.0° to the horizontal. (a) By how much does the ball clear or fall short of clearing the crossbar? (b) Does the ball approach the crossbar while still rising or while falling?

24. A basketball star covers 2.80 m horizontally in a jump to dunk the ball (Fig. P4.24a). His motion through space can be modeled precisely as that of a particle at his *center*

of mass, which we will define in Chapter 9. His center of mass is at elevation 1.02 m when he leaves the floor. It reaches a maximum height of 1.85 m above the floor and is at elevation 0.900 m when he touches down again. Determine (a) his time of flight (his "hang time"), (b) his horizontal and (c) vertical velocity components at the instant of takeoff, and (d) his takeoff angle. (e) For comparison, determine the hang time of a whitetail deer making a jump (Fig. P4.24b) with center-of-mass elevations $y_i = 1.20$ m, $y_{max} = 2.50$ m, and $y_f = 0.700$ m.

Figure P4.24

25. A playground is on the flat roof of a city school, 6.00 m above the street below (Fig. P4.25). The vertical wall of the building is $h = 7.00$ m high, forming a 1-m-high railing around the playground. A ball has fallen to the street below, and a passerby returns it by launching it at an angle of $\theta = 53.0°$ above the horizontal at a point $d = 24.0$ m from the base of the building wall. The ball takes 2.20 s to reach a point vertically above the wall. (a) Find the speed at which the ball was launched. (b) Find the vertical distance by which the ball clears the wall. (c) Find the horizontal distance from the wall to the point on the roof where the ball lands.

Figure P4.25

26. The motion of a human body through space can be modeled as the motion of a particle at the body's center of mass as we will study in Chapter 9. The components of the displacement of an athlete's center of mass from the beginning to the end of a certain jump are described by the equations

$$x_f = 0 + (11.2 \text{ m/s})(\cos 18.5°)t$$
$$0.360 \text{ m} = 0.840 \text{ m} + (11.2 \text{ m/s})(\sin 18.5°)t - \tfrac{1}{2}(9.80 \text{ m/s}^2)t^2$$

where t is in seconds and is the time at which the athlete ends the jump. Identify (a) the athlete's position and (b) his vector velocity at the takeoff point. (c) How far did he jump?

27. A soccer player kicks a rock horizontally off a
W 40.0-m-high cliff into a pool of water. If the player

hears the sound of the splash 3.00 s later, what was the initial speed given to the rock? Assume the speed of sound in air is 343 m/s.

28. A projectile is fired from the top of a cliff of height h above the ocean below. The projectile is fired at an angle θ above the horizontal and with an initial speed v_i. (a) Find a symbolic expression in terms of the variables v_i, g, and θ for the time at which the projectile reaches its maximum height. (b) Using the result of part (a), find an expression for the maximum height h_{max} above the ocean attained by the projectile in terms of h, v_i, g, and θ.

29. A student stands at the
GP edge of a cliff and throws a stone horizontally over the edge with a speed of $v_i = 18.0$ m/s. The cliff is $h = 50.0$ m above a body of water as shown in Figure P4.29. (a) What are the coordinates of the initial position of the stone? (b) What are the components of the initial velocity of the stone? (c) What is the appropriate analysis model for the vertical motion of the stone? (d) What is the appropriate analysis model for the horizontal motion of the stone? (e) Write symbolic equations for the x and y components of the velocity of the stone as a function of time. (f) Write symbolic equations for the position of the stone as a function of time. (g) How long after being released does the stone strike the water below the cliff? (h) With what speed and angle of impact does the stone land?

Figure P4.29

30. The record distance in the sport of throwing cowpats is 81.1 m. This record toss was set by Steve Urner of the United States in 1981. Assuming the initial launch angle was 45° and neglecting air resistance, determine (a) the initial speed of the projectile and (b) the total time interval the projectile was in flight. (c) How would the answers change if the range were the same but the launch angle were greater than 45°? Explain.

31. A boy stands on a diving board and tosses a stone into a swimming pool. The stone is thrown from a height of 2.50 m above the water surface with a velocity of 4.00 m/s at an angle of 60.0° above the horizontal. As the stone strikes the water surface, it immediately slows down to exactly half the speed it had when it struck the water and maintains that speed while in the water. After the stone enters the water, it moves in a straight line in the direction of the velocity it had when it struck the water. If the pool is 3.00 m deep, how much time elapses between when the stone is thrown and when it strikes the bottom of the pool?

32. A home run is hit in such a way that the baseball just
M clears a wall 21.0 m high, located 130 m from home plate. The ball is hit at an angle of 35.0° to the horizontal, and air resistance is negligible. Find (a) the

initial speed of the ball, (b) the time it takes the ball to reach the wall, and (c) the velocity components and the speed of the ball when it reaches the wall. (Assume the ball is hit at a height of 1.00 m above the ground.)

Section 4.4 Analysis Model: Particle in Uniform Circular Motion

Note: Problems 6 and 13 in Chapter 6 can also be assigned with this section.

33. The athlete shown in Figure P4.33 rotates a 1.00-kg discus along a circular path of radius 1.06 m. The maximum speed of the discus is 20.0 m/s. Determine the magnitude of the maximum radial acceleration of the discus.

Figure P4.33

34. In Example 4.6, we found the centripetal acceleration of the Earth as it revolves around the Sun. From information on the endpapers of this book, compute the centripetal acceleration of a point on the surface of the Earth at the equator caused by the rotation of the Earth about its axis.

35. Casting molten metal is important in many industrial processes. *Centrifugal casting* is used for manufacturing pipes, bearings, and many other structures. A variety of sophisticated techniques have been invented, but the basic idea is as illustrated in Figure P4.35. A cylindrical enclosure is rotated rapidly and steadily about a horizontal axis. Molten metal is poured into the rotating cylinder and then cooled, forming the finished product. Turning the cylinder at a high rotation rate forces the solidifying metal strongly to the outside. Any bubbles are displaced toward the axis, so unwanted voids will not be present in the casting. Sometimes it is desirable to form a composite casting, such as for a bearing. Here a strong steel outer surface is poured and then inside it a lining of special low-friction metal. In some applications, a very strong metal is given a coating of corrosion-resistant metal. Centrifugal casting results in strong bonding between the layers.

Figure P4.35

Suppose a copper sleeve of inner radius 2.10 cm and outer radius 2.20 cm is to be cast. To eliminate bubbles and give high structural integrity, the centripetal acceleration of each bit of metal should be at least 100*g*. What rate of rotation is required? State the answer in revolutions per minute.

36. A tire 0.500 m in radius rotates at a constant rate of 200 rev/min. Find the speed and acceleration of a small stone lodged in the tread of the tire (on its outer edge).

37. Review. The 20-*g* centrifuge at NASA's Ames Research **AMT** Center in Mountain View, California, is a horizontal, cylindrical tube 58.0 ft long and is represented in Figure P4.37. Assume an astronaut in training sits in a seat at one end, facing the axis of rotation 29.0 ft away. Determine the rotation rate, in revolutions per second, required to give the astronaut a centripetal acceleration of 20.0*g*.

Figure P4.37

38. An athlete swings a ball, connected to the end of a chain, in a horizontal circle. The athlete is able to rotate the ball at the rate of 8.00 rev/s when the length of the chain is 0.600 m. When he increases the length to 0.900 m, he is able to rotate the ball only 6.00 rev/s. (a) Which rate of rotation gives the greater speed for the ball? (b) What is the centripetal acceleration of the ball at 8.00 rev/s? (c) What is the centripetal acceleration at 6.00 rev/s?

39. The astronaut orbiting the Earth in Figure P4.39 is preparing to dock with a Westar VI satellite. The satellite is in a circular orbit 600 km above the Earth's surface, where the free-fall acceleration is 8.21 m/s². Take the radius of the Earth as 6 400 km. Determine the speed of the satellite and the time interval required to complete one orbit around the Earth, which is the period of the satellite.

Figure P4.39

NASA

Section 4.5 Tangential and Radial Acceleration

40. Figure P4.40 represents the **W** total acceleration of a particle moving clockwise in a circle of radius 2.50 m at a certain instant of time. For that instant, find (a) the radial acceleration of the particle, (b) the speed of the particle, and (c) its tangential acceleration.

$a = 15.0 \text{ m/s}^2$

2.50 m

$\vec{\mathbf{v}}$

$\vec{\mathbf{a}}$

30.0°

Figure P4.40

41. A train slows down as it rounds a sharp horizontal **M** turn, going from 90.0 km/h to 50.0 km/h in the 15.0 s it takes to round the bend. The radius of the curve is 150 m. Compute the acceleration at the moment the train speed reaches 50.0 km/h. Assume the train continues to slow down at this time at the same rate.

42. A ball swings counterclockwise in a vertical circle at the end of a rope 1.50 m long. When the ball is 36.9° past the lowest point on its way up, its total acceleration is $(-22.5\hat{\mathbf{i}} + 20.2\hat{\mathbf{j}})$ m/s². For that instant, (a) sketch a vector diagram showing the components of its acceleration, (b) determine the magnitude of its radial acceleration, and (c) determine the speed and velocity of the ball.

43. (a) Can a particle moving with instantaneous speed 3.00 m/s on a path with radius of curvature 2.00 m have an acceleration of magnitude 6.00 m/s²? (b) Can it have an acceleration of magnitude 4.00 m/s²? In each case, if the answer is yes, explain how it can happen; if the answer is no, explain why not.

Section 4.6 Relative Velocity and Relative Acceleration

44. The pilot of an airplane notes that the compass indicates a heading due west. The airplane's speed relative to the air is 150 km/h. The air is moving in a wind at 30.0 km/h toward the north. Find the velocity of the airplane relative to the ground.

45. An airplane maintains a speed of 630 km/h relative to the air it is flying through as it makes a trip to a city 750 km away to the north. (a) What time interval is required for the trip if the plane flies through a headwind blowing at 35.0 km/h toward the south? (b) What time interval is required if there is a tailwind with the same speed? (c) What time interval is required if there is a crosswind blowing at 35.0 km/h to the east relative to the ground?

46. A moving beltway at an airport has a speed v_1 and a length L. A woman stands on the beltway as it moves from one end to the other, while a man in a hurry to reach his flight walks on the beltway with a speed of v_2 relative to the moving beltway. (a) What time interval is required for the woman to travel the distance L? (b) What time interval is required for the man to travel this distance? (c) A second beltway is located next to the first one. It is identical to the first one but moves in the opposite direction at speed v_1. Just as the man steps onto the beginning of the beltway and begins to walk at speed v_2 relative to his beltway, a child steps on the other end of the adjacent beltway. The child stands at rest relative to this second beltway. How long after stepping on the beltway does the man pass the child?

47. A police car traveling at 95.0 km/h is traveling west, chasing a motorist traveling at 80.0 km/h. (a) What is the velocity of the motorist relative to the police car? (b) What is the velocity of the police car relative to the motorist? (c) If they are originally 250 m apart, in what time interval will the police car overtake the motorist?

48. A car travels due east with a speed of 50.0 km/h. Rain- **M** drops are falling at a constant speed vertically with respect to the Earth. The traces of the rain on the side windows of the car make an angle of 60.0° with the vertical. Find the velocity of the rain with respect to (a) the car and (b) the Earth.

49. A bolt drops from the ceiling of a moving train car that is accelerating northward at a rate of 2.50 m/s². (a) What is the acceleration of the bolt relative to the train car? (b) What is the acceleration of the bolt relative to the Earth? (c) Describe the trajectory of the bolt as seen by an observer inside the train car. (d) Describe the trajectory of the bolt as seen by an observer fixed on the Earth.

50. A river has a steady speed of 0.500 m/s. A student swims upstream a distance of 1.00 km and swims back to the starting point. (a) If the student can swim at a speed of 1.20 m/s in still water, how long does the trip take? (b) How much time is required in still water for the same length swim? (c) Intuitively, why does the swim take longer when there is a current?

51. A river flows with a steady speed v. A student swims upstream a distance d and then back to the starting point. The student can swim at speed c in still water. (a) In terms of d, v, and c, what time interval is required for the round trip? (b) What time interval would be required if the water were still? (c) Which time interval is larger? Explain whether it is always larger.

52. A Coast Guard cutter detects an unidentified ship at a distance of 20.0 km in the direction 15.0° east of north. The ship is traveling at 26.0 km/h on a course at 40.0° east of north. The Coast Guard wishes to send a speedboat to intercept and investigate the vessel. If the speedboat travels at 50.0 km/h, in what direction should it head? Express the direction as a compass bearing with respect to due north.

53. A science student is riding on a flatcar of a train traveling along a straight, horizontal track at a constant speed of 10.0 m/s. The student throws a ball into the air along a path that he judges to make an initial angle of 60.0° with the horizontal and to be in line with the track. The student's professor, who is standing on the ground nearby, observes the ball to rise vertically. How high does she see the ball rise?

54. A farm truck moves due east with a constant velocity of 9.50 m/s on a limitless, horizontal stretch of road. A boy riding on the back of the truck throws a can of soda upward (Fig. P4.54) and catches the projectile at the same location on the truck bed, but 16.0 m farther down the road. (a) In the frame of reference of the truck, at what angle to the vertical does the boy throw the can? (b) What is the initial speed of the can relative to the truck? (c) What is the shape of the can's trajectory as seen by the boy? An observer on the ground watches the boy throw the

Figure P4.54

can and catch it. In this observer's frame of reference, (d) describe the shape of the can's path and (e) determine the initial velocity of the can.

Additional Problems

55. A ball on the end of a string is whirled around in a horizontal circle of radius 0.300 m. The plane of the circle is 1.20 m above the ground. The string breaks and the ball lands 2.00 m (horizontally) away from the point on the ground directly beneath the ball's location when the string breaks. Find the radial acceleration of the ball during its circular motion.

56. A ball is thrown with an initial speed v_i at an angle θ_i with the horizontal. The horizontal range of the ball is R, and the ball reaches a maximum height $R/6$. In terms of R and g, find (a) the time interval during which the ball is in motion, (b) the ball's speed at the peak of its path, (c) the initial vertical component of its velocity, (d) its initial speed, and (e) the angle θ_i. (f) Suppose the ball is thrown at the same initial speed found in (d) but at the angle appropriate for reaching the greatest height that it can. Find this height. (g) Suppose the ball is thrown at the same initial speed but at the angle for greatest possible range. Find this maximum horizontal range.

57. *Why is the following situation impossible?* A normally proportioned adult walks briskly along a straight line in the $+x$ direction, standing straight up and holding his right arm vertical and next to his body so that the arm does not swing. His right hand holds a ball at his side a distance h above the floor. When the ball passes above a point marked as $x = 0$ on the horizontal floor, he opens his fingers to release the ball from rest relative to his hand. The ball strikes the ground for the first time at position $x = 7.00h$.

58. A particle starts from the origin with velocity $5\hat{\mathbf{i}}$ m/s at $t = 0$ and moves in the xy plane with a varying acceleration given by $\vec{\mathbf{a}} = (6\sqrt{t}\,\hat{\mathbf{j}})$, where $\vec{\mathbf{a}}$ is in meters per second squared and t is in seconds. (a) Determine the velocity of the particle as a function of time. (b) Determine the position of the particle as a function of time.

59. *The "Vomit Comet."* In microgravity astronaut training and equipment testing, NASA flies a KC135A aircraft along a parabolic flight path. As shown in Figure P4.59, the aircraft climbs from 24 000 ft to 31 000 ft, where

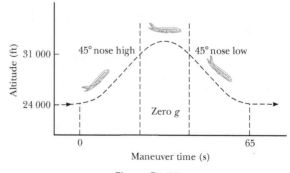

Figure P4.59

it enters a parabolic path with a velocity of 143 m/s nose high at 45.0° and exits with velocity 143 m/s at 45.0° nose low. During this portion of the flight, the aircraft and objects inside its padded cabin are in free fall; astronauts and equipment float freely as if there were no gravity. What are the aircraft's (a) speed and (b) altitude at the top of the maneuver? (c) What is the time interval spent in microgravity?

60. A basketball player is standing on the floor 10.0 m from the basket as in Figure P4.60. The height of the basket is 3.05 m, and he shoots the ball at a 40.0° angle with the horizontal from a height of 2.00 m. (a) What is the acceleration of the basketball at the highest point in its trajectory? (b) At what speed must the player throw the basketball so that the ball goes through the hoop without striking the backboard?

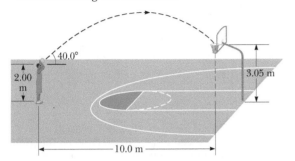

Figure P4.60

61. Lisa in her Lamborghini accelerates at the rate of $(3.00\hat{\mathbf{i}} - 2.00\hat{\mathbf{j}})$ m/s², while Jill in her Jaguar accelerates at $(1.00\hat{\mathbf{i}} + 3.00\hat{\mathbf{j}})$ m/s². They both start from rest at the origin of an xy coordinate system. After 5.00 s, (a) what is Lisa's speed with respect to Jill, (b) how far apart are they, and (c) what is Lisa's acceleration relative to Jill?

62. A boy throws a stone horizontally from the top of a cliff of height h toward the ocean below. The stone strikes the ocean at distance d from the base of the cliff. In terms of h, d, and g, find expressions for (a) the time t at which the stone lands in the ocean, (b) the initial speed of the stone, (c) the speed of the stone immediately before it reaches the ocean, and (d) the direction of the stone's velocity immediately before it reaches the ocean.

63. A flea is at point Ⓐ on a horizontal turntable, 10.0 cm from the center. The turntable is rotating at 33.3 rev/min in the clockwise direction. The flea jumps straight up to a height of 5.00 cm. At takeoff, it gives itself no horizontal velocity relative to the turntable. The flea lands on the turntable at point Ⓑ. Choose the origin of coordinates to be at the center of the turntable and the positive x axis passing through Ⓐ at the moment of takeoff. Then the original position of the flea is $10.0\hat{\mathbf{i}}$ cm. (a) Find the position of point Ⓐ when the flea lands. (b) Find the position of point Ⓑ when the flea lands.

64. Towns A and B in Figure P4.64 are 80.0 km apart. A Ⓜ couple arranges to drive from town A and meet a couple driving from town B at the lake, L. The two couples

leave simultaneously and drive for 2.50 h in the directions shown. Car 1 has a speed of 90.0 km/h. If the cars arrive simultaneously at the lake, what is the speed of car 2?

Figure P4.64

65. A catapult launches a rocket at an angle of 53.0° above the horizontal with an initial speed of 100 m/s. The rocket engine immediately starts a burn, and for 3.00 s the rocket moves along its initial line of motion with an acceleration of 30.0 m/s². Then its engine fails, and the rocket proceeds to move in free fall. Find (a) the maximum altitude reached by the rocket, (b) its total time of flight, and (c) its horizontal range.

66. A cannon with a muzzle speed of 1 000 m/s is used to start an avalanche on a mountain slope. The target is 2 000 m from the cannon horizontally and 800 m above the cannon. At what angle, above the horizontal, should the cannon be fired?

67. *Why is the following situation impossible?* Albert Pujols hits a home run so that the baseball just clears the top row of bleachers, 24.0 m high, located 130 m from home plate. The ball is hit at 41.7 m/s at an angle of 35.0° to the horizontal, and air resistance is negligible.

68. As some molten metal splashes, one droplet flies off to the east with initial velocity v_i at angle θ_i above the horizontal, and another droplet flies off to the west with the same speed at the same angle above the horizontal as shown in Figure P4.68. In terms of v_i and θ_i, find the distance between the two droplets as a function of time.

Figure P4.68

69. An astronaut on the surface of the Moon fires a cannon to launch an experiment package, which leaves the barrel moving horizontally. Assume the free-fall acceleration on the Moon is one-sixth of that on the

Earth. (a) What must the muzzle speed of the package be so that it travels completely around the Moon and returns to its original location? (b) What time interval does this trip around the Moon require?

70. A pendulum with a cord of length $r = 1.00$ m swings in a vertical plane (Fig. P4.70). When the pendulum is in the two horizontal positions $\theta = 90.0°$ and $\theta = 270°$, its speed is 5.00 m/s. Find the magnitude of (a) the radial acceleration and (b) the tangential acceleration for these positions. (c) Draw vector diagrams to determine the direction of the total acceleration for these two positions. (d) Calculate the magnitude and direction of the total acceleration at these two positions.

Figure P4.70

71. A hawk is flying horizontally at 10.0 m/s in a straight
M line, 200 m above the ground. A mouse it has been carrying struggles free from its talons. The hawk continues on its path at the same speed for 2.00 s before attempting to retrieve its prey. To accomplish the retrieval, it dives in a straight line at constant speed and recaptures the mouse 3.00 m above the ground. (a) Assuming no air resistance acts on the mouse, find the diving speed of the hawk. (b) What angle did the hawk make with the horizontal during its descent? (c) For what time interval did the mouse experience free fall?

72. A projectile is launched from the point ($x = 0$, $y = 0$), with velocity $(12.0\hat{\mathbf{i}} + 49.0\hat{\mathbf{j}})$ m/s, at $t = 0$. (a) Make a table listing the projectile's distance $|\vec{\mathbf{r}}|$ from the origin at the end of each second thereafter, for $0 \le t \le 10$ s. Tabulating the x and y coordinates and the components of velocity v_x and v_y will also be useful. (b) Notice that the projectile's distance from its starting point increases with time, goes through a maximum, and starts to decrease. Prove that the distance is a maximum when the position vector is perpendicular to the velocity. *Suggestion:* Argue that if $\vec{\mathbf{v}}$ is not perpendicular to $\vec{\mathbf{r}}$, then $|\vec{\mathbf{r}}|$ must be increasing or decreasing. (c) Determine the magnitude of the maximum displacement. (d) Explain your method for solving part (c).

73. A spring cannon is located at the edge of a table that is 1.20 m above the floor. A steel ball is launched from the cannon with speed v_i at 35.0° above the horizontal. (a) Find the horizontal position of the ball as a function of v_i at the instant it lands on the floor. We write this function as $x(v_i)$. Evaluate x for (b) $v_i = 0.100$ m/s and for (c) $v_i = 100$ m/s. (d) Assume v_i is close to but not equal to zero. Show that one term in the answer to part (a) dominates so that the function $x(v_i)$ reduces to a simpler form. (e) If v_i is very large, what is the approximate form of $x(v_i)$? (f) Describe the overall shape of the graph of the function $x(v_i)$.

74. An outfielder throws a baseball to his catcher in an attempt to throw out a runner at home plate. The ball bounces once before reaching the catcher. Assume the angle at which the bounced ball leaves the ground is the same as the angle at which the outfielder threw it as shown in Figure P4.74, but that the ball's speed after the bounce is one-half of what it was before the bounce. (a) Assume the ball is always thrown with the same initial speed and ignore air resistance. At what angle θ should the fielder throw the ball to make it go the same distance D with one bounce (blue path) as a ball thrown upward at 45.0° with no bounce (green path)? (b) Determine the ratio of the time interval for the one-bounce throw to the flight time for the no-bounce throw.

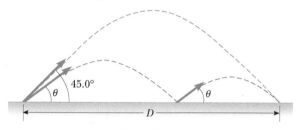

Figure P4.74

75. A World War II bomber flies horizontally over level terrain with a speed of 275 m/s relative to the ground and at an altitude of 3.00 km. The bombardier releases one bomb. (a) How far does the bomb travel horizontally between its release and its impact on the ground? Ignore the effects of air resistance. (b) The pilot maintains the plane's original course, altitude, and speed through a storm of flak. Where is the plane when the bomb hits the ground? (c) The bomb hits the target seen in the telescopic bombsight at the moment of the bomb's release. At what angle from the vertical was the bombsight set?

76. A truck loaded with cannonball watermelons stops suddenly to avoid running over the edge of a washed-out bridge (Fig. P4.76). The quick stop causes a number of melons to fly off the truck. One melon leaves the hood of the truck with an initial speed $v_i = 10.0$ m/s in the horizontal direction. A cross section of the bank has the shape of the bottom half of a parabola, with its vertex at the initial location of the projected watermelon and with the equation $y^2 = 16x$, where x and y are mea-

Figure P4.76 The blue dashed curve shows the parabolic shape of the bank.

sured in meters. What are the x and y coordinates of the melon when it splatters on the bank?

77. A car is parked on a steep incline, making an angle of **M** 37.0° below the horizontal and overlooking the ocean, when its brakes fail and it begins to roll. Starting from rest at $t = 0$, the car rolls down the incline with a constant acceleration of 4.00 m/s², traveling 50.0 m to the edge of a vertical cliff. The cliff is 30.0 m above the ocean. Find (a) the speed of the car when it reaches the edge of the cliff, (b) the time interval elapsed when it arrives there, (c) the velocity of the car when it lands in the ocean, (d) the total time interval the car is in motion, and (e) the position of the car when it lands in the ocean, relative to the base of the cliff.

78. An aging coyote cannot run fast enough to catch a road-runner. He purchases on eBay a set of jet-powered roller skates, which provide a constant horizontal acceleration of 15.0 m/s² (Fig. P4.78). The coyote starts at rest 70.0 m from the edge of a cliff at the instant the roadrunner zips past in the direction of the cliff. (a) Determine the minimum constant speed the roadrunner must have to reach the cliff before the coyote. At the edge of the cliff, the roadrunner escapes by making a sudden turn, while the coyote continues straight ahead. The coyote's skates remain horizontal and continue to operate while he is in flight, so his acceleration while in the air is $(15.0\hat{\mathbf{i}} - 9.80\hat{\mathbf{j}})$ m/s². (b) The cliff is 100 m above the flat floor of the desert. Determine how far from the base of the vertical cliff the coyote lands. (c) Determine the components of the coyote's impact velocity.

Figure P4.78

79. A fisherman sets out upstream on a river. His small boat, powered by an outboard motor, travels at a constant speed v in still water. The water flows at a lower constant speed v_w. The fisherman has traveled upstream for 2.00 km when his ice chest falls out of the boat. He notices that the chest is missing only after he has gone upstream for another 15.0 min. At that point, he turns around and heads back downstream, all the time traveling at the same speed relative to the water. He catches up with the floating ice chest just as he returns to his starting point. How fast is the river flowing? Solve this problem in two ways. (a) First, use the Earth as a reference frame. With respect to the Earth, the boat travels upstream at speed $v - v_w$ and downstream at $v + v_w$. (b) A second much simpler and more elegant solution is obtained by using the water as the reference frame. This approach has important applications in many more complicated problems; examples are calculating the motion of rockets and satellites and analyzing the scattering of subatomic particles from massive targets.

80. Do not hurt yourself; do not strike your hand against anything. Within these limitations, describe what you do to give your hand a large acceleration. Compute an order-of-magnitude estimate of this acceleration, stating the quantities you measure or estimate and their values.

Challenge Problems

81. A skier leaves the ramp of a ski jump with a velocity of $v = 10.0$ m/s at $\theta = 15.0°$ above the horizontal as shown in Figure P4.81. The slope where she will land is inclined downward at $\phi = 50.0°$, and air resistance is negligible. Find (a) the distance from the end of the ramp to where the jumper lands and (b) her velocity components just before the landing. (c) Explain how you think the results might be affected if air resistance were included.

Figure P4.81

82. Two swimmers, Chris and Sarah, start together at the same point on the bank of a wide stream that flows with a speed v. Both move at the same speed c (where $c > v$) relative to the water. Chris swims downstream a distance L and then upstream the same distance. Sarah swims so that her motion relative to the Earth is perpendicular to the banks of the stream. She swims the distance L and then back the same distance, with both swimmers returning to the starting point. In terms of L, c, and v, find the time intervals required (a) for Chris's round trip and (b) for Sarah's round trip. (c) Explain which swimmer returns first.

83. The water in a river flows uniformly at a constant speed of 2.50 m/s between parallel banks 80.0 m apart. You are to deliver a package across the river, but you can swim only at 1.50 m/s. (a) If you choose to minimize the time you spend in the water, in what direction should you head? (b) How far downstream will you be carried? (c) If you choose to minimize the distance downstream that the river carries you, in what direction should you head? (d) How far downstream will you be carried?

84. A person standing at the top of a hemispherical rock of radius R kicks a ball (initially at rest on the top of the rock) to give it horizontal velocity $\vec{\mathbf{v}}_i$ as shown in Figure P4.84. (a) What must be its minimum initial speed

Figure P4.84

if the ball is never to hit the rock after it is kicked?
(b) With this initial speed, how far from the base of the
rock does the ball hit the ground?

85. A dive-bomber has a velocity of 280 m/s at an angle θ
below the horizontal. When the altitude of the aircraft
is 2.15 km, it releases a bomb, which subsequently hits
a target on the ground. The magnitude of the displacement from the point of release of the bomb to the target is 3.25 km. Find the angle θ.

86. A projectile is fired up an incline (incline angle ϕ)
with an initial speed v_i at an angle θ_i with respect to the
horizontal ($\theta_i > \phi$) as shown in Figure P4.86. (a) Show
that the projectile travels a distance d up the incline,
where

$$d = \frac{2v_i^2 \cos\theta_i \sin(\theta_i - \phi)}{g\cos^2\phi}$$

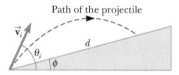

Path of the projectile

Figure P4.86

(b) For what value of θ_i is d a maximum, and what is
that maximum value?

87. A fireworks rocket explodes at height h, the peak of
its vertical trajectory. It throws out burning fragments
in all directions, but all at the same speed v. Pellets
of solidified metal fall to the ground without air resistance. Find the smallest angle that the final velocity of
an impacting fragment makes with the horizontal.

88. In the What If? section of Example 4.5, it was claimed
that the maximum range of a ski jumper occurs for a
launch angle θ given by

$$\theta = 45° - \frac{\phi}{2}$$

where ϕ is the angle the hill makes with the horizontal
in Figure 4.14. Prove this claim by deriving the equation above.

89. An enemy ship is on the east side of a mountain island
as shown in Figure P4.89. The enemy ship has maneu-
vered to within 2 500 m of the 1 800-m-high mountain
peak and can shoot projectiles with an initial speed of
250 m/s. If the western shoreline is horizontally 300 m
from the peak, what are the distances from the western
shore at which a ship can be safe from the bombard-
ment of the enemy ship?

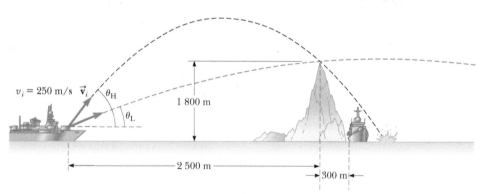

$v_i = 250$ m/s

1 800 m

2 500 m

300 m

Figure P4.89

The Laws of Motion

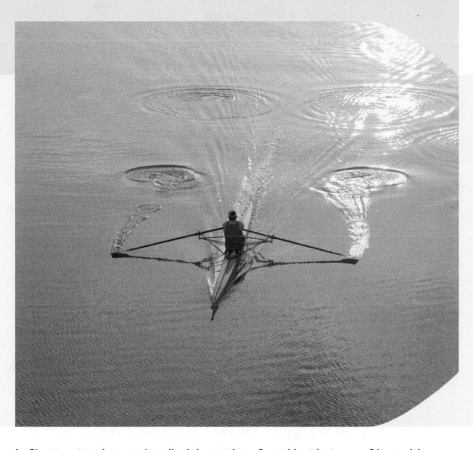

In Chapters 2 and 4, we *described* the motion of an object in terms of its position, velocity, and acceleration without considering what might *influence* that motion. Now we consider that influence: Why does the motion of an object change? What might cause one object to remain at rest and another object to accelerate? Why is it generally easier to move a small object than a large object? The two main factors we need to consider are the *forces* acting on an object and the *mass* of the object. In this chapter, we begin our study of *dynamics* by discussing the three basic laws of motion, which deal with forces and masses and were formulated more than three centuries ago by Isaac Newton.

A person sculls on a calm waterway. The water exerts forces on the oars to accelerate the boat. *(Tetra Images/ Getty Images)*

5.1 The Concept of Force

Everyone has a basic understanding of the concept of force from everyday experience. When you push your empty dinner plate away, you exert a force on it. Similarly, you exert a force on a ball when you throw or kick it. In these examples, the word *force* refers to an interaction with an object by means of muscular activity and some change in the object's velocity. Forces do not always cause motion, however. For example, when you are sitting, a gravitational force acts on your body and yet you remain stationary. As a second example, you can push (in other words, exert a force) on a large boulder and not be able to move it.

What force (if any) causes the Moon to orbit the Earth? Newton answered this and related questions by stating that forces are what cause any change in the velocity of an object. The Moon's velocity changes in direction as it moves in a nearly circular

111

Figure 5.1 Some examples of applied forces. In each case, a force is exerted on the object within the boxed area. Some agent in the environment external to the boxed area exerts a force on the object.

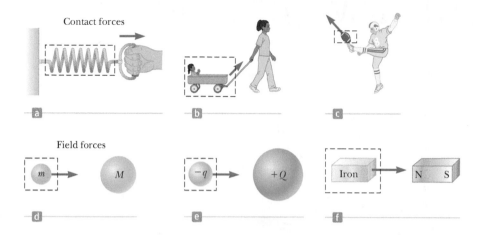

orbit around the Earth. This change in velocity is caused by the gravitational force exerted by the Earth on the Moon.

When a coiled spring is pulled, as in Figure 5.1a, the spring stretches. When a stationary cart is pulled, as in Figure 5.1b, the cart moves. When a football is kicked, as in Figure 5.1c, it is both deformed and set in motion. These situations are all examples of a class of forces called *contact forces*. That is, they involve physical contact between two objects. Other examples of contact forces are the force exerted by gas molecules on the walls of a container and the force exerted by your feet on the floor.

Another class of forces, known as *field forces*, does not involve physical contact between two objects. These forces act through empty space. The gravitational force of attraction between two objects with mass, illustrated in Figure 5.1d, is an example of this class of force. The gravitational force keeps objects bound to the Earth and the planets in orbit around the Sun. Another common field force is the electric force that one electric charge exerts on another (Fig. 5.1e), such as the attractive electric force between an electron and a proton that form a hydrogen atom. A third example of a field force is the force a bar magnet exerts on a piece of iron (Fig. 5.1f).

The distinction between contact forces and field forces is not as sharp as you may have been led to believe by the previous discussion. When examined at the atomic level, all the forces we classify as contact forces turn out to be caused by electric (field) forces of the type illustrated in Figure 5.1e. Nevertheless, in developing models for macroscopic phenomena, it is convenient to use both classifications of forces. The only known *fundamental* forces in nature are all field forces: (1) *gravitational forces* between objects, (2) *electromagnetic forces* between electric charges, (3) *strong forces* between subatomic particles, and (4) *weak forces* that arise in certain radioactive decay processes. In classical physics, we are concerned only with gravitational and electromagnetic forces. We will discuss strong and weak forces in Chapter 46.

The Vector Nature of Force

It is possible to use the deformation of a spring to measure force. Suppose a vertical force is applied to a spring scale that has a fixed upper end as shown in Figure 5.2a. The spring elongates when the force is applied, and a pointer on the scale reads the extension of the spring. We can calibrate the spring by defining a reference force $\vec{\mathbf{F}}_1$ as the force that produces a pointer reading of 1.00 cm. If we now apply a different downward force $\vec{\mathbf{F}}_2$ whose magnitude is twice that of the reference force $\vec{\mathbf{F}}_1$ as seen in Figure 5.2b, the pointer moves to 2.00 cm. Figure 5.2c shows that the combined effect of the two collinear forces is the sum of the effects of the individual forces.

Now suppose the two forces are applied simultaneously with $\vec{\mathbf{F}}_1$ downward and $\vec{\mathbf{F}}_2$ horizontal as illustrated in Figure 5.2d. In this case, the pointer reads 2.24 cm. The single force $\vec{\mathbf{F}}$ that would produce this same reading is the sum of the two vectors $\vec{\mathbf{F}}_1$ and $\vec{\mathbf{F}}_2$ as described in Figure 5.2d. That is, $|\vec{\mathbf{F}}_1| = \sqrt{F_1^2 + F_2^2} = 2.24$ units,

Isaac Newton
English physicist and mathematician
(1642–1727)
Isaac Newton was one of the most brilliant scientists in history. Before the age of 30, he formulated the basic concepts and laws of mechanics, discovered the law of universal gravitation, and invented the mathematical methods of calculus. As a consequence of his theories, Newton was able to explain the motions of the planets, the ebb and flow of the tides, and many special features of the motions of the Moon and the Earth. He also interpreted many fundamental observations concerning the nature of light. His contributions to physical theories dominated scientific thought for two centuries and remain important today.

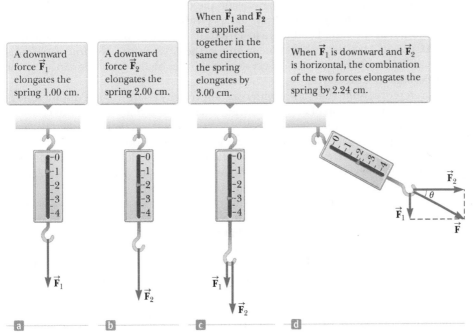

When $\vec{F}_1$ and $\vec{F}_2$ are applied together in the same direction, the spring elongates by 3.00 cm.

A downward force $\vec{F}_1$ elongates the spring 1.00 cm.

A downward force $\vec{F}_2$ elongates the spring 2.00 cm.

When $\vec{F}_1$ is downward and $\vec{F}_2$ is horizontal, the combination of the two forces elongates the spring by 2.24 cm.

Figure 5.2 The vector nature of a force is tested with a spring scale.

and its direction is $\theta = \tan^{-1}(-0.500) = -26.6°$. Because forces have been experimentally verified to behave as vectors, you *must* use the rules of vector addition to obtain the net force on an object.

5.2 Newton's First Law and Inertial Frames

We begin our study of forces by imagining some physical situations involving a puck on a perfectly level air hockey table (Fig. 5.3). You expect that the puck will remain stationary when it is placed gently at rest on the table. Now imagine your air hockey table is located on a train moving with constant velocity along a perfectly smooth track. If the puck is placed on the table, the puck again remains where it is placed. If the train were to accelerate, however, the puck would start moving along the table opposite the direction of the train's acceleration, just as a set of papers on your dashboard falls onto the floor of your car when you step on the accelerator.

As we saw in Section 4.6, a moving object can be observed from any number of reference frames. **Newton's first law of motion,** sometimes called the *law of inertia,* defines a special set of reference frames called *inertial frames.* This law can be stated as follows:

> If an object does not interact with other objects, it is possible to identify a reference frame in which the object has zero acceleration.

Such a reference frame is called an **inertial frame of reference.** When the puck is on the air hockey table located on the ground, you are observing it from an inertial reference frame; there are no horizontal interactions of the puck with any other objects, and you observe it to have zero acceleration in that direction. When you are on the train moving at constant velocity, you are also observing the puck from an inertial reference frame. Any reference frame that moves with constant velocity relative to an inertial frame is itself an inertial frame. When you and the train accelerate, however, you are observing the puck from a **noninertial reference frame** because the train is accelerating relative to the inertial reference frame of the Earth's surface. While the puck appears to be accelerating according to your observations, a reference frame can be identified in which the puck has zero acceleration.

Airflow

Electric blower

Figure 5.3 On an air hockey table, air blown through holes in the surface allows the puck to move almost without friction. If the table is not accelerating, a puck placed on the table will remain at rest.

◀ Newton's first law

◀ Inertial frame of reference

For example, an observer standing outside the train on the ground sees the puck sliding relative to the table but always moving with the same velocity with respect to the ground as the train had before it started to accelerate (because there is almost no friction to "tie" the puck and the train together). Therefore, Newton's first law is still satisfied even though your observations as a rider on the train show an apparent acceleration relative to you.

A reference frame that moves with constant velocity relative to the distant stars is the best approximation of an inertial frame, and for our purposes we can consider the Earth as being such a frame. The Earth is not really an inertial frame because of its orbital motion around the Sun and its rotational motion about its own axis, both of which involve centripetal accelerations. These accelerations are small compared with g, however, and can often be neglected. For this reason, we model the Earth as an inertial frame, along with any other frame attached to it.

Let us assume we are observing an object from an inertial reference frame. (We will return to observations made in noninertial reference frames in Section 6.3.) Before about 1600, scientists believed that the natural state of matter was the state of rest. Observations showed that moving objects eventually stopped moving. Galileo was the first to take a different approach to motion and the natural state of matter. He devised thought experiments and concluded that it is not the nature of an object to stop once set in motion: rather, it is its nature to *resist changes in its motion*. In his words, "Any velocity once imparted to a moving body will be rigidly maintained as long as the external causes of retardation are removed." For example, a spacecraft drifting through empty space with its engine turned off will keep moving forever. It would *not* seek a "natural state" of rest.

Given our discussion of observations made from inertial reference frames, we can pose a more practical statement of Newton's first law of motion:

> In the absence of external forces and when viewed from an inertial reference frame, an object at rest remains at rest and an object in motion continues in motion with a constant velocity (that is, with a constant speed in a straight line).

In other words, **when no force acts on an object, the acceleration of the object is zero.** From the first law, we conclude that any *isolated object* (one that does not interact with its environment) is either at rest or moving with constant velocity. The tendency of an object to resist any attempt to change its velocity is called **inertia.** Given the statement of the first law above, we can conclude that an object that is accelerating must be experiencing a force. In turn, from the first law, we can define **force** as **that which causes a change in motion of an object.**

Quick Quiz 5.1 Which of the following statements is correct? **(a)** It is possible for an object to have motion in the absence of forces on the object. **(b)** It is possible to have forces on an object in the absence of motion of the object. **(c)** Neither statement **(a)** nor statement **(b)** is correct. **(d)** Both statements **(a)** and **(b)** are correct.

5.3 Mass

Imagine playing catch with either a basketball or a bowling ball. Which ball is more likely to keep moving when you try to catch it? Which ball requires more effort to throw it? The bowling ball requires more effort. In the language of physics, we say that the bowling ball is more resistant to changes in its velocity than the basketball. How can we quantify this concept?

Mass is that property of an object that specifies how much resistance an object exhibits to changes in its velocity, and as we learned in Section 1.1, the SI unit of mass is the kilogram. Experiments show that the greater the mass of an object, the less that object accelerates under the action of a given applied force.

To describe mass quantitatively, we conduct experiments in which we compare the accelerations a given force produces on different objects. Suppose a force act-

Pitfall Prevention 5.1
Newton's First Law Newton's first law does *not* say what happens for an object with *zero net force*, that is, multiple forces that cancel; it says what happens *in the absence of external forces*. This subtle but important difference allows us to define force as that which causes a change in the motion. The description of an object under the effect of forces that balance is covered by Newton's second law.

▶ Another statement of Newton's first law

▶ Definition of force

Pitfall Prevention 5.2
Force Is the Cause of Changes in Motion An object can have motion in the absence of forces as described in Newton's first law. Therefore, don't interpret force as the cause of *motion*. Force is the cause of *changes in motion*.

▶ Definition of mass

ing on an object of mass m_1 produces a change in motion of the object that we can quantify with the object's acceleration $\vec{a}_1$, and the *same force* acting on an object of mass m_2 produces an acceleration $\vec{a}_2$. The ratio of the two masses is defined as the *inverse* ratio of the magnitudes of the accelerations produced by the force:

$$\frac{m_1}{m_2} \equiv \frac{a_2}{a_1} \tag{5.1}$$

For example, if a given force acting on a 3-kg object produces an acceleration of 4 m/s², the same force applied to a 6-kg object produces an acceleration of 2 m/s². According to a huge number of similar observations, we conclude that the magnitude of the acceleration of an object is inversely proportional to its mass when acted on by a given force. If one object has a known mass, the mass of the other object can be obtained from acceleration measurements.

Mass is an inherent property of an object and is independent of the object's surroundings and of the method used to measure it. Also, mass is a scalar quantity and thus obeys the rules of ordinary arithmetic. For example, if you combine a 3-kg mass with a 5-kg mass, the total mass is 8 kg. This result can be verified experimentally by comparing the acceleration that a known force gives to several objects separately with the acceleration that the same force gives to the same objects combined as one unit.

Mass should not be confused with weight. Mass and weight are two different quantities. The weight of an object is equal to the magnitude of the gravitational force exerted on the object and varies with location (see Section 5.5). For example, a person weighing 180 lb on the Earth weighs only about 30 lb on the Moon. On the other hand, the mass of an object is the same everywhere: an object having a mass of 2 kg on the Earth also has a mass of 2 kg on the Moon.

◀ **Mass and weight are different quantities**

5.4 Newton's Second Law

Newton's first law explains what happens to an object when no forces act on it: it maintains its original motion; it either remains at rest or moves in a straight line with constant speed. Newton's second law answers the question of what happens to an object when one or more forces act on it.

Imagine performing an experiment in which you push a block of mass m across a frictionless, horizontal surface. When you exert some horizontal force $\vec{F}$ on the block, it moves with some acceleration $\vec{a}$. If you apply a force twice as great on the same block, experimental results show that the acceleration of the block doubles; if you increase the applied force to $3\vec{F}$, the acceleration triples; and so on. From such observations, we conclude that the acceleration of an object is directly proportional to the force acting on it: $\vec{F} \propto \vec{a}$. This idea was first introduced in Section 2.4 when we discussed the direction of the acceleration of an object. We also know from the preceding section that the magnitude of the acceleration of an object is inversely proportional to its mass: $|\vec{a}| \propto 1/m$.

These experimental observations are summarized in **Newton's second law:**

> When viewed from an inertial reference frame, the acceleration of an object is directly proportional to the net force acting on it and inversely proportional to its mass:
>
> $$\vec{a} \propto \frac{\sum \vec{F}}{m}$$

If we choose a proportionality constant of 1, we can relate mass, acceleration, and force through the following mathematical statement of Newton's second law:[1]

$$\sum \vec{F} = m\vec{a} \tag{5.2}$$

Pitfall Prevention 5.3
$m\vec{a}$ **Is Not a Force** Equation 5.2 does *not* say that the product $m\vec{a}$ is a force. All forces on an object are added vectorially to generate the net force on the left side of the equation. This net force is then equated to the product of the mass of the object and the acceleration that results from the net force. Do *not* include an "$m\vec{a}$ force" in your analysis of the forces on an object.

◀ **Newton's second law**

---

[1]Equation 5.2 is valid only when the speed of the object is much less than the speed of light. We treat the relativistic situation in Chapter 39.

In both the textual and mathematical statements of Newton's second law, we have indicated that the acceleration is due to the *net force* $\sum \vec{F}$ acting on an object. The **net force** on an object is the vector sum of all forces acting on the object. (We sometimes refer to the net force as the *total force*, the *resultant force*, or the *unbalanced force*.) In solving a problem using Newton's second law, it is imperative to determine the correct net force on an object. Many forces may be acting on an object, but there is only one acceleration.

Equation 5.2 is a vector expression and hence is equivalent to three component equations:

◀ **Newton's second law: component form**

$$\sum F_x = ma_x \qquad \sum F_y = ma_y \qquad \sum F_z = ma_z \qquad \text{(5.3)}$$

Quick Quiz 5.2 An object experiences no acceleration. Which of the following *cannot* be true for the object? **(a)** A single force acts on the object. **(b)** No forces act on the object. **(c)** Forces act on the object, but the forces cancel.

Quick Quiz 5.3 You push an object, initially at rest, across a frictionless floor with a constant force for a time interval Δt, resulting in a final speed of v for the object. You then repeat the experiment, but with a force that is twice as large. What time interval is now required to reach the same final speed v? **(a)** $4\,\Delta t$ **(b)** $2\,\Delta t$ **(c)** Δt **(d)** $\Delta t/2$ **(e)** $\Delta t/4$

The SI unit of force is the **newton** (N). A force of 1 N is the force that, when acting on an object of mass 1 kg, produces an acceleration of 1 m/s². From this definition and Newton's second law, we see that the newton can be expressed in terms of the following fundamental units of mass, length, and time:

◀ **Definition of the newton**

$$1\ \mathrm{N} \equiv 1\ \mathrm{kg} \cdot \mathrm{m/s^2} \qquad \text{(5.4)}$$

In the U.S. customary system, the unit of force is the **pound** (lb). A force of 1 lb is the force that, when acting on a 1-slug mass,[2] produces an acceleration of 1 ft/s²:

$$1\ \mathrm{lb} \equiv 1\ \mathrm{slug} \cdot \mathrm{ft/s^2}$$

A convenient approximation is $1\ \mathrm{N} \approx \frac{1}{4}\ \mathrm{lb}$.

Example 5.1 | **An Accelerating Hockey Puck** AM

A hockey puck having a mass of 0.30 kg slides on the frictionless, horizontal surface of an ice rink. Two hockey sticks strike the puck simultaneously, exerting the forces on the puck shown in Figure 5.4. The force $\vec{F}_1$ has a magnitude of 5.0 N, and is directed at $\theta = 20°$ below the x axis. The force $\vec{F}_2$ has a magnitude of 8.0 N and its direction is $\phi = 60°$ above the x axis. Determine both the magnitude and the direction of the puck's acceleration.

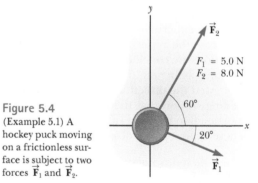

Figure 5.4
(Example 5.1) A hockey puck moving on a frictionless surface is subject to two forces $\vec{F}_1$ and $\vec{F}_2$.

$F_1 = 5.0$ N
$F_2 = 8.0$ N

SOLUTION

Conceptualize Study Figure 5.4. Using your expertise in vector addition from Chapter 3, predict the approximate direction of the net force vector on the puck. The acceleration of the puck will be in the same direction.

Categorize Because we can determine a net force and we want an acceleration, this problem is categorized as one that may be solved using Newton's second law. In Section 5.7, we will formally introduce the *particle under a net force* analysis model to describe a situation such as this one.

Analyze Find the component of the net force acting on the puck in the x direction:

$$\sum F_x = F_{1x} + F_{2x} = F_1 \cos \theta + F_2 \cos \phi$$

---

[2]The *slug* is the unit of mass in the U.S. customary system and is that system's counterpart of the SI unit the *kilogram*. Because most of the calculations in our study of classical mechanics are in SI units, the slug is seldom used in this text.

▶ **5.1** continued

Find the component of the net force acting on the puck in the y direction:

$$\sum F_y = F_{1y} + F_{2y} = F_1 \sin \theta + F_2 \sin \phi$$

Use Newton's second law in component form (Eq. 5.3) to find the x and y components of the puck's acceleration:

$$a_x = \frac{\sum F_x}{m} = \frac{F_1 \cos \theta + F_2 \cos \phi}{m}$$

$$a_y = \frac{\sum F_y}{m} = \frac{F_1 \sin \theta + F_2 \sin \phi}{m}$$

Substitute numerical values:

$$a_x = \frac{(5.0 \text{ N}) \cos(-20°) + (8.0 \text{ N}) \cos(60°)}{0.30 \text{ kg}} = 29 \text{ m/s}^2$$

$$a_y = \frac{(5.0 \text{ N}) \sin(-20°) + (8.0 \text{ N}) \sin(60°)}{0.30 \text{ kg}} = 17 \text{ m/s}^2$$

Find the magnitude of the acceleration:

$$a = \sqrt{(29 \text{ m/s}^2)^2 + (17 \text{ m/s}^2)^2} = \boxed{34 \text{ m/s}^2}$$

Find the direction of the acceleration relative to the positive x axis:

$$\theta = \tan^{-1}\left(\frac{a_y}{a_x}\right) = \tan^{-1}\left(\frac{17}{29}\right) = \boxed{31°}$$

Finalize The vectors in Figure 5.4 can be added graphically to check the reasonableness of our answer. Because the acceleration vector is along the direction of the resultant force, a drawing showing the resultant force vector helps us check the validity of the answer. (Try it!)

WHAT IF? Suppose three hockey sticks strike the puck simultaneously, with two of them exerting the forces shown in Figure 5.4. The result of the three forces is that the hockey puck shows *no* acceleration. What must be the components of the third force?

Answer If there is zero acceleration, the net force acting on the puck must be zero. Therefore, the three forces must cancel. The components of the third force must be of equal magnitude and opposite sign compared to the components of the net force applied by the first two forces so that all the components add to zero. Therefore, $F_{3x} = -\sum F_x = -(0.30 \text{ kg})(29 \text{ m/s}^2) = -8.7 \text{ N}$ and $F_{3y} = -\sum F_y = -(0.30 \text{ kg})(17 \text{ m/s}^2) = -5.2 \text{ N}$.

5.5 The Gravitational Force and Weight

All objects are attracted to the Earth. The attractive force exerted by the Earth on an object is called the **gravitational force** $\vec{F}_g$. This force is directed toward the center of the Earth,[3] and its magnitude is called the **weight** of the object.

We saw in Section 2.6 that a freely falling object experiences an acceleration $\vec{g}$ acting toward the center of the Earth. Applying Newton's second law $\sum \vec{F} = m\vec{a}$ to a freely falling object of mass m, with $\vec{a} = \vec{g}$ and $\sum \vec{F} = \vec{F}_g$, gives

$$\vec{F}_g = m\vec{g} \tag{5.5}$$

Therefore, the weight of an object, being defined as the magnitude of $\vec{F}_g$, is given by

$$F_g = mg \tag{5.6}$$

Because it depends on g, weight varies with geographic location. Because g decreases with increasing distance from the center of the Earth, objects weigh less at higher altitudes than at sea level. For example, a 1 000-kg pallet of bricks used in the construction of the Empire State Building in New York City weighed 9 800 N at street level, but weighed about 1 N less by the time it was lifted from sidewalk level to the top of the building. As another example, suppose a student has a mass

Pitfall Prevention 5.4
"Weight of an Object" We are familiar with the everyday phrase, the "weight of an object." Weight, however, is not an inherent property of an object; rather, it is a measure of the gravitational force between the object and the Earth (or other planet). Therefore, weight is a property of a *system* of items: the object and the Earth.

Pitfall Prevention 5.5
Kilogram Is Not a Unit of Weight You may have seen the "conversion" 1 kg = 2.2 lb. Despite popular statements of weights expressed in kilograms, the kilogram is not a unit of *weight*, it is a unit of *mass*. The conversion statement is not an equality; it is an *equivalence* that is valid only on the Earth's surface.

---

[3]This statement ignores that the mass distribution of the Earth is not perfectly spherical.

The life-support unit strapped to the back of astronaut Harrison Schmitt weighed 300 lb on the Earth and had a mass of 136 kg. During his training, a 50-lb mock-up with a mass of 23 kg was used. Although this strategy effectively simulated the reduced weight the unit would have on the Moon, it did not correctly mimic the unchanging mass. It was more difficult to accelerate the 136-kg unit (perhaps by jumping or twisting suddenly) on the Moon than it was to accelerate the 23-kg unit on the Earth.

NASA/Eugene Cernan

of 70.0 kg. The student's weight in a location where $g = 9.80$ m/s^2 is 686 N (about 150 lb). At the top of a mountain, however, where $g = 9.77$ m/s^2, the student's weight is only 684 N. Therefore, if you want to lose weight without going on a diet, climb a mountain or weigh yourself at 30 000 ft during an airplane flight!

Equation 5.6 quantifies the gravitational force on the object, but notice that this equation does not require the object to be moving. Even for a stationary object or for an object on which several forces act, Equation 5.6 can be used to calculate the magnitude of the gravitational force. The result is a subtle shift in the interpretation of m in the equation. The mass m in Equation 5.6 determines the strength of the gravitational attraction between the object and the Earth. This role is completely different from that previously described for mass, that of measuring the resistance to changes in motion in response to an external force. In that role, mass is also called **inertial mass.** We call m in Equation 5.6 the **gravitational mass.** Even though this quantity is different in behavior from inertial mass, it is one of the experimental conclusions in Newtonian dynamics that gravitational mass and inertial mass have the same value.

Although this discussion has focused on the gravitational force on an object due to the Earth, the concept is generally valid on any planet. The value of g will vary from one planet to the next, but the magnitude of the gravitational force will always be given by the value of mg.

Quick Quiz 5.4 Suppose you are talking by interplanetary telephone to a friend who lives on the Moon. He tells you that he has just won a newton of gold in a contest. Excitedly, you tell him that you entered the Earth version of the same contest and also won a newton of gold! Who is richer? **(a)** You are. **(b)** Your friend is. **(c)** You are equally rich.

Conceptual Example 5.2   How Much Do You Weigh in an Elevator?

You have most likely been in an elevator that accelerates upward as it moves toward a higher floor. In this case, you feel heavier. In fact, if you are standing on a bathroom scale at the time, the scale measures a force having a magnitude that is greater than your weight. Therefore, you have tactile and measured evidence that leads you to believe you are heavier in this situation. *Are* you heavier?

SOLUTION

No; your weight is unchanged. Your experiences are due to your being in a noninertial reference frame. To provide the acceleration upward, the floor or scale must exert on your feet an upward force that is greater in magnitude than your weight. It is this greater force you feel, which you interpret as feeling heavier. The scale reads this upward force, not your weight, and so its reading increases.

5.6 Newton's Third Law

If you press against a corner of this textbook with your fingertip, the book pushes back and makes a small dent in your skin. If you push harder, the book does the same and the dent in your skin is a little larger. This simple activity illustrates that forces are *interactions* between two objects: when your finger pushes on the book, the book pushes back on your finger. This important principle is known as **Newton's third law:**

Newton's third law ▶

If two objects interact, the force $\vec{F}_{12}$ exerted by object 1 on object 2 is equal in magnitude and opposite in direction to the force $\vec{F}_{21}$ exerted by object 2 on object 1:

$$\vec{F}_{12} = -\vec{F}_{21} \qquad (5.7)$$

When it is important to designate forces as interactions between two objects, we will use this subscript notation, where $\vec{\mathbf{F}}_{ab}$ means "the force exerted *by* a *on* b." The third law is illustrated in Figure 5.5. The force that object 1 exerts on object 2 is popularly called the *action force*, and the force of object 2 on object 1 is called the *reaction force*. These italicized terms are not scientific terms; furthermore, either force can be labeled the action or reaction force. We will use these terms for convenience. In all cases, the action and reaction forces act on *different* objects and must be of the same type (gravitational, electrical, etc.). For example, the force acting on a freely falling projectile is the gravitational force exerted by the Earth on the projectile $\vec{\mathbf{F}}_g = \vec{\mathbf{F}}_{Ep}$ (E = Earth, p = projectile), and the magnitude of this force is *mg*. The reaction to this force is the gravitational force exerted by the projectile on the Earth $\vec{\mathbf{F}}_{pE} = -\vec{\mathbf{F}}_{Ep}$. The reaction force $\vec{\mathbf{F}}_{pE}$ must accelerate the Earth toward the projectile just as the action force $\vec{\mathbf{F}}_{Ep}$ accelerates the projectile toward the Earth. Because the Earth has such a large mass, however, its acceleration due to this reaction force is negligibly small.

Consider a computer monitor at rest on a table as in Figure 5.6a. The gravitational force on the monitor is $\vec{\mathbf{F}}_g = \vec{\mathbf{F}}_{Em}$. The reaction to this force is the force $\vec{\mathbf{F}}_{mE} = -\vec{\mathbf{F}}_{Em}$ exerted by the monitor on the Earth. The monitor does not accelerate because it is held up by the table. The table exerts on the monitor an upward force $\vec{\mathbf{n}} = \vec{\mathbf{F}}_{tm}$, called the **normal force.** (*Normal* in this context means *perpendicular.*) In general, whenever an object is in contact with a surface, the surface exerts a normal force on the object. The normal force on the monitor can have any value needed, up to the point of breaking the table. Because the monitor has zero acceleration, Newton's second law applied to the monitor gives us $\sum \vec{\mathbf{F}} = \vec{\mathbf{n}} + m\vec{\mathbf{g}} = 0$, so $n\hat{\mathbf{j}} - mg\hat{\mathbf{j}} = 0$, or $n = mg$. The normal force balances the gravitational force on the monitor, so the net force on the monitor is zero. The reaction force to $\vec{\mathbf{n}}$ is the force exerted by the monitor downward on the table, $\vec{\mathbf{F}}_{mt} = -\vec{\mathbf{F}}_{tm} = -\vec{\mathbf{n}}$.

Notice that the forces acting on the monitor are $\vec{\mathbf{F}}_g$ and $\vec{\mathbf{n}}$ as shown in Figure 5.6b. The two forces $\vec{\mathbf{F}}_{mE}$ and $\vec{\mathbf{F}}_{mt}$ are exerted on objects other than the monitor.

Figure 5.6 illustrates an extremely important step in solving problems involving forces. Figure 5.6a shows many of the forces in the situation: those acting on the monitor, one acting on the table, and one acting on the Earth. Figure 5.6b, by contrast, shows only the forces acting on *one object*, the monitor, and is called a **force diagram** or a *diagram showing the forces on the object.* The important pictorial representation in Figure 5.6c is called a **free-body diagram.** In a free-body diagram, the particle model is used by representing the object as a dot and showing the forces that act on the object as being applied to the dot. When analyzing an object subject to forces, we are interested in the net force acting on one object, which we will model as a particle. Therefore, a free-body diagram helps us isolate only those forces on the object and eliminate the other forces from our analysis.

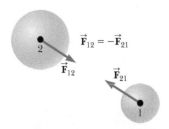

Figure 5.5 Newton's third law. The force $\vec{\mathbf{F}}_{12}$ exerted by object 1 on object 2 is equal in magnitude and opposite in direction to the force $\vec{\mathbf{F}}_{21}$ exerted by object 2 on object 1.

Pitfall Prevention 5.6

n Does Not Always Equal mg In the situation shown in Figure 5.6 and in many others, we find that $n = mg$ (the normal force has the same magnitude as the gravitational force). This result, however, is *not* generally true. If an object is on an incline, if there are applied forces with vertical components, or if there is a vertical acceleration of the system, then $n \neq mg$. *Always* apply Newton's second law to find the relationship between *n* and *mg.*

Pitfall Prevention 5.7

Newton's Third Law Remember that Newton's third-law action and reaction forces act on *different* objects. For example, in Figure 5.6, $\vec{\mathbf{n}} = \vec{\mathbf{F}}_{tm} = -m\vec{\mathbf{g}} = -\vec{\mathbf{F}}_{Em}$. The forces $\vec{\mathbf{n}}$ and $m\vec{\mathbf{g}}$ are equal in magnitude and opposite in direction, but they do not represent an action–reaction pair because both forces act on the *same* object, the monitor.

Pitfall Prevention 5.8

Free-Body Diagrams The *most important* step in solving a problem using Newton's laws is to draw a proper sketch, the free-body diagram. Be sure to draw *only* those forces that act on the object you are isolating. Be sure to draw *all* forces acting on the object, including any field forces, such as the gravitational force.

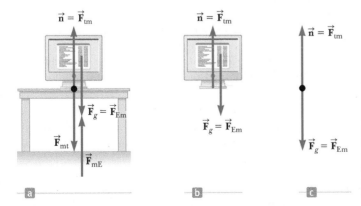

Figure 5.6 (a) When a computer monitor is at rest on a table, the forces acting on the monitor are the normal force $\vec{\mathbf{n}}$ and the gravitational force $\vec{\mathbf{F}}_g$. The reaction to $\vec{\mathbf{n}}$ is the force $\vec{\mathbf{F}}_{mt}$ exerted by the monitor on the table. The reaction to $\vec{\mathbf{F}}_g$ is the force $\vec{\mathbf{F}}_{mE}$ exerted by the monitor on the Earth. (b) A *force diagram* shows the forces on the monitor. (c) A *free-body diagram* shows the monitor as a black dot with the forces acting on it.

Quick Quiz 5.5   **(i)** If a fly collides with the windshield of a fast-moving bus, which experiences an impact force with a larger magnitude? (a) The fly. (b) The bus. (c) The same force is experienced by both. **(ii)** Which experiences the greater acceleration? (a) The fly. (b) The bus. (c) The same acceleration is experienced by both.

Conceptual Example 5.3   **You Push Me and I'll Push You**

A large man and a small boy stand facing each other on frictionless ice. They put their hands together and push against each other so that they move apart.

(A) Who moves away with the higher speed?

SOLUTION

This situation is similar to what we saw in Quick Quiz 5.5. According to Newton's third law, the force exerted by the man on the boy and the force exerted by the boy on the man are a third-law pair of forces, so they must be equal in magnitude. (A bathroom scale placed between their hands would read the same, regardless of which way it faced.) Therefore, the boy, having the smaller mass, experiences the greater acceleration. Both individuals accelerate for the same amount of time, but the greater acceleration of the boy over this time interval results in his moving away from the interaction with the higher speed.

(B) Who moves farther while their hands are in contact?

SOLUTION

Because the boy has the greater acceleration and therefore the greater average velocity, he moves farther than the man during the time interval during which their hands are in contact.

5.7 Analysis Models Using Newton's Second Law

In this section, we discuss two analysis models for solving problems in which objects are either in equilibrium ($\vec{\mathbf{a}} = 0$) or accelerating under the action of constant external forces. Remember that when Newton's laws are applied to an object, we are interested only in external forces that act on the object. If the objects are modeled as particles, we need not worry about rotational motion. For now, we also neglect the effects of friction in those problems involving motion, which is equivalent to stating that the surfaces are *frictionless*. (The friction force is discussed in Section 5.8.)

We usually neglect the mass of any ropes, strings, or cables involved. In this approximation, the magnitude of the force exerted by any element of the rope on the adjacent element is the same for all elements along the rope. In problem statements, the synonymous terms *light* and *of negligible mass* are used to indicate that a mass is to be ignored when you work the problems. When a rope attached to an object is pulling on the object, the rope exerts a force on the object in a direction away from the object, parallel to the rope. The magnitude T of that force is called the **tension** in the rope. Because it is the magnitude of a vector quantity, tension is a scalar quantity.

Analysis Model: The Particle in Equilibrium

If the acceleration of an object modeled as a particle is zero, the object is treated with the **particle in equilibrium** model. In this model, the net force on the object is zero:

$$\sum \vec{\mathbf{F}} = 0 \tag{5.8}$$

Consider a lamp suspended from a light chain fastened to the ceiling as in Figure 5.7a. The force diagram for the lamp (Fig. 5.7b) shows that the forces acting on the lamp are the downward gravitational force $\vec{F}_g$ and the upward force $\vec{T}$ exerted by the chain. Because there are no forces in the x direction, $\Sigma F_x = 0$ provides no helpful information. The condition $\Sigma F_y = 0$ gives

$$\Sigma F_y = T - F_g = 0 \quad \text{or} \quad T = F_g$$

Again, notice that $\vec{T}$ and $\vec{F}_g$ are *not* an action–reaction pair because they act on the same object, the lamp. The reaction force to $\vec{T}$ is a downward force exerted by the lamp on the chain.

Example 5.4 (page 122) shows an application of the particle in equilibrium model.

Analysis Model: The Particle Under a Net Force

If an object experiences an acceleration, its motion can be analyzed with the **particle under a net force** model. The appropriate equation for this model is Newton's second law, Equation 5.2:

$$\sum \vec{F} = m\vec{a} \tag{5.2}$$

Consider a crate being pulled to the right on a frictionless, horizontal floor as in Figure 5.8a. Of course, the floor directly under the boy must have friction; otherwise, his feet would simply slip when he tries to pull on the crate! Suppose you wish to find the acceleration of the crate and the force the floor exerts on it. The forces acting on the crate are illustrated in the free-body diagram in Figure 5.8b. Notice that the horizontal force $\vec{T}$ being applied to the crate acts through the rope. The magnitude of $\vec{T}$ is equal to the tension in the rope. In addition to the force $\vec{T}$, the free-body diagram for the crate includes the gravitational force $\vec{F}_g$ and the normal force $\vec{n}$ exerted by the floor on the crate.

We can now apply Newton's second law in component form to the crate. The only force acting in the x direction is $\vec{T}$. Applying $\Sigma F_x = ma_x$ to the horizontal motion gives

$$\sum F_x = T = ma_x \quad \text{or} \quad a_x = \frac{T}{m}$$

No acceleration occurs in the y direction because the crate moves only horizontally. Therefore, we use the particle in equilibrium model in the y direction. Applying the y component of Equation 5.8 yields

$$\sum F_y = n - F_g = 0 \quad \text{or} \quad n = F_g$$

That is, the normal force has the same magnitude as the gravitational force but acts in the opposite direction.

If $\vec{T}$ is a constant force, the acceleration $a_x = T/m$ also is constant. Hence, the crate is also modeled as a particle under constant acceleration in the x direction, and the equations of kinematics from Chapter 2 can be used to obtain the crate's position x and velocity v_x as functions of time.

Notice from this discussion two concepts that will be important in future problem solving: (1) *In a given problem, it is possible to have different analysis models applied in different directions.* The crate in Figure 5.8 is a particle in equilibrium in the vertical direction and a particle under a net force in the horizontal direction. (2) *It is possible to describe an object by multiple analysis models.* The crate is a particle under a net force in the horizontal direction and is also a particle under constant acceleration in the same direction.

In the situation just described, the magnitude of the normal force $\vec{n}$ is equal to the magnitude of $\vec{F}_g$, but that is not always the case, as noted in Pitfall Prevention 5.6. For example, suppose a book is lying on a table and you push down on the book with a force $\vec{F}$ as in Figure 5.9. Because the book is at rest and therefore not accelerating, $\Sigma F_y = 0$, which gives $n - F_g - F = 0$, or $n = F_g + F = mg + F$. In this situation, the normal force is *greater* than the gravitational force. Other examples in which $n \neq F_g$ are presented later.

Figure 5.7 (a) A lamp suspended from a ceiling by a chain of negligible mass. (b) The forces acting on the lamp are the gravitational force $\vec{F}_g$ and the force $\vec{T}$ exerted by the chain.

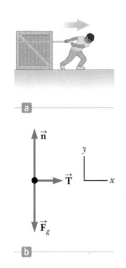

Figure 5.8 (a) A crate being pulled to the right on a frictionless floor. (b) The free-body diagram representing the external forces acting on the crate.

Figure 5.9 When a force $\vec{F}$ pushes vertically downward on another object, the normal force $\vec{n}$ on the object is greater than the gravitational force: $n = F_g + F$.

Several examples below demonstrate the use of the particle under a net force model.

Particle in Equilibrium

Imagine an object that can be modeled as a particle. If it has several forces acting on it so that the forces all cancel, giving a net force of zero, the object will have an acceleration of zero. This condition is mathematically described as

$$\sum \vec{\mathbf{F}} = 0 \qquad (5.8)$$

$$\vec{\mathbf{a}} = 0$$
$$\overset{m}{\underset{\Sigma\vec{\mathbf{F}} = 0}{\longleftrightarrow}}$$

Examples

- a chandelier hanging over a dining room table
- an object moving at terminal speed through a viscous medium (Chapter 6)
- a steel beam in the frame of a building (Chapter 12)
- a boat floating on a body of water (Chapter 14)

Particle Under a Net Force

Imagine an object that can be modeled as a particle. If it has one or more forces acting on it so that there is a net force on the object, it will accelerate in the direction of the net force. The relationship between the net force and the acceleration is

$$\sum \vec{\mathbf{F}} = m\vec{\mathbf{a}} \qquad (5.2)$$

$$\overset{m}{\bullet} \overset{\vec{\mathbf{a}}}{\longrightarrow}$$
$$\underset{\Sigma\vec{\mathbf{F}}}{\longrightarrow}$$

Examples

- a crate pushed across a factory floor
- a falling object acted upon by a gravitational force
- a piston in an automobile engine pushed by hot gases (Chapter 22)
- a charged particle in an electric field (Chapter 23)

Example 5.4   **A Traffic Light at Rest**   AM

A traffic light weighing 122 N hangs from a cable tied to two other cables fastened to a support as in Figure 5.10a. The upper cables make angles of $\theta_1 = 37.0°$ and $\theta_2 = 53.0°$ with the horizontal. These upper cables are not as strong as the vertical cable and will break if the tension in them exceeds 100 N. Does the traffic light remain hanging in this situation, or will one of the cables break?

SOLUTION

Conceptualize Inspect the drawing in Figure 5.10a. Let us assume the cables do not break and nothing is moving.

Categorize If nothing is moving, no part of the system is accelerating. We can now model the light as a *particle in equilibrium* on which the net force is zero. Similarly, the net force on the knot (Fig. 5.10c) is zero, so it is also modeled as a *particle in equilibrium*.

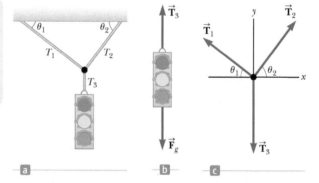

Figure 5.10 (Example 5.4) (a) A traffic light suspended by cables. (b) The forces acting on the traffic light. (c) The free-body diagram for the knot where the three cables are joined.

Analyze We construct a diagram of the forces acting on the traffic light, shown in Figure 5.10b, and a free-body diagram for the knot that holds the three cables together, shown in Figure 5.10c. This knot is a convenient object to choose because all the forces of interest act along lines passing through the knot.

From the particle in equilibrium model, apply Equation 5.8 for the traffic light in the *y* direction:

$$\sum F_y = 0 \quad \rightarrow \quad T_3 - F_g = 0$$
$$T_3 = F_g$$

▶ **5.4** c o n t i n u e d

Choose the coordinate axes as shown in Figure 5.10c and resolve the forces acting on the knot into their components:

Force	x Component	y Component
$\vec{\mathbf{T}}_1$	$-T_1 \cos \theta_1$	$T_1 \sin \theta_1$
$\vec{\mathbf{T}}_2$	$T_2 \cos \theta_2$	$T_2 \sin \theta_2$
$\vec{\mathbf{T}}_3$	0	$-F_g$

Apply the particle in equilibrium model to the knot:

$$(1) \quad \sum F_x = -T_1 \cos \theta_1 + T_2 \cos \theta_2 = 0$$

$$(2) \quad \sum F_y = T_1 \sin \theta_1 + T_2 \sin \theta_2 + (-F_g) = 0$$

Equation (1) shows that the horizontal components of $\vec{\mathbf{T}}_1$ and $\vec{\mathbf{T}}_2$ must be equal in magnitude, and Equation (2) shows that the sum of the vertical components of $\vec{\mathbf{T}}_1$ and $\vec{\mathbf{T}}_2$ must balance the downward force $\vec{\mathbf{T}}_3$, which is equal in magnitude to the weight of the light.

Solve Equation (1) for T_2 in terms of T_1:

$$(3) \quad T_2 = T_1 \left(\frac{\cos \theta_1}{\cos \theta_2} \right)$$

Substitute this value for T_2 into Equation (2):

$$T_1 \sin \theta_1 + T_1 \left(\frac{\cos \theta_1}{\cos \theta_2} \right)(\sin \theta_2) - F_g = 0$$

Solve for T_1:

$$T_1 = \frac{F_g}{\sin \theta_1 + \cos \theta_1 \tan \theta_2}$$

Substitute numerical values:

$$T_1 = \frac{122 \text{ N}}{\sin 37.0° + \cos 37.0° \tan 53.0°} = 73.4 \text{ N}$$

Using Equation (3), solve for T_2:

$$T_2 = (73.4 \text{ N}) \left(\frac{\cos 37.0°}{\cos 53.0°} \right) = 97.4 \text{ N}$$

Both values are less than 100 N (just barely for T_2), so the cables will not break .

. . . .

Finalize Let us finalize this problem by imagining a change in the system, as in the following What If?

WHAT IF? Suppose the two angles in Figure 5.10a are equal. What would be the relationship between T_1 and T_2?

Answer We can argue from the symmetry of the problem that the two tensions T_1 and T_2 would be equal to each other. Mathematically, if the equal angles are called θ, Equation (3) becomes

$$T_2 = T_1 \left(\frac{\cos \theta}{\cos \theta} \right) = T_1$$

which also tells us that the tensions are equal. Without knowing the specific value of θ, we cannot find the values of T_1 and T_2. The tensions will be equal to each other, however, regardless of the value of θ.

Conceptual Example 5.5 **Forces Between Cars in a Train**

Train cars are connected by *couplers*, which are under tension as the locomotive pulls the train. Imagine you are on a train speeding up with a constant acceleration. As you move through the train from the locomotive to the last car, measuring the tension in each set of couplers, does the tension increase, decrease, or stay the same? When the engineer applies the brakes, the couplers are under compression. How does this compression force vary from the locomotive to the last car? (Assume only the brakes on the wheels of the engine are applied.)

SOLUTION

While the train is speeding up, tension decreases from the front of the train to the back. The coupler between the locomotive and the first car must apply enough force to accelerate the rest of the cars. As you move back along the

continued

▶ **5.5** continued

train, each coupler is accelerating less mass behind it. The last coupler has to accelerate only the last car, and so it is under the least tension.

When the brakes are applied, the force again decreases from front to back. The coupler connecting the locomotive to the first car must apply a large force to slow down the rest of the cars, but the final coupler must apply a force large enough to slow down only the last car.

Example 5.6   The Runaway Car AM

A car of mass m is on an icy driveway inclined at an angle θ as in Figure 5.11a.

(A) Find the acceleration of the car, assuming the driveway is frictionless.

SOLUTION

Conceptualize Use Figure 5.11a to conceptualize the situation. From everyday experience, we know that a car on an icy incline will accelerate down the incline. (The same thing happens to a car on a hill with its brakes not set.)

Categorize We categorize the car as a *particle under a net force* because it accelerates. Further-

Figure 5.11 (Example 5.6) (a) A car on a frictionless incline. (b) The free-body diagram for the car. The black dot represents the position of the center of mass of the car. We will learn about center of mass in Chapter 9.

more, this example belongs to a very common category of problems in which an object moves under the influence of gravity on an inclined plane.

Analyze Figure 5.11b shows the free-body diagram for the car. The only forces acting on the car are the normal force $\vec{n}$ exerted by the inclined plane, which acts perpendicular to the plane, and the gravitational force $\vec{F}_g = m\vec{g}$, which acts vertically downward. For problems involving inclined planes, it is convenient to choose the coordinate axes with x along the incline and y perpendicular to it as in Figure 5.11b. With these axes, we represent the gravitational force by a component of magnitude $mg \sin \theta$ along the positive x axis and one of magnitude $mg \cos \theta$ along the negative y axis. Our choice of axes results in the car being modeled as a particle under a net force in the x direction and a particle in equilibrium in the y direction.

Apply these models to the car:

$$(1) \quad \sum F_x = mg \sin \theta = ma_x$$

$$(2) \quad \sum F_y = n - mg \cos \theta = 0$$

Solve Equation (1) for a_x:

$$(3) \quad a_x = \boxed{g \sin \theta}$$

Finalize Note that the acceleration component a_x is independent of the mass of the car! It depends only on the angle of inclination and on g.

From Equation (2), we conclude that the component of $\vec{F}_g$ perpendicular to the incline is balanced by the normal force; that is, $n = mg \cos \theta$. This situation is a case in which the normal force is *not* equal in magnitude to the weight of the object (as discussed in Pitfall Prevention 5.6 on page 119).

It is possible, although inconvenient, to solve the problem with "standard" horizontal and vertical axes. You may want to try it, just for practice.

(B) Suppose the car is released from rest at the top of the incline and the distance from the car's front bumper to the bottom of the incline is d. How long does it take the front bumper to reach the bottom of the hill, and what is the car's speed as it arrives there?

▶ **5.6** continued

$F_{net} = m \cdot a$

SOLUTION

Conceptualize Imagine the car is sliding down the hill and you use a stopwatch to measure the entire time interval until it reaches the bottom.

Categorize This part of the problem belongs to kinematics rather than to dynamics, and Equation (3) shows that the acceleration a_x is constant. Therefore, you should categorize the car in this part of the problem as a particle under constant acceleration.

....

Analyze Defining the initial position of the front bumper as $x_i = 0$ and its final position as $x_f = d$, and recognizing that $v_{xi} = 0$, choose Equation 2.16 from the particle under constant acceleration model, $x_f = x_i + v_{xi}t + \frac{1}{2}a_x t^2$:

$$d = \tfrac{1}{2}a_x t^2$$

Solve for t:

$$(4) \quad t = \sqrt{\frac{2d}{a_x}} = \sqrt{\frac{2d}{g\sin\theta}}$$

Use Equation 2.17, with $v_{xi} = 0$, to find the final velocity of the car:

$$v_{xf}^2 = 2a_x d$$

$$(5) \quad v_{xf} = \sqrt{2a_x d} = \sqrt{2gd\sin\theta}$$

....

Finalize We see from Equations (4) and (5) that the time t at which the car reaches the bottom and its final speed v_{xf} are independent of the car's mass, as was its acceleration. Notice that we have combined techniques from Chapter 2 with new techniques from this chapter in this example. As we learn more techniques in later chapters, this process of combining analysis models and information from several parts of the book will occur more often. In these cases, use the General Problem-Solving Strategy to help you identify what analysis models you will need.

WHAT IF? What previously solved problem does this situation become if $\theta = 90°$?

Answer Imagine θ going to $90°$ in Figure 5.11. The inclined plane becomes vertical, and the car is an object in free fall! Equation (3) becomes

$$a_x = g\sin\theta = g\sin 90° = g$$

which is indeed the free-fall acceleration. (We find $a_x = g$ rather than $a_x = -g$ because we have chosen positive x to be downward in Fig. 5.11.) Notice also that the condition $n = mg\cos\theta$ gives us $n = mg\cos 90° = 0$. That is consistent with the car falling downward *next to* the vertical plane, in which case there is no contact force between the car and the plane.

Example 5.7 **One Block Pushes Another** AM

Two blocks of masses m_1 and m_2, with $m_1 > m_2$, are placed in contact with each other on a frictionless, horizontal surface as in Figure 5.12a. A constant horizontal force $\vec{F}$ is applied to m_1 as shown.

(A) Find the magnitude of the acceleration of the system.

SOLUTION

Conceptualize Conceptualize the situation by using Figure 5.12a and realize that both blocks must experience the *same* acceleration because they are in contact with each other and remain in contact throughout the motion.

Categorize We categorize this problem as one involving a *particle under a net force* because a force is applied to a system of blocks and we are looking for the acceleration of the system.

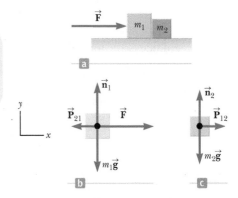

Figure 5.12 (Example 5.7) (a) A force is applied to a block of mass m_1, which pushes on a second block of mass m_2. (b) The forces acting on m_1. (c) The forces acting on m_2.

continued

▶ **5.7** c o n t i n u e d

Analyze First model the combination of two blocks as a single particle under a net force. Apply Newton's second law to the combination in the x direction to find the acceleration:

$$\sum F_x = F = (m_1 + m_2)a_x$$

$$(1) \quad a_x = \frac{F}{m_1 + m_2}$$

Finalize The acceleration given by Equation (1) is the same as that of a single object of mass $m_1 + m_2$ and subject to the same force.

(B) Determine the magnitude of the contact force between the two blocks.

S O L U T I O N

Conceptualize The contact force is internal to the system of two blocks. Therefore, we cannot find this force by modeling the whole system (the two blocks) as a single particle.

Categorize Now consider each of the two blocks individually by categorizing each as a *particle under a net force*.

Analyze We construct a diagram of forces acting on the object for each block as shown in Figures 5.12b and 5.12c, where the contact force is denoted by $\vec{P}$. From Figure 5.12c, we see that the only horizontal force acting on m_2 is the contact force $\vec{P}_{12}$ (the force exerted by m_1 on m_2), which is directed to the right.

Apply Newton's second law to m_2:

$$(2) \quad \sum F_x = P_{12} = m_2 a_x$$

Substitute the value of the acceleration a_x given by Equation (1) into Equation (2):

$$(3) \quad P_{12} = m_2 a_x = \left(\frac{m_2}{m_1 + m_2}\right)F$$

Finalize This result shows that the contact force P_{12} is *less* than the applied force F. The force required to accelerate block 2 alone must be less than the force required to produce the same acceleration for the two-block system.

To finalize further, let us check this expression for P_{12} by considering the forces acting on m_1, shown in Figure 5.12b. The horizontal forces acting on m_1 are the applied force $\vec{F}$ to the right and the contact force $\vec{P}_{21}$ to the left (the force exerted by m_2 on m_1). From Newton's third law, $\vec{P}_{21}$ is the reaction force to $\vec{P}_{12}$, so $P_{21} = P_{12}$.

Apply Newton's second law to m_1:

$$(4) \quad \sum F_x = F - P_{21} = F - P_{12} = m_1 a_x$$

Solve for P_{12} and substitute the value of a_x from Equation (1):

$$P_{12} = F - m_1 a_x = F - m_1\left(\frac{F}{m_1 + m_2}\right) = \left(\frac{m_2}{m_1 + m_2}\right)F$$

This result agrees with Equation (3), as it must.

WHAT IF?   Imagine that the force $\vec{F}$ in Figure 5.12 is applied toward the left on the right-hand block of mass m_2. Is the magnitude of the force $\vec{P}_{12}$ the same as it was when the force was applied toward the right on m_1?

Answer When the force is applied toward the left on m_2, the contact force must accelerate m_1. In the original situation, the contact force accelerates m_2. Because $m_1 > m_2$, more force is required, so the magnitude of $\vec{P}_{12}$ is greater than in the original situation. To see this mathematically, modify Equation (4) appropriately and solve for $\vec{P}_{12}$.

Example 5.8   **Weighing a Fish in an Elevator**   AM

A person weighs a fish of mass m on a spring scale attached to the ceiling of an elevator as illustrated in Figure 5.13.

(A) Show that if the elevator accelerates either upward or downward, the spring scale gives a reading that is different from the weight of the fish.

▶ **5.8** continued

SOLUTION

Conceptualize The reading on the scale is related to the extension of the spring in the scale, which is related to the force on the end of the spring as in Figure 5.2. Imagine that the fish is hanging on a string attached to the end of the spring. In this case, the magnitude of the force exerted on the spring is equal to the tension T in the string. Therefore, we are looking for T. The force $\vec{T}$ pulls down on the string and pulls up on the fish.

Categorize We can categorize this problem by identifying the fish as a *particle in equilibrium* if the elevator is not accelerating or as a *particle under a net force* if the elevator is accelerating.

When the elevator accelerates upward, the spring scale reads a value greater than the weight of the fish.

When the elevator accelerates downward, the spring scale reads a value less than the weight of the fish.

Figure 5.13 (Example 5.8) A fish is weighed on a spring scale in an accelerating elevator car.

Analyze Inspect the diagrams of the forces acting on the fish in Figure 5.13 and notice that the external forces acting on the fish are the downward gravitational force $\vec{F}_g = m\vec{g}$ and the force $\vec{T}$ exerted by the string. If the elevator is either at rest or moving at constant velocity, the fish is a particle in equilibrium, so $\Sigma F_y = T - F_g = 0$ or $T = F_g = mg$. (Remember that the scalar mg is the weight of the fish.)

Now suppose the elevator is moving with an acceleration $\vec{a}$ relative to an observer standing outside the elevator in an inertial frame. The fish is now a particle under a net force.

Apply Newton's second law to the fish:

$$\Sigma F_y = T - mg = ma_y$$

Solve for T:

$$(1) \quad T = ma_y + mg = mg\left(\frac{a_y}{g} + 1\right) = F_g\left(\frac{a_y}{g} + 1\right)$$

where we have chosen upward as the positive y direction. We conclude from Equation (1) that the scale reading T is greater than the fish's weight mg if $\vec{a}$ is upward, so a_y is positive (Fig. 5.13a), and that the reading is less than mg if $\vec{a}$ is downward, so a_y is negative (Fig. 5.13b).

(B) Evaluate the scale readings for a 40.0-N fish if the elevator moves with an acceleration $a_y = \pm 2.00$ m/s².

SOLUTION

Evaluate the scale reading from Equation (1) if $\vec{a}$ is upward:

$$T = (40.0 \text{ N})\left(\frac{2.00 \text{ m/s}^2}{9.80 \text{ m/s}^2} + 1\right) = \boxed{48.2 \text{ N}}$$

Evaluate the scale reading from Equation (1) if $\vec{a}$ is downward:

$$T = (40.0 \text{ N})\left(\frac{-2.00 \text{ m/s}^2}{9.80 \text{ m/s}^2} + 1\right) = \boxed{31.8 \text{ N}}$$

Finalize Take this advice: if you buy a fish in an elevator, make sure the fish is weighed while the elevator is either at rest or accelerating downward! Furthermore, notice that from the information given here, one cannot determine the direction of the velocity of the elevator.

WHAT IF? Suppose the elevator cable breaks and the elevator and its contents are in free fall. What happens to the reading on the scale?

Answer If the elevator falls freely, the fish's acceleration is $a_y = -g$. We see from Equation (1) that the scale reading T is zero in this case; that is, the fish *appears* to be weightless.

Example 5.9   The Atwood Machine   AM

When two objects of unequal mass are hung vertically over a frictionless pulley of negligible mass as in Figure 5.14a, the arrangement is called an *Atwood machine*. The device is sometimes used in the laboratory to determine the value of *g*. Determine the magnitude of the acceleration of the two objects and the tension in the lightweight string.

SOLUTION

Conceptualize Imagine the situation pictured in Figure 5.14a in action: as one object moves upward, the other object moves downward. Because the objects are connected by an inextensible string, their accelerations must be of equal magnitude.

Categorize The objects in the Atwood machine are subject to the gravitational force as well as to the forces exerted by the strings connected to them. Therefore, we can categorize this problem as one involving two *particles under a net force*.

Figure 5.14 (Example 5.9) The Atwood machine. (a) Two objects connected by a massless inextensible string over a frictionless pulley. (b) The free-body diagrams for the two objects.

Analyze The free-body diagrams for the two objects are shown in Figure 5.14b. Two forces act on each object: the upward force $\vec{\mathbf{T}}$ exerted by the string and the downward gravitational force. In problems such as this one in which the pulley is modeled as massless and frictionless, the tension in the string on both sides of the pulley is the same. If the pulley has mass or is subject to friction, the tensions on either side are not the same and the situation requires techniques we will learn in Chapter 10.

We must be very careful with signs in problems such as this one. In Figure 5.14a, notice that if object 1 accelerates upward, object 2 accelerates downward. Therefore, for consistency with signs, if we define the upward direction as positive for object 1, we must define the downward direction as positive for object 2. With this sign convention, both objects accelerate in the same direction as defined by the choice of sign. Furthermore, according to this sign convention, the *y* component of the net force exerted on object 1 is $T - m_1 g$, and the *y* component of the net force exerted on object 2 is $m_2 g - T$.

From the particle under a net force model, apply Newton's second law to object 1:

$$(1) \quad \sum F_y = T - m_1 g = m_1 a_y$$

Apply Newton's second law to object 2:

$$(2) \quad \sum F_y = m_2 g - T = m_2 a_y$$

Add Equation (2) to Equation (1), noticing that *T* cancels:

$$-m_1 g + m_2 g = m_1 a_y + m_2 a_y$$

Solve for the acceleration:

$$(3) \quad a_y = \left(\frac{m_2 - m_1}{m_1 + m_2}\right)g$$

Substitute Equation (3) into Equation (1) to find *T*:

$$(4) \quad T = m_1(g + a_y) = \left(\frac{2m_1 m_2}{m_1 + m_2}\right)g$$

Finalize The acceleration given by Equation (3) can be interpreted as the ratio of the magnitude of the unbalanced force on the system $(m_2 - m_1)g$ to the total mass of the system $(m_1 + m_2)$, as expected from Newton's second law. Notice that the sign of the acceleration depends on the relative masses of the two objects.

WHAT IF? Describe the motion of the system if the objects have equal masses, that is, $m_1 = m_2$.

Answer If we have the same mass on both sides, the system is balanced and should not accelerate. Mathematically, we see that if $m_1 = m_2$, Equation (3) gives us $a_y = 0$.

WHAT IF? What if one of the masses is much larger than the other: $m_1 \gg m_2$?

Answer In the case in which one mass is infinitely larger than the other, we can ignore the effect of the smaller mass. Therefore, the larger mass should simply fall as if the smaller mass were not there. We see that if $m_1 \gg m_2$, Equation (3) gives us $a_y = -g$.

| Example 5.10 | Acceleration of Two Objects Connected by a Cord AM |

A ball of mass m_1 and a block of mass m_2 are attached by a lightweight cord that passes over a frictionless pulley of negligible mass as in Figure 5.15a. The block lies on a frictionless incline of angle θ. Find the magnitude of the acceleration of the two objects and the tension in the cord.

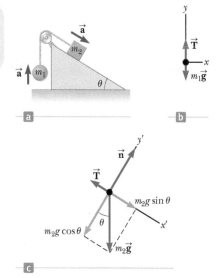

SOLUTION

Conceptualize Imagine the objects in Figure 5.15 in motion. If m_2 moves down the incline, then m_1 moves upward. Because the objects are connected by a cord (which we assume does not stretch), their accelerations have the same magnitude. Notice the normal coordinate axes in Figure 5.15b for the ball and the "tilted" axes for the block in Figure 5.15c.

Categorize We can identify forces on each of the two objects and we are looking for an acceleration, so we categorize the objects as *particles under a net force*. For the block, this model is only valid for the x' direction. In the y' direction, we apply the *particle in equilibrium* model because the block does not accelerate in that direction.

Analyze Consider the free-body diagrams shown in Figures 5.15b and 5.15c.

Figure 5.15 (Example 5.10) (a) Two objects connected by a lightweight cord strung over a frictionless pulley. (b) The free-body diagram for the ball. (c) The free-body diagram for the block. (The incline is frictionless.)

Apply Newton's second law in the y direction to the ball, choosing the upward direction as positive:

$$(1) \quad \sum F_y = T - m_1 g = m_1 a_y = m_1 a$$

For the ball to accelerate upward, it is necessary that $T > m_1 g$. In Equation (1), we replaced a_y with a because the acceleration has only a y component.

For the block, we have chosen the x' axis along the incline as in Figure 5.15c. For consistency with our choice for the ball, we choose the positive x' direction to be down the incline.

Apply the particle under a net force model to the block in the x' direction and the particle in equilibrium model in the y' direction:

$$(2) \quad \sum F_{x'} = m_2 g \sin\theta - T = m_2 a_{x'} = m_2 a$$

$$(3) \quad \sum F_{y'} = n - m_2 g \cos\theta = 0$$

In Equation (2), we replaced $a_{x'}$ with a because the two objects have accelerations of equal magnitude a.

Solve Equation (1) for T:

$$(4) \quad T = m_1(g + a)$$

Substitute this expression for T into Equation (2):

$$m_2 g \sin\theta - m_1(g + a) = m_2 a$$

Solve for a:

$$(5) \quad a = \left(\frac{m_2 \sin\theta - m_1}{m_1 + m_2} \right) g$$

Substitute this expression for a into Equation (4) to find T:

$$(6) \quad T = \left(\frac{m_1 m_2 (\sin\theta + 1)}{m_1 + m_2} \right) g$$

Finalize The block accelerates down the incline only if $m_2 \sin\theta > m_1$. If $m_1 > m_2 \sin\theta$, the acceleration is up the incline for the block and downward for the ball. Also notice that the result for the acceleration, Equation (5), can be interpreted as the magnitude of the net external force acting on the ball–block system divided by the total mass of the system; this result is consistent with Newton's second law.

WHAT IF? What happens in this situation if $\theta = 90°$?

continued

▶ **5.10** continued

Answer If $\theta = 90°$, the inclined plane becomes vertical and there is no interaction between its surface and m_2. Therefore, this problem becomes the Atwood machine of Example 5.9. Letting $\theta \rightarrow 90°$ in Equations (5) and (6) causes them to reduce to Equations (3) and (4) of Example 5.9!

WHAT IF?   What if $m_1 = 0$?

Answer If $m_1 = 0$, then m_2 is simply sliding down an inclined plane without interacting with m_1 through the string. Therefore, this problem becomes the sliding car problem in Example 5.6. Letting $m_1 \rightarrow 0$ in Equation (5) causes it to reduce to Equation (3) of Example 5.6!

5.8 Forces of Friction

When an object is in motion either on a surface or in a viscous medium such as air or water, there is resistance to the motion because the object interacts with its surroundings. We call such resistance a **force of friction.** Forces of friction are very important in our everyday lives. They allow us to walk or run and are necessary for the motion of wheeled vehicles.

Imagine that you are working in your garden and have filled a trash can with yard clippings. You then try to drag the trash can across the surface of your concrete patio as in Figure 5.16a. This surface is *real*, not an idealized, frictionless surface. If we apply an external horizontal force $\vec{F}$ to the trash can, acting to the right, the trash can remains stationary when $\vec{F}$ is small. The force on the trash can that counteracts $\vec{F}$ and keeps it from moving acts toward the left and is called the

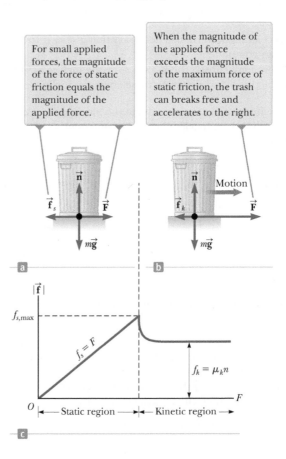

Figure 5.16 (a) and (b) When pulling on a trash can, the direction of the force of friction $\vec{f}$ between the can and a rough surface is opposite the direction of the applied force $\vec{F}$. (c) A graph of friction force versus applied force. Notice that $f_{s,max} > f_k$.

force of static friction $\vec{\mathbf{f}}_s$. As long as the trash can is not moving, $f_s = F$. Therefore, if $\vec{\mathbf{F}}$ is increased, $\vec{\mathbf{f}}_s$ also increases. Likewise, if $\vec{\mathbf{F}}$ decreases, $\vec{\mathbf{f}}_s$ also decreases.

◀ **Force of static friction**

Experiments show that the friction force arises from the nature of the two surfaces: because of their roughness, contact is made only at a few locations where peaks of the material touch. At these locations, the friction force arises in part because one peak physically blocks the motion of a peak from the opposing surface and in part from chemical bonding ("spot welds") of opposing peaks as they come into contact. Although the details of friction are quite complex at the atomic level, this force ultimately involves an electrical interaction between atoms or molecules.

If we increase the magnitude of $\vec{\mathbf{F}}$ as in Figure 5.16b, the trash can eventually slips. When the trash can is on the verge of slipping, f_s has its maximum value $f_{s,max}$ as shown in Figure 5.16c. When F exceeds $f_{s,max}$, the trash can moves and accelerates to the right. We call the friction force for an object in motion the **force of kinetic friction** $\vec{\mathbf{f}}_k$. When the trash can is in motion, the force of kinetic friction on the can is less than $f_{s,max}$ (Fig. 5.16c). The net force $F - f_k$ in the x direction produces an acceleration to the right, according to Newton's second law. If $F = f_k$, the acceleration is zero and the trash can moves to the right with constant speed. If the applied force $\vec{\mathbf{F}}$ is removed from the moving can, the friction force $\vec{\mathbf{f}}_k$ acting to the left provides an acceleration of the trash can in the $-x$ direction and eventually brings it to rest, again consistent with Newton's second law.

◀ **Force of kinetic friction**

Experimentally, we find that, to a good approximation, both $f_{s,max}$ and f_k are proportional to the magnitude of the normal force exerted on an object by the surface. The following descriptions of the force of friction are based on experimental observations and serve as the simplification model we shall use for forces of friction in problem solving:

- The magnitude of the force of static friction between any two surfaces in contact can have the values

$$f_s \leq \mu_s n \tag{5.9}$$

where the dimensionless constant μ_s is called the **coefficient of static friction** and n is the magnitude of the normal force exerted by one surface on the other. The equality in Equation 5.9 holds when the surfaces are on the verge of slipping, that is, when $f_s = f_{s,max} = \mu_s n$. This situation is called *impending motion*. The inequality holds when the surfaces are not on the verge of slipping.

- The magnitude of the force of kinetic friction acting between two surfaces is

$$f_k = \mu_k n \tag{5.10}$$

where μ_k is the **coefficient of kinetic friction.** Although the coefficient of kinetic friction can vary with speed, we shall usually neglect any such variations in this text.

- The values of μ_k and μ_s depend on the nature of the surfaces, but μ_k is generally less than μ_s. Typical values range from around 0.03 to 1.0. Table 5.1 (page 132) lists some reported values.

- The direction of the friction force on an object is parallel to the surface with which the object is in contact and opposite to the actual motion (kinetic friction) or the impending motion (static friction) of the object relative to the surface.

- The coefficients of friction are nearly independent of the area of contact between the surfaces. We might expect that placing an object on the side having the most area might increase the friction force. Although this method provides more points in contact, the weight of the object is spread out over a larger area and the individual points are not pressed together as tightly. Because these effects approximately compensate for each other, the friction force is independent of the area.

Pitfall Prevention 5.9
The Equal Sign Is Used in Limited Situations In Equation 5.9, the equal sign is used *only* in the case in which the surfaces are just about to break free and begin sliding. Do not fall into the common trap of using $f_s = \mu_s n$ in *any* static situation.

Pitfall Prevention 5.10
Friction Equations Equations 5.9 and 5.10 are *not* vector equations. They are relationships between the *magnitudes* of the vectors representing the friction and normal forces. Because the friction and normal forces are perpendicular to each other, the vectors cannot be related by a multiplicative constant.

Pitfall Prevention 5.11
The Direction of the Friction Force Sometimes, an incorrect statement about the friction force between an object and a surface is made—"the friction force on an object is opposite to its motion or impending motion"—rather than the correct phrasing, "the friction force on an object is opposite to its motion or impending motion *relative to the surface.*"

Table 5.1	Coefficients of Friction	
	μ_s	μ_k
Rubber on concrete	1.0	0.8
Steel on steel	0.74	0.57
Aluminum on steel	0.61	0.47
Glass on glass	0.94	0.4
Copper on steel	0.53	0.36
Wood on wood	0.25–0.5	0.2
Waxed wood on wet snow	0.14	0.1
Waxed wood on dry snow	—	0.04
Metal on metal (lubricated)	0.15	0.06
Teflon on Teflon	0.04	0.04
Ice on ice	0.1	0.03
Synovial joints in humans	0.01	0.003

Note: All values are approximate. In some cases, the coefficient of friction can exceed 1.0.

a

b

Figure 5.17 (Quick Quiz 5.7) A father slides his daughter on a sled either by (a) pushing down on her shoulders or (b) pulling up on a rope.

Quick Quiz 5.6 You press your physics textbook flat against a vertical wall with your hand. What is the direction of the friction force exerted by the wall on the book? **(a)** downward **(b)** upward **(c)** out from the wall **(d)** into the wall

Quick Quiz 5.7 You are playing with your daughter in the snow. She sits on a sled and asks you to slide her across a flat, horizontal field. You have a choice of **(a)** pushing her from behind by applying a force downward on her shoulders at 30° below the horizontal (Fig. 5.17a) or **(b)** attaching a rope to the front of the sled and pulling with a force at 30° above the horizontal (Fig. 5.17b). Which would be easier for you and why?

Example 5.11   Experimental Determination of μ_s and μ_k AM

The following is a simple method of measuring coefficients of friction. Suppose a block is placed on a rough surface inclined relative to the horizontal as shown in Figure 5.18. The incline angle is increased until the block starts to move. Show that you can obtain μ_s by measuring the critical angle θ_c at which this slipping just occurs.

SOLUTION

Conceptualize Consider Figure 5.18 and imagine that the block tends to slide down the incline due to the gravitational force. To simulate the situation, place a coin on this book's cover and tilt the book until the coin begins to slide. Notice how this example differs from Example 5.6. When there is no friction on an incline, *any* angle of the incline will cause a stationary object to begin moving. When there is friction, however, there is no movement of the object for angles less than the critical angle.

Categorize The block is subject to various forces. Because we are raising the plane to the angle at which the block is just ready to begin to move but is not moving, we categorize the block as a *particle in equilibrium*.

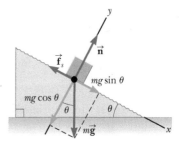

Figure 5.18 (Example 5.11) The external forces exerted on a block lying on a rough incline are the gravitational force $m\vec{g}$, the normal force $\vec{n}$, and the force of friction $\vec{f}_s$. For convenience, the gravitational force is resolved into a component $mg \sin \theta$ along the incline and a component $mg \cos \theta$ perpendicular to the incline.

Analyze The diagram in Figure 5.18 shows the forces on the block: the gravitational force $m\vec{g}$, the normal force $\vec{n}$, and the force of static friction $\vec{f}_s$. We choose x to be parallel to the plane and y perpendicular to it.

From the particle in equilibrium model, apply Equation 5.8 to the block in both the x and y directions:

(1) $\sum F_x = mg \sin \theta - f_s = 0$

(2) $\sum F_y = n - mg \cos \theta = 0$

▶ **5.11** continued

Substitute $mg = n/\cos\theta$ from Equation (2) into Equation (1):

(3) $f_s = mg\sin\theta = \left(\dfrac{n}{\cos\theta}\right)\sin\theta = n\tan\theta$

When the incline angle is increased until the block is on the verge of slipping, the force of static friction has reached its maximum value $\mu_s n$. The angle θ in this situation is the critical angle θ_c. Make these substitutions in Equation (3):

$\mu_s n = n\tan\theta_c$

$\mu_s = \tan\theta_c$

We have shown, as requested, that the coefficient of static friction is related only to the critical angle. For example, if the block just slips at $\theta_c = 20.0°$, we find that $\mu_s = \tan 20.0° = 0.364$.

Finalize Once the block starts to move at $\theta \geq \theta_c$, it accelerates down the incline and the force of friction is $f_k = \mu_k n$. If θ is reduced to a value less than θ_c, however, it may be possible to find an angle θ_c' such that the block moves down the incline with constant speed as a particle in equilibrium again ($a_x = 0$). In this case, use Equations (1) and (2) with f_s replaced by f_k to find μ_k: $\mu_k = \tan\theta_c'$, where $\theta_c' < \theta_c$.

Example 5.12   **The Sliding Hockey Puck**   AM

A hockey puck on a frozen pond is given an initial speed of 20.0 m/s. If the puck always remains on the ice and slides 115 m before coming to rest, determine the coefficient of kinetic friction between the puck and ice.

SOLUTION

Conceptualize Imagine that the puck in Figure 5.19 slides to the right. The kinetic friction force acts to the left and slows the puck, which eventually comes to rest due to that force.

Figure 5.19 (Example 5.12) After the puck is given an initial velocity to the right, the only external forces acting on it are the gravitational force $m\vec{\mathbf{g}}$, the normal force $\vec{\mathbf{n}}$, and the force of kinetic friction $\vec{\mathbf{f}}_k$.

Categorize The forces acting on the puck are identified in Figure 5.19, but the text of the problem provides kinematic variables. Therefore, we categorize the problem in several ways. First, it involves modeling the puck as a *particle under a net force* in the horizontal direction: kinetic friction causes the puck to accelerate. There is no acceleration of the puck in the vertical direction, so we use the *particle in equilibrium* model for that direction. Furthermore, because we model the force of kinetic friction as independent of speed, the acceleration of the puck is constant. So, we can also categorize this problem by modeling the puck as a *particle under constant acceleration*.

Analyze First, let's find the acceleration algebraically in terms of the coefficient of kinetic friction, using Newton's second law. Once we know the acceleration of the puck and the distance it travels, the equations of kinematics can be used to find the numerical value of the coefficient of kinetic friction. The diagram in Figure 5.19 shows the forces on the puck.

Apply the particle under a net force model in the x direction to the puck:

(1) $\sum F_x = -f_k = ma_x$

Apply the particle in equilibrium model in the y direction to the puck:

(2) $\sum F_y = n - mg = 0$

Substitute $n = mg$ from Equation (2) and $f_k = \mu_k n$ into Equation (1):

$-\mu_k n = -\mu_k mg = ma_x$

$a_x = -\mu_k g$

The negative sign means the acceleration is to the left in Figure 5.19. Because the velocity of the puck is to the right, the puck is slowing down. The acceleration is independent of the mass of the puck and is constant because we assume μ_k remains constant.

continued

▶ **5.12** continued

Apply the particle under constant acceleration model to the puck, choosing Equation 2.17 from the model, $v_{xf}^2 = v_{xi}^2 + 2a_x(x_f - x_i)$, with $x_i = 0$ and $v_{xf} = 0$:

$$0 = v_{xi}^2 + 2a_x x_f = v_{xi}^2 - 2\mu_k g x_f$$

Solve for the coefficient of kinetic friction:

$$\mu_k = \frac{v_{xi}^2}{2gx_f}$$

Substitute the numerical values:

$$\mu_k = \frac{(20.0 \text{ m/s})^2}{2(9.80 \text{ m/s}^2)(115 \text{ m})} = \boxed{0.177}$$

Finalize Notice that μ_k is dimensionless, as it should be, and that it has a low value, consistent with an object sliding on ice.

Example 5.13   Acceleration of Two Connected Objects When Friction Is Present   AM

A block of mass m_2 on a rough, horizontal surface is connected to a ball of mass m_1 by a lightweight cord over a lightweight, frictionless pulley as shown in Figure 5.20a. A force of magnitude F at an angle θ with the horizontal is applied to the block as shown, and the block slides to the right. The coefficient of kinetic friction between the block and surface is μ_k. Determine the magnitude of the acceleration of the two objects.

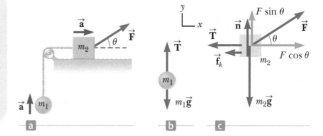

Figure 5.20 (Example 5.13) (a) The external force $\vec{\mathbf{F}}$ applied as shown can cause the block to accelerate to the right. (b, c) Diagrams showing the forces on the two objects, assuming the block accelerates to the right and the ball accelerates upward.

SOLUTION

Conceptualize Imagine what happens as $\vec{\mathbf{F}}$ is applied to the block. Assuming $\vec{\mathbf{F}}$ is large enough to break the block free from static friction but not large enough to lift the block, the block slides to the right and the ball rises.

Categorize We can identify forces and we want an acceleration, so we categorize this problem as one involving two *particles under a net force*, the ball and the block. Because we assume that the block does not rise into the air due to the applied force, we model the block as a *particle in equilibrium* in the vertical direction.

Analyze First draw force diagrams for the two objects as shown in Figures 5.20b and 5.20c. Notice that the string exerts a force of magnitude T on both objects. The applied force $\vec{\mathbf{F}}$ has x and y components $F \cos \theta$ and $F \sin \theta$, respectively. Because the two objects are connected, we can equate the magnitudes of the x component of the acceleration of the block and the y component of the acceleration of the ball and call them both a. Let us assume the motion of the block is to the right.

Apply the particle under a net force model to the block in the horizontal direction:

(1) $\sum F_x = F \cos \theta - f_k - T = m_2 a_x = m_2 a$

Because the block moves only horizontally, apply the particle in equilibrium model to the block in the vertical direction:

(2) $\sum F_y = n + F \sin \theta - m_2 g = 0$

Apply the particle under a net force model to the ball in the vertical direction:

(3) $\sum F_y = T - m_1 g = m_1 a_y = m_1 a$

Solve Equation (2) for n:

$n = m_2 g - F \sin \theta$

Substitute n into $f_k = \mu_k n$ from Equation 5.10:

(4) $f_k = \mu_k (m_2 g - F \sin \theta)$

▶ **5.13** continued

Substitute Equation (4) and the value of T from Equation (3) into Equation (1):

$$F\cos\theta - \mu_k(m_2 g - F\sin\theta) - m_1(a + g) = m_2 a$$

Solve for a:

$$(5) \quad a = \frac{F(\cos\theta + \mu_k \sin\theta) - (m_1 + \mu_k m_2)g}{m_1 + m_2}$$

Finalize The acceleration of the block can be either to the right or to the left depending on the sign of the numerator in Equation (5). If the velocity is to the left, we must reverse the sign of f_k in Equation (1) because the force of kinetic friction must oppose the motion of the block relative to the surface. In this case, the value of a is the same as in Equation (5), with the two plus signs in the numerator changed to minus signs.

What does Equation (5) reduce to if the force $\vec{F}$ is removed and the surface becomes frictionless? Call this expression Equation (6). Does this algebraic expression match your intuition about the physical situation in this case? Now go back to Example 5.10 and let angle θ go to zero in Equation (5) of that example. How does the resulting equation compare with your Equation (6) here in Example 5.13? Should the algebraic expressions compare in this way based on the physical situations?

Summary

Definitions

An **inertial frame of reference** is a frame in which an object that does not interact with other objects experiences zero acceleration. Any frame moving with constant velocity relative to an inertial frame is also an inertial frame.

We define **force** as **that which causes a change in motion of an object.**

Concepts and Principles

Newton's first law states that it is possible to find an inertial frame in which an object that does not interact with other objects experiences zero acceleration, or, equivalently, in the absence of an external force, when viewed from an inertial frame, an object at rest remains at rest and an object in uniform motion in a straight line maintains that motion.

Newton's second law states that the acceleration of an object is directly proportional to the net force acting on it and inversely proportional to its mass.

Newton's third law states that if two objects interact, the force exerted by object 1 on object 2 is equal in magnitude and opposite in direction to the force exerted by object 2 on object 1.

The **gravitational force** exerted on an object is equal to the product of its mass (a scalar quantity) and the free-fall acceleration:

$$\vec{F}_g = m\vec{g} \qquad (5.5)$$

The **weight** of an object is the magnitude of the gravitational force acting on the object:

$$F_g = mg \qquad (5.6)$$

The maximum **force of static friction** $\vec{f}_{s,\max}$ between an object and a surface is proportional to the normal force acting on the object. In general, $f_s \leq \mu_s n$, where μ_s is the **coefficient of static friction** and n is the magnitude of the normal force.

When an object slides over a surface, the magnitude of the **force of kinetic friction** $\vec{f}_k$ is given by $f_k = \mu_k n$, where μ_k is the **coefficient of kinetic friction.**

continued

Analysis Models for Problem Solving

Particle Under a Net Force If a particle of mass m experiences a nonzero net force, its acceleration is related to the net force by Newton's second law:

$$\sum \vec{F} = m\vec{a} \qquad (5.2)$$

Particle in Equilibrium If a particle maintains a constant velocity (so that $\vec{a} = 0$), which could include a velocity of zero, the forces on the particle balance and Newton's second law reduces to

$$\sum \vec{F} = 0 \qquad (5.8)$$

Objective Questions   1. denotes answer available in *Student Solutions Manual/Study Guide*

1. The driver of a speeding empty truck slams on the brakes and skids to a stop through a distance d. On a second trial, the truck carries a load that doubles its mass. What will now be the truck's "skidding distance"? (a) $4d$ (b) $2d$ (c) $\sqrt{2}d$ (d) d (e) $d/2$

2. In Figure OQ5.2, a locomotive has broken through the wall of a train station. During the collision, what can be said about the force exerted by the locomotive on the wall? (a) The force exerted by the locomotive on the wall was larger than the force the wall could exert on the locomotive. (b) The force exerted by the locomotive on the wall was the same in magnitude as the force exerted by the wall on the locomotive. (c) The force exerted by the locomotive on the wall was less than the force exerted by the wall on the locomotive. (d) The wall cannot be said to "exert" a force; after all, it broke.

Figure OQ5.2

3. The third graders are on one side of a schoolyard, and the fourth graders are on the other. They are throwing snowballs at each other. Between them, snowballs of various masses are moving with different velocities as shown in Figure OQ5.3. Rank the snowballs (a) through (e) according to the magnitude of the total force exerted on each one. Ignore air resistance. If two snowballs rank together, make that fact clear.

Figure OQ5.3

4. The driver of a speeding truck slams on the brakes and skids to a stop through a distance d. On another trial, the initial speed of the truck is half as large. What now will be the truck's skidding distance? (a) $2d$ (b) $\sqrt{2}d$ (c) d (d) $d/2$ (e) $d/4$

5. An experiment is performed on a puck on a level air hockey table, where friction is negligible. A constant horizontal force is applied to the puck, and the puck's acceleration is measured. Now the same puck is transported far into outer space, where both friction and gravity are negligible. The same constant force is applied to the puck (through a spring scale that stretches the same amount), and the puck's acceleration (relative to the distant stars) is measured. What is the puck's acceleration in outer space? (a) It is somewhat greater than its acceleration on the Earth. (b) It is the same as its acceleration on the Earth. (c) It is less than its acceleration on the Earth. (d) It is infinite because neither friction nor gravity constrains it. (e) It is very large because acceleration is inversely proportional to weight and the puck's weight is very small but not zero.

6. The manager of a department store is pushing horizontally with a force of magnitude 200 N on a box of shirts. The box is sliding across the horizontal floor with a forward acceleration. Nothing else touches the box. What must be true about the magnitude of the force of kinetic friction acting on the box (choose one)? (a) It is greater than 200 N. (b) It is less than 200 N. (c) It is equal to 200 N. (d) None of those statements is necessarily true.

7. Two objects are connected by a string that passes over a frictionless pulley as in Figure 5.14a, where $m_1 < m_2$ and a_1 and a_2 are the magnitudes of the respective accelerations. Which mathematical statement is true regarding the magnitude of the acceleration a_2 of the mass m_2? (a) $a_2 < g$ (b) $a_2 > g$ (c) $a_2 = g$ (d) $a_2 < a_1$ (e) $a_2 > a_1$

8. An object of mass m is sliding with speed v_i at some instant across a level tabletop, with which its coefficient of kinetic friction is μ. It then moves through a distance d and comes to rest. Which of the following equations for the speed v_i is reasonable? (a) $v_i = \sqrt{-2\mu mgd}$ (b) $v_i = \sqrt{2\mu mgd}$ (c) $v_i = \sqrt{-2\mu gd}$ (d) $v_i = \sqrt{2\mu gd}$ (e) $v_i = \sqrt{2\mu d}$

9. A truck loaded with sand accelerates along a highway. The driving force on the truck remains constant. What happens to the acceleration of the truck if its trailer leaks sand at a constant rate through a hole in its bottom? (a) It decreases at a steady rate. (b) It increases at a steady rate. (c) It increases and then decreases. (d) It decreases and then increases. (e) It remains constant.

10. A large crate of mass m is place on the flatbed of a truck but not tied down. As the truck accelerates forward with acceleration a, the crate remains at rest relative to the truck. What force causes the crate to accelerate? (a) the normal force (b) the gravitational

force (c) the friction force (d) the ma force exerted by the crate (e) No force is required.

11. If an object is in equilibrium, which of the following statements is *not* true? (a) The speed of the object remains constant. (b) The acceleration of the object is zero. (c) The net force acting on the object is zero. (d) The object must be at rest. (e) There are at least two forces acting on the object.

12. A crate remains stationary after it has been placed on a ramp inclined at an angle with the horizontal. Which of the following statements is or are correct about the magnitude of the friction force that acts on the crate? Choose all that are true. (a) It is larger than the weight of the crate. (b) It is equal to $\mu_s n$. (c) It is greater than the component of the gravitational force acting down the ramp. (d) It is equal to the component of the gravitational force acting down the ramp. (e) It is less than the component of the gravitational force acting down the ramp.

13. An object of mass m moves with acceleration $\vec{a}$ down a rough incline. Which of the following forces should appear in a free-body diagram of the object? Choose all correct answers. (a) the gravitational force exerted by the planet (b) $m\vec{a}$ in the direction of motion (c) the normal force exerted by the incline (d) the friction force exerted by the incline (e) the force exerted by the object on the incline

Conceptual Questions   1. denotes answer available in *Student Solutions Manual/Study Guide*

1. If you hold a horizontal metal bar several centimeters above the ground and move it through grass, each leaf of grass bends out of the way. If you increase the speed of the bar, each leaf of grass will bend more quickly. How then does a rotary power lawn mower manage to cut grass? How can it exert enough force on a leaf of grass to shear it off?

2. Your hands are wet, and the restroom towel dispenser is empty. What do you do to get drops of water off your hands? How does the motion of the drops exemplify one of Newton's laws? Which one?

3. In the motion picture *It Happened One Night* (Columbia Pictures, 1934), Clark Gable is standing inside a stationary bus in front of Claudette Colbert, who is seated. The bus suddenly starts moving forward and Clark falls into Claudette's lap. Why did this happen?

4. If a car is traveling due westward with a constant speed of 20 m/s, what is the resultant force acting on it?

5. A passenger sitting in the rear of a bus claims that she was injured when the driver slammed on the brakes, causing a suitcase to come flying toward her from the front of the bus. If you were the judge in this case, what disposition would you make? Why?

6. A child tosses a ball straight up. She says that the ball is moving away from her hand because the ball feels an upward "force of the throw" as well as the gravitational force. (a) Can the "force of the throw" exceed the

gravitational force? How would the ball move if it did? (b) Can the "force of the throw" be equal in magnitude to the gravitational force? Explain. (c) What strength can accurately be attributed to the "force of the throw"? Explain. (d) Why does the ball move away from the child's hand?

7. A person holds a ball in her hand. (a) Identify all the external forces acting on the ball and the Newton's third-law reaction force to each one. (b) If the ball is dropped, what force is exerted on it while it is falling? Identify the reaction force in this case. (Ignore air resistance.)

8. A spherical rubber balloon inflated with air is held stationary, with its opening, on the west side, pinched shut. (a) Describe the forces exerted by the air inside and outside the balloon on sections of the rubber. (b) After the balloon is released, it takes off toward the east, gaining speed rapidly. Explain this motion in terms of the forces now acting on the rubber. (c) Account for the motion of a skyrocket taking off from its launch pad.

9. A rubber ball is dropped onto the floor. What force causes the ball to bounce?

10. Twenty people participate in a tug-of-war. The two teams of ten people are so evenly matched that neither team wins. After the game they notice that a car is stuck in the mud. They attach the tug-of-war rope to the bumper of the car, and all the people pull on the

rope. The heavy car has just moved a couple of decimeters when the rope breaks. Why did the rope break in this situation when it did not break when the same twenty people pulled on it in a tug-of-war?

11. Can an object exert a force on itself? Argue for your answer.

12. When you push on a box with a 200-N force instead of a 50-N force, you can feel that you are making a greater effort. When a table exerts a 200-N normal force instead of one of smaller magnitude, is the table really doing anything differently?

13. A weightlifter stands on a bathroom scale. He pumps a barbell up and down. What happens to the reading on the scale as he does so? **What If?** What if he is strong enough to actually *throw* the barbell upward? How does the reading on the scale vary now?

14. An athlete grips a light rope that passes over a low-friction pulley attached to the ceiling of a gym. A sack of sand precisely equal in weight to the athlete is tied to the other end of the rope. Both the sand and the athlete are initially at rest. The athlete climbs the rope, sometimes speeding up and slowing down as he does so. What happens to the sack of sand? Explain.

15. Suppose you are driving a classic car. Why should you avoid slamming on your brakes when you want to stop in the shortest possible distance? (Many modern cars have antilock brakes that avoid this problem.)

16. In Figure CQ5.16, the light, taut, unstretchable cord B joins block 1 and the larger-mass block 2. Cord A exerts a force on block 1 to make it accelerate forward. (a) How does the magnitude of the force exerted by cord A on block 1 compare with the magnitude of the force exerted by cord B on block 2? Is it larger, smaller, or equal? (b) How does the acceleration of block 1 compare with the acceleration (if any) of block 2? (c) Does cord B exert a force on block 1? If so, is it forward or backward? Is it larger, smaller, or equal in magnitude to the force exerted by cord B on block 2?

Figure CQ5.16

17. Describe two examples in which the force of friction exerted on an object is in the direction of motion of the object.

18. The mayor of a city reprimands some city employees because they will not remove the obvious sags from the cables that support the city traffic lights. What explanation can the employees give? How do you think the case will be settled in mediation?

19. Give reasons for the answers to each of the following questions: (a) Can a normal force be horizontal? (b) Can a normal force be directed vertically downward? (c) Consider a tennis ball in contact with a stationary floor and with nothing else. Can the normal force be different in magnitude from the gravitational force exerted on the ball? (d) Can the force exerted by the floor on the ball be different in magnitude from the force the ball exerts on the floor?

20. Balancing carefully, three boys inch out onto a horizontal tree branch above a pond, each planning to dive in separately. The third boy in line notices that the branch is barely strong enough to support them. He decides to jump straight up and land back on the branch to break it, spilling all three into the pond. When he starts to carry out his plan, at what precise moment does the branch break? Explain. *Suggestion:* Pretend to be the third boy and imitate what he does in slow motion. If you are still unsure, stand on a bathroom scale and repeat the suggestion.

21. Identify action–reaction pairs in the following situations: (a) a man takes a step (b) a snowball hits a girl in the back (c) a baseball player catches a ball (d) a gust of wind strikes a window

22. As shown in Figure CQ5.22, student A, a 55-kg girl, sits on one chair with metal runners, at rest on a classroom floor. Student B, an 80-kg boy, sits on an identical chair. Both students keep their feet off the floor. A rope runs from student A's hands around a light pulley and then over her shoulder to the hands of a teacher standing on the floor behind her. The low-friction axle of the pulley is attached to a second rope held by student B. All ropes run parallel to the chair runners. (a) If student A pulls on her end of the rope, will her chair or will B's chair slide on the floor? Explain why. (b) If instead the teacher pulls on his rope end, which chair slides? Why this one? (c) If student B pulls on his rope, which chair slides? Why? (d) Now the teacher ties his end of the rope to student A's chair. Student A pulls on the end of the rope in her hands. Which chair slides and why?

Student B

Student A

Teacher

Figure CQ5.22

23. A car is moving forward slowly and is speeding up. A student claims that "the car exerts a force on itself" or that "the car's engine exerts a force on the car." (a) Argue that this idea cannot be accurate and that friction exerted by the road is the propulsive force on the car. Make your evidence and reasoning as persuasive as possible. (b) Is it static or kinetic friction? *Suggestions:* Consider a road covered with light gravel. Consider a sharp print of the tire tread on an asphalt road, obtained by coating the tread with dust.

Problems

Section 5.1 The Concept of Force

Section 5.2 Newton's First Law and Inertial Frames

Section 5.3 Mass

Section 5.4 Newton's Second Law

Section 5.5 The Gravitational Force and Weight

Section 5.6 Newton's Third Law

1. A woman weighs 120 lb. Determine (a) her weight in newtons and (b) her mass in kilograms.

2. If a man weighs 900 N on the Earth, what would he weigh on Jupiter, where the free-fall acceleration is 25.9 m/s²?

3. A 3.00-kg object undergoes an acceleration given by $\vec{\mathbf{a}} = (2.00\hat{\mathbf{i}} + 5.00\hat{\mathbf{j}})$ m/s². Find (a) the resultant force acting on the object and (b) the magnitude of the resultant force.

4. A certain orthodontist uses a wire brace to align a patient's crooked tooth as in Figure P5.4. The tension in the wire is adjusted to have a magnitude of 18.0 N. Find the magnitude of the net force exerted by the wire on the crooked tooth.

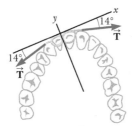

Figure P5.4

5. A toy rocket engine is securely fastened to a large puck that can glide with negligible friction over a horizontal surface, taken as the *xy* plane. The 4.00-kg puck has a velocity of $3.00\hat{\mathbf{i}}$ m/s at one instant. Eight seconds later, its velocity is $(8.00\hat{\mathbf{i}} + 10.00\hat{\mathbf{j}})$ m/s. Assuming the rocket engine exerts a constant horizontal force, find (a) the components of the force and (b) its magnitude.

6. The average speed of a nitrogen molecule in air is about 6.70×10^2 m/s, and its mass is 4.68×10^{-26} kg. (a) If it takes 3.00×10^{-13} s for a nitrogen molecule to hit a wall and rebound with the same speed but moving in the opposite direction, what is the average acceleration of the molecule during this time interval? (b) What average force does the molecule exert on the wall?

7. The distinction between mass and weight was discovered after Jean Richer transported pendulum clocks from Paris, France, to Cayenne, French Guiana, in 1671. He found that they quite systematically ran slower in Cayenne than in Paris. The effect was reversed when the clocks returned to Paris. How much weight would a 90.0 kg person lose in traveling from Paris, where *g* = 9.809 5 m/s², to Cayenne, where *g* = 9.780 8 m/s²? (We will consider how the free-fall acceleration influences the period of a pendulum in Section 15.5.)

8. (a) A car with a mass of 850 kg is moving to the right with a constant speed of 1.44 m/s. What is the total force on the car? (b) What is the total force on the car if it is moving to the left?

9. **Review.** The gravitational force exerted on a baseball is 2.21 N down. A pitcher throws the ball horizontally with velocity 18.0 m/s by uniformly accelerating it along a straight horizontal line for a time interval of 170 ms. The ball starts from rest. (a) Through what distance does it move before its release? (b) What are the magnitude and direction of the force the pitcher exerts on the ball?

10. **Review.** The gravitational force exerted on a baseball is $-F_g\hat{\mathbf{j}}$. A pitcher throws the ball with velocity $v\hat{\mathbf{i}}$ by uniformly accelerating it along a straight horizontal line for a time interval of $\Delta t = t - 0 = t$. (a) Starting from rest, through what distance does the ball move before its release? (b) What force does the pitcher exert on the ball?

11. **Review.** An electron of mass 9.11×10^{-31} kg has an initial speed of 3.00×10^5 m/s. It travels in a straight line, and its speed increases to 7.00×10^5 m/s in a distance of 5.00 cm. Assuming its acceleration is constant, (a) determine the magnitude of the force exerted on the electron and (b) compare this force with the weight of the electron, which we ignored.

12. Besides the gravitational force, a 2.80-kg object is subjected to one other constant force. The object starts from rest and in 1.20 s experiences a displacement of $(4.20\hat{\mathbf{i}} - 3.30\hat{\mathbf{j}})$ m, where the direction of $\hat{\mathbf{j}}$ is the upward vertical direction. Determine the other force.

13. One or more external forces, large enough to be easily measured, are exerted on each object enclosed in a dashed box shown in Figure 5.1. Identify the reaction to each of these forces.

14. A brick of mass M has been placed on a rubber cushion of mass m. Together they are sliding to the right at constant velocity on an ice-covered parking lot. (a) Draw a free-body diagram of the brick and identify each force acting on it. (b) Draw a free-body diagram of the cushion and identify each force acting on it. (c) Identify all of the action–reaction pairs of forces in the brick–cushion–planet system.

15. Two forces, $\vec{F}_1 = (-6.00\hat{i} - 4.00\hat{j})$ N and $\vec{F}_2 = (-3.00\hat{i} + 7.00\hat{j})$ N, act on a particle of mass 2.00 kg that is initially at rest at coordinates $(-2.00$ m, $+4.00$ m). (a) What are the components of the particle's velocity at $t = 10.0$ s? (b) In what direction is the particle moving at $t = 10.0$ s? (c) What displacement does the particle undergo during the first 10.0 s? (d) What are the coordinates of the particle at $t = 10.0$ s?

16. The force exerted by the wind on the sails of a sailboat is 390 N north. The water exerts a force of 180 N east. If the boat (including its crew) has a mass of 270 kg, what are the magnitude and direction of its acceleration?

17. An object of mass m is dropped at $t = 0$ from the roof of a building of height h. While the object is falling, a wind blowing parallel to the face of the building exerts a constant horizontal force F on the object. (a) At what time t does the object strike the ground? Express t in terms of g and h. (b) Find an expression in terms of m and F for the acceleration a_x of the object in the horizontal direction (taken as the positive x direction). (c) How far is the object displaced horizontally before hitting the ground? Answer in terms of m, g, F, and h. (d) Find the magnitude of the object's acceleration while it is falling, using the variables F, m, and g.

18. A force $\vec{F}$ applied to an object of mass m_1 produces an acceleration of 3.00 m/s². The same force applied to a second object of mass m_2 produces an acceleration of 1.00 m/s². (a) What is the value of the ratio m_1/m_2? (b) If m_1 and m_2 are combined into one object, find its acceleration under the action of the force $\vec{F}$.

19. Two forces $\vec{F}_1$ and $\vec{F}_2$ act on a 5.00-kg object. Taking $F_1 = 20.0$ N and $F_2 = 15.0$ N, find the accelerations of the object for the configurations of forces shown in parts (a) and (b) of Figure P5.19.

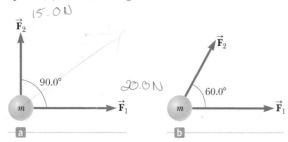

Figure P5.19

20. You stand on the seat of a chair and then hop off. (a) During the time interval you are in flight down to the floor, the Earth moves toward you with an acceleration of what order of magnitude? In your solution, explain your logic. Model the Earth as a perfectly solid object. (b) The Earth moves toward you through a distance of what order of magnitude?

21. A 15.0-lb block rests on the floor. (a) What force does the floor exert on the block? (b) A rope is tied to the block and is run vertically over a pulley. The other end is attached to a free-hanging 10.0-lb object. What now is the force exerted by the floor on the 15.0-lb block? (c) If the 10.0-lb object in part (b) is replaced with a 20.0-lb object, what is the force exerted by the floor on the 15.0-lb block?

22. Review. Three forces acting on an object are given by $\vec{F}_1 = (-2.00\hat{i} + 2.00\hat{j})$ N, and $\vec{F}_2 = (5.00\hat{i} - 3.00\hat{j})$ N, and $\vec{F}_3 = (-45.0\hat{i})$ N. The object experiences an acceleration of magnitude 3.75 m/s². (a) What is the direction of the acceleration? (b) What is the mass of the object? (c) If the object is initially at rest, what is its speed after 10.0 s? (d) What are the velocity components of the object after 10.0 s?

23. A 1 000-kg car is pulling a 300-kg trailer. Together, the car and trailer move forward with an acceleration of 2.15 m/s². Ignore any force of air drag on the car and all friction forces on the trailer. Determine (a) the net force on the car, (b) the net force on the trailer, (c) the force exerted by the trailer on the car, and (d) the resultant force exerted by the car on the road.

24. If a single constant force acts on an object that moves on a straight line, the object's velocity is a linear function of time. The equation $v = v_i + at$ gives its velocity v as a function of time, where a is its constant acceleration. What if velocity is instead a linear function of position? Assume that as a particular object moves through a resistive medium, its speed decreases as described by the equation $v = v_i - kx$, where k is a constant coefficient and x is the position of the object. Find the law describing the total force acting on this object.

Section 5.7 Analysis Models Using Newton's Second Law

25. Review. Figure P5.25 shows a worker poling a boat—a very efficient mode of transportation—across a shallow lake. He pushes parallel to the length of the light pole, exerting a force of magnitude 240 N on the bottom of the lake. Assume the pole lies in the vertical plane containing the keel of the boat. At one moment, the pole makes an angle of 35.0° with the vertical and the water exerts a horizontal drag force of 47.5 N on the boat, opposite to its forward velocity of magnitude 0.857 m/s. The mass of the boat including its cargo and the worker is 370 kg. (a) The water exerts

Figure P5.25

a buoyant force vertically upward on the boat. Find the magnitude of this force. (b) Model the forces as constant over a short interval of time to find the velocity of the boat 0.450 s after the moment described.

26. An iron bolt of mass 65.0 g hangs from a string 35.7 cm long. The top end of the string is fixed. Without touching it, a magnet attracts the bolt so that it remains stationary, but is displaced horizontally 28.0 cm to the right from the previously vertical line of the string. The magnet is located to the right of the bolt and on the same vertical level as the bolt in the final configuration. (a) Draw a free-body diagram of the bolt. (b) Find the tension in the string. (c) Find the magnetic force on the bolt.

27. Figure P5.27 shows the horizontal forces acting on a sailboat moving north at constant velocity, seen from a point straight above its mast. At the particular speed of the sailboat, the water exerts a 220-N drag force on its hull and $\theta = 40.0°$. For each of the situations (a) and (b) described

Figure P5.27

below, write two component equations representing Newton's second law. Then solve the equations for P (the force exerted by the wind on the sail) and for n (the force exerted by the water on the keel). (a) Choose the x direction as east and the y direction as north. (b) Now choose the x direction as $\theta = 40.0°$ north of east and the y direction as $\theta = 40.0°$ west of north. (c) Compare your solutions to parts (a) and (b). Do the results agree? Is one method significantly easier?

28. The systems shown in Figure P5.28 are in equilibrium. If the spring scales are calibrated in newtons, what do they read? Ignore the masses of the pulleys and strings and assume the pulleys and the incline in Figure P5.28d are frictionless.

Figure P5.28

29. Assume the three blocks portrayed in Figure P5.29 move on a frictionless surface and a 42-N force acts as shown on the 3.0-kg block. Determine (a) the acceleration given this system, (b) the tension in the cord connecting the 3.0-kg and the 1.0-kg blocks, and (c) the force exerted by the 1.0-kg block on the 2.0-kg block.

Figure P5.29

30. A block slides down a frictionless plane having an inclination of $\theta = 15.0°$. The block starts from rest at the top, and the length of the incline is 2.00 m. (a) Draw a free-body diagram of the block. Find (b) the acceleration of the block and (c) its speed when it reaches the bottom of the incline.

31. The distance between two telephone poles is 50.0 m. When a 1.00-kg bird lands on the telephone wire midway between the poles, the wire sags 0.200 m. (a) Draw a free-body diagram of the bird. (b) How much tension does the bird produce in the wire? Ignore the weight of the wire.

32. A 3.00-kg object is moving in a plane, with its x and y coordinates given by $x = 5t^2 - 1$ and $y = 3t^3 + 2$, where x and y are in meters and t is in seconds. Find the magnitude of the net force acting on this object at $t = 2.00$ s.

33. A bag of cement weighing 325 N hangs in equilibrium from three wires as suggested in Figure P5.33. Two of the wires make angles $\theta_1 = 60.0°$ and $\theta_2 = 40.0°$ with the horizontal. Assuming the system is in equilibrium, find the tensions T_1, T_2, and T_3 in the wires.

34. A bag of cement whose weight is F_g hangs in equilibrium from three wires as shown in Figure P5.33. Two of the wires make

Figure P5.33
Problems 33 and 34.

angles θ_1 and θ_2 with the horizontal. Assuming the system is in equilibrium, show that the tension in the left-hand wire is

$$T_1 = \frac{F_g \cos \theta_2}{\sin (\theta_1 + \theta_2)}$$

35. Two people pull as hard as they can on horizontal ropes attached to a boat that has a mass of 200 kg. If they pull in the same direction, the boat has an acceleration of 1.52 m/s² to the right. If they pull in opposite directions, the boat has an acceleration of 0.518 m/s² to the left. What is the magnitude of the force each person exerts on the boat? Disregard any other horizontal forces on the boat.

36. Figure P5.36 shows loads hanging from the ceiling of an elevator that is moving at constant velocity. Find the tension in each of the three strands of cord supporting each load.

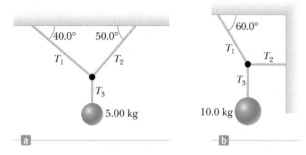

Figure P5.36

37. An object of mass $m = 1.00$ kg is observed to have an acceleration $\vec{\mathbf{a}}$ with a magnitude of 10.0 m/s^2 in a direction 60.0° east of north. Figure P5.37 shows a view of the object from above. The force $\vec{\mathbf{F}}_2$ acting on the object has a magnitude of 5.00 N and is directed north. Determine the magnitude and direction of the one other horizontal force $\vec{\mathbf{F}}_1$ acting on the object.

Figure P5.37

38. A setup similar to the one shown in Figure P5.38 is often used in hospitals to support and apply a horizontal traction force to an injured leg. (a) Determine the force of tension in the rope supporting the leg. (b) What is the traction force exerted to the right on the leg?

Figure P5.38

39. A simple accelerometer is constructed inside a car by suspending an object of mass m from a string of length L that is tied to the car's ceiling. As the car accelerates the string–object system makes a constant angle of θ with the vertical. (a) Assuming that the string mass is negligible compared with m, derive an expression for the car's acceleration in terms of θ and show that it is independent of the mass m and the length L. (b) Determine the acceleration of the car when $\theta = 23.0°$.

40. An object of mass $m_1 = 5.00$ kg **AMT** placed on a frictionless, horizon- **W** tal table is connected to a string that passes over a pulley and then is fastened to a hanging object of mass $m_2 = 9.00$ kg as shown in Figure P5.40. (a) Draw free-body

Figure P5.40
Problems 40, 63, and 87.

diagrams of both objects. Find (b) the magnitude of the acceleration of the objects and (c) the tension in the string.

41. Figure P5.41 shows the speed of a person's body as he does a chin-up. Assume the motion is vertical and the mass of the person's body is 64.0 kg. Determine the force exerted by the chin-up bar on his body at (a) $t = 0$, (b) $t = 0.5$ s, (c) $t = 1.1$ s, and (d) $t = 1.6$ s.

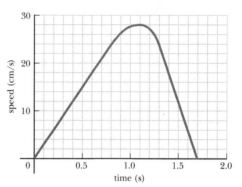

Figure P5.41

42. Two objects are connected by a light string that passes over a frictionless pulley as shown in Figure P5.42. Assume the incline is frictionless and take $m_1 = 2.00$ kg, $m_2 = 6.00$ kg, and $\theta = 55.0°$. (a) Draw free-body diagrams of both objects. Find (b) the magnitude of the acceleration of the objects, (c) the tension in the string, and (d) the speed of each object 2.00 s after it is released from rest.

Figure P5.42

43. Two blocks, each of mass $m = 3.50$ kg, are hung from the ceiling of an elevator as in Figure P5.43. (a) If the elevator moves with an upward acceleration $\vec{\mathbf{a}}$ of magnitude 1.60 m/s^2, find the tensions T_1 and T_2 in the upper and lower strings. (b) If the strings can withstand a maximum tension of 85.0 N, what maximum acceleration can the elevator have before a string breaks?

Figure P5.43
Problems 43 and 44.

44. Two blocks, each of mass m, are hung from the ceiling of an elevator as in Figure P5.43. The elevator has an upward acceleration a. The strings have negligible mass. (a) Find the tensions T_1 and T_2 in the upper and lower strings in terms of m, a, and g. (b) Compare the two tensions and determine which string would break first if a is made sufficiently large. (c) What are the tensions if the cable supporting the elevator breaks?

45. In the system shown in Figure P5.45, a horizontal force **M** $\vec{\mathbf{F}}_x$ acts on an object of mass $m_2 = 8.00$ kg. The horizontal surface is frictionless. Consider the acceleration of the sliding object as a function of F_x. (a) For what values of F_x does the object of mass $m_1 = 2.00$ kg accelerate upward? (b) For what values of F_x is the tension in the cord zero? (c) Plot the acceleration of the m_2 object versus F_x. Include values of F_x from -100 N to $+100$ N.

Figure P5.45

46. An object of mass m_1 hangs from a string that passes over a very light fixed pulley P_1 as shown in Figure P5.46. The string connects to a second very light pulley P_2. A second string passes around this pulley with one end attached to a wall and the other to an object of mass m_2 on a frictionless, horizontal table. (a) If a_1 and a_2 are the accelerations of m_1 and m_2, respectively, what is the relation between these accelerations? Find expressions for (b) the tensions in the strings and (c) the accelerations a_1 and a_2 in terms of the masses m_1 and m_2, and g.

Figure P5.46

47. A block is given an initial velocity of 5.00 m/s up a frictionless incline of angle $\theta = 20.0°$ (Fig. P5.47). How far up the incline does the block slide before coming to rest?

Figure P5.47

48. A car is stuck in the mud. A tow truck pulls on the car with the arrangement shown in Fig. P5.48. The tow cable is under a tension of 2 500 N and pulls downward and to the left on the pin at its upper end. The light pin is held in equilibrium by forces exerted by the two bars A and B. Each bar is a *strut*; that is, each is a bar whose weight is small compared to the forces it exerts and which exerts forces only through hinge pins at its ends. Each strut exerts a force directed parallel to its length. Determine the force of tension or compression in each strut. Proceed as follows. Make a guess as to which way (pushing or pulling) each force

acts on the top pin. Draw a free-body diagram of the pin. Use the condition for equilibrium of the pin to translate the free-body diagram into equations. From the equations calculate the forces exerted by struts A and B. If you obtain a positive answer, you correctly guessed the direction of the force. A negative answer means that the direction should be reversed, but the absolute value correctly gives the magnitude of the force. If a strut pulls on a pin, it is in tension. If it pushes, the strut is in compression. Identify whether each strut is in tension or in compression.

Figure P5.48

49. Two blocks of mass 3.50 kg and 8.00 kg are connected by a massless string that passes over a frictionless pulley (Fig. P5.49). The inclines are frictionless. Find (a) the magnitude of the acceleration of each block and (b) the tension in the string.

Figure P5.49 Problems 49 and 71.

50. In the Atwood machine discussed in Example 5.9 and shown in Figure 5.14a, $m_1 = 2.00$ kg and $m_2 = 7.00$ kg. The masses of the pulley and string are negligible by comparison. The pulley turns without friction, and the string does not stretch. The lighter object is released with a sharp push that sets it into motion at $v_i = 2.40$ m/s downward. (a) How far will m_1 descend below its initial level? (b) Find the velocity of m_1 after 1.80 s.

51. In Example 5.8, we investigated the apparent weight of a fish in an elevator. Now consider a 72.0-kg man standing on a spring scale in an elevator. Starting from rest, the elevator ascends, attaining its maximum speed of 1.20 m/s in 0.800 s. It travels with this constant speed for the next 5.00 s. The elevator then undergoes a uniform acceleration in the negative y direction for 1.50 s and comes to rest. What does the spring scale register (a) before the elevator starts to move, (b) during the first 0.800 s, (c) while the elevator is traveling at constant speed, and (d) during the time interval it is slowing down?

Section 5.8 Forces of Friction

52. Consider a large truck carrying a heavy load, such as steel beams. A significant hazard for the driver is that the load may slide forward, crushing the cab, if the truck stops suddenly in an accident or even in braking. Assume, for example, that a 10 000-kg load sits on the

flatbed of a 20 000-kg truck moving at 12.0 m/s. Assume that the load is not tied down to the truck, but has a coefficient of friction of 0.500 with the flatbed of the truck. (a) Calculate the minimum stopping distance for which the load will not slide forward relative to the truck. (b) Is any piece of data unnecessary for the solution?

53. Review. A rifle bullet with a mass of 12.0 g traveling toward the right at 260 m/s strikes a large bag of sand and penetrates it to a depth of 23.0 cm. Determine the magnitude and direction of the friction force (assumed constant) that acts on the bullet.

54. Review. A car is traveling at 50.0 mi/h on a horizontal highway. (a) If the coefficient of static friction between road and tires on a rainy day is 0.100, what is the minimum distance in which the car will stop? (b) What is the stopping distance when the surface is dry and $\mu_s = 0.600$?

55. A 25.0-kg block is initially at rest on a horizontal surface. A horizontal force of 75.0 N is required to set the block in motion, after which a horizontal force of 60.0 N is required to keep the block moving with constant speed. Find (a) the coefficient of static friction and (b) the coefficient of kinetic friction between the block and the surface.

56. *Why is the following situation impossible?* Your 3.80-kg physics book is placed next to you on the horizontal seat of your car. The coefficient of static friction between the book and the seat is 0.650, and the coefficient of kinetic friction is 0.550. You are traveling forward at 72.0 km/h and brake to a stop with constant acceleration over a distance of 30.0 m. Your physics book remains on the seat rather than sliding forward onto the floor.

57. To determine the coefficients of friction between rubber and various surfaces, a student uses a rubber eraser and an incline. In one experiment, the eraser begins to slip down the incline when the angle of inclination is 36.0° and then moves down the incline with constant speed when the angle is reduced to 30.0°. From these data, determine the coefficients of static and kinetic friction for this experiment.

58. Before 1960, people believed that the maximum attainable coefficient of static friction for an automobile tire on a roadway was $\mu_s = 1$. Around 1962, three companies independently developed racing tires with coefficients of 1.6. This problem shows that tires have improved further since then. The shortest time interval in which a piston-engine car initially at rest has covered a distance of one-quarter mile is about 4.43 s. (a) Assume the car's rear wheels lift the front wheels off the pavement as shown in Figure P5.58. What mini-

Figure P5.58

mum value of μ_s is necessary to achieve the record time? (b) Suppose the driver were able to increase his or her engine power, keeping other things equal. How would this change affect the elapsed time?

59. To meet a U.S. Postal Service requirement, employees' footwear must have a coefficient of static friction of 0.5 or more on a specified tile surface. A typical athletic shoe has a coefficient of static friction of 0.800. In an emergency, what is the minimum time interval in which a person starting from rest can move 3.00 m on the tile surface if she is wearing (a) footwear meeting the Postal Service minimum and (b) a typical athletic shoe?

60. A woman at an airport is towing her 20.0-kg suitcase at constant speed by pulling on a strap at an angle θ above the horizontal (Fig. P5.60). She pulls on the strap with a 35.0-N force, and the friction force on the suitcase is 20.0 N. (a) Draw a free-body diagram of the suitcase. (b) What angle does the strap make with the horizontal? (c) What is the magnitude of the normal force that the ground exerts on the suitcase?

Figure P5.60

61. Review. A 3.00-kg block starts from rest at the top of a 30.0° incline and slides a distance of 2.00 m down the incline in 1.50 s. Find (a) the magnitude of the acceleration of the block, (b) the coefficient of kinetic friction between block and plane, (c) the friction force acting on the block, and (d) the speed of the block after it has slid 2.00 m.

62. The person in Figure P5.62 weighs 170 lb. As seen from the front, each light crutch makes an angle of 22.0° with the vertical. Half of the person's weight is supported by the crutches. The other half is supported by the vertical forces of the ground on the person's feet. Assuming that the person is moving with constant velocity and the force exerted by the ground on the crutches acts along the crutches, determine (a) the smallest possible coefficient of friction between crutches and ground and (b) the magnitude of the compression force in each crutch.

Figure P5.62

63. A 9.00-kg hanging object is connected by a light, inextensible cord over a light, frictionless pulley to a 5.00-kg block that is sliding on a flat table (Fig. P5.40). Taking the coefficient of kinetic friction as 0.200, find the tension in the string.

64. Three objects are connected on a table as shown in Figure P5.64. The coefficient of kinetic friction between the block of mass m_2 and the table is 0.350. The objects have masses of $m_1 = 4.00$ kg, $m_2 = 1.00$ kg, and $m_3 =$

2.00 kg, and the pulleys are frictionless. (a) Draw a free-body diagram of each object. (b) Determine the acceleration of each object, including its direction. (c) Determine the tensions in the two cords. **What If?** (d) If the tabletop were smooth, would the tensions increase, decrease, or remain the same? Explain.

Figure P5.64

65. Two blocks connected by a rope of negligible mass are being dragged by a horizontal force (Fig. P5.65). Suppose $F = 68.0$ N, $m_1 = 12.0$ kg, $m_2 = 18.0$ kg, and the coefficient of kinetic friction between each block and the surface is 0.100. (a) Draw a free-body diagram for each block. Determine (b) the acceleration of the system and (c) the tension T in the rope.

Figure P5.65

66. A block of mass 3.00 kg is pushed up against a wall by a force $\vec{\textbf{P}}$ that makes an angle of $\theta = 50.0°$ with the horizontal as shown in Figure P5.66. The coefficient of static friction between the block and the wall is 0.250. (a) Determine the possible values for the magnitude of $\vec{\textbf{P}}$ that allow the block to remain stationary. (b) Describe what happens if $|\vec{\textbf{P}}|$ has a larger value and what happens if it is smaller. (c) Repeat parts (a) and (b), assuming the force makes an angle of $\theta = 13.0°$ with the horizontal.

Figure P5.66

67. **Review.** One side of the roof of a house slopes up at 37.0°. A roofer kicks a round, flat rock that has been thrown onto the roof by a neighborhood child. The rock slides straight up the incline with an initial speed of 15.0 m/s. The coefficient of kinetic friction between the rock and the roof is 0.400. The rock slides 10.0 m up the roof to its peak. It crosses the ridge and goes into free fall, following a parabolic trajectory above the far side of the roof, with negligible air resistance. Determine the maximum height the rock reaches above the point where it was kicked.

68. **Review.** A Chinook salmon can swim underwater at 3.58 m/s, and it can also jump vertically upward, leaving the water with a speed of 6.26 m/s. A record salmon has length 1.50 m and mass 61.0 kg. Consider the fish swimming straight upward in the water below the surface of a lake. The gravitational force exerted on it is very nearly canceled out by a buoyant force exerted by the water as we will study in Chapter 14. The fish experiences an upward force P exerted by the water on its threshing tail fin and a downward fluid friction force that we model as acting on its front end. Assume

the fluid friction force disappears as soon as the fish's head breaks the water surface and assume the force on its tail is constant. Model the gravitational force as suddenly switching full on when half the length of the fish is out of the water. Find the value of P.

69. **Review.** A magician pulls a tablecloth from under a 200-g mug located 30.0 cm from the edge of the cloth. The cloth exerts a friction force of 0.100 N on the mug, and the cloth is pulled with a constant acceleration of 3.00 m/s². How far does the mug move relative to the horizontal tabletop before the cloth is completely out from under it? Note that the cloth must move more than 30 cm relative to the tabletop during the process.

70. A 5.00-kg block is placed on top of a 10.0-kg block (Fig. P5.70). A horizontal force of 45.0 N is applied to the 10-kg block, and the 5.00-kg block is tied to the wall. The coefficient of kinetic friction between all moving surfaces is 0.200. (a) Draw a free-body diagram for each block and identify the action–reaction forces between the blocks. (b) Determine the tension in the string and the magnitude of the acceleration of the 10.0-kg block.

Figure P5.70

71. The system shown in Figure P5.49 has an acceleration of magnitude 1.50 m/s². Assume that the coefficient of kinetic friction between block and incline is the same for both inclines. Find (a) the coefficient of kinetic friction and (b) the tension in the string.

Additional Problems

72. A black aluminum glider floats on a film of air above a level aluminum air track. Aluminum feels essentially no force in a magnetic field, and air resistance is negligible. A strong magnet is attached to the top of the glider, forming a total mass of 240 g. A piece of scrap iron attached to one end stop on the track attracts the magnet with a force of 0.823 N when the iron and the magnet are separated by 2.50 cm. (a) Find the acceleration of the glider at this instant. (b) The scrap iron is now attached to another green glider, forming total mass 120 g. Find the acceleration of each glider when the gliders are simultaneously released at 2.50-cm separation.

73. A young woman buys an inexpensive used car for stock car racing. It can attain highway speed with an acceleration of 8.40 mi/h · s. By making changes to its engine, she can increase the net horizontal force on the car by 24.0%. With much less expense, she can remove material from the body of the car to decrease its mass by 24.0%. (a) Which of these two changes, if either, will result in the greater increase in the car's acceleration? (b) If she makes both changes, what acceleration can she attain?

74. *Why is the following situation impossible?* A book sits on an inclined plane on the surface of the Earth. The angle

of the plane with the horizontal is 60.0°. The coefficient of kinetic friction between the book and the plane is 0.300. At time $t = 0$, the book is released from rest. The book then slides through a distance of 1.00 m, measured along the plane, in a time interval of 0.483 s.

75. **Review.** A hockey puck struck by a hockey stick is given an initial speed v_i in the positive x direction. The coefficient of kinetic friction between the ice and the puck is μ_k. (a) Obtain an expression for the acceleration of the puck as it slides across the ice. (b) Use the result of part (a) to obtain an expression for the distance d the puck slides. The answer should be in terms of the variables v_i, μ_k, and g only.

76. A 1.00-kg glider on a horizontal air track is pulled by a string at an angle θ. The taut string runs over a pulley and is attached to a hanging object of mass 0.500 kg as shown in Figure P5.76. (a) Show that the speed v_x of the glider and the speed v_y of the hanging object are related by $v_x = uv_y$, where $u = z(z^2 - h_0^2)^{-1/2}$. (b) The glider is released from rest. Show that at that instant the acceleration a_x of the glider and the acceleration a_y of the hanging object are related by $a_x = ua_y$. (c) Find the tension in the string at the instant the glider is released for $h_0 = 80.0$ cm and $\theta = 30.0°$.

Figure P5.76

77. A frictionless plane is 10.0 m long and inclined at 35.0°. A sled starts at the bottom with an initial speed of 5.00 m/s up the incline. When the sled reaches the point at which it momentarily stops, a second sled is released from the top of the incline with an initial speed v_i. Both sleds reach the bottom of the incline at the same moment. (a) Determine the distance that the first sled traveled up the incline. (b) Determine the initial speed of the second sled.

78. A rope with mass m_r is attached to a block with mass m_b as in Figure P5.78. The block rests on a frictionless, horizontal surface. The rope does not stretch. The free end of the rope is pulled to the right with a horizontal force $\vec{F}$. (a) Draw force diagrams for the rope and the block, noting that the tension in the rope is not uniform. (b) Find the acceleration of the system in terms of m_b, m_r, and F. (c) Find the magnitude of the force the rope exerts on the block. (d) What happens

Figure P5.78

to the force on the block as the rope's mass approaches zero? What can you state about the tension in a *light* cord joining a pair of moving objects?

79. Two blocks of masses m_1 and m_2 are placed on a table in contact with each other as discussed in Example 5.7 and shown in Figure 5.12a. The coefficient of kinetic friction between the block of mass m_1 and the table is μ_1, and that between the block of mass m_2 and the table is μ_2. A horizontal force of magnitude F is applied to the block of mass m_1. We wish to find P, the magnitude of the contact force between the blocks. (a) Draw diagrams showing the forces for each block. (b) What is the net force on the system of two blocks? (c) What is the net force acting on m_1? (d) What is the net force acting on m_2? (e) Write Newton's second law in the x direction for each block. (f) Solve the two equations in two unknowns for the acceleration a of the blocks in terms of the masses, the applied force F, the coefficients of friction, and g. (g) Find the magnitude P of the contact force between the blocks in terms of the same quantities.

80. On a single, light, vertical cable that does not stretch, a crane is lifting a 1 207-kg Ferrari and, below it, a 1 461-kg BMW Z8. The Ferrari is moving upward with speed 3.50 m/s and acceleration 1.25 m/s². (a) How do the velocity and acceleration of the BMW compare with those of the Ferrari? (b) Find the tension in the cable between the BMW and the Ferrari. (c) Find the tension in the cable above the Ferrari.

81. An inventive child named Nick wants to reach an apple in a tree without climbing the tree. Sitting in a chair connected to a rope that passes over a frictionless pulley (Fig. P5.81), Nick pulls on the loose end of the rope with such a force that the spring scale reads 250 N. Nick's true weight is 320 N, and the chair weighs 160 N. Nick's feet are not touching the ground. (a) Draw one pair of diagrams showing the forces for Nick and the chair considered as separate systems and another diagram for Nick and the chair considered as one system. (b) Show that the acceleration of the system is *upward* and find its magnitude. (c) Find the force Nick exerts on the chair.

Figure P5.81 Problems 81 and 82.

82. In the situation described in Problem 81 and Figure P5.81, the masses of the rope, spring balance, and pul-

ley are negligible. Nick's feet are not touching the ground. (a) Assume Nick is momentarily at rest when he stops pulling down on the rope and passes the end of the rope to another child, of weight 440 N, who is standing on the ground next to him. The rope does not break. Describe the ensuing motion. (b) Instead, assume Nick is momentarily at rest when he ties the end of the rope to a strong hook projecting from the tree trunk. Explain why this action can make the rope break.

83. In Example 5.7, we pushed on two blocks on a table. Suppose three blocks are in contact with one another on a frictionless, horizontal surface as shown in Figure P5.83. A horizontal force $\vec{F}$ is applied to m_1. Take $m_1 =$ 2.00 kg, $m_2 = 3.00$ kg, $m_3 = 4.00$ kg, and $F = 18.0$ N. (a) Draw a separate free-body diagram for each block. (b) Determine the acceleration of the blocks. (c) Find the *resultant* force on each block. (d) Find the magnitudes of the contact forces between the blocks. (e) You are working on a construction project. A coworker is nailing up plasterboard on one side of a light partition, and you are on the opposite side, providing "backing" by leaning against the wall with your back pushing on it. Every hammer blow makes your back sting. The supervisor helps you put a heavy block of wood between the wall and your back. Using the situation analyzed in parts (a) through (d) as a model, explain how this change works to make your job more comfortable.

Figure P5.83

84. An aluminum block of mass $m_1 = 2.00$ kg and a copper block of mass $m_2 =$ 6.00 kg are connected by a light string over a frictionless pulley. They sit on a steel surface as shown in Figure P5.84, where $\theta =$ 30.0°. (a) When they are released from rest, will they start to move? If they do, determine (b) their acceleration and (c) the tension in the string. If they do not move, determine (d) the sum of the magnitudes of the forces of friction acting on the blocks.

Figure P5.84

85. An object of mass M is held in place by an applied force $\vec{F}$ and a pulley system as shown in Figure P5.85. The pulleys are massless and frictionless. (a) Draw diagrams showing the forces on each pulley. Find (b) the tension in each section of rope, T_1, T_2, T_3, T_4, and T_5 and (c) the magnitude of $\vec{F}$.

86. Any device that allows you to increase the force you exert is a kind of *machine*. Some machines, such as the prybar

Figure P5.85

or the inclined plane, are very simple. Some machines do not even look like machines. For example, your car is stuck in the mud and you can't pull hard enough to get it out. You do, however, have a long cable that you connect taut between your front bumper and the trunk of a stout tree. You now pull sideways on the cable at its midpoint, exerting a force f. Each half of the cable is displaced through a small angle θ from the straight line between the ends of the cable. (a) Deduce an expression for the force acting on the car. (b) Evaluate the cable tension for the case where $\theta = 7.00°$ and $f = 100$ N.

87. Objects with masses $m_1 = 10.0$ kg and $m_2 = 5.00$ kg are M connected by a light string that passes over a frictionless pulley as in Figure P5.40. If, when the system starts from rest, m_2 falls 1.00 m in 1.20 s, determine the coefficient of kinetic friction between m_1 and the table.

88. Consider the three connected objects shown in Figure P5.88. Assume first that the inclined plane is frictionless and that the system is in equilibrium. In terms of m, g, and θ, find (a) the mass M and (b) the tensions T_1 and T_2. Now assume that the value of M is double the value found in part (a). Find (c) the acceleration of each object and (d) the tensions T_1 and T_2. Next, assume that the coefficient of static friction between m and $2m$ and the inclined plane is m_s and that the system is in equilibrium. Find (e) the maximum value of M and (f) the minimum value of M. (g) Compare the values of T_2 when M has its minimum and maximum values.

Figure P5.88

89. A crate of weight F_g is pushed by a force $\vec{P}$ on a horizontal floor as shown in Figure P5.89. The coefficient of static friction is μ_s, and $\vec{P}$ is directed at angle θ below the horizontal. (a) Show that the minimum value of P that will move the crate is given by

Figure P5.89

$$P = \frac{\mu_s F_g \sec \theta}{1 - \mu_s \tan \theta}$$

(b) Find the condition on θ in terms of μ_s for which motion of the crate is impossible for any value of P.

90. A student is asked to measure the acceleration of a glider on a frictionless, inclined plane, using an air track, a stopwatch, and a meterstick. The top of the track is measured to be 1.774 cm higher than the bottom of the track, and the length of the track is $d =$ 127.1 cm. The cart is released from rest at the top of the incline, taken as $x = 0$, and its position x along the incline is measured as a function of time. For x values of 10.0 cm, 20.0 cm, 35.0 cm, 50.0 cm, 75.0 cm, and 100 cm, the measured times at which these positions are reached (averaged over five runs) are 1.02 s, 1.53 s,

2.01 s, 2.64 s, 3.30 s, and 3.75 s, respectively. (a) Construct a graph of x versus t^2, with a best-fit straight line to describe the data. (b) Determine the acceleration of the cart from the slope of this graph. (c) Explain how your answer to part (b) compares with the theoretical value you calculate using $a = g \sin \theta$ as derived in Example 5.6.

91. A flat cushion of mass m is released from rest at the corner of the roof of a building, at height h. A wind blowing along the side of the building exerts a constant horizontal force of magnitude F on the cushion as it drops as shown in Figure P5.91. The air exerts no vertical force. (a) Show that the path of the cushion is a straight line. (b) Does the cushion fall with constant velocity? Explain. (c) If $m = 1.20$ kg, $h = 8.00$ m, and $F = 2.40$ N, how far from the building will the cushion hit the level ground? **What If?** (d) If the cushion is thrown downward with a non-zero speed at the top of the building, what will be the shape of its trajectory? Explain.

Figure P5.91

92. In Figure P5.92, the pulleys and the cord are light, all surfaces are frictionless, and the cord does not stretch. (a) How does the acceleration of block 1 compare with the acceleration of block 2? Explain your reasoning. (b) The mass of block 2 is 1.30 kg. Find its acceleration as it depends on the mass m_1 of block 1.

Figure P5.92

(c) **What If?** What does the result of part (b) predict if m_1 is very much less than 1.30 kg? (d) What does the result of part (b) predict if m_1 approaches infinity? (e) In this last case, what is the tension in the cord? (f) Could you anticipate the answers to parts (c), (d), and (e) without first doing part (b)? Explain.

93. What horizontal force must be applied to a large block of mass M shown in Figure P5.93 so that the tan blocks remain stationary relative to M? Assume all surfaces and the pulley are frictionless. Notice that the force exerted by the string accelerates m_2.

Figure P5.93
Problems 93 and 98.

94. An 8.40-kg object slides down a fixed, frictionless, inclined plane. Use a computer to determine and tabulate (a) the normal force exerted on the object and (b) its acceleration for a series of incline angles (measured from the horizontal) ranging from 0° to 90° in 5° increments. (c) Plot a graph of the normal force and the acceleration as functions of the incline angle. (d) In the limiting cases of 0° and 90°, are your results consistent with the known behavior?

95. A car accelerates down a hill (Fig. P5.95), going from rest to 30.0 m/s in 6.00 s. A toy inside the car hangs by a string from the car's ceiling. The ball in the figure represents the toy, of mass 0.100 kg. The acceleration is such that the string remains perpendicular to the ceiling. Determine (a) the angle θ and (b) the tension in the string.

Figure P5.95

Challenge Problems

96. A time-dependent force, $\vec{F} = (8.00\hat{i} - 4.00t\hat{j})$, where $\vec{F}$ is in newtons and t is in seconds, is exerted on a 2.00-kg object initially at rest. (a) At what time will the object be moving with a speed of 15.0 m/s? (b) How far is the object from its initial position when its speed is 15.0 m/s? (c) Through what total displacement has the object traveled at this moment?

97. The board sandwiched between two other boards in Figure P5.97 weighs 95.5 N. If the coefficient of static friction between the boards is 0.663, what must be the magnitude of the compression forces (assumed horizontal) acting on both sides of the center board to keep it from slipping?

Figure P5.97

98. Initially, the system of objects shown in Figure P5.93 is held motionless. The pulley and all surfaces and wheels are frictionless. Let the force $\vec{F}$ be zero and assume that m_1 can move only vertically. At the instant after the system of objects is released, find (a) the tension T in the string, (b) the acceleration of m_2, (c) the acceleration of M, and (d) the acceleration of m_1. (*Note:* The pulley accelerates along with the cart.)

99. A block of mass 2.20 kg is accelerated across a rough surface by a light cord passing over a small pulley as shown in Figure P5.99. The tension T in the cord is maintained at 10.0 N, and the pulley is 0.100 m above the top of the block. The coefficient of kinetic friction is 0.400. (a) Determine the acceleration of the block when $x = 0.400$ m. (b) Describe the general behavior of the acceleration as the block slides from a location where x is large to $x = 0$. (c) Find the maximum value of the acceleration and the position x for which it occurs. (d) Find the value of x for which the acceleration is zero.

Figure P5.99

100. *Why is the following situation impossible?* A 1.30-kg toaster is not plugged in. The coefficient of static friction between the toaster and a horizontal countertop is 0.350. To make the toaster start moving, you carelessly pull on its electric cord. Unfortunately, the cord has become frayed from your previous similar actions and will break if the tension in the cord exceeds 4.00 N. By pulling on the cord at a particular angle, you successfully start the toaster moving without breaking the cord.

101. **Review.** A block of mass $m = 2.00$ kg is released from rest at $h = 0.500$ m above the surface of a table, at the top of a $\theta = 30.0°$ incline as shown in Figure P5.101. The frictionless incline is fixed on a table of height $H = 2.00$ m. (a) Determine the acceleration of the block as it slides down the incline. (b) What is the velocity of the block as it leaves the incline? (c) How far from the table will the block hit the floor? (d) What time interval elapses between when the block is released and when it hits the floor? (e) Does the mass of the block affect any of the above calculations?

Figure P5.101 Problems 101 and 102.

102. In Figure P5.101, the incline has mass M and is fastened to the stationary horizontal tabletop. The block of mass m is placed near the bottom of the incline and is released with a quick push that sets it sliding upward. The block stops near the top of the incline as shown in the figure and then slides down again, always without friction. Find the force that the tabletop exerts on the incline throughout this motion in terms of m, M, g, and θ.

103. A block of mass $m = 2.00$ kg rests on the left edge of a block of mass $M = 8.00$ kg. The coefficient of kinetic friction between the two blocks is 0.300, and the surface on which the 8.00-kg block rests is frictionless. A constant horizontal force of magnitude $F = 10.0$ N is applied to the 2.00-kg block, setting it in motion as

shown in Figure P5.103a. If the distance L that the leading edge of the smaller block travels on the larger block is 3.00 m, (a) in what time interval will the smaller block make it to the right side of the 8.00-kg block as shown in Figure P5.103b? (*Note:* Both blocks are set into motion when $\vec{F}$ is applied.) (b) How far does the 8.00-kg block move in the process?

Figure P5.103

104. A mobile is formed by supporting four metal butterflies of equal mass m from a string of length L. The points of support are evenly spaced a distance ℓ apart as shown in Figure P5.104. The string forms an angle θ_1 with the ceiling at each endpoint. The center section of string is horizontal. (a) Find the tension in each section of string in terms of θ_1, m, and g. (b) In terms of θ_1, find the angle θ_2 that the sections of string between the outside butterflies and the inside butterflies form with the horizontal. (c) Show that the distance D between the endpoints of the string is

$$D = \frac{L}{5}\left\{2\cos\theta_1 + 2\cos[\tan^{-1}(\tfrac{1}{2}\tan\theta_1)] + 1\right\}$$

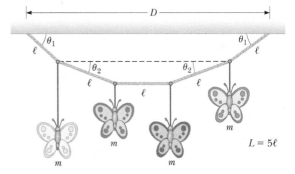

Figure P5.104

Circular Motion and Other Applications of Newton's Laws

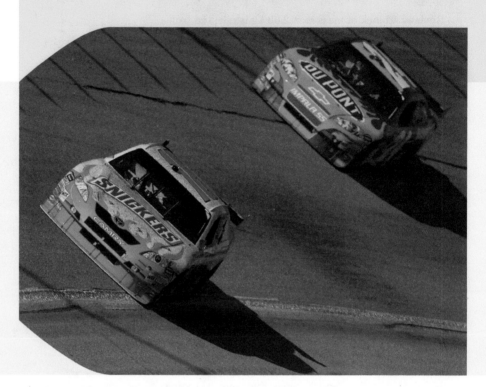

Kyle Busch, driver of the #18 Snickers Toyota, leads Jeff Gordon, driver of the #24 Dupont Chevrolet, during the NASCAR Sprint Cup Series Kobalt Tools 500 at the Atlanta Motor Speedway on March 9, 2008, in Hampton, Georgia. The cars travel on a banked roadway to help them undergo circular motion on the turns. *(Chris Graythen/Getty Images for NASCAR)*

In the preceding chapter, we introduced Newton's laws of motion and incorporated them into two analysis models involving linear motion. Now we discuss motion that is slightly more complicated. For example, we shall apply Newton's laws to objects traveling in circular paths. We shall also discuss motion observed from an accelerating frame of reference and motion of an object through a viscous medium. For the most part, this chapter consists of a series of examples selected to illustrate the application of Newton's laws to a variety of new circumstances.

6.1 Extending the Particle in Uniform Circular Motion Model

In Section 4.4, we discussed the analysis model of a particle in uniform circular motion, in which a particle moves with constant speed v in a circular path having a radius r. The particle experiences an acceleration that has a magnitude

$$a_c = \frac{v^2}{r}$$

150

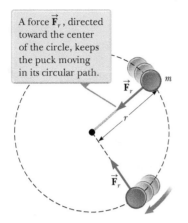

A force $\vec{\mathbf{F}}_r$, directed toward the center of the circle, keeps the puck moving in its circular path.

Figure 6.1 An overhead view of a puck moving in a circular path in a horizontal plane.

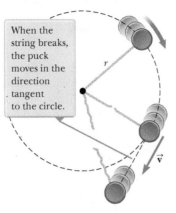

When the string breaks, the puck moves in the direction tangent to the circle.

Figure 6.2 The string holding the puck in its circular path breaks.

23.6 × 9.8 = 231.3 N

The acceleration is called *centripetal acceleration* because $\vec{\mathbf{a}}_c$ is directed toward the center of the circle. Furthermore, $\vec{\mathbf{a}}_c$ is *always* perpendicular to $\vec{\mathbf{v}}$. (If there were a component of acceleration parallel to $\vec{\mathbf{v}}$, the particle's speed would be changing.)

Let us now extend the particle in uniform circular motion model from Section 4.4 by incorporating the concept of force. Consider a puck of mass m that is tied to a string of length r and moves at constant speed in a horizontal, circular path as illustrated in Figure 6.1. Its weight is supported by a frictionless table, and the string is anchored to a peg at the center of the circular path of the puck. Why does the puck move in a circle? According to Newton's first law, the puck would move in a straight line if there were no force on it; the string, however, prevents motion along a straight line by exerting on the puck a radial force $\vec{\mathbf{F}}_r$ that makes it follow the circular path. This force is directed along the string toward the center of the circle as shown in Figure 6.1.

If Newton's second law is applied along the radial direction, the net force causing the centripetal acceleration can be related to the acceleration as follows:

$$\sum F = ma_c = m\frac{v^2}{r} \tag{6.1}$$

◀ **Force causing centripetal acceleration**

A force causing a centripetal acceleration acts toward the center of the circular path and causes a change in the direction of the velocity vector. If that force should vanish, the object would no longer move in its circular path; instead, it would move along a straight-line path tangent to the circle. This idea is illustrated in Figure 6.2 for the puck moving in a circular path at the end of a string in a horizontal plane. If the string breaks at some instant, the puck moves along the straight-line path that is tangent to the circle at the position of the puck at this instant.

Pitfall Prevention 6.1
Direction of Travel When the String Is Cut Study Figure 6.2 very carefully. Many students (wrongly) think that the puck will move *radially* away from the center of the circle when the string is cut. The velocity of the puck is *tangent* to the circle. By Newton's first law, the puck continues to move in the same direction in which it is moving just as the force from the string disappears.

Quick Quiz 6.1 You are riding on a Ferris wheel that is rotating with constant speed. The car in which you are riding always maintains its correct upward orientation; it does not invert. **(i)** What is the direction of the normal force on you from the seat when you are at the top of the wheel? (a) upward (b) downward (c) impossible to determine **(ii)** From the same choices, what is the direction of the net force on you when you are at the top of the wheel?

Analysis Model   Particle in Uniform Circular Motion (Extension)

Imagine a moving object that can be modeled as a particle. If it moves in a circular path of radius r at a constant speed v, it experiences a centripetal acceleration. Because the particle is accelerating, there must be a net force acting on the particle. That force is directed toward the center of the circular path and is given by

$$\sum F = ma_c = m\frac{v^2}{r} \qquad \textbf{(6.1)}$$

Examples

- the tension in a string of constant length acting on a rock twirled in a circle
- the gravitational force acting on a planet traveling around the Sun in a perfectly circular orbit (Chapter 13)
- the magnetic force acting on a charged particle moving in a uniform magnetic field (Chapter 29)
- the electric force acting on an electron in orbit around a nucleus in the Bohr model of the hydrogen atom (Chapter 42)

Example 6.1   The Conical Pendulum   AM

A small ball of mass m is suspended from a string of length L. The ball revolves with constant speed v in a horizontal circle of radius r as shown in Figure 6.3. (Because the string sweeps out the surface of a cone, the system is known as a *conical pendulum*.) Find an expression for v in terms of the geometry in Figure 6.3.

SOLUTION

Conceptualize Imagine the motion of the ball in Figure 6.3a and convince yourself that the string sweeps out a cone and that the ball moves in a horizontal circle.

Categorize The ball in Figure 6.3 does not accelerate vertically. Therefore, we model it as a *particle in equilibrium* in the vertical direction. It experiences a centripetal acceleration in the horizontal direction, so it is modeled as a *particle in uniform circular motion* in this direction.

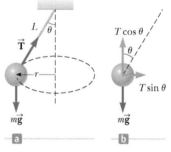

Figure 6.3 (Example 6.1) (a) A conical pendulum. The path of the ball is a horizontal circle. (b) The forces acting on the ball.

Analyze Let θ represent the angle between the string and the vertical. In the diagram of forces acting on the ball in Figure 6.3b, the force $\vec{T}$ exerted by the string on the ball is resolved into a vertical component $T\cos\theta$ and a horizontal component $T\sin\theta$ acting toward the center of the circular path.

Apply the particle in equilibrium model in the vertical direction:

$$\sum F_y = T\cos\theta - mg = 0$$
$$(1) \quad T\cos\theta = mg$$

Use Equation 6.1 from the particle in uniform circular motion model in the horizontal direction:

$$(2) \quad \sum F_x = T\sin\theta = ma_c = \frac{mv^2}{r}$$

Divide Equation (2) by Equation (1) and use $\sin\theta/\cos\theta = \tan\theta$:

$$\tan\theta = \frac{v^2}{rg}$$

Solve for v:

$$v = \sqrt{rg\tan\theta}$$

Incorporate $r = L\sin\theta$ from the geometry in Figure 6.3a:

$$v = \sqrt{Lg\sin\theta\tan\theta}$$

Finalize Notice that the speed is independent of the mass of the ball. Consider what happens when θ goes to 90° so that the string is horizontal. Because the tangent of 90° is infinite, the speed v is infinite, which tells us the string cannot possibly be horizontal. If it were, there would be no vertical component of the force $\vec{T}$ to balance the gravitational force on the ball. That is why we mentioned in regard to Figure 6.1 that the puck's weight in the figure is supported by a frictionless table.

0.2754

$mv^2 = T \qquad \dfrac{T \cdot r}{v^2} = m$

$\dfrac{mv^2}{r} = T$

$m \cdot 20.7^2 = 44.5$

$\dfrac{2.55}{} \qquad m =$

Example 6.2   **How Fast Can It Spin?** AM

A puck of mass 0.500 kg is attached to the end of a cord 1.50 m long. The puck moves in a horizontal circle as shown in Figure 6.1. If the cord can withstand a maximum tension of 50.0 N, what is the maximum speed at which the puck can move before the cord breaks? Assume the string remains horizontal during the motion.

SOLUTION

Conceptualize It makes sense that the stronger the cord, the faster the puck can move before the cord breaks. Also, we expect a more massive puck to break the cord at a lower speed. (Imagine whirling a bowling ball on the cord!)

Categorize Because the puck moves in a circular path, we model it as a *particle in uniform circular motion.*

Analyze Incorporate the tension and the centripetal acceleration into Newton's second law as described by Equation 6.1:

$$T = m \frac{v^2}{r}$$

$231.28 \cdot 0.807$

$\dfrac{}{2.97}$

Solve for v:

q

$$(1) \quad v = \sqrt{\frac{Tr}{m}}$$

Find the maximum speed the puck can have, which corresponds to the maximum tension the string can withstand:

$$v_{max} = \sqrt{\frac{T_{max}r}{m}} = \sqrt{\frac{(50.0\ \text{N})(1.50\ \text{m})}{0.500\ \text{kg}}} = \boxed{12.2\ \text{m/s}}$$

Finalize Equation (1) shows that v increases with T and decreases with larger m, as we expected from our conceptualization of the problem.

WHAT IF? Suppose the puck moves in a circle of larger radius at the same speed v. Is the cord more likely or less likely to break?

Answer The larger radius means that the change in the direction of the velocity vector will be smaller in a given time interval. Therefore, the acceleration is smaller and the required tension in the string is smaller. As a result, the string is less likely to break when the puck travels in a circle of larger radius.

Example 6.3   **What Is the Maximum Speed of the Car?** AM

A 1 500-kg car moving on a flat, horizontal road negotiates a curve as shown in Figure 6.4a. If the radius of the curve is 35.0 m and the coefficient of static friction between the tires and dry pavement is 0.523, find the maximum speed the car can have and still make the turn successfully.

SOLUTION

Conceptualize Imagine that the curved roadway is part of a large circle so that the car is moving in a circular path.

Categorize Based on the Conceptualize step of the problem, we model the car as a *particle in uniform circular motion* in the horizontal direction. The car is not accelerating vertically, so it is modeled as a *particle in equilibrium* in the vertical direction.

Analyze Figure 6.4b shows the forces on the car. The force that enables the car to remain in its circular path is the force of static friction. (It is *static* because no slipping occurs at the point of contact between road and tires. If this force of static friction were zero—for example, if the car were on an icy road—the car would continue in a straight line and slide off the curved road.) The maximum speed v_{max} the car can have around the curve is the speed at which it is on the verge of skidding outward. At this point, the friction force has its maximum value $f_{s,max} = \mu_s n$.

Figure 6.4 (Example 6.3) (a) The force of static friction directed toward the center of the curve keeps the car moving in a circular path. (b) The forces acting on the car.

continued

▶ **6.3** continued

Apply Equation 6.1 from the particle in uniform circular motion model in the radial direction for the maximum speed condition:

$$(1) \quad f_{s,max} = \mu_s n = m\frac{v_{max}^2}{r}$$

Apply the particle in equilibrium model to the car in the vertical direction:

$$\sum F_y = 0 \quad \rightarrow \quad n - mg = 0 \quad \rightarrow \quad n = mg$$

Solve Equation (1) for the maximum speed and substitute for n:

$$(2) \quad v_{max} = \sqrt{\frac{\mu_s n r}{m}} = \sqrt{\frac{\mu_s m g r}{m}} = \sqrt{\mu_s g r}$$

Substitute numerical values:

$$v_{max} = \sqrt{(0.523)(9.80 \text{ m/s}^2)(35.0 \text{ m})} = \boxed{13.4 \text{ m/s}}$$

Finalize This speed is equivalent to 30.0 mi/h. Therefore, if the speed limit on this roadway is higher than 30 mi/h, this roadway could benefit greatly from some banking, as in the next example! Notice that the maximum speed does not depend on the mass of the car, which is why curved highways do not need multiple speed limits to cover the various masses of vehicles using the road.

WHAT IF?   Suppose a car travels this curve on a wet day and begins to skid on the curve when its speed reaches only 8.00 m/s. What can we say about the coefficient of static friction in this case?

Answer The coefficient of static friction between the tires and a wet road should be smaller than that between the tires and a dry road. This expectation is consistent with experience with driving because a skid is more likely on a wet road than a dry road.

To check our suspicion, we can solve Equation (2) for the coefficient of static friction:

$$\mu_s = \frac{v_{max}^2}{gr}$$

Substituting the numerical values gives

$$\mu_s = \frac{v_{max}^2}{gr} = \frac{(8.00 \text{ m/s})^2}{(9.80 \text{ m/s}^2)(35.0 \text{ m})} = 0.187$$

which is indeed smaller than the coefficient of 0.523 for the dry road.

Example 6.4   The Banked Roadway   AM

A civil engineer wishes to redesign the curved roadway in Example 6.3 in such a way that a car will not have to rely on friction to round the curve without skidding. In other words, a car moving at the designated speed can negotiate the curve even when the road is covered with ice. Such a road is usually *banked*, which means that the roadway is tilted toward the inside of the curve as seen in the opening photograph for this chapter. Suppose the designated speed for the road is to be 13.4 m/s (30.0 mi/h) and the radius of the curve is 35.0 m. At what angle should the curve be banked?

SOLUTION

Conceptualize The difference between this example and Example 6.3 is that the car is no longer moving on a flat roadway. Figure 6.5 shows the banked roadway, with the center of the circular path of the car far to the left of the figure. Notice that the horizontal component of the normal force participates in causing the car's centripetal acceleration.

Categorize As in Example 6.3, the car is modeled as a *particle in equilibrium* in the vertical direction and a *particle in uniform circular motion* in the horizontal direction.

Analyze On a level (unbanked) road, the force that causes the centripetal acceleration is the force of static friction between tires and the road as we saw in the preceding example. If the road is banked at an angle θ as in Figure 6.5, however, the

Figure 6.5 (Example 6.4) A car moves into the page and is rounding a curve on a road banked at an angle θ to the horizontal. When friction is neglected, the force that causes the centripetal acceleration and keeps the car moving in its circular path is the horizontal component of the normal force.

▶ 6.4 continued

normal force $\vec{\mathbf{n}}$ has a horizontal component toward the center of the curve. Because the road is to be designed so that the force of static friction is zero, the component $n_x = n \sin \theta$ is the only force that causes the centripetal acceleration.

Write Newton's second law for the car in the radial direction, which is the x direction:

(1) $\sum F_r = n \sin \theta = \dfrac{mv^2}{r}$

Apply the particle in equilibrium model to the car in the vertical direction:

$\sum F_y = n \cos \theta - mg = 0$

(2) $n \cos \theta = mg$

Divide Equation (1) by Equation (2):

(3) $\tan \theta = \dfrac{v^2}{rg}$

Solve for the angle θ:

$\theta = \tan^{-1}\left[\dfrac{(13.4 \text{ m/s})^2}{(35.0 \text{ m})(9.80 \text{ m/s}^2)}\right] = \boxed{27.6°}$

Finalize Equation (3) shows that the banking angle is independent of the mass of the vehicle negotiating the curve. If a car rounds the curve at a speed less than 13.4 m/s, the centripetal acceleration decreases. Therefore, the normal force, which is unchanged, is sufficient to cause *two* accelerations: the lower centripetal acceleration and an acceleration of the car down the inclined roadway. Consequently, an additional friction force parallel to the roadway and upward is needed to keep the car from sliding down the bank (to the left in Fig. 6.5). Similarly, a driver attempting to negotiate the curve at a speed greater than 13.4 m/s has to depend on friction to keep from sliding up the bank (to the right in Fig. 6.5).

WHAT IF? Imagine that this same roadway were built on Mars in the future to connect different colony centers. Could it be traveled at the same speed?

Answer The reduced gravitational force on Mars would mean that the car is not pressed as tightly to the roadway. The reduced normal force results in a smaller component of the normal force toward the center of the circle. This smaller component would not be sufficient to provide the centripetal acceleration associated with the original speed. The centripetal acceleration must be reduced, which can be done by reducing the speed v.

Mathematically, notice that Equation (3) shows that the speed v is proportional to the square root of g for a roadway of fixed radius r banked at a fixed angle θ. Therefore, if g is smaller, as it is on Mars, the speed v with which the roadway can be safely traveled is also smaller.

Example 6.5   Riding the Ferris Wheel   AM

A child of mass m rides on a Ferris wheel as shown in Figure 6.6a. The child moves in a vertical circle of radius 10.0 m at a constant speed of 3.00 m/s.

(A) Determine the force exerted by the seat on the child at the bottom of the ride. Express your answer in terms of the weight of the child, mg.

SOLUTION

Conceptualize Look carefully at Figure 6.6a. Based on experiences you may have had on a Ferris wheel or driving over small hills on a roadway, you would expect to feel lighter at the top of the path. Similarly, you would expect to feel heavier at the bottom of the path. At both the bottom of the path and the top, the normal and gravitational forces on the child act in *opposite* directions. The vector sum of these two forces gives a

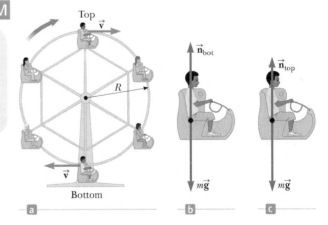

Figure 6.6 (Example 6.5) (a) A child rides on a Ferris wheel. (b) The forces acting on the child at the bottom of the path. (c) The forces acting on the child at the top of the path.

force of constant magnitude that keeps the child moving in a circular path at a constant speed. To yield net force vectors with the same magnitude, the normal force at the bottom must be greater than that at the top.

continued

▶ **6.5** continued

Categorize Because the speed of the child is constant, we can categorize this problem as one involving a *particle* (the child) *in uniform circular motion*, complicated by the gravitational force acting at all times on the child.

Analyze We draw a diagram of forces acting on the child at the bottom of the ride as shown in Figure 6.6b. The only forces acting on him are the downward gravitational force $\vec{\mathbf{F}}_g = m\vec{\mathbf{g}}$ and the upward force $\vec{\mathbf{n}}_{bot}$ exerted by the seat. The net upward force on the child that provides his centripetal acceleration has a magnitude $n_{bot} - mg$.

Using the particle in uniform circular motion model, apply Newton's second law to the child in the radial direction when he is at the bottom of the ride:

$$\sum F = n_{bot} - mg = m\frac{v^2}{r}$$

Solve for the force exerted by the seat on the child:

$$n_{bot} = mg + m\frac{v^2}{r} = mg\left(1 + \frac{v^2}{rg}\right)$$

Substitute numerical values given for the speed and radius:

$$n_{bot} = mg\left[1 + \frac{(3.00 \text{ m/s})^2}{(10.0 \text{ m})(9.80 \text{ m/s}^2)}\right]$$

$$= \boxed{1.09\, mg}$$

Hence, the magnitude of the force $\vec{\mathbf{n}}_{bot}$ exerted by the seat on the child is *greater* than the weight of the child by a factor of 1.09. So, the child experiences an apparent weight that is greater than his true weight by a factor of 1.09.

(B) Determine the force exerted by the seat on the child at the top of the ride.

SOLUTION

Analyze The diagram of forces acting on the child at the top of the ride is shown in Figure 6.6c. The net downward force that provides the centripetal acceleration has a magnitude $mg - n_{top}$.

Apply Newton's second law to the child at this position:

$$\sum F = mg - n_{top} = m\frac{v^2}{r}$$

Solve for the force exerted by the seat on the child:

$$n_{top} = mg - m\frac{v^2}{r} = mg\left(1 - \frac{v^2}{rg}\right)$$

Substitute numerical values:

$$n_{top} = mg\left[1 - \frac{(3.00 \text{ m/s})^2}{(10.0 \text{ m})(9.80 \text{ m/s}^2)}\right]$$

$$= \boxed{0.908\, mg}$$

In this case, the magnitude of the force exerted by the seat on the child is *less* than his true weight by a factor of 0.908, and the child feels lighter.

Finalize The variations in the normal force are consistent with our prediction in the Conceptualize step of the problem.

WHAT IF?   Suppose a defect in the Ferris wheel mechanism causes the speed of the child to increase to 10.0 m/s. What does the child experience at the top of the ride in this case?

Answer If the calculation above is performed with $v = 10.0$ m/s, the magnitude of the normal force at the top of the ride is negative, which is impossible. We interpret it to mean that the required centripetal acceleration of the child is larger than that due to gravity. As a result, the child will lose contact with the seat and will only stay in his circular path if there is a safety bar or a seat belt that provides a downward force on him to keep him in his seat. At the bottom of the ride, the normal force is 2.02 mg, which would be uncomfortable.

6.2 Nonuniform Circular Motion

In Chapter 4, we found that if a particle moves with varying speed in a circular path, there is, in addition to the radial component of acceleration, a tangential component having magnitude $|dv/dt|$. Therefore, the force acting on the particle

Figure 6.7 When the net force acting on a particle moving in a circular path has a tangential component ΣF_t, the particle's speed changes.

> The net force exerted on the particle is the vector sum of the radial force and the tangential force.

must also have a tangential and a radial component. Because the total acceleration is $\vec{a} = \vec{a}_r + \vec{a}_t$, the total force exerted on the particle is $\Sigma \vec{F} = \Sigma \vec{F}_r + \Sigma \vec{F}_t$ as shown in Figure 6.7. (We express the radial and tangential forces as net forces with the summation notation because each force could consist of multiple forces that combine.) The vector $\Sigma \vec{F}_r$ is directed toward the center of the circle and is responsible for the centripetal acceleration. The vector $\Sigma \vec{F}_t$ tangent to the circle is responsible for the tangential acceleration, which represents a change in the particle's speed with time.

Quick Quiz 6.2 A bead slides at constant speed along a curved wire lying on a horizontal surface as shown in Figure 6.8. **(a)** Draw the vectors representing the force exerted by the wire on the bead at points Ⓐ, Ⓑ, and Ⓒ. **(b)** Suppose the bead in Figure 6.8 speeds up with constant tangential acceleration as it moves toward the right. Draw the vectors representing the force on the bead at points Ⓐ, Ⓑ, and Ⓒ.

Figure 6.8 (Quick Quiz 6.2) A bead slides along a curved wire.

Example 6.6 **Keep Your Eye on the Ball** AM

A small sphere of mass m is attached to the end of a cord of length R and set into motion in a *vertical* circle about a fixed point O as illustrated in Figure 6.9. Determine the tangential acceleration of the sphere and the tension in the cord at any instant when the speed of the sphere is v and the cord makes an angle θ with the vertical.

SOLUTION

Conceptualize Compare the motion of the sphere in Figure 6.9 with that of the child in Figure 6.6a associated with Example 6.5. Both objects travel in a circular path. Unlike the child in Example 6.5, however, the speed of the sphere is *not* uniform in this example because, at most points along the path, a tangential component of acceleration arises from the gravitational force exerted on the sphere.

Categorize We model the sphere as a *particle under a net force* and moving in a circular path, but it is not a particle in *uniform* circular motion. We need to use the techniques discussed in this section on nonuniform circular motion.

....

Analyze From the force diagram in Figure 6.9, we see that the only forces acting on the sphere are the gravitational force

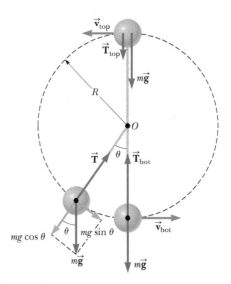

Figure 6.9 (Example 6.6) The forces acting on a sphere of mass m connected to a cord of length R and rotating in a vertical circle centered at O. Forces acting on the sphere are shown when the sphere is at the top and bottom of the circle and at an arbitrary location.

▶ **6.6** continued

$\vec{\mathbf{F}}_g = m\vec{\mathbf{g}}$ exerted by the Earth and the force $\vec{\mathbf{T}}$ exerted by the cord. We resolve $\vec{\mathbf{F}}_g$ into a tangential component $mg\sin\theta$ and a radial component $mg\cos\theta$.

From the particle under a net force model, apply Newton's second law to the sphere in the tangential direction:

$$\sum F_t = mg\sin\theta = ma_t$$
$$a_t = g\sin\theta$$

Apply Newton's second law to the forces acting on the sphere in the radial direction, noting that both $\vec{\mathbf{T}}$ and $\vec{\mathbf{a}}_r$ are directed toward O. As noted in Section 4.5, we can use Equation 4.14 for the centripetal acceleration of a particle even when it moves in a circular path in nonuniform motion:

$$\sum F_r = T - mg\cos\theta = \frac{mv^2}{R}$$
$$T = mg\left(\frac{v^2}{Rg} + \cos\theta\right)$$

. . . .

Finalize Let us evaluate this result at the top and bottom of the circular path (Fig. 6.9):

$$T_{\text{top}} = mg\left(\frac{v_{\text{top}}^2}{Rg} - 1\right) \qquad T_{\text{bot}} = mg\left(\frac{v_{\text{bot}}^2}{Rg} + 1\right)$$

These results have similar mathematical forms as those for the normal forces n_{top} and n_{bot} on the child in Example 6.5, which is consistent with the normal force on the child playing a similar physical role in Example 6.5 as the tension in the string plays in this example. Keep in mind, however, that the normal force $\vec{\mathbf{n}}$ on the child in Example 6.5 is always upward, whereas the force $\vec{\mathbf{T}}$ in this example changes direction because it must always point inward along the string. Also note that v in the expressions above varies for different positions of the sphere, as indicated by the subscripts, whereas v in Example 6.5 is constant.

WHAT IF?   What if the ball is set in motion with a slower speed?

(A) What speed would the ball have as it passes over the top of the circle if the tension in the cord goes to zero instantaneously at this point?

Answer Let us set the tension equal to zero in the expression for T_{top}:

$$0 = mg\left(\frac{v_{\text{top}}^2}{Rg} - 1\right) \quad \rightarrow \quad v_{\text{top}} = \sqrt{gR}$$

(B) What if the ball is set in motion such that the speed at the top is less than this value? What happens?

Answer In this case, the ball never reaches the top of the circle. At some point on the way up, the tension in the string goes to zero and the ball becomes a projectile. It follows a segment of a parabolic path over the top of its motion, rejoining the circular path on the other side when the tension becomes nonzero again.

6.3 Motion in Accelerated Frames

Newton's laws of motion, which we introduced in Chapter 5, describe observations that are made in an inertial frame of reference. In this section, we analyze how Newton's laws are applied by an observer in a noninertial frame of reference, that is, one that is accelerating. For example, recall the discussion of the air hockey table on a train in Section 5.2. The train moving at constant velocity represents an inertial frame. An observer on the train sees the puck at rest remain at rest, and Newton's first law appears to be obeyed. The accelerating train is not an inertial frame. According to you as the observer on this train, there appears to be no force on the puck, yet it accelerates from rest toward the back of the train, appearing to violate Newton's first law. This property is a general property of observations made in noninertial frames: there appear to be unexplained accelerations of objects that are not "fastened" to the frame. Newton's first law is not violated, of course. It only appears to be violated because of observations made from a noninertial frame.

On the accelerating train, as you watch the puck accelerating toward the back of the train, you might conclude based on your belief in Newton's second law that a force has acted on the puck to cause it to accelerate. We call an apparent force such as this one a **fictitious force** because it is not a real force and is due only to observations made in an accelerated reference frame. A fictitious force appears to act on an object in the same way as a real force. Real forces are always interactions between two objects, however, and you cannot identify a second object for a fictitious force. (What second object is interacting with the puck to cause it to accelerate?) In general, simple fictitious forces appear to act in the direction *opposite* that of the acceleration of the noninertial frame. For example, the train accelerates forward and there appears to be a fictitious force causing the puck to slide toward the back of the train.

The train example describes a fictitious force due to a change in the train's speed. Another fictitious force is due to the change in the *direction* of the velocity vector. To understand the motion of a system that is noninertial because of a change in direction, consider a car traveling along a highway at a high speed and approaching a curved exit ramp on the left as shown in Figure 6.10a. As the car takes the sharp left turn on the ramp, a person sitting in the passenger seat leans or slides to the right and hits the door. At that point the force exerted by the door on the passenger keeps her from being ejected from the car. What causes her to move toward the door? A popular but incorrect explanation is that a force acting toward the right in Figure 6.10b pushes the passenger outward from the center of the circular path. Although often called the "centrifugal force," it is a fictitious force. The car represents a noninertial reference frame that has a centripetal acceleration toward the center of its circular path. As a result, the passenger feels an apparent force which is outward from the center of the circular path, or to the right in Figure 6.10b, in the direction opposite that of the acceleration.

Let us address this phenomenon in terms of Newton's laws. Before the car enters the ramp, the passenger is moving in a straight-line path. As the car enters the ramp and travels a curved path, the passenger tends to move along the original straight-line path, which is in accordance with Newton's first law: the natural tendency of an object is to continue moving in a straight line. If a sufficiently large force (toward the center of curvature) acts on the passenger as in Figure 6.10c, however, she moves in a curved path along with the car. This force is the force of friction between her and the car seat. If this friction force is not large enough, the seat follows a curved path while the passenger tends to continue in the straight-line path of the car before the car began the turn. Therefore, from the point of view of an observer in the car, the passenger leans or slides to the right relative to the seat. Eventually, she encounters the door, which provides a force large enough to enable her to follow the same curved path as the car.

Another interesting fictitious force is the "Coriolis force." It is an apparent force caused by changing the radial position of an object in a rotating coordinate system.

For example, suppose you and a friend are on opposite sides of a rotating circular platform and you decide to throw a baseball to your friend. Figure 6.11a on page 160 represents what an observer would see if the ball is viewed while the observer is hovering at rest above the rotating platform. According to this observer, who is in an inertial frame, the ball follows a straight line as it must according to Newton's first law. At $t = 0$ you throw the ball toward your friend, but by the time t_f when the ball has crossed the platform, your friend has moved to a new position and can't catch the ball. Now, however, consider the situation from your friend's viewpoint. Your friend is in a noninertial reference frame because he is undergoing a centripetal acceleration relative to the inertial frame of the Earth's surface. He starts off seeing the baseball coming toward him, but as it crosses the platform, it veers to one side as shown in Figure 6.11b. Therefore, your friend on the rotating platform states that the ball does not obey Newton's first law and claims that a sideways force is causing the ball to follow a curved path. This fictitious force is called the Coriolis force.

Fictitious forces may not be real forces, but they can have real effects. An object on your dashboard *really* slides off if you press the accelerator of your car. As you ride on a merry-go-round, you feel pushed toward the outside as if due to the fictitious "centrifugal force." You are likely to fall over and injure yourself due to the

From the passenger's frame of reference, a force appears to push her toward the right door, but it is a fictitious force.

Fictitious force

Relative to the reference frame of the Earth, the car seat applies a real force (friction) toward the left on the passenger, causing her to change direction along with the rest of the car.

Real force

Figure 6.10 (a) A car approaching a curved exit ramp. What causes a passenger in the front seat to move toward the right-hand door? (b) Passenger's frame of reference. (c) Reference frame of the Earth.

By the time t_f that the ball arrives at the other side of the platform, your friend is no longer there to catch it. According to this observer, the ball follows a straight-line path, consistent with Newton's laws.

From your friend's point of view, the ball veers to one side during its flight. Your friend introduces a fictitious force to explain this deviation from the expected path.

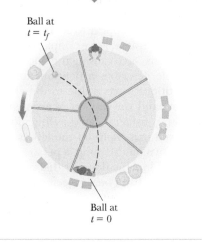

Figure 6.11 You and your friend stand at the edge of a rotating circular platform. You throw the ball at $t = 0$ in the direction of your friend. (a) Overhead view observed by someone in an inertial reference frame attached to the Earth. The ground appears stationary, and the platform rotates clockwise. (b) Overhead view observed by someone in an inertial reference frame attached to the platform. The platform appears stationary, and the ground rotates counterclockwise.

Pitfall Prevention 6.2

Centrifugal Force The commonly heard phrase "centrifugal force" is described as a force pulling *outward* on an object moving in a circular path. If you are feeling a "centrifugal force" on a rotating carnival ride, what is the other object with which you are interacting? You cannot identify another object because it is a fictitious force that occurs when you are in a noninertial reference frame.

Coriolis force if you walk along a radial line while a merry-go-round rotates. (One of the authors did so and suffered a separation of the ligaments from his ribs when he fell over.) The Coriolis force due to the rotation of the Earth is responsible for rotations of hurricanes and for large-scale ocean currents.

Quick Quiz 6.3 Consider the passenger in the car making a left turn in Figure 6.10. Which of the following is correct about forces in the horizontal direction if she is making contact with the right-hand door? **(a)** The passenger is in equilibrium between real forces acting to the right and real forces acting to the left. **(b)** The passenger is subject only to real forces acting to the right. **(c)** The passenger is subject only to real forces acting to the left. **(d)** None of those statements is true.

Example 6.7   Fictitious Forces in Linear Motion   AM

A small sphere of mass m hangs by a cord from the ceiling of a boxcar that is accelerating to the right as shown in Figure 6.12. Both the inertial observer on the ground in Figure 6.12a and the noninertial observer on the train in Figure 6.12b agree that the cord makes an angle θ with respect to the vertical. The noninertial observer claims that a force, which we know to be fictitious, causes the observed deviation of the cord from the vertical. How is the magnitude of this force related to the boxcar's acceleration measured by the inertial observer in Figure 6.12a?

SOLUTION

Conceptualize Place yourself in the role of each of the two observers in Figure 6.12. As the inertial observer on the ground, you see the boxcar accelerating and know that the deviation of the cord is due to this acceleration. As the noninertial observer on the boxcar, imagine that you ignore any effects of the car's motion so that you are not aware of its acceleration. Because you are unaware of this acceleration, you claim that a force is pushing sideways on the sphere to cause the deviation of the cord from the vertical. To make the conceptualization more real, try running from rest while holding a hanging object on a string and notice that the string is at an angle to the vertical while you are accelerating, as if a force is pushing the object backward.

▶ **6.7** continued

Figure 6.12 (Example 6.7) A small sphere suspended from the ceiling of a boxcar accelerating to the right is deflected as shown.

Categorize For the inertial observer, we model the sphere as a *particle under a net force* in the horizontal direction and a *particle in equilibrium* in the vertical direction. For the noninertial observer, the sphere is modeled as a *particle in equilibrium* in both directions.

Analyze According to the inertial observer at rest (Fig. 6.12a), the forces on the sphere are the force $\vec{T}$ exerted by the cord and the gravitational force. The inertial observer concludes that the sphere's acceleration is the same as that of the boxcar and that this acceleration is provided by the horizontal component of $\vec{T}$.

For this observer, apply the particle under a net force and particle in equilibrium models:

Inertial observer $\begin{cases} (1) \quad \sum F_x = T\sin\theta = ma \\ (2) \quad \sum F_y = T\cos\theta - mg = 0 \end{cases}$

According to the noninertial observer riding in the car (Fig. 6.12b), the cord also makes an angle θ with the vertical; to that observer, however, the sphere is at rest and so its acceleration is zero. Therefore, the noninertial observer introduces a force (which we know to be fictitious) in the horizontal direction to balance the horizontal component of $\vec{T}$ and claims that the net force on the sphere is zero.

Apply the particle in equilibrium model for this observer in both directions:

Noninertial observer $\begin{cases} \sum F'_x = T\sin\theta - F_{fictitious} = 0 \\ \sum F'_y = T\cos\theta - mg = 0 \end{cases}$

These expressions are equivalent to Equations (1) and (2) if $F_{fictitious} = ma$, where a is the acceleration according to the inertial observer.

Finalize If we make this substitution in the equation for $\sum F'_x$ above, we obtain the same mathematical results as the inertial observer. The physical interpretation of the cord's deflection, however, differs in the two frames of reference.

WHAT IF? Suppose the inertial observer wants to measure the acceleration of the train by means of the pendulum (the sphere hanging from the cord). How could she do so?

Answer Our intuition tells us that the angle θ the cord makes with the vertical should increase as the acceleration increases. By solving Equations (1) and (2) simultaneously for a, we find that $a = g\tan\theta$. Therefore, the inertial observer can determine the magnitude of the car's acceleration by measuring the angle θ and using that relationship. Because the deflection of the cord from the vertical serves as a measure of acceleration, *a simple pendulum can be used as an accelerometer.*

6.4 Motion in the Presence of Resistive Forces

In Chapter 5, we described the force of kinetic friction exerted on an object moving on some surface. We completely ignored any interaction between the object and the medium through which it moves. Now consider the effect of that medium, which can be either a liquid or a gas. The medium exerts a **resistive force** $\vec{\textbf{R}}$ on the object moving through it. Some examples are the air resistance associated with moving vehicles (sometimes called *air drag*) and the viscous forces that act on objects moving through a liquid. The magnitude of $\vec{\textbf{R}}$ depends on factors such as the speed of the object, and the direction of $\vec{\textbf{R}}$ is always opposite the direction of the object's motion relative to the medium. This direction may or may not be in the direction opposite the object's velocity according to the observer. For example, if a marble is dropped into a bottle of shampoo, the marble moves downward and the resistive force is upward, resisting the falling of the marble. In contrast, imagine the moment at which there is no wind and you are looking at a flag hanging limply on a flagpole. When a breeze begins to blow toward the right, the flag moves toward the right. In this case, the drag force on the flag from the moving air is to the right and the motion of the flag in response is also to the right, the *same* direction as the drag force. Because the air moves toward the right with respect to the flag, the flag moves to the left relative to the air. Therefore, the direction of the drag force is indeed opposite to the direction of the motion of the flag with respect to the air!

The magnitude of the resistive force can depend on speed in a complex way, and here we consider only two simplified models. In the first model, we assume the resistive force is proportional to the velocity of the moving object; this model is valid for objects falling slowly through a liquid and for very small objects, such as dust particles, moving through air. In the second model, we assume a resistive force that is proportional to the square of the speed of the moving object; large objects, such as skydivers moving through air in free fall, experience such a force.

Model 1: Resistive Force Proportional to Object Velocity

If we model the resistive force acting on an object moving through a liquid or gas as proportional to the object's velocity, the resistive force can be expressed as

$$\vec{\textbf{R}} = -b\vec{\textbf{v}} \tag{6.2}$$

where b is a constant whose value depends on the properties of the medium and on the shape and dimensions of the object and $\vec{\textbf{v}}$ is the velocity of the object relative to the medium. The negative sign indicates that $\vec{\textbf{R}}$ is in the opposite direction to $\vec{\textbf{v}}$.

Consider a small sphere of mass m released from rest in a liquid as in Figure 6.13a. Assuming the only forces acting on the sphere are the resistive force $\vec{\textbf{R}} = -b\vec{\textbf{v}}$ and the gravitational force $\vec{\textbf{F}}_g$, let us describe its motion.[1] We model the sphere as a par-

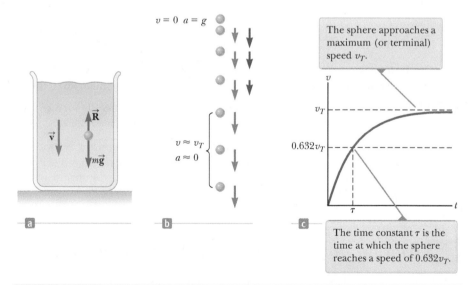

Figure 6.13 (a) A small sphere falling through a liquid. (b) A motion diagram of the sphere as it falls. Velocity vectors (red) and acceleration vectors (violet) are shown for each image after the first one. (c) A speed–time graph for the sphere.

---

[1]A *buoyant force* is also acting on the submerged object. This force is constant, and its magnitude is equal to the weight of the displaced liquid. This force can be modeled by changing the apparent weight of the sphere by a constant factor, so we will ignore the force here. We will discuss buoyant forces in Chapter 14.

ticle under a net force. Applying Newton's second law to the vertical motion of the sphere and choosing the downward direction to be positive, we obtain

$$\sum F_y = ma \rightarrow mg - bv = ma \qquad (6.3)$$

where the acceleration of the sphere is downward. Noting that the acceleration a is equal to dv/dt gives

$$\frac{dv}{dt} = g - \frac{b}{m}v \qquad (6.4)$$

This equation is called a *differential equation*, and the methods of solving it may not be familiar to you as yet. Notice, however, that initially when $v = 0$, the magnitude of the resistive force is also zero and the acceleration of the sphere is simply g. As t increases, the magnitude of the resistive force increases and the acceleration decreases. The acceleration approaches zero when the magnitude of the resistive force approaches the sphere's weight so that the net force on the sphere is zero. In this situation, the speed of the sphere approaches its **terminal speed** v_T.

◀ **Terminal speed**

The terminal speed is obtained from Equation 6.4 by setting $dv/dt = 0$, which gives

$$mg - bv_T = 0 \quad \text{or} \quad v_T = \frac{mg}{b} \qquad (6.5)$$

Because you may not be familiar with differential equations yet, we won't show the details of the process that gives the expression for v for all times t. If $v = 0$ at $t = 0$, this expression is

$$v = \frac{mg}{b}(1 - e^{-bt/m}) = v_T(1 - e^{-t/\tau}) \qquad (6.6)$$

This function is plotted in Figure 6.13c. The symbol e represents the base of the natural logarithm and is also called *Euler's number*: $e = 2.718\ 28$. The **time constant** $\tau = m/b$ (Greek letter tau) is the time at which the sphere released from rest at $t = 0$ reaches 63.2% of its terminal speed; when $t = \tau$, Equation 6.6 yields $v = 0.632v_T$. (The number 0.632 is $1 - e^{-1}$.)

We can check that Equation 6.6 is a solution to Equation 6.4 by direct differentiation:

$$\frac{dv}{dt} = \frac{d}{dt}\left[\frac{mg}{b}(1 - e^{-bt/m})\right] = \frac{mg}{b}\left(0 + \frac{b}{m}e^{-bt/m}\right) = ge^{-bt/m}$$

(See Appendix Table B.4 for the derivative of e raised to some power.) Substituting into Equation 6.4 both this expression for dv/dt and the expression for v given by Equation 6.6 shows that our solution satisfies the differential equation.

Example 6.8   Sphere Falling in Oil   AM

A small sphere of mass 2.00 g is released from rest in a large vessel filled with oil, where it experiences a resistive force proportional to its speed. The sphere reaches a terminal speed of 5.00 cm/s. Determine the time constant τ and the time at which the sphere reaches 90.0% of its terminal speed.

SOLUTION

Conceptualize With the help of Figure 6.13, imagine dropping the sphere into the oil and watching it sink to the bottom of the vessel. If you have some thick shampoo in a clear container, drop a marble in it and observe the motion of the marble.

Categorize We model the sphere as a *particle under a net force*, with one of the forces being a resistive force that depends on the speed of the sphere. This model leads to the result in Equation 6.5.

....

Analyze From Equation 6.5, evaluate the coefficient b: $b = \dfrac{mg}{v_T}$

continued

▶ **6.8** continued

Evaluate the time constant τ:

$$\tau = \frac{m}{b} = m\left(\frac{v_\tau}{mg}\right) = \frac{v_\tau}{g}$$

Substitute numerical values:

$$\tau = \frac{5.00 \text{ cm/s}}{980 \text{ cm/s}^2} = 5.10 \times 10^{-3} \text{ s}$$

Find the time t at which the sphere reaches a speed of $0.900v_T$ by setting $v = 0.900v_T$ in Equation 6.6 and solving for t:

$$0.900v_T = v_T(1 - e^{-t/\tau})$$

$$1 - e^{-t/\tau} = 0.900$$

$$e^{-t/\tau} = 0.100$$

$$-\frac{t}{\tau} = \ln(0.100) = -2.30$$

$$t = 2.30\tau = 2.30(5.10 \times 10^{-3} \text{ s}) = 11.7 \times 10^{-3} \text{ s}$$

$$= \boxed{11.7 \text{ ms}}$$

Finalize The sphere reaches 90.0% of its terminal speed in a very short time interval. You should have also seen this behavior if you performed the activity with the marble and the shampoo. Because of the short time interval required to reach terminal velocity, you may not have noticed the time interval at all. The marble may have appeared to immediately begin moving through the shampoo at a constant velocity.

Model 2: Resistive Force Proportional to Object Speed Squared

For objects moving at high speeds through air, such as airplanes, skydivers, cars, and baseballs, the resistive force is reasonably well modeled as proportional to the square of the speed. In these situations, the magnitude of the resistive force can be expressed as

$$R = \tfrac{1}{2}D\rho Av^2 \tag{6.7}$$

where D is a dimensionless empirical quantity called the *drag coefficient*, ρ is the density of air, and A is the cross-sectional area of the moving object measured in a plane perpendicular to its velocity. The drag coefficient has a value of about 0.5 for spherical objects but can have a value as great as 2 for irregularly shaped objects.

Let us analyze the motion of a falling object subject to an upward air resistive force of magnitude $R = \tfrac{1}{2}D\rho Av^2$. Suppose an object of mass m is released from rest. As Figure 6.14 shows, the object experiences two external forces:[2] the downward gravitational force $\vec{F}_g = m\vec{g}$ and the upward resistive force $\vec{R}$. Hence, the magnitude of the net force is

$$\sum F = mg - \tfrac{1}{2}D\rho Av^2 \tag{6.8}$$

where we have taken downward to be the positive vertical direction. Modeling the object as a particle under a net force, with the net force given by Equation 6.8, we find that the object has a downward acceleration of magnitude

$$a = g - \left(\frac{D\rho A}{2m}\right)v^2 \tag{6.9}$$

We can calculate the terminal speed v_T by noticing that when the gravitational force is balanced by the resistive force, the net force on the object is zero and therefore its acceleration is zero. Setting $a = 0$ in Equation 6.9 gives

$$g - \left(\frac{D\rho A}{2m}\right)v_T^2 = 0$$

Figure 6.14 (a) An object falling through air experiences a resistive force $\vec{R}$ and a gravitational force $\vec{F}_g = m\vec{g}$. (b) The object reaches terminal speed when the net force acting on it is zero, that is, when $\vec{R} = -\vec{F}_g$ or $R = mg$.

---

[2]As with Model 1, there is also an upward buoyant force that we neglect.

	Mass	Cross-Sectional Area	v_T
Table 6.1 **Terminal Speed for Various Objects Falling Through Air**			
Object	(kg)	(m²)	(m/s)
Skydiver	75	0.70	60
Baseball (radius 3.7 cm)	0.145	4.2×10^{-3}	43
Golf ball (radius 2.1 cm)	0.046	1.4×10^{-3}	44
Hailstone (radius 0.50 cm)	4.8×10^{-4}	7.9×10^{-5}	14
Raindrop (radius 0.20 cm)	3.4×10^{-5}	1.3×10^{-5}	9.0

so

$$v_T = \sqrt{\frac{2mg}{D\rho A}} \qquad\qquad (6.10)$$

Table 6.1 lists the terminal speeds for several objects falling through air.

Quick Quiz 6.4 A baseball and a basketball, having the same mass, are dropped through air from rest such that their bottoms are initially at the same height above the ground, on the order of 1 m or more. Which one strikes the ground first? **(a)** The baseball strikes the ground first. **(b)** The basketball strikes the ground first. **(c)** Both strike the ground at the same time.

Conceptual Example 6.9 The Skysurfer

Consider a skysurfer (Fig. 6.15) who jumps from a plane with his feet attached firmly to his surfboard, does some tricks, and then opens his parachute. Describe the forces acting on him during these maneuvers.

SOLUTION

When the surfer first steps out of the plane, he has no vertical velocity. The downward gravitational force causes him to accelerate toward the ground. As his downward speed increases, so does the upward resistive force exerted by the air on his body and the board. This upward force reduces their acceleration, and so their speed increases more slowly. Eventually, they are going so fast that the upward resistive force matches the downward gravitational force. Now the net force is zero and they no longer accelerate, but instead reach their terminal speed. At some point after reaching terminal speed, he opens his parachute, resulting in a drastic increase in the upward resistive force. The net force (and therefore the acceleration) is now upward, in the direction opposite the direction of the velocity. The downward velocity therefore decreases rapidly, and the resistive force on the parachute also decreases. Eventually, the upward resistive force and the downward gravitational force balance each other again and a much smaller terminal speed is reached, permitting a safe landing.

Oliver Furrer/Jupiter Images

Figure 6.15 (Conceptual Example 6.9) A skysurfer.

(Contrary to popular belief, the velocity vector of a skydiver never points upward. You may have seen a video in which a skydiver appears to "rocket" upward once the parachute opens. In fact, what happens is that the skydiver slows down but the person holding the camera continues falling at high speed.)

Example 6.10 Falling Coffee Filters

The dependence of resistive force on the square of the speed is a simplification model. Let's test the model for a specific situation. Imagine an experiment in which we drop a series of bowl-shaped, pleated coffee filters and measure their terminal speeds. Table 6.2 on page 166 presents typical terminal speed data from a real experiment using these coffee filters as

continued

▶ 6.10 continued

they fall through the air. The time constant τ is small, so a dropped filter quickly reaches terminal speed. Each filter has a mass of 1.64 g. When the filters are nested together, they combine in such a way that the front-facing surface area does not increase. Determine the relationship between the resistive force exerted by the air and the speed of the falling filters.

Conceptualize Imagine dropping the coffee filters through the air. (If you have some coffee filters, try dropping them.) Because of the relatively small mass of the coffee filter, you probably won't notice the time interval during which there is an acceleration. The filters will appear to fall at constant velocity immediately upon leaving your hand.

Categorize Because a filter moves at constant velocity, we model it as a *particle in equilibrium*.

Analyze At terminal speed, the upward resistive force on the filter balances the downward gravitational force so that $R = mg$.

Evaluate the magnitude of the resistive force:

$$R = mg = (1.64 \text{ g})\left(\frac{1 \text{ kg}}{1\,000 \text{ g}}\right)(9.80 \text{ m/s}^2) = 0.016\,1 \text{ N}$$

Likewise, two filters nested together experience 0.032 2 N of resistive force, and so forth. These values of resistive force are shown in the far right column of Table 6.2. A graph of the resistive force on the filters as a function of terminal speed is shown in Figure 6.16a. A straight line is not a good fit, indicating that the resistive force is *not* proportional to the speed. The behavior is more clearly seen in Figure 6.16b, in which the resistive force is plotted as a function of the square of the terminal speed. This graph indicates that the resistive force is proportional to the *square* of the speed as suggested by Equation 6.7.

Finalize Here is a good opportunity for you to take some actual data at home on real coffee filters and see if you can reproduce the results shown in Figure 6.16. If you have shampoo and a marble as mentioned in Example 6.8, take data on that system too and see if the resistive force is appropriately modeled as being proportional to the speed.

Table 6.2   **Terminal Speed and Resistive Force for Nested Coffee Filters**

Number of Filters	v_T (m/s)[a]	R (N)
1	1.01	0.016 1
2	1.40	0.032 2
3	1.63	0.048 3
4	2.00	0.064 4
5	2.25	0.080 5
6	2.40	0.096 6
7	2.57	0.112 7
8	2.80	0.128 8
9	3.05	0.144 9
10	3.22	0.161 0

[a]All values of v_T are approximate.

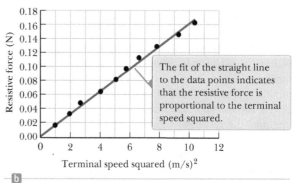

The data points do not lie along a straight line, but instead suggest a curve.

The fit of the straight line to the data points indicates that the resistive force is proportional to the terminal speed squared.

Figure 6.16 (Example 6.10) (a) Relationship between the resistive force acting on falling coffee filters and their terminal speed. (b) Graph relating the resistive force to the square of the terminal speed.

Example 6.11   **Resistive Force Exerted on a Baseball**   AM

A pitcher hurls a 0.145-kg baseball past a batter at 40.2 m/s (= 90 mi/h). Find the resistive force acting on the ball at this speed.

▶ **6.11** continued

SOLUTION

Conceptualize This example is different from the previous ones in that the object is now moving horizontally through the air instead of moving vertically under the influence of gravity and the resistive force. The resistive force causes the ball to slow down, and gravity causes its trajectory to curve downward. We simplify the situation by assuming the velocity vector is exactly horizontal at the instant it is traveling at 40.2 m/s.

Categorize In general, the ball is a *particle under a net force.* Because we are considering only one instant of time, however, we are not concerned about acceleration, so the problem involves only finding the value of one of the forces.

Analyze To determine the drag coefficient D, imagine that we drop the baseball and allow it to reach terminal speed. Solve Equation 6.10 for D:

$$D = \frac{2mg}{v_T^2 \rho A}$$

Use this expression for D in Equation 6.7 to find an expression for the magnitude of the resistive force:

$$R = \tfrac{1}{2} D \rho A v^2 = \frac{1}{2} \left(\frac{2mg}{v_T^2 \rho A} \right) \rho A v^2 = mg \left(\frac{v}{v_T} \right)^2$$

Substitute numerical values, using the terminal speed from Table 6.1:

$$R = (0.145 \text{ kg})(9.80 \text{ m/s}^2) \left(\frac{40.2 \text{ m/s}}{43 \text{ m/s}} \right)^2 = 1.2 \text{ N}$$

Finalize The magnitude of the resistive force is similar in magnitude to the weight of the baseball, which is about 1.4 N. Therefore, air resistance plays a major role in the motion of the ball, as evidenced by the variety of curve balls, floaters, sinkers, and the like thrown by baseball pitchers.

Summary

Concepts and Principles

A particle moving in uniform circular motion has a centripetal acceleration; this acceleration must be provided by a net force directed toward the center of the circular path.

An observer in a noninertial (accelerating) frame of reference introduces **fictitious forces** when applying Newton's second law in that frame.

An object moving through a liquid or gas experiences a speed-dependent **resistive force.** This resistive force is in a direction opposite that of the velocity of the object relative to the medium and generally increases with speed. The magnitude of the resistive force depends on the object's size and shape and on the properties of the medium through which the object is moving. In the limiting case for a falling object, when the magnitude of the resistive force equals the object's weight, the object reaches its **terminal speed.**

Analysis Model for Problem-Solving

Particle in Uniform Circular Motion (Extension) With our new knowledge of forces, we can extend the model of a particle in uniform circular motion, first introduced in Chapter 4. Newton's second law applied to a particle moving in uniform circular motion states that the net force causing the particle to undergo a centripetal acceleration (Eq. 4.14) is related to the acceleration according to

$$\sum F = ma_c = m \frac{v^2}{r} \qquad \text{(6.1)}$$

Objective Questions   **1.** denotes answer available in *Student Solutions Manual/Study Guide*

1. A child is practicing for a BMX race. His speed remains constant as he goes counterclockwise around a level track with two straight sections and two nearly semicircular sections as shown in the aerial view of Figure OQ6.1. (a) Rank the magnitudes of his acceleration at the points A, B, C, D, and E from largest to smallest. If his acceleration is the same size at two points, display that fact in your ranking. If his acceleration is zero, display that fact. (b) What are the directions of his velocity at points A, B, and C? For each point, choose one: north, south, east, west, or nonexistent. (c) What are the directions of his acceleration at points A, B, and C?

Figure OQ6.1

2. Consider a skydiver who has stepped from a helicopter and is falling through air. Before she reaches terminal speed and long before she opens her parachute, does her speed (a) increase, (b) decrease, or (c) stay constant?

3. A door in a hospital has a pneumatic closer that pulls the door shut such that the doorknob moves with constant speed over most of its path. In this part of its motion, (a) does the doorknob experience a centripetal acceleration? (b) Does it experience a tangential acceleration?

4. A pendulum consists of a small object called a bob hanging from a light cord of fixed length, with the top end of the cord fixed, as represented in Figure OQ6.4. The bob moves without friction, swinging equally high on both sides. It moves from its turning point A through point B and reaches its maximum speed at point C. (a) Of these points, is there a point where the bob has nonzero radial acceleration and zero tangential acceleration? If so, which point? What is the direction of its total acceleration at this point? (b) Of these points, is there a point where the bob has nonzero tangential acceleration and zero radial acceleration? If so, which point? What is the direction of its total acceleration at this point? (c) Is there a point where the bob has no acceleration? If so, which point? (d) Is there a point where the bob has both nonzero tangential and radial acceleration? If so, which point? What is the direction of its total acceleration at this point?

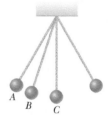

Figure OQ6.4

5. As a raindrop falls through the atmosphere, its speed initially changes as it falls toward the Earth. Before the raindrop reaches its terminal speed, does the magnitude of its acceleration (a) increase, (b) decrease, (c) stay constant at zero, (d) stay constant at 9.80 m/s^2, or (e) stay constant at some other value?

6. An office door is given a sharp push and swings open against a pneumatic device that slows the door down and then reverses its motion. At the moment the door is open the widest, (a) does the doorknob have a centripetal acceleration? (b) Does it have a tangential acceleration?

7. Before takeoff on an airplane, an inquisitive student on the plane dangles an iPod by its earphone wire. It hangs straight down as the plane is at rest waiting to take off. The plane then gains speed rapidly as it moves down the runway. **(i)** Relative to the student's hand, does the iPod (a) shift toward the front of the plane, (b) continue to hang straight down, or (c) shift toward the back of the plane? **(ii)** The speed of the plane increases at a constant rate over a time interval of several seconds. During this interval, does the angle the earphone wire makes with the vertical (a) increase, (b) stay constant, or (c) decrease?

Conceptual Questions   **1.** denotes answer available in *Student Solutions Manual/Study Guide*

1. What forces cause (a) an automobile, (b) a propeller-driven airplane, and (c) a rowboat to move?

2. A falling skydiver reaches terminal speed with her parachute closed. After the parachute is opened, what parameters change to decrease this terminal speed?

3. An object executes circular motion with constant speed whenever a net force of constant magnitude acts perpendicular to the velocity. What happens to the speed if the force is not perpendicular to the velocity?

4. Describe the path of a moving body in the event that (a) its acceleration is constant in magnitude at all times and perpendicular to the velocity, and (b) its acceleration is constant in magnitude at all times and parallel to the velocity.

5. The observer in the accelerating elevator of Example 5.8 would claim that the "weight" of the fish is T, the scale reading, but this answer is obviously wrong. Why does this observation differ from that of a person outside the elevator, at rest with respect to the Earth?

6. If someone told you that astronauts are weightless in orbit because they are beyond the pull of gravity, would you accept the statement? Explain.

7. It has been suggested that rotating cylinders about 20 km in length and 8 km in diameter be placed in space and used as colonies. The purpose of the rotation is to simulate gravity for the inhabitants. Explain this concept for producing an effective imitation of gravity.

8. Consider a small raindrop and a large raindrop falling through the atmosphere. (a) Compare their terminal speeds. (b) What are their accelerations when they reach terminal speed?

9. Why does a pilot tend to black out when pulling out of a steep dive?

10. A pail of water can be whirled in a vertical path such that no water is spilled. Why does the water stay in the pail, even when the pail is above your head?

11. "If the current position and velocity of every particle in the Universe were known, together with the laws describing the forces that particles exert on one another, the whole future of the Universe could be calculated. The future is determinate and preordained. Free will is an illusion." Do you agree with this thesis? Argue for or against it.

Problems

 WebAssign The problems found in this chapter may be assigned online in Enhanced WebAssign

1. straightforward; **2.** intermediate; **3.** challenging

1. full solution available in the *Student Solutions Manual/Study Guide*

AMT Analysis Model tutorial available in Enhanced WebAssign

GP Guided Problem

M Master It tutorial available in Enhanced WebAssign

W Watch It video solution available in Enhanced WebAssign

Section 6.1 Extending the Particle in Uniform Circular Motion Model

1. A light string can support a stationary hanging load of 25.0 kg before breaking. An object of mass $m = 3.00$ kg attached to the string rotates on a frictionless, horizontal table in a circle of radius $r = 0.800$ m, and the other end of the string is held fixed as in Figure P6.1. What range of speeds can the object have before the string breaks?

Figure P6.1

2. Whenever two *Apollo* astronauts were on the surface of the Moon, a third astronaut orbited the Moon. Assume the orbit to be circular and 100 km above the surface of the Moon, where the acceleration due to gravity is 1.52 m/s². The radius of the Moon is 1.70×10^6 m. Determine (a) the astronaut's orbital speed and (b) the period of the orbit.

3. In the Bohr model of the hydrogen atom, an electron moves in a circular path around a proton. The speed of the electron is approximately 2.20×10^6 m/s. Find (a) the force acting on the electron as it revolves in a circular orbit of radius 0.529×10^{-10} m and (b) the centripetal acceleration of the electron.

4. A curve in a road forms part of a horizontal circle. As a car goes around it at constant speed 14.0 m/s, the total horizontal force on the driver has magnitude 130 N. What is the total horizontal force on the driver if the speed on the same curve is 18.0 m/s instead?

5. In a cyclotron (one type of particle accelerator), a deuteron (of mass 2.00 u) reaches a final speed of 10.0% of the speed of light while moving in a circular path of radius 0.480 m. What magnitude of magnetic force is required to maintain the deuteron in a circular path?

6. A car initially traveling eastward turns north by traveling in a circular path at uniform speed as shown in Figure P6.6. The length of the arc *ABC* is 235 m, and the car completes the turn in 36.0 s. (a) What is the acceleration when the car is at *B* located at an angle of 35.0°? Express your answer in terms of the unit vectors $\hat{\mathbf{i}}$ and $\hat{\mathbf{j}}$. Determine (b) the car's average speed and (c) its average acceleration during the 36.0-s interval.

Figure P6.6

7. A space station, in the form of a wheel 120 m in diameter, rotates to provide an "artificial gravity" of 3.00 m/s² for persons who walk around on the inner wall of the outer rim. Find the rate of the wheel's rotation in revolutions per minute that will produce this effect.

8. Consider a conical pendulum (Fig. P6.8) with a bob of mass $m = 80.0$ kg on a string of length $L = 10.0$ m that makes an angle of $\theta = 5.00°$ with the vertical. Determine (a) the horizontal and vertical components of the force exerted by the string on the pendulum and (b) the radial acceleration of the bob.

Figure P6.8

9. A coin placed 30.0 cm from the center of a rotating, horizontal turntable slips when its speed is 50.0 cm/s. (a) What force causes the centripetal acceleration when the coin is stationary relative to the turntable? (b) What is the coefficient of static friction between coin and turntable?

10. *Why is the following situation impossible?* The object of mass $m = 4.00$ kg in Figure P6.10 is attached to a vertical rod by two strings of length $\ell = 2.00$ m. The strings are attached to the rod at points a distance $d = 3.00$ m apart. The object rotates in a horizontal circle at a constant speed of $v = 3.00$ m/s, and the strings remain taut. The rod rotates along with the object so that the strings do not wrap onto the rod. **What If?** Could this situation be possible on another planet?

Figure P6.10

11. A crate of eggs is located in the middle of the flatbed of a pickup truck as the truck negotiates a curve in the flat road. The curve may be regarded as an arc of a circle of radius 35.0 m. If the coefficient of static friction between crate and truck is 0.600, how fast can the truck be moving without the crate sliding?

Section 6.2 Nonuniform Circular Motion

12. A pail of water is rotated in a vertical circle of radius 1.00 m. (a) What two external forces act on the water in the pail? (b) Which of the two forces is most important in causing the water to move in a circle? (c) What is the pail's minimum speed at the top of the circle if no water is to spill out? (d) Assume the pail with the speed found in part (c) were to suddenly disappear at the top of the circle. Describe the subsequent motion of the water. Would it differ from the motion of a projectile?

13. A hawk flies in a horizontal arc of radius 12.0 m at constant speed 4.00 m/s. (a) Find its centripetal acceleration. (b) It continues to fly along the same horizontal arc, but increases its speed at the rate of 1.20 m/s². Find the acceleration (magnitude and direction) in this situation at the moment the hawk's speed is 4.00 m/s.

14. A 40.0-kg child swings in a swing supported by two chains, each 3.00 m long. The tension in each chain at the lowest point is 350 N. Find (a) the child's speed at the lowest point and (b) the force exerted by the seat on the child at the lowest point. (Ignore the mass of the seat.)

15. A child of mass m swings in a swing supported by two chains, each of length R. If the tension in each chain at the lowest point is T, find (a) the child's speed at the lowest point and (b) the force exerted by the seat on the child at the lowest point. (Ignore the mass of the seat.)

16. A roller-coaster car (Fig. P6.16) has a mass of 500 kg when fully loaded with passengers. The path of the coaster from its initial point shown in the figure to point Ⓑ involves only up-and-down motion (as seen by the riders), with no motion to the left or right. (a) If the vehicle has a speed of 20.0 m/s at point Ⓐ, what is the force exerted by the track on the car at this point? (b) What is the maximum speed the vehicle can have at point Ⓑ and still remain on the track? Assume the roller-coaster tracks at points Ⓐ and Ⓑ are parts of vertical circles of radius $r_1 = 10.0$ m and $r_2 = 15.0$ m, respectively.

Figure P6.16 Problems 16 and 38.

17. A roller coaster at the Six Flags Great America amusement park in Gurnee, Illinois, incorporates some clever design technology and some basic physics. Each vertical loop, instead of being circular, is shaped like a teardrop (Fig. P6.17). The cars ride on the inside of the loop at the top, and the speeds are fast enough to ensure the cars remain on the track. The biggest loop is 40.0 m high. Suppose the speed at the top of the loop is 13.0 m/s and the corresponding centripetal acceleration of the riders is $2g$. (a) What is the radius of the arc of the teardrop at the top? (b) If the total mass of a car plus the riders is M, what force does the rail exert on the car at the top? (c) Suppose the roller coaster had a circular loop of radius 20.0 m. If the cars have the same speed, 13.0 m/s at the top, what is the centripetal acceleration of the riders at the top? (d) Comment on the normal force at the top in the situation described in part (c) and on the advantages of having teardrop-shaped loops.

Figure P6.17

Frank Cezus/Getty Images

18. One end of a cord is fixed and a small 0.500-kg object is attached to the other end, where it swings in a section of a vertical circle of radius 2.00 m as shown in Figure P6.18. When $\theta = 20.0°$, the speed of the object is 8.00 m/s. At this instant, find (a) the tension in the string, (b) the tangential and radial components of acceleration, and (c) the total acceleration. (d) Is your answer changed if the object is swinging down toward its

Figure P6.18

lowest point instead of swinging up? (e) Explain your answer to part (d).

19. An adventurous archeologist ($m = 85.0$ kg) tries to cross a river by swinging from a vine. The vine is 10.0 m long, and his speed at the bottom of the swing is 8.00 m/s. The archeologist doesn't know that the vine has a breaking strength of 1 000 N. Does he make it across the river without falling in?

Section 6.3 Motion in Accelerated Frames

20. An object of mass $m = $ 5.00 kg, attached to a spring scale, rests on a frictionless, horizontal surface as shown in Figure P6.20. The spring scale, attached to the front end of a boxcar, reads zero when the

Figure P6.20

car is at rest. (a) Determine the acceleration of the car if the spring scale has a constant reading of 18.0 N when the car is in motion. (b) What constant reading will the spring scale show if the car moves with constant velocity? Describe the forces on the object as observed (c) by someone in the car and (d) by someone at rest outside the car.

21. An object of mass $m = $ 0.500 kg is suspended from the ceiling of an accelerating truck as shown in Figure P6.21. Taking $a = 3.00$ m/s^2, find (a) the angle θ that the string makes with the vertical and (b) the tension T in the string.

Figure P6.21

22. A child lying on her back experiences 55.0 N tension in the muscles on both sides of her neck when she raises her head to look past her toes. Later, sliding feet first down a water slide at terminal speed 5.70 m/s and riding high on the outside wall of a horizontal curve of radius 2.40 m, she raises her head again to look forward past her toes. Find the tension in the muscles on both sides of her neck while she is sliding.

23. A person stands on a scale in an elevator. As the elevator starts, the scale has a constant reading of 591 N. As the elevator later stops, the scale reading is 391 N. Assuming the magnitude of the acceleration is the same during starting and stopping, determine (a) the weight of the person, (b) the person's mass, and (c) the acceleration of the elevator.

24. **Review.** A student, along with her backpack on the floor next to her, are in an elevator that is accelerating upward with acceleration a. The student gives her backpack a quick kick at $t = 0$, imparting to it speed v and causing it to slide across the elevator floor. At time t, the backpack hits the opposite wall a distance L away from the student. Find the coefficient

of kinetic friction μ_k between the backpack and the elevator floor.

25. A small container of water is placed on a turntable inside a microwave oven, at a radius of 12.0 cm from the center. The turntable rotates steadily, turning one revolution in each 7.25 s. What angle does the water surface make with the horizontal?

Section 6.4 Motion in the Presence of Resistive Forces

26. **Review.** (a) Estimate the terminal speed of a wooden sphere (density 0.830 g/cm^3) falling through air, taking its radius as 8.00 cm and its drag coefficient as 0.500. (b) From what height would a freely falling object reach this speed in the absence of air resistance?

27. The mass of a sports car is 1 200 kg. The shape of the body is such that the aerodynamic drag coefficient is 0.250 and the frontal area is 2.20 m^2. Ignoring all other sources of friction, calculate the initial acceleration the car has if it has been traveling at 100 km/h and is now shifted into neutral and allowed to coast.

28. A skydiver of mass 80.0 kg jumps from a slow-moving aircraft and reaches a terminal speed of 50.0 m/s. (a) What is her acceleration when her speed is 30.0 m/s? What is the drag force on the skydiver when her speed is (b) 50.0 m/s and (c) 30.0 m/s?

29. Calculate the force required to pull a copper ball of radius 2.00 cm upward through a fluid at the constant speed 9.00 cm/s. Take the drag force to be proportional to the speed, with proportionality constant 0.950 kg/s. Ignore the buoyant force.

30. A small piece of Styrofoam packing material is dropped from a height of 2.00 m above the ground. Until it reaches terminal speed, the magnitude of its acceleration is given by $a = g - Bv$. After falling 0.500 m, the Styrofoam effectively reaches terminal speed and then takes 5.00 s more to reach the ground. (a) What is the value of the constant B? (b) What is the acceleration at $t = 0$? (c) What is the acceleration when the speed is 0.150 m/s?

31. A small, spherical bead of mass 3.00 g is released from rest at $t = 0$ from a point under the surface of a viscous liquid. The terminal speed is observed to be $v_T = 2.00$ cm/s. Find (a) the value of the constant b that appears in Equation 6.2, (b) the time t at which the bead reaches $0.632v_T$, and (c) the value of the resistive force when the bead reaches terminal speed.

32. At major league baseball games, it is commonplace to flash on the scoreboard a speed for each pitch. This speed is determined with a radar gun aimed by an operator positioned behind home plate. The gun uses the Doppler shift of microwaves reflected from the baseball, an effect we will study in Chapter 39. The gun determines the speed at some particular point on the baseball's path, depending on when the operator pulls the trigger. Because the ball is subject to a drag force due to air proportional to the square of its speed given by $R = kmv^2$, it slows as it travels 18.3 m toward the

plate according to the formula $v = v_i e^{-kx}$. Suppose the ball leaves the pitcher's hand at 90.0 mi/h = 40.2 m/s. Ignore its vertical motion. Use the calculation of R for baseballs from Example 6.11 to determine the speed of the pitch when the ball crosses the plate.

33. Assume the resistive force acting on a speed skater is proportional to the square of the skater's speed v and is given by $f = -kmv^2$, where k is a constant and m is the skater's mass. The skater crosses the finish line of a straight-line race with speed v_i and then slows down by coasting on his skates. Show that the skater's speed at any time t after crossing the finish line is $v(t) = v_i/(1 + ktv_i)$.

34. **Review.** A window washer pulls a rubber squeegee
AMT down a very tall vertical window. The squeegee has mass 160 g and is mounted on the end of a light rod. The coefficient of kinetic friction between the squeegee and the dry glass is 0.900. The window washer presses it against the window with a force having a horizontal component of 4.00 N. (a) If she pulls the squeegee down the window at constant velocity, what vertical force component must she exert? (b) The window washer increases the downward force component by 25.0%, while all other forces remain the same. Find the squeegee's acceleration in this situation. (c) The squeegee is moved into a wet portion of the window, where its motion is resisted by a fluid drag force R proportional to its velocity according to $R = -20.0v$, where R is in newtons and v is in meters per second. Find the terminal velocity that the squeegee approaches, assuming the window washer exerts the same force described in part (b).

35. A motorboat cuts its engine when its speed is 10.0 m/s and then coasts to rest. The equation describing the motion of the motorboat during this period is $v = v_i e^{-ct}$, where v is the speed at time t, v_i is the initial speed at $t = 0$, and c is a constant. At $t = 20.0$ s, the speed is 5.00 m/s. (a) Find the constant c. (b) What is the speed at $t = 40.0$ s? (c) Differentiate the expression for $v(t)$ and thus show that the acceleration of the boat is proportional to the speed at any time.

36. You can feel a force of air drag on your hand if you stretch your arm out of the open window of a speeding car. *Note:* Do not endanger yourself. What is the order of magnitude of this force? In your solution, state the quantities you measure or estimate and their values.

Additional Problems

37. A car travels clockwise at constant speed around a circular section of a horizontal road as shown in the aerial view of Figure P6.37. Find the directions of its velocity and acceleration at (a) position Ⓐ and (b) position Ⓑ.

38. The mass of a roller-coaster car, including its passengers, is 500 kg. Its speed at the bottom of the track in Figure P6.16 is 19 m/s. The radius of this section of the track is

Figure P6.37

$r_1 = 25$ m. Find the force that a seat in the roller-coaster car exerts on a 50-kg passenger at the lowest point.

39. A string under a tension of 50.0 N is used to whirl a rock in a horizontal circle of radius 2.50 m at a speed of 20.4 m/s on a frictionless surface as shown in Figure P6.39. As the string is pulled in, the speed of the rock increases. When the string on the table is 1.00 m long and the speed of the rock is 51.0 m/s, the string breaks. What is the breaking strength, in newtons, of the string?

Figure P6.39

40. Disturbed by speeding cars outside his workplace, Nobel laureate Arthur Holly Compton designed a speed bump (called the "Holly hump") and had it installed. Suppose a 1 800-kg car passes over a hump in a roadway that follows the arc of a circle of radius 20.4 m as shown in Figure P6.40. (a) If the car travels at 30.0 km/h, what force does the road exert on the car as the car passes the highest point of the hump? (b) **What If?** What is the maximum speed the car can have without losing contact with the road as it passes this highest point?

Figure P6.40
Problems 40 and 41.

41. A car of mass m passes over a hump in a road that follows the arc of a circle of radius R as shown in Figure P6.40. (a) If the car travels at a speed v, what force does the road exert on the car as the car passes the highest point of the hump? (b) **What If?** What is the maximum speed the car can have without losing contact with the road as it passes this highest point?

42. A child's toy consists of a small wedge that has an acute angle θ (Fig. P6.42). The sloping side of the wedge is frictionless, and an object of mass m on it remains at constant height if the wedge is spun at a certain constant speed. The wedge is spun by rotating, as an axis, a vertical rod that is firmly attached to the wedge at the bottom end. Show that, when the object sits at rest at a point at distance L up along the wedge, the speed of the object must be $v = (gL \sin \theta)^{1/2}$.

Figure P6.42

43. A seaplane of total mass m lands on a lake with initial speed $v_i \hat{\mathbf{i}}$. The only horizontal force on it is a resistive force on its pontoons from the water. The resistive force is proportional to the velocity of the seaplane: $\vec{\mathbf{R}} = -b\vec{\mathbf{v}}$. Newton's second law applied to the plane is $-bv\hat{\mathbf{i}} = m(dv/dt)\hat{\mathbf{i}}$. From the fundamental theorem

of calculus, this differential equation implies that the speed changes according to

$$\int_{v_i}^{v} \frac{dv}{v} = -\frac{b}{m}\int_{0}^{t} dt$$

(a) Carry out the integration to determine the speed of the seaplane as a function of time. (b) Sketch a graph of the speed as a function of time. (c) Does the seaplane come to a complete stop after a finite interval of time? (d) Does the seaplane travel a finite distance in stopping?

44. An object of mass $m_1 =$ W 4.00 kg is tied to an object of mass $m_2 =$ 3.00 kg with String 1 of length $\ell = 0.500$ m. The combination is swung in a vertical circular path on a second string, String 2, of length $\ell = 0.500$ m. During the motion, the two strings are collinear at all times as shown in Figure P6.44. At the top of its motion, m_2 is traveling at $v = 4.00$ m/s. (a) What is the tension in String 1 at this instant? (b) What is the tension in String 2 at this instant? (c) Which string will break first if the combination is rotated faster and faster?

Figure P6.44

45. A ball of mass $m = 0.275$ kg swings in a vertical circular path on a string $L = 0.850$ m long as in Figure P6.45. (a) What are the forces acting on the ball at any point on the path? (b) Draw force diagrams for the ball when it is at the bottom of the circle and when it is at the top. (c) If its speed is 5.20 m/s at the top of the circle, what is the tension in the string there? (d) If the string breaks when its tension exceeds 22.5 N, what is the maximum speed the ball can have at the bottom before that happens?

Figure P6.45

46. *Why is the following situation impossible?* A mischievous child goes to an amusement park with his family. On one ride, after a severe scolding from his mother, he slips out of his seat and climbs to the top of the ride's structure, which is shaped like a cone with its axis vertical and its sloped sides making an angle of $\theta = 20.0°$ with the horizontal as shown in Figure P6.46. This part

Figure P6.46

of the structure rotates about the vertical central axis when the ride operates. The child sits on the sloped surface at a point $d = 5.32$ m down the sloped side from the center of the cone and pouts. The coefficient of static friction between the boy and the cone is 0.700. The ride operator does not notice that the child has slipped away from his seat and so continues to operate the ride. As a result, the sitting, pouting boy rotates in a circular path at a speed of 3.75 m/s.

47. (a) A luggage carousel at an airport has the form of a section of a large cone, steadily rotating about its vertical axis. Its metallic surface slopes downward toward the outside, making an angle of 20.0° with the horizontal. A piece of luggage having mass 30.0 kg is placed on the carousel at a position 7.46 m measured horizontally from the axis of rotation. The travel bag goes around once in 38.0 s. Calculate the force of static friction exerted by the carousel on the bag. (b) The drive motor is shifted to turn the carousel at a higher constant rate of rotation, and the piece of luggage is bumped to another position, 7.94 m from the axis of rotation. Now going around once in every 34.0 s, the bag is on the verge of slipping down the sloped surface. Calculate the coefficient of static friction between the bag and the carousel.

48. In a home laundry dryer, a cylindrical tub containing wet clothes is rotated steadily about a horizontal axis as shown in Figure P6.48. So that the clothes will dry uniformly, they are made to tumble. The rate of rotation of the smooth-walled tub is chosen so that a small piece of cloth will lose contact with the tub when the cloth is at an angle of $\theta = 68.0°$ above the horizontal. If the radius of the tub is $r = 0.330$ m, what rate of revolution is needed?

Figure P6.48

49. Interpret the graph in Figure 6.16(b), which describes the results for falling coffee filters discussed in Example 6.10. Proceed as follows. (a) Find the slope of the straight line, including its units. (b) From Equation 6.6, $R = \frac{1}{2}D\rho A v^2$, identify the theoretical slope of a graph of resistive force versus squared speed. (c) Set the experimental and theoretical slopes equal to each other and proceed to calculate the drag coefficient of the filters. Model the cross-sectional area of the filters as that of a circle of radius 10.5 cm and take the density of air to be 1.20 kg/m³. (d) Arbitrarily choose the eighth data point on the graph and find its vertical

separation from the line of best fit. Express this scatter as a percentage. (e) In a short paragraph, state what the graph demonstrates and compare it with the theoretical prediction. You will need to make reference to the quantities plotted on the axes, to the shape of the graph line, to the data points, and to the results of parts (c) and (d).

50. A basin surrounding a drain has the shape of a circular cone opening upward, having everywhere an angle of $35.0°$ with the horizontal. A 25.0-g ice cube is set sliding around the cone without friction in a horizontal circle of radius R. (a) Find the speed the ice cube must have as a function of R. (b) Is any piece of data unnecessary for the solution? Suppose R is made two times larger. (c) Will the required speed increase, decrease, or stay constant? If it changes, by what factor? (d) Will the time required for each revolution increase, decrease, or stay constant? If it changes, by what factor? (e) Do the answers to parts (c) and (d) seem contradictory? Explain.

51. A truck is moving with constant acceleration a up a hill that makes an angle ϕ with the horizontal as in Figure P6.51. A small sphere of mass m is suspended from the ceiling of the truck by a light cord. If the pendulum makes a constant angle θ with the perpendicular to the ceiling, what is a?

Figure P6.51

52. The pilot of an airplane executes a loop-the-loop maneuver in a vertical circle. The speed of the airplane is 300 mi/h at the top of the loop and 450 mi/h at the bottom, and the radius of the circle is 1 200 ft. (a) What is the pilot's apparent weight at the lowest point if his true weight is 160 lb? (b) What is his apparent weight at the highest point? (c) **What If?** Describe how the pilot could experience weightlessness if both the radius and the speed can be varied. *Note:* His apparent weight is equal to the magnitude of the force exerted by the seat on his body.

53. **Review.** While learning to drive, you are in a 1 200-kg car moving at 20.0 m/s across a large, vacant, level parking lot. Suddenly you realize you are heading straight toward the brick sidewall of a large supermarket and are in danger of running into it. The pavement can exert a maximum horizontal force of 7 000 N on the car. (a) Explain why you should expect the force to have a well-defined maximum value. (b) Suppose you apply the brakes and do not turn the steering wheel. Find the minimum distance you must be from the wall to avoid a collision. (c) If you do not brake but instead maintain constant speed and turn the steering wheel, what is the minimum distance you must be from the wall to avoid a collision? (d) Of the two methods in parts (b) and (c), which is better for avoiding a collision? Or should you use both the brakes and the steering wheel, or neither? Explain. (e) Does the conclusion in part (d) depend on the numerical values given in this problem, or is it true in general? Explain.

54. A puck of mass m_1 is tied to a string and allowed to revolve in a circle of radius R on a frictionless, horizontal table. The other end of the string passes through a small hole in the center of the table, and an object of mass m_2 is tied to it (Fig. P6.54). The suspended object

Figure P6.54

remains in equilibrium while the puck on the tabletop revolves. Find symbolic expressions for (a) the tension in the string, (b) the radial force acting on the puck, and (c) the speed of the puck. (d) Qualitatively describe what will happen in the motion of the puck if the value of m_2 is increased by placing a small additional load on the puck. (e) Qualitatively describe what will happen in the motion of the puck if the value of m_2 is instead decreased by removing a part of the hanging load.

55. Because the Earth rotates about its axis, a point on the equator experiences a centripetal acceleration of $0.033\ 7\ \text{m/s}^2$, whereas a point at the poles experiences no centripetal acceleration. If a person at the equator has a mass of 75.0 kg, calculate (a) the gravitational force (true weight) on the person and (b) the normal force (apparent weight) on the person. (c) Which force is greater? Assume the Earth is a uniform sphere and take $g = 9.800\ \text{m/s}^2$.

56. Galileo thought about whether acceleration should be defined as the rate of change of velocity over time or as the rate of change in velocity over distance. He chose the former, so let's use the name "vroomosity" for the rate of change of velocity over distance. For motion of a particle on a straight line with constant acceleration, the equation $v = v_i + at$ gives its velocity v as a function of time. Similarly, for a particle's linear motion with constant vroomosity k, the equation $v = v_i + kx$ gives the velocity as a function of the position x if the particle's speed is v_i at $x = 0$. (a) Find the law describing the total force acting on this object of mass m. (b) Describe an example of such a motion or explain why it is unrealistic. Consider (c) the possibility of k positive and (d) the possibility of k negative.

57. Figure P6.57 shows a photo of a swing ride at an amusement park. The structure consists of a horizontal, rotating, circular platform of diameter D from which seats of mass m are suspended at the end of massless chains of length d. When the system rotates at

Figure P6.57

t (s)	d (ft)	t (s)	d (ft)	t (s)	d (ft)
0	0	7	652	14	1 831
1	16	8	808	15	2 005
2	62	9	971	16	2 179
3	138	10	1 138	17	2 353
4	242	11	1 309	18	2 527
5	366	12	1 483	19	2 701
6	504	13	1 657	20	2 875

constant speed, the chains swing outward and make an angle θ with the vertical. Consider such a ride with the following parameters: $D = 8.00$ m, $d = 2.50$ m, $m = 10.0$ kg, and $\theta = 28.0°$. (a) What is the speed of each seat? (b) Draw a diagram of forces acting on the combination of a seat and a 40.0-kg child and (c) find the tension in the chain.

58. **Review.** A piece of putty is initially located at point A
GP on the rim of a grinding wheel rotating at constant angular speed about a horizontal axis. The putty is dislodged from point A when the diameter through A is horizontal. It then rises vertically and returns to A at the instant the wheel completes one revolution. From this information, we wish to find the speed v of the putty when it leaves the wheel and the force holding it to the wheel. (a) What analysis model is appropriate for the motion of the putty as it rises and falls? (b) Use this model to find a symbolic expression for the time interval between when the putty leaves point A and when it arrives back at A, in terms of v and g. (c) What is the appropriate analysis model to describe point A on the wheel? (d) Find the period of the motion of point A in terms of the tangential speed v and the radius R of the wheel. (e) Set the time interval from part (b) equal to the period from part (d) and solve for the speed v of the putty as it leaves the wheel. (f) If the mass of the putty is m, what is the magnitude of the force that held it to the wheel before it was released?

59. An amusement park ride consists of a large vertical cylinder that spins about its axis fast enough that any person inside is held up against the wall when the floor drops away (Fig. P6.59). The coefficient of static friction between person and wall is μ_s, and the radius of the cylinder is R. (a) Show that the maximum period of revolution necessary to keep the person from falling is $T = (4\pi^2 R \mu_s/g)^{1/2}$. (b) If the rate of revolution of the cylinder is made to be somewhat larger, what happens to the magnitude of each one of the forces acting on the person? What happens in the motion of the person? (c) If the rate of revolution of the cylinder is instead made to be somewhat smaller, what happens to the magnitude of each one of the forces acting on the person? How does the motion of the person change?

Figure P6.59

60. Members of a skydiving club were given the following data to use in planning their jumps. In the table, d is the distance fallen from rest by a skydiver in a "free-fall stable spread position" versus the time of fall t. (a) Convert the distances in feet into meters. (b) Graph d (in meters) versus t. (c) Determine the value of the terminal speed v_T by finding the slope of the straight portion of the curve. Use a least-squares fit to determine this slope.

61. A car rounds a banked curve as discussed in Example 6.4 and shown in Figure 6.5. The radius of curvature of the road is R, the banking angle is θ, and the coefficient of static friction is μ_s. (a) Determine the range of speeds the car can have without slipping up or down the road. (b) Find the minimum value for μ_s such that the minimum speed is zero.

62. In Example 6.5, we investigated the forces a child expe-
W riences on a Ferris wheel. Assume the data in that example applies to this problem. What force (magnitude and direction) does the seat exert on a 40.0-kg child when the child is halfway between top and bottom?

63. A model airplane of mass 0.750 kg flies with a speed of
M 35.0 m/s in a horizontal circle at the end of a 60.0-m-long control wire as shown in Figure P6.63a. The forces exerted on the airplane are shown in Figure P6.63b: the tension in the control wire, the gravitational force, and aerodynamic lift that acts at $\theta = 20.0°$ inward from the vertical. Compute the tension in the wire, assuming it makes a constant angle of $\theta = 20.0°$ with the horizontal.

Figure P6.63

64. A student builds and calibrates an accelerometer and uses it to determine the speed of her car around a certain unbanked highway curve. The accelerometer is a plumb bob with a protractor that she attaches to the roof of her car. A friend riding in the car with the student observes that the plumb bob hangs at an angle of 15.0° from the vertical when the car has a speed of 23.0 m/s. (a) What is the centripetal acceleration of the car rounding the curve? (b) What is the radius of the curve? (c) What is the speed of the car if the plumb bob deflection is 9.00° while rounding the same curve?

Challenge Problems

65. A 9.00-kg object starting from rest falls through a viscous medium and experiences a resistive force given by Equation 6.2. The object reaches one half its terminal speed in 5.54 s. (a) Determine the terminal speed. (b) At what time is the speed of the object three-fourths the terminal speed? (c) How far has the object traveled in the first 5.54 s of motion?

66. For $t < 0$, an object of mass m experiences no force and moves in the positive x direction with a constant speed v_i. Beginning at $t = 0$, when the object passes position $x = 0$, it experiences a net resistive force proportional to the square of its speed: $\vec{\mathbf{F}}_{net} = -mkv^2\hat{\mathbf{i}}$, where k is a constant. The speed of the object after $t = 0$ is given by $v = v_i/(1 + kv_i t)$. (a) Find the position x of the object as a function of time. (b) Find the object's velocity as a function of position.

67. A golfer tees off from a location precisely at $\phi_i = 35.0°$ north latitude. He hits the ball due south, with range 285 m. The ball's initial velocity is at $48.0°$ above the horizontal. Suppose air resistance is negligible for the golf ball. (a) For how long is the ball in flight? The cup is due south

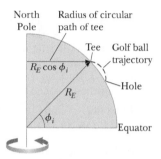

Figure P6.67

of the golfer's location, and the golfer would have a hole-in-one if the Earth were not rotating. The Earth's rotation makes the tee move in a circle of radius $R_E \cos \phi_i = (6.37 \times 10^6 \text{ m}) \cos 35.0°$ as shown in Figure P6.67. The tee completes one revolution each day. (b) Find the eastward speed of the tee relative to the stars. The hole is also moving east, but it is 285 m farther south and thus at a slightly lower latitude ϕ_f. Because the hole moves in a slightly larger circle, its speed must be greater than that of the tee. (c) By how much does the hole's speed exceed that of the tee? During the time interval the ball is in flight, it moves upward and downward as well as southward with the projectile motion you studied in Chapter 4, but it also moves eastward with the speed you found in part (b). The hole moves to the east at a faster speed, however, pulling ahead of the ball with the relative speed

you found in part (c). (d) How far to the west of the hole does the ball land?

68. A single bead can slide with negligible friction on a stiff wire that has been bent into a circular loop of radius 15.0 cm as shown in Figure P6.68. The circle is always in a vertical plane and rotates steadily about its vertical diameter with a period of 0.450 s. The posi-

Figure P6.68

tion of the bead is described by the angle θ that the radial line, from the center of the loop to the bead, makes with the vertical. (a) At what angle up from the bottom of the circle can the bead stay motionless relative to the turning circle? (b) **What If?** Repeat the problem, this time taking the period of the circle's rotation as 0.850 s. (c) Describe how the solution to part (b) is different from the solution to part (a). (d) For any period or loop size, is there always an angle at which the bead can stand still relative to the loop? (e) Are there ever more than two angles? Arnold Arons suggested the idea for this problem.

69. The expression $F = arv + br^2v^2$ gives the magnitude of the resistive force (in newtons) exerted on a sphere of radius r (in meters) by a stream of air moving at speed v (in meters per second), where a and b are constants with appropriate SI units. Their numerical values are $a = 3.10 \times 10^{-4}$ and $b = 0.870$. Using this expression, find the terminal speed for water droplets falling under their own weight in air, taking the following values for the drop radii: (a) 10.0 μm, (b) 100 μm, (c) 1.00 mm. For parts (a) and (c), you can obtain accurate answers without solving a quadratic equation by considering which of the two contributions to the air resistance is dominant and ignoring the lesser contribution.

70. Because of the Earth's rotation, a plumb bob does not hang exactly along a line directed to the center of the Earth. How much does the plumb bob deviate from a radial line at 35.0° north latitude? Assume the Earth is spherical.

Energy of a System

The definitions of quantities such as position, velocity, acceleration, and force and associated principles such as Newton's second law have allowed us to solve a variety of problems. Some problems that could theoretically be solved with Newton's laws, however, are very difficult in practice, but they can be made much simpler with a different approach. Here and in the following chapters, we will investigate this new approach, which will include definitions of quantities that may not be familiar to you. Other quantities may sound familiar, but they may have more specific meanings in physics than in everyday life. We begin this discussion by exploring the notion of *energy*.

The concept of energy is one of the most important topics in science and engineering. In everyday life, we think of energy in terms of fuel for transportation and heating, electricity for lights and appliances, and foods for consumption. These ideas, however, do not truly define energy. They merely tell us that fuels are needed to do a job and that those fuels provide us with something we call energy.

Energy is present in the Universe in various forms. *Every* physical process that occurs in the Universe involves energy and energy transfers or transformations. Unfortunately, despite its extreme importance, energy cannot be easily defined. The variables in previous chapters were relatively concrete; we have everyday experience with velocities and forces, for example. Although we have *experiences* with energy, such as running out of gasoline or losing our electrical service following a violent storm, the *notion* of energy is more abstract.

On a wind farm at the mouth of the River Mersey in Liverpool, England, the moving air does work on the blades of the windmills, causing the blades and the rotor of an electrical generator to rotate. Energy is transferred out of the system of the windmill by means of electricity. *(Christopher Furlong/Getty Images)*

177

The concept of energy can be applied to mechanical systems without resorting to Newton's laws. Furthermore, the energy approach allows us to understand thermal and electrical phenomena in later chapters of the book in terms of the same models that we will develop here in our study of mechanics.

Our analysis models presented in earlier chapters were based on the motion of a *particle* or an object that could be modeled as a particle. We begin our new approach by focusing our attention on a new simplification model, a *system,* and analysis models based on the model of a system. These analysis models will be formally introduced in Chapter 8. In this chapter, we introduce systems and three ways to store energy in a system.

7.1 Systems and Environments

In the system model, we focus our attention on a small portion of the Universe— the **system**—and ignore details of the rest of the Universe outside of the system. A critical skill in applying the system model to problems is *identifying the system.*

A valid system

- may be a single object or particle
- may be a collection of objects or particles
- may be a region of space (such as the interior of an automobile engine combustion cylinder)
- may vary with time in size and shape (such as a rubber ball, which deforms upon striking a wall)

Identifying the need for a system approach to solving a problem (as opposed to a particle approach) is part of the Categorize step in the General Problem-Solving Strategy outlined in Chapter 2. Identifying the particular system is a second part of this step.

No matter what the particular system is in a given problem, we identify a **system boundary,** an imaginary surface (not necessarily coinciding with a physical surface) that divides the Universe into the system and the **environment** surrounding the system.

As an example, imagine a force applied to an object in empty space. We can define the object as the system and its surface as the system boundary. The force applied to it is an influence on the system from the environment that acts across the system boundary. We will see how to analyze this situation from a system approach in a subsequent section of this chapter.

Another example was seen in Example 5.10, where the system can be defined as the combination of the ball, the block, and the cord. The influence from the environment includes the gravitational forces on the ball and the block, the normal and friction forces on the block, and the force exerted by the pulley on the cord. The forces exerted by the cord on the ball and the block are internal to the system and therefore are not included as an influence from the environment.

There are a number of mechanisms by which a system can be influenced by its environment. The first one we shall investigate is *work.*

7.2 Work Done by a Constant Force

Almost all the terms we have used thus far—velocity, acceleration, force, and so on—convey a similar meaning in physics as they do in everyday life. Now, however, we encounter a term whose meaning in physics is distinctly different from its everyday meaning: work.

To understand what work as an influence on a system means to the physicist, consider the situation illustrated in Figure 7.1. A force $\vec{\mathbf{F}}$ is applied to a chalkboard

Figure 7.1 An eraser being pushed along a chalkboard tray by a force acting at different angles with respect to the horizontal direction.

eraser, which we identify as the system, and the eraser slides along the tray. If we want to know how effective the force is in moving the eraser, we must consider not only the magnitude of the force but also its direction. Notice that the finger in Figure 7.1 applies forces in three different directions on the eraser. Assuming the magnitude of the applied force is the same in all three photographs, the push applied in Figure 7.1b does more to move the eraser than the push in Figure 7.1a. On the other hand, Figure 7.1c shows a situation in which the applied force does not move the eraser at all, regardless of how hard it is pushed (unless, of course, we apply a force so great that we break the chalkboard tray!). These results suggest that when analyzing forces to determine the influence they have on the system, we must consider the vector nature of forces. We must also consider the magnitude of the force. Moving a force with a magnitude of $|\vec{\mathbf{F}}| = 2$ N through a displacement represents a greater influence on the system than moving a force of magnitude 1 N through the same displacement. The magnitude of the displacement is also important. Moving the eraser 3 m along the tray represents a greater influence than moving it 2 cm if the same force is used in both cases.

Let us examine the situation in Figure 7.2, where the object (the system) undergoes a displacement along a straight line while acted on by a constant force of magnitude F that makes an angle θ with the direction of the displacement.

> The **work** W done on a system by an agent exerting a constant force on the system is the product of the magnitude F of the force, the magnitude Δr of the displacement of the point of application of the force, and $\cos \theta$, where θ is the angle between the force and displacement vectors:
>
> $$W \equiv F\,\Delta r \cos \theta \qquad (7.1)$$

Notice in Equation 7.1 that work is a scalar, even though it is defined in terms of two vectors, a force $\vec{\mathbf{F}}$ and a displacement $\Delta\vec{\mathbf{r}}$. In Section 7.3, we explore how to combine two vectors to generate a scalar quantity.

Notice also that the displacement in Equation 7.1 is that of *the point of application of the force*. If the force is applied to a particle or a rigid object that can be modeled as a particle, this displacement is the same as that of the particle. For a deformable system, however, these displacements are not the same. For example, imagine pressing in on the sides of a balloon with both hands. The center of the balloon moves through zero displacement. The points of application of the forces from your hands on the sides of the balloon, however, do indeed move through a displacement as the balloon is compressed, and that is the displacement to be used in Equation 7.1. We will see other examples of deformable systems, such as springs and samples of gas contained in a vessel.

As an example of the distinction between the definition of work and our everyday understanding of the word, consider holding a heavy chair at arm's length for 3 min. At the end of this time interval, your tired arms may lead you to think you

Pitfall Prevention 7.2

Work Is Done by . . . on . . . Not only must you identify the system, you must also identify what agent in the environment is doing work on the system. When discussing work, always use the phrase, "the work done by . . . on . . . ." After "by," insert the part of the environment that is interacting directly with the system. After "on," insert the system. For example, "the work done by the hammer on the nail" identifies the nail as the system, and the force from the hammer represents the influence from the environment.

Figure 7.2 An object undergoes a displacement $\Delta\vec{\mathbf{r}}$ under the action of a constant force $\vec{\mathbf{F}}$.

◄ **Work done by a constant force**

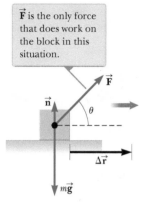

F is the only force that does work on the block in this situation.

Figure 7.3 An object is displaced on a frictionless, horizontal surface. The normal force $\vec{n}$ and the gravitational force $m\vec{g}$ do no work on the object.

Pitfall Prevention 7.3

Cause of the Displacement We can calculate the work done by a force on an object, but that force is *not* necessarily the cause of the object's displacement. For example, if you lift an object, (negative) work is done on the object by the gravitational force, although gravity is not the cause of the object moving upward!

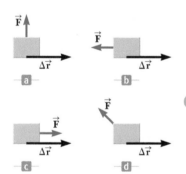

Figure 7.4 (Quick Quiz 7.2) A block is pulled by a force in four different directions. In each case, the displacement of the block is to the right and of the same magnitude.

have done a considerable amount of work on the chair. According to our definition, however, you have done no work on it whatsoever. You exert a force to support the chair, but you do not move it. A force does no work on an object if the force does not move through a displacement. If $\Delta r = 0$, Equation 7.1 gives $W = 0$, which is the situation depicted in Figure 7.1c.

Also notice from Equation 7.1 that the work done by a force on a moving object is zero when the force applied is perpendicular to the displacement of its point of application. That is, if $\theta = 90°$, then $W = 0$ because $\cos 90° = 0$. For example, in Figure 7.3, the work done by the normal force on the object and the work done by the gravitational force on the object are both zero because both forces are perpendicular to the displacement and have zero components along an axis in the direction of $\Delta \vec{r}$.

The sign of the work also depends on the direction of $\vec{F}$ relative to $\Delta \vec{r}$. The work done by the applied force on a system is positive when the projection of $\vec{F}$ onto $\Delta \vec{r}$ is in the same direction as the displacement. For example, when an object is lifted, the work done by the applied force on the object is positive because the direction of that force is upward, in the same direction as the displacement of its point of application. When the projection of $\vec{F}$ onto $\Delta \vec{r}$ is in the direction opposite the displacement, W is negative. For example, as an object is lifted, the work done by the gravitational force on the object is negative. The factor $\cos \theta$ in the definition of W (Eq. 7.1) automatically takes care of the sign.

If an applied force $\vec{F}$ is in the same direction as the displacement $\Delta \vec{r}$, then $\theta = 0$ and $\cos 0 = 1$. In this case, Equation 7.1 gives

$$W = F\Delta r$$

The units of work are those of force multiplied by those of length. Therefore, the SI unit of work is the **newton · meter** ($\text{N} \cdot \text{m} = \text{kg} \cdot \text{m}^2/\text{s}^2$). This combination of units is used so frequently that it has been given a name of its own, the **joule** (J).

An important consideration for a system approach to problems is that **work is an energy transfer.** If W is the work done on a system and W is positive, energy is transferred *to* the system; if W is negative, energy is transferred *from* the system. Therefore, if a system interacts with its environment, this interaction can be described as a transfer of energy across the system boundary. The result is a change in the energy stored in the system. We will learn about the first type of energy storage in Section 7.5, after we investigate more aspects of work.

Quick Quiz 7.1   The gravitational force exerted by the Sun on the Earth holds the Earth in an orbit around the Sun. Let us assume that the orbit is perfectly circular. The work done by this gravitational force during a short time interval in which the Earth moves through a displacement in its orbital path is **(a)** zero **(b)** positive **(c)** negative **(d)** impossible to determine

Quick Quiz 7.2   Figure 7.4 shows four situations in which a force is applied to an object. In all four cases, the force has the same magnitude, and the displacement of the object is to the right and of the same magnitude. Rank the situations in order of the work done by the force on the object, from most positive to most negative.

Example 7.1   **Mr. Clean**

A man cleaning a floor pulls a vacuum cleaner with a force of magnitude $F = 50.0$ N at an angle of $30.0°$ with the horizontal (Fig. 7.5). Calculate the work done by the force on the vacuum cleaner as the vacuum cleaner is displaced 3.00 m to the right.

▶ **7.1** continued

SOLUTION

Conceptualize Figure 7.5 helps conceptualize the situation. Think about an experience in your life in which you pulled an object across the floor with a rope or cord.

Categorize We are asked for the work done on an object by a force and are given the force on the object, the displacement of the object, and the angle between the two vectors, so we categorize this example as a substitution problem. We identify the vacuum cleaner as the system.

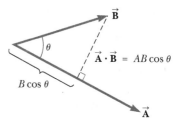

Figure 7.5 (Example 7.1) A vacuum cleaner being pulled at an angle of 30.0° from the horizontal.

Use the definition of work (Eq. 7.1):

$$W = F\,\Delta r \cos\theta = (50.0\ \text{N})(3.00\ \text{m})(\cos 30.0°)$$
$$= \boxed{130\ \text{J}}$$

Notice in this situation that the normal force $\vec{n}$ and the gravitational $\vec{F}_g = m\vec{g}$ do no work on the vacuum cleaner because these forces are perpendicular to the displacements of their points of application. Furthermore, there was no mention of whether there was friction between the vacuum cleaner and the floor. The presence or absence of friction is not important when calculating the work done by the applied force. In addition, this work does not depend on whether the vacuum moved at constant velocity or if it accelerated.

7.3 The Scalar Product of Two Vectors

Because of the way the force and displacement vectors are combined in Equation 7.1, it is helpful to use a convenient mathematical tool called the **scalar product** of two vectors. We write this scalar product of vectors $\vec{A}$ and $\vec{B}$ as $\vec{A} \cdot \vec{B}$. (Because of the dot symbol, the scalar product is often called the **dot product**.)

The scalar product of any two vectors $\vec{A}$ and $\vec{B}$ is defined as a scalar quantity equal to the product of the magnitudes of the two vectors and the cosine of the angle θ between them:

$$\vec{A} \cdot \vec{B} \equiv AB \cos\theta \tag{7.2}$$

◀ **Scalar product of any two vectors $\vec{A}$ and $\vec{B}$**

As is the case with any multiplication, $\vec{A}$ and $\vec{B}$ need not have the same units.

By comparing this definition with Equation 7.1, we can express Equation 7.1 as a scalar product:

$$W = F\,\Delta r \cos\theta = \vec{F} \cdot \Delta\vec{r} \tag{7.3}$$

In other words, $\vec{F} \cdot \Delta\vec{r}$ is a shorthand notation for $F\,\Delta r \cos\theta$.

Before continuing with our discussion of work, let us investigate some properties of the dot product. Figure 7.6 shows two vectors $\vec{A}$ and $\vec{B}$ and the angle θ between them used in the definition of the dot product. In Figure 7.6, $B \cos\theta$ is the projection of $\vec{B}$ onto $\vec{A}$. Therefore, Equation 7.2 means that $\vec{A} \cdot \vec{B}$ is the product of the magnitude of $\vec{A}$ and the projection of $\vec{B}$ onto $\vec{A}$.[1]

From the right-hand side of Equation 7.2, we also see that the scalar product is **commutative**.[2] That is,

$$\vec{A} \cdot \vec{B} = \vec{B} \cdot \vec{A}$$

Pitfall Prevention 7.4

Work Is a Scalar Although Equation 7.3 defines the work in terms of two vectors, *work is a scalar;* there is no direction associated with it. *All* types of energy and energy transfer are scalars. This fact is a major advantage of the energy approach because we don't need vector calculations!

Figure 7.6 The scalar product $\vec{A} \cdot \vec{B}$ equals the magnitude of $\vec{A}$ multiplied by $B \cos\theta$, which is the projection of $\vec{B}$ onto $\vec{A}$.

---

[1]This statement is equivalent to stating that $\vec{A} \cdot \vec{B}$ equals the product of the magnitude of $\vec{B}$ and the projection of $\vec{A}$ onto $\vec{B}$.

[2]In Chapter 11, you will see another way of combining vectors that proves useful in physics and is not commutative.

Finally, the scalar product obeys the **distributive law of multiplication,** so

$$\vec{A} \cdot (\vec{B} + \vec{C}) = \vec{A} \cdot \vec{B} + \vec{A} \cdot \vec{C}$$

The scalar product is simple to evaluate from Equation 7.2 when $\vec{A}$ is either perpendicular or parallel to $\vec{B}$. If $\vec{A}$ is perpendicular to $\vec{B}$ ($\theta = 90°$), then $\vec{A} \cdot \vec{B} = 0$. (The equality $\vec{A} \cdot \vec{B} = 0$ also holds in the more trivial case in which either $\vec{A}$ or $\vec{B}$ is zero.) If vector $\vec{A}$ is parallel to vector $\vec{B}$ and the two point in the same direction ($\theta = 0$), then $\vec{A} \cdot \vec{B} = AB$. If vector $\vec{A}$ is parallel to vector $\vec{B}$ but the two point in opposite directions ($\theta = 180°$), then $\vec{A} \cdot \vec{B} = -AB$. The scalar product is negative when $90° < \theta \leq 180°$.

The unit vectors $\hat{i}$, $\hat{j}$, and $\hat{k}$, which were defined in Chapter 3, lie in the positive x, y, and z directions, respectively, of a right-handed coordinate system. Therefore, it follows from the definition of $\vec{A} \cdot \vec{B}$ that the scalar products of these unit vectors are

◄ Scalar products of unit vectors

$$\hat{i} \cdot \hat{i} = \hat{j} \cdot \hat{j} = \hat{k} \cdot \hat{k} = 1 \tag{7.4}$$

$$\hat{i} \cdot \hat{j} = \hat{i} \cdot \hat{k} = \hat{j} \cdot \hat{k} = 0 \tag{7.5}$$

Equations 3.18 and 3.19 state that two vectors $\vec{A}$ and $\vec{B}$ can be expressed in unit-vector form as

$$\vec{A} = A_x\hat{i} + A_y\hat{j} + A_z\hat{k}$$

$$\vec{B} = B_x\hat{i} + B_y\hat{j} + B_z\hat{k}$$

Using these expressions for the vectors and the information given in Equations 7.4 and 7.5 shows that the scalar product of $\vec{A}$ and $\vec{B}$ reduces to

$$\vec{A} \cdot \vec{B} = A_x B_x + A_y B_y + A_z B_z \tag{7.6}$$

(Details of the derivation are left for you in Problem 7 at the end of the chapter.) In the special case in which $\vec{A} = \vec{B}$, we see that

$$\vec{A} \cdot \vec{A} = A_x^2 + A_y^2 + A_z^2 = A^2$$

Quick Quiz 7.3 Which of the following statements is true about the relationship between the dot product of two vectors and the product of the magnitudes of the vectors? (a) $\vec{A} \cdot \vec{B}$ is larger than AB. (b) $\vec{A} \cdot \vec{B}$ is smaller than AB. (c) $\vec{A} \cdot \vec{B}$ could be larger or smaller than AB, depending on the angle between the vectors. (d) $\vec{A} \cdot \vec{B}$ could be equal to AB.

Example 7.2   **The Scalar Product**

The vectors $\vec{A}$ and $\vec{B}$ are given by $\vec{A} = 2\hat{i} + 3\hat{j}$ and $\vec{B} = -\hat{i} + 2\hat{j}$.

(A) Determine the scalar product $\vec{A} \cdot \vec{B}$.

SOLUTION

Conceptualize There is no physical system to imagine here. Rather, it is purely a mathematical exercise involving two vectors.

Categorize Because we have a definition for the scalar product, we categorize this example as a substitution problem.

Substitute the specific vector expressions for $\vec{A}$ and $\vec{B}$:

$$\vec{A} \cdot \vec{B} = (2\hat{i} + 3\hat{j}) \cdot (-\hat{i} + 2\hat{j})$$

$$= -2\hat{i} \cdot \hat{i} + 2\hat{i} \cdot 2\hat{j} - 3\hat{j} \cdot \hat{i} + 3\hat{j} \cdot 2\hat{j}$$

$$= -2(1) + 4(0) - 3(0) + 6(1) = -2 + 6 = \boxed{4}$$

The same result is obtained when we use Equation 7.6 directly, where $A_x = 2$, $A_y = 3$, $B_x = -1$, and $B_y = 2$.

▶ **7.2** continued

(B) Find the angle θ between $\vec{\mathbf{A}}$ and $\vec{\mathbf{B}}$.

SOLUTION

Evaluate the magnitudes of $\vec{\mathbf{A}}$ and $\vec{\mathbf{B}}$ using the Pythagorean theorem:

$$A = \sqrt{A_x^2 + A_y^2} = \sqrt{(2)^2 + (3)^2} = \sqrt{13}$$

$$B = \sqrt{B_x^2 + B_y^2} = \sqrt{(-1)^2 + (2)^2} = \sqrt{5}$$

Use Equation 7.2 and the result from part (A) to find the angle:

$$\cos \theta = \frac{\vec{\mathbf{A}} \cdot \vec{\mathbf{B}}}{AB} = \frac{4}{\sqrt{13}\sqrt{5}} = \frac{4}{\sqrt{65}}$$

$$\theta = \cos^{-1} \frac{4}{\sqrt{65}} = \boxed{60.3°}$$

Example 7.3   Work Done by a Constant Force

A particle moving in the xy plane undergoes a displacement given by $\Delta \vec{\mathbf{r}} = (2.0\hat{\mathbf{i}} + 3.0\hat{\mathbf{j}})$ m as a constant force $\vec{\mathbf{F}} = (5.0\hat{\mathbf{i}} + 2.0\hat{\mathbf{j}})$ N acts on the particle. Calculate the work done by $\vec{\mathbf{F}}$ on the particle.

SOLUTION

Conceptualize Although this example is a little more physical than the previous one in that it identifies a force and a displacement, it is similar in terms of its mathematical structure.

Categorize Because we are given force and displacement vectors and asked to find the work done by this force on the particle, we categorize this example as a substitution problem.

Substitute the expressions for $\vec{\mathbf{F}}$ and $\Delta \vec{\mathbf{r}}$ into Equation 7.3 and use Equations 7.4 and 7.5:

$$W = \vec{\mathbf{F}} \cdot \Delta \vec{\mathbf{r}} = [(5.0\hat{\mathbf{i}} + 2.0\hat{\mathbf{j}}) \text{ N}] \cdot [(2.0\hat{\mathbf{i}} + 3.0\hat{\mathbf{j}}) \text{ m}]$$

$$= (5.0\hat{\mathbf{i}} \cdot 2.0\hat{\mathbf{i}} + 5.0\hat{\mathbf{i}} \cdot 3.0\hat{\mathbf{j}} + 2.0\hat{\mathbf{j}} \cdot 2.0\hat{\mathbf{i}} + 2.0\hat{\mathbf{j}} \cdot 3.0\hat{\mathbf{j}}) \text{ N} \cdot \text{m}$$

$$= [10 + 0 + 0 + 6] \text{ N} \cdot \text{m} = \boxed{16 \text{ J}}$$

7.4 Work Done by a Varying Force

Consider a particle being displaced along the x axis under the action of a force that varies with position. In such a situation, we cannot use Equation 7.1 to calculate the work done by the force because this relationship applies only when $\vec{\mathbf{F}}$ is constant in magnitude and direction. Figure 7.7a (page 184) shows a varying force applied on a particle that moves from initial position x_i to final position x_f. Imagine a particle undergoing a very small displacement Δx, shown in the figure. The x component F_x of the force is approximately constant over this small interval; for this small displacement, we can approximate the work done on the particle by the force using Equation 7.1 as

$$W \approx F_x \Delta x$$

which is the area of the shaded rectangle in Figure 7.7a. If the F_x versus x curve is divided into a large number of such intervals, the total work done for the displacement from x_i to x_f is approximately equal to the sum of a large number of such terms:

$$W \approx \sum_{x_i}^{x_f} F_x \, \Delta x$$

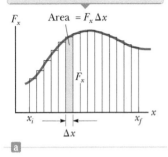

The total work done for the displacement from x_i to x_f is approximately equal to the sum of the areas of all the rectangles.

a

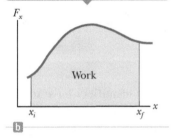

The work done by the component F_x of the varying force as the particle moves from x_i to x_f is *exactly* equal to the area under the curve.

b

Figure 7.7 (a) The work done on a particle by the force component F_x for the small displacement Δx is $F_x \Delta x$, which equals the area of the shaded rectangle. (b) The width Δx of each rectangle is shrunk to zero.

If the size of the small displacements is allowed to approach zero, the number of terms in the sum increases without limit but the value of the sum approaches a definite value equal to the area bounded by the F_x curve and the x axis:

$$\lim_{\Delta x \to 0} \sum_{x_i}^{x_f} F_x \, \Delta x = \int_{x_i}^{x_f} F_x \, dx$$

Therefore, we can express the work done by F_x on the system of the particle as it moves from x_i to x_f as

$$W = \int_{x_i}^{x_f} F_x \, dx \qquad (7.7)$$

This equation reduces to Equation 7.1 when the component $F_x = F \cos \theta$ remains constant.

If more than one force acts on a system *and the system can be modeled as a particle*, the total work done on the system is just the work done by the net force. If we express the net force in the x direction as ΣF_x, the total work, or *net work*, done as the particle moves from x_i to x_f is

$$\sum W = W_{\text{ext}} = \int_{x_i}^{x_f} \left(\sum F_x \right) dx \quad \text{(particle)}$$

For the general case of a net force $\sum \vec{F}$ whose magnitude and direction may both vary, we use the scalar product,

$$\sum W = W_{\text{ext}} = \int \left(\sum \vec{F} \right) \cdot d\vec{r} \quad \text{(particle)} \qquad (7.8)$$

where the integral is calculated over the path that the particle takes through space. The subscript "ext" on work reminds us that the net work is done by an *external* agent on the system. We will use this notation in this chapter as a reminder and to differentiate this work from an *internal* work to be described shortly.

If the system cannot be modeled as a particle (for example, if the system is deformable), we cannot use Equation 7.8 because different forces on the system may move through different displacements. In this case, we must evaluate the work done by each force separately and then add the works algebraically to find the net work done on the system:

$$\sum W = W_{\text{ext}} = \sum_{\text{forces}} \left(\int \vec{F} \cdot d\vec{r} \right) \quad \text{(deformable system)}$$

Example 7.4   Calculating Total Work Done from a Graph

A force acting on a particle varies with x as shown in Figure 7.8. Calculate the work done by the force on the particle as it moves from $x = 0$ to $x = 6.0$ m.

SOLUTION

Conceptualize Imagine a particle subject to the force in Figure 7.8. The force remains constant as the particle moves through the first 4.0 m and then decreases linearly to zero at 6.0 m. In terms of earlier discussions of motion, the particle could be modeled as a particle under constant acceleration for the first 4.0 m because the force is constant. Between 4.0 m and 6.0 m, however, the motion does not fit into one of our earlier analysis models because the acceleration of the particle is changing. If the particle starts from rest, its speed increases throughout the motion, and the particle is always moving in the positive x direction. These details about its speed and direction are not necessary for the calculation of the work done, however.

Categorize Because the force varies during the motion of the particle, we must use the techniques for work done by varying forces. In this case, the graphical representation in Figure 7.8 can be used to evaluate the work done.

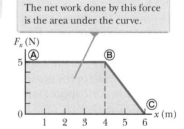

The net work done by this force is the area under the curve.

Figure 7.8 (Example 7.4) The force acting on a particle is constant for the first 4.0 m of motion and then decreases linearly with x from $x_{\text{B}} = 4.0$ m to $x_{\text{C}} = 6.0$ m.

▶ **7.4** continued

Analyze The work done by the force is equal to the area under the curve from $x_{Ⓐ} = 0$ to $x_{Ⓒ} = 6.0$ m. This area is equal to the area of the rectangular section from Ⓐ to Ⓑ plus the area of the triangular section from Ⓑ to Ⓒ.

Evaluate the area of the rectangle: $W_{Ⓐ \text{ to } Ⓑ} = (5.0 \text{ N})(4.0 \text{ m}) = 20 \text{ J}$

Evaluate the area of the triangle: $W_{Ⓑ \text{ to } Ⓒ} = \frac{1}{2}(5.0 \text{ N})(2.0 \text{ m}) = 5.0 \text{ J}$

Find the total work done by the force on the particle: $W_{Ⓐ \text{ to } Ⓒ} = W_{Ⓐ \text{ to } Ⓑ} + W_{Ⓑ \text{ to } Ⓒ} = 20 \text{ J} + 5.0 \text{ J} = \boxed{25 \text{ J}}$

Finalize Because the graph of the force consists of straight lines, we can use rules for finding the areas of simple geometric models to evaluate the total work done in this example. If a force does not vary linearly as in Figure 7.7, such rules cannot be used and the force function must be integrated as in Equation 7.7 or 7.8.

Work Done by a Spring

A model of a common physical system on which the force varies with position is shown in Figure 7.9. The system is a block on a frictionless, horizontal surface and connected to a spring. For many springs, if the spring is either stretched or compressed a small distance from its unstretched (equilibrium) configuration, it exerts on the block a force that can be mathematically modeled as

$$F_s = -kx \tag{7.9}$$ ◀ **Spring force**

where x is the position of the block relative to its equilibrium ($x = 0$) position and k is a positive constant called the **force constant** or the **spring constant** of the spring. In other words, the force required to stretch or compress a spring is proportional to the amount of stretch or compression x. This force law for springs is known as **Hooke's law.** The value of k is a measure of the *stiffness* of the spring. Stiff springs have large k values, and soft springs have small k values. As can be seen from Equation 7.9, the units of k are N/m.

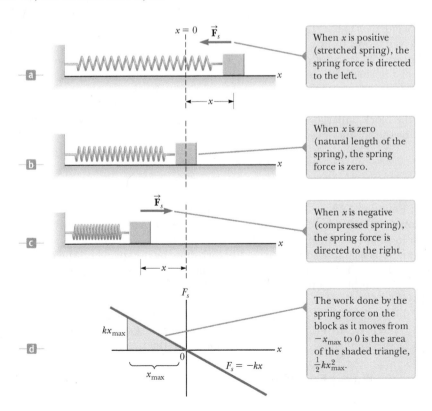

When x is positive (stretched spring), the spring force is directed to the left.

When x is zero (natural length of the spring), the spring force is zero.

When x is negative (compressed spring), the spring force is directed to the right.

The work done by the spring force on the block as it moves from $-x_{max}$ to 0 is the area of the shaded triangle, $\frac{1}{2}kx_{max}^2$.

Figure 7.9 The force exerted by a spring on a block varies with the block's position x relative to the equilibrium position $x = 0$. (a) x is positive. (b) x is zero. (c) x is negative. (d) Graph of F_s versus x for the block–spring system.

The vector form of Equation 7.9 is

$$\vec{\mathbf{F}}_s = F_s \hat{\mathbf{i}} = -kx\hat{\mathbf{i}} \tag{7.10}$$

where we have chosen the x axis to lie along the direction the spring extends or compresses.

The negative sign in Equations 7.9 and 7.10 signifies that the force exerted by the spring is always directed *opposite* the displacement from equilibrium. When $x > 0$ as in Figure 7.9a so that the block is to the right of the equilibrium position, the spring force is directed to the left, in the negative x direction. When $x < 0$ as in Figure 7.9c, the block is to the left of equilibrium and the spring force is directed to the right, in the positive x direction. When $x = 0$ as in Figure 7.9b, the spring is unstretched and $F_s = 0$. Because the spring force always acts toward the equilibrium position ($x = 0$), it is sometimes called a *restoring force*.

If the spring is compressed until the block is at the point $-x_{max}$ and is then released, the block moves from $-x_{max}$ through zero to $+x_{max}$. It then reverses direction, returns to $-x_{max}$, and continues oscillating back and forth. We will study these oscillations in more detail in Chapter 15. For now, let's investigate the work done by the spring on the block over small portions of one oscillation.

Suppose the block has been pushed to the left to a position $-x_{max}$ and is then released. We identify the block as our system and calculate the work W_s done by the spring force on the block as the block moves from $x_i = -x_{max}$ to $x_f = 0$. Applying Equation 7.8 and assuming the block may be modeled as a particle, we obtain

$$W_s = \int \vec{\mathbf{F}}_s \cdot d\vec{\mathbf{r}} = \int_{x_i}^{x_f} (-kx\hat{\mathbf{i}}) \cdot (dx\hat{\mathbf{i}}) = \int_{-x_{max}}^{0} (-kx)\, dx = \tfrac{1}{2}kx_{max}^2 \tag{7.11}$$

where we have used the integral $\int x^n\, dx = x^{n+1}/(n+1)$ with $n = 1$. The work done by the spring force is positive because the force is in the same direction as its displacement (both are to the right). Because the block arrives at $x = 0$ with some speed, it will continue moving until it reaches a position $+x_{max}$. The work done by the spring force on the block as it moves from $x_i = 0$ to $x_f = x_{max}$ is $W_s = -\tfrac{1}{2}kx_{max}^2$. The work is negative because for this part of the motion the spring force is to the left and its displacement is to the right. Therefore, the *net* work done by the spring force on the block as it moves from $x_i = -x_{max}$ to $x_f = x_{max}$ is *zero*.

Figure 7.9d is a plot of F_s versus x. The work calculated in Equation 7.11 is the area of the shaded triangle, corresponding to the displacement from $-x_{max}$ to 0. Because the triangle has base x_{max} and height kx_{max}, its area is $\tfrac{1}{2}kx_{max}^2$, agreeing with the work done by the spring as given by Equation 7.11.

If the block undergoes an arbitrary displacement from $x = x_i$ to $x = x_f$, the work done by the spring force on the block is

Work done by a spring ▶

$$W_s = \int_{x_i}^{x_f} (-kx)\, dx = \tfrac{1}{2}kx_i^2 - \tfrac{1}{2}kx_f^2 \tag{7.12}$$

From Equation 7.12, we see that the work done by the spring force is zero for any motion that ends where it began ($x_i = x_f$). We shall make use of this important result in Chapter 8 when we describe the motion of this system in greater detail.

Equations 7.11 and 7.12 describe the work done by the spring on the block. Now let us consider the work done on the block by an *external agent* as the agent applies a force on the block and the block moves *very slowly* from $x_i = -x_{max}$ to $x_f = 0$ as in Figure 7.10. We can calculate this work by noting that at any value of the position, the *applied force* $\vec{\mathbf{F}}_{app}$ is equal in magnitude and opposite in direction to the spring force $\vec{\mathbf{F}}_s$, so $\vec{\mathbf{F}}_{app} = F_{app}\hat{\mathbf{i}} = -\vec{\mathbf{F}}_s = -(-kx\hat{\mathbf{i}}) = kx\hat{\mathbf{i}}$. Therefore, the work done by this applied force (the external agent) on the system of the block is

$$W_{ext} = \int \vec{\mathbf{F}}_{app} \cdot d\vec{\mathbf{r}} = \int_{x_i}^{x_f} (kx\hat{\mathbf{i}}) \cdot (dx\hat{\mathbf{i}}) = \int_{-x_{max}}^{0} kx\, dx = -\tfrac{1}{2}kx_{max}^2$$

This work is equal to the negative of the work done by the spring force for this displacement (Eq. 7.11). The work is negative because the external agent must push inward on the spring to prevent it from expanding, and this direction is opposite the direction of the displacement of the point of application of the force as the block moves from $-x_{max}$ to 0.

For an arbitrary displacement of the block, the work done on the system by the external agent is

$$W_{ext} = \int_{x_i}^{x_f} kx \, dx = \tfrac{1}{2}kx_f^2 - \tfrac{1}{2}kx_i^2 \qquad (7.13)$$

Notice that this equation is the negative of Equation 7.12.

Quick Quiz 7.4 A dart is inserted into a spring-loaded dart gun by pushing the spring in by a distance x. For the next loading, the spring is compressed a distance $2x$. How much work is required to load the second dart compared with that required to load the first? **(a)** four times as much **(b)** two times as much **(c)** the same **(d)** half as much **(e)** one-fourth as much

If the process of moving the block is carried out very slowly, then $\vec{\mathbf{F}}_{app}$ is equal in magnitude and opposite in direction to $\vec{\mathbf{F}}_s$ at all times.

$x_i = -x_{max} \qquad x_f = 0$

Figure 7.10 A block moves from $x_i = -x_{max}$ to $x_f = 0$ on a frictionless surface as a force $\vec{\mathbf{F}}_{app}$ is applied to the block.

Example 7.5   **Measuring k for a Spring**   AM

A common technique used to measure the force constant of a spring is demonstrated by the setup in Figure 7.11. The spring is hung vertically (Fig. 7.11a), and an object of mass m is attached to its lower end. Under the action of the "load" mg, the spring stretches a distance d from its equilibrium position (Fig. 7.11b).

(A) If a spring is stretched 2.0 cm by a suspended object having a mass of 0.55 kg, what is the force constant of the spring?

SOLUTION

Conceptualize Figure 7.11b shows what happens to the spring when the object is attached to it. Simulate this situation by hanging an object on a rubber band.

Categorize The object in Figure 7.11b is at rest and not accelerating, so it is modeled as a *particle in equilibrium*.

Analyze Because the object is in equilibrium, the net force on it is zero and the upward spring force balances the downward gravitational force $m\vec{\mathbf{g}}$ (Fig. 7.11c).

The elongation d is caused by the weight mg of the attached object.

ⓐ   ⓑ   ⓒ

Figure 7.11 (Example 7.5) Determining the force constant k of a spring.

Apply the particle in equilibrium model to the object: $\vec{\mathbf{F}}_s + m\vec{\mathbf{g}} = 0 \quad \rightarrow \quad F_s - mg = 0 \quad \rightarrow \quad F_s = mg$

Apply Hooke's law to give $F_s = kd$ and solve for k: $k = \dfrac{mg}{d} = \dfrac{(0.55 \text{ kg})(9.80 \text{ m/s}^2)}{2.0 \times 10^{-2} \text{ m}} = \boxed{2.7 \times 10^2 \text{ N/m}}$

(B) How much work is done by the spring on the object as it stretches through this distance?

SOLUTION

Use Equation 7.12 to find the work done by the spring on the object: $W_s = 0 - \tfrac{1}{2}kd^2 = -\tfrac{1}{2}(2.7 \times 10^2 \text{ N/m})(2.0 \times 10^{-2} \text{ m})^2$

$= -5.4 \times 10^{-2} \text{ J}$

Finalize This work is negative because the spring force acts upward on the object, but its point of application (where the spring attaches to the object) moves downward. As the object moves through the 2.0-cm distance, the gravitational force also does work on it. This work is positive because the gravitational force is downward and so is the displacement

continued

of the point of application of this force. Would we expect the work done by the gravitational force, as the applied force in a direction opposite to the spring force, to be the negative of the answer above? Let's find out.

Evaluate the work done by the gravitational force on the object:

$$W = \vec{\mathbf{F}} \cdot \Delta\vec{\mathbf{r}} = (mg)(d)\cos 0 = mgd$$
$$= (0.55 \text{ kg})(9.80 \text{ m/s}^2)(2.0 \times 10^{-2} \text{ m}) = 1.1 \times 10^{-1} \text{ J}$$

If you expected the work done by gravity simply to be that done by the spring with a positive sign, you may be surprised by this result! To understand why that is not the case, we need to explore further, as we do in the next section.

7.5 Kinetic Energy and the Work–Kinetic Energy Theorem

Figure 7.12 An object undergoing a displacement $\Delta\vec{\mathbf{r}} = \Delta x\hat{\mathbf{i}}$ and a change in velocity under the action of a constant net force $\Sigma\,\vec{\mathbf{F}}$.

We have investigated work and identified it as a mechanism for transferring energy into a system. We have stated that work is an influence on a system from the environment, but we have not yet discussed the *result* of this influence on the system. One possible result of doing work on a system is that the system changes its speed. In this section, we investigate this situation and introduce our first type of energy that a system can possess, called *kinetic energy*.

Consider a system consisting of a single object. Figure 7.12 shows a block of mass m moving through a displacement directed to the right under the action of a net force $\Sigma\,\vec{\mathbf{F}}$, also directed to the right. We know from Newton's second law that the block moves with an acceleration $\vec{\mathbf{a}}$. If the block (and therefore the force) moves through a displacement $\Delta\vec{\mathbf{r}} = \Delta x\hat{\mathbf{i}} = (x_f - x_i)\hat{\mathbf{i}}$, the net work done on the block by the external net force $\Sigma\,\vec{\mathbf{F}}$ is

$$W_{\text{ext}} = \int_{x_i}^{x_f} \sum F\, dx \tag{7.14}$$

Using Newton's second law, we substitute for the magnitude of the net force $\Sigma F = ma$ and then perform the following chain-rule manipulations on the integrand:

$$W_{\text{ext}} = \int_{x_i}^{x_f} ma\, dx = \int_{x_i}^{x_f} m\frac{dv}{dt}\, dx = \int_{x_i}^{x_f} m\frac{dv}{dx}\frac{dx}{dt}\, dx = \int_{v_i}^{v_f} mv\, dv$$

$$W_{\text{ext}} = \tfrac{1}{2}mv_f^2 - \tfrac{1}{2}mv_i^2 \tag{7.15}$$

where v_i is the speed of the block at $x = x_i$ and v_f is its speed at x_f.

Equation 7.15 was generated for the specific situation of one-dimensional motion, but it is a general result. It tells us that the work done by the net force on a particle of mass m is equal to the difference between the initial and final values of a quantity $\tfrac{1}{2}mv^2$. This quantity is so important that it has been given a special name, **kinetic energy**:

Kinetic energy ▶

$$K \equiv \tfrac{1}{2}mv^2 \tag{7.16}$$

Kinetic energy represents the energy associated with the motion of the particle. Note that kinetic energy is a scalar quantity and has the same units as work. For example, a 2.0-kg object moving with a speed of 4.0 m/s has a kinetic energy of 16 J. Table 7.1 lists the kinetic energies for various objects.

Equation 7.15 states that the work done on a particle by a net force $\Sigma\,\vec{\mathbf{F}}$ acting on it equals the change in kinetic energy of the particle. It is often convenient to write Equation 7.15 in the form

$$W_{\text{ext}} = K_f - K_i = \Delta K \tag{7.17}$$

Another way to write it is $K_f = K_i + W_{\text{ext}}$, which tells us that the final kinetic energy of an object is equal to its initial kinetic energy plus the change in energy due to the net work done on it.

Table 7.1	Kinetic Energies for Various Objects		
Object	**Mass (kg)**	**Speed (m/s)**	**Kinetic Energy (J)**
Earth orbiting the Sun	5.97×10^{24}	2.98×10^{4}	2.65×10^{33}
Moon orbiting the Earth	7.35×10^{22}	1.02×10^{3}	3.82×10^{28}
Rocket moving at escape speed[a]	500	1.12×10^{4}	3.14×10^{10}
Automobile at 65 mi/h	2 000	29	8.4×10^{5}
Running athlete	70	10	3 500
Stone dropped from 10 m	1.0	14	98
Golf ball at terminal speed	0.046	44	45
Raindrop at terminal speed	3.5×10^{-5}	9.0	1.4×10^{-3}
Oxygen molecule in air	5.3×10^{-26}	500	6.6×10^{-21}

[a]Escape speed is the minimum speed an object must reach near the Earth's surface to move infinitely far away from the Earth.

We have generated Equation 7.17 by imagining doing work on a particle. We could also do work on a deformable system, in which parts of the system move with respect to one another. In this case, we also find that Equation 7.17 is valid as long as the net work is found by adding up the works done by each force and adding, as discussed earlier with regard to Equation 7.8.

Equation 7.17 is an important result known as the **work–kinetic energy theorem:**

> When work is done on a system and the only change in the system is in its speed, the net work done on the system equals the change in kinetic energy of the system, as expressed by Equation 7.17: $W = \Delta K$.

◀ **Work–kinetic energy theorem**

The work–kinetic energy theorem indicates that the speed of a system *increases* if the net work done on it is *positive* because the final kinetic energy is greater than the initial kinetic energy. The speed *decreases* if the net work is *negative* because the final kinetic energy is less than the initial kinetic energy.

Because we have so far only investigated translational motion through space, we arrived at the work–kinetic energy theorem by analyzing situations involving translational motion. Another type of motion is *rotational motion*, in which an object spins about an axis. We will study this type of motion in Chapter 10. The work–kinetic energy theorem is also valid for systems that undergo a change in the rotational speed due to work done on the system. The windmill in the photograph at the beginning of this chapter is an example of work causing rotational motion.

The work–kinetic energy theorem will clarify a result seen earlier in this chapter that may have seemed odd. In Section 7.4, we arrived at a result of zero net work done when we let a spring push a block from $x_i = -x_{max}$ to $x_f = x_{max}$. Notice that because the speed of the block is continually changing, it may seem complicated to analyze this process. The quantity ΔK in the work–kinetic energy theorem, however, only refers to the initial and final points for the speeds; it does not depend on details of the path followed between these points. Therefore, because the speed is zero at both the initial and final points of the motion, the net work done on the block is zero. We will often see this concept of path independence in similar approaches to problems.

Let us also return to the mystery in the Finalize step at the end of Example 7.5. Why was the work done by gravity not just the value of the work done by the spring with a positive sign? Notice that the work done by gravity is larger than the magnitude of the work done by the spring. Therefore, the total work done by all forces on the object is positive. Imagine now how to create the situation in which the *only* forces on the object are the spring force and the gravitational force. You must support the object at the highest point and then remove your hand and let the object fall. If you do so, you know that when the object reaches a position 2.0 cm below your hand, it will be *moving*, which is consistent with Equation 7.17. Positive net

Pitfall Prevention 7.5

Conditions for the Work–Kinetic Energy Theorem The work–kinetic energy theorem is important but limited in its application; it is not a general principle. In many situations, other changes in the system occur besides its speed, and there are other interactions with the environment besides work. A more general principle involving energy is *conservation of energy* in Section 8.1.

Pitfall Prevention 7.6

The Work–Kinetic Energy Theorem: Speed, Not Velocity The work–kinetic energy theorem relates work to a change in the *speed* of a system, not a change in its velocity. For example, if an object is in uniform circular motion, its speed is constant. Even though its velocity is changing, no work is done on the object by the force causing the circular motion.

work is done on the object, and the result is that it has a kinetic energy as it passes through the 2.0-cm point.

The only way to prevent the object from having a kinetic energy after moving through 2.0 cm is to slowly lower it with your hand. Then, however, there is a third force doing work on the object, the normal force from your hand. If this work is calculated and added to that done by the spring force and the gravitational force, the net work done on the object is zero, which is consistent because it is not moving at the 2.0-cm point.

Earlier, we indicated that work can be considered as a mechanism for transferring energy into a system. Equation 7.17 is a mathematical statement of this concept. When work W_{ext} is done on a system, the result is a transfer of energy across the boundary of the system. The result on the system, in the case of Equation 7.17, is a change ΔK in kinetic energy. In the next section, we investigate another type of energy that can be stored in a system as a result of doing work on the system.

Quick Quiz 7.5 A dart is inserted into a spring-loaded dart gun by pushing the spring in by a distance x. For the next loading, the spring is compressed a distance $2x$. How much faster does the second dart leave the gun compared with the first? **(a)** four times as fast **(b)** two times as fast **(c)** the same **(d)** half as fast **(e)** one-fourth as fast

---

Example 7.6   A Block Pulled on a Frictionless Surface   AM

A 6.0-kg block initially at rest is pulled to the right along a frictionless, horizontal surface by a constant horizontal force of magnitude 12 N. Find the block's speed after it has moved through a horizontal distance of 3.0 m.

SOLUTION

Conceptualize Figure 7.13 illustrates this situation. Imagine pulling a toy car across a table with a horizontal rubber band attached to the front of the car. The force is maintained constant by ensuring that the stretched rubber band always has the same length.

Figure 7.13 (Example 7.6) A block pulled to the right on a frictionless surface by a constant horizontal force.

Categorize We could apply the equations of kinematics to determine the answer, but let us practice the energy approach. The block is the system, and three external forces act on the system. The normal force balances the gravitational force on the block, and neither of these vertically acting forces does work on the block because their points of application are horizontally displaced.

Analyze The net external force acting on the block is the horizontal 12-N force.

Use the work–kinetic energy theorem for the block, noting that its initial kinetic energy is zero:

$$W_{ext} = \Delta K = K_f - K_i = \tfrac{1}{2}mv_f^2 - 0 = \tfrac{1}{2}mv_f^2$$

Solve for v_f and use Equation 7.1 for the work done on the block by $\vec{F}$:

$$v_f = \sqrt{\frac{2W_{ext}}{m}} = \sqrt{\frac{2F\,\Delta x}{m}}$$

Substitute numerical values:

$$v_f = \sqrt{\frac{2(12\text{ N})(3.0\text{ m})}{6.0\text{ kg}}} = \boxed{3.5\text{ m/s}}$$

Finalize You should solve this problem again by modeling the block as a *particle under a net force* to find its acceleration and then as a *particle under constant acceleration* to find its final velocity. In Chapter 8, we will see that the energy procedure followed above is an example of the analysis model of the *nonisolated system*.

WHAT IF? Suppose the magnitude of the force in this example is doubled to $F' = 2F$. The 6.0-kg block accelerates to 3.5 m/s due to this applied force while moving through a displacement $\Delta x'$. How does the displacement $\Delta x'$ compare with the original displacement Δx?

▶ **7.6** continued

Answer If we pull harder, the block should accelerate to a given speed in a shorter distance, so we expect that $\Delta x' < \Delta x$. In both cases, the block experiences the same change in kinetic energy ΔK. Mathematically, from the work–kinetic energy theorem, we find that

$$W_{ext} = F'\Delta x' = \Delta K = F\Delta x$$

$$\Delta x' = \frac{F}{F'}\Delta x = \frac{F}{2F}\Delta x = \tfrac{1}{2}\Delta x$$

and the distance is shorter as suggested by our conceptual argument.

Conceptual Example 7.7   **Does the Ramp Lessen the Work Required?**

A man wishes to load a refrigerator onto a truck using a ramp at angle θ as shown in Figure 7.14. He claims that less work would be required to load the truck if the length L of the ramp were increased. Is his claim valid?

SOLUTION

No. Suppose the refrigerator is wheeled on a hand truck up the ramp at constant speed. In this case, for the system of the refrigerator and the hand truck, $\Delta K = 0$. The normal force exerted by the ramp on the system is directed at 90° to the displacement of its point of application and so does no work on the system. Because $\Delta K = 0$, the work–kinetic energy theorem gives

Figure 7.14 (Conceptual Example 7.7) A refrigerator attached to a frictionless, wheeled hand truck is moved up a ramp at constant speed.

$$W_{ext} = W_{by\ man} + W_{by\ gravity} = 0$$

The work done by the gravitational force equals the product of the weight mg of the system, the distance L through which the refrigerator is displaced, and $\cos(\theta + 90°)$. Therefore,

$$W_{by\ man} = -W_{by\ gravity} = -(mg)(L)[\cos(\theta + 90°)]$$
$$= mgL\sin\theta = mgh$$

where $h = L\sin\theta$ is the height of the ramp. Therefore, the man must do the same amount of work mgh on the system *regardless* of the length of the ramp. The work depends only on the height of the ramp. Although less force is required with a longer ramp, the point of application of that force moves through a greater displacement.

7.6 Potential Energy of a System

So far in this chapter, we have defined a system in general, but have focused our attention primarily on single particles or objects under the influence of external forces. Let us now consider systems of two or more particles or objects interacting via a force that is *internal* to the system. The kinetic energy of such a system is the algebraic sum of the kinetic energies of all members of the system. There may be systems, however, in which one object is so massive that it can be modeled as stationary and its kinetic energy can be neglected. For example, if we consider a ball–Earth system as the ball falls to the Earth, the kinetic energy of the system can be considered as just the kinetic energy of the ball. The Earth moves so slowly in this process that we can ignore its kinetic energy. On the other hand, the kinetic energy of a system of two electrons must include the kinetic energies of both particles.

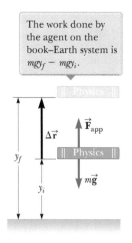

The work done by the agent on the book–Earth system is $mgy_f - mgy_i$.

Figure 7.15 An external agent lifts a book slowly from a height y_i to a height y_f.

Pitfall Prevention 7.7

Potential Energy The phrase *potential energy* does not refer to something that has the potential to become energy. Potential energy *is* energy.

Pitfall Prevention 7.8

Potential Energy Belongs to a System Potential energy is always associated with a *system* of two or more interacting objects. When a small object moves near the surface of the Earth under the influence of gravity, we may sometimes refer to the potential energy "associated with the object" rather than the more proper "associated with the system" because the Earth does not move significantly. We will not, however, refer to the potential energy "of the object" because this wording ignores the role of the Earth.

Gravitational ▶ potential energy

Let us imagine a system consisting of a book and the Earth, interacting via the gravitational force. We do some work on the system by lifting the book slowly from rest through a vertical displacement $\Delta \vec{r} = (y_f - y_i)\hat{j}$ as in Figure 7.15. According to our discussion of work as an energy transfer, this work done on the system must appear as an increase in energy of the system. The book is at rest before we perform the work and is at rest after we perform the work. Therefore, there is no change in the kinetic energy of the system.

Because the energy change of the system is not in the form of kinetic energy, the work-kinetic energy theorem does not apply here and the energy change must appear as some form of energy storage other than kinetic energy. After lifting the book, we could release it and let it fall back to the position y_i. Notice that the book (and therefore, the system) now has kinetic energy and that its source is in the work that was done in lifting the book. While the book was at the highest point, the system had the *potential* to possess kinetic energy, but it did not do so until the book was allowed to fall. Therefore, we call the energy storage mechanism before the book is released **potential energy.** We will find that the potential energy of a system can only be associated with specific types of forces acting between members of a system. The amount of potential energy in the system is determined by the *configuration* of the system. Moving members of the system to different positions or rotating them may change the configuration of the system and therefore its potential energy.

Let us now derive an expression for the potential energy associated with an object at a given location above the surface of the Earth. Consider an external agent lifting an object of mass m from an initial height y_i above the ground to a final height y_f as in Figure 7.15. We assume the lifting is done slowly, with no acceleration, so the applied force from the agent is equal in magnitude to the gravitational force on the object: the object is modeled as a particle in equilibrium moving at constant velocity. The work done by the external agent on the system (object and the Earth) as the object undergoes this upward displacement is given by the product of the upward applied force $\vec{F}_{app}$ and the upward displacement of this force, $\Delta \vec{r} = \Delta y \hat{j}$:

$$W_{ext} = (\vec{F}_{app}) \cdot \Delta \vec{r} = (mg\hat{j}) \cdot [(y_f - y_i)\hat{j}] = mgy_f - mgy_i \qquad (7.18)$$

where this result is the net work done on the system because the applied force is the only force on the system from the environment. (Remember that the gravitational force is *internal* to the system.) Notice the similarity between Equation 7.18 and Equation 7.15. In each equation, the work done on a system equals a difference between the final and initial values of a quantity. In Equation 7.15, the work represents a transfer of energy into the system and the increase in energy of the system is kinetic in form. In Equation 7.18, the work represents a transfer of energy into the system and the system energy appears in a different form, which we have called potential energy.

Therefore, we can identify the quantity mgy as the **gravitational potential energy** U_g of the system of an object of mass m and the Earth:

$$U_g \equiv mgy \qquad (7.19)$$

The units of gravitational potential energy are joules, the same as the units of work and kinetic energy. Potential energy, like work and kinetic energy, is a scalar quantity. Notice that Equation 7.19 is valid only for objects near the surface of the Earth, where g is approximately constant.[3]

Using our definition of gravitational potential energy, Equation 7.18 can now be rewritten as

$$W_{ext} = \Delta U_g \qquad (7.20)$$

which mathematically describes that the net external work done on the system in this situation appears as a change in the gravitational potential energy of the system.

Equation 7.20 is similar in form to the work–kinetic energy theorem, Equation 7.17. In Equation 7.17, work is done on a system and energy appears in the system as

---

[3]The assumption that g is constant is valid as long as the vertical displacement of the object is small compared with the Earth's radius.

kinetic energy, representing *motion* of the members of the system. In Equation 7.20, work is done on the system and energy appears in the system as potential energy, representing a change in the *configuration* of the members of the system.

Gravitational potential energy depends only on the vertical height of the object above the surface of the Earth. The same amount of work must be done on an object–Earth system whether the object is lifted vertically from the Earth or is pushed starting from the same point up a frictionless incline, ending up at the same height. We verified this statement for a specific situation of rolling a refrigerator up a ramp in Conceptual Example 7.7. This statement can be shown to be true in general by calculating the work done on an object by an agent moving the object through a displacement having both vertical and horizontal components:

$$W_{\text{ext}} = (\vec{\mathbf{F}}_{\text{app}}) \cdot \Delta \vec{\mathbf{r}} = (mg\hat{\mathbf{j}}) \cdot [(x_f - x_i)\hat{\mathbf{i}} + (y_f - y_i)\hat{\mathbf{j}}] = mgy_f - mgy_i$$

where there is no term involving x in the final result because $\hat{\mathbf{j}} \cdot \hat{\mathbf{i}} = 0$.

In solving problems, you must choose a reference configuration for which the gravitational potential energy of the system is set equal to some reference value, which is normally zero. The choice of reference configuration is completely arbitrary because the important quantity is the *difference* in potential energy, and this difference is independent of the choice of reference configuration.

It is often convenient to choose as the reference configuration for zero gravitational potential energy the configuration in which an object is at the surface of the Earth, but this choice is not essential. Often, the statement of the problem suggests a convenient configuration to use.

Quick Quiz 7.6 Choose the correct answer. The gravitational potential energy of a system **(a)** is always positive **(b)** is always negative **(c)** can be negative or positive

---

Example 7.8   The Proud Athlete and the Sore Toe

A trophy being shown off by a careless athlete slips from the athlete's hands and drops on his foot. Choosing floor level as the $y = 0$ point of your coordinate system, estimate the change in gravitational potential energy of the trophy–Earth system as the trophy falls. Repeat the calculation, using the top of the athlete's head as the origin of coordinates.

SOLUTION

Conceptualize The trophy changes its vertical position with respect to the surface of the Earth. Associated with this change in position is a change in the gravitational potential energy of the trophy–Earth system.

Categorize We evaluate a change in gravitational potential energy defined in this section, so we categorize this example as a substitution problem. Because there are no numbers provided in the problem statement, it is also an estimation problem.

The problem statement tells us that the reference configuration of the trophy–Earth system corresponding to zero potential energy is when the bottom of the trophy is at the floor. To find the change in potential energy for the system, we need to estimate a few values. Let's say the trophy has a mass of approximately 2 kg, and the top of a person's foot is about 0.05 m above the floor. Also, let's assume the trophy falls from a height of 1.4 m.

Calculate the gravitational potential energy of the trophy–Earth system just before the trophy is released:

$$U_i = mgy_i = (2 \text{ kg})(9.80 \text{ m/s}^2)(1.4 \text{ m}) = 27.4 \text{ J}$$

Calculate the gravitational potential energy of the trophy–Earth system when the trophy reaches the athlete's foot:

$$U_f = mgy_f = (2 \text{ kg})(9.80 \text{ m/s}^2)(0.05 \text{ m}) = 0.98 \text{ J}$$

Evaluate the change in gravitational potential energy of the trophy–Earth system:

$$\Delta U_g = 0.98 \text{ J} - 27.4 \text{ J} = -26.4 \text{ J}$$

continued

▶ **7.8** continued

We should probably keep only two digits because of the roughness of our estimates; therefore, we estimate that the change in gravitational potential energy is −26 J . The system had about 27 J of gravitational potential energy before the trophy began its fall and approximately 1 J of potential energy as the trophy reaches the top of the foot.

The second case presented indicates that the reference configuration of the system for zero potential energy is chosen to be when the trophy is on the athlete's head (even though the trophy is never at this position in its motion). We estimate this position to be 2.0 m above the floor).

Calculate the gravitational potential energy of the trophy–Earth system just before the trophy is released from its position 0.6 m below the athlete's head:

$$U_i = mgy_i = (2 \text{ kg})(9.80 \text{ m/s}^2)(-0.6 \text{ m}) = -11.8 \text{ J}$$

Calculate the gravitational potential energy of the trophy–Earth system when the trophy reaches the athlete's foot located 1.95 m below its initial position:

$$U_f = mgy_f = (2 \text{ kg})(9.80 \text{ m/s}^2)(-1.95 \text{ m}) = -38.2 \text{ J}$$

Evaluate the change in gravitational potential energy of the trophy–Earth system:

$$\Delta U_g = -38.2 \text{ J} - (-11.8 \text{ J}) = -26.4 \text{ J} \approx -26 \text{ J}$$

This value is the same as before, as it must be. The change in potential energy is independent of the choice of configuration of the system representing the zero of potential energy. If we wanted to keep only one digit in our estimates, we could write the final result as 3×10^1 J.

Elastic Potential Energy

Because members of a system can interact with one another by means of different types of forces, it is possible that there are different types of potential energy in a system. We have just become familiar with gravitational potential energy of a system in which members interact via the gravitational force. Let us explore a second type of potential energy that a system can possess.

Consider a system consisting of a block and a spring as shown in Figure 7.16. In Section 7.4, we identified *only* the block as the system. Now we include both the block and the spring in the system and recognize that the spring force is the interaction between the two members of the system. The force that the spring exerts on the block is given by $F_s = -kx$ (Eq. 7.9). The external work done by an applied force F_{app} on the block–spring system is given by Equation 7.13:

$$W_{ext} = \tfrac{1}{2}kx_f^2 - \tfrac{1}{2}kx_i^2 \tag{7.21}$$

In this situation, the initial and final x coordinates of the block are measured from its equilibrium position, $x = 0$. Again (as in the gravitational case, Eq. 7.18) the work done on the system is equal to the difference between the initial and final values of an expression related to the system's configuration. The **elastic potential energy** function associated with the block–spring system is defined by

Elastic potential energy ▶

$$U_s \equiv \tfrac{1}{2}kx^2 \tag{7.22}$$

Equation 7.21 can be expressed as

$$W_{ext} = \Delta U_s \tag{7.23}$$

Compare this equation to Equations 7.17 and 7.20. In all three situations, external work is done on a system and a form of energy storage in the system changes as a result.

The elastic potential energy of the system can be thought of as the energy stored in the deformed spring (one that is either compressed or stretched from its equilibrium position). The elastic potential energy stored in a spring is zero whenever the spring is undeformed ($x = 0$). Energy is stored in the spring only when the spring is

Figure 7.16 A spring on a frictionless, horizontal surface is compressed a distance x_{max} when a block of mass m is pushed against it. The block is then released and the spring pushes it to the right, where the block eventually loses contact with the spring. Parts (a) through (e) show various instants in the process. Energy bar charts on the right of each part of the figure help keep track of the energy in the system.

either stretched or compressed. Because the elastic potential energy is proportional to x^2, we see that U_s is always positive in a deformed spring. Everyday examples of the storage of elastic potential energy can be found in old-style clocks or watches that operate from a wound-up spring and small wind-up toys for children.

Consider Figure 7.16 once again, which shows a spring on a frictionless, horizontal surface. When a block is pushed against the spring by an external agent, the elastic potential energy and the total energy of the system increase as indicated in Figure 7.16b. When the spring is compressed a distance x_{max} (Fig. 7.16c), the elastic potential energy stored in the spring is $\frac{1}{2}kx_{max}^2$. When the block is released from rest, the spring exerts a force on the block and pushes the block to the right. The elastic potential energy of the system decreases, whereas the kinetic energy increases and the total energy remains fixed (Fig. 7.16d). When the spring returns to its original length, the stored elastic potential energy is completely transformed into kinetic energy of the block (Fig. 7.16e).

Figure 7.17 (Quick Quiz 7.7) A ball connected to a massless spring suspended vertically. What forms of potential energy are associated with the system when the ball is displaced downward?

Figure 7.18 (a) A book sliding to the right on a horizontal surface slows down in the presence of a force of kinetic friction acting to the left. (b) An energy bar chart showing the energy in the system of the book and the surface at the initial instant of time. The energy of the system is all kinetic energy. (c) While the book is sliding, the kinetic energy of the system decreases as it is transformed to internal energy. (d) After the book has stopped, the energy of the system is all internal energy.

Quick Quiz 7.7 A ball is connected to a light spring suspended vertically as shown in Figure 7.17. When pulled downward from its equilibrium position and released, the ball oscillates up and down. **(i)** In the system of *the ball, the spring, and the Earth*, what forms of energy are there during the motion? (a) kinetic and elastic potential (b) kinetic and gravitational potential (c) kinetic, elastic potential, and gravitational potential (d) elastic potential and gravitational potential **(ii)** In the system of *the ball and the spring*, what forms of energy are there during the motion? Choose from the same possibilities (a) through (d).

Energy Bar Charts

Figure 7.16 shows an important graphical representation of information related to energy of systems called an **energy bar chart.** The vertical axis represents the amount of energy of a given type in the system. The horizontal axis shows the types of energy in the system. The bar chart in Figure 7.16a shows that the system contains zero energy because the spring is relaxed and the block is not moving. Between Figure 7.16a and Figure 7.16c, the hand does work on the system, compressing the spring and storing elastic potential energy in the system. In Figure 7.16d, the block has been released and is moving to the right while still in contact with the spring. The height of the bar for the elastic potential energy of the system decreases, the kinetic energy bar increases, and the total energy bar remains fixed. In Figure 7.16e, the spring has returned to its relaxed length and the system now contains only kinetic energy associated with the moving block.

Energy bar charts can be a very useful representation for keeping track of the various types of energy in a system. For practice, try making energy bar charts for the book–Earth system in Figure 7.15 when the book is dropped from the higher position. Figure 7.17 associated with Quick Quiz 7.7 shows another system for which drawing an energy bar chart would be a good exercise. We will show energy bar charts in some figures in this chapter. Some figures will not show a bar chart in the text but will include one in animated versions that appear in Enhanced WebAssign.

7.7 Conservative and Nonconservative Forces

We now introduce a third type of energy that a system can possess. Imagine that the book in Figure 7.18a has been accelerated by your hand and is now sliding to the right on the surface of a heavy table and slowing down due to the friction force. Suppose the *surface* is the system. Then the friction force from the sliding book does work on the surface. The force on the surface is to the right and the displacement of the point of application of the force is to the right because the book has moved to the right. The work done on the surface is therefore positive, but the surface is not moving after the book has stopped. Positive work has been done on the surface, yet there is no increase in the surface's kinetic energy or the potential energy of any system. So where is the energy?

From your everyday experience with sliding over surfaces with friction, you can probably guess that the surface will be *warmer* after the book slides over it. The work that was done on the surface has gone into warming the surface rather than increasing its speed or changing the configuration of a system. We call the energy associated with the temperature of a system its **internal energy,** symbolized E_{int}. (We will define internal energy more generally in Chapter 20.) In this case, the work done on the surface does indeed represent energy transferred into the system, but it appears in the system as internal energy rather than kinetic or potential energy.

Now consider the book and the surface in Figure 7.18a together as a system. Initially, the system has kinetic energy because the book is moving. While the book is sliding, the internal energy of the system increases: the book and the surface are warmer than before. When the book stops, the kinetic energy has been completely transformed to internal energy. We can consider the nonconservative force within the system—that is, between the book and the surface—as a *transformation mechanism* for energy. This nonconservative force transforms the kinetic energy of the system into internal energy. Rub your hands together briskly to experience this effect!

Figures 7.18b through 7.18d show energy bar charts for the situation in Figure 7.18a. In Figure 7.18b, the bar chart shows that the system contains kinetic energy at the instant the book is released by your hand. We define the reference amount of internal energy in the system as zero at this instant. Figure 7.18c shows the kinetic energy transforming to internal energy as the book slows down due to the friction force. In Figure 7.18d, after the book has stopped sliding, the kinetic energy is zero, and the system now contains only internal energy E_{int}. Notice that the total energy bar in red has not changed during the process. The amount of internal energy in the system after the book has stopped is equal to the amount of kinetic energy in the system at the initial instant. This equality is described by an important principle called *conservation of energy*. We will explore this principle in Chapter 8.

Now consider in more detail an object moving downward near the surface of the Earth. The work done by the gravitational force on the object does not depend on whether it falls vertically or slides down a sloping incline with friction. All that matters is the change in the object's elevation. The energy transformation to internal energy due to friction on that incline, however, depends very much on the distance the object slides. The longer the incline, the more potential energy is transformed to internal energy. In other words, the path makes no difference when we consider the work done by the gravitational force, but it does make a difference when we consider the energy transformation due to friction forces. We can use this varying dependence on path to classify forces as either *conservative* or *nonconservative*. Of the two forces just mentioned, the gravitational force is conservative and the friction force is nonconservative.

Conservative Forces

Conservative forces have these two equivalent properties:

◀ **Properties of conservative forces**

1. The work done by a conservative force on a particle moving between any two points is independent of the path taken by the particle.
2. The work done by a conservative force on a particle moving through any closed path is zero. (A closed path is one for which the beginning point and the endpoint are identical.)

The gravitational force is one example of a conservative force; the force that an ideal spring exerts on any object attached to the spring is another. The work done by the gravitational force on an object moving between any two points near the Earth's surface is $W_g = -mg\hat{\mathbf{j}} \cdot [(y_f - y_i)\hat{\mathbf{j}}] = mgy_i - mgy_f$. From this equation, notice that W_g depends only on the initial and final y coordinates of the object and hence is independent of the path. Furthermore, W_g is zero when the object moves over any closed path (where $y_i = y_f$).

For the case of the object–spring system, the work W_s done by the spring force is given by $W_s = \frac{1}{2}kx_i^2 - \frac{1}{2}kx_f^2$ (Eq. 7.12). We see that the spring force is conservative because W_s depends only on the initial and final x coordinates of the object and is zero for any closed path.

We can associate a potential energy for a system with a force acting between members of the system, but we can do so only if the force is conservative. In general, the work W_{int} done by a conservative force on an object that is a member of a system as the system changes from one configuration to another is equal to the initial value of the potential energy of the system minus the final value:

$$W_{int} = U_i - U_f = -\Delta U \tag{7.24}$$

The subscript "int" in Equation 7.24 reminds us that the work we are discussing is done by one member of the system on another member and is therefore *internal* to

Pitfall Prevention 7.9
Similar Equation Warning Compare Equation 7.24 with Equation 7.20. These equations are similar except for the negative sign, which is a common source of confusion. Equation 7.20 tells us that positive work done *by an outside agent* on a system causes an increase in the potential energy of the system (with no change in the kinetic or internal energy). Equation 7.24 states that positive work done on a component of a system by a conservative force *internal to the system* causes a decrease in the potential energy of the system.

the system. It is different from the work W_{ext} done *on* the system as a whole by an external agent. As an example, compare Equation 7.24 with the equation for the work done by an external agent on a block–spring system (Eq. 7.23) as the extension of the spring changes.

Nonconservative Forces

A force is **nonconservative** if it does not satisfy properties 1 and 2 above. The work done by a nonconservative force is path-dependent. We define the sum of the kinetic and potential energies of a system as the **mechanical energy** of the system:

$$E_{mech} \equiv K + U \tag{7.25}$$

where K includes the kinetic energy of all moving members of the system and U includes all types of potential energy in the system. For a book falling under the action of the gravitational force, the mechanical energy of the book–Earth system remains fixed; gravitational potential energy transforms to kinetic energy, and the total energy of the system remains constant. Nonconservative forces acting within a system, however, cause a *change* in the mechanical energy of the system. For example, for a book sent sliding on a horizontal surface that is not frictionless (Fig. 7.18a), the mechanical energy of the book–surface system is transformed to internal energy as we discussed earlier. Only part of the book's kinetic energy is transformed to internal energy in the book. The rest appears as internal energy in the surface. (When you trip and slide across a gymnasium floor, not only does the skin on your knees warm up, so does the floor!) Because the force of kinetic friction transforms the mechanical energy of a system into internal energy, it is a nonconservative force.

As an example of the path dependence of the work for a nonconservative force, consider Figure 7.19. Suppose you displace a book between two points on a table. If the book is displaced in a straight line along the blue path between points Ⓐ and Ⓑ in Figure 7.19, you do a certain amount of work against the kinetic friction force to keep the book moving at a constant speed. Now, imagine that you push the book along the brown semicircular path in Figure 7.19. You perform more work against friction along this curved path than along the straight path because the curved path is longer. The work done on the book depends on the path, so the friction force *cannot* be conservative.

The work done in moving the book is greater along the brown path than along the blue path.

Figure 7.19 The work done against the force of kinetic friction depends on the path taken as the book is moved from Ⓐ to Ⓑ.

7.8 Relationship Between Conservative Forces and Potential Energy

In the preceding section, we found that the work done on a member of a system by a conservative force between the members of the system does not depend on the path taken by the moving member. The work depends only on the initial and final coordinates. For such a system, we can define a **potential energy function U** such that the work done within the system by the conservative force equals the negative of the change in the potential energy of the system according to Equation 7.24. Let us imagine a system of particles in which a conservative force $\vec{F}$ acts between the particles. Imagine also that the configuration of the system changes due to the motion of one particle along the x axis. Then we can evaluate the internal work done by this force as the particle moves along the x axis[4] using Equations 7.7 and 7.24:

$$W_{int} = \int_{x_i}^{x_f} F_x \, dx = -\Delta U \tag{7.26}$$

[4]For a general displacement, the work done in two or three dimensions also equals $-\Delta U$, where $U = U(x, y, z)$. We write this equation formally as $W_{int} = \int_i^f \vec{F} \cdot d\vec{r} = U_i - U_f$.

where F_x is the component of $\vec{F}$ in the direction of the displacement. We can also express Equation 7.26 as

$$\Delta U = U_f - U_i = -\int_{x_i}^{x_f} F_x \, dx \qquad (7.27)$$

Therefore, ΔU is negative when F_x and dx are in the same direction, as when an object is lowered in a gravitational field or when a spring pushes an object toward equilibrium.

It is often convenient to establish some particular location x_i of one member of a system as representing a reference configuration and measure all potential energy differences with respect to it. We can then define the potential energy function as

$$U_f(x) = -\int_{x_i}^{x_f} F_x \, dx + U_i \qquad (7.28)$$

The value of U_i is often taken to be zero for the reference configuration. It does not matter what value we assign to U_i because any nonzero value merely shifts $U_f(x)$ by a constant amount and only the *change* in potential energy is physically meaningful.

If the point of application of the force undergoes an infinitesimal displacement dx, we can express the infinitesimal change in the potential energy of the system dU as

$$dU = -F_x \, dx$$

Therefore, the conservative force is related to the potential energy function through the relationship[5]

$$\boxed{F_x = -\frac{dU}{dx}} \qquad (7.29)$$

◀ Relation of force between members of a system to the potential energy of the system

That is, the x component of a conservative force acting on a member within a system equals the negative derivative of the potential energy of the system with respect to x.

We can easily check Equation 7.29 for the two examples already discussed. In the case of the deformed spring, $U_s = \frac{1}{2}kx^2$; therefore,

$$F_s = -\frac{dU_s}{dx} = -\frac{d}{dx}\left(\tfrac{1}{2}kx^2\right) = -kx$$

which corresponds to the restoring force in the spring (Hooke's law). Because the gravitational potential energy function is $U_g = mgy$, it follows from Equation 7.29 that $F_g = -mg$ when we differentiate U_g with respect to y instead of x.

We now see that U is an important function because a conservative force can be derived from it. Furthermore, Equation 7.29 should clarify that adding a constant to the potential energy is unimportant because the derivative of a constant is zero.

Quick Quiz 7.8 What does the slope of a graph of $U(x)$ versus x represent? **(a)** the magnitude of the force on the object **(b)** the negative of the magnitude of the force on the object **(c)** the x component of the force on the object **(d)** the negative of the x component of the force on the object

7.9 Energy Diagrams and Equilibrium of a System

The motion of a system can often be understood qualitatively through a graph of its potential energy versus the position of a member of the system. Consider the potential

---

[5]In three dimensions, the expression is

$$\vec{F} = -\frac{\partial U}{\partial x}\hat{i} - \frac{\partial U}{\partial y}\hat{j} - \frac{\partial U}{\partial z}\hat{k}$$

where $(\partial U/\partial x)$ and so forth are partial derivatives. In the language of vector calculus, $\vec{F}$ equals the negative of the *gradient* of the scalar quantity $U(x, y, z)$.

$U_s = \frac{1}{2}kx^2$

a

The restoring force exerted by the spring always acts toward $x = 0$, the position of stable equilibrium.

b

Figure 7.20 (a) Potential energy as a function of x for the friction-less block–spring system shown in (b). For a given energy E of the system, the block oscillates between the turning points, which have the coordinates $x = \pm x_{max}$.

Pitfall Prevention 7.10

Energy Diagrams A common mistake is to think that potential energy on the graph in an energy diagram represents the height of some object. For example, that is not the case in Figure 7.20, where the block is only moving horizontally.

energy function for a block–spring system, given by $U_s = \frac{1}{2}kx^2$. This function is plotted versus x in Figure 7.20a, where x is the position of the block. The force F_s exerted by the spring on the block is related to U_s through Equation 7.29:

$$F_s = -\frac{dU_s}{dx} = -kx$$

As we saw in Quick Quiz 7.8, the x component of the force is equal to the negative of the slope of the U-versus-x curve. When the block is placed at rest at the equilibrium position of the spring ($x = 0$), where $F_s = 0$, it will remain there unless some external force F_{ext} acts on it. If this external force stretches the spring from equilibrium, x is positive and the slope dU/dx is positive; therefore, the force F_s exerted by the spring is negative and the block accelerates back toward $x = 0$ when released. If the external force compresses the spring, x is negative and the slope is negative; therefore, F_s is positive and again the mass accelerates toward $x = 0$ upon release.

From this analysis, we conclude that the $x = 0$ position for a block–spring system is one of **stable equilibrium.** That is, any movement away from this position results in a force directed back toward $x = 0$. In general, configurations of a system in stable equilibrium correspond to those for which $U(x)$ for the system is a minimum.

If the block in Figure 7.20 is moved to an initial position x_{max} and then released from rest, its total energy initially is the potential energy $\frac{1}{2}kx_{max}^2$ stored in the spring. As the block starts to move, the system acquires kinetic energy and loses potential energy. The block oscillates (moves back and forth) between the two points $x = -x_{max}$ and $x = +x_{max}$, called the *turning points*. In fact, because no energy is transformed to internal energy due to friction, the block oscillates between $-x_{max}$ and $+x_{max}$ forever. (We will discuss these oscillations further in Chapter 15.)

Another simple mechanical system with a configuration of stable equilibrium is a ball rolling about in the bottom of a bowl. Anytime the ball is displaced from its lowest position, it tends to return to that position when released.

Now consider a particle moving along the x axis under the influence of a conservative force F_x, where the U-versus-x curve is as shown in Figure 7.21. Once again, $F_x = 0$ at $x = 0$, and so the particle is in equilibrium at this point. This position, however, is one of **unstable equilibrium** for the following reason. Suppose the particle is displaced to the right ($x > 0$). Because the slope is negative for $x > 0$, $F_x = -dU/dx$ is positive and the particle accelerates away from $x = 0$. If instead the particle is at $x = 0$ and is displaced to the left ($x < 0$), the force is negative because the slope is positive for $x < 0$ and the particle again accelerates away from the equilibrium position. The position $x = 0$ in this situation is one of unstable equilibrium because for any displacement from this point, the force pushes the particle farther away from equilibrium and toward a position of lower potential energy. A pencil balanced on its point is in a position of unstable equilibrium. If the pencil is displaced slightly from its absolutely vertical position and is then released, it will surely fall over. In general, configurations of a system in unstable equilibrium correspond to those for which $U(x)$ for the system is a maximum.

Finally, a configuration called **neutral equilibrium** arises when U is constant over some region. Small displacements of an object from a position in this region produce neither restoring nor disrupting forces. A ball lying on a flat, horizontal surface is an example of an object in neutral equilibrium.

Figure 7.21 A plot of U versus x for a particle that has a position of unstable equilibrium located at $x = 0$. For any finite displacement of the particle, the force on the particle is directed away from $x = 0$.

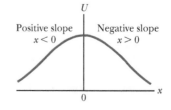

Example 7.9   Force and Energy on an Atomic Scale

The potential energy associated with the force between two neutral atoms in a molecule can be modeled by the Lennard–Jones potential energy function:

$$U(x) = 4\epsilon\left[\left(\frac{\sigma}{x}\right)^{12} - \left(\frac{\sigma}{x}\right)^{6}\right]$$

where x is the separation of the atoms. The function $U(x)$ contains two parameters σ and ϵ that are determined from experiments. Sample values for the interaction between two atoms in a molecule are $\sigma = 0.263$ nm and $\epsilon = 1.51 \times 10^{-22}$ J. Using a spreadsheet or similar tool, graph this function and find the most likely distance between the two atoms.

SOLUTION

Conceptualize We identify the two atoms in the molecule as a system. Based on our understanding that stable molecules exist, we expect to find stable equilibrium when the two atoms are separated by some equilibrium distance.

Categorize Because a potential energy function exists, we categorize the force between the atoms as conservative. For a conservative force, Equation 7.29 describes the relationship between the force and the potential energy function.

Analyze Stable equilibrium exists for a separation distance at which the potential energy of the system of two atoms (the molecule) is a minimum.

Take the derivative of the function $U(x)$:

$$\frac{dU(x)}{dx} = 4\epsilon\frac{d}{dx}\left[\left(\frac{\sigma}{x}\right)^{12} - \left(\frac{\sigma}{x}\right)^{6}\right] = 4\epsilon\left[\frac{-12\sigma^{12}}{x^{13}} + \frac{6\sigma^{6}}{x^{7}}\right]$$

Minimize the function $U(x)$ by setting its derivative equal to zero:

$$4\epsilon\left[\frac{-12\sigma^{12}}{x_{eq}^{13}} + \frac{6\sigma^{6}}{x_{eq}^{7}}\right] = 0 \quad \rightarrow \quad x_{eq} = (2)^{1/6}\sigma$$

Evaluate x_{eq}, the equilibrium separation of the two atoms in the molecule:

$$x_{eq} = (2)^{1/6}(0.263\text{ nm}) = \boxed{2.95 \times 10^{-10}\text{ m}}$$

We graph the Lennard–Jones function on both sides of this critical value to create our energy diagram as shown in Figure 7.22.

Finalize Notice that $U(x)$ is extremely large when the atoms are very close together, is a minimum when the atoms are at their critical separation, and then increases again as the atoms move apart. When $U(x)$ is a minimum, the atoms are in stable equilibrium, indicating that the most likely separation between them occurs at this point.

Figure 7.22 (Example 7.9) Potential energy curve associated with a molecule. The distance x is the separation between the two atoms making up the molecule.

Summary

Definitions

A **system** is most often a single particle, a collection of particles, or a region of space, and may vary in size and shape. A **system boundary** separates the system from the **environment**.

The **work** W done on a system by an agent exerting a constant force $\vec{F}$ on the system is the product of the magnitude Δr of the displacement of the point of application of the force and the component $F\cos\theta$ of the force along the direction of the displacement $\Delta\vec{r}$:

$$W \equiv F\Delta r\cos\theta \qquad (7.1)$$

continued

If a varying force does work on a particle as the particle moves along the *x* axis from x_i to x_f, the work done by the force on the particle is given by

$$W = \int_{x_i}^{x_f} F_x \, dx \qquad (7.7)$$

where F_x is the component of force in the *x* direction.

The **scalar product** (dot product) of two vectors $\vec{A}$ and $\vec{B}$ is defined by the relationship

$$\vec{A} \cdot \vec{B} \equiv AB \cos \theta \qquad (7.2)$$

where the result is a scalar quantity and θ is the angle between the two vectors. The scalar product obeys the commutative and distributive laws.

The **kinetic energy** of a particle of mass *m* moving with a speed *v* is

$$K \equiv \tfrac{1}{2}mv^2 \qquad (7.16)$$

If a particle of mass *m* is at a distance *y* above the Earth's surface, the **gravitational potential energy** of the particle–Earth system is

$$U_g \equiv mgy \qquad (7.19)$$

The **elastic potential energy** stored in a spring of force constant *k* is

$$U_s \equiv \tfrac{1}{2}kx^2 \qquad (7.22)$$

A force is **conservative** if the work it does on a particle that is a member of the system as the particle moves between two points is independent of the path the particle takes between the two points. Furthermore, a force is conservative if the work it does on a particle is zero when the particle moves through an arbitrary closed path and returns to its initial position. A force that does not meet these criteria is said to be **nonconservative.**

The **total mechanical energy of a system** is defined as the sum of the kinetic energy and the potential energy:

$$E_{\text{mech}} \equiv K + U \qquad (7.25)$$

Concepts and Principles

The **work–kinetic energy theorem** states that if work is done on a system by external forces and the only change in the system is in its speed,

$$W_{\text{ext}} = K_f - K_i = \Delta K = \tfrac{1}{2}mv_f^2 - \tfrac{1}{2}mv_i^2 \quad (7.15, 7.17)$$

A **potential energy function** *U* can be associated only with a conservative force. If a conservative force $\vec{F}$ acts between members of a system while one member moves along the *x* axis from x_i to x_f, the change in the potential energy of the system equals the negative of the work done by that force:

$$U_f - U_i = -\int_{x_i}^{x_f} F_x \, dx \qquad (7.27)$$

Systems can be in three types of equilibrium configurations when the net force on a member of the system is zero. Configurations of **stable equilibrium** correspond to those for which $U(x)$ is a minimum.

Configurations of **unstable equilibrium** correspond to those for which $U(x)$ is a maximum.

Neutral equilibrium arises when *U* is constant as a member of the system moves over some region.

Objective Questions   1. denotes answer available in *Student Solutions Manual/Study Guide*

1. Alex and John are loading identical cabinets onto a truck. Alex lifts his cabinet straight up from the ground to the bed of the truck, whereas John slides his cabinet up a rough ramp to the truck. Which statement is correct about the work done on the cabinet–Earth system? (a) Alex and John do the same amount of work. (b) Alex does more work than John. (c) John does more work than Alex. (d) None of those statements is necessarily true because the force of friction is unknown. (e) None of those statements is necessarily true because the angle of the incline is unknown.

2. If the net work done by external forces on a particle is zero, which of the following statements about the particle must be true? (a) Its velocity is zero. (b) Its velocity is decreased. (c) Its velocity is unchanged. (d) Its speed is unchanged. (e) More information is needed.

3. A worker pushes a wheelbarrow with a horizontal force of 50 N on level ground over a distance of 5.0 m. If a friction force of 43 N acts on the wheelbarrow in a direction opposite that of the worker, what work is done on the wheelbarrow by the worker? (a) 250 J (b) 215 J (c) 35 J (d) 10 J (e) None of those answers is correct.

4. A cart is set rolling across a level table, at the same speed on every trial. If it runs into a patch of sand, the cart exerts on the sand an average horizontal force of 6 N and travels a distance of 6 cm through the sand as it comes to a stop. If instead the cart runs into a patch of gravel on which the cart exerts an average horizontal force of 9 N, how far into the gravel will the cart roll before stopping? (a) 9 cm (b) 6 cm (c) 4 cm (d) 3 cm (e) none of those answers

5. Let $\hat{\mathbf{N}}$ represent the direction horizontally north, $\widehat{\mathbf{NE}}$ represent northeast (halfway between north and east), and so on. Each direction specification can be thought of as a unit vector. Rank from the largest to the smallest the following dot products. Note that zero is larger than a negative number. If two quantities are equal, display that fact in your ranking. (a) $\hat{\mathbf{N}} \cdot \hat{\mathbf{N}}$ (b) $\hat{\mathbf{N}} \cdot \widehat{\mathbf{NE}}$ (c) $\hat{\mathbf{N}} \cdot \hat{\mathbf{S}}$ (d) $\hat{\mathbf{N}} \cdot \hat{\mathbf{E}}$ (e) $\widehat{\mathbf{SE}} \cdot \hat{\mathbf{S}}$

6. Is the work required to be done by an external force on an object on a frictionless, horizontal surface to accelerate it from a speed v to a speed $2v$ (a) equal to the work required to accelerate the object from $v = 0$ to v, (b) twice the work required to accelerate the object from $v = 0$ to v, (c) three times the work required to accelerate the object from $v = 0$ to v, (d) four times the work required to accelerate the object from 0 to v, or (e) not known without knowledge of the acceleration?

7. A block of mass m is dropped from the fourth floor of an office building and hits the sidewalk below at speed v. From what floor should the block be dropped to double that impact speed? (a) the sixth floor (b) the eighth floor (c) the tenth floor (d) the twelfth floor (e) the sixteenth floor

8. As a simple pendulum swings back and forth, the forces acting on the suspended object are (a) the gravitational force, (b) the tension in the supporting cord, and (c) air resistance. (i) Which of these forces, if any, does no work on the pendulum at any time? (ii) Which of these forces does negative work on the pendulum at all times during its motion?

9. Bullet 2 has twice the mass of bullet 1. Both are fired so that they have the same speed. If the kinetic energy of bullet 1 is K, is the kinetic energy of bullet 2 (a) $0.25K$, (b) $0.5K$, (c) $0.71K$, (d) K, or (e) $2K$?

10. Figure OQ7.10 shows a light extended spring exerting a force F_s to the left on a block. (i) Does the block exert a force on the spring? Choose every correct answer. (a) No, it doesn't. (b) Yes, it does, to the left. (c) Yes, it does, to the right. (d) Yes, it does, and its magnitude is larger than F_s. (e) Yes, it does, and its magnitude is equal to F_s. (ii) Does the spring exert a force

Figure OQ7.10

on the wall? Choose your answers from the same list (a) through (e).

11. If the speed of a particle is doubled, what happens to its kinetic energy? (a) It becomes four times larger. (b) It becomes two times larger. (c) It becomes $\sqrt{2}$ times larger. (d) It is unchanged. (e) It becomes half as large.

12. Mark and David are loading identical cement blocks onto David's pickup truck. Mark lifts his block straight up from the ground to the truck, whereas David slides his block up a ramp containing frictionless rollers. Which statement is true about the work done on the block–Earth system? (a) Mark does more work than David. (b) Mark and David do the same amount of work. (c) David does more work than Mark. (d) None of those statements is necessarily true because the angle of the incline is unknown. (e) None of those statements is necessarily true because the mass of one block is not given.

13. (i) Rank the gravitational accelerations you would measure for the following falling objects: (a) a 2-kg object 5 cm above the floor, (b) a 2-kg object 120 cm above the floor, (c) a 3-kg object 120 cm above the floor, and (d) a 3-kg object 80 cm above the floor. List the one with the largest magnitude of acceleration first. If any are equal, show their equality in your list. (ii) Rank the gravitational forces on the same four objects, listing the one with the largest magnitude first. (iii) Rank the gravitational potential energies (of the object–Earth system) for the same four objects, largest first, taking $y = 0$ at the floor.

14. A certain spring that obeys Hooke's law is stretched by an external agent. The work done in stretching the spring by 10 cm is 4 J. How much additional work is required to stretch the spring an additional 10 cm? (a) 2 J (b) 4 J (c) 8 J (d) 12 J (e) 16 J

15. A cart is set rolling across a level table, at the same speed on every trial. If it runs into a patch of sand, the cart exerts on the sand an average horizontal force of 6 N and travels a distance of 6 cm through the sand as it comes to a stop. If instead the cart runs into a patch of flour, it rolls an average of 18 cm before stopping. What is the average magnitude of the horizontal force the cart exerts on the flour? (a) 2 N (b) 3 N (c) 6 N (d) 18 N (e) none of those answers

16. An ice cube has been given a push and slides without friction on a level table. Which is correct? (a) It is in stable equilibrium. (b) It is in unstable equilibrium. (c) It is in neutral equilibrium. (d) It is not in equilibrium.

1. Can a normal force do work? If not, why not? If so, give an example.

2. Object 1 pushes on object 2 as the objects move together, like a bulldozer pushing a stone. Assume object 1 does 15.0 J of work on object 2. Does object 2 do work on object 1? Explain your answer. If possible, determine how much work and explain your reasoning.

3. A student has the idea that the total work done on an object is equal to its final kinetic energy. Is this idea true always, sometimes, or never? If it is sometimes true, under what circumstances? If it is always or never true, explain why.

4. (a) For what values of the angle θ between two vectors is their scalar product positive? (b) For what values of θ is their scalar product negative?

5. Can kinetic energy be negative? Explain.

6. Discuss the work done by a pitcher throwing a baseball. What is the approximate distance through which the force acts as the ball is thrown?

7. Discuss whether any work is being done by each of the following agents and, if so, whether the work is positive or negative. (a) a chicken scratching the ground (b) a person studying (c) a crane lifting a bucket of concrete (d) the gravitational force on the bucket in part (c) (e) the leg muscles of a person in the act of sitting down

8. If only one external force acts on a particle, does it necessarily change the particle's (a) kinetic energy? (b) Its velocity?

9. Preparing to clean them, you pop all the removable keys off a computer keyboard. Each key has the shape of a tiny box with one side open. By accident, you spill the keys onto the floor. Explain why many more keys land letter-side down than land open-side down.

10. You are reshelving books in a library. You lift a book from the floor to the top shelf. The kinetic energy of the book on the floor was zero and the kinetic energy of the book on the top shelf is zero, so no change occurs in the kinetic energy, yet you did some work in lifting the book. Is the work–kinetic energy theorem violated? Explain.

11. A certain uniform spring has spring constant k. Now the spring is cut in half. What is the relationship between k and the spring constant k' of each resulting smaller spring? Explain your reasoning.

12. What shape would the graph of U versus x have if a particle were in a region of neutral equilibrium?

13. Does the kinetic energy of an object depend on the frame of reference in which its motion is measured? Provide an example to prove this point.

14. Cite two examples in which a force is exerted on an object without doing any work on the object.

Section 7.2 Work Done by a Constant Force

1. A shopper in a supermarket pushes a cart with a force of 35.0 N directed at an angle of 25.0° below the horizontal. The force is just sufficient to balance various friction forces, so the cart moves at constant speed. (a) Find the work done by the shopper on the cart as she moves down a 50.0-m-long aisle. (b) The shopper goes down the next aisle, pushing horizontally and maintaining the same speed as before. If the friction force doesn't change, would the shopper's applied force be larger, smaller, or the same? (c) What about the work done on the cart by the shopper?

2. A raindrop of mass 3.35×10^{-5} kg falls vertically at constant speed under the influence of gravity and air resistance. Model the drop as a particle. As it falls 100 m, what is the work done on the raindrop (a) by the gravitational force and (b) by air resistance?

3. In 1990, Walter Arfeuille of Belgium lifted a 281.5-kg object through a distance of 17.1 cm using only his teeth. (a) How much work was done on the object by Arfeuille in this lift, assuming the object was lifted at constant speed? (b) What total force was exerted on Arfeuille's teeth during the lift?

4. The record number of boat lifts, including the boat and its ten crew members, was achieved by Sami Heinonen and Juha Räsänen of Sweden in 2000. They lifted a total mass of 653.2 kg approximately 4 in. off the ground a total of 24 times. Estimate the total work done by the two men on the boat in this record lift, ignoring the negative work done by the men when they lowered the boat back to the ground.

5. A block of mass $m =$ 2.50 kg is pushed a distance $d = 2.20$ m along a frictionless, horizontal table by a constant applied force of magnitude $F = 16.0$ N directed at an angle $\theta = 25.0°$ below the horizontal as shown in Figure P7.5. Determine the work done on the block by (a) the applied force, (b) the normal force exerted by the table, (c) the gravitational force, and (d) the net force on the block.

Figure P7.5

6. Spiderman, whose mass is 80.0 kg, is dangling on the free end of a 12.0-m-long rope, the other end of which is fixed to a tree limb above. By repeatedly bending at the waist, he is able to get the rope in motion, eventually getting it to swing enough that he can reach a ledge when the rope makes a 60.0° angle with the vertical. How much work was done by the gravitational force on Spiderman in this maneuver?

Section 7.3 The Scalar Product of Two Vectors

7. For any two vectors $\vec{A}$ and $\vec{B}$, show that $\vec{A} \cdot \vec{B} = A_x B_x + A_y B_y + A_z B_z$. *Suggestions:* Write $\vec{A}$ and $\vec{B}$ in unit-vector form and use Equations 7.4 and 7.5.

8. Vector $\vec{A}$ has a magnitude of 5.00 units, and vector $\vec{B}$ has a magnitude of 9.00 units. The two vectors make an angle of 50.0° with each other. Find $\vec{A} \cdot \vec{B}$.

Note: In Problems 9 through 12, calculate numerical answers to three significant figures as usual.

9. For $\vec{A} = 3\hat{i} + \hat{j} - \hat{k}$, $\vec{B} = -\hat{i} + 2\hat{j} + 5\hat{k}$, and $\vec{C} = 2\hat{j} - 3\hat{k}$, find $\vec{C} \cdot (\vec{A} - \vec{B})$.

10. Find the scalar product of the vectors in Figure P7.10.

Figure P7.10

11. A force $\vec{F} = (6\hat{i} - 2\hat{j})$ N acts on a particle that undergoes a displacement $\Delta\vec{r} = (3\hat{i} + \hat{j})$ m. Find (a) the work done by the force on the particle and (b) the angle between $\vec{F}$ and $\Delta\vec{r}$.

12. Using the definition of the scalar product, find the angles between (a) $\vec{A} = 3\hat{i} - 2\hat{j}$ and $\vec{B} = 4\hat{i} - 4\hat{j}$, (b) $\vec{A} = -2\hat{i} + 4\hat{j}$ and $\vec{B} = 3\hat{i} - 4\hat{j} + 2\hat{k}$, and (c) $\vec{A} = \hat{i} - 2\hat{j} + 2\hat{k}$ and $\vec{B} = 3\hat{j} + 4\hat{k}$.

13. Let $\vec{B} = 5.00$ m at 60.0°. Let the vector $\vec{C}$ have the same magnitude as $\vec{A}$ and a direction angle greater than that of $\vec{A}$ by 25.0°. Let $\vec{A} \cdot \vec{B} = 30.0$ m^2 and $\vec{B} \cdot \vec{C} = 35.0$ m^2. Find the magnitude and direction of $\vec{A}$.

Section 7.4 Work Done by a Varying Force

14. The force acting on a particle varies as shown in Figure P7.14. Find the work done by the force on the particle as it moves (a) from $x = 0$ to $x = 8.00$ m, (b) from $x = 8.00$ m to $x = 10.0$ m, and (c) from $x = 0$ to $x = 10.0$ m.

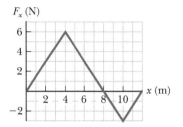

Figure P7.14

15. A particle is subject to a force F_x that varies with position as shown in Figure P7.15. Find the work done by the force on the particle as it moves (a) from $x = 0$ to $x = 5.00$ m, (b) from $x = 5.00$ m to $x = 10.0$ m, and (c) from $x = 10.0$ m to $x = 15.0$ m. (d) What is the total work done by the force over the distance $x = 0$ to $x = 15.0$ m?

Figure P7.15 Problems 15 and 34.

16. In a control system, an accelerometer consists of a 4.70-g object sliding on a calibrated horizontal rail. A low-mass spring attaches the object to a flange at one end of the rail. Grease on the rail makes static friction negligible, but rapidly damps out vibrations of the sliding object. When subject to a steady acceleration of $0.800g$, the object should be at a location 0.500 cm away from its equilibrium position. Find the force constant of the spring required for the calibration to be correct.

17. When a 4.00-kg object is hung vertically on a certain light spring that obeys Hooke's law, the spring stretches 2.50 cm. If the 4.00-kg object is removed, (a) how far will the spring stretch if a 1.50-kg block is hung on it? (b) How much work must an external agent do to stretch the same spring 4.00 cm from its unstretched position?

18. Hooke's law describes a certain light spring of unstretched length 35.0 cm. When one end is attached to the top of a doorframe and a 7.50-kg object is hung from the other end, the length of the spring is 41.5 cm. (a) Find its spring constant. (b) The load and the spring are taken down. Two people pull in opposite directions on the ends of the spring, each with a force of 190 N. Find the length of the spring in this situation.

19. An archer pulls her bowstring back 0.400 m by exerting a force that increases uniformly from zero to 230 N. (a) What is the equivalent spring constant of the bow?

(b) How much work does the archer do on the string in drawing the bow?

20. A light spring with spring constant 1 200 N/m is hung from an elevated support. From its lower end hangs a second light spring, which has spring constant 1 800 N/m. An object of mass 1.50 kg is hung at rest from the lower end of the second spring. (a) Find the total extension distance of the pair of springs. (b) Find the effective spring constant of the pair of springs as a system. We describe these springs as *in series*.

21. A light spring with spring constant k_1 is hung from an elevated support. From its lower end a second light spring is hung, which has spring constant k_2. An object of mass m is hung at rest from the lower end of the second spring. (a) Find the total extension distance of the pair of springs. (b) Find the effective spring constant of the pair of springs as a system.

22. Express the units of the force constant of a spring in SI fundamental units.

23. A cafeteria tray dispenser supports a stack of trays on a shelf that hangs from four identical spiral springs under tension, one near each corner of the shelf. Each tray is rectangular, 45.3 cm by 35.6 cm, 0.450 cm thick, and with mass 580 g. (a) Demonstrate that the top tray in the stack can always be at the same height above the floor, however many trays are in the dispenser. (b) Find the spring constant each spring should have for the dispenser to function in this convenient way. (c) Is any piece of data unnecessary for this determination?

24. A light spring with force constant 3.85 N/m is compressed by 8.00 cm as it is held between a 0.250-kg block on the left and a 0.500-kg block on the right, both resting on a horizontal surface. The spring exerts a force on each block, tending to push the blocks apart. The blocks are simultaneously released from rest. Find the acceleration with which each block starts to move, given that the coefficient of kinetic friction between each block and the surface is (a) 0, (b) 0.100, and (c) 0.462.

25. A small particle of mass m is pulled to the top of a frictionless half-cylinder (of radius R) by a light cord that passes over the top of the cylinder as illustrated in Figure P7.25. (a) Assuming the particle moves at a constant speed, show that $F = mg \cos \theta$. *Note:* If the particle moves at constant speed, the component of its acceleration tangent to the cylinder must be zero at all times. (b) By directly integrating $W = \int \vec{\mathbf{F}} \cdot d\vec{\mathbf{r}}$, find the work done in moving the particle at constant speed from the bottom to the top of the half-cylinder.

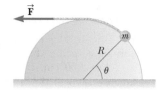

Figure P7.25

26. The force acting on a particle is $F_x = (8x - 16)$, where F is in newtons and x is in meters. (a) Make a plot of this force versus x from $x = 0$ to $x = 3.00$ m. (b) From your graph, find the net work done by this force on the particle as it moves from $x = 0$ to $x = 3.00$ m.

27. When different loads hang on a spring, the spring stretches to different lengths as shown in the following table. (a) Make a graph of the applied force versus the extension of the spring. (b) By least-squares fitting, determine the straight line that best fits the data. (c) To complete part (b), do you want to use all the data points, or should you ignore some of them? Explain. (d) From the slope of the best-fit line, find the spring constant k. (e) If the spring is extended to 105 mm, what force does it exert on the suspended object?

F (N)	2.0	4.0	6.0	8.0	10	12	14	16	18	20	22
L (mm)	15	32	49	64	79	98	112	126	149	175	190

28. A 100-g bullet is fired from a rifle having a barrel 0.600 m long. Choose the origin to be at the location where the bullet begins to move. Then the force (in newtons) exerted by the expanding gas on the bullet is $15\,000 + 10\,000x - 25\,000x^2$, where x is in meters. (a) Determine the work done by the gas on the bullet as the bullet travels the length of the barrel. (b) **What If?** If the barrel is 1.00 m long, how much work is done, and (c) how does this value compare with the work calculated in part (a)?

29. A force $\vec{\mathbf{F}} = (4x\hat{\mathbf{i}} + 3y\hat{\mathbf{j}})$, where $\vec{\mathbf{F}}$ is in newtons and x and y are in meters, acts on an object as the object moves in the x direction from the origin to $x = 5.00$ m. Find the work $W = \int \vec{\mathbf{F}} \cdot d\vec{\mathbf{r}}$ done by the force on the object.

30. **Review.** The graph in Figure P7.30 specifies a functional relationship between the two variables u and v. (a) Find $\int_a^b u\,dv$. (b) Find $\int_b^a u\,dv$. (c) Find $\int_a^b v\,du$.

Figure P7.30

Section 7.5 Kinetic Energy and the Work–Kinetic Energy Theorem

31. A 3.00-kg object has a velocity $(6.00\hat{\mathbf{i}} - 2.00\hat{\mathbf{j}})$ m/s. (a) What is its kinetic energy at this moment? (b) What is the net work done on the object if its velocity changes to $(8.00\hat{\mathbf{i}} + 4.00\hat{\mathbf{j}})$ m/s? (*Note:* From the definition of the dot product, $v^2 = \vec{\mathbf{v}} \cdot \vec{\mathbf{v}}$.)

32. A worker pushing a 35.0-kg wooden crate at a constant speed for 12.0 m along a wood floor does 350 J of work by applying a constant horizontal force of magnitude F on the crate. (a) Determine the value of F. (b) If the worker now applies a force greater than F, describe the subsequent motion of the crate. (c) Describe what would happen to the crate if the applied force is less than F.

33. A 0.600-kg particle has a speed of 2.00 m/s at point Ⓐ and kinetic energy of 7.50 J at point Ⓑ. What is (a) its kinetic energy at Ⓐ, (b) its speed at Ⓑ, and (c) the net work done on the particle by external forces as it moves from Ⓐ to Ⓑ?

34. A 4.00-kg particle is subject to a net force that varies
Ⓦ with position as shown in Figure P7.15. The particle
starts from rest at $x = 0$. What is its speed at (a) $x =$
5.00 m, (b) $x = 10.0$ m, and (c) $x = 15.0$ m?

35. A 2 100-kg pile driver is used to drive a steel I-beam into
Ⓜ the ground. The pile driver falls 5.00 m before coming
into contact with the top of the beam, and it drives the
beam 12.0 cm farther into the ground before coming
to rest. Using energy considerations, calculate the average force the beam exerts on the pile driver while the
pile driver is brought to rest.

36. Review. In an electron microscope, there is an electron
AMT gun that contains two charged metallic plates 2.80 cm
apart. An electric force accelerates each electron in
the beam from rest to 9.60% of the speed of light over
this distance. (a) Determine the kinetic energy of the
electron as it leaves the electron gun. Electrons carry
this energy to a phosphorescent viewing screen where
the microscope's image is formed, making it glow. For
an electron passing between the plates in the electron
gun, determine (b) the magnitude of the constant
electric force acting on the electron, (c) the acceleration of the electron, and (d) the time interval the electron spends between the plates.

37. Review. You can think of the work–kinetic energy the-
GP orem as a second theory of motion, parallel to Newton's laws in describing how outside influences affect
the motion of an object. In this problem, solve parts
(a), (b), and (c) separately from parts (d) and (e) so
you can compare the predictions of the two theories.
A 15.0-g bullet is accelerated from rest to a speed of
780 m/s in a rifle barrel of length 72.0 cm. (a) Find
the kinetic energy of the bullet as it leaves the barrel. (b) Use the work–kinetic energy theorem to find
the net work that is done on the bullet. (c) Use your
result to part (b) to find the magnitude of the average
net force that acted on the bullet while it was in the
barrel. (d) Now model the bullet as a particle under
constant acceleration. Find the constant acceleration
of a bullet that starts from rest and gains a speed of
780 m/s over a distance of 72.0 cm. (e) Modeling the
bullet as a particle under a net force, find the net
force that acted on it during its acceleration. (f) What
conclusion can you draw from comparing your results
of parts (c) and (e)?

38. Review. A 7.80-g bullet moving at 575 m/s strikes the
hand of a superhero, causing the hand to move 5.50 cm
in the direction of the bullet's velocity before stopping.
(a) Use work and energy considerations to find the
average force that stops the bullet. (b) Assuming the
force is constant, determine how much time elapses
between the moment the bullet strikes the hand and
the moment it stops moving.

39. Review. A 5.75-kg object passes through the origin
at time $t = 0$ such that its x component of velocity is
5.00 m/s and its y component of velocity is −3.00 m/s.
(a) What is the kinetic energy of the object at this time?
(b) At a later time $t = 2.00$ s, the particle is located at
$x = 8.50$ m and $y = 5.00$ m. What constant force acted

on the object during this time interval? (c) What is the
speed of the particle at $t = 2.00$ s?

Section 7.6 Potential Energy of a System

40. A 1 000-kg roller coaster car is initially at the top of a
rise, at point Ⓐ. It then moves 135 ft, at an angle of 40.0°
below the horizontal, to a lower point Ⓑ. (a) Choose
the car at point Ⓑ to be the zero configuration for
gravitational potential energy of the roller coaster–
Earth system. Find the potential energy of the system
when the car is at points Ⓐ and Ⓑ, and the change
in potential energy as the car moves between these
points. (b) Repeat part (a), setting the zero configuration with the car at point Ⓐ.

41. A 0.20-kg stone is held 1.3 m above the top edge of a
water well and then dropped into it. The well has a
depth of 5.0 m. Relative to the configuration with the
stone at the top edge of the well, what is the gravitational potential energy of the stone–Earth system
(a) before the stone is released and (b) when it reaches
the bottom of the well? (c) What is the change in gravitational potential energy of the system from release to
reaching the bottom of the well?

42. A 400-N child is in a swing that is attached to a pair
Ⓦ of ropes 2.00 m long. Find the gravitational potential
energy of the child–Earth system relative to the child's
lowest position when (a) the ropes are horizontal,
(b) the ropes make a 30.0° angle with the vertical, and
(c) the child is at the bottom of the circular arc.

Section 7.7 Conservative and Nonconservative Forces

43. A 4.00-kg particle moves
Ⓜ from the origin to position Ⓒ, having coordi-
nates $x = 5.00$ m and $y =$
5.00 m (Fig. P7.43). One
force on the particle is
the gravitational force
acting in the negative y
direction. Using Equa-
tion 7.3, calculate the
work done by the gravi-
tational force on the
particle as it goes from O

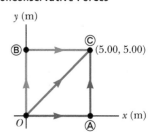

Figure P7.43
Problems 43 through 46.

to Ⓒ along (a) the purple path, (b) the red path, and
(c) the blue path. (d) Your results should all be identical. Why?

44. (a) Suppose a constant force acts on an object. The
force does not vary with time or with the position or
the velocity of the object. Start with the general definition for work done by a force

$$W = \int_i^f \vec{\mathbf{F}} \cdot d\vec{\mathbf{r}}$$

and show that the force is conservative. (b) As a spe-
cial case, suppose the force $\vec{\mathbf{F}} = (3\hat{\mathbf{i}} + 4\hat{\mathbf{j}})$ N acts on a
particle that moves from O to Ⓒ in Figure P7.43. Cal-
culate the work done by $\vec{\mathbf{F}}$ on the particle as it moves
along each one of the three paths shown in the figure

and show that the work done along the three paths is identical.

45. A force acting on a particle moving in the xy plane is given by $\vec{F} = (2y\hat{i} + x^2\hat{j})$, where $\vec{F}$ is in newtons and x and y are in meters. The particle moves from the origin to a final position having coordinates $x = 5.00$ m and $y = 5.00$ m as shown in Figure P7.43. Calculate the work done by $\vec{F}$ on the particle as it moves along (a) the purple path, (b) the red path, and (c) the blue path. (d) Is $\vec{F}$ conservative or nonconservative? (e) Explain your answer to part (d).

46. An object moves in the xy plane in Figure P7.43 and experiences a friction force with constant magnitude 3.00 N, always acting in the direction opposite the object's velocity. Calculate the work that you must do to slide the object at constant speed against the friction force as the object moves along (a) the purple path O to Ⓐ followed by a return purple path to O, (b) the purple path O to Ⓒ followed by a return blue path to O, and (c) the blue path O to Ⓒ followed by a return blue path to O. (d) Each of your three answers should be nonzero. What is the significance of this observation?

Section 7.8 Relationship Between Conservative Forces and Potential Energy

47. The potential energy of a system of two particles separated by a distance r is given by $U(r) = A/r$, where A is a constant. Find the radial force $\vec{F}_r$ that each particle exerts on the other.

48. *Why is the following situation impossible?* A librarian lifts a book from the ground to a high shelf, doing 20.0 J of work in the lifting process. As he turns his back, the book falls off the shelf back to the ground. The gravitational force from the Earth on the book does 20.0 J of work on the book while it falls. Because the work done was 20.0 J + 20.0 J = 40.0 J, the book hits the ground with 40.0 J of kinetic energy.

49. A potential energy function for a system in which a two-dimensional force acts is of the form $U = 3x^3y - 7x$. Find the force that acts at the point (x, y).

50. A single conservative force acting on a particle within a system varies as $\vec{F} = (-Ax + Bx^2)\hat{i}$, where A and B are constants, $\vec{F}$ is in newtons, and x is in meters. (a) Calculate the potential energy function $U(x)$ associated with this force for the system, taking $U = 0$ at $x = 0$. Find (b) the change in potential energy and (c) the change in kinetic energy of the system as the particle moves from $x = 2.00$ m to $x = 3.00$ m.

51. A single conservative force acts on a 5.00-kg particle within a system due to its interaction with the rest of the system. The equation $F_x = 2x + 4$ describes the force, where F_x is in newtons and x is in meters. As the particle moves along the x axis from $x = 1.00$ m to $x = 5.00$ m, calculate (a) the work done by this force on the particle, (b) the change in the potential energy of the system, and (c) the kinetic energy the particle has at $x = 5.00$ m if its speed is 3.00 m/s at $x = 1.00$ m.

Section 7.9 Energy Diagrams and Equilibrium of a System

52. For the potential energy curve shown in Figure P7.52, (a) determine whether the force F_x is positive, negative, or zero at the five points indicated. (b) Indicate points of stable, unstable, and neutral equilibrium. (c) Sketch the curve for F_x versus x from $x = 0$ to $x = 9.5$ m.

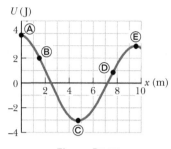

Figure P7.52

53. A right circular cone can theoretically be balanced on a horizontal surface in three different ways. Sketch these three equilibrium configurations and identify them as positions of stable, unstable, or neutral equilibrium.

Additional Problems

54. The potential energy function for a system of particles is given by $U(x) = -x^3 + 2x^2 + 3x$, where x is the position of one particle in the system. (a) Determine the force F_x on the particle as a function of x. (b) For what values of x is the force equal to zero? (c) Plot $U(x)$ versus x and F_x versus x and indicate points of stable and unstable equilibrium.

55. **Review.** A baseball outfielder throws a 0.150-kg baseball at a speed of 40.0 m/s and an initial angle of 30.0° to the horizontal. What is the kinetic energy of the baseball at the highest point of its trajectory?

56. A particle moves along the x axis from $x = 12.8$ m to $x = 23.7$ m under the influence of a force

$$F = \frac{375}{x^3 + 3.75x}$$

where F is in newtons and x is in meters. Using numerical integration, determine the work done by this force on the particle during this displacement. Your result should be accurate to within 2%.

57. Two identical steel balls, each of diameter 25.4 mm and moving in opposite directions at 5 m/s, run into each other head-on and bounce apart. Prior to the collision, one of the balls is squeezed in a vise while precise measurements are made of the resulting amount of compression. The results show that Hooke's law is a fair model of the ball's elastic behavior. For one datum, a force of 16 kN exerted by each jaw of the vise results in a 0.2-mm reduction in the diameter. The diameter returns to its original value when the force is removed. (a) Modeling the ball as a spring, find its spring constant. (b) Does the interaction of the balls during the collision last only for an instant or for a nonzero time interval? State your evidence. (c) Compute an estimate for the kinetic energy of each of the balls before they collide. (d) Compute an estimate for the maximum amount of compression each ball undergoes when the balls collide. (e) Compute an order-of-magnitude estimate for the time interval for which the balls are in contact. (In Chapter 15, you will learn to calculate the contact time interval precisely.)

58. When an object is displaced by an amount x from stable equilibrium, a restoring force acts on it, tending to return the object to its equilibrium position. The magnitude of the restoring force can be a complicated function of x. In such cases, we can generally imagine the force function $F(x)$ to be expressed as a power series in x as $F(x) = -(k_1 x + k_2 x^2 + k_3 x^3 + \cdots)$. The first term here is just Hooke's law, which describes the force exerted by a simple spring for small displacements. For small excursions from equilibrium, we generally ignore the higher-order terms, but in some cases it may be desirable to keep the second term as well. If we model the restoring force as $F = -(k_1 x + k_2 x^2)$, how much work is done on an object in displacing it from $x = 0$ to $x = x_{max}$ by an applied force $-F$?

59. A 6 000-kg freight car rolls along rails with negligible friction. The car is brought to rest by a combination of two coiled springs as illustrated in Figure P7.59. Both springs are described by Hooke's law and have spring constants $k_1 = 1\ 600$ N/m and $k_2 = 3\ 400$ N/m. After the first spring compresses a distance of 30.0 cm, the second spring acts with the first to increase the force as additional compression occurs as shown in the graph. The car comes to rest 50.0 cm after first contacting the two-spring system. Find the car's initial speed.

Figure P7.59

60. *Why is the following situation impossible?* In a new casino, a supersized pinball machine is introduced. Casino advertising boasts that a professional basketball player can lie on top of the machine and his head and feet will not hang off the edge! The ball launcher in the machine sends metal balls up one side of the machine and then into play. The spring in the launcher (Fig. P7.60) has a force constant of 1.20 N/cm. The surface on which the ball moves is inclined $\theta = 10.0°$ with respect to the horizontal. The spring is initially compressed its maximum distance $d = 5.00$ cm. A

ball of mass 100 g is projected into play by releasing the plunger. Casino visitors find the play of the giant machine quite exciting.

Figure P7.60

61. **Review.** Two constant forces act on an object of mass $m = 5.00$ kg moving in the xy plane as shown in Figure P7.61. Force $\vec{F}_1$ is 25.0 N at 35.0°, and force $\vec{F}_2$ is 42.0 N at 150°. At time $t = 0$, the object is at the origin and has velocity $(4.00\hat{i} + 2.50\hat{j})$ m/s. (a) Express the two forces in unit-vector notation. Use unit-vector notation for your other answers. (b) Find the total force exerted on the object. (c) Find the object's acceleration. Now, considering the instant $t = 3.00$ s, find (d) the object's velocity, (e) its position, (f) its kinetic energy from $\frac{1}{2}mv_f^2$, and (g) its kinetic energy from $\frac{1}{2}mv_i^2 + \sum \vec{F} \cdot \Delta \vec{r}$. (h) What conclusion can you draw by comparing the answers to parts (f) and (g)?

Figure P7.61

62. The spring constant of an automotive suspension spring increases with increasing load due to a spring coil that is widest at the bottom, smoothly tapering to a smaller diameter near the top. The result is a softer ride on normal road surfaces from the wider coils, but the car does not bottom out on bumps because when the lower coils collapse, the stiffer coils near the top absorb the load. For such springs, the force exerted by the spring can be empirically found to be given by $F = ax^b$. For a tapered spiral spring that compresses 12.9 cm with a 1 000-N load and 31.5 cm with a 5 000-N load, (a) evaluate the constants a and b in the empirical equation for F and (b) find the work needed to compress the spring 25.0 cm.

63. An inclined plane of angle $\theta = 20.0°$ has a spring of force constant $k = 500$ N/m fastened securely at the bottom so that the spring is parallel to the surface as shown in Figure P7.63. A block of mass $m = 2.50$ kg is placed on the

Figure P7.63
Problems 63 and 64.

plane at a distance $d = 0.300$ m from the spring. From this position, the block is projected downward toward the spring with speed $v = 0.750$ m/s. By what distance is the spring compressed when the block momentarily comes to rest?

64. An inclined plane of angle θ has a spring of force constant k fastened securely at the bottom so that the

spring is parallel to the surface. A block of mass m is placed on the plane at a distance d from the spring. From this position, the block is projected downward toward the spring with speed v as shown in Figure P7.63. By what distance is the spring compressed when the block momentarily comes to rest?

65. (a) Take $U = 5$ for a system with a particle at position $x = 0$ and calculate the potential energy of the system as a function of the particle position x. The force on the particle is given by $(8e^{-2x})\hat{i}$. (b) Explain whether the force is conservative or nonconservative and how you can tell.

Challenge Problems

66. A particle of mass $m = 1.18$ kg is attached between two identical springs on a frictionless, horizontal tabletop. Both springs have spring constant k and are initially unstressed, and the particle is at $x = 0$. (a) The particle is pulled a distance x along a direction perpendicular to the initial configuration of the springs as shown in Figure P7.66. Show that the force exerted by the springs on the particle is

$$\vec{F} = -2kx\left(1 - \frac{L}{\sqrt{x^2 + L^2}}\right)\hat{i}$$

(b) Show that the potential energy of the system is

$$U(x) = kx^2 + 2kL(L - \sqrt{x^2 + L^2})$$

(c) Make a plot of $U(x)$ versus x and identify all equilibrium points. Assume $L = 1.20$ m and $k = 40.0$ N/m. (d) If the particle is pulled 0.500 m to the right and then released, what is its speed when it reaches $x = 0$?

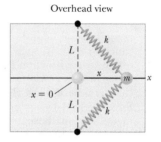

Figure P7.66

67. **Review.** A light spring has unstressed length 15.5 cm. It is described by Hooke's law with spring constant 4.30 N/m. One end of the horizontal spring is held on a fixed vertical axle, and the other end is attached to a puck of mass m that can move without friction over a horizontal surface. The puck is set into motion in a circle with a period of 1.30 s. (a) Find the extension of the spring x as it depends on m. Evaluate x for (b) $m = 0.070\ 0$ kg, (c) $m = 0.140$ kg, (d) $m = 0.180$ kg, and (e) $m = 0.190$ kg. (f) Describe the pattern of variation of x as it depends on m.

Conservation of Energy

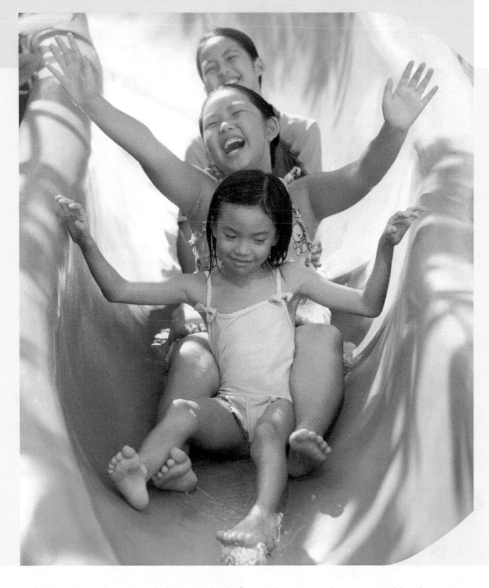

In Chapter 7, we introduced three methods for storing energy in a system: kinetic energy, associated with movement of members of the system; potential energy, determined by the configuration of the system; and internal energy, which is related to the temperature of the system.

We now consider analyzing physical situations using the energy approach for two types of systems: *nonisolated* and *isolated* systems. For nonisolated systems, we shall investigate ways that energy can cross the boundary of the system, resulting in a change in the system's total energy. This analysis leads to a critically important principle called *conservation of energy.* The conservation of energy principle extends well beyond physics and can be applied to biological organisms, technological systems, and engineering situations.

In isolated systems, energy does not cross the boundary of the system. For these systems, the total energy of the system is constant. If no nonconservative forces act within the system, we can use *conservation of mechanical energy* to solve a variety of problems.

Three youngsters enjoy the transformation of potential energy to kinetic energy on a waterslide. We can analyze processes such as these with the techniques developed in this chapter.
(Jade Lee/Asia Images/Getty Images)

211

Situations involving the transformation of mechanical energy to internal energy due to nonconservative forces require special handling. We investigate the procedures for these types of problems.

Finally, we recognize that energy can cross the boundary of a system at different rates. We describe the rate of energy transfer with the quantity *power*.

8.1 Analysis Model: Nonisolated System (Energy)

As we have seen, an object, modeled as a particle, can be acted on by various forces, resulting in a change in its kinetic energy according to the work–kinetic energy theorem from Chapter 7. If we choose the object as the system, this very simple situation is the first example of a *nonisolated system*, for which energy crosses the boundary of the system during some time interval due to an interaction with the environment. This scenario is common in physics problems. If a system does not interact with its environment, it is an *isolated system*, which we will study in Section 8.2.

The work–kinetic energy theorem is our first example of an energy equation appropriate for a nonisolated system. In the case of that theorem, the interaction of the system with its environment is the work done by the external force, and the quantity in the system that changes is the kinetic energy.

So far, we have seen only one way to transfer energy into a system: work. We mention below a few other ways to transfer energy into or out of a system. The details of these processes will be studied in other sections of the book. We illustrate mechanisms to transfer energy in Figure 8.1 and summarize them as follows.

Work, as we have learned in Chapter 7, is a method of transferring energy to a system by applying a force to the system such that the point of application of the force undergoes a displacement (Fig. 8.1a).

Energy is transferred to the block by *work*.

Energy leaves the radio from the speaker by *mechanical waves*.

Energy transfers to the handle of the spoon by *heat*.

Energy enters the automobile gas tank by *matter transfer*.

Energy enters the hair dryer by *electrical transmission*.

Energy leaves the light-bulb by *electromagnetic radiation*.

Figure 8.1 Energy transfer mechanisms. In each case, the system into which or from which energy is transferred is indicated.

Mechanical waves (Chapters 16–18) are a means of transferring energy by allowing a disturbance to propagate through air or another medium. It is the method by which energy (which you detect as sound) leaves the system of your clock radio through the loudspeaker and enters your ears to stimulate the hearing process (Fig. 8.1b). Other examples of mechanical waves are seismic waves and ocean waves.

Heat (Chapter 20) is a mechanism of energy transfer that is driven by a temperature difference between a system and its environment. For example, imagine dividing a metal spoon into two parts: the handle, which we identify as the system, and the portion submerged in a cup of coffee, which is part of the environment (Fig. 8.1c). The handle of the spoon becomes hot because fast-moving electrons and atoms in the submerged portion bump into slower ones in the nearby part of the handle. These particles move faster because of the collisions and bump into the next group of slow particles. Therefore, the internal energy of the spoon handle rises from energy transfer due to this collision process.

Matter transfer (Chapter 20) involves situations in which matter physically crosses the boundary of a system, carrying energy with it. Examples include filling your automobile tank with gasoline (Fig. 8.1d) and carrying energy to the rooms of your home by circulating warm air from the furnace, a process called *convection*.

Electrical transmission (Chapters 27 and 28) involves energy transfer into or out of a system by means of electric currents. It is how energy transfers into your hair dryer (Fig. 8.1e), home theater system, or any other electrical device.

Electromagnetic radiation (Chapter 34) refers to electromagnetic waves such as light (Fig. 8.1f), microwaves, and radio waves crossing the boundary of a system. Examples of this method of transfer include cooking a baked potato in your microwave oven and energy traveling from the Sun to the Earth by light through space.[1]

A central feature of the energy approach is the notion that we can neither create nor destroy energy, that energy is always *conserved*. This feature has been tested in countless experiments, and no experiment has ever shown this statement to be incorrect. Therefore, **if the total amount of energy in a system changes, it can *only* be because energy has crossed the boundary of the system by a transfer mechanism such as one of the methods listed above.**

Energy is one of several quantities in physics that are conserved. We will see other conserved quantities in subsequent chapters. There are many physical quantities that do not obey a conservation principle. For example, there is no conservation of force principle or conservation of velocity principle. Similarly, in areas other than physical quantities, such as in everyday life, some quantities are conserved and some are not. For example, the money in the system of your bank account is a conserved quantity. The only way the account balance changes is if money crosses the boundary of the system by deposits or withdrawals. On the other hand, the number of people in the system of a country is not conserved. Although people indeed cross the boundary of the system, which changes the total population, the population can also change by people dying and by giving birth to new babies. Even if no people cross the system boundary, the births and deaths will change the number of people in the system. There is no equivalent in the concept of energy to dying or giving birth. The general statement of the principle of **conservation of energy** can be described mathematically with the **conservation of energy equation** as follows:

$$\Delta E_{\text{system}} = \sum T \qquad\qquad (8.1)$$

◀ Conservation of energy

where E_{system} is the total energy of the system, including all methods of energy storage (kinetic, potential, and internal), and T (for *transfer*) is the amount of energy transferred across the system boundary by some mechanism. Two of our transfer mechanisms have well-established symbolic notations. For work, $T_{\text{work}} = W$ as discussed in Chapter 7, and for heat, $T_{\text{heat}} = Q$ as defined in Chapter 20. (Now that we

Pitfall Prevention 8.1
Heat Is Not a Form of Energy
The word *heat* is one of the most misused words in our popular language. Heat is a method of *transferring* energy, *not* a form of storing energy. Therefore, phrases such as "heat content," "the heat of the summer," and "the heat escaped" all represent uses of this word that are inconsistent with our physics definition. See Chapter 20.

---

[1]Electromagnetic radiation and work done by field forces are the only energy transfer mechanisms that do not require molecules of the environment to be available at the system boundary. Therefore, systems surrounded by a vacuum (such as planets) can only exchange energy with the environment by means of these two possibilities.

are familiar with work, we can simplify the appearance of equations by letting the simple symbol W represent the external work W_{ext} on a system. For internal work, we will *always* use W_{int} to differentiate it from W.) The other four members of our list do not have established symbols, so we will call them T_{MW} (mechanical waves), T_{MT} (matter transfer), T_{ET} (electrical transmission), and T_{ER} (electromagnetic radiation).

The full expansion of Equation 8.1 is

$$\Delta K + \Delta U + \Delta E_{int} = W + Q + T_{MW} + T_{MT} + T_{ET} + T_{ER} \qquad \textbf{(8.2)}$$

which is the primary mathematical representation of the energy version of the analysis model of the **nonisolated system.** (We will see other versions of the nonisolated system model, involving linear momentum and angular momentum, in later chapters.) In most cases, Equation 8.2 reduces to a much simpler one because some of the terms are zero for the specific situation. If, for a given system, all terms on the right side of the conservation of energy equation are zero, the system is an *isolated system*, which we study in the next section.

The conservation of energy equation is no more complicated in theory than the process of balancing your checking account statement. If your account is the system, the change in the account balance for a given month is the sum of all the transfers: deposits, withdrawals, fees, interest, and checks written. You may find it useful to think of energy as the *currency of nature!*

Suppose a force is applied to a nonisolated system and the point of application of the force moves through a displacement. Then suppose the only effect on the system is to change its speed. In this case, the only transfer mechanism is work (so that the right side of Eq. 8.2 reduces to just W) and the only kind of energy in the system that changes is the kinetic energy (so that the left side of Eq. 8.2 reduces to just ΔK). Equation 8.2 then becomes

$$\Delta K = W$$

which is the work–kinetic energy theorem. This theorem is a special case of the more general principle of conservation of energy. We shall see several more special cases in future chapters.

Quick Quiz 8.1 By what transfer mechanisms does energy enter and leave **(a)** your television set? **(b)** Your gasoline-powered lawn mower? **(c)** Your hand-cranked pencil sharpener?

Quick Quiz 8.2 Consider a block sliding over a horizontal surface with friction. Ignore any sound the sliding might make. **(i)** If the system is the *block*, this system is (a) isolated (b) nonisolated (c) impossible to determine **(ii)** If the system is the *surface*, describe the system from the same set of choices. **(iii)** If the system is the *block and the surface*, describe the system from the same set of choices.

Analysis Model   **Nonisolated System (Energy)**

Imagine you have identified a system to be analyzed and have defined a system boundary. Energy can exist in the system in three forms: kinetic, potential, and internal. The total of that energy can be changed when energy crosses the system boundary by any of six transfer methods shown in the diagram here. The total change in the energy in the system is equal to the total amount of energy that has crossed the system boundary. The mathematical statement of that concept is expressed in the **conservation of energy equation:**

$$\Delta E_{system} = \Sigma T \qquad \textbf{(8.1)}$$

The full expansion of Equation 8.1 shows the specific types of energy storage and transfer:

$$\Delta K + \Delta U + \Delta E_{int} = W + Q + T_{MW} + T_{MT} + T_{ET} + T_{ER} \tag{8.2}$$

For a specific problem, this equation is generally reduced to a smaller number of terms by eliminating the terms that are equal to zero because they are not appropriate to the situation.

Examples:

- a force does work on a system of a single object, changing its speed: the work–kinetic energy theorem, $W = \Delta K$
- a gas contained in a vessel has work done on it and experiences a transfer of energy by heat, resulting in a change in its temperature: the first law of thermodynamics, $\Delta E_{int} = W + Q$ (Chapter 20)
- an incandescent light bulb is turned on, with energy entering the filament by electricity, causing its temperature to increase, and leaving by light: $\Delta E_{int} = T_{ET} + T_{ER}$ (Chapter 27)
- a photon enters a metal, causing an electron to be ejected from the metal: the photoelectric effect, $\Delta K + \Delta U = T_{ER}$ (Chapter 40)

8.2 Analysis Model: Isolated System (Energy)

In this section, we study another very common scenario in physics problems: a system is chosen such that no energy crosses the system boundary by any method. We begin by considering a gravitational situation. Think about the book–Earth system in Figure 7.15 in the preceding chapter. After we have lifted the book, there is gravitational potential energy stored in the system, which can be calculated from the work done by the external agent on the system, using $W = \Delta U_g$. (Check to see that this equation, which we've seen before, is contained within Eq. 8.2 above.)

Let us now shift our focus to the work done *on the book alone* by the gravitational force (Fig. 8.2) as the book falls back to its original height. As the book falls from y_i to y_f, the work done by the gravitational force on the book is

$$W_{on\ book} = (m\vec{\mathbf{g}}) \cdot \Delta\vec{\mathbf{r}} = (-mg\hat{\mathbf{j}}) \cdot [(y_f - y_i)\hat{\mathbf{j}}] = mgy_i - mgy_f \tag{8.3}$$

From the work–kinetic energy theorem of Chapter 7, the work done on the book is equal to the change in the kinetic energy of the book:

$$W_{on\ book} = \Delta K_{book}$$

We can equate these two expressions for the work done on the book:

$$\Delta K_{book} = mgy_i - mgy_f \tag{8.4}$$

Let us now relate each side of this equation to the *system* of the book and the Earth. For the right-hand side,

$$mgy_i - mgy_f = -(mgy_f - mgy_i) = -\Delta U_g$$

where $U_g = mgy$ is the gravitational potential energy of the system. For the left-hand side of Equation 8.4, because the book is the only part of the system that is moving, we see that $\Delta K_{book} = \Delta K$, where K is the kinetic energy of the system. Therefore, with each side of Equation 8.4 replaced with its system equivalent, the equation becomes

$$\Delta K = -\Delta U_g \tag{8.5}$$

This equation can be manipulated to provide a very important general result for solving problems. First, we move the change in potential energy to the left side of the equation:

$$\Delta K + \Delta U_g = 0$$

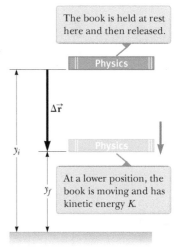

The book is held at rest here and then released.

At a lower position, the book is moving and has kinetic energy K.

Figure 8.2 A book is released from rest and falls due to work done by the gravitational force on the book.

The left side represents a sum of changes of the energy stored in the system. The right-hand side is zero because there are no transfers of energy across the boundary of the system; the book–Earth system is *isolated* from the environment. We developed this equation for a gravitational system, but it can be shown to be valid for a system with any type of potential energy. Therefore, for an isolated system,

$$\Delta K + \Delta U = 0 \tag{8.6}$$

(Check to see that this equation is contained within Eq. 8.2.)

We defined in Chapter 7 the sum of the kinetic and potential energies of a system as its mechanical energy:

◄ **Mechanical energy of a system**

$$E_{mech} \equiv K + U \tag{8.7}$$

where U represents the total of *all* types of potential energy. Because the system under consideration is isolated, Equations 8.6 and 8.7 tell us that the mechanical energy of the system is conserved:

◄ **The mechanical energy of an isolated system with no nonconservative forces acting is conserved.**

$$\Delta E_{mech} = 0 \tag{8.8}$$

Equation 8.8 is a statement of **conservation of mechanical energy** for an isolated system with no nonconservative forces acting. The mechanical energy in such a system is conserved: the sum of the kinetic and potential energies remains constant:

Let us now write the changes in energy in Equation 8.6 explicitly:

$$(K_f - K_i) + (U_f - U_i) = 0$$

$$K_f + U_f = K_i + U_i \tag{8.9}$$

For the gravitational situation of the falling book, Equation 8.9 can be written as

$$\tfrac{1}{2}mv_f^2 + mgy_f = \tfrac{1}{2}mv_i^2 + mgy_i$$

As the book falls to the Earth, the book–Earth system loses potential energy and gains kinetic energy such that the total of the two types of energy always remains constant: $E_{total,i} = E_{total,f}$.

If there are nonconservative forces acting within the system, mechanical energy is transformed to internal energy as discussed in Section 7.7. If nonconservative forces act in an isolated system, the total energy of the system is conserved although the mechanical energy is not. In that case, we can express the conservation of energy of the system as

◄ **The total energy of an isolated system is conserved.**

Figure 8.3 (Quick Quiz 8.4)
Three identical balls are thrown with the same initial speed from the top of a building.

$$\Delta E_{system} = 0 \tag{8.10}$$

where E_{system} includes all kinetic, potential, and internal energies. This equation is the most general statement of the energy version of the **isolated system** model. It is equivalent to Equation 8.2 with all terms on the right-hand side equal to zero.

Quick Quiz 8.3 A rock of mass m is dropped to the ground from a height h. A second rock, with mass $2m$, is dropped from the same height. When the second rock strikes the ground, what is its kinetic energy? **(a)** twice that of the first rock **(b)** four times that of the first rock **(c)** the same as that of the first rock **(d)** half as much as that of the first rock **(e)** impossible to determine

Quick Quiz 8.4 Three identical balls are thrown from the top of a building, all with the same initial speed. As shown in Figure 8.3, the first is thrown horizontally, the second at some angle above the horizontal, and the third at some angle below the horizontal. Neglecting air resistance, rank the speeds of the balls at the instant each hits the ground.

Analysis Model Isolated System (Energy)

Imagine you have identified a system to be analyzed and have defined a system boundary. Energy can exist in the system in three forms: kinetic, potential, and internal. Imagine also a situation in which no energy crosses the boundary of the system by any method. Then, the system is isolated; energy transforms from one form to another and Equation 8.2 becomes

System boundary

Kinetic energy
Potential energy
Internal energy

The total amount of energy in the system is constant. Energy transforms among the three possible types.

$$\Delta E_{\text{system}} = 0 \qquad (8.10)$$

If no nonconservative forces act within the isolated system, the mechanical energy of the system is conserved, so

$$\Delta E_{\text{mech}} = 0 \qquad (8.8)$$

Examples:

- an object is in free-fall; gravitational potential energy transforms to kinetic energy: $\Delta K + \Delta U = 0$
- a basketball rolling across a gym floor comes to rest; kinetic energy transforms to internal energy: $\Delta K + \Delta E_{\text{int}} = 0$
- a pendulum is raised and released with an initial speed; its motion eventually stops due to air resistance; gravitational potential energy and kinetic energy transform to internal energy, $\Delta K + \Delta U + \Delta E_{\text{int}} = 0$ (Chapter 15)
- a battery is connected to a resistor; chemical potential energy in the battery transforms to internal energy in the resistor: $\Delta U + \Delta E_{\text{int}} = 0$ (Chapter 27)

Problem-Solving Strategy Isolated and Nonisolated Systems with No Nonconservative Forces: Conservation of Energy

Many problems in physics can be solved using the principle of conservation of energy. The following procedure should be used when you apply this principle:

1. Conceptualize. Study the physical situation carefully and form a mental representation of what is happening. As you become more proficient working energy problems, you will begin to be comfortable imagining the types of energy that are changing in the system and the types of energy transfers occurring across the system boundary.

2. Categorize. Define your system, which may consist of more than one object and may or may not include springs or other possibilities for storing potential energy. Identify the time interval over which you will analyze the energy changes in the problem. Determine if any energy transfers occur across the boundary of your system during this time interval. If so, use the nonisolated system model, $\Delta E_{\text{system}} = \Sigma T$, from Section 8.1. If not, use the isolated system model, $\Delta E_{\text{system}} = 0$.

Determine whether any nonconservative forces are present within the system. If so, use the techniques of Sections 8.3 and 8.4. If not, use the principle of conservation of energy as outlined below.

3. Analyze. Choose configurations to represent the initial and final conditions of the system based on your choice of time interval. For each object that changes elevation, select a reference position for the object that defines the zero configuration of gravitational potential energy for the system. For an object on a spring, the zero configuration for elastic potential energy is when the object is at its equilibrium position. If there is more than one conservative force, write an expression for the potential energy associated with each force.

Begin with Equation 8.2 and retain only those terms in the equation that are appropriate for the situation in the problem. Express each change of energy stored in the system as the final value minus the initial value. Substitute appropriate expressions for each initial and final value of energy storage on the left side of the equation and for the energy transfers on the right side of the equation. Solve for the unknown quantity.

continued

▶ **Problem-Solving Strategy** continued

4. Finalize. Make sure your results are consistent with your mental representation. Also make sure the values of your results are reasonable and consistent with connections to everyday experience.

Example 8.1   **Ball in Free Fall**   AM

A ball of mass m is dropped from a height h above the ground as shown in Figure 8.4.

(A) Neglecting air resistance, determine the speed of the ball when it is at a height y above the ground. Choose the system as the ball and the Earth.

SOLUTION

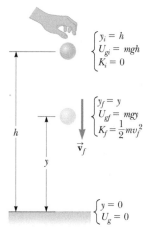

Conceptualize Figure 8.4 and our everyday experience with falling objects allow us to conceptualize the situation. Although we can readily solve this problem with the techniques of Chapter 2, let us practice an energy approach.

Categorize As suggested in the problem, we identify the system as the ball and the Earth. Because there is neither air resistance nor any other interaction between the system and the environment, the system is isolated and we use the *isolated system* model. The only force between members of the system is the gravitational force, which is conservative.

Figure 8.4 (Example 8.1) A ball is dropped from a height h above the ground. Initially, the total energy of the ball–Earth system is gravitational potential energy, equal to mgh relative to the ground. At the position y, the total energy is the sum of the kinetic and potential energies.

Analyze Because the system is isolated and there are no nonconservative forces acting within the system, we apply the principle of conservation of mechanical energy to the ball–Earth system. At the instant the ball is released, its kinetic energy is $K_i = 0$ and the gravitational potential energy of the system is $U_{gi} = mgh$. When the ball is at a position y above the ground, its kinetic energy is $K_f = \frac{1}{2}mv_f^2$ and the potential energy relative to the ground is $U_{gf} = mgy$.

Write the appropriate reduction of Equation 8.2, noting that the only types of energy in the system that change are kinetic energy and gravitational potential energy:

$$\Delta K + \Delta U_g = 0$$

Substitute for the energies:

$$\left(\tfrac{1}{2}mv_f^2 - 0\right) + \left(mgy - mgh\right) = 0$$

Solve for v_f:

$$v_f^2 = 2g(h - y) \quad \rightarrow \quad v_f = \boxed{\sqrt{2g(h - y)}}$$

The speed is always positive. If you had been asked to find the ball's velocity, you would use the negative value of the square root as the y component to indicate the downward motion.

(B) Find the speed of the ball again at height y by choosing the ball as the system.

SOLUTION

Categorize In this case, the only type of energy in the system that changes is kinetic energy. A single object that can be modeled as a particle cannot possess potential energy. The effect of gravity is to do work on the ball across the boundary of the system. We use the *nonisolated system* model.

Analyze Write the appropriate reduction of Equation 8.2:

$$\Delta K = W$$

Substitute for the initial and final kinetic energies and the work:

$$\left(\tfrac{1}{2}mv_f^2 - 0\right) = \vec{\mathbf{F}}_g \cdot \Delta \vec{\mathbf{r}} = -mg\hat{\mathbf{j}} \cdot \Delta y\hat{\mathbf{j}}$$
$$= -mg\Delta y = -mg(y - h) = mg(h - y)$$

Solve for v_f:

$$v_f^2 = 2g(h - y) \quad \rightarrow \quad v_f = \boxed{\sqrt{2g(h - y)}}$$

▶ **8.1** continued

Finalize The final result is the same, regardless of the choice of system. In your future problem solving, keep in mind that the choice of system is yours to make. Sometimes the problem is much easier to solve if a judicious choice is made as to the system to analyze.

WHAT IF? What if the ball were thrown downward from its highest position with a speed v_i? What would its speed be at height y?

Answer If the ball is thrown downward initially, we would expect its speed at height y to be larger than if simply dropped. Make your choice of system, either the ball alone or the ball and the Earth. You should find that either choice gives you the following result:

$$v_f = \sqrt{v_i^2 + 2g(h - y)}$$

[handwritten notes: $W_+ = \Delta K$ → force of gravity → spring force → other forces]

Example 8.2 A Grand Entrance AM

You are designing an apparatus to support an actor of mass 65.0 kg who is to "fly" down to the stage during the performance of a play. You attach the actor's harness to a 130-kg sandbag by means of a lightweight steel cable running smoothly over two frictionless pulleys as in Figure 8.5a. You need 3.00 m of cable between the harness and the nearest pulley so that the pulley can be hidden behind a curtain. For the apparatus to work successfully, the sandbag must never lift above the floor as the actor swings from above the stage to the floor. Let us call the initial angle that the actor's cable makes with the vertical θ. What is the maximum value θ can have before the sandbag lifts off the floor?

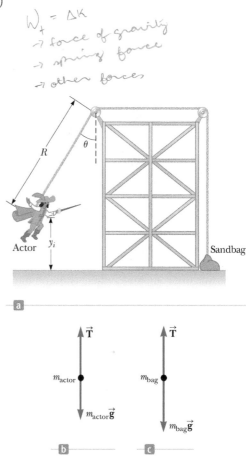

SOLUTION

Conceptualize We must use several concepts to solve this problem. Imagine what happens as the actor approaches the bottom of the swing. At the bottom, the cable is vertical and must support his weight as well as provide centripetal acceleration of his body in the upward direction. At this point in his swing, the tension in the cable is the highest and the sandbag is most likely to lift off the floor.

Categorize Looking first at the swinging of the actor from the initial point to the lowest point, we model the actor and the Earth as an *isolated system*. We ignore air resistance, so there are no nonconservative forces acting. You might initially be tempted to model the system as nonisolated because of the interaction of the system with the cable, which is in the environment. The force applied to the actor by the cable, however, is always perpendicular to each element of the displacement of the actor and hence does no work. Therefore, in terms of energy transfers across the boundary, the system is isolated.

Figure 8.5 (Example 8.2) (a) An actor uses some clever staging to make his entrance. (b) The free-body diagram for the actor at the bottom of the circular path. (c) The free-body diagram for the sandbag if the normal force from the floor goes to zero.

Analyze We first find the actor's speed as he arrives at the floor as a function of the initial angle θ and the radius R of the circular path through which he swings.

From the isolated system model, make the appropriate reduction of Equation 8.2 for the actor–Earth system:

$$\Delta K + \Delta U_g = 0$$

continued

▶ **8.2** continued

Let y_i be the initial height of the actor above the floor and v_f be his speed at the instant before he lands. (Notice that $K_i = 0$ because the actor starts from rest and that $U_f = 0$ because we define the configuration of the actor at the floor as having a gravitational potential energy of zero.)

(1) $(\frac{1}{2}m_{actor}v_f^2 - 0) + (0 - m_{actor}gy_i) = 0$

From the geometry in Figure 8.5a, notice that $y_f = 0$, so $y_i = R - R\cos\theta = R(1 - \cos\theta)$. Use this relationship in Equation (1) and solve for v_f^2:

(2) $v_f^2 = 2gR(1 - \cos\theta)$

- -

Categorize Next, focus on the instant the actor is at the lowest point. Because the tension in the cable is transferred as a force applied to the sandbag, we model the actor at this instant as a *particle under a net force*. Because the actor moves along a circular arc, he experiences at the bottom of the swing a centripetal acceleration of v_f^2/r directed upward.

- -

Analyze Apply Newton's second law from the particle under a net force model to the actor at the bottom of his path, using the free-body diagram in Figure 8.5b as a guide, and recognizing the acceleration as centripetal:

$$\sum F_y = T - m_{actor}g = m_{actor}\frac{v_f^2}{R}$$

(3) $T = m_{actor}g + m_{actor}\dfrac{v_f^2}{R}$

- -

Categorize Finally, notice that the sandbag lifts off the floor when the upward force exerted on it by the cable exceeds the gravitational force acting on it; the normal force from the floor is zero when that happens. We do *not*, however, want the sandbag to lift off the floor. The sandbag must remain at rest, so we model it as a *particle in equilibrium*.

- -

Analyze A force T of the magnitude given by Equation (3) is transmitted by the cable to the sandbag. If the sandbag remains at rest but is just ready to be lifted off the floor if any more force were applied by the cable, the normal force on it becomes zero and the particle in equilibrium model tells us that $T = m_{bag}g$ as in Figure 8.5c.

Substitute this condition and Equation (2) into Equation (3):

$$m_{bag}g = m_{actor}g + m_{actor}\frac{2gR(1 - \cos\theta)}{R}$$

Solve for $\cos\theta$ and substitute the given parameters:

$$\cos\theta = \frac{3m_{actor} - m_{bag}}{2m_{actor}} = \frac{3(65.0\text{ kg}) - 130\text{ kg}}{2(65.0\text{ kg})} = 0.500$$

$$\theta = \boxed{60.0°}$$

- -

Finalize Here we had to combine several analysis models from different areas of our study. Notice that the length R of the cable from the actor's harness to the leftmost pulley did not appear in the final algebraic equation for $\cos\theta$. Therefore, the final answer is independent of R.

Example 8.3   **The Spring-Loaded Popgun**   AM

The launching mechanism of a popgun consists of a trigger-released spring (Fig. 8.6a). The spring is compressed to a position $y_Ⓐ$, and the trigger is fired. The projectile of mass m rises to a position $y_Ⓒ$ above the position at which it leaves the spring, indicated in Figure 8.6b as position $y_Ⓑ = 0$. Consider a firing of the gun for which $m = 35.0$ g, $y_Ⓐ = -0.120$ m, and $y_Ⓒ = 20.0$ m.

(A) Neglecting all resistive forces, determine the spring constant.

SOLUTION

Conceptualize Imagine the process illustrated in parts (a) and (b) of Figure 8.6. The projectile starts from rest at Ⓐ, speeds up as the spring pushes upward on it, leaves the spring at Ⓑ, and then slows down as the gravitational force pulls downward on it, eventually coming to rest at point Ⓒ.

▶ 8.3 continued

Figure 8.6 (Example 8.3)
A spring-loaded popgun (a) before firing and (b) when the spring extends to its relaxed length. (c) An energy bar chart for the popgun–projectile–Earth system before the popgun is loaded. The energy in the system is zero. (d) The popgun is loaded by means of an external agent doing work on the system to push the spring downward. Therefore the system is nonisolated during this process. After the popgun is loaded, elastic potential energy is stored in the spring and the gravitational potential energy of the system is lower because the projectile is below point Ⓑ. (e) as the projectile passes through point Ⓑ, all of the energy of the isolated system is kinetic. (f) When the projectile reaches point Ⓒ, all of the energy of the isolated system is gravitational potential.

Categorize We identify the system as the projectile, the spring, and the Earth. We ignore both air resistance on the projectile and friction in the gun, so we model the system as isolated with no nonconservative forces acting.

Analyze Because the projectile starts from rest, its initial kinetic energy is zero. We choose the zero configuration for the gravitational potential energy of the system to be when the projectile leaves the spring at Ⓑ. For this configuration, the elastic potential energy is also zero.

After the gun is fired, the projectile rises to a maximum height $y_Ⓒ$. The final kinetic energy of the projectile is zero.

From the isolated system model, write a conservation of mechanical energy equation for the system between configurations when the projectile is at points Ⓐ and Ⓒ:

$$(1) \quad \Delta K + \Delta U_g + \Delta U_s = 0$$

Substitute for the initial and final energies:

$$(0 - 0) + (mgy_Ⓒ - mgy_Ⓐ) + (0 - \tfrac{1}{2}kx^2) = 0$$

Solve for k:

$$k = \frac{2mg(y_Ⓒ - y_Ⓐ)}{x^2}$$

Substitute numerical values:

$$k = \frac{2(0.035\,0\ \text{kg})(9.80\ \text{m/s}^2)[20.0\ \text{m} - (-0.120\ \text{m})]}{(0.120\ \text{m})^2} = \boxed{958\ \text{N/m}}$$

(B) Find the speed of the projectile as it moves through the equilibrium position Ⓑ of the spring as shown in Figure 8.6b.

SOLUTION

Analyze The energy of the system as the projectile moves through the equilibrium position of the spring includes only the kinetic energy of the projectile $\tfrac{1}{2}mv_Ⓑ^2$. Both types of potential energy are equal to zero for this configuration of the system.

continued

▶ **8.3** continued

Write Equation (1) again for the system between points Ⓐ and Ⓑ:

$$\Delta K + \Delta U_g + \Delta U_s = 0$$

Substitute for the initial and final energies:

$$(\tfrac{1}{2}mv_Ⓑ^2 - 0) + (0 - mgy_Ⓐ) + (0 - \tfrac{1}{2}kx^2) = 0$$

Solve for $v_Ⓑ$:

$$v_Ⓑ = \sqrt{\frac{kx^2}{m} + 2gy_Ⓐ}$$

Substitute numerical values:

$$v_Ⓑ = \sqrt{\frac{(958\ \text{N/m})(0.120\ \text{m})^2}{(0.035\ 0\ \text{kg})} + 2(9.80\ \text{m/s}^2)(-0.120\ \text{m})} = \boxed{19.8\ \text{m/s}}$$

Finalize This example is the first one we have seen in which we must include two different types of potential energy. Notice in part (A) that we never needed to consider anything about the speed of the ball between points Ⓐ and Ⓒ, which is part of the power of the energy approach: changes in kinetic and potential energy depend only on the initial and final values, not on what happens between the configurations corresponding to these values.

8.3 Situations Involving Kinetic Friction

Consider again the book in Figure 7.18a sliding to the right on the surface of a heavy table and slowing down due to the friction force. Work is done by the friction force on the book because there is a force and a displacement. Keep in mind, however, that our equations for work involve the displacement *of the point of application of the force*. A simple model of the friction force between the book and the surface is shown in Figure 8.7a. We have represented the entire friction force between the book and surface as being due to two identical teeth that have been spot-welded together.[2] One tooth projects upward from the surface, the other downward from the book, and they are welded at the points where they touch. The friction force acts at the junction of the two teeth. Imagine that the book slides a small distance d to the right as in Figure 8.7b. Because the teeth are modeled as identical, the junction of the teeth moves to the right by a distance $d/2$. Therefore, the displacement of the point of application of the friction force is $d/2$, but the displacement of the book is d!

In reality, the friction force is spread out over the entire contact area of an object sliding on a surface, so the force is not localized at a point. In addition, because the magnitudes of the friction forces at various points are constantly changing as individual spot welds occur, the surface and the book deform locally, and so on, the displacement of the point of application of the friction force is not at all the same as the displacement of the book. In fact, the displacement of the point of application of the friction force is not calculable and so neither is the work done by the friction force.

The work–kinetic energy theorem is valid for a particle or an object that can be modeled as a particle. When a friction force acts, however, we cannot calculate the work done by friction. For such situations, Newton's second law is still valid for the system even though the work–kinetic energy theorem is not. The case of a nondeformable object like our book sliding on the surface[3] can be handled in a relatively straightforward way.

Starting from a situation in which forces, including friction, are applied to the book, we can follow a similar procedure to that done in developing Equation 7.17. Let us start by writing Equation 7.8 for all forces on an object other than friction:

$$\sum W_{\text{other forces}} = \int \left(\sum \vec{F}_{\text{other forces}}\right) \cdot d\vec{r} \tag{8.11}$$

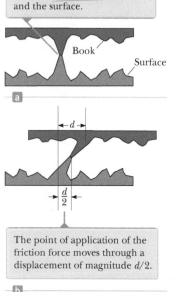

The entire friction force is modeled to be applied at the interface between two identical teeth projecting from the book and the surface.

Book

Surface

a

The point of application of the friction force moves through a displacement of magnitude $d/2$.

b

Figure 8.7 (a) A simplified model of friction between a book and a surface. (b) The book is moved to the right by a distance d.

[2]Figure 8.7 and its discussion are inspired by a classic article on friction: B. A. Sherwood and W. H. Bernard, "Work and heat transfer in the presence of sliding friction," *American Journal of Physics*, **52**:1001, 1984.

[3]The overall shape of the book remains the same, which is why we say it is nondeformable. On a microscopic level, however, there is deformation of the book's face as it slides over the surface.

The $d\vec{\mathbf{r}}$ in this equation is the displacement of the object because for forces other than friction, under the assumption that these forces do not deform the object, this displacement is the same as the displacement of the point of application of the forces. To each side of Equation 8.11 let us add the integral of the scalar product of the force of kinetic friction and $d\vec{\mathbf{r}}$. In doing so, we are not defining this quantity as work! We are simply saying that it is a quantity that can be calculated mathematically and will turn out to be useful to us in what follows.

$$\sum W_{\text{other forces}} + \int \vec{\mathbf{f}}_k \cdot d\vec{\mathbf{r}} = \int \left(\sum \vec{\mathbf{F}}_{\text{other forces}} \right) \cdot d\vec{\mathbf{r}} + \int \vec{\mathbf{f}}_k \cdot d\vec{\mathbf{r}}$$

$$= \int \left(\sum \vec{\mathbf{F}}_{\text{other forces}} + \vec{\mathbf{f}}_k \right) \cdot d\vec{\mathbf{r}}$$

The integrand on the right side of this equation is the net force $\sum \vec{\mathbf{F}}$ on the object, so

$$\sum W_{\text{other forces}} + \int \vec{\mathbf{f}}_k \cdot d\vec{\mathbf{r}} = \int \sum \vec{\mathbf{F}} \cdot d\vec{\mathbf{r}}$$

Incorporating Newton's second law $\sum \vec{\mathbf{F}} = m\vec{\mathbf{a}}$ gives

$$\sum W_{\text{other forces}} + \int \vec{\mathbf{f}}_k \cdot d\vec{\mathbf{r}} = \int m\vec{\mathbf{a}} \cdot d\vec{\mathbf{r}} = \int m \frac{d\vec{\mathbf{v}}}{dt} \cdot d\vec{\mathbf{r}} = \int_{t_i}^{t_f} m \frac{d\vec{\mathbf{v}}}{dt} \cdot \vec{\mathbf{v}} \, dt \quad \textbf{(8.12)}$$

where we have used Equation 4.3 to rewrite $d\vec{\mathbf{r}}$ as $\vec{\mathbf{v}} \, dt$. The scalar product obeys the product rule for differentiation (See Eq. B.30 in Appendix B.6), so the derivative of the scalar product of $\vec{\mathbf{v}}$ with itself can be written

$$\frac{d}{dt}(\vec{\mathbf{v}} \cdot \vec{\mathbf{v}}) = \frac{d\vec{\mathbf{v}}}{dt} \cdot \vec{\mathbf{v}} + \vec{\mathbf{v}} \cdot \frac{d\vec{\mathbf{v}}}{dt} = 2\frac{d\vec{\mathbf{v}}}{dt} \cdot \vec{\mathbf{v}}$$

We used the commutative property of the scalar product to justify the final expression in this equation. Consequently,

$$\frac{d\vec{\mathbf{v}}}{dt} \cdot \vec{\mathbf{v}} = \tfrac{1}{2} \frac{d}{dt}(\vec{\mathbf{v}} \cdot \vec{\mathbf{v}}) = \tfrac{1}{2} \frac{dv^2}{dt}$$

Substituting this result into Equation 8.12 gives

$$\sum W_{\text{other forces}} + \int \vec{\mathbf{f}}_k \cdot d\vec{\mathbf{r}} = \int_{t_i}^{t_f} m\left(\tfrac{1}{2} \frac{dv^2}{dt} \right) dt = \tfrac{1}{2}m \int_{v_i}^{v_f} d(v^2) = \tfrac{1}{2}mv_f^2 - \tfrac{1}{2}mv_i^2 = \Delta K$$

Looking at the left side of this equation, notice that in the inertial frame of the surface, $\vec{\mathbf{f}}_k$ and $d\vec{\mathbf{r}}$ will be in opposite directions for every increment $d\vec{\mathbf{r}}$ of the path followed by the object. Therefore, $\vec{\mathbf{f}}_k \cdot d\vec{\mathbf{r}} = -f_k \, dr$. The previous expression now becomes

$$\sum W_{\text{other forces}} - \int f_k \, dr = \Delta K$$

In our model for friction, the magnitude of the kinetic friction force is constant, so f_k can be brought out of the integral. The remaining integral $\int dr$ is simply the sum of increments of length along the path, which is the total path length d. Therefore,

$$\sum W_{\text{other forces}} - f_k d = \Delta K \quad \textbf{(8.13)}$$

Equation 8.13 can be used when a friction force acts on an object. The change in kinetic energy is equal to the work done by all forces other than friction minus a term $f_k d$ associated with the friction force.

Considering the sliding book situation again, let's identify the larger system of the book *and* the surface as the book slows down under the influence of a friction force alone. There is no work done across the boundary of this system by other forces because the system does not interact with the environment. There are no other types of energy transfer occurring across the boundary of the system, assuming we ignore the inevitable sound the sliding book makes! In this case, Equation 8.2 becomes

$$\Delta E_{\text{system}} = \Delta K + \Delta E_{\text{int}} = 0$$

The change in kinetic energy of this book–surface system is the same as the change in kinetic energy of the book alone because the book is the only part of the system that is moving. Therefore, incorporating Equation 8.13 with no work done by other forces gives

$$-f_k d + \Delta E_{int} = 0$$

Change in internal energy ▶
due to a constant friction
force within the system

$$\Delta E_{int} = f_k d \qquad (8.14)$$

Equation 8.14 tells us that the increase in internal energy of the system is equal to the product of the friction force and the path length through which the block moves. In summary, a friction force transforms kinetic energy in a system to internal energy. If work is done on the system by forces other than friction, Equation 8.13, with the help of Equation 8.14, can be written as

$$\sum W_{\text{other forces}} = W = \Delta K + \Delta E_{int} \qquad (8.15)$$

which is a reduced form of Equation 8.2 and represents the nonisolated system model for a system within which a nonconservative force acts.

Quick Quiz 8.5 You are traveling along a freeway at 65 mi/h. Your car has kinetic energy. You suddenly skid to a stop because of congestion in traffic. Where is the kinetic energy your car once had? **(a)** It is all in internal energy in the road. **(b)** It is all in internal energy in the tires. **(c)** Some of it has transformed to internal energy and some of it transferred away by mechanical waves. **(d)** It is all transferred away from your car by various mechanisms.

Example 8.4   **A Block Pulled on a Rough Surface**   **AM**

A 6.0-kg block initially at rest is pulled to the right along a horizontal surface by a constant horizontal force of 12 N.

(A) Find the speed of the block after it has moved 3.0 m if the surfaces in contact have a coefficient of kinetic friction of 0.15.

SOLUTION

Conceptualize This example is similar to Example 7.6 (page 190), but modified so that the surface is no longer frictionless. The rough surface applies a friction force on the block opposite to the applied force. As a result, we expect the speed to be lower than that found in Example 7.6.

Figure 8.8 (Example 8.4) (a) A block pulled to the right on a rough surface by a constant horizontal force. (b) The applied force is at an angle θ to the horizontal.

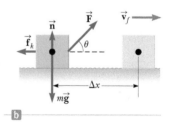

Categorize The block is pulled by a force and the surface is rough, so the block and the surface are modeled as a *nonisolated system* with a nonconservative force acting.

· ·

Analyze Figure 8.8a illustrates this situation. Neither the normal force nor the gravitational force does work on the system because their points of application are displaced horizontally.

Find the work done on the system by the applied force just as in Example 7.6:

$$\sum W_{\text{other forces}} = W_F = F \Delta x$$

Apply the *particle in equilibrium* model to the block in the vertical direction:

$$\sum F_y = 0 \rightarrow n - mg = 0 \rightarrow n = mg$$

Find the magnitude of the friction force:

$$f_k = \mu_k n = \mu_k mg = (0.15)(6.0 \text{ kg})(9.80 \text{ m/s}^2) = 8.82 \text{ N}$$

▶ **8.4** c o n t i n u e d

Substitute the energies into Equation 8.15 and solve for the final speed of the block:

$$F\Delta x = \Delta K + \Delta E_{\text{int}} = (\tfrac{1}{2}mv_f^2 - 0) + f_k d$$

$$v_f = \sqrt{\frac{2}{m}(-f_k d + F\Delta x)}$$

Substitute numerical values:

$$v_f = \sqrt{\frac{2}{6.0\text{ kg}}[-(8.82\text{ N})(3.0\text{ m}) + (12\text{ N})(3.0\text{ m})]} = \boxed{1.8\text{ m/s}}$$

Finalize As expected, this value is less than the 3.5 m/s found in the case of the block sliding on a frictionless surface (see Example 7.6). The difference in kinetic energies between the block in Example 7.6 and the block in this example is equal to the increase in internal energy of the block–surface system in this example.

(B) Suppose the force $\vec{F}$ is applied at an angle θ as shown in Figure 8.8b. At what angle should the force be applied to achieve the largest possible speed after the block has moved 3.0 m to the right?

S O L U T I O N

Conceptualize You might guess that $\theta = 0$ would give the largest speed because the force would have the largest component possible in the direction parallel to the surface. Think about $\vec{F}$ applied at an arbitrary nonzero angle, however. Although the horizontal component of the force would be reduced, the vertical component of the force would reduce the normal force, in turn reducing the force of friction, which suggests that the speed could be maximized by pulling at an angle other than $\theta = 0$.

Categorize As in part (A), we model the block and the surface as a *nonisolated system* with a nonconservative force acting.

Analyze Find the work done by the applied force, noting that $\Delta x = d$ because the path followed by the block is a straight line:

$$(1)\quad \sum W_{\text{other forces}} = W_F = F\Delta x \cos\theta = Fd\cos\theta$$

Apply the particle in equilibrium model to the block in the vertical direction:

$$\sum F_y = n + F\sin\theta - mg = 0$$

Solve for n:

$$(2)\quad n = mg - F\sin\theta$$

Use Equation 8.15 to find the final kinetic energy for this situation:

$$W_F = \Delta K + \Delta E_{\text{int}} = (K_f - 0) + f_k d \rightarrow K_f = W_F - f_k d$$

Substitute the results in Equations (1) and (2):

$$K_f = Fd\cos\theta - \mu_k n d = Fd\cos\theta - \mu_k(mg - F\sin\theta)d$$

Maximizing the speed is equivalent to maximizing the final kinetic energy. Consequently, differentiate K_f with respect to θ and set the result equal to zero:

$$\frac{dK_f}{d\theta} = -Fd\sin\theta - \mu_k(0 - F\cos\theta)d = 0$$

$$-\sin\theta + \mu_k\cos\theta = 0$$

$$\tan\theta = \mu_k$$

Evaluate θ for $\mu_k = 0.15$:

$$\theta = \tan^{-1}(\mu_k) = \tan^{-1}(0.15) = \boxed{8.5°}$$

Finalize Notice that the angle at which the speed of the block is a maximum is indeed not $\theta = 0$. When the angle exceeds 8.5°, the horizontal component of the applied force is too small to be compensated by the reduced friction force and the speed of the block begins to decrease from its maximum value.

Conceptual Example 8.5   Useful Physics for Safer Driving

A car traveling at an initial speed v slides a distance d to a halt after its brakes lock. If the car's initial speed is instead $2v$ at the moment the brakes lock, estimate the distance it slides.

continued

▶ 8.5 continued

SOLUTION

Let us assume the force of kinetic friction between the car and the road surface is constant and the same for both speeds. According to Equation 8.13, the friction force multiplied by the distance d is equal to the initial kinetic energy of the car (because $K_f = 0$ and there is no work done by other forces). If the speed is doubled, as it is in this example, the kinetic energy is quadrupled. For a given friction force, the distance traveled is four times as great when the initial speed is doubled, and so the estimated distance the car slides is $4d$.

Example 8.6   A Block–Spring System   AM

A block of mass 1.6 kg is attached to a horizontal spring that has a force constant of 1 000 N/m as shown in Figure 8.9a. The spring is compressed 2.0 cm and is then released from rest as in Figure 8.9b.

(A) Calculate the speed of the block as it passes through the equilibrium position $x = 0$ if the surface is frictionless.

SOLUTION

Conceptualize This situation has been discussed before, and it is easy to visualize the block being pushed to the right by the spring and moving with some speed at $x = 0$.

Categorize We identify the system as the block and model the block as a *nonisolated system*.

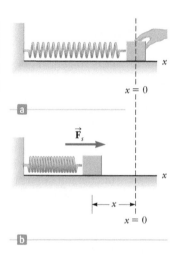

Figure 8.9 (Example 8.6) (a) A block attached to a spring is pushed inward from an initial position $x = 0$ by an external agent. (b) At position x, the block is released from rest and the spring pushes it to the right.

Analyze In this situation, the block starts with $v_i = 0$ at $x_i = -2.0$ cm, and we want to find v_f at $x_f = 0$.

Use Equation 7.11 to find the work done by the spring on the system with $x_{max} = x_i$:

$$W_s = \tfrac{1}{2}kx_{max}^2$$

Work is done on the block, and its speed changes. The conservation of energy equation, Equation 8.2, reduces to the work–kinetic energy theorem. Use that theorem to find the speed at $x = 0$:

$$W_s = \tfrac{1}{2}mv_f^2 - \tfrac{1}{2}mv_i^2$$

$$v_f = \sqrt{v_i^2 + \frac{2}{m} W_s} = \sqrt{v_i^2 + \frac{2}{m}\left(\tfrac{1}{2}kx_{max}^2\right)}$$

Substitute numerical values:

$$v_f = \sqrt{0 + \frac{2}{1.6 \text{ kg}}\left[\tfrac{1}{2}(1\,000 \text{ N/m})(0.020 \text{ m})^2\right]} = \boxed{0.50 \text{ m/s}}$$

Finalize Although this problem could have been solved in Chapter 7, it is presented here to provide contrast with the following part (B), which requires the techniques of this chapter.

(B) Calculate the speed of the block as it passes through the equilibrium position if a constant friction force of 4.0 N retards its motion from the moment it is released.

SOLUTION

Conceptualize The correct answer must be less than that found in part (A) because the friction force retards the motion.

Categorize We identify the system as the block and the surface, a *nonisolated system* because of the work done by the spring. There is a nonconservative force acting within the system: the friction between the block and the surface.

▶ **8.6** continued

Analyze Write Equation 8.15:

$$W_s = \Delta K + \Delta E_{int} = (\tfrac{1}{2}mv_f^2 - 0) + f_k d$$

Solve for v_f:

$$v_f = \sqrt{\frac{2}{m}(W_s - f_k d)}$$

Substitute for the work done by the spring:

$$v_f = \sqrt{\frac{2}{m}(\tfrac{1}{2}kx_{max}^2 - f_k d)}$$

Substitute numerical values:

$$v_f = \sqrt{\frac{2}{1.6\ kg}[\tfrac{1}{2}(1\ 000\ N/m)(0.020\ m)^2 - (4.0\ N)(0.020\ m)]} = 0.39\ m/s$$

Finalize As expected, this value is less than the 0.50 m/s found in part (A).

WHAT IF? What if the friction force were increased to 10.0 N? What is the block's speed at $x = 0$?

Answer In this case, the value of $f_k d$ as the block moves to $x = 0$ is

$$f_k d = (10.0\ N)(0.020\ m) = 0.20\ J$$

which is equal in magnitude to the kinetic energy at $x = 0$ for the frictionless case. (Verify it!). Therefore, all the kinetic energy has been transformed to internal energy by friction when the block arrives at $x = 0$, and its speed at this point is $v = 0$.

In this situation as well as that in part (B), the speed of the block reaches a maximum at some position other than $x = 0$. Problem 53 asks you to locate these positions.

8.4 Changes in Mechanical Energy for Nonconservative Forces

Consider the book sliding across the surface in the preceding section. As the book moves through a distance d, the only force in the horizontal direction is the force of kinetic friction. This force causes a change $-f_k d$ in the kinetic energy of the book as described by Equation 8.13.

Now, however, suppose the book is part of a system that also exhibits a change in potential energy. In this case, $-f_k d$ is the amount by which the *mechanical* energy of the system changes because of the force of kinetic friction. For example, if the book moves on an incline that is not frictionless, there is a change in both the kinetic energy and the gravitational potential energy of the book–Earth system. Consequently,

$$\Delta E_{mech} = \Delta K + \Delta U_g = -f_k d = -\Delta E_{int}$$

In general, if a nonconservative force acts within an isolated system,

$$\Delta K + \Delta U + \Delta E_{int} = 0 \qquad (8.16)$$

where ΔU is the change in all forms of potential energy. We recognize Equation 8.16 as Equation 8.2 with no transfers of energy across the boundary of the system.

If the system in which nonconservative forces act is nonisolated and the external influence on the system is by means of work, the generalization of Equation 8.13 is

$$\sum W_{other\ forces} - f_k d = \Delta E_{mech}$$

This equation, with the help of Equations 8.7 and 8.14, can be written as

$$\sum W_{other\ forces} = W = \Delta K + \Delta U + \Delta E_{int} \qquad (8.17)$$

This reduced form of Equation 8.2 represents the nonisolated system model for a system that possesses potential energy and within which a nonconservative force acts.

| Example 8.7 | Crate Sliding Down a Ramp   AM |

A 3.00-kg crate slides down a ramp. The ramp is 1.00 m in length and inclined at an angle of 30.0° as shown in Figure 8.10. The crate starts from rest at the top, experiences a constant friction force of magnitude 5.00 N, and continues to move a short distance on the horizontal floor after it leaves the ramp.

(A) Use energy methods to determine the speed of the crate at the bottom of the ramp.

SOLUTION

Conceptualize Imagine the crate sliding down the ramp in Figure 8.10. The larger the friction force, the more slowly the crate will slide.

Categorize We identify the crate, the surface, and the Earth as an *isolated system* with a nonconservative force acting.

Figure 8.10 (Example 8.7) A crate slides down a ramp under the influence of gravity. The potential energy of the system decreases, whereas the kinetic energy increases.

Analyze Because $v_i = 0$, the initial kinetic energy of the system when the crate is at the top of the ramp is zero. If the y coordinate is measured from the bottom of the ramp (the final position of the crate, for which we choose the gravitational potential energy of the system to be zero) with the upward direction being positive, then $y_i = 0.500$ m.

Write the conservation of energy equation (Eq. 8.2) for this system:

$$\Delta K + \Delta U + \Delta E_{\text{int}} = 0$$

Substitute for the energies:

$$(\tfrac{1}{2}mv_f^2 - 0) + (0 - mgy_i) + f_k d = 0$$

Solve for v_f:

$$(1) \quad v_f = \sqrt{\frac{2}{m}(mgy_i - f_k d)}$$

Substitute numerical values:

$$v_f = \sqrt{\frac{2}{3.00 \text{ kg}}[(3.00 \text{ kg})(9.80 \text{ m/s}^2)(0.500 \text{ m}) - (5.00 \text{ N})(1.00 \text{ m})]} = \boxed{2.54 \text{ m/s}}$$

(B) How far does the crate slide on the horizontal floor if it continues to experience a friction force of magnitude 5.00 N?

SOLUTION

Analyze This part of the problem is handled in exactly the same way as part (A), but in this case we can consider the mechanical energy of the system to consist only of kinetic energy because the potential energy of the system remains fixed.

Write the conservation of energy equation for this situation:

$$\Delta K + \Delta E_{\text{int}} = 0$$

Substitute for the energies:

$$(0 - \tfrac{1}{2}mv_i^2) + f_k d = 0$$

Solve for the distance d and substitute numerical values:

$$d = \frac{mv_i^2}{2f_k} = \frac{(3.00 \text{ kg})(2.54 \text{ m/s})^2}{2(5.00 \text{ N})} = \boxed{1.94 \text{ m}}$$

Finalize For comparison, you may want to calculate the speed of the crate at the bottom of the ramp in the case in which the ramp is frictionless. Also notice that the increase in internal energy of the system as the crate slides down the ramp is $f_k d = (5.00 \text{ N})(1.00 \text{ m}) = 5.00$ J. This energy is shared between the crate and the surface, each of which is a bit warmer than before.

Also notice that the distance d the object slides on the horizontal surface is infinite if the surface is frictionless. Is that consistent with your conceptualization of the situation?

WHAT IF? A cautious worker decides that the speed of the crate when it arrives at the bottom of the ramp may be so large that its contents may be damaged. Therefore, he replaces the ramp with a longer one such that the new

▶ **8.7** continued

ramp makes an angle of 25.0° with the ground. Does this new ramp reduce the speed of the crate as it reaches the ground?

Answer Because the ramp is longer, the friction force acts over a longer distance and transforms more of the mechanical energy into internal energy. The result is a reduction in the kinetic energy of the crate, and we expect a lower speed as it reaches the ground.

Find the length d of the new ramp:

$$\sin 25.0° = \frac{0.500 \text{ m}}{d} \quad \rightarrow \quad d = \frac{0.500 \text{ m}}{\sin 25.0°} = 1.18 \text{ m}$$

Find v_f from Equation (1) in part (A):

$$v_f = \sqrt{\frac{2}{3.00 \text{ kg}}[(3.00 \text{ kg})(9.80 \text{ m/s}^2)(0.500 \text{ m}) - (5.00 \text{ N})(1.18 \text{ m})]} = 2.42 \text{ m/s}$$

The final speed is indeed lower than in the higher-angle case.

Example 8.8   Block–Spring Collision   AM

A block having a mass of 0.80 kg is given an initial velocity $v_Ⓐ = 1.2$ m/s to the right and collides with a spring whose mass is negligible and whose force constant is $k = 50$ N/m as shown in Figure 8.11.

(A) Assuming the surface to be frictionless, calculate the maximum compression of the spring after the collision.

SOLUTION

Conceptualize The various parts of Figure 8.11 help us imagine what the block will do in this situation. All motion takes place in a horizontal plane, so we do not need to consider changes in gravitational potential energy.

Categorize We identify the system to be the block and the spring and model it as an *isolated system* with no nonconservative forces acting.

Figure 8.11 (Example 8.8) A block sliding on a frictionless, horizontal surface collides with a light spring. (a) Initially, the mechanical energy is all kinetic energy. (b) The mechanical energy is the sum of the kinetic energy of the block and the elastic potential energy in the spring. (c) The energy is entirely potential energy. (d) The energy is transformed back to the kinetic energy of the block. The total energy of the system remains constant throughout the motion.

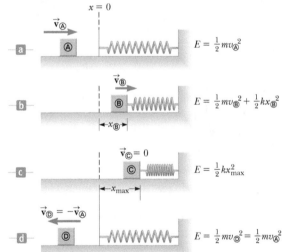

Analyze Before the collision, when the block is at Ⓐ, it has kinetic energy and the spring is uncompressed, so the elastic potential energy stored in the system is zero. Therefore, the total mechanical energy of the system before the collision is just $\frac{1}{2}mv_Ⓐ^2$. After the collision, when the block is at Ⓒ, the spring is fully compressed; now the block is at rest and so has zero kinetic energy. The elastic potential energy stored in the system, however, has its maximum value $\frac{1}{2}kx^2 = \frac{1}{2}kx_{\text{max}}^2$, where the origin of coordinates $x = 0$ is chosen to be the equilibrium position of the spring and x_{max} is the maximum compression of the spring, which in this case happens to be $x_Ⓒ$. The total mechanical energy of the system is conserved because no nonconservative forces act on objects within the isolated system.

Write the conservation of energy equation for this situation:

$$\Delta K + \Delta U = 0$$

Substitute for the energies:

$$\left(0 - \tfrac{1}{2}mv_Ⓐ^2\right) + \left(\tfrac{1}{2}kx_{\text{max}}^2 - 0\right) = 0$$

Solve for x_{max} and evaluate:

$$x_{\text{max}} = \sqrt{\frac{m}{k}}\, v_Ⓐ = \sqrt{\frac{0.80 \text{ kg}}{50 \text{ N/m}}}(1.2 \text{ m/s}) = \boxed{0.15 \text{ m}}$$

continued

▶ **8.8** continued

(B) Suppose a constant force of kinetic friction acts between the block and the surface, with $\mu_k = 0.50$. If the speed of the block at the moment it collides with the spring is $v_{\circledA} = 1.2$ m/s, what is the maximum compression $x_{\circledcirc}$ in the spring?

SOLUTION

Conceptualize Because of the friction force, we expect the compression of the spring to be smaller than in part (A) because some of the block's kinetic energy is transformed to internal energy in the block and the surface.

Categorize We identify the system as the block, the surface, and the spring. This is an *isolated system* but now involves a nonconservative force.

Analyze In this case, the mechanical energy $E_{\text{mech}} = K + U_s$ of the system is *not* conserved because a friction force acts on the block. From the *particle in equilibrium* model in the vertical direction, we see that $n = mg$.

Evaluate the magnitude of the friction force:

$$f_k = \mu_k n = \mu_k mg$$

Write the conservation of energy equation for this situation:

$$\Delta K + \Delta U + \Delta E_{\text{int}} = 0$$

Substitute the initial and final energies:

$$(0 - \tfrac{1}{2}mv_{\circledA}^2) + (\tfrac{1}{2}kx_{\circledcirc}^2 - 0) + \mu_k mgx_{\circledcirc} = 0$$

Rearrange the terms into a qaudratic equation:

$$kx_{\circledcirc}^2 + 2\mu_k mgx_{\circledcirc} - mv_{\circledA}^2 = 0$$

Substitute numerical values:

$$50x_{\circledcirc}^2 + 2(0.50)(0.80)(9.80)x_{\circledcirc} - (0.80)(1.2)^2 = 0$$

$$50x_{\circledcirc}^2 + 7.84x_{\circledcirc} - 1.15 = 0$$

Solving the quadratic equation for $x_{\circledcirc}$ gives $x_{\circledcirc} = 0.092$ m and $x_{\circledcirc} = -0.25$ m. The physically meaningful root is $x_{\circledcirc} = \boxed{0.092 \text{ m}}$.

Finalize The negative root does not apply to this situation because the block must be to the right of the origin (positive value of x) when it comes to rest. Notice that the value of 0.092 m is less than the distance obtained in the frictionless case of part (A) as we expected.

Example 8.9   **Connected Blocks in Motion**   `AM`

Two blocks are connected by a light string that passes over a frictionless pulley as shown in Figure 8.12. The block of mass m_1 lies on a horizontal surface and is connected to a spring of force constant k. The system is released from rest when the spring is unstretched. If the hanging block of mass m_2 falls a distance h before coming to rest, calculate the coefficient of kinetic friction between the block of mass m_1 and the surface.

Figure 8.12 (Example 8.9) As the hanging block moves from its highest elevation to its lowest, the system loses gravitational potential energy but gains elastic potential energy in the spring. Some mechanical energy is transformed to internal energy because of friction between the sliding block and the surface.

SOLUTION

Conceptualize The key word *rest* appears twice in the problem statement. This word suggests that the configurations of the system associated with rest are good candidates for the initial and final configurations because the kinetic energy of the system is zero for these configurations.

Categorize In this situation, the system consists of the two blocks, the spring, the surface, and the Earth. This is an *isolated system* with a nonconservative force acting. We also model the sliding block as a *particle in equilibrium* in the vertical direction, leading to $n = m_1 g$.

Analyze We need to consider two forms of potential energy for the system, gravitational and elastic: $\Delta U_g = U_{gf} - U_{gi}$ is the change in the system's gravitational potential energy, and $\Delta U_s = U_{sf} - U_{si}$ is the change in the system's elastic potential energy. The change in the gravitational potential energy of the system is associated with only the falling block

▶ **8.9** continued

because the vertical coordinate of the horizontally sliding block does not change. The initial and final kinetic energies of the system are zero, so $\Delta K = 0$.

Write the appropriate reduction of Equation 8.2:

$$(1)\quad \Delta U_g + \Delta U_s + \Delta E_{\text{int}} = 0$$

Substitute for the energies, noting that as the hanging block falls a distance h, the horizontally moving block moves the same distance h to the right, and the spring stretches by a distance h:

$$(0 - m_2gh) + (\tfrac{1}{2}kh^2 - 0) + f_k h = 0$$

Substitute for the friction force:

$$-m_2gh + \tfrac{1}{2}kh^2 + \mu_k m_1 gh = 0$$

Solve for μ_k:

$$\mu_k = \frac{m_2g - \tfrac{1}{2}kh}{m_1g}$$

Finalize This setup represents a method of measuring the coefficient of kinetic friction between an object and some surface. Notice how we have solved the examples in this chapter using the energy approach. We begin with Equation 8.2 and then tailor it to the physical situation. This process may include deleting terms, such as the kinetic energy term and all terms on the right-hand side of Equation 8.2 in this example. It can also include expanding terms, such as rewriting ΔU due to two types of potential energy in this example.

Conceptual Example 8.10   **Interpreting the Energy Bars**

The energy bar charts in Figure 8.13 show three instants in the motion of the system in Figure 8.12 and described in Example 8.9. For each bar chart, identify the configuration of the system that corresponds to the chart.

SOLUTION

In Figure 8.13a, there is no kinetic energy in the system. Therefore, nothing in the system is moving. The bar chart shows that the system contains only gravitational potential energy and no internal energy yet, which corresponds to the configuration with the darker blocks in Figure 8.12 and represents the instant just after the system is released.

In Figure 8.13b, the system contains four types of energy. The height of the gravitational potential energy bar is at 50%, which tells us that the hanging block has moved halfway between its position corresponding to Figure 8.13a and the position defined as $y = 0$. Therefore, in this configuration, the hanging block is between the dark and light images of the hanging block in Figure 8.12. The system has gained kinetic energy because the blocks are moving, elastic potential energy because the spring is stretching, and internal energy because of friction between the block of mass m_1 and the surface.

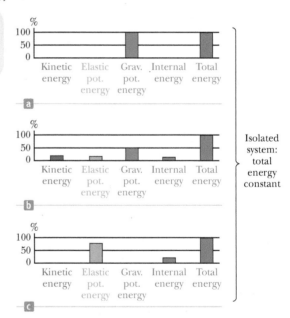

Figure 8.13 (Conceptual Example 8.10) Three energy bar charts are shown for the system in Figure 8.12.

In Figure 8.13c, the height of the gravitational potential energy bar is zero, telling us that the hanging block is at $y = 0$. In addition, the height of the kinetic energy bar is zero, indicating that the blocks have stopped moving momentarily. Therefore, the configuration of the system is that shown by the light images of the blocks in Figure 8.12. The height of the elastic potential energy bar is high because the spring is stretched its maximum amount. The height of the internal energy bar is higher than in Figure 8.13b because the block of mass m_1 has continued to slide over the surface after the configuration shown in Figure 8.13b.

8.5 Power

Consider Conceptual Example 7.7 again, which involved rolling a refrigerator up a ramp into a truck. Suppose the man is not convinced the work is the same regardless of the ramp's length and sets up a long ramp with a gentle rise. Although he does the same amount of work as someone using a shorter ramp, he takes longer to do the work because he has to move the refrigerator over a greater distance. Although the work done on both ramps is the same, there is *something* different about the tasks: the *time interval* during which the work is done.

The time rate of energy transfer is called the **instantaneous power** P and is defined as

◀ Definition of power

$$P \equiv \frac{dE}{dt} \tag{8.18}$$

We will focus on work as the energy transfer method in this discussion, but keep in mind that the notion of power is valid for *any* means of energy transfer discussed in Section 8.1. If an external force is applied to an object (which we model as a particle) and if the work done by this force on the object in the time interval Δt is W, the **average power** during this interval is

$$P_{avg} = \frac{W}{\Delta t}$$

Therefore, in Conceptual Example 7.7, although the same work is done in rolling the refrigerator up both ramps, less power is required for the longer ramp.

In a manner similar to the way we approached the definition of velocity and acceleration, the instantaneous power is the limiting value of the average power as Δt approaches zero:

$$P = \lim_{\Delta t \to 0} \frac{W}{\Delta t} = \frac{dW}{dt}$$

where we have represented the infinitesimal value of the work done by dW. We find from Equation 7.3 that $dW = \vec{\mathbf{F}} \cdot d\vec{\mathbf{r}}$. Therefore, the instantaneous power can be written

$$P = \frac{dW}{dt} = \vec{\mathbf{F}} \cdot \frac{d\vec{\mathbf{r}}}{dt} = \vec{\mathbf{F}} \cdot \vec{\mathbf{v}} \tag{8.19}$$

where $\vec{\mathbf{v}} = d\vec{\mathbf{r}}/dt$.

The SI unit of power is joules per second (J/s), also called the **watt** (W) after James Watt:

◀ The watt

$$1 \text{ W} = 1 \text{ J/s} = 1 \text{ kg} \cdot \text{m}^2/\text{s}^3$$

A unit of power in the U.S. customary system is the **horsepower** (hp):

$$1 \text{ hp} = 746 \text{ W}$$

A unit of energy (or work) can now be defined in terms of the unit of power. One **kilowatt-hour** (kWh) is the energy transferred in 1 h at the constant rate of 1 kW = 1 000 J/s. The amount of energy represented by 1 kWh is

$$1 \text{ kWh} = (10^3 \text{ W})(3\,600 \text{ s}) = 3.60 \times 10^6 \text{ J}$$

A kilowatt-hour is a unit of energy, not power. When you pay your electric bill, you are buying energy, and the amount of energy transferred by electrical transmission into a home during the period represented by the electric bill is usually expressed in kilowatt-hours. For example, your bill may state that you used 900 kWh of energy during a month and that you are being charged at the rate of 10¢ per kilowatt-hour. Your obligation is then $90 for this amount of energy. As another example, suppose an electric bulb is rated at 100 W. In 1.00 h of operation, it would have energy transferred to it by electrical transmission in the amount of (0.100 kW)(1.00 h) = 0.100 kWh = 3.60×10^5 J.

Pitfall Prevention 8.3

W, *W*, and watts Do not confuse the symbol W for the watt with the italic symbol *W* for work. Also, remember that the watt already represents a rate of energy transfer, so "watts per second" does not make sense. The watt is *the same as* a joule per second.

Example 8.11 Power Delivered by an Elevator Motor AM

An elevator car (Fig. 8.14a) has a mass of 1 600 kg and is carrying passengers having a combined mass of 200 kg. A constant friction force of 4 000 N retards its motion.

(A) How much power must a motor deliver to lift the elevator car and its passengers at a constant speed of 3.00 m/s?

SOLUTION

Conceptualize The motor must supply the force of magnitude T that pulls the elevator car upward.

Categorize The friction force increases the power necessary to lift the elevator. The problem states that the speed of the elevator is constant, which tells us that $a = 0$. We model the elevator as a *particle in equilibrium*.

Analyze The free-body diagram in Figure 8.14b specifies the upward direction as positive. The *total* mass M of the system (car plus passengers) is equal to 1 800 kg.

Figure 8.14 (Example 8.11) (a) The motor exerts an upward force $\vec{T}$ on the elevator car. The magnitude of this force is the total tension T in the cables connecting the car and motor. The downward forces acting on the car are a friction force $\vec{f}$ and the gravitational force $\vec{F}_g = M\vec{g}$. (b) The free-body diagram for the elevator car.

Using the particle in equilibrium model, apply Newton's second law to the car:

$$\sum F_y = T - f - Mg = 0$$

Solve for T:

$$T = Mg + f$$

Use Equation 8.19 and that $\vec{T}$ is in the same direction as $\vec{v}$ to find the power:

$$P = \vec{T} \cdot \vec{v} = Tv = (Mg + f)v$$

Substitute numerical values:

$$P = [(1\ 800\ \text{kg})(9.80\ \text{m/s}^2) + (4\ 000\ \text{N})](3.00\ \text{m/s}) = 6.49 \times 10^4\ \text{W}$$

(B) What power must the motor deliver at the instant the speed of the elevator is v if the motor is designed to provide the elevator car with an upward acceleration of 1.00 m/s²?

SOLUTION

Conceptualize In this case, the motor must supply the force of magnitude T that pulls the elevator car upward with an increasing speed. We expect that more power will be required to do that than in part (A) because the motor must now perform the additional task of accelerating the car.

Categorize In this case, we model the elevator car as a *particle under a net force* because it is accelerating.

Analyze Using the particle under a net force model, apply Newton's second law to the car:

$$\sum F_y = T - f - Mg = Ma$$

Solve for T:

$$T = M(a + g) + f$$

Use Equation 8.19 to obtain the required power:

$$P = Tv = [M(a + g) + f]v$$

Substitute numerical values:

$$P = [(1\ 800\ \text{kg})(1.00\ \text{m/s}^2 + 9.80\ \text{m/s}^2) + 4\ 000\ \text{N}]v$$

$$= (2.34 \times 10^4)v$$

where v is the instantaneous speed of the car in meters per second and P is in watts.

Finalize To compare with part (A), let $v = 3.00$ m/s, giving a power of

$$P = (2.34 \times 10^4\ \text{N})(3.00\ \text{m/s}) = 7.02 \times 10^4\ \text{W}$$

which is larger than the power found in part (A), as expected.

Summary

Definitions

A **nonisolated system** is one for which energy crosses the boundary of the system. An **isolated system** is one for which no energy crosses the boundary of the system.

The **instantaneous power** P is defined as the time rate of energy transfer:

$$P \equiv \frac{dE}{dt} \qquad (8.18)$$

Concepts and Principles

For a nonisolated system, we can equate the change in the total energy stored in the system to the sum of all the transfers of energy across the system boundary, which is a statement of **conservation of energy**. For an isolated system, the total energy is constant.

If a friction force of magnitude f_k acts over a distance d within a system, the change in internal energy of the system is

$$\Delta E_{\text{int}} = f_k d \qquad (8.14)$$

Analysis Models for Problem Solving

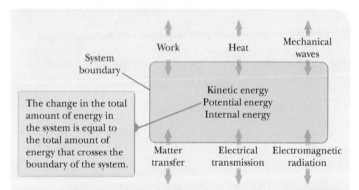

Nonisolated System (Energy). The most general statement describing the behavior of a nonisolated system is the **conservation of energy equation**:

$$\Delta E_{\text{system}} = \Sigma T \qquad (8.1)$$

Including the types of energy storage and energy transfer that we have discussed gives

$$\Delta K + \Delta U + \Delta E_{\text{int}} = W + Q + T_{\text{MW}} + T_{\text{MT}} + T_{\text{ET}} + T_{\text{ER}} \qquad (8.2)$$

For a specific problem, this equation is generally reduced to a smaller number of terms by eliminating the terms that are not appropriate to the situation.

Isolated System (Energy). The total energy of an isolated system is conserved, so

$$\Delta E_{\text{system}} = 0 \qquad (8.10)$$

which can be written as

$$\Delta K + \Delta U + \Delta E_{\text{int}} = 0 \qquad (8.16)$$

If no nonconservative forces act within the isolated system, the mechanical energy of the system is conserved, so

$$\Delta E_{\text{mech}} = 0 \qquad (8.8)$$

which can be written as

$$\Delta K + \Delta U = 0 \qquad (8.6)$$

1. You hold a slingshot at arm's length, pull the light elastic band back to your chin, and release it to launch a pebble horizontally with speed 200 cm/s. With the same procedure, you fire a bean with speed 600 cm/s. What is the ratio of the mass of the bean to the mass of the pebble? (a) $\frac{1}{9}$ (b) $\frac{1}{3}$ (c) 1 (d) 3 (e) 9

2. Two children stand on a platform at the top of a curving slide next to a backyard swimming pool. At the same moment the smaller child hops off to jump straight down into the pool, the bigger child releases herself at the top of the frictionless slide. **(i)** Upon reaching the water, the kinetic energy of the smaller child compared with that of the larger child is (a) greater (b) less (c) equal. **(ii)** Upon reaching the water, the speed of the smaller child compared with that of the larger child is (a) greater (b) less (c) equal. **(iii)** During their motions from the platform to the water, the average acceleration of the smaller child compared with that of the larger child is (a) greater (b) less (c) equal.

3. At the bottom of an air track tilted at angle θ, a glider of mass m is given a push to make it coast a distance d up the slope as it slows down and stops. Then the glider comes back down the track to its starting point. Now the experiment is repeated with the same original speed but with a second identical glider set on top of the first. The airflow from the track is strong enough to support the stacked pair of gliders so that the combination moves over the track with negligible friction. Static friction holds the second glider stationary relative to the first glider throughout the motion. The coefficient of static friction between the two gliders is μ_s. What is the change in mechanical energy of the two-glider–Earth system in the up- and down-slope motion after the pair of gliders is released? Choose one. (a) $-2\mu_s mg$ (b) $-2mgd\cos\theta$ (c) $-2\mu_s mgd\cos\theta$ (d) 0 (e) $+2\mu_s mgd\cos\theta$

4. An athlete jumping vertically on a trampoline leaves the surface with a velocity of 8.5 m/s upward. What maximum height does she reach? (a) 13 m (b) 2.3 m (c) 3.7 m (d) 0.27 m (e) The answer can't be determined because the mass of the athlete isn't given.

5. Answer yes or no to each of the following questions. (a) Can an object–Earth system have kinetic energy and not gravitational potential energy? (b) Can it have gravitational potential energy and not kinetic energy? (c) Can it have both types of energy at the same moment? (d) Can it have neither?

6. In a laboratory model of cars skidding to a stop, data are measured for four trials using two blocks. The blocks have identical masses but different coefficients of kinetic friction with a table: $\mu_k = 0.2$ and 0.8. Each block is launched with speed $v_i = 1$ m/s and slides across the level table as the block comes to rest. This process represents the first two trials. For the next two trials, the procedure is repeated but the blocks are launched with speed $v_i = 2$ m/s. Rank the four trials (a) through (d) according to the stopping distance from largest to smallest. If the stopping distance is the same in two cases, give them equal rank. (a) $v_i = 1$ m/s, $\mu_k = 0.2$ (b) $v_i = 1$ m/s, $\mu_k = 0.8$ (c) $v_i = 2$ m/s, $\mu_k = 0.2$ (d) $v_i = 2$ m/s, $\mu_k = 0.8$

7. What average power is generated by a 70.0-kg mountain climber who climbs a summit of height 325 m in 95.0 min? (a) 39.1 W (b) 54.6 W (c) 25.5 W (d) 67.0 W (e) 88.4 W

8. A ball of clay falls freely to the hard floor. It does not bounce noticeably, and it very quickly comes to rest. What, then, has happened to the energy the ball had while it was falling? (a) It has been used up in producing the downward motion. (b) It has been transformed back into potential energy. (c) It has been transferred into the ball by heat. (d) It is in the ball and floor (and walls) as energy of invisible molecular motion. (e) Most of it went into sound.

9. A pile driver drives posts into the ground by repeatedly dropping a heavy object on them. Assume the object is dropped from the same height each time. By what factor does the energy of the pile driver–Earth system change when the mass of the object being dropped is doubled? (a) $\frac{1}{2}$ (b) 1; the energy is the same (c) 2 (d) 4

1. One person drops a ball from the top of a building while another person at the bottom observes its motion. Will these two people agree (a) on the value of the gravitational potential energy of the ball–Earth system? (b) On the change in potential energy? (c) On the kinetic energy of the ball at some point in its motion?

2. A car salesperson claims that a 300-hp engine is a necessary option in a compact car, in place of the conventional 130-hp engine. Suppose you intend to drive the car within speed limits (≤ 65 mi/h) on flat terrain. How would you counter this sales pitch?

3. Does everything have energy? Give the reasoning for your answer.

4. You ride a bicycle. In what sense is your bicycle solar-powered?

5. A bowling ball is suspended from the ceiling of a lecture hall by a strong cord. The ball is drawn away from its equilibrium position and released from rest at the

tip of the demonstrator's nose as shown in Figure CQ8.5. The demonstrator remains stationary. (a) Explain why the ball does not strike her on its return swing. (b) Would this demonstrator be safe if the ball were given a push from its starting position at her nose?

Figure CQ8.5

6. Can a force of static friction do work? If not, why not? If so, give an example.

7. In the general conservation of energy equation, state which terms predominate in describing each of the following devices and processes. For a process going on continuously, you may consider what happens in a 10-s time interval. State which terms in the equation represent original and final forms of energy, which would be inputs, and which outputs. (a) a slingshot firing a pebble (b) a fire burning (c) a portable radio operating (d) a car braking to a stop (e) the surface of the Sun shining visibly (f) a person jumping up onto a chair

8. Consider the energy transfers and transformations listed below in parts (a) through (e). For each part, **(i)** describe human-made devices designed to produce each of the energy transfers or transformations and, **(ii)** whenever possible, describe a natural process in which the energy transfer or transformation occurs. Give details to defend your choices, such as identifying the system and identifying other output energy if the device or natural process has limited efficiency. (a) Chemical potential energy transforms into internal energy. (b) Energy transferred by electrical transmission becomes gravitational potential energy. (c) Elastic potential energy transfers out of a system by heat. (d) Energy transferred by mechanical waves does work on a system. (e) Energy carried by electromagnetic waves becomes kinetic energy in a system.

9. A block is connected to a spring that is suspended from the ceiling. Assuming air resistance is ignored, describe the energy transformations that occur within the system consisting of the block, the Earth, and the spring when the block is set into vertical motion.

10. In Chapter 7, the work–kinetic energy theorem, $W = \Delta K$, was introduced. This equation states that work done on a system appears as a change in kinetic energy. It is a special-case equation, valid if there are no changes in any other type of energy such as potential or internal. Give two or three examples in which work is done on a system but the change in energy of the system is not a change in kinetic energy.

Problems

The problems found in this chapter may be assigned online in Enhanced WebAssign

1. straightforward; **2.** intermediate; **3.** challenging

1. full solution available in the *Student Solutions Manual/Study Guide*

AMT Analysis Model tutorial available in Enhanced WebAssign

GP Guided Problem

M Master It tutorial available in Enhanced WebAssign

W Watch It video solution available in Enhanced WebAssign

Section 8.1 Analysis Model: Nonisolated System (Energy)

1. For each of the following systems and time intervals, write the appropriate version of Equation 8.2, the conservation of energy equation. (a) the heating coils in your toaster during the first five seconds after you turn the toaster on (b) your automobile from just before you fill it with gasoline until you pull away from the gas station at speed v (c) your body while you sit quietly and eat a peanut butter and jelly sandwich for lunch (d) your home during five minutes of a sunny afternoon while the temperature in the home remains fixed

2. A ball of mass m falls from a height h to the floor. (a) Write the appropriate version of Equation 8.2 for the system of the ball and the Earth and use it to calculate the speed of the ball just before it strikes the Earth. (b) Write the appropriate version of Equation 8.2 for the system of the ball and use it to calculate the speed of the ball just before it strikes the Earth.

Section 8.2 Analysis Model: Isolated System (Energy)

3. **W** A block of mass 0.250 kg is placed on top of a light, vertical spring of force constant 5 000 N/m and pushed downward so that the spring is compressed by 0.100 m. After the block is released from rest, it travels upward and then leaves the spring. To what maximum height above the point of release does it rise?

4. **W** A 20.0-kg cannonball is fired from a cannon with muzzle speed of 1 000 m/s at an angle of 37.0° with the horizontal. A second ball is fired at an angle of 90.0°. Use the isolated system model to find (a) the maximum height reached by each ball and (b) the total mechanical energy of the ball–Earth system at the maximum height for each ball. Let $y = 0$ at the cannon.

5. **AMT M** **Review.** A bead slides without friction around a loop-the-loop (Fig. P8.5). The bead is released from rest at a height $h = 3.50R$. (a) What

Figure P8.5

is its speed at point Ⓐ? (b) How large is the normal force on the bead at point Ⓐ if its mass is 5.00 g?

6. A block of mass $m = 5.00$ kg is released from point Ⓐ Ⓦ and slides on the frictionless track shown in Figure P8.6. Determine (a) the block's speed at points Ⓑ and Ⓒ and (b) the net work done by the gravitational force on the block as it moves from point Ⓐ to point Ⓒ.

Figure P8.6

7. Two objects are connected Ⓜ by a light string passing over a light, frictionless pulley as shown in Figure P8.7. The object of mass $m_1 = 5.00$ kg is released from rest at a height $h = 4.00$ m above the table. Using the isolated system model, (a) determine the speed of the object of mass $m_2 = 3.00$ kg just as the 5.00-kg object hits the table and (b) find the maximum height above the table to which the 3.00-kg object rises.

Figure P8.7
Problems 7 and 8.

8. Two objects are connected by a light string passing over a light, frictionless pulley as shown in Figure P8.7. The object of mass m_1 is released from rest at height h above the table. Using the isolated system model, (a) determine the speed of m_2 just as m_1 hits the table and (b) find the maximum height above the table to which m_2 rises.

9. A light, rigid rod is 77.0 cm long. Its top end is pivoted on a frictionless, horizontal axle. The rod hangs straight down at rest with a small, massive ball attached to its bottom end. You strike the ball, suddenly giving it a horizontal velocity so that it swings around in a full circle. What minimum speed at the bottom is required to make the ball go over the top of the circle?

10. At 11:00 a.m. on September 7, 2001, more than one million British schoolchildren jumped up and down for one minute to simulate an earthquake. (a) Find the energy stored in the children's bodies that was converted into internal energy in the ground and their bodies and propagated into the ground by seismic waves during the experiment. Assume 1 050 000 children of average mass 36.0 kg jumped 12 times each, raising their centers of mass by 25.0 cm each time and briefly resting between one jump and the next. (b) Of the energy that propagated into the ground, most pro-

duced high-frequency "microtremor" vibrations that were rapidly damped and did not travel far. Assume 0.01% of the total energy was carried away by long-range seismic waves. The magnitude of an earthquake on the Richter scale is given by

$$M = \frac{\log E - 4.8}{1.5}$$

where E is the seismic wave energy in joules. According to this model, what was the magnitude of the demonstration quake?

11. **Review.** The system shown in Figure P8.11 consists of a light, inextensible cord, light, frictionless pulleys, and blocks of equal mass. Notice that block B is attached to one of the pulleys. The system is initially held at rest so that the blocks are at the same height above the ground. The blocks are then released. Find the speed of block A at the moment the vertical separation of the blocks is h.

Figure P8.11

Section 8.3 Situations Involving Kinetic Friction

12. A sled of mass m is given a kick on a frozen pond. The kick imparts to the sled an initial speed of 2.00 m/s. The coefficient of kinetic friction between sled and ice is 0.100. Use energy considerations to find the distance the sled moves before it stops.

13. A sled of mass m is given a kick on a frozen pond. The kick imparts to the sled an initial speed of v. The coefficient of kinetic friction between sled and ice is μ_k. Use energy considerations to find the distance the sled moves before it stops.

14. A crate of mass 10.0 kg is pulled up a rough incline with Ⓜ an initial speed of 1.50 m/s. The pulling force is 100 N parallel to the incline, which makes an angle of 20.0° with the horizontal. The coefficient of kinetic friction is 0.400, and the crate is pulled 5.00 m. (a) How much work is done by the gravitational force on the crate? (b) Determine the increase in internal energy of the crate–incline system owing to friction. (c) How much work is done by the 100-N force on the crate? (d) What is the change in kinetic energy of the crate? (e) What is the speed of the crate after being pulled 5.00 m?

15. A block of mass $m = 2.00$ kg Ⓦ is attached to a spring of force constant $k = 500$ N/m as shown in Figure P8.15. The block is pulled to a position $x_i = 5.00$ cm to the right of equilibrium and released from rest. Find the speed the block has as it passes

Figure P8.15

through equilibrium if (a) the horizontal surface is frictionless and (b) the coefficient of friction between block and surface is $\mu_k = 0.350$.

16. A 40.0-kg box initially at rest is pushed 5.00 m along a rough, horizontal floor with a constant applied horizontal force of 130 N. The coefficient of friction

between box and floor is 0.300. Find (a) the work done by the applied force, (b) the increase in internal energy in the box–floor system as a result of friction, (c) the work done by the normal force, (d) the work done by the gravitational force, (e) the change in kinetic energy of the box, and (f) the final speed of the box.

17. A smooth circular hoop with a radius of 0.500 m is placed flat on the floor. A 0.400-kg particle slides around the inside edge of the hoop. The particle is given an initial speed of 8.00 m/s. After one revolution, its speed has dropped to 6.00 m/s because of friction with the floor. (a) Find the energy transformed from mechanical to internal in the particle–hoop–floor system as a result of friction in one revolution. (b) What is the total number of revolutions the particle makes before stopping? Assume the friction force remains constant during the entire motion.

Section 8.4 Changes in Mechanical Energy for Nonconservative Forces

18. At time t_i, the kinetic energy of a particle is 30.0 J and the potential energy of the system to which it belongs is 10.0 J. At some later time t_f, the kinetic energy of the particle is 18.0 J. (a) If only conservative forces act on the particle, what are the potential energy and the total energy of the system at time t_f? (b) If the potential energy of the system at time t_f is 5.00 J, are any nonconservative forces acting on the particle? (c) Explain your answer to part (b).

19. A boy in a wheelchair (total mass 47.0 kg) has speed 1.40 m/s at the crest of a slope 2.60 m high and 12.4 m long. At the bottom of the slope his speed is 6.20 m/s. Assume air resistance and rolling resistance can be modeled as a constant friction force of 41.0 N. Find the work he did in pushing forward on his wheels during the downhill ride.

20. As shown in Figure P8.20, a green bead of mass 25 g slides along a straight wire. The length of the wire from point Ⓐ to point Ⓑ is 0.600 m, and point Ⓐ is 0.200 m higher than point Ⓑ. A constant friction force of magnitude 0.025 0 N acts on the bead. (a) If the bead is released from rest at point Ⓐ, what is its speed at point Ⓑ? (b) A red bead of mass 25 g slides along a curved wire, subject to a friction force with the same constant magnitude as that on the green bead. If the green and red beads are released simultaneously from rest at point Ⓐ, which bead reaches point Ⓑ with a higher speed? Explain.

Figure P8.20

21. A toy cannon uses a spring to project a 5.30-g soft rubber ball. The spring is originally compressed by 5.00 cm and has a force constant of 8.00 N/m. When the cannon is fired, the ball moves 15.0 cm through the horizontal barrel of the cannon, and the barrel exerts a constant friction force of 0.032 0 N on the ball. (a) With what speed does the projectile leave the barrel

of the cannon? (b) At what point does the ball have maximum speed? (c) What is this maximum speed?

22. The coefficient of friction between the block of mass $m_1 = 3.00$ kg and the surface in Figure P8.22 is $\mu_k = 0.400$. The system starts from rest. What is the speed of the ball of mass $m_2 = 5.00$ kg when it has fallen a distance $h = 1.50$ m?

Figure P8.22

23. A 5.00-kg block is set into motion up an inclined plane with an initial speed of $v_i = 8.00$ m/s (Fig. P8.23). The block comes to rest after traveling $d = 3.00$ m along the plane, which is inclined at an angle of $\theta = 30.0°$ to the horizontal. For this motion, determine (a) the change in the block's kinetic energy, (b) the change in the potential energy of the block–Earth system, and (c) the friction force exerted on the block (assumed to be constant). (d) What is the coefficient of kinetic friction?

Figure P8.23

24. A 1.50-kg object is held 1.20 m above a relaxed massless, vertical spring with a force constant of 320 N/m. The object is dropped onto the spring. (a) How far does the object compress the spring? (b) **What If?** Repeat part (a), but this time assume a constant air-resistance force of 0.700 N acts on the object during its motion. (c) **What If?** How far does the object compress the spring if the same experiment is performed on the Moon, where $g = 1.63$ m/s^2 and air resistance is neglected?

25. A 200-g block is pressed against a spring of force constant 1.40 kN/m until the block compresses the spring 10.0 cm. The spring rests at the bottom of a ramp inclined at 60.0° to the horizontal. Using energy considerations, determine how far up the incline the block moves from its initial position before it stops (a) if the ramp exerts no friction force on the block and (b) if the coefficient of kinetic friction is 0.400.

26. An 80.0-kg skydiver jumps out of a balloon at an altitude of 1 000 m and opens his parachute at an altitude of 200 m. (a) Assuming the total retarding force on the skydiver is constant at 50.0 N with the parachute closed and constant at 3 600 N with the parachute open, find the speed of the skydiver when he lands on the ground. (b) Do you think the skydiver will be injured? Explain. (c) At what height should the parachute be opened so that the final speed of the skydiver when he hits the ground is 5.00 m/s? (d) How realistic is the assumption that the total retarding force is constant? Explain.

27. A child of mass m starts from rest and slides without friction from a height h along a slide next to a pool (Fig. P8.27). She is launched from a height $h/5$ into the air over the pool. We wish to find the maximum height she reaches above the water in her projectile motion. (a) Is the child–Earth system isolated or

nonisolated? Why? (b) Is there a nonconservative force acting within the system? (c) Define the configuration of the system when the child is at the water level as having zero gravitational potential energy. Express the total energy of the system when the child is at the top of the waterslide. (d) Express the total energy of the system when the child is at the launching point. (e) Express the total energy of the system when the child is at the highest point in her projectile motion. (f) From parts (c) and (d), determine her initial speed v_i at the launch point in terms of g and h. (g) From parts (d), (e), and (f), determine her maximum airborne height y_{max} in terms of h and the launch angle θ. (h) Would your answers be the same if the waterslide were not frictionless? Explain.

Figure P8.27

Section 8.5 **Power**

28. Sewage at a certain pumping station is raised vertically by 5.49 m at the rate of 1 890 000 liters each day. The sewage, of density 1 050 kg/m^3, enters and leaves the pump at atmospheric pressure and through pipes of equal diameter. (a) Find the output mechanical power of the lift station. (b) Assume an electric motor continuously operating with average power 5.90 kW runs the pump. Find its efficiency.

29. An 820-N Marine in basic training climbs a 12.0-m vertical rope at a constant speed in 8.00 s. What is his power output?

30. The electric motor of a model train accelerates the train from rest to 0.620 m/s in 21.0 ms. The total mass of the train is 875 g. (a) Find the minimum power delivered to the train by electrical transmission from the metal rails during the acceleration. (b) Why is it the minimum power?

31. When an automobile moves with constant speed down a highway, most of the power developed by the engine is used to compensate for the energy transformations due to friction forces exerted on the car by the air and the road. If the power developed by an engine is 175 hp, estimate the total friction force acting on the car when it is moving at a speed of 29 m/s. One horsepower equals 746 W.

32. A certain rain cloud at an altitude of 1.75 km contains 3.20×10^7 kg of water vapor. How long would it take a 2.70-kW pump to raise the same amount of water from the Earth's surface to the cloud's position?

33. An energy-efficient lightbulb, taking in 28.0 W of power, can produce the same level of brightness as a conventional lightbulb operating at power 100 W. The lifetime of the energy-efficient bulb is 10 000 h and its purchase price is $4.50, whereas the conventional bulb has a lifetime of 750 h and costs $0.42. Determine the total savings obtained by using one energy-efficient bulb over its lifetime as opposed to using conventional bulbs over the same time interval. Assume an energy cost of $0.200 per kilowatt-hour.

34. An electric scooter has a battery capable of supplying 120 Wh of energy. If friction forces and other losses account for 60.0% of the energy usage, what altitude change can a rider achieve when driving in hilly terrain if the rider and scooter have a combined weight of 890 N?

35. Make an order-of-magnitude estimate of the power a car engine contributes to speeding the car up to highway speed. In your solution, state the physical quantities you take as data and the values you measure or estimate for them. The mass of a vehicle is often given in the owner's manual.

36. An older-model car accelerates from 0 to speed v in a time interval of Δt. A newer, more powerful sports car accelerates from 0 to $2v$ in the same time period. Assuming the energy coming from the engine appears only as kinetic energy of the cars, compare the power of the two cars.

37. For saving energy, bicycling and walking are far more efficient means of transportation than is travel by automobile. For example, when riding at 10.0 mi/h, a cyclist uses food energy at a rate of about 400 kcal/h above what he would use if merely sitting still. (In exercise physiology, power is often measured in kcal/h rather than in watts. Here 1 kcal = 1 nutritionist's Calorie = 4 186 J.) Walking at 3.00 mi/h requires about 220 kcal/h. It is interesting to compare these values with the energy consumption required for travel by car. Gasoline yields about 1.30×10^8 J/gal. Find the fuel economy in equivalent miles per gallon for a person (a) walking and (b) bicycling.

38. A 650-kg elevator starts from rest. It moves upward for 3.00 s with constant acceleration until it reaches its cruising speed of 1.75 m/s. (a) What is the average power of the elevator motor during this time interval? (b) How does this power compare with the motor power when the elevator moves at its cruising speed?

39. A 3.50-kN piano is lifted by three workers at constant speed to an apartment 25.0 m above the street using a pulley system fastened to the roof of the building. Each worker is able to deliver 165 W of power, and the pulley system is 75.0% efficient (so that 25.0% of the mechanical energy is transformed to other forms due to friction in the pulley). Neglecting the mass of the pulley, find the time required to lift the piano from the street to the apartment.

40. Energy is conventionally measured in Calories as well as in joules. One Calorie in nutrition is one kilocalorie, defined as 1 kcal = 4 186 J. Metabolizing 1 g of fat can release 9.00 kcal. A student decides to try to lose weight by exercising. He plans to run up and down the stairs in a football stadium as fast as he can and as many times as necessary. To evaluate the program, suppose he runs up a flight of 80 steps, each 0.150 m high, in 65.0 s. For simplicity, ignore the energy he uses in coming down (which is small). Assume a typical efficiency for human muscles is 20.0%. This statement means that when your body converts 100 J from metabolizing fat, 20 J goes into doing mechanical work (here, climbing stairs). The remainder goes into extra internal energy. Assume the student's mass is 75.0 kg. (a) How many times must the student run the flight of stairs to lose 1.00 kg of fat? (b) What is his average power output, in watts and in horsepower, as he runs up the stairs? (c) Is this activity in itself a practical way to lose weight?

41. A loaded ore car has a mass of 950 kg and rolls on rails with negligible friction. It starts from rest and is pulled up a mine shaft by a cable connected to a winch. The shaft is inclined at 30.0° above the horizontal. The car accelerates uniformly to a speed of 2.20 m/s in 12.0 s and then continues at constant speed. (a) What power must the winch motor provide when the car is moving at constant speed? (b) What maximum power must the winch motor provide? (c) What total energy has transferred out of the motor by work by the time the car moves off the end of the track, which is of length 1 250 m?

Additional Problems

42. Make an order-of-magnitude estimate of your power output as you climb stairs. In your solution, state the physical quantities you take as data and the values you measure or estimate for them. Do you consider your peak power or your sustainable power?

43. A small block of mass $m = 200$ g is released from rest at point Ⓐ along the horizontal diameter on the inside of a frictionless, hemispherical bowl of radius $R = 30.0$ cm (Fig. P8.43). Calculate (a) the gravitational potential energy of the block–Earth system when the block is at point Ⓐ relative to point Ⓑ, (b) the kinetic energy of the block at point Ⓑ, (c) its speed at point Ⓑ, and (d) its kinetic energy and the potential energy when the block is at point Ⓒ.

Figure P8.43 Problems 43 and 44.

44. **What If?** The block of mass $m = 200$ g described in Problem 43 (Fig. P8.43) is released from rest at point Ⓐ, and the surface of the bowl is rough. The block's speed at point Ⓑ is 1.50 m/s. (a) What is its kinetic energy at point Ⓑ? (b) How much mechanical energy is transformed into internal energy as the block moves from point Ⓐ to point Ⓑ? (c) Is it possible to determine the coefficient of friction from these results in any simple manner? (d) Explain your answer to part (c).

45. **Review.** A boy starts at rest and slides down a frictionless slide as in Figure P8.45. The bottom of the track is a height h above the ground. The boy then leaves the track horizontally, striking the ground at a distance d as shown. Using energy methods, determine the initial height H of the boy above the ground in terms of h and d.

Figure P8.45

46. **Review.** As shown in Figure P8.46, a light string that does not stretch changes from horizontal to vertical as it passes over the edge of a table. The string connects m_1, a 3.50-kg block originally at rest on the horizontal table at a height $h = 1.20$ m above the floor, to m_2, a hanging 1.90-kg block originally a distance $d = 0.900$ m above the floor. Neither the surface of the table nor its edge exerts a force of kinetic friction. The blocks start to move from rest. The sliding block m_1 is projected horizontally after reaching the edge of the table. The hanging block m_2 stops without bouncing when it strikes the floor. Consider the two blocks plus the Earth as the system. (a) Find the speed at which m_1 leaves the edge of the table. (b) Find the impact speed of m_1 on the floor. (c) What is the shortest length of the string so that it does not go taut while m_1 is in flight? (d) Is the energy of the system when it is released from rest equal to the energy of the system just before m_1 strikes the ground? (e) Why or why not?

Figure P8.46

47. A 4.00-kg particle moves along the x axis. Its position varies with time according to $x = t + 2.0t^3$, where x is in meters and t is in seconds. Find (a) the kinetic energy of the particle at any time t, (b) the acceleration of the particle and the force acting on it at time t, (c) the power being delivered to the particle at time t, and (d) the work done on the particle in the interval $t = 0$ to $t = 2.00$ s.

48. *Why is the following situation impossible?* A softball pitcher has a strange technique: she begins with her hand at rest at the highest point she can reach and then quickly rotates her arm backward so that the ball moves through a half-circle path. She releases the ball when her hand reaches the bottom of the path. The pitcher maintains a component of force on the 0.180-kg ball of constant magnitude 12.0 N in the direction of motion around the complete path. As the ball arrives

at the bottom of the path, it leaves her hand with a speed of 25.0 m/s.

49. A skateboarder with his board can be modeled as a particle of mass 76.0 kg, located at his center of mass (which we will study in Chapter 9). As shown in Figure P8.49, the skateboarder starts from rest in a crouching position at one lip of a half-pipe (point Ⓐ). The half-pipe is one half of a cylinder of radius 6.80 m with its axis horizontal. On his descent, the skateboarder moves without friction so that his center of mass moves through one quarter of a circle of radius 6.30 m. (a) Find his speed at the bottom of the half-pipe (point Ⓑ). (b) Immediately after passing point Ⓑ, he stands up and raises his arms, lifting his center of mass from 0.500 m to 0.950 m above the concrete (point Ⓒ). Next, the skateboarder glides upward with his center of mass moving in a quarter circle of radius 5.85 m. His body is horizontal when he passes point Ⓓ, the far lip of the half-pipe. As he passes through point Ⓓ, the speed of the skateboarder is 5.14 m/s. How much chemical potential energy in the body of the skateboarder was converted to mechanical energy in the skateboarder–Earth system when he stood up at point Ⓑ? (c) How high above point Ⓓ does he rise? *Caution:* Do not try this stunt yourself without the required skill and protective equipment.

Figure P8.49

50. Heedless of danger, a child leaps onto a pile of old mattresses to use them as a trampoline. His motion between two particular points is described by the energy conservation equation

$$\tfrac{1}{2}(46.0 \text{ kg})(2.40 \text{ m/s})^2 + (46.0 \text{ kg})(9.80 \text{ m/s}^2)(2.80 \text{ m} + x) = \tfrac{1}{2}(1.94 \times 10^4 \text{ N/m})x^2$$

(a) Solve the equation for x. (b) Compose the statement of a problem, including data, for which this equation gives the solution. (c) Add the two values of x obtained in part (a) and divide by 2. (d) What is the significance of the resulting value in part (c)?

51. Jonathan is riding a bicycle and encounters a hill of height 7.30 m. At the base of the hill, he is traveling at 6.00 m/s. When he reaches the top of the hill, he is traveling at 1.00 m/s. Jonathan and his bicycle together have a mass of 85.0 kg. Ignore friction in the bicycle mechanism and between the bicycle tires and the road. (a) What is the total external work done on the system of Jonathan and the bicycle between the time he starts up the hill and the time he reaches the top? (b) What is the change in potential energy stored in Jonathan's body during this process? (c) How much work does

Jonathan do on the bicycle pedals within the Jonathan–bicycle–Earth system during this process?

52. Jonathan is riding a bicycle and encounters a hill of height h. At the base of the hill, he is traveling at a speed v_i. When he reaches the top of the hill, he is traveling at a speed v_f. Jonathan and his bicycle together have a mass m. Ignore friction in the bicycle mechanism and between the bicycle tires and the road. (a) What is the total external work done on the system of Jonathan and the bicycle between the time he starts up the hill and the time he reaches the top? (b) What is the change in potential energy stored in Jonathan's body during this process? (c) How much work does Jonathan do on the bicycle pedals within the Jonathan-bicycle–Earth system during this process?

53. Consider the block–spring–surface system in part (B) of Example 8.6. (a) Using an energy approach, find the position x of the block at which its speed is a maximum. (b) In the **What If?** section of this example, we explored the effects of an increased friction force of 10.0 N. At what position of the block does its maximum speed occur in this situation?

54. As it plows a parking lot, a snowplow pushes an ever-growing pile of snow in front of it. Suppose a car moving through the air is similarly modeled as a cylinder of area A pushing a growing disk of air in front of it. The origi-

Figure P8.54

nally stationary air is set into motion at the constant speed v of the cylinder as shown in Figure P8.54. In a time interval Δt, a new disk of air of mass Δm must be moved a distance $v \Delta t$ and hence must be given a kinetic energy $\tfrac{1}{2}(\Delta m)v^2$. Using this model, show that the car's power loss owing to air resistance is $\tfrac{1}{2}\rho A v^3$ and that the resistive force acting on the car is $\tfrac{1}{2}\rho A v^2$, where ρ is the density of air. Compare this result with the empirical expression $\tfrac{1}{2}D\rho A v^2$ for the resistive force.

55. A wind turbine on a wind farm turns in response to a force of high-speed air resistance, $R = \tfrac{1}{2}D\rho A v^2$. The power available is $P = Rv = \tfrac{1}{2}D\rho\pi r^2 v^3$, where v is the wind speed and we have assumed a circular face for the wind turbine of radius r. Take the drag coefficient as $D = 1.00$ and the density of air from the front endpaper. For a wind turbine having $r = 1.50$ m, calculate the power available with (a) $v = 8.00$ m/s and (b) $v = 24.0$ m/s. The power delivered to the generator is limited by the efficiency of the system, about 25%. For comparison, a large American home uses about 2 kW of electric power.

56. Consider the popgun in Example 8.3. Suppose the projectile mass, compression distance, and spring constant remain the same as given or calculated in the example. Suppose, however, there is a friction force of magnitude 2.00 N acting on the projectile as it rubs against the interior of the barrel. The vertical length from point Ⓐ to the end of the barrel is 0.600 m.

(a) After the spring is compressed and the popgun fired, to what height does the projectile rise above point ⑧? (b) Draw four energy bar charts for this situation, analogous to those in Figures 8.6c–d.

57. As the driver steps on the gas pedal, a car of mass 1 160 kg accelerates from rest. During the first few seconds of motion, the car's acceleration increases with time according to the expression

$$a = 1.16t - 0.210t^2 + 0.240t^3$$

where t is in seconds and a is in m/s². (a) What is the change in kinetic energy of the car during the interval from $t = 0$ to $t = 2.50$ s? (b) What is the minimum average power output of the engine over this time interval? (c) Why is the value in part (b) described as the *minimum* value?

58. **Review.** *Why is the following situation impossible?* A new high-speed roller coaster is claimed to be so safe that the passengers do not need to wear seat belts or any other restraining device. The coaster is designed with a vertical circular section over which the coaster travels on the inside of the circle so that the passengers are upside down for a short time interval. The radius of the circular section is 12.0 m, and the coaster enters the bottom of the circular section at a speed of 22.0 m/s. Assume the coaster moves without friction on the track and model the coaster as a particle.

59. A horizontal spring attached to a wall has a force constant of $k = 850$ N/m. A block of mass $m = 1.00$ kg is attached to the spring and rests on a frictionless, horizontal surface as in Figure P8.59. (a) The block is pulled to a position $x_i = 6.00$ cm from equilibrium and released. Find the elastic potential energy stored in the spring when the block is 6.00 cm from equilibrium and when the block passes through equilibrium. (b) Find the speed of the block as it passes through the equilibrium point. (c) What is the speed of the block when it is at a position $x_i/2 = 3.00$ cm? (d) Why isn't the answer to part (c) half the answer to part (b)?

$x = 0$ $x = x_i/2$ $x = x_i$

Figure P8.59

60. More than 2 300 years ago, the Greek teacher Aristotle wrote the first book called *Physics*. Put into more precise terminology, this passage is from the end of its Section Eta:

Let P be the power of an agent causing motion; w, the load moved; d, the distance covered; and Δt, the time interval required. Then (1) a power equal to P will in an interval of time equal to Δt move $w/2$ a distance $2d$; or (2) it will move $w/2$ the given distance d in the time interval $\Delta t/2$. Also, if (3) the given power P moves the given

load w a distance $d/2$ in time interval $\Delta t/2$, then (4) $P/2$ will move $w/2$ the given distance d in the given time interval Δt.

(a) Show that Aristotle's proportions are included in the equation $P\Delta t = bwd$, where b is a proportionality constant. (b) Show that our theory of motion includes this part of Aristotle's theory as one special case. In particular, describe a situation in which it is true, derive the equation representing Aristotle's proportions, and determine the proportionality constant.

61. A child's pogo stick (Fig. P8.61) stores energy in a spring with a force constant of 2.50×10^4 N/m. At position ⓐ ($x_ⓐ = -0.100$ m), the spring compression is a maximum and the child is momentarily at rest. At position ⑧ ($x_ⓑ = 0$), the spring is relaxed and the child is moving upward. At position ©, the child is again momentarily at rest at the top of the jump. The combined mass of child and pogo stick is 25.0 kg. Although the boy must lean forward to remain balanced, the angle is small, so let's assume the pogo stick is vertical. Also assume the boy does not bend his legs during the motion. (a) Calculate the total energy of the child–stick–Earth system, taking both gravitational and elastic potential energies as zero for $x = 0$. (b) Determine $x_©$. (c) Calculate the speed of the child at $x = 0$. (d) Determine the value of x for which the kinetic energy of the system is a maximum. (e) Calculate the child's maximum upward speed.

Figure P8.61

62. A 1.00-kg object slides [W] to the right on a surface having a coefficient of kinetic friction 0.250 (Fig. P8.62a). The object has a speed of $v_i = 3.00$ m/s when it makes contact with a light spring (Fig. P8.62b) that has a force constant of 50.0 N/m. The object comes to rest after the spring has been compressed a distance d (Fig. P8.62c). The object is then forced toward the left by the spring (Fig. P8.62d) and continues to move in that direction beyond the spring's unstretched position. Finally, the object comes to rest a distance D to the left of the unstretched spring (Fig. P8.62e). Find (a) the distance of compression d, (b) the speed v at the unstretched position when the object is moving to the left (Fig. P8.62d), and (c) the distance D where the object comes to rest.

Figure P8.62

63. A 10.0-kg block is released from rest at point Ⓐ in Figure P8.63. The track is frictionless except for the portion between points Ⓑ and Ⓒ, which has a length of 6.00 m. The block travels down the track, hits a spring of force constant 2 250 N/m, and compresses the spring 0.300 m from its equilibrium position before coming to rest momentarily. Determine the coefficient of kinetic friction between the block and the rough surface between points Ⓑ and Ⓒ.

Figure P8.63

64. A block of mass $m_1 = 20.0$ kg is connected to a block of mass $m_2 = 30.0$ kg by a massless string that passes over a light, frictionless pulley. The 30.0-kg block is connected to a spring that has negligible mass and a force constant of $k = 250$ N/m as shown in Figure P8.64. The spring is unstretched when the system is as shown in the figure, and the incline is frictionless. The 20.0-kg block is pulled a distance $h = 20.0$ cm down the incline of angle $\theta = 40.0°$ and released from rest. Find the speed of each block when the spring is again unstretched.

Figure P8.64

65. A block of mass 0.500 kg is pushed against a horizontal spring of negligible mass until the spring is compressed a distance x (Fig. P8.65). The force constant of the spring is 450 N/m. When it is released, the block travels along a frictionless, horizontal surface to point Ⓐ, the bottom of a vertical circular track of radius $R = 1.00$ m, and continues to move up the track. The block's speed at the bottom of the track is $v_Ⓐ = 12.0$ m/s, and the block experiences an average friction force of 7.00 N while sliding up the track. (a) What is x? (b) If the block were to reach the top of the track, what would be its speed at that point? (c) Does the block actually reach the top of the track, or does it fall off before reaching the top?

Figure P8.65

66. Review. As a prank, someone has balanced a pumpkin at the highest point of a grain silo. The silo is topped with a hemispherical cap that is frictionless when wet. The line from the center of curvature of the cap to the pumpkin makes an angle $\theta_i = 0°$ with the vertical. While you happen to be standing nearby in the middle of a rainy night, a breath of wind makes the pumpkin start sliding downward from rest. It loses contact with the cap when the line from the center of the hemisphere to the pumpkin makes a certain angle with the vertical. What is this angle?

67. Review. The mass of a car is 1 500 kg. The shape of the car's body is such that its aerodynamic drag coefficient is $D = 0.330$ and its frontal area is 2.50 m². Assuming the drag force is proportional to v^2 and ignoring other sources of friction, calculate the power required to maintain a speed of 100 km/h as the car climbs a long hill sloping at 3.20°.

68. A pendulum, comprising a light string of length L and a small sphere, swings in the vertical plane. The string hits a peg located a distance d below the point of suspension (Fig. P8.68). (a) Show that if the sphere is released from a height below that of the peg, it will return to this height after the string strikes the peg. (b) Show that if the pendulum is released from rest at the horizontal position ($\theta = 90°$) and is to swing in a complete circle centered on the peg, the minimum value of d must be $3L/5$.

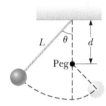

Figure P8.68

69. A block of mass M rests on a table. It is fastened to the lower end of a light, vertical spring. The upper end of the spring is fastened to a block of mass m. The upper block is pushed down by an additional force $3mg$, so the spring compression is $4mg/k$. In this configuration, the upper block is released from rest. The spring lifts the lower block off the table. In terms of m, what is the greatest possible value for M?

70. Review. *Why is the following situation impossible?* An athlete tests her hand strength by having an assistant hang weights from her belt as she hangs onto a horizontal bar with her hands. When the weights hanging on her belt have increased to 80% of her body weight, her hands can no longer support her and she drops to the floor. Frustrated at not meeting her hand-strength goal, she decides to swing on a trapeze. The trapeze consists of a bar suspended by two parallel ropes, each of length ℓ, allowing performers to swing in a vertical circular arc (Fig. P8.70). The athlete holds the bar and steps off an elevated platform, starting from rest with the ropes at an angle $\theta_i = 60.0°$ with respect to the vertical. As she swings several times back and forth in a circular arc, she forgets her frustration related to the hand-strength test. Assume the size of the

Figure P8.70

performer's body is small compared to the length ℓ and air resistance is negligible.

71. While running, a person transforms about 0.600 J of chemical energy to mechanical energy per step per kilogram of body mass. If a 60.0-kg runner transforms energy at a rate of 70.0 W during a race, how fast is the person running? Assume that a running step is 1.50 m long.

72. A roller-coaster car shown in Figure P8.72 is released from rest from a height h and then moves freely with negligible friction. The roller-coaster track includes a circular loop of radius R in a vertical plane. (a) First suppose the car barely makes it around the loop; at the top of the loop, the riders are upside down and feel weightless. Find the required height h of the release point above the bottom of the loop in terms of R. (b) Now assume the release point is at or above the minimum required height. Show that the normal force on the car at the bottom of the loop exceeds the normal force at the top of the loop by six times the car's weight. The normal force on each rider follows the same rule. Such a large normal force is dangerous and very uncomfortable for the riders. Roller coasters are therefore not built with circular loops in vertical planes. Figure P6.17 (page 170) shows an actual design.

Figure P8.72

73. A ball whirls around in a *vertical* circle at the end of a string. The other end of the string is fixed at the center of the circle. Assuming the total energy of the ball–Earth system remains constant, show that the tension in the string at the bottom is greater than the tension at the top by six times the ball's weight.

74. An airplane of mass 1.50×10^4 kg is in level flight, initially moving at 60.0 m/s. The resistive force exerted by air on the airplane has a magnitude of 4.0×10^4 N. By Newton's third law, if the engines exert a force on the exhaust gases to expel them out of the back of the engine, the exhaust gases exert a force on the engines in the direction of the airplane's travel. This force is called thrust, and the value of the thrust in this situation is 7.50×10^4 N. (a) Is the work done by the exhaust gases on the airplane during some time interval equal to the change in the airplane's kinetic energy? Explain. (b) Find the speed of the airplane after it has traveled 5.0×10^2 m.

75. Consider the block–spring collision discussed in Example 8.8. (a) For the situation in part (B), in which the surface exerts a friction force on the block, show that the block never arrives back at $x = 0$. (b) What is

the maximum value of the coefficient of friction that would allow the block to return to $x = 0$?

76. In bicycling for aerobic exercise, a woman wants her heart rate to be between 136 and 166 beats per minute. Assume that her heart rate is directly proportional to her mechanical power output within the range relevant here. Ignore all forces on the woman–bicycle system except for static friction forward on the drive wheel of the bicycle and an air resistance force proportional to the square of her speed. When her speed is 22.0 km/h, her heart rate is 90.0 beats per minute. In what range should her speed be so that her heart rate will be in the range she wants?

77. **Review.** In 1887 in Bridgeport, Connecticut, C. J. Belknap built the water slide shown in Figure P8.77. A rider on a small sled, of total mass 80.0 kg, pushed off to start at the top of the slide (point Ⓐ) with a speed of 2.50 m/s. The chute was 9.76 m high at the top and 54.3 m long. Along its length, 725 small wheels made friction negligible. Upon leaving the chute horizontally at its bottom end (point Ⓒ), the rider skimmed across the water of Long Island Sound for as much as 50 m, "skipping along like a flat pebble," before at last coming to rest and swimming ashore, pulling his sled after him. (a) Find the speed of the sled and rider at point Ⓒ. (b) Model the force of water friction as a constant retarding force acting on a particle. Find the magnitude of the friction force the water exerts on the sled. (c) Find the magnitude of the force the chute exerts on the sled at point Ⓑ. (d) At point Ⓒ, the chute is horizontal but curving in the vertical plane. Assume its radius of curvature is 20.0 m. Find the force the chute exerts on the sled at point Ⓒ.

Figure P8.77

78. In a needle biopsy, a narrow strip of tissue is extracted from a patient using a hollow needle. Rather than being pushed by hand, to ensure a clean cut the needle can be fired into the patient's body by a spring. Assume that the needle has mass 5.60 g, the light spring has

force constant 375 N/m, and the spring is originally compressed 8.10 cm to project the needle horizontally without friction. After the needle leaves the spring, the tip of the needle moves through 2.40 cm of skin and soft tissue, which exerts on it a resistive force of 7.60 N. Next, the needle cuts 3.50 cm into an organ, which exerts on it a backward force of 9.20 N. Find (a) the maximum speed of the needle and (b) the speed at which the flange on the back end of the needle runs into a stop that is set to limit the penetration to 5.90 cm.

Challenge Problems

79. **Review.** A uniform board of length L is sliding along a smooth, frictionless, horizontal plane as shown in Figure P8.79a. The board then slides across the boundary with a rough horizontal surface. The coefficient of kinetic friction between the board and the second surface is μ_k. (a) Find the acceleration of the board at the moment its front end has traveled a distance x beyond the boundary. (b) The board stops at the moment its back end reaches the boundary as shown in Figure P8.79b. Find the initial speed v of the board.

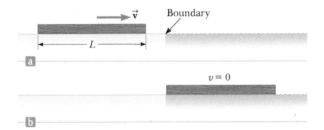

Figure P8.79

80. Starting from rest, a 64.0-kg person bungee jumps from a tethered hot-air balloon 65.0 m above the ground. The bungee cord has negligible mass and unstretched length 25.8 m. One end is tied to the basket of the balloon and the other end to a harness around the person's body. The cord is modeled as a spring that obeys Hooke's law with a spring constant of 81.0 N/m, and the person's body is modeled as a particle. The hot-air balloon does not move. (a) Express the gravitational potential energy of the person–Earth system as a function of the person's variable height y above the ground. (b) Express the elastic potential energy of the cord as a function of y. (c) Express the total potential energy of the person–cord–Earth system as a function of y. (d) Plot a graph of the gravitational, elastic, and total potential energies as functions of y. (e) Assume air resistance is negligible. Determine the minimum height of the person above the ground during his plunge. (f) Does the potential energy graph show any equilibrium position or positions? If so, at what elevations? Are they stable or unstable? (g) Determine the jumper's maximum speed.

81. Jane, whose mass is 50.0 kg, needs to swing across a river (having width D) filled with person-eating crocodiles to save Tarzan from danger. She must swing into

a wind exerting constant horizontal force $\vec{F}$, on a vine having length L and initially making an angle θ with the vertical (Fig. P8.81). Take $D = 50.0$ m, $F = 110$ N, $L = 40.0$ m, and $\theta = 50.0°$. (a) With what minimum speed must Jane begin her swing to just make it to the other side? (b) Once the rescue is complete, Tarzan and Jane must swing back across the river. With what minimum speed must they begin their swing? Assume Tarzan has a mass of 80.0 kg.

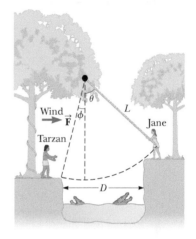

Figure P8.81

82. A ball of mass $m = 300$ g is connected by a strong string of length $L = 80.0$ cm to a pivot and held in place with the string vertical. A wind exerts constant force F to the right on the ball as shown in Figure P8.82. The ball is released from rest. The wind makes it swing up to attain maximum height H above its starting point before it swings down again. (a) Find H as a function of F. Evaluate H for (b) $F = 1.00$ N and (c) $F = 10.0$ N. How does H behave (d) as F approaches zero and (e) as F approaches infinity? (f) Now consider the equilibrium height of the ball with the wind blowing. Determine it as a function of F. Evaluate the equilibrium height for (g) $F = 10$ N and (h) F going to infinity.

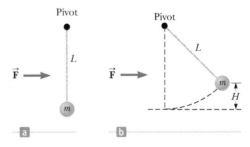

Figure P8.82

83. **What If?** Consider the roller coaster described in Problem 58. Because of some friction between the coaster and the track, the coaster enters the circular section at a speed of 15.0 m/s rather than the 22.0 m/s in Problem 58. Is this situation *more* or *less* dangerous for the passengers than that in Problem 58? Assume the circular section is still frictionless.

84. A uniform chain of length 8.00 m initially lies stretched out on a horizontal table. (a) Assuming the coefficient of static friction between chain and table is 0.600, show that the chain will begin to slide off the table if at least 3.00 m of it hangs over the edge of the table. (b) Determine the speed of the chain as its last link leaves the table, given that the coefficient of kinetic friction between the chain and the table is 0.400.

85. A daredevil plans to bungee jump from a balloon 65.0 m above the ground. He will use a uniform elastic cord, tied to a harness around his body, to stop his fall at a point 10.0 m above the ground. Model his body as a particle and the cord as having negligible mass and obeying Hooke's law. In a preliminary test he finds that when hanging at rest from a 5.00-m length of the cord, his body weight stretches it by 1.50 m. He will drop from rest at the point where the top end of a longer section of the cord is attached to the stationary balloon. (a) What length of cord should he use? (b) What maximum acceleration will he experience?

Linear Momentum and Collisions

Consider what happens when two cars collide as in the opening photograph for this chapter. Both cars change their motion from having a very large velocity to being at rest because of the collision. Because each car experiences a large change in velocity over a very short time interval, the average force on it is very large. By Newton's third law, each of the cars experiences a force of the same magnitude. By Newton's second law, the results of those forces on the motion of the car depends on the mass of the car.

One of the main objectives of this chapter is to enable you to understand and analyze such events in a simple way. First, we introduce the concept of *momentum*, which is useful for describing objects in motion. The momentum of an object is related to both its mass and its velocity. The concept of momentum leads us to a second conservation law, that of conservation of momentum. In turn, we identify new momentum versions of analysis models for isolated and nonisolated system. These models are especially useful for treating problems that involve collisions between objects and for analyzing rocket propulsion. This chapter also introduces the concept of the center of mass of a system of particles. We find that the motion of a system of particles can be described by the motion of one particle located at the center of mass that represents the entire system.

The concept of momentum allows the analysis of car collisions even without detailed knowledge of the forces involved. Such analysis can determine the relative velocity of the cars before the collision, and in addition aid engineers in designing safer vehicles. (The English translation of the German text on the side of the trailer in the background is: "Pit stop for your vehicle.") *(AP Photos/Keystone/ Regina Kuehne)*

9.1 Linear Momentum

In Chapter 8, we studied situations that are difficult to analyze with Newton's laws. We were able to solve problems involving these situations by identifying a system and applying a conservation principle, conservation of energy. Let us consider another situation and see if we can solve it with the models we have developed so far:

A 60-kg archer stands at rest on frictionless ice and fires a 0.030-kg arrow horizontally at 85 m/s. With what velocity does the archer move across the ice after firing the arrow?

From Newton's third law, we know that the force that the bow exerts on the arrow is paired with a force in the opposite direction on the bow (and the archer). This force causes the archer to slide backward on the ice with the speed requested in the problem. We cannot determine this speed using motion models such as the particle under constant acceleration because we don't have any information about the acceleration of the archer. We cannot use force models such as the particle under a net force because we don't know anything about forces in this situation. Energy models are of no help because we know nothing about the work done in pulling the bowstring back or the elastic potential energy in the system related to the taut bowstring.

Despite our inability to solve the archer problem using models learned so far, this problem is very simple to solve if we introduce a new quantity that describes motion, *linear momentum*. To generate this new quantity, consider an isolated system of two particles (Fig. 9.1) with masses m_1 and m_2 moving with velocities $\vec{\mathbf{v}}_1$ and $\vec{\mathbf{v}}_2$ at an instant of time. Because the system is isolated, the only force on one particle is that from the other particle. If a force from particle 1 (for example, a gravitational force) acts on particle 2, there must be a second force—equal in magnitude but opposite in direction—that particle 2 exerts on particle 1. That is, the forces on the particles form a Newton's third law action–reaction pair, and $\vec{\mathbf{F}}_{12} = -\vec{\mathbf{F}}_{21}$. We can express this condition as

$$\vec{\mathbf{F}}_{21} + \vec{\mathbf{F}}_{12} = 0$$

From a system point of view, this equation says that if we add up the forces on the particles in an isolated system, the sum is zero.

Let us further analyze this situation by incorporating Newton's second law. At the instant shown in Figure 9.1, the interacting particles in the system have accelerations corresponding to the forces on them. Therefore, replacing the force on each particle with $m\vec{\mathbf{a}}$ for the particle gives

$$m_1\vec{\mathbf{a}}_1 + m_2\vec{\mathbf{a}}_2 = 0$$

Now we replace each acceleration with its definition from Equation 4.5:

$$m_1\frac{d\vec{\mathbf{v}}_1}{dt} + m_2\frac{d\vec{\mathbf{v}}_2}{dt} = 0$$

If the masses m_1 and m_2 are constant, we can bring them inside the derivative operation, which gives

$$\frac{d(m_1\vec{\mathbf{v}}_1)}{dt} + \frac{d(m_2\vec{\mathbf{v}}_2)}{dt} = 0$$

$$\frac{d}{dt}(m_1\vec{\mathbf{v}}_1 + m_2\vec{\mathbf{v}}_2) = 0 \qquad (9.1)$$

Notice that the derivative of the sum $m_1\vec{\mathbf{v}}_1 + m_2\vec{\mathbf{v}}_2$ with respect to time is zero. Consequently, this sum must be constant. We learn from this discussion that the quantity $m\vec{\mathbf{v}}$ for a particle is important in that the sum of these quantities for an isolated system of particles is conserved. We call this quantity *linear momentum*:

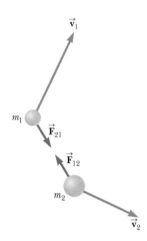

Figure 9.1 Two particles interact with each other. According to Newton's third law, we must have $\vec{\mathbf{F}}_{12} = -\vec{\mathbf{F}}_{21}$.

Definition of linear ▶
momentum of a particle

The **linear momentum** of a particle or an object that can be modeled as a particle of mass m moving with a velocity $\vec{\mathbf{v}}$ is defined to be the product of the mass and velocity of the particle:

$$\vec{\mathbf{p}} \equiv m\vec{\mathbf{v}} \qquad (9.2)$$

Linear momentum is a vector quantity because it equals the product of a scalar quantity m and a vector quantity $\vec{\mathbf{v}}$. Its direction is along $\vec{\mathbf{v}}$, it has dimensions ML/T, and its SI unit is kg · m/s.

If a particle is moving in an arbitrary direction, $\vec{\mathbf{p}}$ has three components, and Equation 9.2 is equivalent to the component equations

$$p_x = mv_x \qquad p_y = mv_y \qquad p_z = mv_z$$

As you can see from its definition, the concept of momentum[1] provides a quantitative distinction between heavy and light particles moving at the same velocity. For example, the momentum of a bowling ball is much greater than that of a tennis ball moving at the same speed. Newton called the product $m\vec{\mathbf{v}}$ *quantity of motion;* this term is perhaps a more graphic description than our present-day word *momentum*, which comes from the Latin word for movement.

We have seen another quantity, kinetic energy, that is a combination of mass and speed. It would be a legitimate question to ask why we need another quantity, momentum, based on mass and velocity. There are clear differences between kinetic energy and momentum. First, kinetic energy is a scalar, whereas momentum is a vector. Consider a system of two equal-mass particles heading toward each other along a line with equal speeds. There is kinetic energy associated with this system because members of the system are moving. Because of the vector nature of momentum, however, the momentum of this system is zero. A second major difference is that kinetic energy can transform to other types of energy, such as potential energy or internal energy. There is only one type of linear momentum, so we see no such transformations when using a momentum approach to a problem. These differences are sufficient to make models based on momentum separate from those based on energy, providing an independent tool to use in solving problems.

Using Newton's second law of motion, we can relate the linear momentum of a particle to the resultant force acting on the particle. We start with Newton's second law and substitute the definition of acceleration:

$$\sum \vec{\mathbf{F}} = m\vec{\mathbf{a}} = m\,\frac{d\vec{\mathbf{v}}}{dt}$$

In Newton's second law, the mass m is assumed to be constant. Therefore, we can bring m inside the derivative operation to give us

$$\sum \vec{\mathbf{F}} = \frac{d(m\vec{\mathbf{v}})}{dt} = \frac{d\vec{\mathbf{p}}}{dt} \qquad (9.3)$$

◀ Newton's second law for a particle

This equation shows that **the time rate of change of the linear momentum of a particle is equal to the net force acting on the particle.** In Chapter 5, we identified force as that which causes a change in the motion of an object (Section 5.2). In Newton's second law (Eq. 5.2), we used acceleration $\vec{\mathbf{a}}$ to represent the change in motion. We see now in Equation 9.3 that we can use the derivative of momentum $\vec{\mathbf{p}}$ with respect to time to represent the change in motion.

This alternative form of Newton's second law is the form in which Newton presented the law, and it is actually more general than the form introduced in Chapter 5. In addition to situations in which the velocity vector varies with time, we can use Equation 9.3 to study phenomena in which the mass changes. For example, the mass of a rocket changes as fuel is burned and ejected from the rocket. We cannot use $\sum \vec{\mathbf{F}} = m\vec{\mathbf{a}}$ to analyze rocket propulsion; we must use a momentum approach, as we will show in Section 9.9.

---

[1]In this chapter, the terms *momentum* and *linear momentum* have the same meaning. Later, in Chapter 11, we shall use the term *angular momentum* for a different quantity when dealing with rotational motion.

Quick Quiz 9.1 Two objects have equal kinetic energies. How do the magnitudes of their momenta compare? **(a)** $p_1 < p_2$ **(b)** $p_1 = p_2$ **(c)** $p_1 > p_2$ **(d)** not enough information to tell

Quick Quiz 9.2 Your physical education teacher throws a baseball to you at a certain speed and you catch it. The teacher is next going to throw you a medicine ball whose mass is ten times the mass of the baseball. You are given the following choices: You can have the medicine ball thrown with **(a)** the same speed as the baseball, **(b)** the same momentum, or **(c)** the same kinetic energy. Rank these choices from easiest to hardest to catch.

9.2 Analysis Model: Isolated System (Momentum)

Using the definition of momentum, Equation 9.1 can be written

$$\frac{d}{dt}(\vec{\mathbf{p}}_1 + \vec{\mathbf{p}}_2) = 0$$

Because the time derivative of the total momentum $\vec{\mathbf{p}}_{tot} = \vec{\mathbf{p}}_1 + \vec{\mathbf{p}}_2$ is *zero*, we conclude that the *total* momentum of the isolated system of the two particles in Figure 9.1 must remain constant:

$$\vec{\mathbf{p}}_{tot} = \text{constant} \tag{9.4}$$

or, equivalently, over some time interval,

$$\Delta\vec{\mathbf{p}}_{tot} = 0 \tag{9.5}$$

Equation 9.5 can be written as

$$\vec{\mathbf{p}}_{1i} + \vec{\mathbf{p}}_{2i} = \vec{\mathbf{p}}_{1f} + \vec{\mathbf{p}}_{2f}$$

where $\vec{\mathbf{p}}_{1i}$ and $\vec{\mathbf{p}}_{2i}$ are the initial values and $\vec{\mathbf{p}}_{1f}$ and $\vec{\mathbf{p}}_{2f}$ are the final values of the momenta for the two particles for the time interval during which the particles interact. This equation in component form demonstrates that the total momenta in the *x*, *y*, and *z* directions are all independently conserved:

$$p_{1ix} + p_{2ix} = p_{1fx} + p_{2fx} \qquad p_{1iy} + p_{2iy} = p_{1fy} + p_{2fy} \qquad p_{1iz} + p_{2iz} = p_{1fz} + p_{2fz} \tag{9.6}$$

Equation 9.5 is the mathematical statement of a new analysis model, the **isolated system (momentum)**. It can be extended to any number of particles in an isolated system as we show in Section 9.7. We studied the energy version of the isolated system model in Chapter 8 ($\Delta E_{system} = 0$) and now we have a momentum version. In general, Equation 9.5 can be stated in words as follows:

The momentum version of the ▶ isolated system model

> Whenever two or more particles in an isolated system interact, the total momentum of the system does not change.

This statement tells us that the total momentum of an isolated system at all times equals its initial momentum.

Notice that we have made no statement concerning the type of forces acting on the particles of the system. Furthermore, we have not specified whether the forces are conservative or nonconservative. We have also not indicated whether or not the forces are constant. The only requirement is that the forces must be *internal* to the system. This single requirement should give you a hint about the power of this new model.

Pitfall Prevention 9.1
Momentum of an Isolated *System* Is Conserved Although the momentum of an isolated *system* is conserved, the momentum of one *particle* within an isolated system is not necessarily conserved because other particles in the system may be interacting with it. Avoid applying conservation of momentum to a single particle.

Analysis Model | Isolated System (Momentum)

Imagine you have identified a system to be analyzed and have defined a system boundary. If there are no external forces on the system, the system is *isolated*. In that case, the total momentum of the system, which is the vector sum of the momenta of all members of the system, is conserved:

$$\Delta \vec{\mathbf{p}}_{tot} = 0 \qquad (9.5)$$

Examples:

- a cue ball strikes another ball on a pool table
- a spacecraft fires its rockets and moves faster through space
- molecules in a gas at a specific temperature move about and strike each other (Chapter 21)
- an incoming particle strikes a nucleus, creating a new nucleus and a different outgoing particle (Chapter 44)
- an electron and a positron annihilate to form two outgoing photons (Chapter 46)

System boundary

Momentum

If no external forces act on the system, the total momentum of the system is constant.

Example 9.1 | The Archer AM

Let us consider the situation proposed at the beginning of Section 9.1. A 60-kg archer stands at rest on frictionless ice and fires a 0.030-kg arrow horizontally at 85 m/s (Fig. 9.2). With what velocity does the archer move across the ice after firing the arrow?

SOLUTION

Conceptualize You may have conceptualized this problem already when it was introduced at the beginning of Section 9.1. Imagine the arrow being fired one way and the archer recoiling in the opposite direction.

Categorize As discussed in Section 9.1, we cannot solve this problem with models based on motion, force, or energy. Nonetheless, we *can* solve this problem very easily with an approach involving momentum.

Let us take the system to consist of the archer (including the bow) and the arrow. The system is not isolated because the gravitational force and the normal force from the ice act on the system. These forces, however, are vertical and perpendicular to the motion of the system. There are no external forces in the horizontal direction, and we can apply the *isolated system (momentum)* model in terms of momentum components in this direction.

Figure 9.2 (Example 9.1) An archer fires an arrow horizontally to the right. Because he is standing on frictionless ice, he will begin to slide to the left across the ice.

Analyze The total horizontal momentum of the system before the arrow is fired is zero because nothing in the system is moving. Therefore, the total horizontal momentum of the system after the arrow is fired must also be zero. We choose the direction of firing of the arrow as the positive *x* direction. Identifying the archer as particle 1 and the arrow as particle 2, we have $m_1 = 60$ kg, $m_2 = 0.030$ kg, and $\vec{\mathbf{v}}_{2f} = 85\hat{\mathbf{i}}$ m/s.

Using the isolated system (momentum) model, begin with Equation 9.5:

$$\Delta \vec{\mathbf{p}} = 0 \quad \rightarrow \quad \vec{\mathbf{p}}_f - \vec{\mathbf{p}}_i = 0 \quad \rightarrow \quad \vec{\mathbf{p}}_f = \vec{\mathbf{p}}_i \quad \rightarrow \quad m_1\vec{\mathbf{v}}_{1f} + m_2\vec{\mathbf{v}}_{2f} = 0$$

Solve this equation for $\vec{\mathbf{v}}_{1f}$ and substitute numerical values:

$$\vec{\mathbf{v}}_{1f} = -\frac{m_2}{m_1}\vec{\mathbf{v}}_{2f} = -\left(\frac{0.030 \text{ kg}}{60 \text{ kg}}\right)(85\hat{\mathbf{i}} \text{ m/s}) = -0.042\hat{\mathbf{i}} \text{ m/s}$$

Finalize The negative sign for $\vec{\mathbf{v}}_{1f}$ indicates that the archer is moving to the left in Figure 9.2 after the arrow is fired, in the direction opposite the direction of motion of the arrow, in accordance with Newton's third law. Because the archer

continued

▶ **9.1** continued

is much more massive than the arrow, his acceleration and consequent velocity are much smaller than the acceleration and velocity of the arrow. Notice that this problem sounds very simple, but we could not solve it with models based on motion, force, or energy. Our new momentum model, however, shows us that it not only *sounds* simple, it *is* simple!

WHAT IF? What if the arrow were fired in a direction that makes an angle θ with the horizontal? How will that change the recoil velocity of the archer?

Answer The recoil velocity should decrease in magnitude because only a component of the velocity of the arrow is in the x direction. Conservation of momentum in the x direction gives

$$m_1 v_{1f} + m_2 v_{2f} \cos \theta = 0$$

leading to

$$v_{1f} = -\frac{m_2}{m_1} v_{2f} \cos \theta$$

For $\theta = 0$, $\cos \theta = 1$ and the final velocity of the archer reduces to the value when the arrow is fired horizontally. For nonzero values of θ, the cosine function is less than 1 and the recoil velocity is less than the value calculated for $\theta = 0$. If $\theta = 90°$, then $\cos \theta = 0$ and $v_{1f} = 0$, so there is no recoil velocity. In this case, the archer is simply pushed downward harder against the ice as the arrow is fired.

Example 9.2   Can We Really Ignore the Kinetic Energy of the Earth?   **AM**

In Section 7.6, we claimed that we can ignore the kinetic energy of the Earth when considering the energy of a system consisting of the Earth and a dropped ball. Verify this claim.

SOLUTION

Conceptualize Imagine dropping a ball at the surface of the Earth. From your point of view, the ball falls while the Earth remains stationary. By Newton's third law, however, the Earth experiences an upward force and therefore an upward acceleration while the ball falls. In the calculation below, we will show that this motion is extremely small and can be ignored.

Categorize We identify the system as the ball and the Earth. We assume there are no forces on the system from outer space, so the system is isolated. Let's use the *momentum* version of the *isolated system* model.

Analyze We begin by setting up a ratio of the kinetic energy of the Earth to that of the ball. We identify v_E and v_b as the speeds of the Earth and the ball, respectively, after the ball has fallen through some distance.

Use the definition of kinetic energy to set up this ratio:

$$(1) \quad \frac{K_E}{K_b} = \frac{\frac{1}{2} m_E v_E^2}{\frac{1}{2} m_b v_b^2} = \left(\frac{m_E}{m_b}\right)\left(\frac{v_E}{v_b}\right)^2$$

Apply the isolated system (momentum) model, recognizing that the initial momentum of the system is zero:

$$\Delta \vec{\mathbf{p}} = 0 \quad \rightarrow \quad p_i = p_f \quad \rightarrow \quad 0 = m_b v_b + m_E v_E$$

Solve the equation for the ratio of speeds:

$$\frac{v_E}{v_b} = -\frac{m_b}{m_E}$$

Substitute this expression for v_E/v_b in Equation (1):

$$\frac{K_E}{K_b} = \left(\frac{m_E}{m_b}\right)\left(-\frac{m_b}{m_E}\right)^2 = \frac{m_b}{m_E}$$

Substitute order-of-magnitude numbers for the masses:

$$\frac{K_E}{K_b} = \frac{m_b}{m_E} \sim \frac{1 \text{ kg}}{10^{25} \text{ kg}} \sim 10^{-25}$$

Finalize The kinetic energy of the Earth is a very small fraction of the kinetic energy of the ball, so we are justified in ignoring it in the kinetic energy of the system.

9.3 Analysis Model: Nonisolated System (Momentum)

According to Equation 9.3, the momentum of a particle changes if a net force acts on the particle. The same can be said about a net force applied to a system as we will show explicitly in Section 9.7: the momentum of a system changes if a net force from the environment acts on the system. This may sound similar to our discussion of energy in Chapter 8: the energy of a system changes if energy crosses the boundary of the system to or from the environment. In this section, we consider a *nonisolated system*. For energy considerations, a system is nonisolated if energy transfers across the boundary of the system by any of the means listed in Section 8.1. For momentum considerations, a system is nonisolated if a net force acts on the system for a time interval. In this case, we can imagine momentum being transferred to the system from the environment by means of the net force. Knowing the change in momentum caused by a force is useful in solving some types of problems. To build a better understanding of this important concept, let us assume a net force $\sum \vec{F}$ acts on a particle and this force may vary with time. According to Newton's second law, in the form expressed in Equation 9.3, $\sum \vec{F} = d\vec{p}/dt$, we can write

$$d\vec{p} = \sum \vec{F} \, dt \tag{9.7}$$

We can integrate[2] this expression to find the change in the momentum of a particle when the force acts over some time interval. If the momentum of the particle changes from $\vec{p}_i$ at time t_i to $\vec{p}_f$ at time t_f, integrating Equation 9.7 gives

$$\Delta \vec{p} = \vec{p}_f - \vec{p}_i = \int_{t_i}^{t_f} \sum \vec{F} \, dt \tag{9.8}$$

To evaluate the integral, we need to know how the net force varies with time. The quantity on the right side of this equation is a vector called the **impulse** of the net force $\sum \vec{F}$ acting on a particle over the time interval $\Delta t = t_f - t_i$:

$$\vec{I} \equiv \int_{t_i}^{t_f} \sum \vec{F} \, dt \tag{9.9}$$

◄ Impulse of a force

From its definition, we see that impulse $\vec{I}$ is a vector quantity having a magnitude equal to the area under the force–time curve as described in Figure 9.3a. It is assumed the force varies in time in the general manner shown in the figure and is nonzero in the time interval $\Delta t = t_f - t_i$. The direction of the impulse vector is the same as the direction of the change in momentum. Impulse has the dimensions of momentum, that is, ML/T. Impulse is *not* a property of a particle; rather, it is a measure of the degree to which an external force changes the particle's momentum.

Because the net force imparting an impulse to a particle can generally vary in time, it is convenient to define a time-averaged net force:

$$\left(\sum \vec{F} \right)_{\text{avg}} \equiv \frac{1}{\Delta t} \int_{t_i}^{t_f} \sum \vec{F} \, dt \tag{9.10}$$

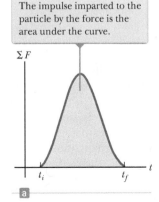

The impulse imparted to the particle by the force is the area under the curve.

ΣF

t_i t_f

a

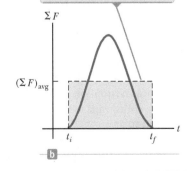

The time-averaged net force gives the same impulse to a particle as does the time-varying force in (a).

ΣF

$(\Sigma F)_{\text{avg}}$

t_i t_f

b

Figure 9.3 (a) A net force acting on a particle may vary in time. (b) The value of the constant force $(\Sigma F)_{\text{avg}}$ (horizontal dashed line) is chosen so that the area $(\Sigma F)_{\text{avg}} \Delta t$ of the rectangle is the same as the area under the curve in (a).

---

[2]Here we are integrating force with respect to time. Compare this strategy with our efforts in Chapter 7, where we integrated force with respect to position to find the work done by the force.

where $\Delta t = t_f - t_i$. (This equation is an application of the mean value theorem of calculus.) Therefore, we can express Equation 9.9 as

$$\vec{\mathbf{I}} = \left(\sum \vec{\mathbf{F}} \right)_{\text{avg}} \Delta t \qquad (9.11)$$

This time-averaged force, shown in Figure 9.3b, can be interpreted as the constant force that would give to the particle in the time interval Δt the same impulse that the time-varying force gives over this same interval.

In principle, if $\sum \vec{\mathbf{F}}$ is known as a function of time, the impulse can be calculated from Equation 9.9. The calculation becomes especially simple if the force acting on the particle is constant. In this case, $\left(\sum \vec{\mathbf{F}} \right)_{\text{avg}} = \sum \vec{\mathbf{F}}$, where $\sum \vec{\mathbf{F}}$ is the constant net force, and Equation 9.11 becomes

$$\vec{\mathbf{I}} = \sum \vec{\mathbf{F}} \, \Delta t \qquad (9.12)$$

Combining Equations 9.8 and 9.9 gives us an important statement known as the **impulse–momentum theorem:**

◀ Impulse–momentum theorem for a particle

> The change in the momentum of a particle is equal to the impulse of the net force acting on the particle:
>
> $$\Delta \vec{\mathbf{p}} = \vec{\mathbf{I}} \qquad (9.13)$$

This statement is equivalent to Newton's second law. When we say that an impulse is given to a particle, we mean that momentum is transferred from an external agent to that particle. Equation 9.13 is identical in form to the conservation of energy equation, Equation 8.1, and its full expansion, Equation 8.2. Equation 9.13 is the most general statement of the principle of **conservation of momentum** and is called the **conservation of momentum equation.** In the case of a momentum approach, isolated systems tend to appear in problems more often than nonisolated systems, so, in practice, the conservation of momentum equation is often identified as the special case of Equation 9.5.

The left side of Equation 9.13 represents the change in the momentum of the system, which in this case is a single particle. The right side is a measure of how much momentum crosses the boundary of the system due to the net force being applied to the system. Equation 9.13 is the mathematical statement of a new analysis model, the **nonisolated system (momentum)** model. Although this equation is similar in form to Equation 8.1, there are several differences in its application to problems. First, Equation 9.13 is a vector equation, whereas Equation 8.1 is a scalar equation. Therefore, directions are important for Equation 9.13. Second, there is only one type of momentum and therefore only one way to store momentum in a system. In contrast, as we see from Equation 8.2, there are three ways to store energy in a system: kinetic, potential, and internal. Third, there is only one way to transfer momentum into a system: by the application of a force on the system over a time interval. Equation 8.2 shows six ways we have identified as transferring energy into a system. Therefore, there is no expansion of Equation 9.13 analogous to Equation 8.2.

In many physical situations, we shall use what is called the **impulse approximation,** in which we assume one of the forces exerted on a particle acts for a short time but is much greater than any other force present. In this case, the net force $\sum \vec{\mathbf{F}}$ in Equation 9.9 is replaced with a single force $\vec{\mathbf{F}}$ to find the impulse on the particle. This approximation is especially useful in treating collisions in which the duration of the collision is very short. When this approximation is made, the single force is referred to as an *impulsive force.* For example, when a baseball is struck with a bat, the time of the collision is about 0.01 s and the average force that the bat exerts on the ball in this time is typically several thousand newtons. Because this contact force is much greater than the magnitude of the gravitational force, the impulse approximation justifies our ignoring the gravitational forces exerted on

Air bags in automobiles have saved countless lives in accidents. The air bag increases the time interval during which the passenger is brought to rest, thereby decreasing the force on (and resultant injury to) the passenger.

© David Woods/CORBIS

the ball and bat during the collision. When we use this approximation, it is important to remember that $\vec{p}_i$ and $\vec{p}_f$ represent the momenta *immediately* before and after the collision, respectively. Therefore, in any situation in which it is proper to use the impulse approximation, the particle moves very little during the collision.

Quick Quiz 9.3 Two objects are at rest on a frictionless surface. Object 1 has a greater mass than object 2. **(i)** When a constant force is applied to object 1, it accelerates through a distance d in a straight line. The force is removed from object 1 and is applied to object 2. At the moment when object 2 has accelerated through the same distance d, which statements are true? (a) $p_1 < p_2$ (b) $p_1 = p_2$ (c) $p_1 > p_2$ (d) $K_1 < K_2$ (e) $K_1 = K_2$ (f) $K_1 > K_2$ **(ii)** When a force is applied to object 1, it accelerates for a time interval Δt. The force is removed from object 1 and is applied to object 2. From the same list of choices, which statements are true after object 2 has accelerated for the same time interval Δt?

Quick Quiz 9.4 Rank an automobile dashboard, seat belt, and air bag, each used alone in separate collisions from the same speed, in terms of (a) the impulse and (b) the average force each delivers to a front-seat passenger, from greatest to least.

Analysis Model   Nonisolated System (Momentum)

Imagine you have identified a system to be analyzed and have defined a system boundary. If external forces are applied on the system, the system is *nonisolated*. In that case, the change in the total momentum of the system is equal to the impulse on the system, a statement known as the **impulse–momentum theorem**:

$$\Delta\vec{p} = \vec{I} \qquad\qquad (9.13)$$

Examples:

- a baseball is struck by a bat
- a spool sitting on a table is pulled by a string (Example 10.14 in Chapter 10)
- a gas molecule strikes the wall of the container holding the gas (Chapter 21)
- photons strike an absorbing surface and exert pressure on the surface (Chapter 34)

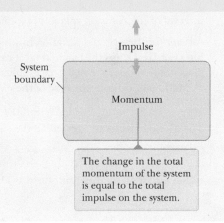

The change in the total momentum of the system is equal to the total impulse on the system.

Example 9.3   How Good Are the Bumpers?   AM

In a particular crash test, a car of mass 1 500 kg collides with a wall as shown in Figure 9.4. The initial and final velocities of the car are $\vec{v}_i = -15.0\,\hat{i}$ m/s and $\vec{v}_f = 2.60\,\hat{i}$ m/s, respectively. If the collision lasts 0.150 s, find the impulse caused by the collision and the average net force exerted on the car.

SOLUTION

Conceptualize The collision time is short, so we can imagine the car being brought to rest very rapidly and then moving back in the opposite direction with a reduced speed.

Categorize Let us assume the net force exerted on the car by the wall and friction from the ground is large compared with other forces on the car (such as

Figure 9.4 (Example 9.3) (a) This car's momentum changes as a result of its collision with the wall. (b) In a crash test, much of the car's initial kinetic energy is transformed into energy associated with the damage to the car.

continued

▶ 9.3 continued

air resistance). Furthermore, the gravitational force and the normal force exerted by the road on the car are perpendicular to the motion and therefore do not affect the horizontal momentum. Therefore, we categorize the problem as one in which we can apply the impulse approximation in the horizontal direction. We also see that the car's momentum changes due to an impulse from the environment. Therefore, we can apply the *nonisolated system (momentum)* model.

Analyze

Use Equation 9.13 to find the impulse on the car:

$$\vec{\mathbf{I}} = \Delta\vec{\mathbf{p}} = \vec{\mathbf{p}}_f - \vec{\mathbf{p}}_i = m\vec{\mathbf{v}}_f - m\vec{\mathbf{v}}_i = m(\vec{\mathbf{v}}_f - \vec{\mathbf{v}}_i)$$

$$= (1\,500\ \text{kg})[2.60\hat{\mathbf{i}}\ \text{m/s} - (-15.0\hat{\mathbf{i}}\ \text{m/s})] = \boxed{2.64 \times 10^4\hat{\mathbf{i}}\ \text{kg} \cdot \text{m/s}}$$

Use Equation 9.11 to evaluate the average net force exerted on the car:

$$\left(\sum\vec{\mathbf{F}}\right)_{\text{avg}} = \frac{\vec{\mathbf{I}}}{\Delta t} = \frac{2.64 \times 10^4\hat{\mathbf{i}}\ \text{kg} \cdot \text{m/s}}{0.150\ \text{s}} = \boxed{1.76 \times 10^5\hat{\mathbf{i}}\ \text{N}}$$

Finalize The net force found above is a combination of the normal force on the car from the wall and any friction force between the tires and the ground as the front of the car crumples. If the brakes are not operating while the crash occurs and the crumpling metal does not interfere with the free rotation of the tires, this friction force could be relatively small due to the freely rotating wheels. Notice that the signs of the velocities in this example indicate the reversal of directions. What would the mathematics be describing if both the initial and final velocities had the same sign?

WHAT IF? What if the car did not rebound from the wall? Suppose the final velocity of the car is zero and the time interval of the collision remains at 0.150 s. Would that represent a larger or a smaller net force on the car?

Answer In the original situation in which the car rebounds, the net force on the car does two things during the time interval: (1) it stops the car, and (2) it causes the car to move away from the wall at 2.60 m/s after the collision. If the car does not rebound, the net force is only doing the first of these steps—stopping the car—which requires a *smaller* force.

Mathematically, in the case of the car that does not rebound, the impulse is

$$\vec{\mathbf{I}} = \Delta\vec{\mathbf{p}} = \vec{\mathbf{p}}_f - \vec{\mathbf{p}}_i = 0 - (1\,500\ \text{kg})(-15.0\hat{\mathbf{i}}\ \text{m/s}) = 2.25 \times 10^4\hat{\mathbf{i}}\ \text{kg} \cdot \text{m/s}$$

The average net force exerted on the car is

$$\left(\sum\vec{\mathbf{F}}\right)_{\text{avg}} = \frac{\vec{\mathbf{I}}}{\Delta t} = \frac{2.25 \times 10^4\hat{\mathbf{i}}\ \text{kg} \cdot \text{m/s}}{0.150\ \text{s}} = 1.50 \times 10^5\hat{\mathbf{i}}\ \text{N}$$

which is indeed smaller than the previously calculated value, as was argued conceptually.

9.4 Collisions in One Dimension

In this section, we use the isolated system (momentum) model to describe what happens when two particles collide. The term **collision** represents an event during which two particles come close to each other and interact by means of forces. The interaction forces are assumed to be much greater than any external forces present, so we can use the impulse approximation.

A collision may involve physical contact between two macroscopic objects as described in Figure 9.5a, but the notion of what is meant by a collision must be generalized because "physical contact" on a submicroscopic scale is ill-defined and hence meaningless. To understand this concept, consider a collision on an atomic scale (Fig. 9.5b) such as the collision of a proton with an alpha particle (the nucleus of a helium atom). Because the particles are both positively charged, they repel each other due to the strong electrostatic force between them at close separations and never come into "physical contact."

When two particles of masses m_1 and m_2 collide as shown in Figure 9.5, the impulsive forces may vary in time in complicated ways, such as that shown in Figure 9.3. Regardless of the complexity of the time behavior of the impulsive force, however, this force is internal to the system of two particles. Therefore, the two particles form an isolated system and the momentum of the system must be conserved in *any* collision.

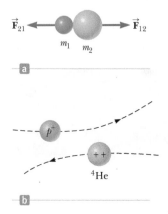

Figure 9.5 (a) The collision between two objects as the result of direct contact. (b) The "collision" between two charged particles.

In contrast, the total kinetic energy of the system of particles may or may not be conserved, depending on the type of collision. In fact, collisions are categorized as being either *elastic* or *inelastic* depending on whether or not kinetic energy is conserved.

An **elastic collision** between two objects is one in which the total kinetic energy (as well as total momentum) of the system is the same before and after the collision. Collisions between certain objects in the macroscopic world, such as billiard balls, are only *approximately* elastic because some deformation and loss of kinetic energy take place. For example, you can hear a billiard ball collision, so you know that some of the energy is being transferred away from the system by sound. An elastic collision must be perfectly silent! *Truly* elastic collisions occur between atomic and subatomic particles. These collisions are described by the isolated system model for both energy and momentum. Furthermore, there must be no transformation of kinetic energy into other types of energy within the system.

An **inelastic collision** is one in which the total kinetic energy of the system is not the same before and after the collision (even though the momentum of the system is conserved). Inelastic collisions are of two types. When the objects stick together after they collide, as happens when a meteorite collides with the Earth, the collision is called **perfectly inelastic.** When the colliding objects do not stick together but some kinetic energy is transformed or transferred away, as in the case of a rubber ball colliding with a hard surface, the collision is called **inelastic** (with no modifying adverb). When the rubber ball collides with the hard surface, some of the ball's kinetic energy is transformed when the ball is deformed while it is in contact with the surface. Inelastic collisions are described by the momentum version of the isolated system model. The system could be isolated for energy, with kinetic energy transformed to potential or internal energy. If the system is nonisolated, there could be energy leaving the system by some means. In this latter case, there could also be some transformation of energy within the system. In either of these cases, the kinetic energy of the system changes.

In the remainder of this section, we investigate the mathematical details for collisions in one dimension and consider the two extreme cases, perfectly inelastic and elastic collisions.

Perfectly Inelastic Collisions

Consider two particles of masses m_1 and m_2 moving with initial velocities $\vec{\mathbf{v}}_{1i}$ and $\vec{\mathbf{v}}_{2i}$ along the same straight line as shown in Figure 9.6. The two particles collide head-on, stick together, and then move with some common velocity $\vec{\mathbf{v}}_f$ after the collision. Because the momentum of an isolated system is conserved in *any* collision, we can say that the total momentum before the collision equals the total momentum of the composite system after the collision:

$$\Delta \vec{\mathbf{p}} = 0 \quad \rightarrow \quad \vec{\mathbf{p}}_i = \vec{\mathbf{p}}_f \quad \rightarrow \quad m_1\vec{\mathbf{v}}_{1i} + m_2\vec{\mathbf{v}}_{2i} = (m_1 + m_2)\vec{\mathbf{v}}_f \qquad (9.14)$$

Solving for the final velocity gives

$$\vec{\mathbf{v}}_f = \frac{m_1\vec{\mathbf{v}}_{1i} + m_2\vec{\mathbf{v}}_{2i}}{m_1 + m_2} \qquad (9.15)$$

Elastic Collisions

Consider two particles of masses m_1 and m_2 moving with initial velocities $\vec{\mathbf{v}}_{1i}$ and $\vec{\mathbf{v}}_{2i}$ along the same straight line as shown in Figure 9.7 on page 258. The two particles collide head-on and then leave the collision site with different velocities, $\vec{\mathbf{v}}_{1f}$ and $\vec{\mathbf{v}}_{2f}$. In an elastic collision, both the momentum and kinetic energy of the system are conserved. Therefore, considering velocities along the horizontal direction in Figure 9.7, we have

$$p_i = p_f \quad \rightarrow \quad m_1 v_{1i} + m_2 v_{2i} = m_1 v_{1f} + m_2 v_{2f} \qquad (9.16)$$

$$K_i = K_f \quad \rightarrow \quad \tfrac{1}{2}m_1 v_{1i}^{\,2} + \tfrac{1}{2}m_2 v_{2i}^{\,2} = \tfrac{1}{2}m_1 v_{1f}^{\,2} + \tfrac{1}{2}m_2 v_{2f}^{\,2} \qquad (9.17)$$

Before the collision, the particles move separately.

After the collision, the particles move together.

Figure 9.6 Schematic representation of a perfectly inelastic head-on collision between two particles.

Before the collision, the particles move separately.

$\vec{\mathbf{v}}_{1i}$ $\vec{\mathbf{v}}_{2i}$

m_1 m_2

a

After the collision, the particles continue to move separately with new velocities.

$\vec{\mathbf{v}}_{1f}$ $\vec{\mathbf{v}}_{2f}$

b

Figure 9.7 Schematic representation of an elastic head-on collision between two particles.

Pitfall Prevention 9.3

Not a General Equation Equation 9.20 can only be used in a very *specific* situation, a one-dimensional, elastic collision between two objects. The *general* concept is conservation of momentum (and conservation of kinetic energy if the collision is elastic) for an isolated system.

Because all velocities in Figure 9.7 are either to the left or right, they can be represented by the corresponding speeds along with algebraic signs indicating direction. We shall indicate v as positive if a particle moves to the right and negative if it moves to the left.

In a typical problem involving elastic collisions, there are two unknown quantities, and Equations 9.16 and 9.17 can be solved simultaneously to find them. An alternative approach, however—one that involves a little mathematical manipulation of Equation 9.17—often simplifies this process. To see how, let us cancel the factor $\frac{1}{2}$ in Equation 9.17 and rewrite it by gathering terms with subscript 1 on the left and 2 on the right:

$$m_1(v_{1i}^2 - v_{1f}^2) = m_2(v_{2f}^2 - v_{2i}^2)$$

Factoring both sides of this equation gives

$$m_1(v_{1i} - v_{1f})(v_{1i} + v_{1f}) = m_2(v_{2f} - v_{2i})(v_{2f} + v_{2i}) \tag{9.18}$$

Next, let us separate the terms containing m_1 and m_2 in Equation 9.16 in a similar way to obtain

$$m_1(v_{1i} - v_{1f}) = m_2(v_{2f} - v_{2i}) \tag{9.19}$$

To obtain our final result, we divide Equation 9.18 by Equation 9.19 and obtain

$$v_{1i} + v_{1f} = v_{2f} + v_{2i}$$

Now rearrange terms once again so as to have initial quantities on the left and final quantities on the right:

$$v_{1i} - v_{2i} = -(v_{1f} - v_{2f}) \tag{9.20}$$

This equation, in combination with Equation 9.16, can be used to solve problems dealing with elastic collisions. This pair of equations (Eqs. 9.16 and 9.20) is easier to handle than the pair of Equations 9.16 and 9.17 because there are no quadratic terms like there are in Equation 9.17. According to Equation 9.20, the *relative* velocity of the two particles before the collision, $v_{1i} - v_{2i}$, equals the negative of their relative velocity after the collision, $-(v_{1f} - v_{2f})$.

Suppose the masses and initial velocities of both particles are known. Equations 9.16 and 9.20 can be solved for the final velocities in terms of the initial velocities because there are two equations and two unknowns:

$$v_{1f} = \left(\frac{m_1 - m_2}{m_1 + m_2}\right)v_{1i} + \left(\frac{2m_2}{m_1 + m_2}\right)v_{2i} \tag{9.21}$$

$$v_{2f} = \left(\frac{2m_1}{m_1 + m_2}\right)v_{1i} + \left(\frac{m_2 - m_1}{m_1 + m_2}\right)v_{2i} \tag{9.22}$$

It is important to use the appropriate signs for v_{1i} and v_{2i} in Equations 9.21 and 9.22.

Let us consider some special cases. If $m_1 = m_2$, Equations 9.21 and 9.22 show that $v_{1f} = v_{2i}$ and $v_{2f} = v_{1i}$, which means that the particles exchange velocities if they have equal masses. That is approximately what one observes in head-on billiard ball collisions: the cue ball stops and the struck ball moves away from the collision with the same velocity the cue ball had.

If particle 2 is initially at rest, then $v_{2i} = 0$, and Equations 9.21 and 9.22 become

◀ **Elastic collision: particle 2 initially at rest**

$$v_{1f} = \left(\frac{m_1 - m_2}{m_1 + m_2}\right)v_{1i} \tag{9.23}$$

$$v_{2f} = \left(\frac{2m_1}{m_1 + m_2}\right)v_{1i} \tag{9.24}$$

If m_1 is much greater than m_2 and $v_{2i} = 0$, we see from Equations 9.23 and 9.24 that $v_{1f} \approx v_{1i}$ and $v_{2f} \approx 2v_{1i}$. That is, when a very heavy particle collides head-on with a very light one that is initially at rest, the heavy particle continues its motion unaltered after the collision and the light particle rebounds with a speed equal to about twice the initial speed of the heavy particle. An example of such a collision is that of a moving heavy atom, such as uranium, striking a light atom, such as hydrogen.

If m_2 is much greater than m_1 and particle 2 is initially at rest, then $v_{1f} \approx -v_{1i}$ and $v_{2f} \approx 0$. That is, when a very light particle collides head-on with a very heavy particle that is initially at rest, the light particle has its velocity reversed and the heavy one remains approximately at rest. For example, imagine what happens when you throw a table tennis ball at a bowling ball as in Quick Quiz 9.6 below.

Quick Quiz 9.5 In a perfectly inelastic one-dimensional collision between two moving objects, what condition alone is necessary so that the final kinetic energy of the system is zero after the collision? **(a)** The objects must have initial momenta with the same magnitude but opposite directions. **(b)** The objects must have the same mass. **(c)** The objects must have the same initial velocity. **(d)** The objects must have the same initial speed, with velocity vectors in opposite directions.

Quick Quiz 9.6 A table-tennis ball is thrown at a stationary bowling ball. The table-tennis ball makes a one-dimensional elastic collision and bounces back along the same line. Compared with the bowling ball after the collision, does the table-tennis ball have **(a)** a larger magnitude of momentum and more kinetic energy, **(b)** a smaller magnitude of momentum and more kinetic energy, **(c)** a larger magnitude of momentum and less kinetic energy, **(d)** a smaller magnitude of momentum and less kinetic energy, or **(e)** the same magnitude of momentum and the same kinetic energy?

Problem-Solving Strategy   One-Dimensional Collisions

You should use the following approach when solving collision problems in one dimension:

1. Conceptualize. Imagine the collision occurring in your mind. Draw simple diagrams of the particles before and after the collision and include appropriate velocity vectors. At first, you may have to guess at the directions of the final velocity vectors.

2. Categorize. Is the system of particles isolated? If so, use the isolated system (momentum) model. Further categorize the collision as elastic, inelastic, or perfectly inelastic.

3. Analyze. Set up the appropriate mathematical representation for the problem. If the collision is perfectly inelastic, use Equation 9.15. If the collision is elastic, use Equations 9.16 and 9.20. If the collision is inelastic, use Equation 9.16. To find the final velocities in this case, you will need some additional information.

4. Finalize. Once you have determined your result, check to see if your answers are consistent with the mental and pictorial representations and that your results are realistic.

Example 9.4   The Executive Stress Reliever

An ingenious device that illustrates conservation of momentum and kinetic energy is shown in Figure 9.8 on page 260. It consists of five identical hard balls supported by strings of equal lengths. When ball 1 is pulled out and released, after the almost-elastic collision between it and ball 2, ball 1 stops and ball 5 moves out as shown in Figure 9.8b. If balls 1 and 2 are pulled out and released, they stop after the collision and balls 4 and 5 swing out, and so forth. Is it ever possible that when ball 1 is released, it stops after the collision and balls 4 and 5 will swing out on the opposite side and travel with half the speed of ball 1 as in Figure 9.8c?

continued

▶ **9.4** continued

SOLUTION

Conceptualize With the help of Figure 9.8c, imagine one ball coming in from the left and two balls exiting the collision on the right. That is the phenomenon we want to test to see if it could ever happen.

Categorize Because of the very short time interval between the arrival of the ball from the left and the departure of the ball(s) from the right, we can use the impulse approximation to ignore the gravitational forces on the balls and model the five balls as an *isolated system* in terms of both *momentum* and *energy*. Because the balls are hard, we can categorize the collisions between them as elastic for purposes of calculation.

Figure 9.8 (Example 9.4) (a) An executive stress reliever. (b) If one ball swings down, we see one ball swing out at the other end. (c) Is it possible for one ball to swing down and two balls to leave the other end with half the speed of the first ball? In (b) and (c), the velocity vectors shown represent those of the balls immediately before and immediately after the collision.

Analyze Let's consider the situation shown in Figure 9.8c. The momentum of the system before the collision is mv, where m is the mass of ball 1 and v is its speed immediately before the collision. After the collision, we imagine that ball 1 stops and balls 4 and 5 swing out, each moving with speed $v/2$. The total momentum of the system after the collision would be $m(v/2) + m(v/2) = mv$. Therefore, the momentum of the system is conserved in the situation shown in Figure 9.8c!

The kinetic energy of the system immediately before the collision is $K_i = \frac{1}{2}mv^2$ and that after the collision is $K_f = \frac{1}{2}m(v/2)^2 + \frac{1}{2}m(v/2)^2 = \frac{1}{4}mv^2$. That shows that the kinetic energy of the system is *not* conserved, which is inconsistent with our assumption that the collisions are elastic.

Finalize Our analysis shows that it is *not* possible for balls 4 and 5 to swing out when only ball 1 is released. The only way to conserve both momentum and kinetic energy of the system is for one ball to move out when one ball is released, two balls to move out when two are released, and so on.

WHAT IF? Consider what would happen if balls 4 and 5 are glued together. Now what happens when ball 1 is pulled out and released?

Answer In this situation, balls 4 and 5 *must* move together as a single object after the collision. We have argued that both momentum and energy of the system cannot be conserved in this case. We assumed, however, ball 1 stopped after striking ball 2. What if we do not make this assumption? Consider the conservation equations with the assumption that ball 1 moves after the collision. For conservation of momentum,

$$p_i = p_f$$
$$mv_{1i} = mv_{1f} + 2mv_{4,5}$$

where $v_{4,5}$ refers to the final speed of the ball 4–ball 5 combination. Conservation of kinetic energy gives us

$$K_i = K_f$$
$$\tfrac{1}{2}mv_{1i}^2 = \tfrac{1}{2}mv_{1f}^2 + \tfrac{1}{2}(2m)v_{4,5}^2$$

Combining these equations gives

$$v_{4,5} = \tfrac{2}{3}v_{1i} \qquad v_{1f} = -\tfrac{1}{3}v_{1i}$$

Therefore, balls 4 and 5 move together as one object after the collision while ball 1 bounces back from the collision with one third of its original speed.

Example 9.5 | Carry Collision Insurance!

An 1 800-kg car stopped at a traffic light is struck from the rear by a 900-kg car. The two cars become entangled, moving along the same path as that of the originally moving car. If the smaller car were moving at 20.0 m/s before the collision, what is the velocity of the entangled cars after the collision?

SOLUTION

Conceptualize This kind of collision is easily visualized, and one can predict that after the collision both cars will be moving in the same direction as that of the initially moving car. Because the initially moving car has only half the mass of the stationary car, we expect the final velocity of the cars to be relatively small.

Categorize We identify the two cars as an *isolated system* in terms of *momentum* in the horizontal direction and apply the impulse approximation during the short time interval of the collision. The phrase "become entangled" tells us to categorize the collision as perfectly inelastic.

Analyze The magnitude of the total momentum of the system before the collision is equal to that of the smaller car because the larger car is initially at rest.

Use the isolated system model for momentum:

$$\Delta \vec{\mathbf{p}} = 0 \quad \rightarrow \quad p_i = p_f \quad \rightarrow \quad m_1 v_i = (m_1 + m_2) v_f$$

Solve for v_f and substitute numerical values:

$$v_f = \frac{m_1 v_i}{m_1 + m_2} = \frac{(900 \text{ kg})(20.0 \text{ m/s})}{900 \text{ kg} + 1\,800 \text{ kg}} = 6.67 \text{ m/s}$$

Finalize Because the final velocity is positive, the direction of the final velocity of the combination is the same as the velocity of the initially moving car as predicted. The speed of the combination is also much lower than the initial speed of the moving car.

WHAT IF? Suppose we reverse the masses of the cars. What if a stationary 900-kg car is struck by a moving 1 800-kg car? Is the final speed the same as before?

Answer Intuitively, we can guess that the final speed of the combination is higher than 6.67 m/s if the initially moving car is the more massive car. Mathematically, that should be the case because the system has a larger momentum if the initially moving car is the more massive one. Solving for the new final velocity, we find

$$v_f = \frac{m_1 v_i}{m_1 + m_2} = \frac{(1\,800 \text{ kg})(20.0 \text{ m/s})}{1\,800 \text{ kg} + 900 \text{ kg}} = 13.3 \text{ m/s}$$

which is two times greater than the previous final velocity.

Example 9.6 | The Ballistic Pendulum

The ballistic pendulum (Fig. 9.9, page 262) is an apparatus used to measure the speed of a fast-moving projectile such as a bullet. A projectile of mass m_1 is fired into a large block of wood of mass m_2 suspended from some light wires. The projectile embeds in the block, and the entire system swings through a height h. How can we determine the speed of the projectile from a measurement of h?

SOLUTION

Conceptualize Figure 9.9a helps conceptualize the situation. Run the animation in your mind: the projectile enters the pendulum, which swings up to some height at which it momentarily comes to rest.

Categorize The projectile and the block form an *isolated system* in terms of *momentum* if we identify configuration A as immediately before the collision and configuration B as immediately after the collision. Because the projectile imbeds in the block, we can categorize the collision between them as perfectly inelastic.

Analyze To analyze the collision, we use Equation 9.15, which gives the speed of the system immediately after the collision when we assume the impulse approximation.

continued

▶ **9.6** continued

Figure 9.9 (Example 9.6) (a) Diagram of a ballistic pendulum. Notice that $\vec{v}_{1A}$ is the velocity of the projectile immediately before the collision and $\vec{v}_B$ is the velocity of the projectile–block system immediately after the perfectly inelastic collision. (b) Multiflash photograph of a ballistic pendulum used in the laboratory.

Noting that $v_{2A} = 0$, solve Equation 9.15 for v_B: (1) $v_B = \dfrac{m_1 v_{1A}}{m_1 + m_2}$

....

Categorize For the process during which the projectile–block combination swings upward to height h (ending at a configuration we'll call C), we focus on a *different* system, that of the projectile, the block, and the Earth. We categorize this part of the problem as one involving an *isolated system* for *energy* with no nonconservative forces acting.

....

Analyze Write an expression for the total kinetic energy of the system immediately after the collision: (2) $K_B = \frac{1}{2}(m_1 + m_2)v_B^2$

Substitute the value of v_B from Equation (1) into Equation (2): $K_B = \dfrac{m_1^2 v_{1A}^2}{2(m_1 + m_2)}$

This kinetic energy of the system immediately after the collision is *less* than the initial kinetic energy of the projectile as is expected in an inelastic collision.

We define the gravitational potential energy of the system for configuration B to be zero. Therefore, $U_B = 0$, whereas $U_C = (m_1 + m_2)gh$.

Apply the isolated system model to the system: $\Delta K + \Delta U = 0 \quad \rightarrow \quad (K_C - K_B) + (U_C - U_B) = 0$

Substitute the energies: $\left(0 - \dfrac{m_1^2 v_{1A}^2}{2(m_1 + m_2)}\right) + [(m_1 + m_2)gh - 0] = 0$

Solve for v_{1A}: $v_{1A} = \left(\dfrac{m_1 + m_2}{m_1}\right)\sqrt{2gh}$

....

Finalize We had to solve this problem in two steps. Each step involved a different system and a different analysis model: isolated system (momentum) for the first step and isolated system (energy) for the second. Because the collision was assumed to be perfectly inelastic, some mechanical energy was transformed to internal energy during the collision. Therefore, it would have been *incorrect* to apply the isolated system (energy) model to the entire process by equating the initial kinetic energy of the incoming projectile with the final gravitational potential energy of the projectile–block–Earth combination.

Example 9.7   A Two-Body Collision with a Spring   AM

A block of mass $m_1 = 1.60$ kg initially moving to the right with a speed of 4.00 m/s on a frictionless, horizontal track collides with a light spring attached to a second block of mass $m_2 = 2.10$ kg initially moving to the left with a speed of 2.50 m/s as shown in Figure 9.10a. The spring constant is 600 N/m.

▶ **9.7** continued

(A) Find the velocities of the two blocks after the collision.

SOLUTION

Conceptualize With the help of Figure 9.10a, run an animation of the collision in your mind. Figure 9.10b shows an instant during the collision when the spring is compressed. Eventually, block 1 and the spring will again separate, so the system will look like Figure 9.10a again but with different velocity vectors for the two blocks.

Categorize Because the spring force is conservative, kinetic energy in the system of two blocks and the spring is not transformed to internal energy during the

Figure 9.10 (Example 9.7) A moving block approaches a second moving block that is attached to a spring.

compression of the spring. Ignoring any sound made when the block hits the spring, we can categorize the collision as being elastic and the two blocks and the spring as an *isolated system* for both *energy* and *momentum*.

Analyze Because momentum of the system is conserved, apply Equation 9.16:

(1) $m_1 v_{1i} + m_2 v_{2i} = m_1 v_{1f} + m_2 v_{2f}$

Because the collision is elastic, apply Equation 9.20:

(2) $v_{1i} - v_{2i} = -(v_{1f} - v_{2f})$

Multiply Equation (2) by m_1:

(3) $m_1 v_{1i} - m_1 v_{2i} = -m_1 v_{1f} + m_1 v_{2f}$

Add Equations (1) and (3):

$2m_1 v_{1i} + (m_2 - m_1)v_{2i} = (m_1 + m_2)v_{2f}$

Solve for v_{2f}:

$$v_{2f} = \frac{2m_1 v_{1i} + (m_2 - m_1)v_{2i}}{m_1 + m_2}$$

Substitute numerical values:

$$v_{2f} = \frac{2(1.60 \text{ kg})(4.00 \text{ m/s}) + (2.10 \text{ kg} - 1.60 \text{ kg})(-2.50 \text{ m/s})}{1.60 \text{ kg} + 2.10 \text{ kg}} = 3.12 \text{ m/s}$$

Solve Equation (2) for v_{1f} and substitute numerical values:

$v_{1f} = v_{2f} - v_{1i} + v_{2i} = 3.12 \text{ m/s} - 4.00 \text{ m/s} + (-2.50 \text{ m/s}) = -3.38 \text{ m/s}$

(B) Determine the velocity of block 2 during the collision, at the instant block 1 is moving to the right with a velocity of +3.00 m/s as in Figure 9.10b.

SOLUTION

Conceptualize Focus your attention now on Figure 9.10b, which represents the final configuration of the system for the time interval of interest.

Categorize Because the momentum and mechanical energy of the *isolated system* of two blocks and the spring are conserved *throughout* the collision, the collision can be categorized as elastic for *any* final instant of time. Let us now choose the final instant to be when block 1 is moving with a velocity of +3.00 m/s.

Analyze Apply Equation 9.16:

$m_1 v_{1i} + m_2 v_{2i} = m_1 v_{1f} + m_2 v_{2f}$

Solve for v_{2f}:

$$v_{2f} = \frac{m_1 v_{1i} + m_2 v_{2i} - m_1 v_{1f}}{m_2}$$

Substitute numerical values:

$$v_{2f} = \frac{(1.60 \text{ kg})(4.00 \text{ m/s}) + (2.10 \text{ kg})(-2.50 \text{ m/s}) - (1.60 \text{ kg})(3.00 \text{ m/s})}{2.10 \text{ kg}}$$

$= -1.74 \text{ m/s}$

continued

▶ **9.7** continued

Finalize The negative value for v_{2f} means that block 2 is still moving to the left at the instant we are considering.

(C) Determine the distance the spring is compressed at that instant.

SOLUTION

Conceptualize Once again, focus on the configuration of the system shown in Figure 9.10b.

Categorize For the system of the spring and two blocks, no friction or other nonconservative forces act within the system. Therefore, we categorize the system as an *isolated system* in terms of *energy* with no nonconservative forces acting. The system also remains an *isolated system* in terms of *momentum*.

Analyze We choose the initial configuration of the system to be that existing immediately before block 1 strikes the spring and the final configuration to be that when block 1 is moving to the right at 3.00 m/s.

Write the appropriate reduction of Equation 8.2:

$$\Delta K + \Delta U = 0$$

Evaluate the energies, recognizing that two objects in the system have kinetic energy and that the potential energy is elastic:

$$[(\tfrac{1}{2}m_1 v_{1f}^2 + \tfrac{1}{2}m_2 v_{2f}^2) - (\tfrac{1}{2}m_1 v_{1i}^2 + \tfrac{1}{2}m_2 v_{2i}^2)] + (\tfrac{1}{2}kx^2 - 0) = 0$$

Solve for x^2:

$$x^2 = \tfrac{1}{k}[m_1(v_{1i}^2 - v_{1f}^2) + m_2(v_{2i}^2 - v_{2f}^2)]$$

Substitute numerical values:

$$x^2 = \left(\frac{1}{600 \text{ N/m}}\right)\{(1.60 \text{ kg})[(4.00 \text{ m/s})^2 - (3.00 \text{ m/s})^2] + (2.10 \text{ kg})[(2.50 \text{ m/s})^2 - (1.74 \text{ m/s})^2]\}$$

$$\rightarrow \; x = \boxed{0.173 \text{ m}}$$

Finalize This answer is not the maximum compression of the spring because the two blocks are still moving toward each other at the instant shown in Figure 9.10b. Can you determine the maximum compression of the spring?

9.5 Collisions in Two Dimensions

Before the collision

After the collision

Figure 9.11 An elastic, glancing collision between two particles.

In Section 9.2, we showed that the momentum of a system of two particles is conserved when the system is isolated. For any collision of two particles, this result implies that the momentum in each of the directions x, y, and z is conserved. An important subset of collisions takes place in a plane. The game of billiards is a familiar example involving multiple collisions of objects moving on a two-dimensional surface. For such two-dimensional collisions, we obtain two component equations for conservation of momentum:

$$m_1 v_{1ix} + m_2 v_{2ix} = m_1 v_{1fx} + m_2 v_{2fx}$$

$$m_1 v_{1iy} + m_2 v_{2iy} = m_1 v_{1fy} + m_2 v_{2fy}$$

where the three subscripts on the velocity components in these equations represent, respectively, the identification of the object (1, 2), initial and final values (i, f), and the velocity component (x, y).

Let us consider a specific two-dimensional problem in which particle 1 of mass m_1 collides with particle 2 of mass m_2 initially at rest as in Figure 9.11. After the collision (Fig. 9.11b), particle 1 moves at an angle θ with respect to the horizontal and particle 2 moves at an angle ϕ with respect to the horizontal. This event is called a *glancing* collision. Applying the law of conservation of momentum in component form and noting that the initial y component of the momentum of the two-particle system is zero gives

$$\Delta p_x = 0 \;\;\rightarrow\;\; p_{ix} = p_{fx} \;\;\rightarrow\;\; m_1 v_{1i} = m_1 v_{1f}\cos\theta + m_2 v_{2f}\cos\phi \qquad \textbf{(9.25)}$$

$$\Delta p_y = 0 \;\;\rightarrow\;\; p_{iy} = p_{fy} \;\;\rightarrow\;\; 0 = m_1 v_{1f}\sin\theta - m_2 v_{2f}\sin\phi \qquad \textbf{(9.26)}$$

where the minus sign in Equation 9.26 is included because after the collision particle 2 has a y component of velocity that is downward. (The symbols v in these particular equations are speeds, not velocity components. The direction of the component vector is indicated explicitly with plus or minus signs.) We now have two independent equations. As long as no more than two of the seven quantities in Equations 9.25 and 9.26 are unknown, we can solve the problem.

If the collision is elastic, we can also use Equation 9.17 (conservation of kinetic energy) with $v_{2i} = 0$:

$$K_i = K_f \quad \rightarrow \quad \tfrac{1}{2}m_1 v_{1i}^2 = \tfrac{1}{2}m_1 v_{1f}^2 + \tfrac{1}{2}m_2 v_{2f}^2 \qquad \textbf{(9.27)}$$

Knowing the initial speed of particle 1 and both masses, we are left with four unknowns (v_{1f}, v_{2f}, θ, and ϕ). Because we have only three equations, one of the four remaining quantities must be given to determine the motion after the elastic collision from conservation principles alone.

If the collision is inelastic, kinetic energy is *not* conserved and Equation 9.27 does *not* apply.

Pitfall Prevention 9.4
Don't Use Equation 9.20 Equation 9.20, relating the initial and final relative velocities of two colliding objects, is only valid for one-dimensional elastic collisions. Do not use this equation when analyzing two-dimensional collisions.

Problem-Solving Strategy   Two-Dimensional Collisions

The following procedure is recommended when dealing with problems involving collisions between two particles in two dimensions.

1. Conceptualize. Imagine the collisions occurring and predict the approximate directions in which the particles will move after the collision. Set up a coordinate system and define your velocities in terms of that system. It is convenient to have the x axis coincide with one of the initial velocities. Sketch the coordinate system, draw and label all velocity vectors, and include all the given information.

2. Categorize. Is the system of particles truly isolated? If so, categorize the collision as elastic, inelastic, or perfectly inelastic.

3. Analyze. Write expressions for the x and y components of the momentum of each object before and after the collision. Remember to include the appropriate signs for the components of the velocity vectors and pay careful attention to signs throughout the calculation.

Apply the isolated system model for momentum $\Delta \vec{p} = 0$. When applied in each direction, this equation will generally reduce to $p_{ix} = p_{fx}$ and $p_{iy} = p_{fy}$, where each of these terms refer to the sum of the momenta of all objects in the system. Write expressions for the *total* momentum in the x direction *before* and *after* the collision and equate the two. Repeat this procedure for the total momentum in the y direction.

Proceed to solve the momentum equations for the unknown quantities. If the collision is inelastic, kinetic energy is *not* conserved and additional information is probably required. If the collision is perfectly inelastic, the final velocities of the two objects are equal.

If the collision is elastic, kinetic energy is conserved and you can equate the total kinetic energy of the system before the collision to the total kinetic energy after the collision, providing an additional relationship between the velocity magnitudes.

4. Finalize. Once you have determined your result, check to see if your answers are consistent with the mental and pictorial representations and that your results are realistic.

Example 9.8   Collision at an Intersection

A 1 500-kg car traveling east with a speed of 25.0 m/s collides at an intersection with a 2 500-kg truck traveling north at a speed of 20.0 m/s as shown in Figure 9.12 on page 266. Find the direction and magnitude of the velocity of the wreckage after the collision, assuming the vehicles stick together after the collision.

continued

▶ **9.8** continued

SOLUTION

Conceptualize Figure 9.12 should help you conceptualize the situation before and after the collision. Let us choose east to be along the positive x direction and north to be along the positive y direction.

Categorize Because we consider moments immediately before and immediately after the collision as defining our time interval, we ignore the small effect that friction would have on the wheels of the vehicles and model the two vehicles as an *isolated system* in terms of *momentum*. We also ignore the vehicles' sizes and model them as particles. The collision is perfectly inelastic because the car and the truck stick together after the collision.

Analyze Before the collision, the only object having momentum in the x direction is the car. Therefore, the magnitude of the total initial momentum of the system (car plus truck) in the x direction is that of only the car. Similarly, the total initial momentum of the system in the y direction is that of the truck. After the collision, let us assume the wreckage moves at an angle θ with respect to the x axis with speed v_f.

Figure 9.12 (Example 9.8) An eastbound car colliding with a northbound truck.

Apply the isolated system model for momentum in the x direction:
$$\Delta p_x = 0 \quad \rightarrow \quad \sum p_{xi} = \sum p_{xf} \quad \rightarrow \quad (1) \quad m_1 v_{1i} = (m_1 + m_2) v_f \cos \theta$$

Apply the isolated system model for momentum in the y direction:
$$\Delta p_y = 0 \quad \rightarrow \quad \sum p_{yi} = \sum p_{yf} \quad \rightarrow \quad (2) \quad m_2 v_{2i} = (m_1 + m_2) v_f \sin \theta$$

Divide Equation (2) by Equation (1):
$$\frac{m_2 v_{2i}}{m_1 v_{1i}} = \frac{\sin \theta}{\cos \theta} = \tan \theta$$

Solve for θ and substitute numerical values:
$$\theta = \tan^{-1}\left(\frac{m_2 v_{2i}}{m_1 v_{1i}}\right) = \tan^{-1}\left[\frac{(2\ 500 \text{ kg})(20.0 \text{ m/s})}{(1\ 500 \text{ kg})(25.0 \text{ m/s})}\right] = \boxed{53.1°}$$

Use Equation (2) to find the value of v_f and substitute numerical values:
$$v_f = \frac{m_2 v_{2i}}{(m_1 + m_2) \sin \theta} = \frac{(2\ 500 \text{ kg})(20.0 \text{ m/s})}{(1\ 500 \text{ kg} + 2\ 500 \text{ kg}) \sin 53.1°} = \boxed{15.6 \text{ m/s}}$$

Finalize Notice that the angle θ is qualitatively in agreement with Figure 9.12. Also notice that the final speed of the combination is less than the initial speeds of the two cars. This result is consistent with the kinetic energy of the system being reduced in an inelastic collision. It might help if you draw the momentum vectors of each vehicle before the collision and the two vehicles together after the collision.

| Example 9.9 | **Proton–Proton Collision** | |

A proton collides elastically with another proton that is initially at rest. The incoming proton has an initial speed of 3.50×10^5 m/s and makes a glancing collision with the second proton as in Figure 9.11. (At close separations, the protons exert a repulsive electrostatic force on each other.) After the collision, one proton moves off at an angle of 37.0° to the original direction of motion and the second deflects at an angle of ϕ to the same axis. Find the final speeds of the two protons and the angle ϕ.

SOLUTION

Conceptualize This collision is like that shown in Figure 9.11, which will help you conceptualize the behavior of the system. We define the x axis to be along the direction of the velocity vector of the initially moving proton.

Categorize The pair of protons form an *isolated system*. Both momentum and kinetic energy of the system are conserved in this glancing elastic collision.

▶ **9.9** continued

Analyze Using the isolated system model for both momentum and energy for a two-dimensional elastic collision, set up the mathematical representation with Equations 9.25 through 9.27:

$(1) \quad v_{1i} = v_{1f} \cos \theta + v_{2f} \cos \phi$

$(2) \quad 0 = v_{1f} \sin \theta - v_{2f} \sin \phi$

$(3) \quad v_{1i}^2 = v_{1f}^2 + v_{2f}^2$

Rearrange Equations (1) and (2):

$v_{2f} \cos \phi = v_{1i} - v_{1f} \cos \theta$

$v_{2f} \sin \phi = v_{1f} \sin \theta$

Square these two equations and add them:

$v_{2f}^2 \cos^2 \phi + v_{2f}^2 \sin^2 \phi =$
$\quad v_{1i}^2 - 2v_{1i}v_{1f} \cos \theta + v_{1f}^2 \cos^2 \theta + v_{1f}^2 \sin^2 \theta$

Incorporate that the sum of the squares of sine and cosine for *any* angle is equal to 1:

$(4) \quad v_{2f}^2 = v_{1i}^2 - 2v_{1i}v_{1f} \cos \theta + v_{1f}^2$

Substitute Equation (4) into Equation (3):

$v_{1f}^2 + (v_{1i}^2 - 2v_{1i}v_{1f} \cos \theta + v_{1f}^2) = v_{1i}^2$

$(5) \quad v_{1f}^2 - v_{1i}v_{1f} \cos \theta = 0$

One possible solution of Equation (5) is $v_{1f} = 0$, which corresponds to a head-on, one-dimensional collision in which the first proton stops and the second continues with the same speed in the same direction. That is not the solution we want.

Divide both sides of Equation (5) by v_{1f} and solve for the remaining factor of v_{1f}:

$v_{1f} = v_{1i} \cos \theta = (3.50 \times 10^5 \text{ m/s}) \cos 37.0° = \boxed{2.80 \times 10^5 \text{ m/s}}$

Use Equation (3) to find v_{2f}:

$v_{2f} = \sqrt{v_{1i}^2 - v_{1f}^2} = \sqrt{(3.50 \times 10^5 \text{ m/s})^2 - (2.80 \times 10^5 \text{ m/s})^2}$

$\quad = \boxed{2.11 \times 10^5 \text{ m/s}}$

Use Equation (2) to find ϕ:

$(2) \quad \phi = \sin^{-1} \left(\frac{v_{1f} \sin \theta}{v_{2f}} \right) = \sin^{-1} \left[\frac{(2.80 \times 10^5 \text{ m/s}) \sin 37.0°}{(2.11 \times 10^5 \text{ m/s})} \right]$

$\quad = \boxed{53.0°}$

Finalize It is interesting that $\theta + \phi = 90°$. This result is *not* accidental. Whenever two objects of equal mass collide elastically in a glancing collision and one of them is initially at rest, their final velocities are perpendicular to each other.

9.6 The Center of Mass

In this section, we describe the overall motion of a system in terms of a special point called the **center of mass** of the system. The system can be either a small number of particles or an extended, continuous object, such as a gymnast leaping through the air. We shall see that the translational motion of the center of mass of the system is the same as if all the mass of the system were concentrated at that point. That is, the system moves as if the net external force were applied to a single particle located at the center of mass. This model, the *particle model*, was introduced in Chapter 2. This behavior is independent of other motion, such as rotation or vibration of the system or deformation of the system (for instance, when a gymnast folds her body).

Consider a system consisting of a pair of particles that have different masses and are connected by a light, rigid rod (Fig. 9.13 on page 268). The position of the center of mass of a system can be described as being the *average position* of the system's mass. The center of mass of the system is located somewhere on the line joining the two particles and is closer to the particle having the larger mass. If a single force is applied at a point on the rod above the center of mass, the system rotates clockwise (see Fig. 9.13a). If the force is applied at a point on the rod below the center of mass, the system rotates counterclockwise (see Fig. 9.13b). If the force

The system rotates clockwise when a force is applied above the center of mass.

CM

a

The system rotates counter-clockwise when a force is applied below the center of mass.

CM

b

The system moves in the direction of the force without rotating when a force is applied at the center of mass.

CM

c

Figure 9.13 A force is applied to a system of two particles of unequal mass connected by a light, rigid rod.

Figure 9.14 The center of mass of two particles of unequal mass on the x axis is located at $x_{\rm CM}$, a point between the particles, closer to the one having the larger mass.

is applied at the center of mass, the system moves in the direction of the force without rotating (see Fig. 9.13c). The center of mass of an object can be located with this procedure.

The center of mass of the pair of particles described in Figure 9.14 is located on the x axis and lies somewhere between the particles. Its x coordinate is given by

$$x_{\rm CM} \equiv \frac{m_1 x_1 + m_2 x_2}{m_1 + m_2} \tag{9.28}$$

For example, if $x_1 = 0$, $x_2 = d$, and $m_2 = 2m_1$, we find that $x_{\rm CM} = \frac{2}{3}d$. That is, the center of mass lies closer to the more massive particle. If the two masses are equal, the center of mass lies midway between the particles.

We can extend this concept to a system of many particles with masses m_i in three dimensions. The x coordinate of the center of mass of n particles is defined to be

$$x_{\rm CM} \equiv \frac{m_1 x_1 + m_2 x_2 + m_3 x_3 + \cdots + m_n x_n}{m_1 + m_2 + m_3 + \cdots + m_n} = \frac{\sum\limits_i m_i x_i}{\sum\limits_i m_i} = \frac{\sum\limits_i m_i x_i}{M} = \frac{1}{M}\sum\limits_i m_i x_i \tag{9.29}$$

where x_i is the x coordinate of the ith particle and the total mass is $M \equiv \sum_i m_i$ where the sum runs over all n particles. The y and z coordinates of the center of mass are similarly defined by the equations

$$y_{\rm CM} \equiv \frac{1}{M}\sum\limits_i m_i y_i \quad \text{and} \quad z_{\rm CM} \equiv \frac{1}{M}\sum\limits_i m_i z_i \tag{9.30}$$

The center of mass can be located in three dimensions by its position vector $\vec{\mathbf{r}}_{\rm CM}$. The components of this vector are $x_{\rm CM}$, $y_{\rm CM}$, and $z_{\rm CM}$, defined in Equations 9.29 and 9.30. Therefore,

$$\vec{\mathbf{r}}_{\rm CM} = x_{\rm CM}\hat{\mathbf{i}} + y_{\rm CM}\hat{\mathbf{j}} + z_{\rm CM}\hat{\mathbf{k}} = \frac{1}{M}\sum\limits_i m_i x_i\hat{\mathbf{i}} + \frac{1}{M}\sum\limits_i m_i y_i\hat{\mathbf{j}} + \frac{1}{M}\sum\limits_i m_i z_i\hat{\mathbf{k}}$$

$$\vec{\mathbf{r}}_{\rm CM} \equiv \frac{1}{M}\sum\limits_i m_i\vec{\mathbf{r}}_i \tag{9.31}$$

where $\vec{\mathbf{r}}_i$ is the position vector of the ith particle, defined by

$$\vec{\mathbf{r}}_i \equiv x_i\hat{\mathbf{i}} + y_i\hat{\mathbf{j}} + z_i\hat{\mathbf{k}}$$

Although locating the center of mass for an extended, continuous object is somewhat more cumbersome than locating the center of mass of a small number of particles, the basic ideas we have discussed still apply. Think of an extended object as a system containing a large number of small mass elements such as the cube in Figure 9.15. Because the separation between elements is very small, the object can be considered to have a continuous mass distribution. By dividing the object into elements of mass Δm_i with coordinates x_i, y_i, z_i, we see that the x coordinate of the center of mass is approximately

$$x_{\rm CM} \approx \frac{1}{M}\sum\limits_i x_i\,\Delta m_i$$

with similar expressions for $y_{\rm CM}$ and $z_{\rm CM}$. If we let the number of elements n approach infinity, the size of each element approaches zero and $x_{\rm CM}$ is given precisely. In this limit, we replace the sum by an integral and Δm_i by the differential element dm:

$$x_{\rm CM} = \lim_{\Delta m_i \to 0}\frac{1}{M}\sum\limits_i x_i\,\Delta m_i = \frac{1}{M}\int x\,dm \tag{9.32}$$

Likewise, for $y_{\rm CM}$ and $z_{\rm CM}$ we obtain

$$y_{\rm CM} = \frac{1}{M}\int y\,dm \quad \text{and} \quad z_{\rm CM} = \frac{1}{M}\int z\,dm \tag{9.33}$$

We can express the vector position of the center of mass of an extended object in the form

$$\vec{\mathbf{r}}_{CM} = \frac{1}{M} \int \vec{\mathbf{r}} \, dm \qquad (9.34)$$

which is equivalent to the three expressions given by Equations 9.32 and 9.33.

The center of mass of any symmetric object of uniform density lies on an axis of symmetry and on any plane of symmetry. For example, the center of mass of a uniform rod lies in the rod, midway between its ends. The center of mass of a sphere or a cube lies at its geometric center.

Because an extended object is a continuous distribution of mass, each small mass element is acted upon by the gravitational force. The net effect of all these forces is equivalent to the effect of a single force $M\vec{\mathbf{g}}$ acting through a special point, called the **center of gravity.** If $\vec{\mathbf{g}}$ is constant over the mass distribution, the center of gravity coincides with the center of mass. If an extended object is pivoted at its center of gravity, it balances in any orientation.

The center of gravity of an irregularly shaped object such as a wrench can be determined by suspending the object first from one point and then from another. In Figure 9.16, a wrench is hung from point A and a vertical line AB (which can be established with a plumb bob) is drawn when the wrench has stopped swinging. The wrench is then hung from point C, and a second vertical line CD is drawn. The center of gravity is halfway through the thickness of the wrench, under the intersection of these two lines. In general, if the wrench is hung freely from any point, the vertical line through this point must pass through the center of gravity.

Quick **Quiz** 9.7 A baseball bat of uniform density is cut at the location of its center of mass as shown in Figure 9.17. Which piece has the smaller mass? **(a)** the piece on the right **(b)** the piece on the left **(c)** both pieces have the same mass **(d)** impossible to determine

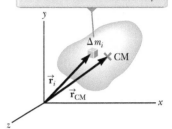

Figure 9.15 The center of mass is located at the vector position $\vec{\mathbf{r}}_{CM}$, which has coordinates x_{CM}, y_{CM}, and z_{CM}.

An extended object can be considered to be a distribution of small elements of mass Δm_i.

Figure 9.17 (Quick Quiz 9.7) A baseball bat cut at the location of its center of mass.

The wrench is hung freely first from point A and then from point C.

The intersection of the two lines AB and CD locates the center of gravity.

Figure 9.16 An experimental technique for determining the center of gravity of a wrench.

Example 9.10   **The Center of Mass of Three Particles**

A system consists of three particles located as shown in Figure 9.18. Find the center of mass of the system. The masses of the particles are $m_1 = m_2 = 1.0$ kg and $m_3 = 2.0$ kg.

SOLUTION

Conceptualize Figure 9.18 shows the three masses. Your intuition should tell you that the center of mass is located somewhere in the region between the blue particle and the pair of tan particles as shown in the figure.

Figure 9.18 (Example 9.10) Two particles are located on the x axis, and a single particle is located on the y axis as shown. The vector indicates the location of the system's center of mass.

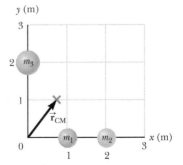

Categorize We categorize this example as a substitution problem because we will be using the equations for the center of mass developed in this section.

continued

▶ 9.10 c o n t i n u e d

Use the defining equations for
the coordinates of the center of
mass and notice that $z_{CM} = 0$:

$$x_{CM} = \frac{1}{M}\sum_i m_i x_i = \frac{m_1 x_1 + m_2 x_2 + m_3 x_3}{m_1 + m_2 + m_3}$$

$$= \frac{(1.0\ \text{kg})(1.0\ \text{m}) + (1.0\ \text{kg})(2.0\ \text{m}) + (2.0\ \text{kg})(0)}{1.0\ \text{kg} + 1.0\ \text{kg} + 2.0\ \text{kg}} = \frac{3.0\ \text{kg}\cdot\text{m}}{4.0\ \text{kg}} = 0.75\ \text{m}$$

$$y_{CM} = \frac{1}{M}\sum_i m_i y_i = \frac{m_1 y_1 + m_2 y_2 + m_3 y_3}{m_1 + m_2 + m_3}$$

$$= \frac{(1.0\ \text{kg})(0) + (1.0\ \text{kg})(0) + (2.0\ \text{kg})(2.0\ \text{m})}{4.0\ \text{kg}} = \frac{4.0\ \text{kg}\cdot\text{m}}{4.0\ \text{kg}} = 1.0\ \text{m}$$

Write the position vector of the
center of mass:

$$\vec{\mathbf{r}}_{CM} \equiv x_{CM}\hat{\mathbf{i}} + y_{CM}\hat{\mathbf{j}} = (0.75\hat{\mathbf{i}} + 1.0\hat{\mathbf{j}})\ \text{m}$$

Example 9.11   The Center of Mass of a Rod

(A) Show that the center of mass of a rod of mass M and length L lies midway
between its ends, assuming the rod has a uniform mass per unit length.

SOLUTION

Conceptualize The rod is shown aligned along the x axis in Figure 9.19, so $y_{CM} = z_{CM} = 0$. What is your prediction of the value of x_{CM}?

Figure 9.19 (Example 9.11) The
geometry used to find the center
of mass of a uniform rod.

Categorize We categorize this example as an analysis problem because we need
to divide the rod into small mass elements to perform the integration in Equation 9.32.

Analyze The mass per unit length (this quantity is called the *linear mass density*) can be written as $\lambda = M/L$ for the uniform rod. If the rod is divided into elements of length dx, the mass of each element is $dm = \lambda\ dx$.

Use Equation 9.32 to find an expression for x_{CM}:

$$x_{CM} = \frac{1}{M}\int x\ dm = \frac{1}{M}\int_0^L x\lambda\ dx = \frac{\lambda}{M}\frac{x^2}{2}\Big|_0^L = \frac{\lambda L^2}{2M}$$

Substitute $\lambda = M/L$:

$$x_{CM} = \frac{L^2}{2M}\left(\frac{M}{L}\right) = \tfrac{1}{2}L$$

One can also use symmetry arguments to obtain the same result.

(B) Suppose a rod is *nonuniform* such that its mass per unit length varies linearly with x according to the expression
$\lambda = \alpha x$, where α is a constant. Find the x coordinate of the center of mass as a fraction of L.

SOLUTION

Conceptualize Because the mass per unit length is not constant in this case but is proportional to x, elements of the
rod to the right are more massive than elements near the left end of the rod.

Categorize This problem is categorized similarly to part (A), with the added twist that the linear mass density is not
constant.

Analyze In this case, we replace dm in Equation 9.32 by $\lambda\ dx$, where $\lambda = \alpha x$.

Use Equation 9.32 to find an expression for x_{CM}:

$$x_{CM} = \frac{1}{M}\int x\ dm = \frac{1}{M}\int_0^L x\lambda\ dx = \frac{1}{M}\int_0^L x\alpha x\ dx$$

$$= \frac{\alpha}{M}\int_0^L x^2\ dx = \frac{\alpha L^3}{3M}$$

▶ **9.11** continued

Find the total mass of the rod:

$$M = \int dm = \int_0^L \lambda \, dx = \int_0^L \alpha x \, dx = \frac{\alpha L^2}{2}$$

Substitute M into the expression for x_{CM}:

$$x_{CM} = \frac{\alpha L^3}{3\alpha L^2/2} = \tfrac{2}{3}L$$

Finalize Notice that the center of mass in part (B) is farther to the right than that in part (A). That result is reasonable because the elements of the rod become more massive as one moves to the right along the rod in part (B).

Example 9.12   **The Center of Mass of a Right Triangle**

You have been asked to hang a metal sign from a single vertical string. The sign has the triangular shape shown in Figure 9.20a. The bottom of the sign is to be parallel to the ground. At what distance from the left end of the sign should you attach the support string?

SOLUTION

Conceptualize Figure 9.20a shows the sign hanging from the string. The string must be attached at a point directly above the center of gravity of the sign, which is the same as its center of mass because it is in a uniform gravitational field.

Categorize As in the case of Example 9.11, we categorize this example as an analysis problem because it is necessary to identify infinitesimal mass elements of the sign to perform the integration in Equation 9.32.

Analyze We assume the triangular sign has a uniform density and total mass M. Because the sign is a continuous distribution of mass, we must use the integral expression in Equation 9.32 to find the x coordinate of the center of mass.

We divide the triangle into narrow strips of width dx and height y as shown in Figure 9.20b, where y is the height of the hypotenuse of the triangle above the x axis for a given value of x. The mass of each strip is the product of the volume of the strip and the density ρ of the material from which the sign is made: $dm = \rho y t \, dx$, where t is the thickness of the metal sign. The density of the material is the total mass of the sign divided by its total volume (area of the triangle times thickness).

Figure 9.20 (Example 9.12) (a) A triangular sign to be hung from a single string. (b) Geometric construction for locating the center of mass.

Evaluate dm:

$$dm = \rho y t \, dx = \left(\frac{M}{\frac{1}{2}abt}\right) y t \, dx = \frac{2My}{ab} \, dx$$

Use Equation 9.32 to find the x coordinate of the center of mass:

$$(1) \quad x_{CM} = \frac{1}{M} \int x \, dm = \frac{1}{M} \int_0^a x \, \frac{2My}{ab} \, dx = \frac{2}{ab} \int_0^a xy \, dx$$

To proceed further and evaluate the integral, we must express y in terms of x. The line representing the hypotenuse of the triangle in Figure 9.20b has a slope of b/a and passes through the origin, so the equation of this line is $y = (b/a)x$.

Substitute for y in Equation (1):

$$x_{CM} = \frac{2}{ab} \int_0^a x\left(\frac{b}{a}x\right) dx = \frac{2}{a^2} \int_0^a x^2 \, dx = \frac{2}{a^2}\left[\frac{x^3}{3}\right]_0^a$$

$$= \tfrac{2}{3}a$$

Therefore, the string must be attached to the sign at a distance two-thirds of the length of the bottom edge from the left end.

continued

Finalize This answer is identical to that in part (B) of Example 9.11. For the triangular sign, the linear increase in height y with position x means that elements in the sign increase in mass linearly along the x axis, just like the linear increase in mass density in Example 9.11. We could also find the y coordinate of the center of mass of the sign, but that is not needed to determine where the string should be attached. You might try cutting a right triangle out of cardboard and hanging it from a string so that the long base is horizontal. Does the string need to be attached at $\frac{2}{3}a$?

9.7 Systems of Many Particles

Consider a system of two or more particles for which we have identified the center of mass. We can begin to understand the physical significance and utility of the center of mass concept by taking the time derivative of the position vector for the center of mass given by Equation 9.31. From Section 4.1, we know that the time derivative of a position vector is by definition the velocity vector. Assuming M remains constant for a system of particles—that is, no particles enter or leave the system—we obtain the following expression for the **velocity of the center of mass** of the system:

◀ **Velocity of the center of mass of a system of particles**

$$\vec{\mathbf{v}}_{CM} = \frac{d\vec{\mathbf{r}}_{CM}}{dt} = \frac{1}{M}\sum_i m_i \frac{d\vec{\mathbf{r}}_i}{dt} = \frac{1}{M}\sum_i m_i \vec{\mathbf{v}}_i \tag{9.35}$$

where $\vec{\mathbf{v}}_i$ is the velocity of the ith particle. Rearranging Equation 9.35 gives

◀ **Total momentum of a system of particles**

$$M\vec{\mathbf{v}}_{CM} = \sum_i m_i \vec{\mathbf{v}}_i = \sum_i \vec{\mathbf{p}}_i = \vec{\mathbf{p}}_{tot} \tag{9.36}$$

Therefore, the total linear momentum of the system equals the total mass multiplied by the velocity of the center of mass. In other words, the total linear momentum of the system is equal to that of a single particle of mass M moving with a velocity $\vec{\mathbf{v}}_{CM}$.

Differentiating Equation 9.35 with respect to time, we obtain the **acceleration of the center of mass** of the system:

◀ **Acceleration of the center of mass of a system of particles**

$$\vec{\mathbf{a}}_{CM} = \frac{d\vec{\mathbf{v}}_{CM}}{dt} = \frac{1}{M}\sum_i m_i \frac{d\vec{\mathbf{v}}_i}{dt} = \frac{1}{M}\sum_i m_i \vec{\mathbf{a}}_i \tag{9.37}$$

Rearranging this expression and using Newton's second law gives

$$M\vec{\mathbf{a}}_{CM} = \sum_i m_i \vec{\mathbf{a}}_i = \sum_i \vec{\mathbf{F}}_i \tag{9.38}$$

where $\vec{\mathbf{F}}_i$ is the net force on particle i.

The forces on any particle in the system may include both external forces (from outside the system) and internal forces (from within the system). By Newton's third law, however, the internal force exerted by particle 1 on particle 2, for example, is equal in magnitude and opposite in direction to the internal force exerted by particle 2 on particle 1. Therefore, when we sum over all internal force vectors in Equation 9.38, they cancel in pairs and we find that the net force on the system is caused *only* by external forces. We can then write Equation 9.38 in the form

◀ **Newton's second law for a system of particles**

$$\sum \vec{\mathbf{F}}_{ext} = M\vec{\mathbf{a}}_{CM} \tag{9.39}$$

That is, the net external force on a system of particles equals the total mass of the system multiplied by the acceleration of the center of mass. Comparing Equation 9.39 with Newton's second law for a single particle, we see that the particle model we have used in several chapters can be described in terms of the center of mass:

The center of mass of a system of particles having combined mass M moves like an equivalent particle of mass M would move under the influence of the net external force on the system.

Let us integrate Equation 9.39 over a finite time interval:

$$\int \sum \vec{\mathbf{F}}_{\text{ext}} \, dt = \int M\vec{\mathbf{a}}_{\text{CM}} \, dt = \int M\frac{d\vec{\mathbf{v}}_{\text{CM}}}{dt} \, dt = M \int d\vec{\mathbf{v}}_{\text{CM}} = M\Delta\vec{\mathbf{v}}_{\text{CM}}$$

Notice that this equation can be written as

$$\Delta\vec{\mathbf{p}}_{\text{tot}} = \vec{\mathbf{I}} \qquad (9.40)$$

◀ Impulse–momentum theorem for a system of particles

where $\vec{\mathbf{I}}$ is the impulse imparted to the system by external forces and $\vec{\mathbf{p}}_{\text{tot}}$ is the momentum of the system. Equation 9.40 is the generalization of the impulse–momentum theorem for a particle (Eq. 9.13) to a system of many particles. It is also the mathematical representation of the nonisolated system (momentum) model for a system of many particles.

Finally, if the net external force on a system is zero so that the system is isolated, it follows from Equation 9.39 that

$$M\vec{\mathbf{a}}_{\text{CM}} = M\frac{d\vec{\mathbf{v}}_{\text{CM}}}{dt} = 0$$

Therefore, the isolated system model for momentum for a system of many particles is described by

$$\Delta\vec{\mathbf{p}}_{\text{tot}} = 0 \qquad (9.41)$$

which can be rewritten as

$$M\vec{\mathbf{v}}_{\text{CM}} = \vec{\mathbf{p}}_{\text{tot}} = \text{constant} \quad \left(\text{when } \sum \vec{\mathbf{F}}_{\text{ext}} = 0\right) \qquad (9.42)$$

That is, the total linear momentum of a system of particles is conserved if no net external force is acting on the system. It follows that for an isolated system of particles, both the total momentum and the velocity of the center of mass are constant in time. This statement is a generalization of the isolated system (momentum) model for a many-particle system.

Suppose the center of mass of an isolated system consisting of two or more members is at rest. The center of mass of the system remains at rest if there is no net force on the system. For example, consider a system of a swimmer standing on a raft, with the system initially at rest. When the swimmer dives horizontally off the raft, the raft moves in the direction opposite that of the swimmer and the center of mass of the system remains at rest (if we neglect friction between raft and water). Furthermore, the linear momentum of the diver is equal in magnitude to that of the raft, but opposite in direction.

> **Quick Quiz 9.8** A cruise ship is moving at constant speed through the water. The vacationers on the ship are eager to arrive at their next destination. They decide to try to speed up the cruise ship by gathering at the bow (the front) and running together toward the stern (the back) of the ship. **(i)** While they are running toward the stern, is the speed of the ship (a) higher than it was before, (b) unchanged, (c) lower than it was before, or (d) impossible to determine? **(ii)** The vacationers stop running when they reach the stern of the ship. After they have all stopped running, is the speed of the ship (a) higher than it was before they started running, (b) unchanged from what it was before they started running, (c) lower than it was before they started running, or (d) impossible to determine?

Conceptual Example 9.13 **Exploding Projectile**

A projectile fired into the air suddenly explodes into several fragments (Fig. 9.21 on page 274).

(A) What can be said about the motion of the center of mass of the system made up of all the fragments after the explosion?

continued

▶ **9.13** continued

SOLUTION

Neglecting air resistance, the only external force on the projectile is the gravitational force. Therefore, if the projectile did not explode, it would continue to move along the parabolic path indicated by the dashed line in Figure 9.21. Because the forces caused by the explosion are internal, they do not affect the motion of the center of mass of the system (the fragments). Therefore, after the explosion, the center of mass of the fragments follows the same parabolic path the projectile would have followed if no explosion had occurred.

(B) If the projectile did not explode, it would land at a distance R from its launch point. Suppose the projectile explodes and splits into two pieces of equal mass. One piece lands at a distance $2R$ to the right of the launch point. Where does the other piece land?

Figure 9.21 (Conceptual Example 9.13) When a projectile explodes into several fragments, the center of mass of the system made up of all the fragments follows the same parabolic path the projectile would have taken had there been no explosion.

SOLUTION

As discussed in part (A), the center of mass of the two-piece system lands at a distance R from the launch point. One of the pieces lands at a farther distance R from the landing point (or a distance $2R$ from the launch point), to the right in Figure 9.21. Because the two pieces have the same mass, the other piece must land a distance R to the left of the landing point in Figure 9.21, which places this piece right back at the launch point!

Example 9.14   The Exploding Rocket   AM

A rocket is fired vertically upward. At the instant it reaches an altitude of 1 000 m and a speed of $v_i = 300$ m/s, it explodes into three fragments having equal mass. One fragment moves upward with a speed of $v_1 = 450$ m/s following the explosion. The second fragment has a speed of $v_2 = 240$ m/s and is moving east right after the explosion. What is the velocity of the third fragment immediately after the explosion?

SOLUTION

Conceptualize Picture the explosion in your mind, with one piece going upward and a second piece moving horizontally toward the east. Do you have an intuitive feeling about the direction in which the third piece moves?

Categorize This example is a two-dimensional problem because we have two fragments moving in perpendicular directions after the explosion as well as a third fragment moving in an unknown direction in the plane defined by the velocity vectors of the other two fragments. We assume the time interval of the explosion is very short, so we use the impulse approximation in which we ignore the gravitational force and air resistance. Because the forces of the explosion are internal to the system (the rocket), the rocket is an *isolated system* in terms of *momentum*. Therefore, the total momentum $\vec{\mathbf{p}}_i$ of the rocket immediately before the explosion must equal the total momentum $\vec{\mathbf{p}}_f$ of the fragments immediately after the explosion.

· ·

Analyze Because the three fragments have equal mass, the mass of each fragment is $M/3$, where M is the total mass of the rocket. We will let $\vec{\mathbf{v}}_3$ represent the unknown velocity of the third fragment.

Use the isolated system (momentum) model to equate the initial and final momenta of the system and express the momenta in terms of masses and velocities:
$$\Delta\vec{\mathbf{p}} = 0 \quad \rightarrow \quad \vec{\mathbf{p}}_i = \vec{\mathbf{p}}_f \quad \rightarrow \quad M\vec{\mathbf{v}}_i = \frac{M}{3}\vec{\mathbf{v}}_1 + \frac{M}{3}\vec{\mathbf{v}}_2 + \frac{M}{3}\vec{\mathbf{v}}_3$$

Solve for $\vec{\mathbf{v}}_3$:
$$\vec{\mathbf{v}}_3 = 3\vec{\mathbf{v}}_i - \vec{\mathbf{v}}_1 - \vec{\mathbf{v}}_2$$

Substitute the numerical values:
$$\vec{\mathbf{v}}_3 = 3(300\hat{\mathbf{j}}\text{ m/s}) - (450\hat{\mathbf{j}}\text{ m/s}) - (240\hat{\mathbf{i}}\text{ m/s}) = \boxed{(-240\hat{\mathbf{i}} + 450\hat{\mathbf{j}})\text{ m/s}}$$

· ·

Finalize Notice that this event is the reverse of a perfectly inelastic collision. There is one object before the collision and three objects afterward. Imagine running a movie of the event backward: the three objects would come together and become a single object. In a perfectly inelastic collision, the kinetic energy of the system decreases. If you were

▶ **9.14** continued

to calculate the kinetic energy before and after the event in this example, you would find that the kinetic energy of the system increases. (Try it!) This increase in kinetic energy comes from the potential energy stored in whatever fuel exploded to cause the breakup of the rocket.

9.8 Deformable Systems

So far in our discussion of mechanics, we have analyzed the motion of particles or nondeformable systems that can be modeled as particles. The discussion in Section 9.7 can be applied to an analysis of the motion of deformable systems. For example, suppose you stand on a skateboard and push off a wall, setting yourself in motion away from the wall. Your body has deformed during this event: your arms were bent before the event, and they straightened out while you pushed off the wall. How would we describe this event?

The force from the wall on your hands moves through no displacement; the force is always located at the interface between the wall and your hands. Therefore, the force does no work on the system, which is you and your skateboard. Pushing off the wall, however, does indeed result in a change in the kinetic energy of the system. If you try to use the work–kinetic energy theorem, $W = \Delta K$, to describe this event, you will notice that the left side of the equation is zero but the right side is not zero. The work–kinetic energy theorem is not valid for this event and is often not valid for systems that are deformable.

To analyze the motion of deformable systems, we appeal to Equation 8.2, the conservation of energy equation, and Equation 9.40, the impulse–momentum theorem. For the example of you pushing off the wall on your skateboard, identifying the system as you and the skateboard, Equation 8.2 gives

$$\Delta E_{\text{system}} = \sum T \quad \rightarrow \quad \Delta K + \Delta U = 0$$

where ΔK is the change in kinetic energy, which is related to the increased speed of the system, and ΔU is the decrease in potential energy stored in the body from previous meals. This equation tells us that the system transformed potential energy into kinetic energy by virtue of the muscular exertion necessary to push off the wall. Notice that the system is isolated in terms of energy but nonisolated in terms of momentum.

Applying Equation 9.40 to the system in this situation gives us

$$\Delta \vec{\mathbf{p}}_{\text{tot}} = \vec{\mathbf{I}} \quad \rightarrow \quad m \Delta \vec{\mathbf{v}} = \int \vec{\mathbf{F}}_{\text{wall}} \, dt$$

where $\vec{\mathbf{F}}_{\text{wall}}$ is the force exerted by the wall on your hands, m is the mass of you and the skateboard, and $\Delta \vec{\mathbf{v}}$ is the change in the velocity of the system during the event. To evaluate the right side of this equation, we would need to know how the force from the wall varies in time. In general, this process might be complicated. In the case of constant forces, or well-behaved forces, however, the integral on the right side of the equation can be evaluated.

Example 9.15 Pushing on a Spring[3] AM

As shown in Figure 9.22a (page 276), two blocks are at rest on a frictionless, level table. Both blocks have the same mass m, and they are connected by a spring of negligible mass. The separation distance of the blocks when the spring is relaxed is L. During a time interval Δt, a constant force of magnitude F is applied horizontally to the left block,

---

[3]Example 9.15 was inspired in part by C. E. Mungan, "A primer on work–energy relationships for introductory physics," *The Physics Teacher* **43**:10, 2005.

continued

▶ **9.15** continued

moving it through a distance x_1 as shown in Figure 9.22b. During this time inter-val, the right block moves through a distance x_2. At the end of this time interval, the force F is removed.

(A) Find the resulting speed $\vec{v}_{CM}$ of the center of mass of the system.

Figure 9.22 (Example 9.15) (a) Two blocks of equal mass are connected by a spring. (b) The left block is pushed with a constant force of magnitude F and moves a distance x_1 during some time inter-val. During this same time interval, the right block moves through a distance x_2.

SOLUTION

Conceptualize Imagine what happens as you push on the left block. It begins to move to the right in Figure 9.22, and the spring begins to compress. As a result, the spring pushes to the right on the right block, which begins to move to the right. At any given time, the blocks are generally moving with different velocities. As the cen-ter of mass of the system moves to the right with a constant speed after the force is removed, the two blocks oscillate back and forth with respect to the center of mass.

Categorize We apply three analysis models in this problem: the deformable sys-tem of two blocks and a spring is modeled as a *nonisolated system* in terms of *energy* because work is being done on it by the applied force. It is also modeled as a *noniso-lated system* in terms of *momentum* because of the force acting on the system during a time interval. Because the applied force on the system is constant, the acceleration of its center of mass is constant and the center of mass is modeled as a *particle under constant acceleration*.

Analyze Using the nonisolated system (momentum) model, we apply the impulse–momentum theorem to the system of two blocks, recognizing that the force F is constant during the time interval Δt while the force is applied.

Write Equation 9.40 for the system:

$$\Delta p_x = I_x \quad \rightarrow \quad (2m)(v_{CM} - 0) = F \Delta t$$

$$(1) \quad 2mv_{CM} = F \Delta t$$

During the time interval Δt, the center of mass of the sys-tem moves a distance $\frac{1}{2}(x_1 + x_2)$. Use this fact to express the time interval in terms of $v_{CM,avg}$:

$$\Delta t = \frac{\frac{1}{2}(x_1 + x_2)}{v_{CM,avg}}$$

Because the center of mass is modeled as a particle under constant acceleration, the average velocity of the center of mass is the average of the initial velocity, which is zero, and the final velocity v_{CM}:

$$\Delta t = \frac{\frac{1}{2}(x_1 + x_2)}{\frac{1}{2}(0 + v_{CM})} = \frac{(x_1 + x_2)}{v_{CM}}$$

Substitute this expression into Equation (1):

$$2mv_{CM} = F \frac{(x_1 + x_2)}{v_{CM}}$$

Solve for v_{CM}:

$$v_{CM} = \sqrt{F \frac{(x_1 + x_2)}{2m}}$$

(B) Find the total energy of the system associated with vibration relative to its center of mass after the force F is removed.

SOLUTION

Analyze The vibrational energy is all the energy of the system other than the kinetic energy associated with transla-tional motion of the center of mass. To find the vibrational energy, we apply the conservation of energy equation. The kinetic energy of the system can be expressed as $K = K_{CM} + K_{vib}$, where K_{vib} is the kinetic energy of the blocks relative to the center of mass due to their vibration. The potential energy of the system is U_{vib}, which is the potential energy stored in the spring when the separation of the blocks is some value other than L.

From the nonisolated system (energy) model, express Equation 8.2 for this system:

$$(2) \quad \Delta K_{CM} + \Delta K_{vib} + \Delta U_{vib} = W$$

▶ **9.15** continued

Express Equation (2) in an alternate form, noting that $K_{vib} + U_{vib} = E_{vib}$:

$$\Delta K_{CM} + \Delta E_{vib} = W$$

The initial values of the kinetic energy of the center of mass and the vibrational energy of the system are zero. Use this fact and substitute for the work done on the system by the force F:

$$K_{CM} + E_{vib} = W = Fx_1$$

Solve for the vibrational energy and use the result from part (A):

$$E_{vib} = Fx_1 - K_{CM} = Fx_1 - \tfrac{1}{2}(2m)v_{CM}^2 = F\,\frac{(x_1 - x_2)}{2}$$

Finalize Neither of the two answers in this example depends on the spring length, the spring constant, or the time interval. Notice also that the magnitude x_1 of the displacement of the point of application of the applied force is different from the magnitude $\tfrac{1}{2}(x_1 + x_2)$ of the displacement of the center of mass of the system. This difference reminds us that the displacement in the definition of work (Eq. 7.1) is that of the point of application of the force.

9.9 Rocket Propulsion

When ordinary vehicles such as cars are propelled, the driving force for the motion is friction. In the case of the car, the driving force is the force exerted by the road on the car. We can model the car as a nonisolated system in terms of momentum. An impulse is applied to the car from the roadway, and the result is a change in the momentum of the car as described by Equation 9.40.

A rocket moving in space, however, has no road to push against. The rocket is an isolated system in terms of momentum. Therefore, the source of the propulsion of a rocket must be something other than an external force. The operation of a rocket depends on the law of conservation of linear momentum as applied to an isolated system, where the system is the rocket plus its ejected fuel.

Rocket propulsion can be understood by first considering our archer standing on frictionless ice in Example 9.1. Imagine the archer fires several arrows horizontally. For each arrow fired, the archer receives a compensating momentum in the opposite direction. As more arrows are fired, the archer moves faster and faster across the ice. In addition to this analysis in terms of momentum, we can also understand this phenomenon in terms of Newton's second and third laws. Every time the bow pushes an arrow forward, the arrow pushes the bow (and the archer) backward, and these forces result in an acceleration of the archer.

In a similar manner, as a rocket moves in free space, its linear momentum changes when some of its mass is ejected in the form of exhaust gases. Because the gases are given momentum when they are ejected out of the engine, the rocket receives a compensating momentum in the opposite direction. Therefore, the rocket is accelerated as a result of the "push," or thrust, from the exhaust gases. In free space, the center of mass of the system (rocket plus expelled gases) moves uniformly, independent of the propulsion process.[4]

Suppose at some time t the magnitude of the momentum of a rocket plus its fuel is $(M + \Delta m)v$, where v is the speed of the rocket relative to the Earth (Fig. 9.23a). Over a short time interval Δt, the rocket ejects fuel of mass Δm. At the end of this interval, the rocket's mass is M and its speed is $v + \Delta v$, where Δv is the change in speed of the rocket (Fig. 9.23b). If the fuel is ejected with a speed v_e relative to

Courtesy of NASA

The force from a nitrogen-propelled hand-controlled device allows an astronaut to move about freely in space without restrictive tethers, using the thrust force from the expelled nitrogen.

Figure 9.23 Rocket propulsion. (a) The initial mass of the rocket plus all its fuel is $M + \Delta m$ at a time t, and its speed is v. (b) At a time $t + \Delta t$, the rocket's mass has been reduced to M and an amount of fuel Δm has been ejected. The rocket's speed increases by an amount Δv.

---

[4]The rocket and the archer represent cases of the reverse of a perfectly inelastic collision: momentum is conserved, but the kinetic energy of the rocket–exhaust gas system increases (at the expense of chemical potential energy in the fuel), as does the kinetic energy of the archer–arrow system (at the expense of potential energy from the archer's previous meals).

the rocket (the subscript e stands for *exhaust*, and v_e is usually called the *exhaust speed*), the velocity of the fuel relative to the Earth is $v - v_e$. Because the system of the rocket and the ejected fuel is isolated, we apply the isolated system model for momentum and obtain

$$\Delta p = 0 \quad \rightarrow \quad p_i = p_f \quad \rightarrow \quad (M + \Delta m)v = M(v + \Delta v) + \Delta m(v - v_e)$$

Simplifying this expression gives

$$M \Delta v = v_e \Delta m$$

If we now take the limit as Δt goes to zero, we let $\Delta v \rightarrow dv$ and $\Delta m \rightarrow dm$. Furthermore, the increase in the exhaust mass dm corresponds to an equal decrease in the rocket mass, so $dm = -dM$. Notice that dM is negative because it represents a decrease in mass, so $-dM$ is a positive number. Using this fact gives

$$M\,dv = v_e\,dm = -v_e\,dM \tag{9.43}$$

Now divide the equation by M and integrate, taking the initial mass of the rocket plus fuel to be M_i and the final mass of the rocket plus its remaining fuel to be M_f. The result is

$$\int_{v_i}^{v_f} dv = -v_e \int_{M_i}^{M_f} \frac{dM}{M}$$

◀ **Expression for rocket propulsion**

$$v_f - v_i = v_e \ln\left(\frac{M_i}{M_f}\right) \tag{9.44}$$

which is the basic expression for rocket propulsion. First, Equation 9.44 tells us that the increase in rocket speed is proportional to the exhaust speed v_e of the ejected gases. Therefore, the exhaust speed should be very high. Second, the increase in rocket speed is proportional to the natural logarithm of the ratio M_i/M_f. Therefore, this ratio should be as large as possible; that is, the mass of the rocket without its fuel should be as small as possible and the rocket should carry as much fuel as possible.

The **thrust** on the rocket is the force exerted on it by the ejected exhaust gases. We obtain the following expression for the thrust from Newton's second law and Equation 9.43:

$$\text{Thrust} = M\frac{dv}{dt} = \left| v_e \frac{dM}{dt} \right| \tag{9.45}$$

This expression shows that the thrust increases as the exhaust speed increases and as the rate of change of mass (called the *burn rate*) increases.

Example 9.16   **Fighting a Fire**

Two firefighters must apply a total force of 600 N to steady a hose that is discharging water at the rate of 3 600 L/min. Estimate the speed of the water as it exits the nozzle.

SOLUTION

Conceptualize As the water leaves the hose, it acts in a way similar to the gases being ejected from a rocket engine. As a result, a force (thrust) acts on the firefighters in a direction opposite the direction of motion of the water. In this case, we want the end of the hose to be modeled as a particle in equilibrium rather than to accelerate as in the case of the rocket. Consequently, the firefighters must apply a force of magnitude equal to the thrust in the opposite direction to keep the end of the hose stationary.

Categorize This example is a substitution problem in which we use given values in an equation derived in this section. The water exits at 3 600 L/min, which is 60 L/s. Knowing that 1 L of water has a mass of 1 kg, we estimate that about 60 kg of water leaves the nozzle each second.

▶ **9.16** continued

Use Equation 9.45 for the thrust:

$$\text{Thrust} = \left| v_e \frac{dM}{dt} \right|$$

Solve for the exhaust speed:

$$v_e = \frac{\text{Thrust}}{|dM/dt|}$$

Substitute numerical values:

$$v_e = \frac{600 \text{ N}}{60 \text{ kg/s}} = \boxed{10 \text{ m/s}}$$

Example 9.17 A Rocket in Space

A rocket moving in space, far from all other objects, has a speed of 3.0×10^3 m/s relative to the Earth. Its engines are turned on, and fuel is ejected in a direction opposite the rocket's motion at a speed of 5.0×10^3 m/s relative to the rocket.

(A) What is the speed of the rocket relative to the Earth once the rocket's mass is reduced to half its mass before ignition?

SOLUTION

Conceptualize Figure 9.23 shows the situation in this problem. From the discussion in this section and scenes from science fiction movies, we can easily imagine the rocket accelerating to a higher speed as the engine operates.

Categorize This problem is a substitution problem in which we use given values in the equations derived in this section.

Solve Equation 9.44 for the final velocity and substitute the known values:

$$v_f = v_i + v_e \ln\left(\frac{M_i}{M_f}\right)$$

$$= 3.0 \times 10^3 \text{ m/s} + (5.0 \times 10^3 \text{ m/s}) \ln\left(\frac{M_i}{0.50M_i}\right)$$

$$= \boxed{6.5 \times 10^3 \text{ m/s}}$$

(B) What is the thrust on the rocket if it burns fuel at the rate of 50 kg/s?

SOLUTION

Use Equation 9.45, noting that $dM/dt = 50$ kg/s:

$$\text{Thrust} = \left| v_e \frac{dM}{dt} \right| = (5.0 \times 10^3 \text{ m/s})(50 \text{ kg/s}) = \boxed{2.5 \times 10^5 \text{ N}}$$

Summary

Definitions

The **linear momentum** $\vec{p}$ of a particle of mass m moving with a velocity $\vec{v}$ is

$$\vec{p} \equiv m\vec{v} \qquad (9.2)$$

The **impulse** imparted to a particle by a net force $\sum \vec{F}$ is equal to the time integral of the force:

$$\vec{I} \equiv \int_{t_i}^{t_f} \sum \vec{F} \, dt \qquad (9.9)$$

continued

An **inelastic collision** is one for which the total kinetic energy of the system of colliding particles is not conserved. A **perfectly inelastic collision** is one in which the colliding particles stick together after the collision. An **elastic collision** is one in which the kinetic energy of the system is conserved.

The position vector of the **center of mass** of a system of particles is defined as

$$\vec{\mathbf{r}}_{CM} \equiv \frac{1}{M} \sum_i m_i \vec{\mathbf{r}}_i \qquad (9.31)$$

where $M = \Sigma_i\, m_i$ is the total mass of the system and $\vec{\mathbf{r}}_i$ is the position vector of the ith particle.

Concepts and Principles

The position vector of the center of mass of an extended object can be obtained from the integral expression

$$\vec{\mathbf{r}}_{CM} = \frac{1}{M} \int \vec{\mathbf{r}}\; dm \qquad (9.34)$$

The velocity of the center of mass for a system of particles is

$$\vec{\mathbf{v}}_{CM} = \frac{1}{M} \sum_i m_i \vec{\mathbf{v}}_i \qquad (9.35)$$

The total momentum of a system of particles equals the total mass multiplied by the velocity of the center of mass.

Newton's second law applied to a system of particles is

$$\sum \vec{\mathbf{F}}_{ext} = M \vec{\mathbf{a}}_{CM} \qquad (9.39)$$

where $\vec{\mathbf{a}}_{CM}$ is the acceleration of the center of mass and the sum is over all external forces. The center of mass moves like an imaginary particle of mass M under the influence of the resultant external force on the system.

Analysis Models for Problem Solving

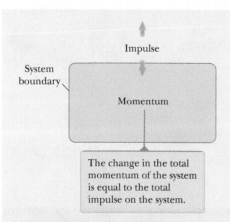

The change in the total momentum of the system is equal to the total impulse on the system.

Nonisolated System (Momentum). If a system interacts with its environment in the sense that there is an external force on the system, the behavior of the system is described by the **impulse–momentum theorem:**

$$\Delta \vec{\mathbf{p}}_{tot} = \vec{\mathbf{I}} \qquad (9.40)$$

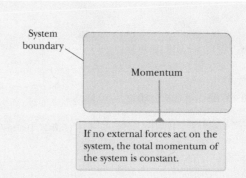

If no external forces act on the system, the total momentum of the system is constant.

Isolated System (Momentum). The total momentum of an isolated system (no external forces) is conserved regardless of the nature of the forces between the members of the system:

$$\Delta \vec{\mathbf{p}}_{tot} = 0 \qquad (9.41)$$

The system may be isolated in terms of momentum but nonisolated in terms of energy, as in the case of inelastic collisions.

1. You are standing on a saucer-shaped sled at rest in the middle of a frictionless ice rink. Your lab partner throws you a heavy Frisbee. You take different actions in successive experimental trials. Rank the following situations according to your final speed from largest to smallest. If your final speed is the same in two cases, give them equal rank. (a) You catch the Frisbee and hold onto it. (b) You catch the Frisbee and throw it back to your partner. (c) You bobble the catch, just touching the Frisbee so that it continues in its original direction more slowly. (d) You catch the Frisbee and throw it so that it moves vertically upward above your head. (e) You catch the Frisbee and set it down so that it remains at rest on the ice.

2. A boxcar at a rail yard is set into motion at the top of a hump. The car rolls down quietly and without friction onto a straight, level track where it couples with a flatcar of smaller mass, originally at rest, so that the two cars then roll together without friction. Consider the two cars as a system from the moment of release of the boxcar until both are rolling together. Answer the following questions yes or no. (a) Is mechanical energy of the system conserved? (b) Is momentum of the system conserved? Next, consider only the process of the boxcar gaining speed as it rolls down the hump. For the boxcar and the Earth as a system, (c) is mechanical energy conserved? (d) Is momentum conserved? Finally, consider the two cars as a system as the boxcar is slowing down in the coupling process. (e) Is mechanical energy of this system conserved? (f) Is momentum of this system conserved?

3. A massive tractor is rolling down a country road. In a perfectly inelastic collision, a small sports car runs into the machine from behind. **(i)** Which vehicle experiences a change in momentum of larger magnitude? (a) The car does. (b) The tractor does. (c) Their momentum changes are the same size. (d) It could be either vehicle. **(ii)** Which vehicle experiences a larger change in kinetic energy? (a) The car does. (b) The tractor does. (c) Their kinetic energy changes are the same size. (d) It could be either vehicle.

4. A 2-kg object moving to the right with a speed of 4 m/s makes a head-on, elastic collision with a 1-kg object that is initially at rest. The velocity of the 1-kg object after the collision is (a) greater than 4 m/s, (b) less than 4 m/s, (c) equal to 4 m/s, (d) zero, or (e) impossible to say based on the information provided. ·

5. A 5-kg cart moving to the right with a speed of 6 m/s collides with a concrete wall and rebounds with a speed of 2 m/s. What is the change in momentum of the cart? (a) 0 (b) 40 kg · m/s (c) −40 kg · m/s (d) −30 kg · m/s (e) −10 kg · m/s

6. A 57.0-g tennis ball is traveling straight at a player at 21.0 m/s. The player volleys the ball straight back at 25.0 m/s. If the ball remains in contact with the racket for 0.060 0 s, what average force acts on the ball? (a) 22.6 N (b) 32.5 N (c) 43.7 N (d) 72.1 N (e) 102 N

7. The momentum of an object is increased by a factor of 4 in magnitude. By what factor is its kinetic energy changed? (a) 16 (b) 8 (c) 4 (d) 2 (e) 1

8. The kinetic energy of an object is increased by a factor of 4. By what factor is the magnitude of its momentum changed? (a) 16 (b) 8 (c) 4 (d) 2 (e) 1

9. If two particles have equal momenta, are their kinetic energies equal? (a) yes, always (b) no, never (c) no, except when their speeds are the same (d) yes, as long as they move along parallel lines

10. If two particles have equal kinetic energies, are their momenta equal? (a) yes, always (b) no, never (c) yes, as long as their masses are equal (d) yes, if both their masses and directions of motion are the same (e) yes, as long as they move along parallel lines

11. A 10.0-g bullet is fired into a 200-g block of wood at rest on a horizontal surface. After impact, the block slides 8.00 m before coming to rest. If the coefficient of friction between the block and the surface is 0.400, what is the speed of the bullet before impact? (a) 106 m/s (b) 166 m/s (c) 226 m/s (d) 286 m/s (e) none of those answers is correct

12. Two particles of different mass start from rest. The same net force acts on both of them as they move over equal distances. How do their final kinetic energies compare? (a) The particle of larger mass has more kinetic energy. (b) The particle of smaller mass has more kinetic energy. (c) The particles have equal kinetic energies. (d) Either particle might have more kinetic energy.

13. Two particles of different mass start from rest. The same net force acts on both of them as they move over equal distances. How do the magnitudes of their final momenta compare? (a) The particle of larger mass has more momentum. (b) The particle of smaller mass has more momentum. (c) The particles have equal momenta. (d) Either particle might have more momentum.

14. A basketball is tossed up into the air, falls freely, and bounces from the wooden floor. From the moment after the player releases it until the ball reaches the top of its bounce, what is the smallest system for which momentum is conserved? (a) the ball (b) the ball plus player (c) the ball plus floor (d) the ball plus the Earth (e) momentum is not conserved for any system

15. A 3-kg object moving to the right on a frictionless, horizontal surface with a speed of 2 m/s collides head-on and sticks to a 2-kg object that is initially moving to the left with a speed of 4 m/s. After the collision, which statement is true? (a) The kinetic energy of the system is 20 J. (b) The momentum of the system is 14 kg · m/s. (c) The kinetic energy of the system is greater than 5 J but less than 20 J. (d) The momentum of the system is −2 kg · m/s. (e) The momentum of the system is less than the momentum of the system before the collision.

16. A ball is suspended by a string that is tied to a fixed point above a wooden block standing on end. The ball is pulled back as shown in Figure OQ9.16 and released. In trial A, the ball rebounds elastically from the block. In trial B, two-sided tape causes the ball to stick to the block. In which case is the ball more likely to knock the block over? (a) It is more likely in trial A. (b) It is more likely in trial B. (c) It makes no difference. (d) It could be either case, depending on other factors.

Figure OQ9.16

17. A car of mass m traveling at speed v crashes into the rear of a truck of mass $2m$ that is at rest and in neutral at an intersection. If the collision is perfectly inelastic, what is the speed of the combined car and truck after the collision? (a) v (b) $v/2$ (c) $v/3$ (d) $2v$ (e) None of those answers is correct.

18. A head-on, elastic collision occurs between two billiard balls of equal mass. If a red ball is traveling to the right with speed v and a blue ball is traveling to the left with speed $3v$ before the collision, what statement is true concerning their velocities subsequent to the collision? Neglect any effects of spin. (a) The red ball travels to the left with speed v, while the blue ball travels to the right with speed $3v$. (b) The red ball travels to the left with speed v, while the blue ball continues to move to the left with a speed $2v$. (c) The red ball travels to the left with speed $3v$, while the blue ball travels to the right with speed v. (d) Their final velocities cannot be determined because momentum is not conserved in the collision. (e) The velocities cannot be determined without knowing the mass of each ball.

Conceptual Questions

1. denotes answer available in *Student Solutions Manual/Study Guide*

1. An airbag in an automobile inflates when a collision occurs, which protects the passenger from serious injury (see the photo on page 254). Why does the airbag soften the blow? Discuss the physics involved in this dramatic photograph.

2. In golf, novice players are often advised to be sure to "follow through" with their swing. Why does this advice make the ball travel a longer distance? If a shot is taken near the green, very little follow-through is required. Why?

3. An open box slides across a frictionless, icy surface of a frozen lake. What happens to the speed of the box as water from a rain shower falls vertically downward into the box? Explain.

4. While in motion, a pitched baseball carries kinetic energy and momentum. (a) Can we say that it carries a force that it can exert on any object it strikes? (b) Can the baseball deliver more kinetic energy to the bat and batter than the ball carries initially? (c) Can the baseball deliver to the bat and batter more momentum than the ball carries initially? Explain each of your answers.

5. You are standing perfectly still and then take a step forward. Before the step, your momentum was zero, but afterward you have some momentum. Is the principle of conservation of momentum violated in this case? Explain your answer.

6. A sharpshooter fires a rifle while standing with the butt of the gun against her shoulder. If the forward momentum of a bullet is the same as the backward momentum of the gun, why isn't it as dangerous to be hit by the gun as by the bullet?

7. Two students hold a large bed sheet vertically between them. A third student, who happens to be the star pitcher on the school baseball team, throws a raw egg at the center of the sheet. Explain why the egg does not break when it hits the sheet, regardless of its initial speed.

8. A juggler juggles three balls in a continuous cycle. Any one ball is in contact with one of his hands for one fifth of the time. (a) Describe the motion of the center of mass of the three balls. (b) What average force does the juggler exert on one ball while he is touching it?

9. (a) Does the center of mass of a rocket in free space accelerate? Explain. (b) Can the speed of a rocket exceed the exhaust speed of the fuel? Explain.

10. On the subject of the following positions, state your own view and argue to support it. (a) The best theory of motion is that force causes acceleration. (b) The true measure of a force's effectiveness is the work it does, and the best theory of motion is that work done on an object changes its energy. (c) The true measure of a force's effect is impulse, and the best theory of motion is that impulse imparted to an object changes its momentum.

11. Does a larger net force exerted on an object always produce a larger change in the momentum of the object compared with a smaller net force? Explain.

12. Does a larger net force always produce a larger change in kinetic energy than a smaller net force? Explain.

13. A bomb, initially at rest, explodes into several pieces. (a) Is linear momentum of the system (the bomb before the explosion, the pieces after the explosion) conserved? Explain. (b) Is kinetic energy of the system conserved? Explain.

Problems

WebAssign The problems found in this chapter may be assigned online in Enhanced WebAssign

1. straightforward; 2. intermediate;
3. challenging

1. full solution available in the *Student Solutions Manual/Study Guide*

AMT Analysis Model tutorial available in Enhanced WebAssign

GP Guided Problem

M Master It tutorial available in Enhanced WebAssign

W Watch It video solution available in Enhanced WebAssign

Section 9.1 Linear Momentum

1. A particle of mass m moves with momentum of magnitude p. (a) Show that the kinetic energy of the particle is $K = p^2/2m$. (b) Express the magnitude of the particle's momentum in terms of its kinetic energy and mass.

2. An object has a kinetic energy of 275 J and a momentum of magnitude 25.0 kg · m/s. Find the speed and mass of the object.

3. At one instant, a 17.5-kg sled is moving over a horizontal surface of snow at 3.50 m/s. After 8.75 s has elapsed, the sled stops. Use a momentum approach to find the average friction force acting on the sled while it was moving.

4. A 3.00-kg particle has a velocity of $(3.00\hat{i} - 4.00\hat{j})$ m/s. (a) Find its x and y components of momentum. (b) Find the magnitude and direction of its momentum.

5. A baseball approaches home plate at a speed of 45.0 m/s, moving horizontally just before being hit by a bat. The batter hits a pop-up such that after hitting the bat, the baseball is moving at 55.0 m/s straight up. The ball has a mass of 145 g and is in contact with the bat for 2.00 ms. What is the average vector force the ball exerts on the bat during their interaction?

Section 9.2 Analysis Model: Isolated System (Momentum)

6. A 45.0-kg girl is standing on a 150-kg plank. Both are originally at rest on a frozen lake that constitutes a frictionless, flat surface. The girl begins to walk along the plank at a constant velocity of $1.50\hat{i}$ m/s relative to the plank. (a) What is the velocity of the plank relative to the ice surface? (b) What is the girl's velocity relative to the ice surface?

7. A girl of mass m_g is standing on a plank of mass m_p. Both are originally at rest on a frozen lake that constitutes a frictionless, flat surface. The girl begins to walk along the plank at a constant velocity v_{gp} to the right relative to the plank. (The subscript gp denotes the girl relative to plank.) (a) What is the velocity v_{pi} of the plank relative to the surface of the ice? (b) What is the girl's velocity v_{gi} relative to the ice surface?

8. A 65.0-kg boy and his 40.0-kg sister, both wearing roller blades, face each other at rest. The girl pushes the boy hard, sending him backward with velocity 2.90 m/s toward the west. Ignore friction. (a) Describe the subsequent motion of the girl. (b) How much potential energy in the girl's body is converted into mechanical energy of the boy–girl system? (c) Is the momentum of the boy–girl system conserved in the pushing-apart process? If so, explain how that is possible considering (d) there are large forces acting and (e) there is no motion beforehand and plenty of motion afterward.

9. In research in cardiology and exercise physiology, it is often important to know the mass of blood pumped by a person's heart in one stroke. This information can be obtained by means of a *ballistocardiograph*. The instrument works as follows. The subject lies on a horizontal pallet floating on a film of air. Friction on the pallet is negligible. Initially, the momentum of the system is zero. When the heart beats, it expels a mass m of blood into the aorta with speed v, and the body and platform move in the opposite direction with speed V. The blood velocity can be determined independently (e.g., by observing the Doppler shift of ultrasound). Assume that it is 50.0 cm/s in one typical trial. The mass of the subject plus the pallet is 54.0 kg. The pallet moves 6.00×10^{-5} m in 0.160 s after one heartbeat. Calculate the mass of blood that leaves the heart. Assume that the mass of blood is negligible compared with the total mass of the person. (This simplified example illustrates the principle of ballistocardiography, but in practice a more sophisticated model of heart function is used.)

10. When you jump straight up as high as you can, what is the order of magnitude of the maximum recoil speed that you give to the Earth? Model the Earth as a perfectly solid object. In your solution, state the physical quantities you take as data and the values you measure or estimate for them.

11. Two blocks of masses m and $3m$ are placed on a frictionless, horizontal surface. A light spring is attached to the more massive block, and the blocks are pushed together with the spring between them (Fig. P9.11). A cord initially holding the blocks together is burned; after that happens, the block of mass $3m$ moves to the right with a speed of 2.00 m/s. (a) What is the velocity of the block of mass m? (b) Find the system's original elastic potential energy, taking $m = 0.350$ kg. (c) Is the original energy

Figure P9.11

in the spring or in the cord? (d) Explain your answer to part (c). (e) Is the momentum of the system conserved in the bursting-apart process? Explain how that is possible considering (f) there are large forces acting and (g) there is no motion beforehand and plenty of motion afterward?

Section 9.3 Analysis Model: Nonisolated System (Momentum)

12. A man claims that he can hold onto a 12.0-kg child in a head-on collision as long as he has his seat belt on. Consider this man in a collision in which he is in one of two identical cars each traveling toward the other at 60.0 mi/h relative to the ground. The car in which he rides is brought to rest in 0.10 s. (a) Find the magnitude of the average force needed to hold onto the child. (b) Based on your result to part (a), is the man's claim valid? (c) What does the answer to this problem say about laws requiring the use of proper safety devices such as seat belts and special toddler seats?

13. An estimated force–time curve for a baseball struck by a bat is shown in Figure P9.13. From this curve, determine (a) the magnitude of the impulse delivered to the ball and (b) the average force exerted on the ball.

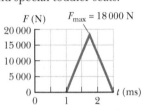

Figure P9.13

14. **Review.** After a 0.300-kg rubber ball is dropped from a height of 1.75 m, it bounces off a concrete floor and rebounds to a height of 1.50 m. (a) Determine the magnitude and direction of the impulse delivered to the ball by the floor. (b) Estimate the time the ball is in contact with the floor and use this estimate to calculate the average force the floor exerts on the ball.

15. A glider of mass m is free to slide along a horizontal air track. It is pushed against a launcher at one end of the track. Model the launcher as a light spring of force constant k compressed by a distance x. The glider is released from rest. (a) Show that the glider attains a speed of $v = x(k/m)^{1/2}$. (b) Show that the magnitude of the impulse imparted to the glider is given by the expression $I = x(km)^{1/2}$. (c) Is more work done on a cart with a large or a small mass?

16. In a slow-pitch softball game, a 0.200-kg softball crosses the plate at 15.0 m/s at an angle of 45.0° below the horizontal. The batter hits the ball toward center field, giving it a velocity of 40.0 m/s at 30.0° above the horizontal. (a) Determine the impulse delivered to the ball. (b) If the force on the ball increases linearly for 4.00 ms, holds constant for 20.0 ms, and then decreases linearly to zero in another 4.00 ms, what is the maximum force on the ball?

17. The front 1.20 m of a 1 400-kg car is designed as a "crumple zone" that collapses to absorb the shock of a collision. If a car traveling 25.0 m/s stops uniformly in 1.20 m, (a) how long does the collision last, (b) what is the magnitude of the average force on the car, and

(c) what is the acceleration of the car? Express the acceleration as a multiple of the acceleration due to gravity.

18. A tennis player receives a shot with the ball (0.060 0 kg) traveling horizontally at 20.0 m/s and returns the shot with the ball traveling horizontally at 40.0 m/s in the opposite direction. (a) What is the impulse delivered to the ball by the tennis racket? (b) Some work is done on the system of the ball and some energy appears in the ball as an increase in internal energy during the collision between the ball and the racket. What is the sum $W - \Delta E_{int}$ for the ball?

19. The magnitude of the net force exerted in the x direction on a 2.50-kg particle varies in time as shown in Figure P9.19. Find (a) the impulse of the force over the 5.00-s time interval, (b) the final velocity the particle attains if it is originally at rest, (c) its final velocity if its original velocity is $-2.00\hat{i}$ m/s, and (d) the average force exerted on the particle for the time interval between 0 and 5.00 s.

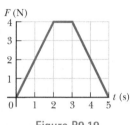

Figure P9.19

20. **Review.** A *force platform* is a tool used to analyze the performance of athletes by measuring the vertical force the athlete exerts on the ground as a function of time. Starting from rest, a 65.0-kg athlete jumps down onto the platform from a height of 0.600 m. While she is in contact with the platform during the time interval $0 < t < 0.800$ s, the force she exerts on it is described by the function

$$F = 9\ 200t - 11\ 500t^2$$

where F is in newtons and t is in seconds. (a) What impulse did the athlete receive from the platform? (b) With what speed did she reach the platform? (c) With what speed did she leave it? (d) To what height did she jump upon leaving the platform?

21. Water falls without splashing at a rate of 0.250 L/s from a height of 2.60 m into a 0.750-kg bucket on a scale. If the bucket is originally empty, what does the scale read in newtons 3.00 s after water starts to accumulate in it?

Section 9.4 Collisions in One Dimension

22. A 1 200-kg car traveling initially at $v_{Ci} = 25.0$ m/s in an easterly direction crashes into the back of a 9 000-kg truck moving in the same direction at $v_{Ti} = 20.0$ m/s (Fig. P9.22). The velocity of the car immediately after the collision is $v_{Cf} = 18.0$ m/s to the east. (a) What is the velocity of the truck immediately after the colli-

Before   After

Figure P9.22

sion? (b) What is the change in mechanical energy of the car–truck system in the collision? (c) Account for this change in mechanical energy.

23. A 10.0-g bullet is fired into a stationary block of wood [W] having mass $m = 5.00$ kg. The bullet imbeds into the block. The speed of the bullet-plus-wood combination immediately after the collision is 0.600 m/s. What was the original speed of the bullet?

24. A car of mass m moving at a speed v_1 collides and couples with the back of a truck of mass $2m$ moving initially in the same direction as the car at a lower speed v_2. (a) What is the speed v_f of the two vehicles immediately after the collision? (b) What is the change in kinetic energy of the car–truck system in the collision?

25. A railroad car of mass 2.50×10^4 kg is moving with a speed of 4.00 m/s. It collides and couples with three other coupled railroad cars, each of the same mass as the single car and moving in the same direction with an initial speed of 2.00 m/s. (a) What is the speed of the four cars after the collision? (b) How much mechanical energy is lost in the collision?

26. Four railroad cars, each of mass 2.50×10^4 kg, are coupled together and coasting along horizontal tracks at speed v_i toward the south. A very strong but foolish movie actor, riding on the second car, uncouples the front car and gives it a big push, increasing its speed to 4.00 m/s southward. The remaining three cars continue moving south, now at 2.00 m/s. (a) Find the initial speed of the four cars. (b) By how much did the potential energy within the body of the actor change? (c) State the relationship between the process described here and the process in Problem 25.

27. A neutron in a nuclear reactor makes an elastic, head-[M] on collision with the nucleus of a carbon atom initially at rest. (a) What fraction of the neutron's kinetic energy is transferred to the carbon nucleus? (b) The initial kinetic energy of the neutron is 1.60×10^{-13} J. Find its final kinetic energy and the kinetic energy of the carbon nucleus after the collision. (The mass of the carbon nucleus is nearly 12.0 times the mass of the neutron.)

28. A 7.00-g bullet, when fired from a gun into a 1.00-kg block of wood held in a vise, penetrates the block to a depth of 8.00 cm. This block of wood is next placed on a frictionless horizontal surface, and a second 7.00-g bullet is fired from the gun into the block. To what depth will the bullet penetrate the block in this case?

29. A tennis ball of mass 57.0 g is held [M] just above a basketball of mass 590 g. With their centers vertically aligned, both balls are released from rest at the same time, to fall through a distance of 1.20 m, as shown in Figure P9.29. (a) Find the magnitude of the downward velocity with which the basketball reaches the ground. (b) Assume that an elastic collision with the ground instantaneously reverses the velocity of the basketball while the tennis ball is still moving down. Next, the two balls meet in an elastic collision. To what height does the tennis ball rebound?

Figure P9.29

30. As shown in Figure P9.30, a bullet of mass m and speed v passes completely through a pendulum bob of mass M. The bullet emerges with a speed of $v/2$. The pendulum bob is suspended by a stiff rod (*not* a string) of length ℓ and negligible mass. What is the minimum value of v such that the pendulum bob will barely swing through a complete vertical circle?

Figure P9.30

31. A 12.0-g wad of sticky clay is hurled horizontally at a [AMT] 100-g wooden block initially at rest on a horizontal sur-[M] face. The clay sticks to the block. After impact, the block slides 7.50 m before coming to rest. If the coefficient of friction between the block and the surface is 0.650, what was the speed of the clay immediately before impact?

32. A wad of sticky clay of mass m is hurled horizontally at a wooden block of mass M initially at rest on a horizontal surface. The clay sticks to the block. After impact, the block slides a distance d before coming to rest. If the coefficient of friction between the block and the surface is μ, what was the speed of the clay immediately before impact?

33. Two blocks are free to slide along the frictionless, [AMT] wooden track shown in Figure P9.33. The block of [W] mass $m_1 = 5.00$ kg is released from the position shown, at height $h = 5.00$ m above the flat part of the track. Protruding from its front end is the north pole of a strong magnet, which repels the north pole of an identical magnet embedded in the back end of the block of mass $m_2 = 10.0$ kg, initially at rest. The two blocks never touch. Calculate the maximum height to which m_1 rises after the elastic collision.

Figure P9.33

34. (a) Three carts of masses $m_1 = 4.00$ kg, $m_2 = 10.0$ kg, and $m_3 = 3.00$ kg move on a frictionless, horizontal track with speeds of $v_1 = 5.00$ m/s to the right, $v_2 = 3.00$ m/s to the right, and $v_3 = 4.00$ m/s to the left as shown in Figure P9.34. Velcro couplers make the carts stick together after colliding. Find the final velocity of the train of three carts. (b) **What If?** Does your answer in part (a) require that all the carts collide and stick

Figure P9.34

together at the same moment? What if they collide in a different order?

Section 9.5 Collisions in Two Dimensions

35. A 0.300-kg puck, initially at rest on a horizontal, frictionless surface, is struck by a 0.200-kg puck moving initially along the x axis with a speed of 2.00 m/s. After the collision, the 0.200-kg puck has a speed of 1.00 m/s at an angle of $\theta = 53.0°$ to the positive x axis (see Figure 9.11). (a) Determine the velocity of the 0.300-kg puck after the collision. (b) Find the fraction of kinetic energy transferred away or transformed to other forms of energy in the collision.

36. Two automobiles of equal mass approach an intersection. One vehicle is traveling with speed 13.0 m/s toward the east, and the other is traveling north with speed v_{2i}. Neither driver sees the other. The vehicles collide in the intersection and stick together, leaving parallel skid marks at an angle of 55.0° north of east. The speed limit for both roads is 35 mi/h, and the driver of the northward-moving vehicle claims he was within the speed limit when the collision occurred. Is he telling the truth? Explain your reasoning.

37. An object of mass 3.00 kg, moving with an initial velocity of $5.00\hat{i}$ m/s, collides with and sticks to an object of mass 2.00 kg with an initial velocity of $-3.00\hat{j}$ m/s. Find the final velocity of the composite object.

38. Two shuffleboard disks of equal mass, one orange and the other yellow, are involved in an elastic, glancing collision. The yellow disk is initially at rest and is struck by the orange disk moving with a speed of 5.00 m/s. After the collision, the orange disk moves along a direction that makes an angle of 37.0° with its initial direction of motion. The velocities of the two disks are perpendicular after the collision. Determine the final speed of each disk.

39. Two shuffleboard disks of equal mass, one orange and the other yellow, are involved in an elastic, glancing collision. The yellow disk is initially at rest and is struck by the orange disk moving with a speed v_i. After the collision, the orange disk moves along a direction that makes an angle θ with its initial direction of motion. The velocities of the two disks are perpendicular after the collision. Determine the final speed of each disk.

40. A proton, moving with a velocity of $v_i\hat{i}$, collides elastically with another proton that is initially at rest. Assuming that the two protons have equal speeds after the collision, find (a) the speed of each proton after the collision in terms of v_i and (b) the direction of the velocity vectors after the collision.

41. A billiard ball moving at 5.00 m/s strikes a stationary ball of the same mass. After the collision, the first ball moves at 4.33 m/s at an angle of 30.0° with respect to the original line of motion. Assuming an elastic collision (and ignoring friction and rotational motion), find the struck ball's velocity after the collision.

42. A 90.0-kg fullback running east with a speed of 5.00 m/s is tackled by a 95.0-kg opponent running north with a speed of 3.00 m/s. (a) Explain why the successful tackle constitutes a perfectly inelastic collision. (b) Calculate the velocity of the players immediately after the tackle. (c) Determine the mechanical energy that disappears as a result of the collision. Account for the missing energy.

43. An unstable atomic nucleus of mass 17.0×10^{-27} kg initially at rest disintegrates into three particles. One of the particles, of mass 5.00×10^{-27} kg, moves in the y direction with a speed of 6.00×10^6 m/s. Another particle, of mass 8.40×10^{-27} kg, moves in the x direction with a speed of 4.00×10^6 m/s. Find (a) the velocity of the third particle and (b) the total kinetic energy increase in the process.

44. The mass of the blue puck in Figure P9.44 is 20.0% greater than the mass of the green puck. Before colliding, the pucks approach each other with momenta of equal magnitudes and opposite directions, and the green puck has an initial speed of 10.0 m/s. Find the speeds the pucks have after the collision if half the kinetic energy of the system becomes internal energy during the collision.

Figure P9.44

Section 9.6 The Center of Mass

45. Four objects are situated along the y axis as follows: a 2.00-kg object is at +3.00 m, a 3.00-kg object is at +2.50 m, a 2.50-kg object is at the origin, and a 4.00-kg object is at −0.500 m. Where is the center of mass of these objects?

46. The mass of the Earth is 5.97×10^{24} kg, and the mass of the Moon is 7.35×10^{22} kg. The distance of separation, measured between their centers, is 3.84×10^8 m. Locate the center of mass of the Earth–Moon system as measured from the center of the Earth.

47. Explorers in the jungle find an ancient monument in the shape of a large isosceles triangle as shown in Figure P9.47. The monument is made from tens of thousands of small stone blocks of density 3 800 kg/m³. The monument is 15.7 m high and 64.8 m wide at its base and is everywhere 3.60 m thick from front to back. Before the monument was built many years ago, all the stone blocks lay on the ground. How much work did laborers do on the blocks to put them in position while building the entire monument? *Note:* The gravitational potential energy of an object–Earth system is given by $U_g = Mgy_{CM}$, where M is the total mass of the object and y_{CM} is the elevation of its center of mass above the chosen reference level.

Figure P9.47

48. A uniform piece of sheet metal is shaped as shown in Figure P9.48. Compute the x and y coordinates of the center of mass of the piece.

Figure P9.48

49. A rod of length 30.0 cm has linear density (mass per length) given by

$$\lambda = 50.0 + 20.0x$$

where x is the distance from one end, measured in meters, and λ is in grams/meter. (a) What is the mass of the rod? (b) How far from the $x = 0$ end is its center of mass?

50. A water molecule consists of an oxygen atom with two hydrogen atoms bound to it (Fig. P9.50). The angle between the two bonds is 106°. If the bonds are 0.100 nm long, where is the center of mass of the molecule?

Figure P9.50

Section 9.7 Systems of Many Particles

51. A 2.00-kg particle has a velocity $(2.00\hat{i} - 3.00\hat{j})$ m/s, and a 3.00-kg particle has a velocity $(1.00\hat{i} + 6.00\hat{j})$ m/s. Find (a) the velocity of the center of mass and (b) the total momentum of the system.

52. Consider a system of two particles in the xy plane: $m_1 = 2.00$ kg is at the location $\vec{r}_1 = (1.00\hat{i} + 2.00\hat{j})$ m and has a velocity of $(3.00\hat{i} + 0.500\hat{j})$ m/s; $m_2 = 3.00$ kg is at $\vec{r}_2 = (-4.00\hat{i} - 3.00\hat{j})$ m and has velocity $(3.00\hat{i} - 2.00\hat{j})$ m/s. (a) Plot these particles on a grid or graph paper. Draw their position vectors and show their velocities. (b) Find the position of the center of mass of the system and mark it on the grid. (c) Determine the velocity of the center of mass and also show it on the diagram. (d) What is the total linear momentum of the system?

53. Romeo (77.0 kg) entertains Juliet (55.0 kg) by playing his guitar from the rear of their boat at rest in still water, 2.70 m away from Juliet, who is in the front of the boat. After the serenade, Juliet carefully moves to the rear of the boat (away from shore) to plant a kiss on Romeo's cheek. How far does the 80.0-kg boat move toward the shore it is facing?

54. The vector position of a 3.50-g particle moving in the xy plane varies in time according to $\vec{r}_1 = (3\hat{i} + 3\hat{j})t + 2\hat{j}t^2$, where t is in seconds and $\vec{r}$ is in centimeters. At the same time, the vector position of a 5.50 g particle varies as $\vec{r}_2 = 3\hat{i} - 2\hat{i}t^2 - 6\hat{j}t$. At $t = 2.50$ s, determine (a) the vector position of the center of mass, (b) the linear momentum of the system, (c) the velocity of the center of mass, (d) the acceleration of the center of mass, and (e) the net force exerted on the two-particle system.

55. A ball of mass 0.200 kg with a velocity of $1.50\hat{i}$ m/s meets a ball of mass 0.300 kg with a velocity of $-0.400\hat{i}$ m/s in a head-on, elastic collision. (a) Find their velocities after the collision. (b) Find the velocity of their center of mass before and after the collision.

Section 9.8 Deformable Systems

56. For a technology project, a student has built a vehicle, of total mass 6.00 kg, that moves itself. As shown in Figure P9.56, it runs on four light wheels. A reel is attached to one of the axles, and a cord originally wound on the reel goes up over a pulley attached to the vehicle to support an elevated load. After the vehicle is released from rest, the load descends very slowly, unwinding the cord to turn the axle and make the vehicle move forward (to the left in Fig. P9.56). Friction is negligible in the pulley and axle bearings. The wheels do not slip on the floor. The reel has been constructed with a conical shape so that the load descends at a constant low speed while the vehicle moves horizontally across the floor with constant acceleration, reaching a final velocity of $3.00\hat{i}$ m/s. (a) Does the floor impart impulse to the vehicle? If so, how much? (b) Does the floor do work on the vehicle? If so, how much? (c) Does it make sense to say that the final momentum of the vehicle came from the floor? If not, where did it come from? (d) Does it make sense to say that the final kinetic energy of the vehicle came from the floor? If not, where did it come from? (e) Can we say that one particular force causes the forward acceleration of the vehicle? What does cause it?

Figure P9.56

57. A particle is suspended from a post on top of a cart by a light string of length L as shown in Figure P9.57a. The cart and particle are initially moving to the right at constant speed v_i, with the string vertical. The cart suddenly comes to rest when it runs into and sticks to a bumper as shown in Figure P9.57b. The suspended particle swings through an angle θ. (a) Show that the original speed of the cart can be computed from $v_i = \sqrt{2gL(1 - \cos\theta)}$. (b) If the bumper is still exerting a horizontal force on the cart when the hanging particle is at its maximum angle *forward* from the vertical, at what moment does the bumper *stop* exerting a horizontal force?

Figure P9.57

58. A 60.0-kg person bends his knees and then jumps straight up. After his feet leave the floor, his motion is

unaffected by air resistance and his center of mass rises by a maximum of 15.0 cm. Model the floor as completely solid and motionless. (a) Does the floor impart impulse to the person? (b) Does the floor do work on the person? (c) With what momentum does the person leave the floor? (d) Does it make sense to say that this momentum came from the floor? Explain. (e) With what kinetic energy does the person leave the floor? (f) Does it make sense to say that this energy came from the floor? Explain.

59. Figure P9.59a shows an overhead view of the initial configuration of two pucks of mass m on frictionless ice. The pucks are tied together with a string of length ℓ and negligible mass. At time $t = 0$, a constant force of magnitude F begins to pull to the right on the center point of the string. At time t, the moving pucks strike each other and stick together. At this time, the force has moved through a distance d, and the pucks have attained a speed v (Fig. P9.59b). (a) What is v in terms of F, d, ℓ, and m? (b) How much of the energy transferred into the system by work done by the force has been transformed to internal energy?

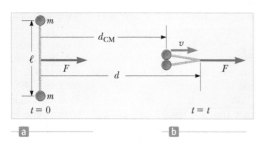

Figure P9.59

Section 9.9 Rocket Propulsion

60. A model rocket engine has an average thrust of 5.26 N. It has an initial mass of 25.5 g, which includes fuel mass of 12.7 g. The duration of its burn is 1.90 s. (a) What is the average exhaust speed of the engine? (b) This engine is placed in a rocket body of mass 53.5 g. What is the final velocity of the rocket if it were to be fired from rest in outer space by an astronaut on a spacewalk? Assume the fuel burns at a constant rate.

61. A garden hose is held as shown in Figure P9.61. The hose is originally full of motionless water. What additional force is necessary to hold the nozzle stationary after the water flow is turned on if the discharge rate is 0.600 kg/s with a speed of 25.0 m/s?

Figure P9.61

62. **Review.** The first stage of a Saturn V space vehicle consumed fuel and oxidizer at the rate of 1.50×10^4 kg/s with an exhaust speed of 2.60×10^3 m/s. (a) Calculate the thrust produced by this engine. (b) Find the acceleration the vehicle had just as it lifted off the launch

pad on the Earth, taking the vehicle's initial mass as 3.00×10^6 kg.

63. A rocket for use in deep space is to be capable of boosting a total load (payload plus rocket frame and engine) of 3.00 metric tons to a speed of 10 000 m/s. (a) It has an engine and fuel designed to produce an exhaust speed of 2 000 m/s. How much fuel plus oxidizer is required? (b) If a different fuel and engine design could give an exhaust speed of 5 000 m/s, what amount of fuel and oxidizer would be required for the same task? (c) Noting that the exhaust speed in part (b) is 2.50 times higher than that in part (a), explain why the required fuel mass is not simply smaller by a factor of 2.50.

64. A rocket has total mass $M_i = 360$ kg, including $M_f = 330$ kg of fuel and oxidizer. In interstellar space, it starts from rest at the position $x = 0$, turns on its engine at time $t = 0$, and puts out exhaust with relative speed $v_e = 1\,500$ m/s at the constant rate $k = 2.50$ kg/s. The fuel will last for a burn time of $T_b = M_f/k = 330$ kg/(2.5 kg/s) = 132 s. (a) Show that during the burn the velocity of the rocket as a function of time is given by

$$v(t) = -v_e \ln\left(1 - \frac{kt}{M_i}\right)$$

(b) Make a graph of the velocity of the rocket as a function of time for times running from 0 to 132 s. (c) Show that the acceleration of the rocket is

$$a(t) = \frac{kv_e}{M_i - kt}$$

(d) Graph the acceleration as a function of time. (e) Show that the position of the rocket is

$$x(t) = v_e\left(\frac{M_i}{k} - t\right)\ln\left(1 - \frac{kt}{M_i}\right) + v_e t$$

(f) Graph the position during the burn as a function of time.

Additional Problems

65. A ball of mass m is thrown straight up into the air with an initial speed v_i. Find the momentum of the ball (a) at its maximum height and (b) halfway to its maximum height.

66. An amateur skater of mass M is trapped in the middle of an ice rink and is unable to return to the side where there is no ice. Every motion she makes causes her to slip on the ice and remain in the same spot. She decides to try to return to safety by throwing her gloves of mass m in the direction opposite the safe side. (a) She throws the gloves as hard as she can, and they leave her hand with a horizontal velocity $\vec{v}_{gloves}$. Explain whether or not she moves. If she does move, calculate her velocity $\vec{v}_{girl}$ relative to the Earth after she throws the gloves. (b) Discuss her motion from the point of view of the forces acting on her.

67. A 3.00-kg steel ball strikes a wall with a speed of 10.0 m/s at an angle of $\theta = 60.0°$ with the surface. It bounces off with the same speed and angle (Fig. P9.67). If the

ball is in contact with the wall for 0.200 s, what is the average force exerted by the wall on the ball?

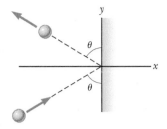

Figure P9.67

68. (a) Figure P9.68 shows three points in the operation of the ballistic pendulum discussed in Example 9.6 (and shown in Fig. 9.9b). The projectile approaches the pendulum in Figure P9.68a. Figure P9.68b shows the situation just after the projectile is captured in the pendulum. In Figure P9.68c, the pendulum arm has swung upward and come to rest at a height h above its initial position. Prove that the ratio of the kinetic energy of the projectile–pendulum system immediately after the collision to the kinetic energy immediately before is $m_1/(m_1 + m_2)$. (b) What is the ratio of the momentum of the system immediately after the collision to the momentum immediately before? (c) A student believes that such a large decrease in mechanical energy must be accompanied by at least a small decrease in momentum. How would you convince this student of the truth?

Figure P9.68 Problems 68 and 86. (a) A metal ball moves toward the pendulum. (b) The ball is captured by the pendulum. (c) The ball–pendulum combination swings up through a height h before coming to rest.

69. **Review.** A 60.0-kg person running at an initial speed of 4.00 m/s jumps onto a 120-kg cart initially at rest (Fig. P9.69). The person slides on the cart's top surface and finally comes to rest relative to the cart. The coefficient of kinetic friction between the person and the cart is 0.400. Friction between the cart and ground can be ignored. (a) Find the final velocity of the person and cart relative to the ground. (b) Find the friction force acting on the person while he is sliding across the top surface of the cart. (c) How long does the friction force act on the person? (d) Find the change in momentum of the person and the change in momentum of the cart. (e) Determine the displacement of the person relative to the ground while he is sliding on the cart. (f) Determine the displacement of the cart relative to the ground while the person is sliding. (g) Find the change in kinetic energy of the person. (h) Find the change in kinetic energy of the cart. (i) Explain why the answers to (g) and (h) differ. (What kind of collision is this one, and what accounts for the loss of mechanical energy?)

Figure P9.69

70. A cannon is rigidly attached to a carriage, which can move along horizontal rails but is connected to a post by a large spring, initially unstretched and with force constant $k = 2.00 \times 10^4$ N/m, as shown in Figure P9.70. The cannon fires a 200-kg projectile at a velocity of 125 m/s directed 45.0° above the horizontal. (a) Assuming that the mass of the cannon and its carriage is 5 000 kg, find the recoil speed of the cannon. (b) Determine the maximum extension of the spring. (c) Find the maximum force the spring exerts on the carriage. (d) Consider the system consisting of the cannon, carriage, and projectile. Is the momentum of this system conserved during the firing? Why or why not?

Figure P9.70

71. A 1.25-kg wooden block rests on a table over a large hole as in Figure P9.71. A 5.00-g bullet with an initial velocity v_i is fired upward into the bottom of the block and remains in the block after the collision. The block and bullet rise to a maximum height of 22.0 cm. (a) Describe how you would find the initial velocity of the bullet using ideas you have learned in this chapter. (b) Calculate the initial velocity of the bullet from the information provided.

Figure P9.71
Problems 71 and 72.

72. A wooden block of mass M rests on a table over a large hole as in Figure 9.71. A bullet of mass m with an initial velocity of v_i is fired upward into the bottom of the block and remains in the block after the collision. The block and bullet rise to a maximum height of h. (a) Describe how you would find the initial velocity of the bullet using ideas you have learned in this chapter. (b) Find an expression for the initial velocity of the bullet.

73. Two particles with masses m and $3m$ are moving toward each other along the x axis with the same initial speeds v_i. The particle with mass m is traveling to the left, and particle with mass $3m$ is traveling to the right. They

undergo a head-on elastic collision, and each rebounds along the same line as it approached. Find the final speeds of the particles.

74. Pursued by ferocious wolves, you are in a sleigh with no horses, gliding without friction across an ice-covered lake. You take an action described by the equations

$$(270 \text{ kg})(7.50 \text{ m/s})\hat{\mathbf{i}} = (15.0 \text{ kg})(-v_{1f}\hat{\mathbf{i}}) + (255 \text{ kg})(v_{2f}\hat{\mathbf{i}})$$
$$v_{1f} + v_{2f} = 8.00 \text{ m/s}$$

(a) Complete the statement of the problem, giving the data and identifying the unknowns. (b) Find the values of v_{1f} and v_{2f}. (c) Find the amount of energy that has been transformed from potential energy stored in your body to kinetic energy of the system.

75. Two gliders are set in motion on a horizontal air track. A spring of force constant k is attached to the back end of the second glider. As shown in Figure P9.75, the first glider, of mass m_1, moves to the right with speed v_1, and the second glider, of mass m_2, moves more slowly to the right with speed v_2. When m_1 collides with the spring attached to m_2, the spring compresses by a distance x_{max}, and the gliders then move apart again. In terms of v_1, v_2, m_1, m_2, and k, find (a) the speed v at maximum compression, (b) the maximum compression x_{max}, and (c) the velocity of each glider after m_1 has lost contact with the spring.

Figure P9.75

76. *Why is the following situation impossible?* An astronaut, together with the equipment he carries, has a mass of 150 kg. He is taking a space walk outside his spacecraft, which is drifting through space with a constant velocity. The astronaut accidentally pushes against the spacecraft and begins moving away at 20.0 m/s, relative to the spacecraft, without a tether. To return, he takes equipment off his space suit and throws it in the direction away from the spacecraft. Because of his bulky space suit, he can throw equipment at a maximum speed of 5.00 m/s relative to himself. After throwing enough equipment, he starts moving back to the spacecraft and is able to grab onto it and climb inside.

77. Two blocks of masses $m_1 = 2.00$ kg and $m_2 = 4.00$ kg are released from rest at a height of $h = 5.00$ m on a frictionless track as shown in Figure P9.77. When they

Figure P9.77

meet on the level portion of the track, they undergo a head-on, elastic collision. Determine the maximum heights to which m_1 and m_2 rise on the curved portion of the track after the collision.

78. **Review.** A metal cannonball of mass m rests next to a tree at the very edge of a cliff 36.0 m above the surface of the ocean. In an effort to knock the cannonball off the cliff, some children tie one end of a rope around a stone of mass 80.0 kg and the other end to a tree limb just above the cannonball. They tighten the rope so that the stone just clears the ground and hangs next to the cannonball. The children manage to swing the stone back until it is held at rest 1.80 m above the ground. The children release the stone, which then swings down and makes a head-on, elastic collision with the cannonball, projecting it horizontally off the cliff. The cannonball lands in the ocean a horizontal distance R away from its initial position. (a) Find the horizontal component R of the cannonball's displacement as it depends on m. (b) What is the maximum possible value for R, and (c) to what value of m does it correspond? (d) For the stone–cannonball–Earth system, is mechanical energy conserved throughout the process? Is this principle sufficient to solve the entire problem? Explain. (e) **What if?** Show that R does not depend on the value of the gravitational acceleration. Is this result remarkable? State how one might make sense of it.

79. A 0.400-kg blue bead slides on a frictionless, curved wire, starting from rest at point Ⓐ in Figure P9.79, where $h =$ 1.50 m. At point Ⓑ, the blue bead collides elastically with a 0.600-kg green bead at rest. Find the maximum height the green bead rises as it moves up the wire.

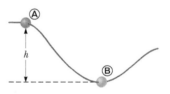

Figure P9.79

80. A small block of mass $m_1 = 0.500$ kg is released from rest at the top of a frictionless, curve-shaped wedge of mass $m_2 = 3.00$ kg, which sits on a frictionless, horizontal surface as shown in Figure P9.80a. When the block leaves the wedge, its velocity is measured to be 4.00 m/s to the right as shown in Figure P9.80b. (a) What is the velocity of the wedge after the block reaches the horizontal surface? (b) What is the height h of the wedge?

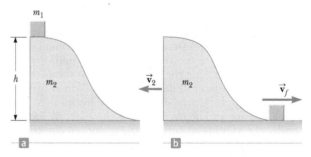

Figure P9.80

81. **Review.** A bullet of mass $m = 8.00$ g is fired into a block
[M] of mass $M = 250$ g that is initially at rest at the edge
of a table of height $h = 1.00$ m (Fig. P9.81). The bullet
remains in the block, and after the impact the block
lands $d = 2.00$ m from the bottom of the table. Determine the initial speed of the bullet.

Figure P9.81 Problems 81 and 82.

82. **Review.** A bullet of mass m is fired into a block of mass
M initially at rest at the edge of a frictionless table of
height h (Fig. P9.81). The bullet remains in the block,
and after impact the block lands a distance d from the
bottom of the table. Determine the initial speed of the
bullet.

83. A 0.500-kg sphere moving with a velocity given by
$(2.00\hat{i} - 3.00\hat{j} + 1.00\hat{k})$ m/s strikes another sphere
of mass 1.50 kg moving with an initial velocity of
$(-1.00\hat{i} + 2.00\hat{j} - 3.00\hat{k})$ m/s. (a) The velocity of
the 0.500-kg sphere after the collision is $(-1.00\hat{i} +$
$3.00\hat{j} - 8.00\hat{k})$ m/s. Find the final velocity of the 1.50-kg
sphere and identify the kind of collision (elastic,
inelastic, or perfectly inelastic). (b) Now assume the
velocity of the 0.500-kg sphere after the collision is
$(-0.250\hat{i} + 0.750\hat{j} - 2.00\hat{k})$ m/s. Find the final velocity
of the 1.50-kg sphere and identify the kind of col-
lision. (c) **What If?** Take the velocity of the 0.500-kg
sphere after the collision as $(-1.00\hat{i} + 3.00\hat{j} + a\hat{k})$ m/s.
Find the value of a and the velocity of the 1.50-kg
sphere after an elastic collision.

84. A 75.0-kg firefighter slides down a pole while a constant
friction force of 300 N retards her motion. A horizontal
20.0-kg platform is supported by a spring at the bottom
of the pole to cushion the fall. The firefighter starts
from rest 4.00 m above the platform, and the spring
constant is 4 000 N/m. Find (a) the firefighter's speed
just before she collides with the platform and (b) the
maximum distance the spring is compressed. Assume
the friction force acts during the entire motion.

85. George of the Jungle, with mass m, swings on a light
vine hanging from a stationary tree branch. A second
vine of equal length hangs from the same point, and a
gorilla of larger mass M swings in the opposite direction on it. Both vines are horizontal when the primates
start from rest at the same moment. George and the
gorilla meet at the lowest point of their swings. Each is
afraid that one vine will break, so they grab each other
and hang on. They swing upward together, reaching a
point where the vines make an angle of 35.0° with the
vertical. Find the value of the ratio m/M.

86. **Review.** A student performs a ballistic pendulum
experiment using an apparatus similar to that discussed in Example 9.6 and shown in Figure P9.68. She
obtains the following average data: $h = 8.68$ cm, projec-

tile mass $m_1 = 68.8$ g, and pendulum mass $m_2 = 263$ g.
(a) Determine the initial speed v_{1A} of the projectile.
(b) The second part of her experiment is to obtain v_{1A}
by firing the same projectile horizontally (with the pendulum removed from the path) and measuring its final
horizontal position x and distance of fall y (Fig. P9.86).
What numerical value does she obtain for v_{1A} based on
her measured values of $x = 257$ cm and $y = 85.3$ cm?
(c) What factors might account for the difference in
this value compared with that obtained in part (a)?

Figure P9.86

87. **Review.** A light spring of force constant 3.85 N/m is
compressed by 8.00 cm and held between a 0.250-kg
block on the left and a 0.500-kg block on the right.
Both blocks are at rest on a horizontal surface. The
blocks are released simultaneously so that the spring
tends to push them apart. Find the maximum velocity
each block attains if the coefficient of kinetic friction
between each block and the surface is (a) 0, (b) 0.100,
and (c) 0.462. Assume the coefficient of static friction
is greater than the coefficient of kinetic friction in
every case.

88. Consider as a system the Sun with the Earth in a circular orbit around it. Find the magnitude of the change
in the velocity of the Sun relative to the center of mass
of the system over a six-month period. Ignore the influence of other celestial objects. You may obtain the necessary astronomical data from the endpapers of the
book.

89. A 5.00-g bullet mov-
[AMT] ing with an initial
[M] speed of $v_i = 400$ m/s
is fired into and passes
through a 1.00-kg
block as shown in Fig-
ure P9.89. The block,
initially at rest on a
frictionless, horizontal
surface, is connected
to a spring with force
constant 900 N/m.

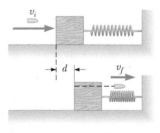

Figure P9.89

The block moves $d = 5.00$ cm to the right after impact
before being brought to rest by the spring. Find (a) the
speed at which the bullet emerges from the block and
(b) the amount of initial kinetic energy of the bullet
that is converted into internal energy in the bullet–
block system during the collision.

90. **Review.** There are (one can say) three coequal theo-
[GP] ries of motion for a single particle: Newton's second
law, stating that the total force on the particle causes its

acceleration; the work–kinetic energy theorem, stating that the total work on the particle causes its change in kinetic energy; and the impulse–momentum theorem, stating that the total impulse on the particle causes its change in momentum. In this problem, you compare predictions of the three theories in one particular case. A 3.00-kg object has velocity $7.00\hat{\mathbf{j}}$ m/s. Then, a constant net force $12.0\hat{\mathbf{i}}$ N acts on the object for 5.00 s. (a) Calculate the object's final velocity, using the impulse–momentum theorem. (b) Calculate its acceleration from $\vec{\mathbf{a}} = (\vec{\mathbf{v}}_f - \vec{\mathbf{v}}_i)/\Delta t$. (c) Calculate its acceleration from $\vec{\mathbf{a}} = \sum \vec{\mathbf{F}}/m$. (d) Find the object's vector displacement from $\Delta\vec{\mathbf{r}} = \vec{\mathbf{v}}_i t + \frac{1}{2}\vec{\mathbf{a}}t^2$. (e) Find the work done on the object from $W = \vec{\mathbf{F}} \cdot \Delta\vec{\mathbf{r}}$. (f) Find the final kinetic energy from $\frac{1}{2}mv_f^2 = \frac{1}{2}m\vec{\mathbf{v}}_f \cdot \vec{\mathbf{v}}_f$. (g) Find the final kinetic energy from $\frac{1}{2}mv_i^2 + W$. (h) State the result of comparing the answers to parts (b) and (c), and the answers to parts (f) and (g).

91. A 2.00-g particle moving at 8.00 m/s makes a perfectly
M elastic head-on collision with a resting 1.00-g object. (a) Find the speed of each particle after the collision. (b) Find the speed of each particle after the collision if the stationary particle has a mass of 10.0 g. (c) Find the final kinetic energy of the incident 2.00-g particle in the situations described in parts (a) and (b). In which case does the incident particle lose more kinetic energy?

Challenge Problems

92. In the 1968 Olympic games, University of Oregon jumper Dick Fosbury introduced a new technique of high jumping called the "Fosbury flop." It contributed to raising the world record by about 30 cm and is currently used by nearly every world-class jumper. In this technique, the jumper goes over the bar face-up while arching her back as much as possible as shown in Figure P9.92a. This action places her center of mass outside her body, below her back. As her body goes over the bar, her center of mass passes below the bar. Because a given energy input implies a certain elevation for her center of mass, the action of arching her back means that her body is higher than if her back were straight. As a model, consider the jumper as a thin uniform rod of length L. When the rod is straight, its center of mass is at its center. Now bend the rod in a circular arc so that it subtends an angle of 90.0° at the center of the arc as shown in Figure P9.92b. In this configuration, how far outside the rod is the center of mass?

Figure P9.92

93. Two particles with masses m and $3m$ are moving toward each other along the x axis with the same initial speeds

v_i. Particle m is traveling to the left, and particle $3m$ is traveling to the right. They undergo an elastic glancing collision such that particle m is moving in the negative y direction after the collision at a right angle from its initial direction. (a) Find the final speeds of the two particles in terms of v_i. (b) What is the angle θ at which the particle $3m$ is scattered?

94. Sand from a stationary hopper falls onto a moving conveyor belt at the rate of 5.00 kg/s as shown in Figure P9.94. The conveyor belt is supported by frictionless rollers and moves at a constant speed of $v = 0.750$ m/s under the action of a constant horizontal external force $\vec{\mathbf{F}}_{ext}$ supplied by the motor that drives the belt. Find (a) the sand's rate of change of momentum in the horizontal direction, (b) the force of friction exerted by the belt on the sand, (c) the external force $\vec{\mathbf{F}}_{ext}$, (d) the work done by $\vec{\mathbf{F}}_{ext}$ in 1 s, and (e) the kinetic energy acquired by the falling sand each second due to the change in its horizontal motion. (f) Why are the answers to parts (d) and (e) different?

Figure P9.94

95. On a horizontal air track, a glider of mass m carries a Γ-shaped post. The post supports a small dense sphere, also of mass m, hanging just above the top of the glider on a cord of length L. The glider and sphere are initially at rest with the cord vertical. (Figure P9.57 shows a cart and a sphere similarly connected.) A constant horizontal force of magnitude F is applied to the glider, moving it through displacement x_1; then the force is removed. During the time interval when the force is applied, the sphere moves through a displacement with horizontal component x_2. (a) Find the horizontal component of the velocity of the center of mass of the glider–sphere system when the force is removed. (b) After the force is removed, the glider continues to move on the track and the sphere swings back and forth, both without friction. Find an expression for the largest angle the cord makes with the vertical.

96. **Review.** A chain of length L and total mass M is released from rest with its lower end just touching the top of a table as shown in Figure P9.96a. Find the force exerted by the table on the chain after the chain has fallen through a distance x as shown in Figure P9.96b. (Assume each link comes to rest the instant it reaches the table.)

Figure P9.96

Rotation of a Rigid Object About a Fixed Axis

When an extended object such as a wheel rotates about its axis, the motion cannot be analyzed by modeling the object as a particle because at any given time different parts of the object have different linear velocities and linear accelerations. We can, however, analyze the motion of an extended object by modeling it as a system of many particles, each of which has its own linear velocity and linear acceleration as discussed in Section 9.7.

In dealing with a rotating object, analysis is greatly simplified by assuming the object is rigid. A **rigid object** is one that is nondeformable; that is, the relative locations of all particles of which the object is composed remain constant. All real objects are deformable to some extent; our rigid-object model, however, is useful in many situations in which deformation is negligible. We have developed analysis models based on particles and systems. In this chapter, we introduce another class of analysis models based on the rigid-object model.

The Malaysian pastime of *gasing* involves the spinning of tops that can have masses up to 5 kg. Professional spinners can spin their tops so that they might rotate for more than an hour before stopping. We will study the rotational motion of objects such as these tops in this chapter. *(Courtesy Tourism Malaysia)*

10.1 Angular Position, Velocity, and Acceleration

We will develop our understanding of rotational motion in a manner parallel to that used for translational motion in previous chapters. We began in Chapter 2 by

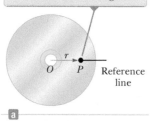

To define angular position for the disc, a fixed reference line is chosen. A particle at P is located at a distance r from the rotation axis through O.

a

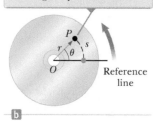

As the disc rotates, a particle at P moves through an arc length s on a circular path of radius r. The angular position of P is θ.

b

Figure 10.1 A compact disc rotating about a fixed axis through O perpendicular to the plane of the figure.

Pitfall Prevention 10.1
Remember the Radian In rotational equations, you *must* use angles expressed in radians. Don't fall into the trap of using angles measured in degrees in rotational equations.

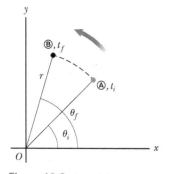

Figure 10.2 A particle on a rotating rigid object moves from Ⓐ to Ⓑ along the arc of a circle. In the time interval $\Delta t = t_f - t_i$, the radial line of length r moves through an angular displacement $\Delta\theta = \theta_f - \theta_i$.

Average angular speed ▶

defining kinematic variables: position, velocity, and acceleration. We do the same here for rotational motion.

Figure 10.1 illustrates an overhead view of a rotating compact disc, or CD. The disc rotates about a fixed axis perpendicular to the plane of the figure and passing through the center of the disc at O. A small element of the disc modeled as a particle at P is at a fixed distance r from the origin and rotates about it in a circle of radius r. (In fact, *every* element of the disc undergoes circular motion about O.) It is convenient to represent the position of P with its polar coordinates (r, θ), where r is the distance from the origin to P and θ is measured *counterclockwise* from some reference line fixed in space as shown in Figure 10.1a. In this representation, the angle θ changes in time while r remains constant. As the particle moves along the circle from the reference line, which is at angle $\theta = 0$, it moves through an arc of length s as in Figure 10.1b. The arc length s is related to the angle θ through the relationship

$$s = r\theta \tag{10.1a}$$

$$\theta = \frac{s}{r} \tag{10.1b}$$

Because θ is the ratio of an arc length and the radius of the circle, it is a pure number. Usually, however, we give θ the artificial unit **radian** (rad), where one radian is the angle subtended by an arc length equal to the radius of the arc. Because the circumference of a circle is $2\pi r$, it follows from Equation 10.1b that $360°$ corresponds to an angle of $(2\pi r/r)$ rad $= 2\pi$ rad. Hence, 1 rad $= 360°/2\pi \approx 57.3°$. To convert an angle in degrees to an angle in radians, we use that π rad $= 180°$, so

$$\theta(\text{rad}) = \frac{\pi}{180°}\theta(\text{deg})$$

For example, $60°$ equals $\pi/3$ rad and $45°$ equals $\pi/4$ rad.

Because the disc in Figure 10.1 is a rigid object, as the particle moves through an angle θ from the reference line, every other particle on the object rotates through the same angle θ. Therefore, we can associate the angle θ with the entire rigid object as well as with an individual particle, which allows us to define the *angular position* of a rigid object in its rotational motion. We choose a reference line on the object, such as a line connecting O and a chosen particle on the object. The **angular position** of the rigid object is the angle θ between this reference line on the object and the fixed reference line in space, which is often chosen as the x axis. Such identification is similar to the way we define the position of an object in translational motion as the distance x between the object and the reference position, which is the origin, $x = 0$. Therefore, the angle θ plays the same role in rotational motion that the position x does in translational motion.

As the particle in question on our rigid object travels from position Ⓐ to position Ⓑ in a time interval Δt as in Figure 10.2, the reference line fixed to the object sweeps out an angle $\Delta\theta = \theta_f - \theta_i$. This quantity $\Delta\theta$ is defined as the **angular displacement** of the rigid object:

$$\Delta\theta \equiv \theta_f - \theta_i$$

The rate at which this angular displacement occurs can vary. If the rigid object spins rapidly, this displacement can occur in a short time interval. If it rotates slowly, this displacement occurs in a longer time interval. These different rotation rates can be quantified by defining the **average angular speed** ω_{avg} (Greek letter omega) as the ratio of the angular displacement of a rigid object to the time interval Δt during which the displacement occurs:

$$\omega_{\text{avg}} \equiv \frac{\theta_f - \theta_i}{t_f - t_i} = \frac{\Delta\theta}{\Delta t} \tag{10.2}$$

In analogy to translational speed, the **instantaneous angular speed** ω is defined as the limit of the average angular speed as Δt approaches zero:

$$\omega \equiv \lim_{\Delta t \to 0} \frac{\Delta \theta}{\Delta t} = \frac{d\theta}{dt} \tag{10.3}$$

◀ Instantaneous angular speed

Angular speed has units of radians per second (rad/s), which can be written as s^{-1} because radians are not dimensional. We take ω to be positive when θ is increasing (counterclockwise motion in Fig. 10.2) and negative when θ is decreasing (clockwise motion in Fig. 10.2).

Quick Quiz 10.1 A rigid object rotates in a counterclockwise sense around a fixed axis. Each of the following pairs of quantities represents an initial angular position and a final angular position of the rigid object. **(i)** Which of the sets can *only* occur if the rigid object rotates through more than 180°? (a) 3 rad, 6 rad (b) −1 rad, 1 rad (c) 1 rad, 5 rad **(ii)** Suppose the change in angular position for each of these pairs of values occurs in 1 s. Which choice represents the lowest average angular speed?

If the instantaneous angular speed of an object changes from ω_i to ω_f in the time interval Δt, the object has an angular acceleration. The **average angular acceleration** α_{avg} (Greek letter alpha) of a rotating rigid object is defined as the ratio of the change in the angular speed to the time interval Δt during which the change in the angular speed occurs:

$$\alpha_{\text{avg}} \equiv \frac{\omega_f - \omega_i}{t_f - t_i} = \frac{\Delta \omega}{\Delta t} \tag{10.4}$$

◀ Average angular acceleration

In analogy to translational acceleration, the **instantaneous angular acceleration** is defined as the limit of the average angular acceleration as Δt approaches zero:

$$\alpha \equiv \lim_{\Delta t \to 0} \frac{\Delta \omega}{\Delta t} = \frac{d\omega}{dt} \tag{10.5}$$

◀ Instantaneous angular acceleration

Angular acceleration has units of radians per second squared (rad/s²), or simply s^{-2}. Notice that α is positive when a rigid object rotating counterclockwise is speeding up or when a rigid object rotating clockwise is slowing down during some time interval.

When a rigid object is rotating about a *fixed* axis, every particle on the object rotates through the same angle in a given time interval and has the same angular speed and the same angular acceleration. Therefore, like the angular position θ, the quantities ω and α characterize the rotational motion of the entire rigid object as well as individual particles in the object.

Angular position (θ), angular speed (ω), and angular acceleration (α) are analogous to translational position (x), translational speed (v), and translational acceleration (a). The variables θ, ω, and α differ dimensionally from the variables x, v, and a only by a factor having the unit of length. (See Section 10.3.)

We have not specified any direction for angular speed and angular acceleration. Strictly speaking, ω and α are the magnitudes of the angular velocity and the angular acceleration vectors[1] $\vec{\omega}$ and $\vec{\alpha}$, respectively, and they should always be positive. Because we are considering rotation about a fixed axis, however, we can use nonvector notation and indicate the vectors' directions by assigning a positive or negative sign to ω and α as discussed earlier with regard to Equations 10.3 and 10.5. For rotation about a fixed axis, the only direction that uniquely specifies the rotational motion is the direction along the axis of rotation. Therefore, the directions of $\vec{\omega}$ and $\vec{\alpha}$ are along this axis. If a particle rotates in the xy plane as in Figure 10.2, the

Pitfall Prevention 10.2
Specify Your Axis In solving rotation problems, you must specify an axis of rotation. This new feature does not exist in our study of translational motion. The choice is arbitrary, but once you make it, you must maintain that choice consistently throughout the problem. In some problems, the physical situation suggests a natural axis, such as one along the axle of an automobile wheel. In other problems, there may not be an obvious choice, and you must exercise judgment.

---

[1]Although we do not verify it here, the instantaneous angular velocity and instantaneous angular acceleration are vector quantities, but the corresponding average values are not because angular displacements do not add as vector quantities for finite rotations.

Figure 10.3 The right-hand rule for determining the direction of the angular velocity vector.

direction of $\vec{\omega}$ for the particle is out of the plane of the diagram when the rotation is counterclockwise and into the plane of the diagram when the rotation is clockwise. To illustrate this convention, it is convenient to use the *right-hand rule* demonstrated in Figure 10.3. When the four fingers of the right hand are wrapped in the direction of rotation, the extended right thumb points in the direction of $\vec{\omega}$. The direction of $\vec{\alpha}$ follows from its definition $\vec{\alpha} \equiv d\vec{\omega}/dt$. It is in the same direction as $\vec{\omega}$ if the angular speed is increasing in time, and it is antiparallel to $\vec{\omega}$ if the angular speed is decreasing in time.

10.2 Analysis Model: Rigid Object Under Constant Angular Acceleration

In our study of translational motion, after introducing the kinematic variables, we considered the special case of a particle under constant acceleration. We follow the same procedure here for a rigid object under constant angular acceleration.

Imagine a rigid object such as the CD in Figure 10.1 rotates about a fixed axis and has a constant angular acceleration. In parallel with our analysis model of the particle under constant acceleration, we generate a new analysis model for rotational motion called the **rigid object under constant angular acceleration.** We develop kinematic relationships for this model in this section. Writing Equation 10.5 in the form $d\omega = \alpha\, dt$ and integrating from $t_i = 0$ to $t_f = t$ gives

◀ **Rotational kinematic equations**

$$\omega_f = \omega_i + \alpha t \quad \text{(for constant } \alpha) \tag{10.6}$$

where ω_i is the angular speed of the rigid object at time $t = 0$. Equation 10.6 allows us to find the angular speed ω_f of the object at any later time t. Substituting Equation 10.6 into Equation 10.3 and integrating once more, we obtain

$$\theta_f = \theta_i + \omega_i t + \tfrac{1}{2}\alpha t^2 \quad \text{(for constant } \alpha) \tag{10.7}$$

where θ_i is the angular position of the rigid object at time $t = 0$. Equation 10.7 allows us to find the angular position θ_f of the object at any later time t. Eliminating t from Equations 10.6 and 10.7 gives

$$\omega_f^2 = \omega_i^2 + 2\alpha(\theta_f - \theta_i) \quad \text{(for constant } \alpha) \tag{10.8}$$

This equation allows us to find the angular speed ω_f of the rigid object for any value of its angular position θ_f. If we eliminate α between Equations 10.6 and 10.7, we obtain

$$\theta_f = \theta_i + \tfrac{1}{2}(\omega_i + \omega_f)t \quad \text{(for constant } \alpha) \tag{10.9}$$

Notice that these kinematic expressions for the rigid object under constant angular acceleration are of the same mathematical form as those for a particle under constant acceleration (Chapter 2). They can be generated from the equations for translational motion by making the substitutions $x \rightarrow \theta$, $v \rightarrow \omega$, and $a \rightarrow \alpha$. Table 10.1 compares the kinematic equations for the rigid object under constant angular acceleration and particle under constant acceleration models.

> **Pitfall Prevention 10.3**
> **Just Like Translation?** Equations 10.6 to 10.9 and Table 10.1 might suggest that rotational kinematics is just like translational kinematics. That is almost true, with two key differences. (1) In rotational kinematics, you must specify a rotation axis (per Pitfall Prevention 10.2). (2) In rotational motion, the object keeps returning to its original orientation; therefore, you may be asked for the number of revolutions made by a rigid object. This concept has no analog in translational motion.

Quick **Quiz 10.2** Consider again the pairs of angular positions for the rigid object in Quick Quiz 10.1. If the object starts from rest at the initial angular position, moves counterclockwise with constant angular acceleration, and arrives at the final angular position with the same angular speed in all three cases, for which choice is the angular acceleration the highest?

Table 10.1   **Kinematic Equations for Rotational and Translational Motion**

Rigid Object Under Constant Angular Acceleration		Particle Under Constant Acceleration	
$\omega_f = \omega_i + \alpha t$	(10.6)	$v_f = v_i + at$	(2.13)
$\theta_f = \theta_i + \omega_i t + \tfrac{1}{2}\alpha t^2$	(10.7)	$x_f = x_i + v_i t + \tfrac{1}{2}at^2$	(2.16)
$\omega_f^2 = \omega_i^2 + 2\alpha(\theta_f - \theta_i)$	(10.8)	$v_f^2 = v_i^2 + 2a(x_f - x_i)$	(2.17)
$\theta_f = \theta_i + \tfrac{1}{2}(\omega_i + \omega_f)t$	(10.9)	$x_f = x_i + \tfrac{1}{2}(v_i + v_f)t$	(2.15)

Analysis Model   Rigid Object Under Constant Angular Acceleration

Imagine an object that undergoes a spinning motion such that its angular acceleration is constant. The equations describing its angular position and angular speed are analogous to those for the particle under constant acceleration model:

$\alpha = $ constant

$$\omega_f = \omega_i + \alpha t \qquad \textbf{(10.6)}$$

$$\theta_f = \theta_i + \omega_i t + \tfrac{1}{2}\alpha t^2 \qquad \textbf{(10.7)}$$

$$\omega_f^2 = \omega_i^2 + 2\alpha(\theta_f - \theta_i) \qquad \textbf{(10.8)}$$

$$\theta_f = \theta_i + \tfrac{1}{2}(\omega_i + \omega_f)t \qquad \textbf{(10.9)}$$

Examples:

- during its spin cycle, the tub of a clothes washer begins from rest and accelerates up to its final spin speed
- a workshop grinding wheel is turned off and comes to rest under the action of a constant friction force in the bearings of the wheel
- a gyroscope is powered up and approaches its operating speed (Chapter 11)
- the crankshaft of a diesel engine changes to a higher angular speed (Chapter 22)

Example 10.1   Rotating Wheel   AM

A wheel rotates with a constant angular acceleration of 3.50 rad/s².

(A) If the angular speed of the wheel is 2.00 rad/s at $t_i = 0$, through what angular displacement does the wheel rotate in 2.00 s?

SOLUTION

Conceptualize Look again at Figure 10.1. Imagine that the compact disc rotates with its angular speed increasing at a constant rate. You start your stopwatch when the disc is rotating at 2.00 rad/s. This mental image is a model for the motion of the wheel in this example.

Categorize The phrase "with a constant angular acceleration" tells us to apply the *rigid object under constant angular acceleration* model to the wheel.

Analyze From the rigid object under constant angular acceleration model, choose Equation 10.7 and rearrange it so that it expresses the angular displacement of the wheel:

$$\Delta\theta = \theta_f - \theta_i = \omega_i t + \tfrac{1}{2}\alpha t^2$$

Substitute the known values to find the angular displacement at $t = 2.00$ s:

$$\Delta\theta = (2.00 \text{ rad/s})(2.00 \text{ s}) + \tfrac{1}{2}(3.50 \text{ rad/s}^2)(2.00 \text{ s})^2$$
$$= \boxed{11.0 \text{ rad}} = (11.0 \text{ rad})(180°/\pi \text{ rad}) = \boxed{630°}$$

(B) Through how many revolutions has the wheel turned during this time interval?

SOLUTION

Multiply the angular displacement found in part (A) by a conversion factor to find the number of revolutions:

$$\Delta\theta = 630°\left(\frac{1 \text{ rev}}{360°}\right) = \boxed{1.75 \text{ rev}}$$

(C) What is the angular speed of the wheel at $t = 2.00$ s?

SOLUTION

Use Equation 10.6 from the rigid object under constant angular acceleration model to find the angular speed at $t = 2.00$ s:

$$\omega_f = \omega_i + \alpha t = 2.00 \text{ rad/s} + (3.50 \text{ rad/s}^2)(2.00 \text{ s})$$
$$= \boxed{9.00 \text{ rad/s}}$$

Finalize We could also obtain this result using Equation 10.8 and the results of part (A). (Try it!)

WHAT IF? Suppose a particle moves along a straight line with a constant acceleration of 3.50 m/s². If the velocity of the particle is 2.00 m/s at $t_i = 0$, through what displacement does the particle move in 2.00 s? What is the velocity of the particle at $t = 2.00$ s? *continued*

▶ **10.1** continued

Answer Notice that these questions are translational analogs to parts (A) and (C) of the original problem. The mathematical solution follows exactly the same form. For the displacement, from the particle under constant acceleration model,

$$\Delta x = x_f - x_i = v_i t + \tfrac{1}{2}at^2$$

$$= (2.00 \text{ m/s})(2.00 \text{ s}) + \tfrac{1}{2}(3.50 \text{ m/s}^2)(2.00 \text{ s})^2 = 11.0 \text{ m}$$

and for the velocity,

$$v_f = v_i + at = 2.00 \text{ m/s} + (3.50 \text{ m/s}^2)(2.00 \text{ s}) = 9.00 \text{ m/s}$$

There is no translational analog to part (B) because translational motion under constant acceleration is not repetitive.

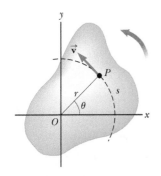

Figure 10.4 As a rigid object rotates about the fixed axis (the *z* axis) through *O*, the point *P* has a tangential velocity $\vec{v}$ that is always tangent to the circular path of radius *r*.

10.3 Angular and Translational Quantities

In this section, we derive some useful relationships between the angular speed and acceleration of a rotating rigid object and the translational speed and acceleration of a point in the object. To do so, we must keep in mind that when a rigid object rotates about a fixed axis as in Figure 10.4, every particle of the object moves in a circle whose center is on the axis of rotation.

Because point *P* in Figure 10.4 moves in a circle, the translational velocity vector $\vec{v}$ is always tangent to the circular path and hence is called *tangential velocity*. The magnitude of the tangential velocity of the point *P* is by definition the tangential speed $v = ds/dt$, where *s* is the distance traveled by this point measured along the circular path. Recalling that $s = r\theta$ (Eq. 10.1a) and noting that *r* is constant, we obtain

$$v = \frac{ds}{dt} = r\frac{d\theta}{dt}$$

Because $d\theta/dt = \omega$ (see Eq. 10.3), it follows that

◀ **Relation between tangential velocity and angular velocity**

$$v = r\omega \qquad\qquad (10.10)$$

As we saw in Equation 4.17, the tangential speed of a point on a rotating rigid object equals the perpendicular distance of that point from the axis of rotation multiplied by the angular speed. Therefore, although every point on the rigid object has the same *angular* speed, not every point has the same *tangential* speed because *r* is not the same for all points on the object. Equation 10.10 shows that the tangential speed of a point on the rotating object increases as one moves outward from the center of rotation, as we would intuitively expect. For example, the outer end of a swinging golf club moves much faster than a point near the handle.

We can relate the angular acceleration of the rotating rigid object to the tangential acceleration of the point *P* by taking the time derivative of *v*:

$$a_t = \frac{dv}{dt} = r\frac{d\omega}{dt}$$

◀ **Relation between tangential acceleration and angular acceleration**

$$a_t = r\alpha \qquad\qquad (10.11)$$

That is, the tangential component of the translational acceleration of a point on a rotating rigid object equals the point's perpendicular distance from the axis of rotation multiplied by the angular acceleration.

In Section 4.4, we found that a point moving in a circular path undergoes a radial acceleration a_r directed toward the center of rotation and whose magnitude is that of the centripetal acceleration v^2/r (Fig. 10.5). Because $v = r\omega$ for a point

P on a rotating object, we can express the centripetal acceleration at that point in terms of angular speed as we did in Equation 4.18:

$$a_c = \frac{v^2}{r} = r\omega^2 \qquad \textbf{(10.12)}$$

The total acceleration vector at the point is $\vec{\mathbf{a}} = \vec{\mathbf{a}}_t + \vec{\mathbf{a}}_r$, where the magnitude of $\vec{\mathbf{a}}_r$ is the centripetal acceleration a_c. Because $\vec{\mathbf{a}}$ is a vector having a radial and a tangential component, the magnitude of $\vec{\mathbf{a}}$ at the point P on the rotating rigid object is

$$a = \sqrt{a_t^2 + a_r^2} = \sqrt{r^2\alpha^2 + r^2\omega^4} = r\sqrt{\alpha^2 + \omega^4} \qquad \textbf{(10.13)}$$

Quick Quiz 10.3 Ethan and Joseph are riding on a merry-go-round. Ethan rides on a horse at the outer rim of the circular platform, twice as far from the center of the circular platform as Joseph, who rides on an inner horse. **(i)** When the merry-go-round is rotating at a constant angular speed, what is Ethan's angular speed? (a) twice Joseph's (b) the same as Joseph's (c) half of Joseph's (d) impossible to determine **(ii)** When the merry-go-round is rotating at a constant angular speed, describe Ethan's tangential speed from the same list of choices.

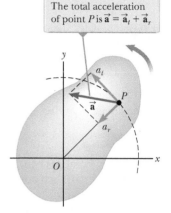

Figure 10.5 As a rigid object rotates about a fixed axis (the z axis) through O, the point P experiences a tangential component of translational acceleration a_t and a radial component of translational acceleration a_r.

Example 10.2   CD Player   AM

On a compact disc (Fig. 10.6), audio information is stored digitally in a series of pits and flat areas on the surface of the disc. The alternations between pits and flat areas on the surface represent binary ones and zeros to be read by the CD player and converted back to sound waves. The pits and flat areas are detected by a system consisting of a laser and lenses. The length of a string of ones and zeros representing one piece of information is the same everywhere on the disc, whether the information is near the center of the disc or near its outer edge. So that this length of ones and zeros always passes by the laser–lens system in the same time interval, the tangential speed of the disc surface at the location of the lens must be constant. According to Equation 10.10, the angular speed must therefore vary as the laser–lens system moves radially along the disc. In a typical CD player, the constant speed of the surface at the point of the laser–lens system is 1.3 m/s.

(A) Find the angular speed of the disc in revolutions per minute when information is being read from the innermost first track ($r = 23$ mm) and the outermost final track ($r = 58$ mm).

SOLUTION

Conceptualize Figure 10.6 shows a photograph of a compact disc. Trace your finger around the circle marked "23 mm" and mentally estimate the time interval to go around the circle once. Now trace your finger around the circle marked "58 mm," moving your finger across the surface of the page at the same speed as you did when tracing the smaller circle. Notice how much longer in time it takes your finger to go around the larger circle. If your finger represents the laser reading the disc, you can see that the disc rotates once in a longer time interval when the laser reads the information in the outer circle. Therefore, the disc must rotate more slowly when the laser is reading information from this part of the disc.

Categorize This part of the example is categorized as a simple substitution problem. In later parts, we will need to identify analysis models.

Figure 10.6 (Example 10.2) A compact disc.

Use Equation 10.10 to find the angular speed that gives the required tangential speed at the position of the inner track:

$$\omega_i = \frac{v}{r_i} = \frac{1.3 \text{ m/s}}{2.3 \times 10^{-2} \text{ m}} = 57 \text{ rad/s}$$

$$= (57 \text{ rad/s})\left(\frac{1 \text{ rev}}{2\pi \text{ rad}}\right)\left(\frac{60 \text{ s}}{1 \text{ min}}\right) = 5.4 \times 10^2 \text{ rev/min}$$

continued

▶ 10.2 continued

Do the same for the outer track:

$$\omega_f = \frac{v}{r_f} = \frac{1.3 \text{ m/s}}{5.8 \times 10^{-2} \text{ m}} = 22 \text{ rad/s} = \boxed{2.1 \times 10^2 \text{ rev/min}}$$

The CD player adjusts the angular speed ω of the disc within this range so that information moves past the objective lens at a constant rate.

(B) The maximum playing time of a standard music disc is 74 min and 33 s. How many revolutions does the disc make during that time?

SOLUTION

Categorize From part (A), the angular speed decreases as the disc plays. Let us assume it decreases steadily, with α constant. We can then apply the *rigid object under constant angular acceleration* model to the disc.

....

Analyze If $t = 0$ is the instant the disc begins rotating, with angular speed of 57 rad/s, the final value of the time t is (74 min)(60 s/min) + 33 s = 4 473 s. We are looking for the angular displacement $\Delta\theta$ during this time interval.

Use Equation 10.9 to find the angular displacement of the disc at $t = 4$ 473 s:

$$\Delta\theta = \theta_f - \theta_i = \tfrac{1}{2}(\omega_i + \omega_f)t$$
$$= \tfrac{1}{2}(57 \text{ rad/s} + 22 \text{ rad/s})(4 \text{ 473 s}) = 1.8 \times 10^5 \text{ rad}$$

Convert this angular displacement to revolutions:

$$\Delta\theta = (1.8 \times 10^5 \text{ rad})\left(\frac{1 \text{ rev}}{2\pi \text{ rad}}\right) = \boxed{2.8 \times 10^4 \text{ rev}}$$

(C) What is the angular acceleration of the compact disc over the 4 473-s time interval?

SOLUTION

Categorize We again model the disc as a *rigid object under constant angular acceleration*. In this case, Equation 10.6 gives the value of the constant angular acceleration. Another approach is to use Equation 10.4 to find the average angular acceleration. In this case, we are not assuming the angular acceleration is constant. The answer is the same from both equations; only the interpretation of the result is different.

....

Analyze Use Equation 10.6 to find the angular acceleration:

$$\alpha = \frac{\omega_f - \omega_i}{t} = \frac{22 \text{ rad/s} - 57 \text{ rad/s}}{4 \text{ 473 s}} = \boxed{-7.6 \times 10^{-3} \text{ rad/s}^2}$$

....

Finalize The disc experiences a very gradual decrease in its rotation rate, as expected from the long time interval required for the angular speed to change from the initial value to the final value. In reality, the angular acceleration of the disc is not constant. Problem 90 allows you to explore the actual time behavior of the angular acceleration.

The component $F \sin \phi$ tends to rotate the wrench about an axis through O.

Figure 10.7 The force $\vec{\mathbf{F}}$ has a greater rotating tendency about an axis through O as F increases and as the moment arm d increases.

10.4 Torque

In our study of translational motion, after investigating the description of motion, we studied the cause of changes in motion: force. We follow the same plan here: What is the cause of changes in rotational motion?

Imagine trying to rotate a door by applying a force of magnitude F perpendicular to the door surface near the hinges and then at various distances from the hinges. You will achieve a more rapid rate of rotation for the door by applying the force near the doorknob than by applying it near the hinges.

When a force is exerted on a rigid object pivoted about an axis, the object tends to rotate about that axis. The tendency of a force to rotate an object about some axis is measured by a quantity called **torque** $\vec{\tau}$ (Greek letter tau). Torque is a vector, but we will consider only its magnitude here; we will explore its vector nature in Chapter 11.

Consider the wrench in Figure 10.7 that we wish to rotate around an axis that is perpendicular to the page and passes through the center of the bolt. The applied force $\vec{F}$ acts at an angle ϕ to the horizontal. We define the magnitude of the torque associated with the force $\vec{F}$ around the axis passing through O by the expression

$$\tau \equiv rF \sin \phi = Fd \qquad (10.14)$$

where r is the distance between the rotation axis and the point of application of $\vec{F}$, and d is the perpendicular distance from the rotation axis to the line of action of $\vec{F}$. (The *line of action* of a force is an imaginary line extending out both ends of the vector representing the force. The dashed line extending from the tail of $\vec{F}$ in Fig. 10.7 is part of the line of action of $\vec{F}$.) From the right triangle in Figure 10.7 that has the wrench as its hypotenuse, we see that $d = r \sin \phi$. The quantity d is called the **moment arm** (or *lever arm*) of $\vec{F}$.

In Figure 10.7, the only component of $\vec{F}$ that tends to cause rotation of the wrench around an axis through O is $F \sin \phi$, the component perpendicular to a line drawn from the rotation axis to the point of application of the force. The horizontal component $F \cos \phi$, because its line of action passes through O, has no tendency to produce rotation about an axis passing through O. From the definition of torque, the rotating tendency increases as F increases and as d increases, which explains why it is easier to rotate a door if we push at the doorknob rather than at a point close to the hinges. We also want to apply our push as closely perpendicular to the door as we can so that ϕ is close to 90°. Pushing sideways on the doorknob ($\phi = 0$) will not cause the door to rotate.

If two or more forces act on a rigid object as in Figure 10.8, each tends to produce rotation about the axis through O. In this example, $\vec{F}_2$ tends to rotate the object clockwise and $\vec{F}_1$ tends to rotate it counterclockwise. We use the convention that the sign of the torque resulting from a force is positive if the turning tendency of the force is counterclockwise and negative if the turning tendency is clockwise. For Example, in Figure 10.8, the torque resulting from $\vec{F}_1$, which has a moment arm d_1, is positive and equal to $+F_1 d_1$; the torque from $\vec{F}_2$ is negative and equal to $-F_2 d_2$. Hence, the *net* torque about an axis through O is

$$\sum \tau = \tau_1 + \tau_2 = F_1 d_1 - F_2 d_2$$

Torque should not be confused with force. Forces can cause a change in translational motion as described by Newton's second law. Forces can also cause a change in rotational motion, but the effectiveness of the forces in causing this change depends on both the magnitudes of the forces and the moment arms of the forces, in the combination we call *torque*. Torque has units of force times length—newton meters (N · m) in SI units—and should be reported in these units. Do not confuse torque and work, which have the same units but are very different concepts.

Quick Quiz 10.4 (i) If you are trying to loosen a stubborn screw from a piece of wood with a screwdriver and fail, should you find a screwdriver for which the handle is (a) longer or (b) fatter? (ii) If you are trying to loosen a stubborn bolt from a piece of metal with a wrench and fail, should you find a wrench for which the handle is (a) longer or (b) fatter?

◄ **Moment arm**

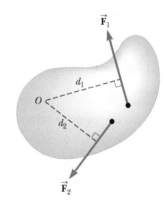

Figure 10.8 The force $\vec{F}_1$ tends to rotate the object counterclockwise about an axis through O, and $\vec{F}_2$ tends to rotate it clockwise.

Example 10.3 The Net Torque on a Cylinder

A one-piece cylinder is shaped as shown in Figure 10.9, with a core section protruding from the larger drum. The cylinder is free to rotate about the central z axis shown in the drawing. A rope wrapped around the drum, which has radius R_1, exerts a force $\vec{T}_1$ to the right on the cylinder. A rope wrapped around the core, which has radius R_2, exerts a force $\vec{T}_2$ downward on the cylinder.

(A) What is the net torque acting on the cylinder about the rotation axis (which is the z axis in Fig. 10.9)?

continued

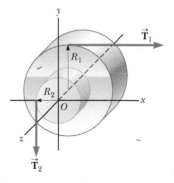

Figure 10.9 (Example 10.3) A solid cylinder pivoted about the z axis through O. The moment arm of $\vec{T}_1$ is R_1, and the moment arm of $\vec{T}_2$ is R_2.

▶ **10.3** continued

SOLUTION

Conceptualize Imagine that the cylinder in Figure 10.9 is a shaft in a machine. The force $\vec{\mathbf{T}}_1$ could be applied by a drive belt wrapped around the drum. The force $\vec{\mathbf{T}}_2$ could be applied by a friction brake at the surface of the core.

Categorize This example is a substitution problem in which we evaluate the net torque using Equation 10.14.

The torque due to $\vec{\mathbf{T}}_1$ about the rotation axis is $-R_1T_1$. (The sign is negative because the torque tends to produce clockwise rotation.) The torque due to $\vec{\mathbf{T}}_2$ is $+R_2T_2$. (The sign is positive because the torque tends to produce counterclockwise rotation of the cylinder.)

Evaluate the net torque about the rotation axis: $\sum \tau = \tau_1 + \tau_2 = \boxed{R_2T_2 - R_1T_1}$

As a quick check, notice that if the two forces are of equal magnitude, the net torque is negative because $R_1 > R_2$. Starting from rest with both forces of equal magnitude acting on it, the cylinder would rotate clockwise because $\vec{\mathbf{T}}_1$ would be more effective at turning it than would $\vec{\mathbf{T}}_2$.

(B) Suppose $T_1 = 5.0$ N, $R_1 = 1.0$ m, $T_2 = 15$ N, and $R_2 = 0.50$ m. What is the net torque about the rotation axis, and which way does the cylinder rotate starting from rest?

SOLUTION

Substitute the given values: $\sum \tau = (0.50 \text{ m})(15 \text{ N}) - (1.0 \text{ m})(5.0 \text{ N}) = \boxed{2.5 \text{ N} \cdot \text{m}}$

Because this net torque is positive, the cylinder begins to rotate in the counterclockwise direction.

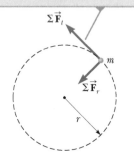

The tangential force on the particle results in a torque on the particle about an axis through the center of the circle.

Figure 10.10 A particle rotating in a circle under the influence of a tangential net force $\sum \vec{\mathbf{F}}_t$. A radial net force $\sum \vec{\mathbf{F}}_r$ also must be present to maintain the circular motion.

10.5 Analysis Model: Rigid Object Under a Net Torque

In Chapter 5, we learned that a net force on an object causes an acceleration of the object and that the acceleration is proportional to the net force. These facts are the basis of the particle under a net force model whose mathematical representation is Newton's second law. In this section, we show the rotational analog of Newton's second law: the angular acceleration of a rigid object rotating about a fixed axis is proportional to the net torque acting about that axis. Before discussing the more complex case of rigid-object rotation, however, it is instructive first to discuss the case of a particle moving in a circular path about some fixed point under the influence of an external force.

Consider a particle of mass m rotating in a circle of radius r under the influence of a tangential net force $\sum \vec{\mathbf{F}}_t$ and a radial net force $\sum \vec{\mathbf{F}}_r$ as shown in Figure 10.10. The radial net force causes the particle to move in the circular path with a centripetal acceleration. The tangential force provides a tangential acceleration $\vec{\mathbf{a}}_t$, and

$$\sum F_t = ma_t$$

The magnitude of the net torque due to $\sum \vec{\mathbf{F}}_t$ on the particle about an axis perpendicular to the page through the center of the circle is

$$\sum \tau = \sum F_t r = (ma_t)r$$

Because the tangential acceleration is related to the angular acceleration through the relationship $a_t = r\alpha$ (Eq. 10.11), the net torque can be expressed as

$$\sum \tau = (mr\alpha)r = (mr^2)\alpha \tag{10.15}$$

Let us denote the quantity mr^2 with the symbol I for now. We will say more about this quantity below. Using this notation, Equation 10.15 can be written as

$$\sum \tau = I\alpha \tag{10.16}$$

That is, the net torque acting on the particle is proportional to its angular acceleration. Notice that $\sum \tau = I\alpha$ has the same mathematical form as Newton's second law of motion, $\sum F = ma$.

Now let us extend this discussion to a rigid object of arbitrary shape rotating about a fixed axis passing through a point O as in Figure 10.11. The object can be regarded as a collection of particles of mass m_i. If we impose a Cartesian coordinate system on the object, each particle rotates in a circle about the origin and each has a tangential acceleration a_i produced by an external tangential force of magnitude F_i. For any given particle, we know from Newton's second law that

$$F_i = m_i a_i$$

The external torque $\vec{\tau}_i$ associated with the force $\vec{F}_i$ acts about the origin and its magnitude is given by

$$\tau_i = r_i F_i = r_i m_i a_i$$

Because $a_i = r_i \alpha$, the expression for τ_i becomes

$$\tau_i = m_i r_i^2 \alpha$$

Although each particle in the rigid object may have a different translational acceleration a_i, they all have the *same* angular acceleration α. With that in mind, we can add the torques on all of the particles making up the rigid object to obtain the net torque on the object about an axis through O due to all external forces:

$$\sum \tau_{\text{ext}} = \sum_i \tau_i = \sum_i m_i r_i^2 \alpha = \left(\sum_i m_i r_i^2 \right) \alpha \qquad \textbf{(10.17)}$$

where α can be taken outside the summation because it is common to all particles. Calling the quantity in parentheses I, the expression for $\sum \tau_{\text{ext}}$ becomes

$$\sum \tau_{\text{ext}} = I\alpha \qquad \textbf{(10.18)}$$

This equation for a rigid object is the same as that found for a particle moving in a circular path (Eq. 10.16). The net torque about the rotation axis is proportional to the angular acceleration of the object, with the proportionality factor being I, a quantity that we have yet to describe fully. Equation 10.18 is the mathematical representation of the analysis model of a **rigid object under a net torque,** the rotational analog to the particle under a net force.

Let us now address the quantity I, defined as follows:

$$I = \sum_i m_i r_i^2 \qquad \textbf{(10.19)}$$

This quantity is called the **moment of inertia** of the object, and depends on the masses of the particles making up the object and their distances from the rotation axis. Notice that Equation 10.19 reduces to $I = mr^2$ for a single particle, consistent with our use of the notation I that we used in going from Equation 10.15 to Equation 10.16. Note that moment of inertia has units of kg · m² in SI units.

Equation 10.18 has the same form as Newton's second law for a system of particles as expressed in Equation 9.39:

$$\sum \vec{F}_{\text{ext}} = M\vec{a}_{\text{CM}}$$

Consequently, the moment of inertia I must play the same role in rotational motion as the role that mass plays in translational motion: the moment of inertia is the resistance to changes in rotational motion. This resistance depends not only on the mass of the object, but also on how the mass is distributed around the rotation axis. Table 10.2 on page 304 gives the moments of inertia[2] for a number of objects about specific axes. The moments of inertia of rigid objects with simple geometry (high symmetry) are relatively easy to calculate provided the rotation axis coincides with an axis of symmetry, as we show in the next section.

---

[2]Civil engineers use moment of inertia to characterize the elastic properties (rigidity) of such structures as loaded beams. Hence, it is often useful even in a nonrotational context.

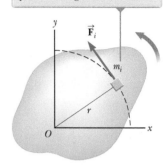

The particle of mass m_i of the rigid object experiences a torque in the same way that the particle in Figure 10.10 does.

Figure 10.11 A rigid object rotating about an axis through O. Each particle of mass m_i rotates about the axis with the same angular acceleration α.

◀ Torque on a rigid object is proportional to angular acceleration

Pitfall Prevention 10.5
No Single Moment of Inertia
There is one major difference between mass and moment of inertia. Mass is an inherent property of an object. The moment of inertia of an object depends on your choice of rotation axis. Therefore, there is no single value of the moment of inertia for an object. There is a *minimum* value of the moment of inertia, which is that calculated about an axis passing through the center of mass of the object.

Table 10.2   **Moments of Inertia of Homogeneous Rigid Objects with Different Geometries**

Hoop or thin cylindrical shell
$I_{CM} = MR^2$

Hollow cylinder
$I_{CM} = \frac{1}{2}M(R_1^2 + R_2^2)$

Solid cylinder or disk
$I_{CM} = \frac{1}{2}MR^2$

Rectangular plate
$I_{CM} = \frac{1}{12}M(a^2 + b^2)$

Long, thin rod with rotation axis through center
$I_{CM} = \frac{1}{12}ML^2$

Long, thin rod with rotation axis through end
$I = \frac{1}{3}ML^2$

Solid sphere
$I_{CM} = \frac{2}{5}MR^2$

Thin spherical shell
$I_{CM} = \frac{2}{3}MR^2$

Ⓠuick Quiz 10.5 You turn off your electric drill and find that the time interval for the rotating bit to come to rest due to frictional torque in the drill is Δt. You replace the bit with a larger one that results in a doubling of the moment of inertia of the drill's entire rotating mechanism. When this larger bit is rotated at the same angular speed as the first and the drill is turned off, the frictional torque remains the same as that for the previous situation. What is the time interval for this second bit to come to rest? **(a)** $4\Delta t$ **(b)** $2\Delta t$ **(c)** Δt **(d)** $0.5\Delta t$ **(e)** $0.25\Delta t$ **(f)** impossible to determine

---

Analysis Model   **Rigid Object Under a Net Torque**

Imagine you are analyzing the motion of an object that is free to rotate about a fixed axis. The cause of changes in rotational motion of this object is torque applied to the object and, in parallel to Newton's second law for translation motion, the torque is equal to the product of the moment of inertia of the object and the angular acceleration:

$$\sum \tau_{ext} = I\alpha \qquad (10.18)$$

The torque, the moment of inertia, and the angular acceleration must all be evaluated around the same rotation axis.

Analysis Model **Rigid Object Under a Net Torque** *(continued)*

Examples:

- a bicycle chain around the sprocket of a bicycle causes the rear wheel of the bicycle to rotate
- an electric dipole moment in an electric field rotates due to the electric force from the field (Chapter 23)
- a magnetic dipole moment in a magnetic field rotates due to the magnetic force from the field (Chapter 30)
- the armature of a motor rotates due to the torque exerted by a surrounding magnetic field (Chapter 31)

Example 10.4 **Rotating Rod** AM

A uniform rod of length L and mass M is attached at one end to a frictionless pivot and is free to rotate about the pivot in the vertical plane as in Figure 10.12. The rod is released from rest in the horizontal position. What are the initial angular acceleration of the rod and the initial translational acceleration of its right end?

Figure 10.12 (Example 10.4) A rod is free to rotate around a pivot at the left end. The gravitational force on the rod acts at its center of mass.

SOLUTION

Conceptualize Imagine what happens to the rod in Figure 10.12 when it is released. It rotates clockwise around the pivot at the left end.

Categorize The rod is categorized as a *rigid object under a net torque*. The torque is due only to the gravitational force on the rod if the rotation axis is chosen to pass through the pivot in Figure 10.12. We *cannot* categorize the rod as a rigid object under constant angular acceleration because the torque exerted on the rod and therefore the angular acceleration of the rod vary with its angular position.

Analyze The only force contributing to the torque about an axis through the pivot is the gravitational force $M\vec{g}$ exerted on the rod. (The force exerted by the pivot on the rod has zero torque about the pivot because its moment arm is zero.) To compute the torque on the rod, we assume the gravitational force acts at the center of mass of the rod as shown in Figure 10.12.

Write an expression for the magnitude of the net external torque due to the gravitational force about an axis through the pivot:

$$\sum \tau_{\text{ext}} = Mg\left(\frac{L}{2}\right)$$

Use Equation 10.18 to obtain the angular acceleration of the rod, using the moment of inertia for the rod from Table 10.2:

$$(1) \quad \alpha = \frac{\sum \tau_{\text{ext}}}{I} = \frac{Mg(L/2)}{\frac{1}{3}ML^2} = \frac{3g}{2L}$$

Use Equation 10.11 with $r = L$ to find the initial translational acceleration of the right end of the rod:

$$a_t = L\alpha = \frac{3}{2}g$$

Finalize These values are the *initial* values of the angular and translational accelerations. Once the rod begins to rotate, the gravitational force is no longer perpendicular to the rod and the values of the two accelerations decrease, going to zero at the moment the rod passes through the vertical orientation.

WHAT IF? What if we were to place a penny on the end of the rod and then release the rod? Would the penny stay in contact with the rod?

Answer The result for the initial acceleration of a point on the end of the rod shows that $a_t > g$. An unsupported penny falls at acceleration g. So, if we place a penny on the end of the rod and then release the rod, the end of the rod falls faster than the penny does! The penny does not stay in contact with the rod. (Try this with a penny and a meterstick!)

The question now is to find the location on the rod at which we can place a penny that *will* stay in contact as both begin to fall. To find the translational acceleration of an arbitrary point on the rod at a distance $r < L$ from the pivot point, we combine Equation (1) with Equation 10.11:

$$a_t = r\alpha = \frac{3g}{2L}r$$

continued

▶ **10.4** continued

For the penny to stay in contact with the rod, the limiting case is that the translational acceleration must be equal to that due to gravity:

$$a_t = g = \frac{3g}{2L} r$$

$$r = \frac{2}{3}L$$

Therefore, a penny placed closer to the pivot than two-thirds of the length of the rod stays in contact with the falling rod, but a penny farther out than this point loses contact.

Conceptual Example 10.5   **Falling Smokestacks and Tumbling Blocks**

When a tall smokestack falls over, it often breaks somewhere along its length before it hits the ground as shown in Figure 10.13. Why?

SOLUTION

As the smokestack rotates around its base, each higher portion of the smokestack falls with a larger tangential acceleration than the portion below it according to Equation 10.11. The angular acceleration increases as the smokestack tips farther. Eventually, higher portions of the smokestack experience an acceleration greater than the acceleration that could result from gravity alone; this situation is similar to that described in Example 10.4. That can happen only if these portions are being pulled downward by a force in addition to the gravitational force. The force that causes that to occur is the shear force from lower portions of the smokestack. Eventually, the shear force that provides this acceleration is greater than the smoke-stack can withstand, and the smokestack breaks. The same thing happens with a tall tower of children's toy blocks. Borrow some blocks from a child and build such a tower. Push it over and watch it come apart at some point before it strikes the floor.

Kevin Spreekmeester/AGE fotostock

Figure 10.13 (Conceptual Example 10.5) A falling smoke-stack breaks at some point along its length.

Example 10.6   **Angular Acceleration of a Wheel**   `AM`

A wheel of radius R, mass M, and moment of inertia I is mounted on a frictionless, horizontal axle as in Figure 10.14. A light cord wrapped around the wheel supports an object of mass m. When the wheel is released, the object accelerates downward, the cord unwraps off the wheel, and the wheel rotates with an angular acceleration. Find expressions for the angular acceleration of the wheel, the translational acceleration of the object, and the tension in the cord.

SOLUTION

Conceptualize Imagine that the object is a bucket in an old-fashioned water well. It is tied to a cord that passes around a cylinder equipped with a crank for raising the bucket. After the bucket has been raised, the system is released and the bucket accelerates downward while the cord unwinds off the cylinder.

Categorize We apply two analysis models here. The object is modeled as a *particle under a net force*. The wheel is modeled as a *rigid object under a net torque*.

Figure 10.14 (Example 10.6) An object hangs from a cord wrapped around a wheel.

Analyze The magnitude of the torque acting on the wheel about its axis of rotation is $\tau = TR$, where T is the force exerted by the cord on the rim of the wheel. (The gravitational force exerted by the Earth on the wheel and the

▶ **10.6** continued

normal force exerted by the axle on the wheel both pass through the axis of rotation and therefore produce no torque.)

From the rigid object under a net torque model, write Equation 10.18:

$$\sum \tau_{ext} = I\alpha$$

Solve for α and substitute the net torque:

$$(1) \quad \alpha = \frac{\sum \tau_{ext}}{I} = \frac{TR}{I}$$

From the particle under a net force model, apply Newton's second law to the motion of the object, taking the downward direction to be positive:

$$\sum F_y = mg - T = ma$$

Solve for the acceleration a:

$$(2) \quad a = \frac{mg - T}{m}$$

Equations (1) and (2) have three unknowns: α, a, and T. Because the object and wheel are connected by a cord that does not slip, the translational acceleration of the suspended object is equal to the tangential acceleration of a point on the wheel's rim. Therefore, the angular acceleration α of the wheel and the translational acceleration of the object are related by $a = R\alpha$.

Use this fact together with Equations (1) and (2):

$$(3) \quad a = R\alpha = \frac{TR^2}{I} = \frac{mg - T}{m}$$

Solve for the tension T:

$$(4) \quad T = \frac{mg}{1 + (mR^2/I)}$$

Substitute Equation (4) into Equation (2) and solve for a:

$$(5) \quad a = \frac{g}{1 + (I/mR^2)}$$

Use $a = R\alpha$ and Equation (5) to solve for α:

$$\alpha = \frac{a}{R} = \frac{g}{R + (I/mR)}$$

Finalize We finalize this problem by imagining the behavior of the system in some extreme limits.

WHAT IF? What if the wheel were to become very massive so that I becomes very large? What happens to the acceleration a of the object and the tension T?

Answer If the wheel becomes infinitely massive, we can imagine that the object of mass m will simply hang from the cord without causing the wheel to rotate.

We can show that mathematically by taking the limit $I \to \infty$. Equation (5) then becomes

$$a = \frac{g}{1 + (I/mR^2)} \to 0$$

which agrees with our conceptual conclusion that the object will hang at rest. Also, Equation (4) becomes

$$T = \frac{mg}{1 + (mR^2/I)} \to mg$$

which is consistent because the object simply hangs at rest in equilibrium between the gravitational force and the tension in the string.

10.6 Calculation of Moments of Inertia

The moment of inertia of a system of discrete particles can be calculated in a straightforward way with Equation 10.19. We can evaluate the moment of inertia of a continuous rigid object by imagining the object to be divided into many small elements, each of which has mass Δm_i. We use the definition $I = \sum_i r_i^2 \Delta m_i$

and take the limit of this sum as $\Delta m_i \rightarrow 0$. In this limit, the sum becomes an integral over the volume of the object:

Moment of inertia ▶
of a rigid object

$$I = \lim_{\Delta m_i \rightarrow 0} \sum_i r_i^2 \, \Delta m_i = \int r^2 \, dm \qquad (10.20)$$

It is usually easier to calculate moments of inertia in terms of the volume of the elements rather than their mass, and we can easily make that change by using Equation 1.1, $\rho \equiv m/V$, where ρ is the density of the object and V is its volume. From this equation, the mass of a small element is $dm = \rho \, dV$. Substituting this result into Equation 10.20 gives

$$I = \int \rho r^2 \, dV \qquad (10.21)$$

If the object is homogeneous, ρ is constant and the integral can be evaluated for a known geometry. If ρ is not constant, its variation with position must be known to complete the integration.

The density given by $\rho = m/V$ sometimes is referred to as *volumetric mass density* because it represents mass per unit volume. Often we use other ways of expressing density. For instance, when dealing with a sheet of uniform thickness t, we can define a *surface mass density* $\sigma = \rho t$, which represents *mass per unit area*. Finally, when mass is distributed along a rod of uniform cross-sectional area A, we sometimes use *linear mass density* $\lambda = M/L = \rho A$, which is the *mass per unit length*.

Example 10.7 Uniform Rigid Rod

Calculate the moment of inertia of a uniform thin rod of length L and mass M (Fig. 10.15) about an axis perpendicular to the rod (the y' axis) and passing through its center of mass.

SOLUTION

Conceptualize Imagine twirling the rod in Figure 10.15 with your fingers around its midpoint. If you have a meterstick handy, use it to simulate the spinning of a thin rod and feel the resistance it offers to being spun.

Figure 10.15 (Example 10.7) A uniform rigid rod of length L. The moment of inertia about the y' axis is less than that about the y axis. The latter axis is examined in Example 10.9.

Categorize This example is a substitution problem, using the definition of moment of inertia in Equation 10.20. As with any integration problem, the solution involves reducing the integrand to a single variable.

The shaded length element dx' in Figure 10.15 has a mass dm equal to the mass per unit length λ multiplied by dx'.

Express dm in terms of dx':

$$dm = \lambda \, dx' = \frac{M}{L} \, dx'$$

Substitute this expression into Equation 10.20, with $r^2 = (x')^2$:

$$I_{y'} = \int r^2 \, dm = \int_{-L/2}^{L/2} (x')^2 \frac{M}{L} \, dx' = \frac{M}{L} \int_{-L/2}^{L/2} (x')^2 \, dx'$$

$$= \frac{M}{L} \left[\frac{(x')^3}{3} \right]_{-L/2}^{L/2} = \frac{1}{12} M L^2$$

Check this result in Table 10.2.

Example 10.8 Uniform Solid Cylinder

A uniform solid cylinder has a radius R, mass M, and length L. Calculate its moment of inertia about its central axis (the z axis in Fig. 10.16).

▶ **10.8** continued

Figure 10.16 (Example 10.8) Calculating I about the z axis for a uniform solid cylinder.

Conceptualize To simulate this situation, imagine twirling a can of frozen juice around its central axis. Don't twirl a nonfrozen can of vegetable soup; it is not a rigid object! The liquid is able to move relative to the metal can.

Categorize This example is a substitution problem, using the definition of moment of inertia. As with Example 10.7, we must reduce the integrand to a single variable.

It is convenient to divide the cylinder into many cylindrical shells, each having radius r, thickness dr, and length L as shown in Figure 10.16. The density of the cylinder is ρ. The volume dV of each shell is its cross-sectional area multiplied by its length: $dV = L\,dA = L(2\pi r)\,dr$.

Express dm in terms of dr:

$$dm = \rho\,dV = \rho L(2\pi r)\,dr$$

Substitute this expression into Equation 10.20:

$$I_z = \int r^2\,dm = \int r^2[\rho L(2\pi r)\,dr] = 2\pi\rho L \int_0^R r^3\,dr = \tfrac{1}{2}\pi\rho L R^4$$

Use the total volume $\pi R^2 L$ of the cylinder to express its density:

$$\rho = \frac{M}{V} = \frac{M}{\pi R^2 L}$$

Substitute this value into the expression for I_z:

$$I_z = \tfrac{1}{2}\pi\left(\frac{M}{\pi R^2 L}\right)LR^4 = \tfrac{1}{2}MR^2$$

Check this result in Table 10.2.

WHAT IF? What if the length of the cylinder in Figure 10.16 is increased to $2L$, while the mass M and radius R are held fixed? How does that change the moment of inertia of the cylinder?

Answer Notice that the result for the moment of inertia of a cylinder does not depend on L, the length of the cylinder. It applies equally well to a long cylinder and a flat disk having the same mass M and radius R. Therefore, the moment of inertia of the cylinder is not affected by how the mass is distributed along its length.

The calculation of moments of inertia of an object about an arbitrary axis can be cumbersome, even for a highly symmetric object. Fortunately, use of an important theorem, called the **parallel-axis theorem,** often simplifies the calculation.

To generate the parallel-axis theorem, suppose the object in Figure 10.17a on page 310 rotates about the z axis. The moment of inertia does not depend on how the mass is distributed along the z axis; as we found in Example 10.8, the moment of inertia of a cylinder is independent of its length. Imagine collapsing the three-dimensional object into a planar object as in Figure 10.17b. In this imaginary process, all mass moves parallel to the z axis until it lies in the xy plane. The coordinates of the object's center of mass are now x_{CM}, y_{CM}, and $z_{CM} = 0$. Let the mass element dm have coordinates $(x, y, 0)$ as shown in the view down the z axis in Figure 10.17c. Because this element is a distance $r = \sqrt{x^2 + y^2}$ from the z axis, the moment of inertia of the entire object about the z axis is

$$I = \int r^2\,dm = \int (x^2 + y^2)\,dm$$

We can relate the coordinates x, y of the mass element dm to the coordinates of this same element located in a coordinate system having the object's center of mass as its origin. If the coordinates of the center of mass are x_{CM}, y_{CM}, and $z_{CM} = 0$ in the original coordinate system centered on O, we see from Figure 10.17c that

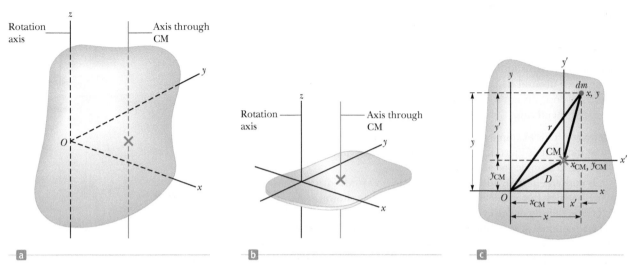

Figure 10.17 (a) An arbitrarily shaped rigid object. The origin of the coordinate system is not at the center of mass of the object. Imagine the object rotating about the z axis. (b) All mass elements of the object are collapsed parallel to the z axis to form a planar object. (c) An arbitrary mass element dm is indicated in blue in this view down the z axis. The parallel axis theorem can be used with the geometry shown to determine the moment of inertia of the original object around the z axis.

the relationships between the unprimed and primed coordinates are $x = x' + x_{CM}$, $y = y' + y_{CM}$, and $z = z' = 0$. Therefore,

$$I = \int \left[(x' + x_{CM})^2 + (y' + y_{CM})^2 \right] dm$$

$$= \int \left[(x')^2 + (y')^2 \right] dm + 2x_{CM} \int x' \, dm + 2y_{CM} \int y' dm + \left(x_{CM}{}^2 + y_{CM}{}^2 \right) \int dm$$

The first integral is, by definition, the moment of inertia I_{CM} about an axis that is parallel to the z axis and passes through the center of mass. The second two integrals are zero because, by definition of the center of mass, $\int x' \, dm = \int y' \, dm = 0$. The last integral is simply MD^2 because $\int dm = M$ and $D^2 = x_{CM}{}^2 + y_{CM}{}^2$. Therefore, we conclude that

Parallel-axis theorem ▶

$$I = I_{CM} + MD^2 \tag{10.22}$$

Example 10.9 **Applying the Parallel-Axis Theorem**

Consider once again the uniform rigid rod of mass M and length L shown in Figure 10.15. Find the moment of inertia of the rod about an axis perpendicular to the rod through one end (the y axis in Fig. 10.15).

SOLUTION

Conceptualize Imagine twirling the rod around an endpoint rather than the midpoint. If you have a meterstick handy, try it and notice the degree of difficulty in rotating it around the end compared with rotating it around the center.

Categorize This example is a substitution problem, involving the parallel-axis theorem.

Intuitively, we expect the moment of inertia to be greater than the result $I_{CM} = \frac{1}{12}ML^2$ from Example 10.7 because there is mass up to a distance of L away from the rotation axis, whereas the farthest distance in Example 10.7 was only $L/2$. The distance between the center-of-mass axis and the y axis is $D = L/2$.

▶ **10.9** continued

Use the parallel-axis theorem:

$$I = I_{CM} + MD^2 = \tfrac{1}{12}ML^2 + M\left(\frac{L}{2}\right)^2 = \boxed{\tfrac{1}{3}ML^2}$$

Check this result in Table 10.2.

10.7 Rotational Kinetic Energy

After investigating the role of forces in our study of translational motion, we turned our attention to approaches involving energy. We do the same thing in our current study of rotational motion.

In Chapter 7, we defined the kinetic energy of an object as the energy associated with its motion through space. An object rotating about a fixed axis remains stationary in space, so there is no kinetic energy associated with translational motion. The individual particles making up the rotating object, however, are moving through space; they follow circular paths. Consequently, there is kinetic energy associated with rotational motion.

Let us consider an object as a system of particles and assume it rotates about a fixed z axis with an angular speed ω. Figure 10.18 shows the rotating object and identifies one particle on the object located at a distance r_i from the rotation axis. If the mass of the ith particle is m_i and its tangential speed is v_i, its kinetic energy is

$$K_i = \tfrac{1}{2}m_i v_i^2$$

To proceed further, recall that although every particle in the rigid object has the same angular speed ω, the individual tangential speeds depend on the distance r_i from the axis of rotation according to Equation 10.10. The *total* kinetic energy of the rotating rigid object is the sum of the kinetic energies of the individual particles:

$$K_R = \sum_i K_i = \sum_i \tfrac{1}{2}m_i v_i^2 = \tfrac{1}{2}\sum_i m_i r_i^2 \omega^2$$

We can write this expression in the form

$$K_R = \tfrac{1}{2}\left(\sum_i m_i r_i^2\right)\omega^2 \tag{10.23}$$

where we have factored ω^2 from the sum because it is common to every particle. We recognize the quantity in parentheses as the moment of inertia of the object, introduced in Section 10.5.

Therefore, Equation 10.23 can be written

$$K_R = \tfrac{1}{2}I\omega^2 \tag{10.24}$$

◀ Rotational kinetic energy

Although we commonly refer to the quantity $\tfrac{1}{2}I\omega^2$ as **rotational kinetic energy,** it is not a new form of energy. It is ordinary kinetic energy because it is derived from a sum over individual kinetic energies of the particles contained in the rigid object. The mathematical form of the kinetic energy given by Equation 10.24 is convenient when we are dealing with rotational motion, provided we know how to calculate I.

Quick Quiz 10.6 A section of hollow pipe and a solid cylinder have the same radius, mass, and length. They both rotate about their long central axes with the same angular speed. Which object has the higher rotational kinetic energy? **(a)** The hollow pipe does. **(b)** The solid cylinder does. **(c)** They have the same rotational kinetic energy. **(d)** It is impossible to determine.

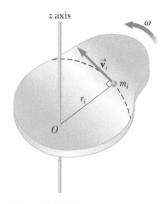

Figure 10.18 A rigid object rotating about the z axis with angular speed ω. The kinetic energy of the particle of mass m_i is $\tfrac{1}{2}m_i v_i^2$. The total kinetic energy of the object is called its rotational kinetic energy.

Example 10.10   An Unusual Baton

Four tiny spheres are fastened to the ends of two rods of negligible mass lying in the xy plane to form an unusual baton (Fig. 10.19). We shall assume the radii of the spheres are small compared with the dimensions of the rods.

(A) If the system rotates about the y axis (Fig. 10.19a) with an angular speed ω, find the moment of inertia and the rotational kinetic energy of the system about this axis.

Figure 10.19 (Example 10.10) Four spheres form an unusual baton. (a) The baton is rotated about the y axis. (b) The baton is rotated about the z axis.

SOLUTION

Conceptualize Figure 10.19 is a pictorial representation that helps conceptualize the system of spheres and how it spins. Model the spheres as particles.

Categorize This example is a substitution problem because it is a straightforward application of the definitions discussed in this section.

Apply Equation 10.19 to the system:

$$I_y = \sum_i m_i r_i^2 = Ma^2 + Ma^2 = \boxed{2Ma^2}$$

Evaluate the rotational kinetic energy using Equation 10.24:

$$K_R = \tfrac{1}{2}I_y\omega^2 = \tfrac{1}{2}(2Ma^2)\omega^2 = \boxed{Ma^2\omega^2}$$

That the two spheres of mass m do not enter into this result makes sense because they have no motion about the axis of rotation; hence, they have no rotational kinetic energy. By similar logic, we expect the moment of inertia about the x axis to be $I_x = 2mb^2$ with a rotational kinetic energy about that axis of $K_R = mb^2\omega^2$.

(B) Suppose the system rotates in the xy plane about an axis (the z axis) through the center of the baton (Fig. 10.19b). Calculate the moment of inertia and rotational kinetic energy about this axis.

SOLUTION

Apply Equation 10.19 for this new rotation axis:

$$I_z = \sum_i m_i r_i^2 = Ma^2 + Ma^2 + mb^2 + mb^2 = \boxed{2Ma^2 + 2mb^2}$$

Evaluate the rotational kinetic energy using Equation 10.24:

$$K_R = \tfrac{1}{2}I_z\omega^2 = \tfrac{1}{2}(2Ma^2 + 2mb^2)\omega^2 = \boxed{(Ma^2 + mb^2)\omega^2}$$

Comparing the results for parts (A) and (B), we conclude that the moment of inertia and therefore the rotational kinetic energy associated with a given angular speed depend on the axis of rotation. In part (B), we expect the result to include all four spheres and distances because all four spheres are rotating in the xy plane. Based on the work–kinetic energy theorem, the smaller rotational kinetic energy in part (A) than in part (B) indicates it would require less work to set the system into rotation about the y axis than about the z axis.

WHAT IF?   What if the mass M is much larger than m? How do the answers to parts (A) and (B) compare?

Answer If $M \gg m$, then m can be neglected and the moment of inertia and the rotational kinetic energy in part (B) become

$$I_z = 2Ma^2 \quad \text{and} \quad K_R = Ma^2\omega^2$$

which are the same as the answers in part (A). If the masses m of the two tan spheres in Figure 10.19 are negligible, these spheres can be removed from the figure and rotations about the y and z axes are equivalent.

10.8 Energy Considerations in Rotational Motion

Having introduced rotational kinetic energy in Section 10.7, let us now see how an energy approach can be useful in solving rotational problems. We begin by considering the relationship between the torque acting on a rigid object and its resulting rotational motion so as to generate expressions for power and a rotational analog to the work–kinetic energy theorem. Consider the rigid object pivoted at O in Figure 10.20. Suppose a single external force $\vec{\mathbf{F}}$ is applied at P, where $\vec{\mathbf{F}}$ lies in the plane of the page. The work done on the object by $\vec{\mathbf{F}}$ as its point of application rotates through an infinitesimal distance $ds = r\,d\theta$ is

$$dW = \vec{\mathbf{F}} \cdot d\vec{\mathbf{s}} = (F\sin\phi)r\,d\theta$$

where $F\sin\phi$ is the tangential component of $\vec{\mathbf{F}}$, or, in other words, the component of the force along the displacement. Notice that the radial component vector of $\vec{\mathbf{F}}$ does no work on the object because it is perpendicular to the displacement of the point of application of $\vec{\mathbf{F}}$.

Because the magnitude of the torque due to $\vec{\mathbf{F}}$ about an axis through O is defined as $rF\sin\phi$ by Equation 10.14, we can write the work done for the infinitesimal rotation as

$$dW = \tau\,d\theta \tag{10.25}$$

The rate at which work is being done by $\vec{\mathbf{F}}$ as the object rotates about the fixed axis through the angle $d\theta$ in a time interval dt is

$$\frac{dW}{dt} = \tau\frac{d\theta}{dt}$$

Because dW/dt is the instantaneous power P (see Section 8.5) delivered by the force and $d\theta/dt = \omega$, this expression reduces to

$$P = \frac{dW}{dt} = \tau\omega \tag{10.26}$$

◀ **Power delivered to a rotating rigid object**

This equation is analogous to $P = Fv$ in the case of translational motion, and Equation 10.25 is analogous to $dW = F_x\,dx$.

In studying translational motion, we have seen that models based on an energy approach can be extremely useful in describing a system's behavior. From what we learned of translational motion, we expect that when a symmetric object rotates about a fixed axis, the work done by external forces equals the change in the rotational energy of the object.

To prove that fact, let us begin with the rigid object under a net torque model, whose mathematical representation is $\sum \tau_{\text{ext}} = I\alpha$. Using the chain rule from calculus, we can express the net torque as

$$\sum\tau_{\text{ext}} = I\alpha = I\frac{d\omega}{dt} = I\frac{d\omega}{d\theta}\frac{d\theta}{dt} = I\frac{d\omega}{d\theta}\omega$$

Rearranging this expression and noting that $\sum\tau_{\text{ext}}\,d\theta = dW$ gives

$$\sum\tau_{\text{ext}}\,d\theta = dW = I\omega\,d\omega$$

Integrating this expression, we obtain for the work W done by the net external force acting on a rotating system

$$W = \int_{\omega_i}^{\omega_f} I\omega\,d\omega = \tfrac{1}{2}I\omega_f^2 - \tfrac{1}{2}I\omega_i^2 \tag{10.27}$$

◀ **Work–kinetic energy theorem for rotational motion**

where the angular speed changes from ω_i to ω_f. Equation 10.27 is the **work–kinetic energy theorem for rotational motion.** Similar to the work–kinetic energy theorem in translational motion (Section 7.5), this theorem states that the net work done by external forces in rotating a symmetric rigid object about a fixed axis equals the change in the object's rotational energy.

This theorem is a form of the nonisolated system (energy) model discussed in Chapter 8. Work is done on the system of the rigid object, which represents a transfer of energy across the boundary of the system that appears as an increase in the object's rotational kinetic energy.

Figure 10.20 A rigid object rotates about an axis through O under the action of an external force $\vec{\mathbf{F}}$ applied at P.

Table 10.3	Useful Equations in Rotational and Translational Motion	
Rotational Motion About a Fixed Axis		**Translational Motion**
Angular speed $\omega = d\theta/dt$		Translational speed $v = dx/dt$
Angular acceleration $\alpha = d\omega/dt$		Translational acceleration $a = dv/dt$
Net torque $\Sigma\tau_{\text{ext}} = I\alpha$		Net force $\Sigma F = ma$
If $\alpha = $ constant $\begin{cases} \omega_f = \omega_i + \alpha t \\ \theta_f = \theta_i + \omega_i t + \frac{1}{2}\alpha t^2 \\ \omega_f^2 = \omega_i^2 + 2\alpha(\theta_f - \theta_i) \end{cases}$		If $a = $ constant $\begin{cases} v_f = v_i + at \\ x_f = x_i + v_i t + \frac{1}{2}at^2 \\ v_f^2 = v_i^2 + 2a(x_f - x_i) \end{cases}$
Work $W = \displaystyle\int_{\theta_i}^{\theta_f} \tau\, d\theta$		Work $W = \displaystyle\int_{x_i}^{x_f} F_x\, dx$
Rotational kinetic energy $K_R = \frac{1}{2}I\omega^2$		Kinetic energy $K = \frac{1}{2}mv^2$
Power $P = \tau\omega$		Power $P = Fv$
Angular momentum $L = I\omega$		Linear momentum $p = mv$
Net torque $\Sigma\tau = dL/dt$		Net force $\Sigma F = dp/dt$

In general, we can combine this theorem with the translational form of the work–kinetic energy theorem from Chapter 7. Therefore, the net work done by external forces on an object is the change in its *total* kinetic energy, which is the sum of the translational and rotational kinetic energies. For example, when a pitcher throws a baseball, the work done by the pitcher's hands appears as kinetic energy associated with the ball moving through space as well as rotational kinetic energy associated with the spinning of the ball.

In addition to the work–kinetic energy theorem, other energy principles can also be applied to rotational situations. For example, if a system involving rotating objects is isolated and no nonconservative forces act within the system, the isolated system model and the principle of conservation of mechanical energy can be used to analyze the system as in Example 10.11 below. In general, Equation 8.2, the conservation of energy equation, applies to rotational situations, with the recognition that the change in kinetic energy ΔK will include changes in both translational and rotational kinetic energies.

Finally, in some situations an energy approach does not provide enough information to solve the problem and it must be combined with a momentum approach. Such a case is illustrated in Example 10.14 in Section 10.9.

Table 10.3 lists the various equations we have discussed pertaining to rotational motion together with the analogous expressions for translational motion. Notice the similar mathematical forms of the equations. The last two equations in the left-hand column of Table 10.3, involving angular momentum L, are discussed in Chapter 11 and are included here only for the sake of completeness.

Example 10.11 Rotating Rod Revisited AM

A uniform rod of length L and mass M is free to rotate on a frictionless pin passing through one end (Fig 10.21). The rod is released from rest in the horizontal position.

(A) What is its angular speed when the rod reaches its lowest position?

SOLUTION

Conceptualize Consider Figure 10.21 and imagine the rod rotating downward through a quarter turn about the pivot at the left end. Also look back at Example 10.8. This physical situation is the same.

Categorize As mentioned in Example 10.4, the angular acceleration of the rod is not constant. Therefore, the kinematic equations for rotation (Section 10.2) can-

Figure 10.21 (Example 10.11)
A uniform rigid rod pivoted at O rotates in a vertical plane under the action of the gravitational force.

▶ **10.11** continued

not be used to solve this example. We categorize the system of the rod and the Earth as an *isolated system* in terms of *energy* with no nonconservative forces acting and use the principle of conservation of mechanical energy.

Analyze We choose the configuration in which the rod is hanging straight down as the reference configuration for gravitational potential energy and assign a value of zero for this configuration. When the rod is in the horizontal position, it has no rotational kinetic energy. The potential energy of the system in this configuration relative to the reference configuration is $MgL/2$ because the center of mass of the rod is at a height $L/2$ higher than its position in the reference configuration. When the rod reaches its lowest position, the energy of the system is entirely rotational energy $\frac{1}{2}I\omega^2$, where I is the moment of inertia of the rod about an axis passing through the pivot.

Using the isolated system (energy) model, write an appropriate reduction of Equation 8.2:

$$\Delta K + \Delta U = 0$$

Substitute for each of the final and initial energies:

$$\left(\tfrac{1}{2}I\omega^2 - 0\right) + \left(0 - \tfrac{1}{2}MgL\right) = 0$$

Solve for ω and use $I = \frac{1}{3}ML^2$ (see Table 10.2) for the rod:

$$\omega = \sqrt{\frac{MgL}{I}} = \sqrt{\frac{MgL}{\frac{1}{3}ML^2}} = \boxed{\sqrt{\frac{3g}{L}}}$$

(B) Determine the tangential speed of the center of mass and the tangential speed of the lowest point on the rod when it is in the vertical position.

SOLUTION

Use Equation 10.10 and the result from part (A):

$$v_{CM} = r\omega = \frac{L}{2}\omega = \boxed{\tfrac{1}{2}\sqrt{3gL}}$$

Because r for the lowest point on the rod is twice what it is for the center of mass, the lowest point has a tangential speed twice that of the center of mass:

$$v = 2v_{CM} = \boxed{\sqrt{3gL}}$$

Finalize The initial configuration in this example is the same as that in Example 10.4. In Example 10.4, however, we could only find the initial angular acceleration of the rod. Applying an energy approach in the current example allows us to find additional information, the angular speed of the rod at the lowest point. Convince yourself that you could find the angular speed of the rod at any angular position by knowing the location of the center of mass at this position.

WHAT IF? What if we want to find the angular speed of the rod when the angle it makes with the horizontal is 45.0°? Because this angle is half of 90.0°, for which we solved the problem above, is the angular speed at this configuration half the answer in the calculation above, that is, $\frac{1}{2}\sqrt{3g/L}$?

Answer Imagine the rod in Figure 10.21 at the 45.0° position. Use a pencil or a ruler to represent the rod at this position. Notice that the center of mass has dropped through more than half of the distance $L/2$ in this configuration. Therefore, more than half of the initial gravitational potential energy has been transformed to rotational kinetic energy. So, we should not expect the value of the angular speed to be as simple as proposed above.

Note that the center of mass of the rod drops through a distance of $0.500L$ as the rod reaches the vertical configuration. When the rod is at 45.0° to the horizontal, we can show that the center of mass of the rod drops through a distance of $0.354L$. Continuing the calculation, we find that the angular speed of the rod at this configuration is $0.841\sqrt{3g/L}$, (not $\frac{1}{2}\sqrt{3g/L}$).

Example 10.12 **Energy and the Atwood Machine**

Two blocks having different masses m_1 and m_2 are connected by a string passing over a pulley as shown in Figure 10.22 on page 316. The pulley has a radius R and moment of inertia I about its axis of rotation. The string does not slip on the pulley, and the system is released from rest. Find the translational speeds of the blocks after block 2 descends through a distance h and find the angular speed of the pulley at this time.

continued

▶ **10.12** continued

Conceptualize We have already seen examples involving the Atwood machine, so the motion of the objects in Figure 10.22 should be easy to visualize.

Categorize Because the string does not slip, the pulley rotates about the axle. We can neglect friction in the axle because the axle's radius is small relative to that of the pulley. Hence, the frictional torque is much smaller than the net torque applied by the two blocks provided that their masses are significantly different. Consequently, the system consisting of the two blocks, the pulley, and the Earth is an *isolated system* in terms of *energy* with no nonconservative forces acting; therefore, the mechanical energy of the system is conserved.

Figure 10.22 (Example 10.12) An Atwood machine with a massive pulley.

Analyze We define the zero configuration for gravitational potential energy as that which exists when the system is released. From Figure 10.22, we see that the descent of block 2 is associated with a decrease in system potential energy and that the rise of block 1 represents an increase in potential energy.

Using the isolated system (energy) model, write an appropriate reduction of the conservation of energy equation:

$$\Delta K + \Delta U = 0$$

Substitute for each of the energies:

$$\left[\left(\tfrac{1}{2}m_1 v_f^2 + \tfrac{1}{2}m_2 v_f^2 + \tfrac{1}{2}I\omega_f^2\right) - 0\right] + \left[(m_1 gh - m_2 gh) - 0\right] = 0$$

Use $v_f = R\omega_f$ to substitute for ω_f:

$$\tfrac{1}{2}m_1 v_f^2 + \tfrac{1}{2}m_2 v_f^2 + \tfrac{1}{2}I\frac{v_f^2}{R^2} = m_2 gh - m_1 gh$$

$$\tfrac{1}{2}\left(m_1 + m_2 + \frac{I}{R^2}\right)v_f^2 = (m_2 - m_1)gh$$

Solve for v_f:

$$(1) \quad v_f = \left[\frac{2(m_2 - m_1)gh}{m_1 + m_2 + I/R^2}\right]^{1/2}$$

Use $v_f = R\omega_f$ to solve for ω_f:

$$\omega_f = \frac{v_f}{R} = \frac{1}{R}\left[\frac{2(m_2 - m_1)gh}{m_1 + m_2 + I/R^2}\right]^{1/2}$$

Finalize Each block can be modeled as a *particle under constant acceleration* because it experiences a constant net force. Think about what you would need to do to use Equation (1) to find the acceleration of one of the blocks. Then imagine the pulley becoming massless and determine the acceleration of a block. How does this result compare with the result of Example 5.9?

10.9 Rolling Motion of a Rigid Object

In this section, we treat the motion of a rigid object rolling along a flat surface. In general, such motion is complex. For example, suppose a cylinder is rolling on a straight path such that the axis of rotation remains parallel to its initial orientation in space. As Figure 10.23 shows, a point on the rim of the cylinder moves in a complex path called a *cycloid*. We can simplify matters, however, by focusing on the center of mass rather than on a point on the rim of the rolling object. As shown in Figure 10.23, the center of mass moves in a straight line. If an object such as a cylinder rolls without slipping on the surface (called *pure rolling motion*), a simple relationship exists between its rotational and translational motions.

Consider a uniform cylinder of radius R rolling without slipping on a horizontal surface (Fig. 10.24). As the cylinder rotates through an angle θ, its center of mass

One light source at the center of a rolling cylinder and another at one point on the rim illustrate the different paths these two points take.

The center moves in a straight line (green line).

The point on the rim moves in the path called a cycloid (red curve).

Figure 10.23 Two points on a rolling object take different paths through space.

moves a linear distance $s = R\theta$ (see Eq. 10.1a). Therefore, the translational speed of the center of mass for pure rolling motion is given by

$$v_{CM} = \frac{ds}{dt} = R\frac{d\theta}{dt} = R\omega \qquad (10.28)$$

where ω is the angular speed of the cylinder. Equation 10.28 holds whenever a cylinder or sphere rolls without slipping and is the **condition for pure rolling motion.** The magnitude of the linear acceleration of the center of mass for pure rolling motion is

$$a_{CM} = \frac{dv_{CM}}{dt} = R\frac{d\omega}{dt} = R\alpha \qquad (10.29)$$

where α is the angular acceleration of the cylinder.

Imagine that you are moving along with a rolling object at speed v_{CM}, staying in a frame of reference at rest with respect to the center of mass of the object. As you observe the object, you will see the object in pure rotation around its center of mass. Figure 10.25a shows the velocities of points at the top, center, and bottom of the object as observed by you. In addition to these velocities, every point on the object moves in the same direction with speed v_{CM} relative to the surface on which it rolls. Figure 10.25b shows these velocities for a nonrotating object. In the reference frame at rest with respect to the surface, the velocity of a given point on the object is the sum of the velocities shown in Figures 10.25a and 10.25b. Figure 10.25c shows the results of adding these velocities.

Notice that the contact point between the surface and object in Figure 10.25c has a translational speed of zero. At this instant, the rolling object is moving in exactly the same way as if the surface were removed and the object were pivoted at point P and spun about an axis passing through P. We can express the total kinetic energy of this imagined spinning object as

$$K = \tfrac{1}{2}I_P\omega^2 \qquad (10.30)$$

where I_P is the moment of inertia about a rotation axis through P.

Figure 10.24 For pure rolling motion, as the cylinder rotates through an angle θ its center moves a linear distance $s = R\theta$.

Pitfall Prevention 10.6
Equation 10.28 Looks Familiar Equation 10.28 looks very similar to Equation 10.10, so be sure to be clear on the difference. Equation 10.10 gives the *tangential* speed of a point on a *rotating* object located a distance r from a fixed rotation axis if the object is rotating with angular speed ω. Equation 10.28 gives the *translational* speed of the center of mass of a *rolling* object of radius R rotating with angular speed ω.

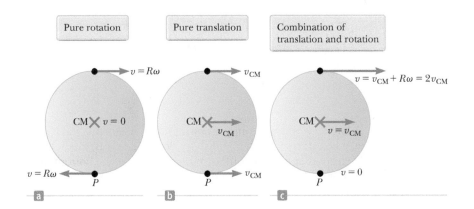

Pure rotation

Pure translation

Combination of translation and rotation

Figure 10.25 The motion of a rolling object can be modeled as a combination of pure translation and pure rotation.

Because the motion of the imagined spinning object is the same at this instant as our actual rolling object, Equation 10.30 also gives the kinetic energy of the rolling object. Applying the parallel-axis theorem, we can substitute $I_P = I_{CM} + MR^2$ into Equation 10.30 to obtain

$$K = \tfrac{1}{2}I_{CM}\omega^2 + \tfrac{1}{2}MR^2\omega^2$$

Using $v_{CM} = R\omega$, this equation can be expressed as

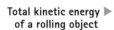

Total kinetic energy of a rolling object ▶

$$K = \tfrac{1}{2}I_{CM}\omega^2 + \tfrac{1}{2}Mv_{CM}^2 \qquad \textbf{(10.31)}$$

The term $\tfrac{1}{2}I_{CM}\omega^2$ represents the rotational kinetic energy of the object about its center of mass, and the term $\tfrac{1}{2}Mv_{CM}^2$ represents the kinetic energy the object would have if it were just translating through space without rotating. Therefore, the total kinetic energy of a rolling object is the sum of the rotational kinetic energy *about* the center of mass and the translational kinetic energy *of* the center of mass. This statement is consistent with the situation illustrated in Figure 10.25, which shows that the velocity of a point on the object is the sum of the velocity of the center of mass and the tangential velocity around the center of mass.

Energy methods can be used to treat a class of problems concerning the rolling motion of an object on a rough incline. For example, consider Figure 10.26, which shows a sphere rolling without slipping after being released from rest at the top of the incline. Accelerated rolling motion is possible only if a friction force is present between the sphere and the incline to produce a net torque about the center of mass. Despite the presence of friction, no loss of mechanical energy occurs because the contact point is at rest relative to the surface at any instant. (On the other hand, if the sphere were to slip, mechanical energy of the sphere–incline–Earth system would decrease due to the nonconservative force of kinetic friction.)

In reality, *rolling friction* causes mechanical energy to transform to internal energy. Rolling friction is due to deformations of the surface and the rolling object. For example, automobile tires flex as they roll on a roadway, representing a transformation of mechanical energy to internal energy. The roadway also deforms a small amount, representing additional rolling friction. In our problem-solving models, we ignore rolling friction unless stated otherwise.

Using $v_{CM} = R\omega$ for pure rolling motion, we can express Equation 10.31 as

$$K = \tfrac{1}{2}I_{CM}\left(\frac{v_{CM}}{R}\right)^2 + \tfrac{1}{2}Mv_{CM}^2$$

$$K = \tfrac{1}{2}\left(\frac{I_{CM}}{R^2} + M\right)v_{CM}^2 \qquad \textbf{(10.32)}$$

For the sphere–Earth system in Figure 10.26, we define the zero configuration of gravitational potential energy to be when the sphere is at the bottom of the incline. Therefore, Equation 8.2 gives

$$\Delta K + \Delta U = 0$$

$$\left[\tfrac{1}{2}\left(\frac{I_{CM}}{R^2} + M\right)v_{CM}^2 - 0\right] + (0 - Mgh) = 0$$

$$v_{CM} = \left[\frac{2gh}{1 + (I_{CM}/MR^2)}\right]^{1/2} \qquad \textbf{(10.33)}$$

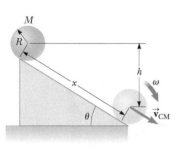

Figure 10.26 A sphere rolling down an incline. Mechanical energy of the sphere–Earth system is conserved if no slipping occurs.

Quick **Q**uiz 10.7 A ball rolls without slipping down incline A, starting from rest. At the same time, a box starts from rest and slides down incline B, which is identical to incline A except that it is frictionless. Which arrives at the bottom first? **(a)** The ball arrives first. **(b)** The box arrives first. **(c)** Both arrive at the same time. **(d)** It is impossible to determine.

Example 10.13 Sphere Rolling Down an Incline AM

For the solid sphere shown in Figure 10.26, calculate the translational speed of the center of mass at the bottom of the incline and the magnitude of the translational acceleration of the center of mass.

SOLUTION

Conceptualize Imagine rolling the sphere down the incline. Compare it in your mind to a book sliding down a frictionless incline. You probably have experience with objects rolling down inclines and may be tempted to think that the sphere would move down the incline faster than the book. You do *not*, however, have experience with objects sliding down *frictionless* inclines! So, which object will reach the bottom first? (See Quick Quiz 10.7.)

Categorize We model the sphere and the Earth as an *isolated system* in terms of *energy* with no nonconservative forces acting. This model is the one that led to Equation 10.33, so we can use that result.

Analyze Evaluate the speed of the center of mass of the sphere from Equation 10.33:

$$(1) \quad v_{CM} = \left[\frac{2gh}{1 + (\frac{2}{5}MR^2 / MR^2)} \right]^{1/2} = (\tfrac{10}{7}gh)^{1/2}$$

This result is less than $\sqrt{2gh}$, which is the speed an object would have if it simply slid down the incline without rotating. (Eliminate the rotation by setting $I_{CM} = 0$ in Eq. 10.33.)

To calculate the translational acceleration of the center of mass, notice that the vertical displacement of the sphere is related to the distance x it moves along the incline through the relationship $h = x \sin \theta$.

Use this relationship to rewrite Equation (1):

$$v_{CM}{}^2 = \tfrac{10}{7}gx \sin \theta$$

Write Equation 2.17 for an object starting from rest and moving through a distance x under constant acceleration:

$$v_{CM}{}^2 = 2a_{CM}x$$

Equate the preceding two expressions to find a_{CM}:

$$a_{CM} = \tfrac{5}{7}g \sin \theta$$

Finalize Both the speed and the acceleration of the center of mass are *independent* of the mass and the radius of the sphere. That is, all homogeneous solid spheres experience the same speed and acceleration on a given incline. Try to verify this statement experimentally with balls of different sizes, such as a marble and a croquet ball.

If we were to repeat the acceleration calculation for a hollow sphere, a solid cylinder, or a hoop, we would obtain similar results in which only the factor in front of $g \sin \theta$ would differ. The constant factors that appear in the expressions for v_{CM} and a_{CM} depend only on the moment of inertia about the center of mass for the specific object. In all cases, the acceleration of the center of mass is *less* than $g \sin \theta$, the value the acceleration would have if the incline were frictionless and no rolling occurred.

Example 10.14 Pulling on a Spool[3] AM

A cylindrically symmetric spool of mass m and radius R sits at rest on a horizontal table with friction (Fig. 10.27). With your hand on a light string wrapped around the axle of radius r, you pull on the spool with a constant horizontal force of magnitude T to the right. As a result, the spool rolls without slipping a distance L along the table with no rolling friction.

(A) Find the final translational speed of the center of mass of the spool.

SOLUTION

Conceptualize Use Figure 10.27 to visualize the motion of the spool when you pull the string. For the spool to roll through a distance L, notice that your hand on the string must pull through a distance *different* from L.

Figure 10.27 (Example 10.14) A spool rests on a horizontal table. A string is wrapped around the axle and is pulled to the right by a hand.

continued

[3]Example 10.14 was inspired in part by C. E. Mungan, "A primer on work–energy relationships for introductory physics," *The Physics Teacher*, **43**:10, 2005.

▶ **10.14** c o n t i n u e d

Categorize The spool is a *rigid object under a net torque,* but the net torque includes that due to the friction force at the bottom of the spool, about which we know nothing. Therefore, an approach based on the rigid object under a net torque model will not be successful. Work is done by your hand on the spool and string, which form a nonisolated system in terms of energy. Let's see if an approach based on the *nonisolated system (energy)* model is fruitful.

....

Analyze The only type of energy that changes in the system is the kinetic energy of the spool. There is no rolling friction, so there is no change in internal energy. The only way that energy crosses the system's boundary is by the work done by your hand on the string. No work is done by the static force of friction on the bottom of the spool (to the left in Fig. 10.27) because the point of application of the force moves through no displacement.

Write the appropriate reduction of the conservation of energy equation, Equation 8.2:

$$(1) \quad W = \Delta K = \Delta K_{\text{trans}} + \Delta K_{\text{rot}}$$

where W is the work done on the string by your hand. To find this work, we need to find the displacement of your hand during the process.

We first find the length of string that has unwound off the spool. If the spool rolls through a distance L, the total angle through which it rotates is $\theta = L/R$. The axle also rotates through this angle.

Use Equation 10.1a to find the total arc length through which the axle turns:

$$\ell = r\theta = \frac{r}{R} L$$

This result also gives the length of string pulled off the axle. Your hand will move through this distance *plus* the distance L through which the spool moves. Therefore, the magnitude of the displacement of the point of application of the force applied by your hand is $\ell + L = L(1 + r/R)$.

Evaluate the work done by your hand on the string:

$$(2) \quad W = TL\left(1 + \frac{r}{R}\right)$$

Substitute Equation (2) into Equation (1):

$$TL\left(1 + \frac{r}{R}\right) = \tfrac{1}{2}mv_{\text{CM}}^2 + \tfrac{1}{2}I\omega^2$$

where I is the moment of inertia of the spool about its center of mass and v_{CM} and ω are the final values after the wheel rolls through the distance L.

Apply the nonslip rolling condition $\omega = v_{\text{CM}}/R$:

$$TL\left(1 + \frac{r}{R}\right) = \tfrac{1}{2}mv_{\text{CM}}^2 + \tfrac{1}{2}I\frac{v_{\text{CM}}^2}{R^2}$$

Solve for v_{CM}:

$$(3) \quad v_{\text{CM}} = \sqrt{\frac{2TL(1 + r/R)}{m(1 + I/mR^2)}}$$

(B) Find the value of the friction force f.

S O L U T I O N

Categorize Because the friction force does no work, we cannot evaluate it from an energy approach. We model the spool as a *nonisolated system,* but this time in terms of *momentum.* The string applies a force across the boundary of the system, resulting in an impulse on the system. Because the forces on the spool are constant, we can model the spool's center of mass as a *particle under constant acceleration.*

....

Analyze Write the impulse–momentum theorem (Eq. 9.40) for the spool:

$$m(v_{\text{CM}} - 0) = (T - f)\Delta t$$
$$(4) \quad mv_{\text{CM}} = (T - f)\Delta t$$

For a particle under constant acceleration starting from rest, Equation 2.14 tells us that the average velocity of the center of mass is half the final velocity.

Use Equation 2.2 to find the time interval for the center of mass of the spool to move a distance L from rest to a final speed v_{CM}:

$$(5) \quad \Delta t = \frac{L}{v_{\text{CM,avg}}} = \frac{2L}{v_{\text{CM}}}$$

▶ **10.14** continued

Substitute Equation (5) into Equation (4):

$$mv_{CM} = (T - f)\frac{2L}{v_{CM}}$$

Solve for the friction force f:

$$f = T - \frac{mv_{CM}^2}{2L}$$

Substitute v_{CM} from Equation (3):

$$f = T - \frac{m}{2L}\left[\frac{2TL(1 + r/R)}{m(1 + I/mR^2)}\right]$$

$$= T - T\frac{(1 + r/R)}{(1 + I/mR^2)} = T\left[\frac{I - mrR}{I + mR^2}\right]$$

Finalize Notice that we could use the impulse–momentum theorem for the translational motion of the spool while ignoring that the spool is rotating! This fact demonstrates the power of our growing list of approaches to solving problems.

Summary

Definitions

The **angular position** of a rigid object is defined as the angle θ between a reference line attached to the object and a reference line fixed in space. The **angular displacement** of a particle moving in a circular path or a rigid object rotating about a fixed axis is $\Delta\theta \equiv \theta_f - \theta_i$.

The **instantaneous angular speed** of a particle moving in a circular path or of a rigid object rotating about a fixed axis is

$$\omega \equiv \frac{d\theta}{dt} \tag{10.3}$$

The **instantaneous angular acceleration** of a particle moving in a circular path or of a rigid object rotating about a fixed axis is

$$\alpha \equiv \frac{d\omega}{dt} \tag{10.5}$$

When a rigid object rotates about a fixed axis, every part of the object has the same angular speed and the same angular acceleration.

The magnitude of the **torque** associated with a force $\vec{F}$ acting on an object at a distance r from the rotation axis is

$$\tau = rF\sin\phi = Fd \tag{10.14}$$

where ϕ is the angle between the position vector of the point of application of the force and the force vector, and d is the moment arm of the force, which is the perpendicular distance from the rotation axis to the line of action of the force.

The **moment of inertia of a system of particles** is defined as

$$I \equiv \sum_i m_i r_i^2 \tag{10.19}$$

where m_i is the mass of the ith particle and r_i is its distance from the rotation axis.

Concepts and Principles

When a rigid object rotates about a fixed axis, the angular position, angular speed, and angular acceleration are related to the translational position, translational speed, and translational acceleration through the relationships

$$s = r\theta \tag{10.1a}$$

$$v = r\omega \tag{10.10}$$

$$a_t = r\alpha \tag{10.11}$$

If a rigid object rotates about a fixed axis with angular speed ω, its **rotational kinetic energy** can be written

$$K_R = \tfrac{1}{2}I\omega^2 \tag{10.24}$$

where I is the moment of inertia of the object about the axis of rotation.

The **moment of inertia of a rigid object** is

$$I = \int r^2\, dm \tag{10.20}$$

where r is the distance from the mass element dm to the axis of rotation.

continued

The rate at which work is done by an external force in rotating a rigid object about a fixed axis, or the **power** delivered, is

$$P = \tau\omega \qquad \textbf{(10.26)}$$

If work is done on a rigid object and the only result of the work is rotation about a fixed axis, the net work done by external forces in rotating the object equals the change in the rotational kinetic energy of the object:

$$W = \tfrac{1}{2}I\omega_f^2 - \tfrac{1}{2}I\omega_i^2 \qquad \textbf{(10.27)}$$

The **total kinetic energy** of a rigid object rolling on a rough surface without slipping equals the rotational kinetic energy about its center of mass plus the translational kinetic energy of the center of mass:

$$K = \tfrac{1}{2}I_{CM}\omega^2 + \tfrac{1}{2}Mv_{CM}^2 \qquad \textbf{(10.31)}$$

Analysis Models for Problem Solving

Rigid Object Under Constant Angular Acceleration. If a rigid object rotates about a fixed axis under constant angular acceleration, one can apply equations of kinematics that are analogous to those for translational motion of a particle under constant acceleration:

α = constant

$$\omega_f = \omega_i + \alpha t \qquad \textbf{(10.6)}$$

$$\theta_f = \theta_i + \omega_i t + \tfrac{1}{2}\alpha t^2 \qquad \textbf{(10.7)}$$

$$\omega_f^2 = \omega_i^2 + 2\alpha(\theta_f - \theta_i) \qquad \textbf{(10.8)}$$

$$\theta_f = \theta_i + \tfrac{1}{2}(\omega_i + \omega_f)t \qquad \textbf{(10.9)}$$

Rigid Object Under a Net Torque. If a rigid object free to rotate about a fixed axis has a net external torque acting on it, the object undergoes an angular acceleration α, where

$$\sum \tau_{ext} = I\alpha \qquad \textbf{(10.18)}$$

This equation is the rotational analog to Newton's second law in the particle under a net force model.

Objective Questions

1. denotes answer available in *Student Solutions Manual/Study Guide*

1. A cyclist rides a bicycle with a wheel radius of 0.500 m across campus. A piece of plastic on the front rim makes a clicking sound every time it passes through the fork. If the cyclist counts 320 clicks between her apartment and the cafeteria, how far has she traveled? (a) 0.50 km (b) 0.80 km (c) 1.0 km (d) 1.5 km (e) 1.8 km

2. Consider an object on a rotating disk a distance r from its center, held in place on the disk by static friction. Which of the following statements is *not* true concerning this object? (a) If the angular speed is constant, the object must have constant tangential speed. (b) If the angular speed is constant, the object is not accelerated. (c) The object has a tangential acceleration only if the disk has an angular acceleration. (d) If the disk has an angular acceleration, the object has both a centripetal acceleration and a tangential acceleration. (e) The object always has a centripetal acceleration except when the angular speed is zero.

3. A wheel is rotating about a fixed axis with constant angular acceleration 3 rad/s². At different moments, its angular speed is −2 rad/s, 0, and +2 rad/s. For a point on the rim of the wheel, consider at these moments the magnitude of the tangential component of acceleration and the magnitude of the radial component of acceleration. Rank the following five items from largest to smallest: (a) $|a_t|$ when $\omega = -2$ rad/s, (b) $|a_r|$ when

$\omega = -2$ rad/s, (c) $|a_r|$ when $\omega = 0$, (d) $|a_t|$ when $\omega = 2$ rad/s, and (e) $|a_r|$ when $\omega = 2$ rad/s. If two items are equal, show them as equal in your ranking. If a quantity is equal to zero, show that fact in your ranking.

4. A grindstone increases in angular speed from 4.00 rad/s to 12.00 rad/s in 4.00 s. Through what angle does it turn during that time interval if the angular acceleration is constant? (a) 8.00 rad (b) 12.0 rad (c) 16.0 rad (d) 32.0 rad (e) 64.0 rad

5. Suppose a car's standard tires are replaced with tires 1.30 times larger in diameter. **(i)** Will the car's speedometer reading be (a) 1.69 times too high, (b) 1.30 times too high, (c) accurate, (d) 1.30 times too low, (e) 1.69 times too low, or (f) inaccurate by an unpredictable factor? **(ii)** Will the car's fuel economy in miles per gallon or km/L appear to be (a) 1.69 times better, (b) 1.30 times better, (c) essentially the same, (d) 1.30 times worse, or (e) 1.69 times worse?

6. Figure OQ10.6 shows a system of four particles joined by light, rigid rods. Assume $a = b$ and M is larger than m. About which of the coordinate axes does the system have **(i)** the smallest and **(ii)** the largest moment of inertia? (a) the x axis (b) the y axis (c) the z axis. (d) The moment of inertia has the same small value for two axes. (e) The moment of inertia is the same for all three axes.

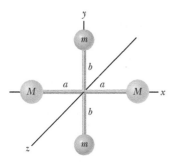

Figure OQ10.6

7. As shown in Figure OQ10.7, a cord is wrapped onto a cylindrical reel mounted on a fixed, frictionless, horizontal axle. When does the reel have a greater magnitude of angular acceleration? (a) When the cord is pulled down with a constant force of 50 N. (b) When an object of weight 50 N is hung from the cord and released. (c) The angular accelerations in parts (a) and (b) are equal. (d) It is impossible to determine.

Figure OQ10.7 Objective Question 7 and Conceptual Question 4.

8. A constant net torque is exerted on an object. Which of the following quantities for the object cannot be constant? Choose all that apply. (a) angular position (b) angular velocity (c) angular acceleration (d) moment of inertia (e) kinetic energy

9. A basketball rolls across a classroom floor without slipping, with its center of mass moving at a certain speed. A block of ice of the same mass is set sliding across the floor with the same speed along a parallel line. Which object has more **(i)** kinetic energy and **(ii)** momentum? (a) The basketball does. (b) The ice does. (c) The two quantities are equal. **(iii)** The two objects encounter a ramp sloping upward. Which object will travel farther up the ramp? (a) The basketball will. (b) The ice will. (c) They will travel equally far up the ramp.

10. A toy airplane hangs from the ceiling at the bottom end of a string. You turn the airplane many times to wind up the string clockwise and release it. The airplane starts to spin counterclockwise, slowly at first and then faster and faster. Take counterclockwise as the positive sense and assume friction is negligible. When the string is entirely unwound, the airplane has its maximum rate of rotation. **(i)** At this moment, is its angular acceleration (a) positive, (b) negative, or (c) zero? **(ii)** The airplane continues to spin, winding the string counterclockwise as it slows down. At the moment it momentarily stops, is its angular acceleration (a) positive, (b) negative, or (c) zero?

11. A solid aluminum sphere of radius R has moment of inertia I about an axis through its center. Will the moment of inertia about a central axis of a solid aluminum sphere of radius $2R$ be (a) $2I$, (b) $4I$, (c) $8I$, (d) $16I$, or (e) $32I$?

Conceptual Questions   1. denotes answer available in *Student Solutions Manual/Study Guide*

1. Is it possible to change the translational kinetic energy of an object without changing its rotational energy?

2. Must an object be rotating to have a nonzero moment of inertia?

3. Suppose just two external forces act on a stationary, rigid object and the two forces are equal in magnitude and opposite in direction. Under what condition does the object start to rotate?

4. Explain how you might use the apparatus described in Figure OQ10.7 to determine the moment of inertia of the wheel. *Note:* If the wheel does not have a uniform mass density, the moment of inertia is not necessarily equal to $\frac{1}{2}MR^2$.

5. Using the results from Example 10.6, how would you calculate the angular speed of the wheel and the linear speed of the hanging object at $t = 2$ s, assuming the system is released from rest at $t = 0$?

6. Explain why changing the axis of rotation of an object changes its moment of inertia.

7. Suppose you have two eggs, one hard-boiled and the other uncooked. You wish to determine which is the hard-boiled egg without breaking the eggs, which

can be done by spinning the two eggs on the floor and comparing the rotational motions. (a) Which egg spins faster? (b) Which egg rotates more uniformly? (c) Which egg begins spinning again after being stopped and then immediately released? Explain your answers to parts (a), (b), and (c).

8. Suppose you set your textbook sliding across a gymnasium floor with a certain initial speed. It quickly stops moving because of a friction force exerted on it by the floor. Next, you start a basketball rolling with the same initial speed. It keeps rolling from one end of the gym to the other. (a) Why does the basketball roll so far? (b) Does friction significantly affect the basketball's motion?

9. (a) What is the angular speed of the second hand of an analog clock? (b) What is the direction of $\vec{\omega}$ as you view a clock hanging on a vertical wall? (c) What is the magnitude of the angular acceleration vector $\vec{\alpha}$ of the second hand?

10. One blade of a pair of scissors rotates counterclockwise in the xy plane. (a) What is the direction of $\vec{\omega}$ for the blade? (b) What is the direction of $\vec{\alpha}$ if the magnitude of the angular velocity is decreasing in time?

11. If you see an object rotating, is there necessarily a net torque acting on it?

12. If a small sphere of mass M were placed at the end of the rod in Figure 10.21, would the result for ω be greater than, less than, or equal to the value obtained in Example 10.11?

13. Three objects of uniform density—a solid sphere, a solid cylinder, and a hollow cylinder—are placed at the top of an incline (Fig. CQ10.13). They are all released from rest at the same elevation and roll without slipping. (a) Which object reaches the bottom first? (b) Which reaches it last? *Note:* The result is independent of the masses and the radii of the objects. (Try this activity at home!)

Figure CQ10.13

14. Which of the entries in Table 10.2 applies to finding the moment of inertia (a) of a long, straight sewer pipe rotating about its axis of symmetry? (b) Of an embroidery hoop rotating about an axis through its center and perpendicular to its plane? (c) Of a uniform door turning on its hinges? (d) Of a coin turning about an axis through its center and perpendicular to its faces?

15. Figure CQ10.15 shows a side view of a child's tricycle with rubber tires on a horizontal concrete sidewalk. If a string were attached to the upper pedal on the far side and pulled forward horizontally, the tricycle would start to roll forward. (a) Instead, assume a string is attached to the lower pedal on the near side and pulled forward horizontally as shown by A. Will the tricycle start to roll? If so, which way? Answer the same questions if (b) the string is pulled forward and upward as shown by B, (c) if the string is pulled straight down as shown by C, and (d) if the string is pulled forward and downward as shown by D. (e) **What If?** Suppose the string is instead attached to the rim of the front wheel and pulled upward and backward as shown by E. Which way does the tricycle roll? (f) Explain a pattern of reasoning, based on the figure, that makes it easy to answer questions such as these. What physical quantity must you evaluate?

Figure CQ10.15

16. A person balances a meterstick in a horizontal position on the extended index fingers of her right and left hands. She slowly brings the two fingers together. The stick remains balanced, and the two fingers always meet at the 50-cm mark regardless of their original positions. (Try it!) Explain why that occurs.

Problems

Section 10.1 Angular Position, Velocity, and Acceleration

1. (a) Find the angular speed of the Earth's rotation about its axis. (b) How does this rotation affect the shape of the Earth?

2. A potter's wheel moves uniformly from rest to an angular speed of 1.00 rev/s in 30.0 s. (a) Find its average angular acceleration in radians per second per second. (b) Would doubling the angular acceleration during the given period have doubled the final angular speed?

3. During a certain time interval, the angular position of a swinging door is described by $\theta = 5.00 + 10.0t + 2.00t^2$, where θ is in radians and t is in seconds. Determine the angular position, angular speed, and angular acceleration of the door (a) at $t = 0$ and (b) at $t = 3.00$ s.

4. A bar on a hinge starts from rest and rotates with an angular acceleration $\alpha = 10 + 6t$, where α is in rad/s^2 and t is in seconds. Determine the angle in radians through which the bar turns in the first 4.00 s.

Section 10.2 Analysis Model: Rigid Object Under Constant Angular Acceleration

5. A wheel starts from rest and rotates with constant angular acceleration to reach an angular speed of 12.0 rad/s in 3.00 s. Find (a) the magnitude of the angular acceleration of the wheel and (b) the angle in radians through which it rotates in this time interval.

6. A centrifuge in a medical laboratory rotates at an angular speed of 3 600 rev/min. When switched off, it rotates through 50.0 revolutions before coming to rest. Find the constant angular acceleration of the centrifuge.

7. An electric motor rotating a workshop grinding wheel at 1.00×10^2 rev/min is switched off. Assume the wheel has a constant negative angular acceleration of magnitude 2.00 rad/s². (a) How long does it take the grinding wheel to stop? (b) Through how many radians has the wheel turned during the time interval found in part (a)?

8. A machine part rotates at an angular speed of 0.060 rad/s; its speed is then increased to 2.2 rad/s at an angular acceleration of 0.70 rad/s². (a) Find the angle through which the part rotates before reaching this final speed. (b) If both the initial and final angular speeds are doubled and the angular acceleration remains the same, by what factor is the angular displacement changed? Why?

9. A dentist's drill starts from rest. After 3.20 s of constant angular acceleration, it turns at a rate of 2.51×10^4 rev/min. (a) Find the drill's angular acceleration. (b) Determine the angle (in radians) through which the drill rotates during this period.

10. *Why is the following situation impossible?* Starting from rest, a disk rotates around a fixed axis through an angle of 50.0 rad in a time interval of 10.0 s. The angular acceleration of the disk is constant during the entire motion, and its final angular speed is 8.00 rad/s.

11. A rotating wheel requires 3.00 s to rotate through 37.0 revolutions. Its angular speed at the end of the 3.00-s interval is 98.0 rad/s. What is the constant angular acceleration of the wheel?

12. The tub of a washer goes into its spin cycle, starting from rest and gaining angular speed steadily for 8.00 s, at which time it is turning at 5.00 rev/s. At this point, the person doing the laundry opens the lid, and a safety switch turns off the washer. The tub smoothly slows to rest in 12.0 s. Through how many revolutions does the tub turn while it is in motion?

13. A spinning wheel is slowed down by a brake, giving it a constant angular acceleration of −5.60 rad/s². During a 4.20-s time interval, the wheel rotates through 62.4 rad. What is the angular speed of the wheel at the end of the 4.20-s interval?

14. **Review.** Consider a tall building located on the Earth's equator. As the Earth rotates, a person on the top floor of the building moves faster than someone on the ground with respect to an inertial reference frame because the person on the ground is closer to the Earth's axis. Consequently, if an object is dropped from the top floor to the ground a distance *h* below, it lands east of the point vertically below where it was dropped. (a) How far to the east will the object land? Express your answer in terms of *h*, *g*, and the angular speed ω of the Earth. Ignore air resistance and assume the free-fall acceleration is constant over this range of heights. (b) Evaluate the eastward displacement for *h* = 50.0 m. (c) In your judgment,

were we justified in ignoring this aspect of the *Coriolis effect* in our previous study of free fall? (d) Suppose the angular speed of the Earth were to decrease due to tidal friction with constant angular acceleration. Would the eastward displacement of the dropped object increase or decrease compared with that in part (b)?

Section 10.3 **Angular and Translational Quantities**

15. A racing car travels on a circular track of radius 250 m. Assuming the car moves with a constant speed of 45.0 m/s, find (a) its angular speed and (b) the magnitude and direction of its acceleration.

16. Make an order-of-magnitude estimate of the number of revolutions through which a typical automobile tire turns in one year. State the quantities you measure or estimate and their values.

17. A discus thrower (Fig. P4.33, page 104) accelerates a discus from rest to a speed of 25.0 m/s by whirling it through 1.25 rev. Assume the discus moves on the arc of a circle 1.00 m in radius. (a) Calculate the final angular speed of the discus. (b) Determine the magnitude of the angular acceleration of the discus, assuming it to be constant. (c) Calculate the time interval required for the discus to accelerate from rest to 25.0 m/s.

18. Figure P10.18 shows the drive train of a bicycle that has wheels 67.3 cm in diameter and pedal cranks 17.5 cm long. The cyclist pedals at a steady cadence of 76.0 rev/min. The chain engages with a front sprocket 15.2 cm in diameter and a rear sprocket 7.00 cm in diameter. Calculate (a) the speed of a link of the chain relative to the bicycle frame, (b) the angular speed of the bicycle wheels, and (c) the speed of the bicycle relative to the road. (d) What pieces of data, if any, are not necessary for the calculations?

Rear sprocket

Pedal crank
Front sprocket

Chain

Figure P10.18

19. A wheel 2.00 m in diameter lies in a vertical plane and rotates about its central axis with a constant angular acceleration of 4.00 rad/s². The wheel starts at rest at *t* = 0, and the radius vector of a certain point *P* on the rim makes an angle of 57.3° with the horizontal at this time. At *t* = 2.00 s, find (a) the angular speed of the wheel and, for point *P*, (b) the tangential speed, (c) the total acceleration, and (d) the angular position.

20. A car accelerates uniformly from rest and reaches a speed of 22.0 m/s in 9.00 s. Assuming the diameter of a tire is 58.0 cm, (a) find the number of revolutions the tire makes during this motion, assuming that no slipping occurs. (b) What is the final angular speed of a tire in revolutions per second?

21. A disk 8.00 cm in radius rotates at a constant rate of
[M] 1 200 rev/min about its central axis. Determine (a) its angular speed in radians per second, (b) the tangential speed at a point 3.00 cm from its center, (c) the radial acceleration of a point on the rim, and (d) the total distance a point on the rim moves in 2.00 s.

22. A straight ladder is leaning against the wall of a house. The ladder has rails 4.90 m long, joined by rungs 0.410 m long. Its bottom end is on solid but sloping ground so that the top of the ladder is 0.690 m to the left of where it should be, and the ladder is unsafe to climb. You want to put a flat rock under one foot of the ladder to compensate for the slope of the ground. (a) What should be the thickness of the rock? (b) Does using ideas from this chapter make it easier to explain the solution to part (a)? Explain your answer.

23. A car traveling on a flat (unbanked), circular track
[W] accelerates uniformly from rest with a tangential acceleration of 1.70 m/s². The car makes it one-quarter of the way around the circle before it skids off the track. From these data, determine the coefficient of static friction between the car and the track.

24. A car traveling on a flat (unbanked), circular track accelerates uniformly from rest with a tangential acceleration of a. The car makes it one-quarter of the way around the circle before it skids off the track. From these data, determine the coefficient of static friction between the car and the track.

25. In a manufacturing process, a large, cylindrical roller is used to flatten material fed beneath it. The diameter of the roller is 1.00 m, and, while being driven into rotation around a fixed axis, its angular position is expressed as

$$\theta = 2.50t^2 - 0.600t^3$$

where θ is in radians and t is in seconds. (a) Find the maximum angular speed of the roller. (b) What is the maximum tangential speed of a point on the rim of the roller? (c) At what time t should the driving force be removed from the roller so that the roller does not reverse its direction of rotation? (d) Through how many rotations has the roller turned between $t = 0$ and the time found in part (c)?

26. Review. A small object with mass 4.00 kg moves counterclockwise with constant angular speed 1.50 rad/s in a circle of radius 3.00 m centered at the origin. It starts at the point with position vector $3.00\hat{i}$ m. It then undergoes an angular displacement of 9.00 rad. (a) What is its new position vector? Use unit-vector notation for all vector answers. (b) In what quadrant is the particle located, and what angle does its position vector make with the positive x axis? (c) What is its velocity? (d) In what direction is it moving? (e) What is its acceleration? (f) Make a sketch of its position, velocity, and acceleration vectors. (g) What total force is exerted on the object?

Section 10.4 Torque

27. Find the net torque on the wheel in Figure P10.27 about
[M] the axle through O, taking $a = 10.0$ cm and $b = 25.0$ cm.

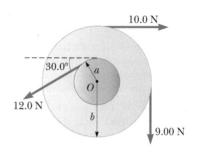

Figure P10.27

28. The fishing pole in Figure P10.28 makes an angle of
[W] 20.0° with the horizontal. What is the torque exerted by the fish about an axis perpendicular to the page and passing through the angler's hand if the fish pulls on the fishing line with a force $\vec{F} = 100$ N at an angle 37.0° below the horizontal? The force is applied at a point 2.00 m from the angler's hands.

Figure P10.28

Section 10.5 Analysis Model: Rigid Object Under a Net Torque

29. An electric motor turns a flywheel through a drive belt that joins a pulley on the motor and a pulley that is rigidly attached to the flywheel as shown in Figure P10.29. The flywheel is a solid disk with a mass of 80.0 kg and a radius $R = 0.625$ m. It turns on a frictionless axle. Its pulley has much smaller mass and a radius of $r = 0.230$ m. The tension T_u in the upper (taut) segment of the belt is 135 N, and the flywheel has a clockwise angular acceleration of 1.67 rad/s². Find the tension in the lower (slack) segment of the belt.

Figure P10.29

30. A grinding wheel is in the form of a uniform solid disk
[AMT] of radius 7.00 cm and mass 2.00 kg. It starts from rest
[W] and accelerates uniformly under the action of the constant torque of 0.600 N · m that the motor exerts on the wheel. (a) How long does the wheel take to reach its final operating speed of 1 200 rev/min? (b) Through how many revolutions does it turn while accelerating?

31. A 150-kg merry-go-round in the shape of a uniform, solid, horizontal disk of radius 1.50 m is set in motion by wrapping a rope about the rim of the disk and pulling on the rope. What constant force must be exerted on the rope to bring the merry-go-round from rest to an angular speed of 0.500 rev/s in 2.00 s?

32. Review. A block of mass m_1 = 2.00 kg and a block of mass m_2 = 6.00 kg are connected by a massless string over a pulley in the shape of a solid disk having radius R = 0.250 m and mass M = 10.0 kg. The fixed, wedge-shaped ramp makes an angle of θ = 30.0° as shown in Figure P10.32. The coefficient of kinetic friction is 0.360 for both blocks. (a) Draw force diagrams of both blocks and of the pulley. Determine (b) the acceleration of the two blocks and (c) the tensions in the string on both sides of the pulley.

Figure P10.32

33. A model airplane with mass 0.750 kg is tethered to the ground by a wire so that it flies in a horizontal circle 30.0 m in radius. The airplane engine provides a net thrust of 0.800 N perpendicular to the tethering wire. (a) Find the torque the net thrust produces about the center of the circle. (b) Find the angular acceleration of the airplane. (c) Find the translational acceleration of the airplane tangent to its flight path.

34. A disk having moment of inertia 100 kg · m² is free to rotate without friction, starting from rest, about a fixed axis through its center. A tangential force whose magnitude can range from F = 0 to F = 50.0 N can be applied at any distance ranging from R = 0 to R = 3.00 m from the axis of rotation. (a) Find a pair of values of F and R that cause the disk to complete 2.00 rev in 10.0 s. (b) Is your answer for part (a) a unique answer? How many answers exist?

35. The combination of an applied force and a friction force produces a constant total torque of 36.0 N · m on a wheel rotating about a fixed axis. The applied force acts for 6.00 s. During this time, the angular speed of the wheel increases from 0 to 10.0 rad/s. The applied force is then removed, and the wheel comes to rest in 60.0 s. Find (a) the moment of inertia of the wheel, (b) the magnitude of the torque due to friction, and (c) the total number of revolutions of the wheel during the entire interval of 66.0 s.

36. Review. Consider the system shown in Figure P10.36 with m_1 = 20.0 kg, m_2 = 12.5 kg, R = 0.200 m, and the mass of the pulley M = 5.00 kg.

Figure P10.36

Object m_2 is resting on the floor, and object m_1 is 4.00 m above the floor when it is released from rest. The pulley axis is frictionless. The cord is light, does not stretch, and does not slip on the pulley. (a) Calculate the time interval required for m_1 to hit the floor. (b) How would your answer change if the pulley were massless?

37. A potter's wheel—a thick stone disk of radius 0.500 m and mass 100 kg—is freely rotating at 50.0 rev/min. The potter can stop the wheel in 6.00 s by pressing a wet rag against the rim and exerting a radially inward force of 70.0 N. Find the effective coefficient of kinetic friction between wheel and rag.

Section 10.6 Calculation of Moments of Inertia

38. Imagine that you stand tall and turn about a vertical axis through the top of your head and the point halfway between your ankles. Compute an order-of-magnitude estimate for the moment of inertia of your body for this rotation. In your solution, state the quantities you measure or estimate and their values.

39. A uniform, thin, solid door has height 2.20 m, width 0.870 m, and mass 23.0 kg. (a) Find its moment of inertia for rotation on its hinges. (b) Is any piece of data unnecessary?

40. Two balls with masses M and m are connected by a rigid rod of length L and negligible mass as shown in Figure P10.40. For an axis perpendicular to the rod, (a) show that the system has the minimum moment of inertia when the axis passes through the center of mass. (b) Show that this moment of inertia is $I = \mu L^2$, where $\mu = mM/(m + M)$.

Figure P10.40

41. Figure P10.41 shows a side view of a car tire before it is mounted on a wheel. Model it as having two sidewalls of uniform thickness 0.635 cm and a tread wall of uniform thickness 2.50 cm and width 20.0 cm. Assume the rubber has uniform density 1.10×10^3 kg/m³. Find its moment of inertia about an axis perpendicular to the page through its center.

Figure P10.41

42. Following the procedure used in Example 10.7, prove that the moment of inertia about the y axis of the rigid rod in Figure 10.15 is $\frac{1}{3}ML^2$.

43. Three identical thin rods, each of length L and mass m, are welded perpendicular to one another as shown in Figure P10.43. The assembly is rotated about an axis that passes through the end of one rod and is parallel to another. Determine the moment of inertia of this structure about this axis.

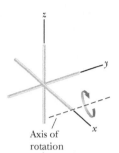

Axis of rotation

Figure P10.43

Section 10.7 Rotational Kinetic Energy

44. Rigid rods of negligible mass lying along the y axis connect three particles (Fig. P10.44). The system rotates about the x axis with an angular speed of 2.00 rad/s. Find (a) the moment of inertia about the x axis, (b) the total rotational kinetic energy evaluated from $\frac{1}{2}I\omega^2$, (c) the tangential speed of each particle, and (d) the total kinetic energy evaluated from $\sum \frac{1}{2}m_i v_i^2$. (e) Compare the answers for kinetic energy in parts (a) and (b).

4.00 kg $y = 3.00$ m

O

2.00 kg $y = -2.00$ m

3.00 kg $y = -4.00$ m

Figure P10.44

45. The four particles in Figure P10.45 are connected by rigid rods of negligible mass. The origin is at the center of the rectangle. The system rotates in the xy plane about the z axis with an angular speed of 6.00 rad/s. Calculate (a) the moment of inertia of the system about the z axis and (b) the rotational kinetic energy of the system.

3.00 kg   2.00 kg

6.00 m

O x

2.00 kg

4.00 kg

←4.00 m→

Figure P10.45

46. Many machines employ cams for various purposes, such as opening and closing valves. In Figure P10.46, the cam is a circular disk of radius R with a hole of diameter R cut through it. As shown in the figure, the

←R→

←$2R$→

Figure P10.46

hole does not pass through the center of the disk. The cam with the hole cut out has mass M. The cam is mounted on a uniform, solid, cylindrical shaft of diameter R and also of mass M. What is the kinetic energy of the cam–shaft combination when it is rotating with angular speed ω about the shaft's axis?

47. A *war-wolf* or *trebuchet* is a device used during the Middle Ages to throw rocks at castles and now sometimes used to fling large vegetables and pianos as a sport. A simple trebuchet is shown in Figure P10.47. Model it as a stiff rod of negligible mass, 3.00 m long, joining particles of mass $m_1 = 0.120$ kg and $m_2 = 60.0$ kg at its ends. It can turn on a frictionless, horizontal axle perpendicular to the rod and 14.0 cm from the large-mass particle. The operator releases the trebuchet from rest in a horizontal orientation. (a) Find the maximum speed that the small-mass object attains. (b) While the small-mass object is gaining speed, does it move with constant acceleration? (c) Does it move with constant tangential acceleration? (d) Does the trebuchet move with constant angular acceleration? (e) Does it have constant momentum? (f) Does the trebuchet–Earth system have constant mechanical energy?

m_1 m_2

←—— 3.00 m ——→

Figure P10.47

Section 10.8 Energy Considerations in Rotational Motion

48. A horizontal 800-N merry-go-round is a solid disk of radius 1.50 m and is started from rest by a constant horizontal force of 50.0 N applied tangentially to the edge of the disk. Find the kinetic energy of the disk after 3.00 s.

49. Big Ben, the nickname for the clock in Elizabeth Tower (named after the Queen in 2012) in London, has an hour hand 2.70 m long with a mass of 60.0 kg and a minute hand 4.50 m long with a mass of 100 kg (Fig. P10.49). Calculate the total rotational kinetic energy of the two hands about the axis of rotation. (You may

Figure P10.49 Problems 49 and 72.

model the hands as long, thin rods rotated about one end. Assume the hour and minute hands are rotating at a constant rate of one revolution per 12 hours and 60 minutes, respectively.)

50. Consider two objects with $m_1 >$ m_2 connected by a light string that passes over a pulley having a moment of inertia of I about its axis of rotation as shown in Figure P10.50. The string does not slip on the pulley or stretch. The pulley turns without friction. The two objects are released from rest separated by a vertical distance $2h$. (a) Use the principle of conservation of energy to find the translational speeds of the objects as they pass each other. (b) Find the angular speed of the pulley at this time.

Figure P10.50

51. The top in Figure P10.51 has a moment of inertia of 4.00×10^{-4} kg · m² and is initially at rest. It is free to rotate about the stationary axis AA'. A string, wrapped around a peg along the axis of the top, is pulled in such a manner as to maintain a constant tension of 5.57 N. If the string does not slip while it is unwound from the peg, what is the angular speed of the top after 80.0 cm of string has been pulled off the peg?

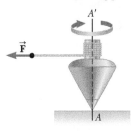

Figure P10.51

52. *Why is the following situation impossible?* In a large city with an air-pollution problem, a bus has no combustion engine. It runs over its citywide route on energy drawn from a large, rapidly rotating flywheel under the floor of the bus. The flywheel is spun up to its maximum rotation rate of 3 000 rev/min by an electric motor at the bus terminal. Every time the bus speeds up, the flywheel slows down slightly. The bus is equipped with regenerative braking so that the flywheel can speed up when the bus slows down. The flywheel is a uniform solid cylinder with mass 1 200 kg and radius 0.500 m. The bus body does work against air resistance and rolling resistance at the average rate of 25.0 hp as it travels its route with an average speed of 35.0 km/h.

53. In Figure P10.53, the hanging object has a mass of $m_1 =$ 0.420 kg; the sliding block has a mass of $m_2 =$ 0.850 kg;

Figure P10.53

and the pulley is a hollow cylinder with a mass of $M =$ 0.350 kg, an inner radius of $R_1 = 0.020\,0$ m, and an outer radius of $R_2 = 0.030\,0$ m. Assume the mass of the spokes is negligible. The coefficient of kinetic friction between the block and the horizontal surface is $\mu_k =$ 0.250. The pulley turns without friction on its axle. The light cord does not stretch and does not slip on the pulley. The block has a velocity of $v_i = 0.820$ m/s toward the pulley when it passes a reference point on the table. (a) Use energy methods to predict its speed after it has moved to a second point, 0.700 m away. (b) Find the angular speed of the pulley at the same moment.

54. **Review.** A thin, cylindrical rod $\ell = 24.0$ cm long with mass $m = 1.20$ kg has a ball of diameter $d =$ 8.00 cm and mass $M =$ 2.00 kg attached to one end. The arrangement is originally vertical and stationary, with the ball at the top as shown in Figure P10.54. The combination is free to pivot about the bottom end of the rod after being given a slight nudge. (a) After the combination rotates through 90 degrees, what is its rotational kinetic energy? (b) What is the angular speed of the rod and ball? (c) What is the linear speed of the center of mass of the ball? (d) How does it compare with the speed had the ball fallen freely through the same distance of 28 cm?

Figure P10.54

55. **Review.** An object with a mass of $m = 5.10$ kg is attached to the free end of a light string wrapped around a reel of radius $R = 0.250$ m and mass $M =$ 3.00 kg. The reel is a solid disk, free to rotate in a vertical plane about the horizontal axis passing through its center as shown in Figure P10.55. The suspended object is released from rest 6.00 m above the floor. Determine (a) the tension in the string, (b) the acceleration of the object, and (c) the speed with which the object hits the floor. (d) Verify your answer to part (c) by using the isolated system (energy) model.

Figure P10.55

56. This problem describes one experimental method for determining the moment of inertia of an irregularly shaped object such as the payload for a satellite. Figure P10.56 shows a counterweight of mass m suspended by a cord wound around a spool of radius r, forming part of a turntable supporting the object. The turntable can rotate without friction. When the counterweight is released from rest, it descends through a distance h, acquiring a speed v. Show that the moment of inertia I of the rotating apparatus (including the turntable) is $mr^2(2gh/v^2 - 1)$.

Figure P10.56

57. A uniform solid disk of radius R and mass M is free to rotate on a frictionless pivot through a point on its rim (Fig. P10.57). If the disk is released from rest in the position shown by the copper-colored circle, (a) what is the speed of its center of mass when the disk reaches the position indicated by the dashed circle? (b) What is the speed of the lowest point on the disk in the dashed position? (c) **What If?** Repeat part (a) using a uniform hoop.

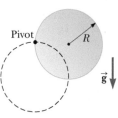

Figure P10.57

58. The head of a grass string trimmer has 100 g of cord wound in a light, cylindrical spool with inside diameter 3.00 cm and outside diameter 18.0 cm as shown in Figure P10.58. The cord has a linear density of 10.0 g/m. A single strand of the cord extends 16.0 cm from the outer edge of the spool. (a) When switched on, the trimmer speeds up from 0 to 2 500 rev/min in 0.215 s. What average power is delivered to the head by the trimmer motor while it is accelerating? (b) When the trimmer is cutting grass, it spins at 2 000 rev/min and the grass exerts an average tangential force of 7.65 N on the outer end of the cord, which is still at a radial distance of 16.0 cm from the outer edge of the spool. What is the power delivered to the head under load?

Figure P10.58

Section 10.9 Rolling Motion of a Rigid Object

59. A cylinder of mass 10.0 kg rolls without slipping on a horizontal surface. At a certain instant, its center of mass has a speed of 10.0 m/s. Determine (a) the translational kinetic energy of its center of mass, (b) the rotational kinetic energy about its center of mass, and (c) its total energy.

60. A solid sphere is released from height h from the top of an incline making an angle θ with the horizontal. Calculate the speed of the sphere when it reaches the bottom of the incline (a) in the case that it rolls without slipping and (b) in the case that it slides frictionlessly without rolling. (c) Compare the time intervals required to reach the bottom in cases (a) and (b).

61. (a) Determine the acceleration of the center of mass of a uniform solid disk rolling down an incline making angle θ with the horizontal. (b) Compare the acceleration found in part (a) with that of a uniform hoop. (c) What is the minimum coefficient of friction required to maintain pure rolling motion for the disk?

62. A smooth cube of mass m and edge length r slides with speed v on a horizontal surface with negligible friction. The cube then moves up a smooth incline that makes an angle θ with the horizontal. A cylinder of mass m and radius r rolls without slipping with its center of mass moving with speed v and encounters an incline of the same angle of inclination but with sufficient friction that the cylinder continues to roll without slipping. (a) Which object will go the greater distance up the incline? (b) Find the difference between the maximum distances the objects travel up the incline. (c) Explain what accounts for this difference in distances traveled.

63. A uniform solid disk and a uniform hoop are placed side by side at the top of an incline of height h. (a) If they are released from rest and roll without slipping, which object reaches the bottom first? (b) Verify your answer by calculating their speeds when they reach the bottom in terms of h.

64. A tennis ball is a hollow sphere with a thin wall. It is set rolling without slipping at 4.03 m/s on a horizontal section of a track as shown in Figure P10.64. It rolls around the inside of a vertical circular loop of radius $r = 45.0$ cm. As the ball nears the bottom of the loop, the shape of the track deviates from a perfect circle so that the ball leaves the track at a point $h = 20.0$ cm below the horizontal section. (a) Find the ball's speed at the top of the loop. (b) Demonstrate that the ball will not fall from the track at the top of the loop. (c) Find the ball's speed as it leaves the track at the bottom. (d) **What If?** Suppose that static friction between ball and track were

Figure P10.64

negligible so that the ball slid instead of rolling. Would its speed then be higher, lower, or the same at the top of the loop? (e) Explain your answer to part (d).

65. A metal can containing condensed mushroom soup has mass 215 g, height 10.8 cm, and diameter 6.38 cm. It is placed at rest on its side at the top of a 3.00-m-long incline that is at 25.0° to the horizontal and is then released to roll straight down. It reaches the bottom of the incline after 1.50 s. (a) Assuming mechanical energy conservation, calculate the moment of inertia of the can. (b) Which pieces of data, if any, are unnecessary for calculating the solution? (c) Why can't the moment of inertia be calculated from $I = \frac{1}{2}mr^2$ for the cylindrical can?

Additional Problems

66. As shown in Figure 10.13 on page 306, toppling chimneys often break apart in midfall because the mortar between the bricks cannot withstand much shear stress. As the chimney begins to fall, shear forces must act on the topmost sections to accelerate them tangentially so that they can keep up with the rotation of the lower part of the stack. For simplicity, let us model the chimney as a uniform rod of length ℓ pivoted at the lower end. The rod starts at rest in a vertical position (with the frictionless pivot at the bottom) and falls over under the influence of gravity. What fraction of the length of the rod has a tangential acceleration greater than $g \sin \theta$, where θ is the angle the chimney makes with the vertical axis?

67. **Review.** A 4.00-m length of light nylon cord is wound around a uniform cylindrical spool of radius 0.500 m and mass 1.00 kg. The spool is mounted on a frictionless axle and is initially at rest. The cord is pulled from the spool with a constant acceleration of magnitude 2.50 m/s². (a) How much work has been done on the spool when it reaches an angular speed of 8.00 rad/s? (b) How long does it take the spool to reach this angular speed? (c) How much cord is left on the spool when it reaches this angular speed?

68. An elevator system in a tall building consists of a 800-kg car and a 950-kg counterweight joined by a light cable of constant length that passes over a pulley of mass 280 kg. The pulley, called a sheave, is a solid cylinder of radius 0.700 m turning on a horizontal axle. The cable does not slip on the sheave. A number n of people, each of mass 80.0 kg, are riding in the elevator car, moving upward at 3.00 m/s and approaching the floor where the car should stop. As an energy-conservation measure, a computer disconnects the elevator motor at just the right moment so that the sheave–car–counterweight system then coasts freely without friction and comes to rest at the floor desired. There it is caught by a simple latch rather than by a massive brake. (a) Determine the distance d the car coasts upward as a function of n. Evaluate the distance for (b) $n = 2$, (c) $n = 12$, and (d) $n = 0$. (e) For what integer values of n does the expression in part (a) apply? (f) Explain your answer to part (e). (g) If an infinite number of people could fit on the elevator, what is the value of d?

69. A shaft is turning at 65.0 rad/s at time $t = 0$. Thereafter, its angular acceleration is given by

$$\alpha = -10.0 - 5.00t$$

where α is in rad/s² and t is in seconds. (a) Find the angular speed of the shaft at $t = 3.00$ s. (b) Through what angle does it turn between $t = 0$ and $t = 3.00$ s?

70. A shaft is turning at angular speed ω at time $t = 0$. Thereafter, its angular acceleration is given by

$$\alpha = A + Bt$$

(a) Find the angular speed of the shaft at time t. (b) Through what angle does it turn between $t = 0$ and t?

71. **Review.** A mixing beater consists of three thin rods, each 10.0 cm long. The rods diverge from a central hub, separated from each other by 120°, and all turn in the same plane. A ball is attached to the end of each rod. Each ball has cross-sectional area 4.00 cm² and is so shaped that it has a drag coefficient of 0.600. Calculate the power input required to spin the beater at 1 000 rev/min (a) in air and (b) in water.

72. The hour hand and the minute hand of Big Ben, the Elizabeth Tower clock in London, are 2.70 m and 4.50 m long and have masses of 60.0 kg and 100 kg, respectively (see Fig. P10.49). (a) Determine the total torque due to the weight of these hands about the axis of rotation when the time reads (i) 3:00, (ii) 5:15, (iii) 6:00, (iv) 8:20, and (v) 9:45. (You may model the hands as long, thin, uniform rods.) (b) Determine all times when the total torque about the axis of rotation is zero. Determine the times to the nearest second, solving a transcendental equation numerically.

73. A long, uniform rod of length L and mass M is pivoted about a frictionless, horizontal pin through one end. The rod is nudged from rest in a vertical position as shown in Figure P10.73. At the instant the rod is horizontal, find (a) its angular speed, (b) the magnitude of its angular acceleration, (c) the x and y components of the acceleration of its center of mass, and (d) the components of the reaction force at the pivot.

Figure P10.73

74. A bicycle is turned upside down while its owner repairs a flat tire on the rear wheel. A friend spins the front wheel, of radius 0.381 m, and observes that drops of water fly off tangentially in an upward direction when the drops are at the same level as the center of the wheel. She measures the height reached by drops moving vertically (Fig. P10.74 on page 332). A drop

that breaks loose from the tire on one turn rises $h = 54.0$ cm above the tangent point. A drop that breaks loose on the next turn rises 51.0 cm above the tangent point. The height to which the drops rise decreases because the angular speed of the wheel decreases. From this information, determine the magnitude of the average angular acceleration of the wheel.

Figure P10.74 Problems 74 and 75.

75. A bicycle is turned upside down while its owner repairs a flat tire on the rear wheel. A friend spins the front wheel, of radius R, and observes that drops of water fly off tangentially in an upward direction when the drops are at the same level as the center of the wheel. She measures the height reached by drops moving vertically (Fig. P10.74). A drop that breaks loose from the tire on one turn rises a distance h_1 above the tangent point. A drop that breaks loose on the next turn rises a distance $h_2 < h_1$ above the tangent point. The height to which the drops rise decreases because the angular speed of the wheel decreases. From this information, determine the magnitude of the average angular acceleration of the wheel.

76. (a) What is the rotational kinetic energy of the Earth about its spin axis? Model the Earth as a uniform sphere and use data from the endpapers of this book. (b) The rotational kinetic energy of the Earth is decreasing steadily because of tidal friction. Assuming the rotational period decreases by 10.0 μs each year, find the change in one day.

77. **Review.** As shown in Figure P10.77, two blocks are con- **GP** nected by a string of negligible mass passing over a pulley of radius $r = 0.250$ m and moment of inertia I. The block on the frictionless incline is moving with a constant acceleration of magnitude $a = 2.00$ m/s^2. From this information, we wish to find the moment of inertia of the pulley. (a) What analysis model is appropriate for the blocks? (b) What analysis model is appropriate

for the pulley? (c) From the analysis model in part (a), find the tension T_1. (d) Similarly, find the tension T_2. (e) From the analysis model in part (b), find a symbolic expression for the moment of inertia of the pulley in terms of the tensions T_1 and T_2, the pulley radius r, and the acceleration a. (f) Find the numerical value of the moment of inertia of the pulley.

78. **Review.** A string is wound around a uniform disk of radius R and mass M. The disk is released from rest with the string vertical and its top end tied to a fixed bar (Fig. P10.78). Show that (a) the tension in the string is one third of the weight of the disk, (b) the magnitude of the acceleration of the center of mass is $2g/3$, and (c) the speed of the center of mass is $(4gh/3)^{1/2}$ after the disk has descended through distance h. (d) Verify your answer to part (c) using the energy approach.

Figure P10.78

79. The reel shown in Figure P10.79 has radius R and moment of inertia I. One end of the block of mass m is connected to a spring of force constant k, and the other end is fastened to a cord wrapped around the reel. The reel axle and the incline are frictionless. The reel is wound counterclockwise so that the spring stretches a distance d from its unstretched position and the reel is then released from rest. Find the angular speed of the reel when the spring is again unstretched.

Figure P10.79

80. A common demonstration, illustrated in Figure P10.80, consists of a ball resting at one end of a uniform board of length ℓ that is hinged at the other end and elevated at an angle θ. A light cup is attached to the board at r_c so that it will catch the ball when the support stick is removed suddenly. (a) Show that the ball will lag behind the falling board when θ is less than 35.3°.

Figure P10.80

Figure P10.77

(b) Assuming the board is 1.00 m long and is supported at this limiting angle, show that the cup must be 18.4 cm from the moving end.

81. A uniform solid sphere of radius r is placed on the inside surface of a hemispherical bowl with radius R. The sphere is released from rest at an angle θ to the vertical and rolls without slipping (Fig. P10.81). Determine the angular speed of the sphere when it reaches the bottom of the bowl.

Figure P10.81

82. **Review.** A spool of wire of mass M and radius R is unwound under a constant force $\vec{F}$ (Fig. P10.82). Assuming the spool is a uniform, solid cylinder that doesn't slip, show that (a) the acceleration of the center of mass is $4\vec{F}/3M$ and (b) the force of friction is to the *right* and equal in magnitude to $F/3$. (c) If the cylinder starts from rest and rolls without slipping, what is the speed of its center of mass after it has rolled through a distance d?

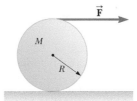

Figure P10.82

83. A solid sphere of mass m and radius r rolls without slipping along the track shown in Figure P10.83. It starts from rest with the lowest point of the sphere at height h above the bottom of the loop of radius R, much larger than r. (a) What is the minimum value of h (in terms of R) such that the sphere completes the loop? (b) What are the force components on the sphere at the point P if $h = 3R$?

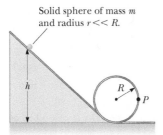

Figure P10.83

84. A thin rod of mass 0.630 kg and length 1.24 m is at rest, hanging vertically from a strong, fixed hinge at its top end. Suddenly, a horizontal impulsive force $14.7\hat{i}$ N is applied to it. (a) Suppose the force acts at the bottom end of the rod. Find the acceleration of its center of mass and (b) the horizontal force the hinge exerts. (c) Suppose the force acts at the midpoint of the rod. Find the acceleration of this point and (d) the horizontal hinge reaction force. (e) Where can the impulse be applied so that the hinge will exert no horizontal force? This point is called the *center of percussion*.

85. A thin rod of length h and mass M is held vertically with its lower end resting on a frictionless, horizontal surface. The rod is then released to fall freely. (a) Determine the speed of its center of mass just before it hits the horizontal surface. (b) **What If?** Now suppose the rod has a fixed pivot at its lower end. Determine the speed of the rod's center of mass just before it hits the surface.

86. **Review.** A clown balances a small spherical grape at the top of his bald head, which also has the shape of a sphere. After drawing sufficient applause, the grape starts from rest and rolls down without slipping. It will leave contact with the clown's scalp when the radial line joining it to the center of curvature makes what angle with the vertical?

Challenge Problems

87. A plank with a mass $M = 6.00$ kg rests on top of two identical, solid, cylindrical rollers that have $R = 5.00$ cm and $m = 2.00$ kg (Fig. P10.87). The plank is pulled by a constant horizontal force $\vec{F}$ of magnitude 6.00 N applied to the end of the plank and perpendicular to the axes of the cylinders (which are parallel). The cylinders roll without slipping on a flat surface. There is also no slipping between the cylinders and the plank. (a) Find the initial acceleration of the plank at the moment the rollers are equidistant from the ends of the plank. (b) Find the acceleration of the rollers at this moment. (c) What friction forces are acting at this moment?

Figure P10.87

88. As a gasoline engine operates, a flywheel turning with the crankshaft stores energy after each fuel explosion, providing the energy required to compress the next charge of fuel and air. For the engine of a certain lawn tractor, suppose a flywheel must be no more than 18.0 cm in diameter. Its thickness, measured along its axis of rotation, must be no larger than 8.00 cm. The flywheel must release energy 60.0 J when its angular speed drops from 800 rev/min to 600 rev/min. Design a sturdy steel (density 7.85×10^3 kg/m^3) flywheel to meet these requirements with the smallest mass you can reasonably attain. Specify the shape and mass of the flywheel.

89. As a result of friction, the angular speed of a wheel changes with time according to

$$\frac{d\theta}{dt} = \omega_0 e^{-\sigma t}$$

where ω_0 and σ are constants. The angular speed changes from 3.50 rad/s at $t = 0$ to 2.00 rad/s at $t = 9.30$ s. (a) Use this information to determine σ and ω_0. Then determine (b) the magnitude of the angular acceleration at $t = 3.00$ s, (c) the number of revolutions the wheel makes in the first 2.50 s, and (d) the number of revolutions it makes before coming to rest.

90. To find the total angular displacement during the playing time of the compact disc in part (B) of Example 10.2, the disc was modeled as a rigid object under constant angular acceleration. In reality, the angular acceleration of a disc is not constant. In this problem, let us explore the actual time dependence of the angular acceleration. (a) Assume the track on the disc is a spiral such that adjacent loops of the track are separated by a small distance h. Show that the radius r of a given portion of the track is given by

$$r = r_i + \frac{h\theta}{2\pi}$$

where r_i is the radius of the innermost portion of the track and θ is the angle through which the disc turns to arrive at the location of the track of radius r. (b) Show that the rate of change of the angle θ is given by

$$\frac{d\theta}{dt} = \frac{v}{r_i + (h\theta/2\pi)}$$

where v is the constant speed with which the disc surface passes the laser. (c) From the result in part (b), use integration to find an expression for the angle θ as a function of time. (d) From the result in part (c), use differentiation to find the angular acceleration of the disc as a function of time.

91. A spool of thread consists of a cylinder of radius R_1 with end caps of radius R_2 as depicted in the end view shown in Figure P10.91. The mass of the spool, including the thread, is m, and its moment of inertia about an axis through its center is I. The spool is placed on a rough, horizontal surface so that it rolls without slipping when a force $\vec{T}$ acting to the right is applied to the free end of the thread. (a) Show that the magnitude of the friction force exerted by the surface on the spool is given by

$$f = \left(\frac{I + mR_1R_2}{I + mR_2^{\,2}}\right) T$$

(b) Determine the direction of the force of friction.

Figure P10.91

92. A cord is wrapped around a pulley that is shaped like a disk of mass m and radius r. The cord's free end is connected to a block of mass M. The block starts from rest and then slides down an incline that makes an angle θ with the horizontal as shown in Figure P10.92. The coefficient of kinetic friction between block and incline is μ. (a) Use energy methods to show that the block's speed as a function of position d down the incline is

$$v = \sqrt{\frac{4Mgd(\sin\theta - \mu\cos\theta)}{m + 2M}}$$

(b) Find the magnitude of the acceleration of the block in terms of μ, m, M, g, and θ.

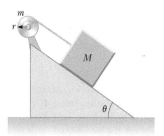

Figure P10.92

93. A merry-go-round is stationary. A dog is running around the merry-go-round on the ground just outside its circumference, moving with a constant angular speed of 0.750 rad/s. The dog does not change his pace when he sees what he has been looking for: a bone resting on the edge of the merry-go-round one-third of a revolution in front of him. At the instant the dog sees the bone ($t = 0$), the merry-go-round begins to move in the direction the dog is running, with a constant angular acceleration of 0.015 0 rad/s². (a) At what time will the dog first reach the bone? (b) The confused dog keeps running and passes the bone. How long after the merry-go-round starts to turn do the dog and the bone draw even with each other for the second time?

94. A uniform, hollow, cylindrical spool has inside radius $R/2$, outside radius R, and mass M (Fig. P10.94). It is mounted so that it rotates on a fixed, horizontal axle. A counterweight of mass m is connected to the end of a string wound around the spool. The counterweight falls from rest at $t = 0$ to a position y at time t. Show that the torque due to the friction forces between spool and axle is

Figure P10.94

$$\tau_f = R\left[m\left(g - \frac{2y}{t^2}\right) - M\frac{5y}{4t^2}\right]$$

Angular Momentum

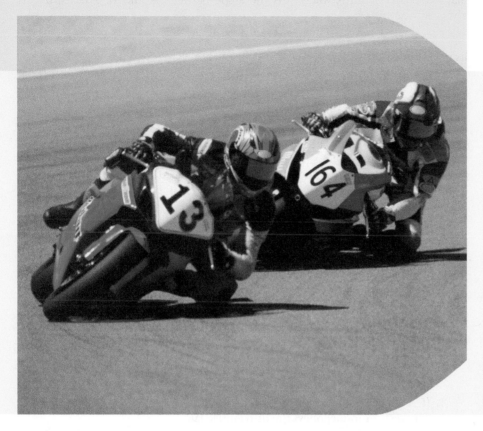

The central topic of this chapter is angular momentum, a quantity that plays a key role in rotational dynamics. In analogy to the principle of conservation of linear momentum, there is also a principle of conservation of angular momentum. The angular momentum of an isolated system is constant. For angular momentum, an isolated system is one for which no external torques act on the system. If a net external torque acts on a system, it is nonisolated. Like the law of conservation of linear momentum, the law of conservation of angular momentum is a fundamental law of physics, equally valid for relativistic and quantum systems.

Two motorcycle racers lean precariously into a turn around a racetrack. The analysis of such a leaning turn is based on principles associated with angular momentum. *(Stuart Westmorland/The Image Bank/ Getty Images)*

11.1 The Vector Product and Torque

An important consideration in defining angular momentum is the process of multiplying two vectors by means of the operation called the *vector product*. We will introduce the vector product by considering the vector nature of torque.

Consider a force $\vec{F}$ acting on a particle located at point P and described by the vector position $\vec{r}$ (Fig. 11.1 on page 336). As we saw in Section 10.6, the *magnitude* of the torque due to this force about an axis through the origin is $rF\sin\phi$, where ϕ is the angle between $\vec{r}$ and $\vec{F}$. The axis about which $\vec{F}$ tends to produce rotation is perpendicular to the plane formed by $\vec{r}$ and $\vec{F}$.

The torque vector $\vec{\tau}$ is related to the two vectors $\vec{r}$ and $\vec{F}$. We can establish a mathematical relationship between $\vec{\tau}$, $\vec{r}$, and $\vec{F}$ using a mathematical operation called the **vector product:**

$$\vec{\tau} \equiv \vec{r} \times \vec{F} \tag{11.1}$$

335

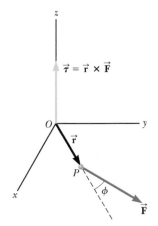

Figure 11.1 The torque vector $\vec{\tau}$ lies in a direction perpendicular to the plane formed by the position vector $\vec{r}$ and the applied force vector $\vec{F}$. In the situation shown, $\vec{r}$ and $\vec{F}$ lie in the xy plane, so the torque is along the z axis.

▶ Properties of the vector product

The direction of $\vec{C}$ is perpendicular to the plane formed by $\vec{A}$ and $\vec{B}$, and its direction is determined by the right-hand rule.

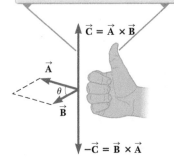

Figure 11.2 The vector product $\vec{A} \times \vec{B}$ is a third vector $\vec{C}$ having a magnitude $AB \sin \theta$ equal to the area of the parallelogram shown.

▶ Cross products of unit vectors

Pitfall Prevention 11.1

The Vector Product Is a Vector Remember that the result of taking a vector product between two vectors is *a third vector*. Equation 11.3 gives only the magnitude of this vector.

We now give a formal definition of the vector product. Given any two vectors $\vec{A}$ and $\vec{B}$, the vector product $\vec{A} \times \vec{B}$ is defined as a third vector $\vec{C}$, which has a magnitude of $AB \sin \theta$, where θ is the angle between $\vec{A}$ and $\vec{B}$. That is, if $\vec{C}$ is given by

$$\vec{C} = \vec{A} \times \vec{B} \tag{11.2}$$

its magnitude is

$$C = AB \sin \theta \tag{11.3}$$

The quantity $AB \sin \theta$ is equal to the area of the parallelogram formed by $\vec{A}$ and $\vec{B}$ as shown in Figure 11.2. The *direction* of $\vec{C}$ is perpendicular to the plane formed by $\vec{A}$ and $\vec{B}$, and the best way to determine this direction is to use the right-hand rule illustrated in Figure 11.2. The four fingers of the right hand are pointed along $\vec{A}$ and then "wrapped" in the direction that would rotate $\vec{A}$ into $\vec{B}$ through the angle θ. The direction of the upright thumb is the direction of $\vec{A} \times \vec{B} = \vec{C}$. Because of the notation, $\vec{A} \times \vec{B}$ is often read "$\vec{A}$ cross $\vec{B}$," so the vector product is also called the **cross product.**

Some properties of the vector product that follow from its definition are as follows:

1. Unlike the scalar product, the vector product is *not* commutative. Instead, the order in which the two vectors are multiplied in a vector product is important:

$$\vec{A} \times \vec{B} = -\vec{B} \times \vec{A} \tag{11.4}$$

Therefore, if you change the order of the vectors in a vector product, you must change the sign. You can easily verify this relationship with the right-hand rule.

2. If $\vec{A}$ is parallel to $\vec{B}$ ($\theta = 0$ or $180°$), then $\vec{A} \times \vec{B} = 0$; therefore, it follows that $\vec{A} \times \vec{A} = 0$.

3. If $\vec{A}$ is perpendicular to $\vec{B}$, then $|\vec{A} \times \vec{B}| = AB$.

4. The vector product obeys the distributive law:

$$\vec{A} \times (\vec{B} + \vec{C}) = \vec{A} \times \vec{B} + \vec{A} \times \vec{C} \tag{11.5}$$

5. The derivative of the vector product with respect to some variable such as t is

$$\frac{d}{dt}(\vec{A} \times \vec{B}) = \frac{d\vec{A}}{dt} \times \vec{B} + \vec{A} \times \frac{d\vec{B}}{dt} \tag{11.6}$$

where it is important to preserve the multiplicative order of the terms on the right side in view of Equation 11.4.

It is left as an exercise (Problem 4) to show from Equations 11.3 and 11.4 and from the definition of unit vectors that the cross products of the unit vectors $\hat{i}$, $\hat{j}$, and $\hat{k}$ obey the following rules:

$$\hat{i} \times \hat{i} = \hat{j} \times \hat{j} = \hat{k} \times \hat{k} = 0 \tag{11.7a}$$

$$\hat{i} \times \hat{j} = -\hat{j} \times \hat{i} = \hat{k} \tag{11.7b}$$

$$\hat{j} \times \hat{k} = -\hat{k} \times \hat{j} = \hat{i} \tag{11.7c}$$

$$\hat{k} \times \hat{i} = -\hat{i} \times \hat{k} = \hat{j} \tag{11.7d}$$

Signs are interchangeable in cross products. For example, $\vec{A} \times (-\vec{B}) = -\vec{A} \times \vec{B}$ and $\hat{i} \times (-\hat{j}) = -\hat{i} \times \hat{j}$.

The cross product of any two vectors $\vec{A}$ and $\vec{B}$ can be expressed in the following determinant form:

$$\vec{A} \times \vec{B} = \begin{vmatrix} \hat{i} & \hat{j} & \hat{k} \\ A_x & A_y & A_z \\ B_x & B_y & B_z \end{vmatrix} = \begin{vmatrix} A_y & A_z \\ B_y & B_z \end{vmatrix} \hat{i} + \begin{vmatrix} A_z & A_x \\ B_z & B_x \end{vmatrix} \hat{j} + \begin{vmatrix} A_x & A_y \\ B_x & B_y \end{vmatrix} \hat{k}$$

Expanding these determinants gives the result

$$\vec{A} \times \vec{B} = (A_y B_z - A_z B_y)\,\hat{i} + (A_z B_x - A_x B_z)\,\hat{j} + (A_x B_y - A_y B_x)\hat{k} \quad (11.8)$$

Given the definition of the cross product, we can now assign a direction to the torque vector. If the force lies in the xy plane as in Figure 11.1, the torque $\vec{\tau}$ is represented by a vector parallel to the z axis. The force in Figure 11.1 creates a torque that tends to rotate the particle counterclockwise about the z axis; the direction of $\vec{\tau}$ is toward increasing z, and $\vec{\tau}$ is therefore in the positive z direction. If we reversed the direction of $\vec{F}$ in Figure 11.1, $\vec{\tau}$ would be in the negative z direction.

Quick Quiz 11.1 Which of the following statements about the relationship between the magnitude of the cross product of two vectors and the product of the magnitudes of the vectors is true? (a) $|\vec{A} \times \vec{B}|$ is larger than AB. (b) $|\vec{A} \times \vec{B}|$ is smaller than AB. (c) $|\vec{A} \times \vec{B}|$ could be larger or smaller than AB, depending on the angle between the vectors. (d) $|\vec{A} \times \vec{B}|$ could be equal to AB.

Example 11.1   The Vector Product

Two vectors lying in the xy plane are given by the equations $\vec{A} = 2\hat{i} + 3\hat{j}$ and $\vec{B} = -\hat{i} + 2\hat{j}$. Find $\vec{A} \times \vec{B}$ and verify that $\vec{A} \times \vec{B} = -\vec{B} \times \vec{A}$.

SOLUTION

Conceptualize Given the unit-vector notations of the vectors, think about the directions the vectors point in space. Draw them on graph paper and imagine the parallelogram shown in Figure 11.2 for these vectors.

Categorize Because we use the definition of the cross product discussed in this section, we categorize this example as a substitution problem.

Write the cross product of the two vectors:

$$\vec{A} \times \vec{B} = (2\hat{i} + 3\hat{j}) \times (-\hat{i} + 2\hat{j})$$

Perform the multiplication:

$$\vec{A} \times \vec{B} = 2\hat{i} \times (-\hat{i}) + 2\hat{i} \times 2\hat{j} + 3\hat{j} \times (-\hat{i}) + 3\hat{j} \times 2\hat{j}$$

Use Equations 11.7a through 11.7d to evaluate the various terms:

$$\vec{A} \times \vec{B} = 0 + 4\hat{k} + 3\hat{k} + 0 = \boxed{7\hat{k}}$$

To verify that $\vec{A} \times \vec{B} = -\vec{B} \times \vec{A}$, evaluate $\vec{B} \times \vec{A}$:

$$\vec{B} \times \vec{A} = (-\hat{i} + 2\hat{j}) \times (2\hat{i} + 3\hat{j})$$

Perform the multiplication:

$$\vec{B} \times \vec{A} = (-\hat{i}) \times 2\hat{i} + (-\hat{i}) \times 3\hat{j} + 2\hat{j} \times 2\hat{i} + 2\hat{j} \times 3\hat{j}$$

Use Equations 11.7a through 11.7d to evaluate the various terms:

$$\vec{B} \times \vec{A} = 0 - 3\hat{k} - 4\hat{k} + 0 = \boxed{-7\hat{k}}$$

Therefore, $\vec{A} \times \vec{B} = -\vec{B} \times \vec{A}$. As an alternative method for finding $\vec{A} \times \vec{B}$, you could use Equation 11.8. Try it!

Example 11.2   The Torque Vector

A force of $\vec{F} = (2.00\hat{i} + 3.00\hat{j})$ N is applied to an object that is pivoted about a fixed axis aligned along the z coordinate axis. The force is applied at a point located at $\vec{r} = (4.00\hat{i} + 5.00\hat{j})$ m. Find the torque $\vec{\tau}$ applied to the object.

SOLUTION

Conceptualize Given the unit-vector notations, think about the directions of the force and position vectors. If this force were applied at this position, in what direction would an object pivoted at the origin turn?

continued

▶ **11.2** c o n t i n u e d

Categorize Because we use the definition of the cross product discussed in this section, we categorize this example as a substitution problem.

Set up the torque vector using Equation 11.1:

$$\vec{\boldsymbol{\tau}} = \vec{\mathbf{r}} \times \vec{\mathbf{F}} = [(4.00\,\hat{\mathbf{i}} + 5.00\,\hat{\mathbf{j}})\ \mathrm{m}] \times [(2.00\,\hat{\mathbf{i}} + 3.00\,\hat{\mathbf{j}})\ \mathrm{N}]$$

Perform the multiplication:

$$\vec{\boldsymbol{\tau}} = [(4.00)(2.00)\,\hat{\mathbf{i}} \times \hat{\mathbf{i}} + (4.00)(3.00)\,\hat{\mathbf{i}} \times \hat{\mathbf{j}}$$
$$+ (5.00)(2.00)\,\hat{\mathbf{j}} \times \hat{\mathbf{i}} + (5.00)(3.00)\,\hat{\mathbf{j}} \times \hat{\mathbf{j}}]\ \mathrm{N} \cdot \mathrm{m}$$

Use Equations 11.7a through 11.7d to evaluate the various terms:

$$\vec{\boldsymbol{\tau}} = [0 + 12.0\,\hat{\mathbf{k}} - 10.0\,\hat{\mathbf{k}} + 0]\ \mathrm{N} \cdot \mathrm{m} = \boxed{2.0\hat{\mathbf{k}}\ \mathrm{N} \cdot \mathrm{m}}$$

Notice that both $\vec{\mathbf{r}}$ and $\vec{\mathbf{F}}$ are in the xy plane. As expected, the torque vector is perpendicular to this plane, having only a z component. We have followed the rules for significant figures discussed in Section 1.6, which lead to an answer with two significant figures. We have lost some precision because we ended up subtracting two numbers that are close.

11.2 Analysis Model: Nonisolated System (Angular Momentum)

Figure 11.3 As the skater passes the pole, she grabs hold of it, which causes her to swing around the pole rapidly in a circular path.

Imagine a rigid pole sticking up through the ice on a frozen pond (Fig. 11.3). A skater glides rapidly toward the pole, aiming a little to the side so that she does not hit it. As she passes the pole, she reaches out to her side and grabs it, an action that causes her to move in a circular path around the pole. Just as the idea of linear momentum helps us analyze translational motion, a rotational analog—*angular momentum*—helps us analyze the motion of this skater and other objects undergoing rotational motion.

In Chapter 9, we developed the mathematical form of linear momentum and then proceeded to show how this new quantity was valuable in problem solving. We will follow a similar procedure for angular momentum.

Consider a particle of mass m located at the vector position $\vec{\mathbf{r}}$ and moving with linear momentum $\vec{\mathbf{p}}$ as in Figure 11.4. In describing translational motion, we found that the net force on the particle equals the time rate of change of its linear momentum, $\sum \vec{\mathbf{F}} = d\vec{\mathbf{p}}/dt$ (see Eq. 9.3). Let us take the cross product of each side of Equation 9.3 with $\vec{\mathbf{r}}$, which gives the net torque on the particle on the left side of the equation:

$$\vec{\mathbf{r}} \times \sum \vec{\mathbf{F}} = \sum \vec{\boldsymbol{\tau}} = \vec{\mathbf{r}} \times \frac{d\vec{\mathbf{p}}}{dt}$$

Now let's add to the right side the term $(d\vec{\mathbf{r}}/dt) \times \vec{\mathbf{p}}$, which is zero because $d\vec{\mathbf{r}}/dt = \vec{\mathbf{v}}$ and $\vec{\mathbf{v}}$ and $\vec{\mathbf{p}}$ are parallel. Therefore,

$$\sum \vec{\boldsymbol{\tau}} = \vec{\mathbf{r}} \times \frac{d\vec{\mathbf{p}}}{dt} + \frac{d\vec{\mathbf{r}}}{dt} \times \vec{\mathbf{p}}$$

We recognize the right side of this equation as the derivative of $\vec{\mathbf{r}} \times \vec{\mathbf{p}}$ (see Eq. 11.6). Therefore,

$$\sum \vec{\boldsymbol{\tau}} = \frac{d(\vec{\mathbf{r}} \times \vec{\mathbf{p}})}{dt} \tag{11.9}$$

which looks very similar in form to Equation 9.3, $\sum \vec{\mathbf{F}} = d\vec{\mathbf{p}}/dt$. Because torque plays the same role in rotational motion that force plays in translational motion, this result suggests that the combination $\vec{\mathbf{r}} \times \vec{\mathbf{p}}$ should play the same role in rotational motion that $\vec{\mathbf{p}}$ plays in translational motion. We call this combination the *angular momentum* of the particle:

> The instantaneous **angular momentum** $\vec{\mathbf{L}}$ of a particle relative to an axis through the origin O is defined by the cross product of the particle's instantaneous position vector $\vec{\mathbf{r}}$ and its instantaneous linear momentum $\vec{\mathbf{p}}$:
>
> $$\vec{\mathbf{L}} \equiv \vec{\mathbf{r}} \times \vec{\mathbf{p}} \qquad (11.10)$$

◀ **Angular momentum of a particle**

We can now write Equation 11.9 as

$$\sum \vec{\boldsymbol{\tau}} = \frac{d\vec{\mathbf{L}}}{dt} \qquad (11.11)$$

which is the rotational analog of Newton's second law, $\sum \vec{\mathbf{F}} = d\vec{\mathbf{p}}/dt$. Torque causes the angular momentum $\vec{\mathbf{L}}$ to change just as force causes linear momentum $\vec{\mathbf{p}}$ to change.

Notice that Equation 11.11 is valid only if $\sum \vec{\boldsymbol{\tau}}$ and $\vec{\mathbf{L}}$ are measured about the same axis. Furthermore, the expression is valid for any axis fixed in an inertial frame.

The SI unit of angular momentum is kg $\cdot$ m^2/s. Notice also that both the magnitude and the direction of $\vec{\mathbf{L}}$ depend on the choice of axis. Following the right-hand rule, we see that the direction of $\vec{\mathbf{L}}$ is perpendicular to the plane formed by $\vec{\mathbf{r}}$ and $\vec{\mathbf{p}}$. In Figure 11.4, $\vec{\mathbf{r}}$ and $\vec{\mathbf{p}}$ are in the xy plane, so $\vec{\mathbf{L}}$ points in the z direction. Because $\vec{\mathbf{p}} = m\vec{\mathbf{v}}$, the magnitude of $\vec{\mathbf{L}}$ is

$$L = mvr \sin \phi \qquad (11.12)$$

where ϕ is the angle between $\vec{\mathbf{r}}$ and $\vec{\mathbf{p}}$. It follows that L is zero when $\vec{\mathbf{r}}$ is parallel to $\vec{\mathbf{p}}$ ($\phi = 0$ or $180°$). In other words, when the translational velocity of the particle is along a line that passes through the axis, the particle has zero angular momentum with respect to the axis. On the other hand, if $\vec{\mathbf{r}}$ is perpendicular to $\vec{\mathbf{p}}$ ($\phi = 90°$), then $L = mvr$. At that instant, the particle moves exactly as if it were on the rim of a wheel rotating about the axis in a plane defined by $\vec{\mathbf{r}}$ and $\vec{\mathbf{p}}$.

The angular momentum $\vec{\mathbf{L}}$ of a particle about an axis is a vector perpendicular to both the particle's position $\vec{\mathbf{r}}$ relative to the axis and its momentum $\vec{\mathbf{p}}$.

Figure 11.4 The angular momentum $\vec{\mathbf{L}}$ of a particle is a vector given by $\vec{\mathbf{L}} = \vec{\mathbf{r}} \times \vec{\mathbf{p}}$.

Quick Quiz 11.2 Recall the skater described at the beginning of this section. Let her mass be m. **(i)** What would be her angular momentum relative to the pole at the instant she is a distance d from the pole if she were skating directly toward it at speed v? (a) zero (b) mvd (c) impossible to determine **(ii)** What would be her angular momentum relative to the pole at the instant she is a distance d from the pole if she were skating at speed v along a straight path that is a perpendicular distance a from the pole? (a) zero (b) mvd (c) mva (d) impossible to determine

Pitfall Prevention 11.2

Is Rotation Necessary for Angular Momentum? We can define angular momentum even if the particle is not moving in a circular path. A particle moving in a straight line has angular momentum about any axis displaced from the path of the particle.

Example 11.3 | Angular Momentum of a Particle in Circular Motion

A particle moves in the xy plane in a circular path of radius r as shown in Figure 11.5. Find the magnitude and direction of its angular momentum relative to an axis through O when its velocity is $\vec{\mathbf{v}}$.

SOLUTION

Conceptualize The linear momentum of the particle is always changing in direction (but not in magnitude). You might therefore be tempted to conclude that the angular momentum of the particle is always changing. In this situation, however, that is not the case. Let's see why.

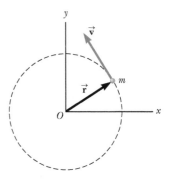

Figure 11.5 (Example 11.3) A particle moving in a circle of radius r has an angular momentum about an axis through O that has magnitude mvr. The vector $\vec{\mathbf{L}} = \vec{\mathbf{r}} \times \vec{\mathbf{p}}$ points *out* of the page.

continued

▶ **11.3** continued

Categorize We use the definition of the angular momentum of a particle discussed in this section, so we categorize this example as a substitution problem.

Use Equation 11.12 to evaluate the magnitude of $\vec{\mathbf{L}}$:

$$L = mvr \sin 90° = \boxed{mvr}$$

This value of L is constant because all three factors on the right are constant. The direction of $\vec{\mathbf{L}}$ also is constant, even though the direction of $\vec{\mathbf{p}} = m\vec{\mathbf{v}}$ keeps changing. To verify this statement, apply the right-hand rule to find the direction of $\vec{\mathbf{L}} = \vec{\mathbf{r}} \times \vec{\mathbf{p}} = m\vec{\mathbf{r}} \times \vec{\mathbf{v}}$ in Figure 11.5. Your thumb points out of the page, so that is the direction of $\vec{\mathbf{L}}$. Hence, we can write the vector expression $\vec{\mathbf{L}} = (mvr)\hat{\mathbf{k}}$. If the particle were to move clockwise, $\vec{\mathbf{L}}$ would point downward and into the page and $\vec{\mathbf{L}} = -(mvr)\hat{\mathbf{k}}$. A particle in uniform circular motion has a constant angular momentum about an axis through the center of its path.

Angular Momentum of a System of Particles

Using the techniques of Section 9.7, we can show that Newton's second law for a system of particles is

$$\sum \vec{\mathbf{F}}_{ext} = \frac{d\vec{\mathbf{p}}_{tot}}{dt}$$

This equation states that the net external force on a system of particles is equal to the time rate of change of the total linear momentum of the system. Let's see if a similar statement can be made for rotational motion. The total angular momentum of a system of particles about some axis is defined as the vector sum of the angular momenta of the individual particles:

$$\vec{\mathbf{L}}_{tot} = \vec{\mathbf{L}}_1 + \vec{\mathbf{L}}_2 + \cdots + \vec{\mathbf{L}}_n = \sum_i \vec{\mathbf{L}}_i$$

where the vector sum is over all n particles in the system.

Differentiating this equation with respect to time gives

$$\frac{d\vec{\mathbf{L}}_{tot}}{dt} = \sum_i \frac{d\vec{\mathbf{L}}_i}{dt} = \sum_i \vec{\boldsymbol{\tau}}_i$$

where we have used Equation 11.11 to replace the time rate of change of the angular momentum of each particle with the net torque on the particle.

The torques acting on the particles of the system are those associated with internal forces between particles and those associated with external forces. The net torque associated with all internal forces, however, is zero. Recall that Newton's third law tells us that internal forces between particles of the system are equal in magnitude and opposite in direction. If we assume these forces lie along the line of separation of each pair of particles, the total torque around some axis passing through an origin O due to each action–reaction force pair is zero (that is, the moment arm d from O to the line of action of the forces is equal for both particles, and the forces are in opposite directions). In the summation, therefore, the net internal torque is zero. We conclude that the total angular momentum of a system can vary with time only if a net external torque is acting on the system:

The net external torque on a ▶
system equals the time rate
of change of angular momentum of the system

$$\sum \vec{\boldsymbol{\tau}}_{ext} = \frac{d\vec{\mathbf{L}}_{tot}}{dt} \tag{11.13}$$

This equation is indeed the rotational analog of $\sum \vec{\mathbf{F}}_{ext} = d\vec{\mathbf{p}}_{tot}/dt$ for a system of particles. Equation 11.13 is the mathematical representation of the **angular momentum version of the nonisolated system model.** If a system is nonisolated in the sense that there is a net torque on it, the torque is equal to the time rate of change of angular momentum.

Although we do not prove it here, this statement is true regardless of the motion of the center of mass. It applies even if the center of mass is accelerating, provided the torque and angular momentum are evaluated relative to an axis through the center of mass.

Equation 11.13 can be rearranged and integrated to give

$$\Delta \vec{\mathbf{L}}_{tot} = \int \left(\sum \vec{\tau}_{ext} \right) dt$$

This equation represents the *angular impulse–angular momentum theorem*. Compare this equation to the translational version, Equation 9.40.

Analysis Model   Nonisolated System (Angular Momentum)

Imagine a system that rotates about an axis. If there is a net external torque acting on the system, the time rate of change of the angular momentum of the system is equal to the net external torque:

$$\sum \vec{\tau}_{ext} = \frac{d\vec{\mathbf{L}}_{tot}}{dt} \qquad (11.13)$$

Examples:

- a flywheel in an automobile engine increases its angular momentum when the engine applies torque to it
- the tub of a washing machine decreases in angular momentum due to frictional torque after the machine is turned off
- the axis of the Earth undergoes a precessional motion due to the torque exerted on the Earth by the gravitational force from the Sun
- the armature of a motor increases its angular momentum due to the torque exerted by a surrounding magnetic field (Chapter 31)

The rate of change in the angular momentum of the nonisolated system is equal to the net external torque on the system.

Example 11.4   A System of Objects   AM

A sphere of mass m_1 and a block of mass m_2 are connected by a light cord that passes over a pulley as shown in Figure 11.6. The radius of the pulley is R, and the mass of the thin rim is M. The spokes of the pulley have negligible mass. The block slides on a frictionless, horizontal surface. Find an expression for the linear acceleration of the two objects, using the concepts of angular momentum and torque.

SOLUTION

Conceptualize When the system is released, the block slides to the left, the sphere drops downward, and the pulley rotates counterclockwise. This situation is similar to problems we have solved earlier except that now we want to use an angular momentum approach.

Figure 11.6 (Example 11.4) When the system is released, the sphere moves downward and the block moves to the left.

Categorize We identify the block, pulley, and sphere as a *nonisolated system* for *angular momentum*, subject to the external torque due to the gravitational force on the sphere. We shall calculate the angular momentum about an axis that coincides with the axle of the pulley. The angular momentum of the system includes that of two objects moving translationally (the sphere and the block) and one object undergoing pure rotation (the pulley).

Analyze At any instant of time, the sphere and the block have a common speed v, so the angular momentum of the sphere about the pulley axle is m_1vR and that of the block is m_2vR. At the same instant, all points on the rim of the pulley also move with speed v, so the angular momentum of the pulley is MvR.

Now let's address the total external torque acting on the system about the pulley axle. Because it has a moment arm of zero, the force exerted by the axle on the pulley does not contribute to the torque. Furthermore, the normal force

continued

▶ **11.4** continued

acting on the block is balanced by the gravitational force $m_2\vec{\mathbf{g}}$, so these forces do not contribute to the torque. The gravitational force $m_1\vec{\mathbf{g}}$ acting on the sphere produces a torque about the axle equal in magnitude to m_1gR, where R is the moment arm of the force about the axle. This result is the total external torque about the pulley axle; that is, $\sum \tau_{ext} = m_1gR$.

Write an expression for the total angular momentum of the system:

$$(1)\quad L = m_1vR + m_2vR + MvR = (m_1 + m_2 + M)vR$$

Substitute this expression and the total external torque into Equation 11.13, the mathematical representation of the nonisolated system model for angular momentum:

$$\sum \tau_{ext} = \frac{dL}{dt}$$

$$m_1gR = \frac{d}{dt}[(m_1 + m_2 + M)vR]$$

$$(2)\quad m_1gR = (m_1 + m_2 + M)R\frac{dv}{dt}$$

Recognizing that $dv/dt = a$, solve Equation (2) for a:

$$(3)\quad a = \frac{m_1g}{m_1 + m_2 + M}$$

Finalize When we evaluated the net torque about the axle, we did not include the forces that the cord exerts on the objects because these forces are internal to the system under consideration. Instead, we analyzed the system as a whole. Only *external* torques contribute to the change in the system's angular momentum. Let $M \to 0$ in Equation (3) and call the result Equation A. Now go back to Equation (5) in Example 5.10, let $\theta \to 0$, and call the result Equation B. Do Equations A and B match? Looking at Figures 5.15 and 11.6 in these limits, *should* the two equations match?

11.3 Angular Momentum of a Rotating Rigid Object

Figure 11.7 When a rigid object rotates about an axis, the angular momentum $\vec{\mathbf{L}}$ is in the same direction as the angular velocity $\vec{\boldsymbol{\omega}}$ according to the expression $\vec{\mathbf{L}} = I\vec{\boldsymbol{\omega}}$.

In Example 11.4, we considered the angular momentum of a deformable system of particles. Let us now restrict our attention to a nondeformable system, a rigid object. Consider a rigid object rotating about a fixed axis that coincides with the z axis of a coordinate system as shown in Figure 11.7. Let's determine the angular momentum of this object. Each *particle* of the object rotates in the xy plane about the z axis with an angular speed ω. The magnitude of the angular momentum of a particle of mass m_i about the z axis is $m_iv_ir_i$. Because $v_i = r_i\omega$ (Eq. 10.10), we can express the magnitude of the angular momentum of this particle as

$$L_i = m_ir_i^2\omega$$

The vector $\vec{\mathbf{L}}_i$ for this particle is directed along the z axis, as is the vector $\vec{\boldsymbol{\omega}}$.

We can now find the angular momentum (which in this situation has only a z component) of the whole object by taking the sum of L_i over all particles:

$$L_z = \sum_i L_i = \sum_i m_ir_i^2\omega = \left(\sum_i m_ir_i^2\right)\omega$$

$$L_z = I\omega \tag{11.14}$$

where we have recognized $\sum_i m_ir_i^2$ as the moment of inertia I of the object about the z axis (Eq. 10.19). Notice that Equation 11.14 is mathematically similar in form to Equation 9.2 for linear momentum: $\vec{\mathbf{p}} = m\vec{\mathbf{v}}$.

Now let's differentiate Equation 11.14 with respect to time, noting that I is constant for a rigid object:

$$\frac{dL_z}{dt} = I\frac{d\omega}{dt} = I\alpha \tag{11.15}$$

where α is the angular acceleration relative to the axis of rotation. Because dL_z/dt is equal to the net external torque (see Eq. 11.13), we can express Equation 11.15 as

$$\sum \tau_{\text{ext}} = I\alpha \qquad \text{(11.16)}$$

◀ Rotational form of Newton's second law

That is, the net external torque acting on a rigid object rotating about a fixed axis equals the moment of inertia about the rotation axis multiplied by the object's angular acceleration relative to that axis. This result is the same as Equation 10.18, which was derived using a force approach, but we derived Equation 11.16 using the concept of angular momentum. As we saw in Section 10.7, Equation 11.16 is the mathematical representation of the rigid object under a net torque analysis model. This equation is also valid for a rigid object rotating about a moving axis, provided the moving axis (1) passes through the center of mass and (2) is a symmetry axis.

If a symmetrical object rotates about a fixed axis passing through its center of mass, you can write Equation 11.14 in vector form as $\vec{L} = I\vec{\omega}$, where $\vec{L}$ is the total angular momentum of the object measured with respect to the axis of rotation. Furthermore, the expression is valid for any object, regardless of its symmetry, if $\vec{L}$ stands for the component of angular momentum along the axis of rotation.[1]

Quick Quiz 11.3 A solid sphere and a hollow sphere have the same mass and radius. They are rotating with the same angular speed. Which one has the higher angular momentum? **(a)** the solid sphere **(b)** the hollow sphere **(c)** both have the same angular momentum **(d)** impossible to determine

---

Example 11.5 | Bowling Ball

Estimate the magnitude of the angular momentum of a bowling ball spinning at 10 rev/s as shown in Figure 11.8.

SOLUTION

Conceptualize Imagine spinning a bowling ball on the smooth floor of a bowling alley. Because a bowling ball is relatively heavy, the angular momentum should be relatively large.

Categorize We evaluate the angular momentum using Equation 11.14, so we categorize this example as a substitution problem.

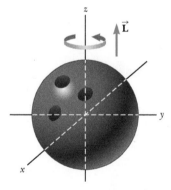

Figure 11.8 (Example 11.5) A bowling ball that rotates about the z axis in the direction shown has an angular momentum $\vec{L}$ in the positive z direction. If the direction of rotation is reversed, then $\vec{L}$ points in the negative z direction.

We start by making some estimates of the relevant physical parameters and model the ball as a uniform solid sphere. A typical bowling ball might have a mass of 7.0 kg and a radius of 12 cm.

Evaluate the moment of inertia of the ball about an axis through its center from Table 10.2:

$$I = \tfrac{2}{5}MR^2 = \tfrac{2}{5}(7.0 \text{ kg})(0.12 \text{ m})^2 = 0.040 \text{ kg} \cdot \text{m}^2$$

Evaluate the magnitude of the angular momentum from Equation 11.14:

$$L_z = I\omega = (0.040 \text{ kg} \cdot \text{m}^2)(10 \text{ rev/s})(2\pi \text{ rad/rev}) = 2.53 \text{ kg} \cdot \text{m}^2/\text{s}$$

Because of the roughness of our estimates, we should keep only one significant figure, so $L_z = \boxed{3 \text{ kg} \cdot \text{m}^2/\text{s}}$.

---

[1]In general, the expression $\vec{L} = I\vec{\omega}$ is not always valid. If a rigid object rotates about an *arbitrary* axis, then $\vec{L}$ and $\vec{\omega}$ may point in different directions. In this case, the moment of inertia cannot be treated as a scalar. Strictly speaking, $\vec{L} = I\vec{\omega}$ applies only to rigid objects of any shape that rotate about one of three mutually perpendicular axes (called *principal axes*) through the center of mass. This concept is discussed in more advanced texts on mechanics.

Example 11.6   The Seesaw   AM

A father of mass m_f and his daughter of mass m_d sit on opposite ends of a seesaw at equal distances from the pivot at the center (Fig. 11.9). The seesaw is modeled as a rigid rod of mass M and length ℓ and is pivoted without friction. At a given moment, the combination rotates in a vertical plane with an angular speed ω.

(A) Find an expression for the magnitude of the system's angular momentum.

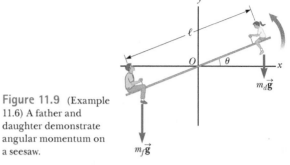

Figure 11.9 (Example 11.6) A father and daughter demonstrate angular momentum on a seesaw.

SOLUTION

Conceptualize Identify the z axis through O as the axis of rotation in Figure 11.9. The rotating system has angular momentum about that axis.

Categorize Ignore any movement of arms or legs of the father and daughter and model them both as particles. The system is therefore modeled as a rigid object. This first part of the example is categorized as a substitution problem.

The moment of inertia of the system equals the sum of the moments of inertia of the three components: the seesaw and the two individuals. We can refer to Table 10.2 to obtain the expression for the moment of inertia of the rod and use the particle expression $I = mr^2$ for each person.

Find the total moment of inertia of the system about the z axis through O:

$$I = \tfrac{1}{12}M\ell^2 + m_f\left(\frac{\ell}{2}\right)^2 + m_d\left(\frac{\ell}{2}\right)^2 = \frac{\ell^2}{4}\left(\frac{M}{3} + m_f + m_d\right)$$

Find the magnitude of the angular momentum of the system:

$$L = I\omega = \frac{\ell^2}{4}\left(\frac{M}{3} + m_f + m_d\right)\omega$$

(B) Find an expression for the magnitude of the angular acceleration of the system when the seesaw makes an angle θ with the horizontal.

SOLUTION

Conceptualize Generally, fathers are more massive than daughters, so the system is not in equilibrium and has an angular acceleration. We expect the angular acceleration to be positive in Figure 11.9.

Categorize The combination of the board, father, and daughter is a *rigid object under a net torque* because of the external torque associated with the gravitational forces on the father and daughter. We again identify the axis of rotation as the z axis in Figure 11.9.

....

Analyze To find the angular acceleration of the system at any angle θ, we first calculate the net torque on the system and then use $\Sigma \tau_{ext} = I\alpha$ from the rigid object under a net torque model to obtain an expression for α.

Evaluate the torque due to the gravitational force on the father:

$$\tau_f = m_f g \frac{\ell}{2} \cos\theta \quad (\vec{\tau}_f \text{ out of page})$$

Evaluate the torque due to the gravitational force on the daughter:

$$\tau_d = -m_d g \frac{\ell}{2} \cos\theta \quad (\vec{\tau}_d \text{ into page})$$

Evaluate the net external torque exerted on the system:

$$\sum \tau_{ext} = \tau_f + \tau_d = \tfrac{1}{2}(m_f - m_d)g\ell \cos\theta$$

Use Equation 11.16 and I from part (A) to find α:

$$\alpha = \frac{\sum \tau_{ext}}{I} = \frac{2(m_f - m_d)g \cos\theta}{\ell\,[(M/3) + m_f + m_d]}$$

....

Finalize For a father more massive than his daughter, the angular acceleration is positive as expected. If the seesaw begins in a horizontal orientation ($\theta = 0$) and is released, the rotation is counterclockwise in Figure 11.9 and the father's end of the seesaw drops, which is consistent with everyday experience.

WHAT IF? Imagine the father moves inward on the seesaw to a distance d from the pivot to try to balance the two sides. What is the angular acceleration of the system in this case when it is released from an arbitrary angle θ?

▶ **11.6** continued

Answer The angular acceleration of the system should decrease if the system is more balanced.

Find the total moment of inertia about the z axis through O for the modified system:	$I = \frac{1}{12}M\ell^2 + m_f d^2 + m_d\left(\frac{\ell}{2}\right)^2 = \frac{\ell^2}{4}\left(\frac{M}{3} + m_d\right) + m_f d^2$
Find the net torque exerted on the system about an axis through O:	$\sum \tau_{ext} = \tau_f + \tau_d = m_f g d \cos\theta - \frac{1}{2}m_d g\ell \cos\theta$
Find the new angular acceleration of the system:	$\alpha = \dfrac{\sum \tau_{ext}}{I} = \dfrac{(m_f d - \frac{1}{2}m_d\ell)g\cos\theta}{(\ell^2/4)\left[(M/3) + m_d\right] + m_f d^2}$

The seesaw is balanced when the angular acceleration is zero. In this situation, both father and daughter can push off the ground and rise to the highest possible point.

Find the required position of the father by setting $\alpha = 0$:

$$\alpha = \frac{(m_f d - \frac{1}{2}m_d\ell)g\cos\theta}{(\ell^2/4)[(M/3) + m_d] + m_f d^2} = 0$$

$$m_f d - \tfrac{1}{2}m_d\ell = 0 \quad \rightarrow \quad d = \left(\frac{m_d}{m_f}\right)\frac{\ell}{2}$$

In the rare case that the father and daughter have the same mass, the father is located at the end of the seesaw, $d = \ell/2$.

11.4 Analysis Model: Isolated System (Angular Momentum)

In Chapter 9, we found that the total linear momentum of a system of particles remains constant if the system is isolated, that is, if the net external force acting on the system is zero. We have an analogous conservation law in rotational motion:

> The total angular momentum of a system is constant in both magnitude and direction if the net external torque acting on the system is zero, that is, if the system is isolated.

◀ **Conservation of angular momentum**

This statement is often called[2] the principle of **conservation of angular momentum** and is the basis of the **angular momentum version of the isolated system model**. This principle follows directly from Equation 11.13, which indicates that if

$$\sum \vec{\tau}_{ext} = \frac{d\vec{L}_{tot}}{dt} = 0 \tag{11.17}$$

then

$$\Delta \vec{L}_{tot} = 0 \tag{11.18}$$

Equation 11.18 can be written as

$$\vec{L}_{tot} = \text{constant} \quad \text{or} \quad \vec{L}_i = \vec{L}_f$$

For an isolated system consisting of a small number of particles, we write this conservation law as $\vec{L}_{tot} = \sum \vec{L}_n = \text{constant}$, where the index n denotes the nth particle in the system.

If an isolated rotating system is deformable so that its mass undergoes redistribution in some way, the system's moment of inertia changes. Because the magnitude of the angular momentum of the system is $L = I\omega$ (Eq. 11.14), conservation

[2]The most general conservation of angular momentum equation is Equation 11.13, which describes how the system interacts with its environment.

of angular momentum requires that the product of I and ω must remain constant. Therefore, a change in I for an isolated system requires a change in ω. In this case, we can express the principle of conservation of angular momentum as

$$I_i \omega_i = I_f \omega_f = \text{constant} \tag{11.19}$$

This expression is valid both for rotation about a fixed axis and for rotation about an axis through the center of mass of a moving system as long as that axis remains fixed in direction. We require only that the net external torque be zero.

Many examples demonstrate conservation of angular momentum for a deformable system. You may have observed a figure skater spinning in the finale of a program (Fig. 11.10). The angular speed of the skater is large when his hands and feet are close to the trunk of his body. (Notice the skater's hair!) Ignoring friction between skater and ice, there are no external torques on the skater. The moment of inertia of his body increases as his hands and feet are moved away from his body at the finish of the spin. According to the isolated system model for angular momentum, his angular speed must decrease. In a similar way, when divers or acrobats wish to make several somersaults, they pull their hands and feet close to their bodies to rotate at a higher rate. In these cases, the external force due to gravity acts through the center of mass and hence exerts no torque about an axis through this point. Therefore, the angular momentum about the center of mass must be conserved; that is, $I_i \omega_i = I_f \omega_f$. For example, when divers wish to double their angular speed, they must reduce their moment of inertia to half its initial value.

In Equation 11.18, we have a third version of the isolated system model. We can now state that the energy, linear momentum, and angular momentum of an isolated system are all constant:

$$\Delta E_{\text{system}} = 0 \quad \text{(if there are no energy transfers across the system boundary)}$$

$$\Delta \vec{\mathbf{p}}_{\text{tot}} = 0 \quad \text{(if the net external force on the system is zero)}$$

$$\Delta \vec{\mathbf{L}}_{\text{tot}} = 0 \quad \text{(if the net external torque on the system is zero)}$$

A system may be isolated in terms of one of these quantities but not in terms of another. If a system is nonisolated in terms of momentum or angular momentum, it will often be nonisolated also in terms of energy because the system has a net force or torque on it and the net force or torque will do work on the system. We can, however, identify systems that are nonisolated in terms of energy but isolated in terms of momentum. For example, imagine pushing inward on a balloon (the system) between your hands. Work is done in compressing the balloon, so the system is nonisolated in terms of energy, but there is zero net force on the system, so the system is isolated in terms of momentum. A similar statement could be made about twisting the ends of a long, springy piece of metal with both hands. Work is done on the metal (the system), so energy is stored in the nonisolated system as elastic potential energy, but the net torque on the system is zero. Therefore, the system is isolated in terms of angular momentum. Other examples are collisions of macroscopic objects, which represent isolated systems in terms of momentum but nonisolated systems in terms of energy because of the output of energy from the system by mechanical waves (sound).

Quick Quiz 11.4 A competitive diver leaves the diving board and falls toward the water with her body straight and rotating slowly. She pulls her arms and legs into a tight tuck position. What happens to her rotational kinetic energy? **(a)** It increases. **(b)** It decreases. **(c)** It stays the same. **(d)** It is impossible to determine.

When his arms and legs are close to his body, the skater's moment of inertia is small and his angular speed is large.

Clive Rose/Getty Images

To slow down for the finish of his spin, the skater moves his arms and legs outward, increasing his moment of inertia.

Al Bello/Getty Images

Figure 11.10 Angular momentum is conserved as Russian gold medalist Evgeni Plushenko performs during the Turin 2006 Winter Olympic Games.

Analysis Model   Isolated System (Angular Momentum)

Imagine a system rotates about an axis. If there is no net external torque on the system, there is no change in the angular momentum of the system:

$$\Delta \vec{L}_{tot} = 0 \qquad \textbf{(11.18)}$$

Applying this law of conservation of angular momentum to a system whose moment of inertia changes gives

$$I_i \omega_i = I_f \omega_f = \text{constant} \qquad \textbf{(11.19)}$$

System boundary

Angular momentum

The angular momentum of the isolated system is constant.

Examples:

- after a supernova explosion, the core of a star collapses to a small radius and spins at a much higher rate
- the square of the orbital period of a planet is proportional to the cube of its semimajor axis; Kepler's third law (Chapter 13)
- in atomic transitions, selection rules on the quantum numbers must be obeyed in order to conserve angular momentum (Chapter 42)
- in beta decay of a radioactive nucleus, a neutrino must be emitted in order to conserve angular momentum (Chapter 44)

Example 11.7   Formation of a Neutron Star

A star rotates with a period of 30 days about an axis through its center. The period is the time interval required for a point on the star's equator to make one complete revolution around the axis of rotation. After the star undergoes a supernova explosion, the stellar core, which had a radius of 1.0×10^4 km, collapses into a neutron star of radius 3.0 km. Determine the period of rotation of the neutron star.

SOLUTION

Conceptualize The change in the neutron star's motion is similar to that of the skater described earlier, but in the reverse direction. As the mass of the star moves closer to the rotation axis, we expect the star to spin faster.

Categorize Let us assume that during the collapse of the stellar core, (1) no external torque acts on it, (2) it remains spherical with the same relative mass distribution, and (3) its mass remains constant. We categorize the star as an *isolated system* in terms of *angular momentum*. We do not know the mass distribution of the star, but we have assumed the distribution is symmetric, so the moment of inertia can be expressed as kMR^2, where k is some numerical constant. (From Table 10.2, for example, we see that $k = \frac{2}{5}$ for a solid sphere and $k = \frac{2}{3}$ for a spherical shell.)

Analyze Let's use the symbol T for the period, with T_i being the initial period of the star and T_f being the period of the neutron star. The star's angular speed is given by $\omega = 2\pi/T$.

From the isolated system model for angular momentum, write Equation 11.19 for the star:

$$I_i \omega_i = I_f \omega_f$$

Use $\omega = 2\pi/T$ to rewrite this equation in terms of the initial and final periods:

$$I_i \left(\frac{2\pi}{T_i}\right) = I_f \left(\frac{2\pi}{T_f}\right)$$

Substitute the moments of inertia in the preceding equation:

$$kMR_i^{\,2}\left(\frac{2\pi}{T_i}\right) = kMR_f^{\,2}\left(\frac{2\pi}{T_f}\right)$$

Solve for the final period of the star:

$$T_f = \left(\frac{R_f}{R_i}\right)^2 T_i$$

Substitute numerical values:

$$T_f = \left(\frac{3.0 \text{ km}}{1.0 \times 10^4 \text{ km}}\right)^2 (30 \text{ days}) = 2.7 \times 10^{-6} \text{ days} = \boxed{0.23 \text{ s}}$$

Finalize The neutron star does indeed rotate faster after it collapses, as predicted. It moves very fast, in fact, rotating about four times each second!

Example 11.8 The Merry-Go-Round

A horizontal platform in the shape of a circular disk rotates freely in a horizontal plane about a frictionless, vertical axle (Fig. 11.11). The platform has a mass $M = 100$ kg and a radius $R = 2.0$ m. A student whose mass is $m = 60$ kg walks slowly from the rim of the disk toward its center. If the angular speed of the system is 2.0 rad/s when the student is at the rim, what is the angular speed when she reaches a point $r = 0.50$ m from the center?

Figure 11.11 (Example 11.8) As the student walks toward the center of the rotating platform, the angular speed of the system increases because the angular momentum of the system remains constant.

SOLUTION

Conceptualize The speed change here is similar to those of the spinning skater and the neutron star in preceding discussions. This problem is different because part of the moment of inertia of the system changes (that of the student) while part remains fixed (that of the platform).

Categorize Because the platform rotates on a frictionless axle, we identify the system of the student and the platform as an *isolated system* in terms of *angular momentum*.

Analyze Let us denote the moment of inertia of the platform as I_p and that of the student as I_s. We model the student as a particle.

Find the initial moment of inertia I_i of the system (student plus platform) about the axis of rotation:

$$I_i = I_{pi} + I_{si} = \tfrac{1}{2}MR^2 + mR^2$$

Find the moment of inertia of the system when the student walks to the position $r < R$:

$$I_f = I_{pf} + I_{sf} = \tfrac{1}{2}MR^2 + mr^2$$

Write Equation 11.19 for the system:

$$I_i\omega_i = I_f\omega_f$$

Substitute the moments of inertia:

$$(\tfrac{1}{2}MR^2 + mR^2)\omega_i = (\tfrac{1}{2}MR^2 + mr^2)\omega_f$$

Solve for the final angular speed:

$$\omega_f = \left(\frac{\tfrac{1}{2}MR^2 + mR^2}{\tfrac{1}{2}MR^2 + mr^2}\right)\omega_i$$

Substitute numerical values:

$$\omega_f = \left[\frac{\tfrac{1}{2}(100\text{ kg})(2.0\text{ m})^2 + (60\text{ kg})(2.0\text{ m})^2}{\tfrac{1}{2}(100\text{ kg})(2.0\text{ m})^2 + (60\text{ kg})(0.50\text{ m})^2}\right](2.0\text{ rad/s}) = 4.1\text{ rad/s}$$

Finalize As expected, the angular speed increases. The fastest that this system could spin would be when the student moves to the center of the platform. Do this calculation to show that this maximum angular speed is 4.4 rad/s. Notice that the activity described in this problem is dangerous as discussed with regard to the Coriolis force in Section 6.3.

WHAT IF? What if you measured the kinetic energy of the system before and after the student walks inward? Are the initial kinetic energy and the final kinetic energy the same?

Answer You may be tempted to say yes because the system is isolated. Remember, however, that energy can be transformed among several forms, so we have to handle an energy question carefully.

Find the initial kinetic energy:

$$K_i = \tfrac{1}{2}I_i\omega_i^2 = \tfrac{1}{2}(440\text{ kg} \cdot \text{m}^2)(2.0\text{ rad/s})^2 = 880\text{ J}$$

Find the final kinetic energy:

$$K_f = \tfrac{1}{2}I_f\omega_f^2 = \tfrac{1}{2}(215\text{ kg} \cdot \text{m}^2)(4.1\text{ rad/s})^2 = 1.80 \times 10^3\text{ J}$$

Therefore, the kinetic energy of the system *increases*. The student must perform muscular activity to move herself closer to the center of rotation, so this extra kinetic energy comes from potential energy stored in the student's body from previous meals. The system is isolated in terms of energy, but a transformation process within the system changes potential energy to kinetic energy.

| Example 11.9 | Disk and Stick Collision | AM |

Before

A 2.0-kg disk traveling at 3.0 m/s strikes a 1.0-kg stick of length 4.0 m that is lying flat on nearly frictionless ice as shown in the overhead view of Figure 11.12a. The disk strikes at the endpoint of the stick, at a distance $r = 2.0$ m from the stick's center. Assume the collision is elastic and the disk does not deviate from its original line of motion. Find the translational speed of the disk, the translational speed of the stick, and the angular speed of the stick after the collision. The moment of inertia of the stick about its center of mass is 1.33 kg · m².

SOLUTION

Conceptualize Examine Figure 11.12a and imagine what happens after the disk hits the stick. Figure 11.12b shows what you might expect: the disk continues to move at a slower speed, and the stick is in both translational and rotational motion. We assume the disk does not deviate from its original line of motion because the force exerted by the stick on the disk is parallel to the original path of the disk.

Figure 11.12 (Example 11.9) Overhead view of a disk striking a stick in an elastic collision. (a) Before the collision, the disk moves toward the stick. (b) The collision causes the stick to rotate and move to the right.

After

Categorize Because the ice is frictionless, the disk and stick form an *isolated system* in terms of *momentum* and *angular momentum*. Ignoring the sound made in the collision, we also model the system as an *isolated system* in terms of *energy*. In addition, because the collision is assumed to be elastic, the kinetic energy of the system is constant.

Analyze First notice that we have three unknowns, so we need three equations to solve simultaneously.

Apply the isolated system model for momentum to the system and then rearrange the result:

$$\Delta \vec{p}_{tot} = 0 \rightarrow (m_d v_{df} + m_s v_s) - m_d v_{di} = 0$$
$$(1) \quad m_d(v_{di} - v_{df}) = m_s v_s$$

Apply the isolated system model for angular momentum to the system and rearrange the result. Use an axis passing through the center of the stick as the rotation axis so that the path of the disk is a distance $r = 2.0$ m from the rotation axis:

$$\Delta \vec{L}_{tot} = 0 \rightarrow (-r m_d v_{df} + I\omega) - (-r m_d v_{di}) = 0$$
$$(2) \quad -r m_d(v_{di} - v_{df}) = I\omega$$

Apply the isolated system model for energy to the system, rearrange the equation, and factor the combination of terms related to the disk:

$$\Delta K = 0 \rightarrow (\tfrac{1}{2} m_d v_{df}^2 + \tfrac{1}{2} m_s v_s^2 + \tfrac{1}{2} I\omega^2) - \tfrac{1}{2} m_d v_{di}^2 = 0$$
$$(3) \quad m_d(v_{di} - v_{df})(v_{di} + v_{df}) = m_s v_s^2 + I\omega^2$$

Multiply Equation (1) by r and add to Equation (2):

$$r m_d(v_{di} - v_{df}) = r m_s v_s$$
$$-r m_d(v_{di} - v_{df}) = I\omega$$
$$0 = r m_s v_s + I\omega$$

Solve for ω:

$$(4) \quad \omega = -\frac{r m_s v_s}{I}$$

Divide Equation (3) by Equation (1):

$$\frac{m_d(v_{di} - v_{df})(v_{di} + v_{df})}{m_d(v_{di} - v_{df})} = \frac{m_s v_s^2 + I\omega^2}{m_s v_s}$$
$$(5) \quad v_{di} + v_{df} = v_s + \frac{I\omega^2}{m_s v_s}$$

Substitute Equation (4) into Equation (5):

$$(6) \quad v_{di} + v_{df} = v_s\left(1 + \frac{r^2 m_s}{I}\right)$$

Substitute v_{df} from Equation (1) into Equation (6):

$$v_{di} + \left(v_{di} - \frac{m_s}{m_d} v_s\right) = v_s\left(1 + \frac{r^2 m_s}{I}\right)$$

continued

▶ **11.9** continued

Solve for v_s and substitute numerical values:

$$v_s = \frac{2v_{di}}{1 + (m_s/m_d) + (r^2 m_s/I)}$$

$$= \frac{2(3.0 \text{ m/s})}{1 + (1.0 \text{ kg}/2.0 \text{ kg}) + [(2.0 \text{ m})^2(1.0 \text{ kg})/1.33 \text{ kg} \cdot \text{m}^2]} = \boxed{1.3 \text{ m/s}}$$

Substitute numerical values into Equation (4):

$$\omega = -\frac{(2.0 \text{ m})(1.0 \text{ kg})(1.3 \text{ m/s})}{1.33 \text{ kg} \cdot \text{m}^2} = \boxed{-2.0 \text{ rad/s}}$$

Solve Equation (1) for v_{df} and substitute numerical values:

$$v_{df} = v_{di} - \frac{m_s}{m_d} v_s = 3.0 \text{ m/s} - \frac{1.0 \text{ kg}}{2.0 \text{ kg}}(1.3 \text{ m/s}) = \boxed{2.3 \text{ m/s}}$$

Finalize These values seem reasonable. The disk is moving more slowly after the collision than it was before the collision, and the stick has a small translational speed. Table 11.1 summarizes the initial and final values of variables for the disk and the stick, and it verifies the conservation of linear momentum, angular momentum, and kinetic energy for the isolated system.

Table 11.1   Comparison of Values in Example 11.9 Before and After the Collision

	v (m/s)	ω (rad/s)	p (kg · m/s)	L (kg · m²/s)	K_{trans} (J)	K_{rot} (J)
Before						
Disk	3.0	—	6.0	−12	9.0	—
Stick	0	0	0	0	0	0
Total for system	—	—	6.0	−12	9.0	0
After						
Disk	2.3	—	4.7	−9.3	5.4	—
Stick	1.3	−2.0	1.3	−2.7	0.9	2.7
Total for system	—	—	6.0	−12	6.3	2.7

Note: Linear momentum, angular momentum, and total kinetic energy of the system are all conserved.

11.5 The Motion of Gyroscopes and Tops

An unusual and fascinating type of motion you have probably observed is that of a top spinning about its axis of symmetry as shown in Figure 11.13a. If the top spins rapidly, the symmetry axis rotates about the z axis, sweeping out a cone (see Fig. 11.13b). The motion of the symmetry axis about the vertical—known as **precessional motion**—is usually slow relative to the spinning motion of the top.

It is quite natural to wonder why the top does not fall over. Because the center of mass is not directly above the pivot point O, a net torque is acting on the top about an axis passing through O, a torque resulting from the gravitational force $M\vec{\mathbf{g}}$. The top would certainly fall over if it were not spinning. Because it is spinning, however, it has an angular momentum $\vec{\mathbf{L}}$ directed along its symmetry axis. We shall show that this symmetry axis moves about the z axis (precessional motion occurs) because the torque produces a change in the *direction* of the symmetry axis. This illustration is an excellent example of the importance of the vector nature of angular momentum.

The essential features of precessional motion can be illustrated by considering the simple gyroscope shown in Figure 11.14a. The two forces acting on the gyroscope are shown in Figure 11.14b: the downward gravitational force $M\vec{\mathbf{g}}$ and the normal force $\vec{\mathbf{n}}$ acting upward at the pivot point O. The normal force produces no torque about an axis passing through the pivot because its moment arm through that point is zero. The gravitational force, however, produces a torque $\vec{\boldsymbol{\tau}} = \vec{\mathbf{r}} \times M\vec{\mathbf{g}}$ about an axis passing through O, where the direction of $\vec{\boldsymbol{\tau}}$ is perpendicular to the plane formed by $\vec{\mathbf{r}}$ and $M\vec{\mathbf{g}}$. By necessity, the vector $\vec{\boldsymbol{\tau}}$ lies in a horizontal xy plane perpendicular to the angular momentum vector. The net torque and angular momentum of the gyroscope are related through Equation 11.13:

$$\sum \vec{\tau}_{ext} = \frac{d\vec{L}}{dt}$$

This expression shows that in the infinitesimal time interval dt, the nonzero torque produces a change in angular momentum $d\vec{L}$, a change that is in the same direction as $\vec{\tau}$. Therefore, like the torque vector, $d\vec{L}$ must also be perpendicular to $\vec{L}$. Figure 11.14c illustrates the resulting precessional motion of the symmetry axis of the gyroscope. In a time interval dt, the change in angular momentum is $d\vec{L} = \vec{L}_f - \vec{L}_i = \vec{\tau}\, dt$. Because $d\vec{L}$ is perpendicular to $\vec{L}$, the magnitude of $\vec{L}$ does not change ($|\vec{L}_i| = |\vec{L}_f|$). Rather, what is changing is the *direction* of $\vec{L}$. Because the change in angular momentum $d\vec{L}$ is in the direction of $\vec{\tau}$, which lies in the xy plane, the gyroscope undergoes precessional motion.

To simplify the description of the system, we assume the total angular momentum of the precessing wheel is the sum of the angular momentum $I\vec{\omega}$ due to the spinning and the angular momentum due to the motion of the center of mass about the pivot. In our treatment, we shall neglect the contribution from the center-of-mass motion and take the total angular momentum to be simply $I\vec{\omega}$. In practice, this approximation is good if $\vec{\omega}$ is made very large.

The vector diagram in Figure 11.14c shows that in the time interval dt, the angular momentum vector rotates through an angle $d\phi$, which is also the angle through which the gyroscope axle rotates. From the vector triangle formed by the vectors $\vec{L}_i$, $\vec{L}_f$, and $d\vec{L}$, we see that

$$d\phi = \frac{dL}{L} = \frac{\sum \tau_{ext}\, dt}{L} = \frac{(Mgr_{CM})\, dt}{L}$$

Dividing through by dt and using the relationship $L = I\omega$, we find that the rate at which the axle rotates about the vertical axis is

$$\omega_p = \frac{d\phi}{dt} = \frac{Mgr_{CM}}{I\omega} \qquad (11.20)$$

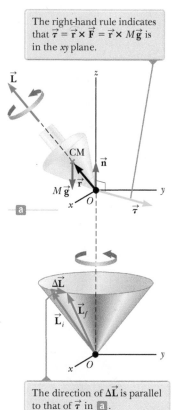

The right-hand rule indicates that $\vec{\tau} = \vec{r} \times \vec{F} = \vec{r} \times M\vec{g}$ is in the xy plane.

a

The direction of $\Delta\vec{L}$ is parallel to that of $\vec{\tau}$ in **a**.

b

Figure 11.13 Precessional motion of a top spinning about its symmetry axis. (a) The only external forces acting on the top are the normal force $\vec{n}$ and the gravitational force $M\vec{g}$. The direction of the angular momentum $\vec{L}$ is along the axis of symmetry. (b) Because $\vec{L}_f = \Delta\vec{L} + \vec{L}_i$, the top precesses about the z axis.

The gravitational force $M\vec{g}$ in the negative z direction produces a torque on the gyroscope in the positive y direction about the pivot.

The torque results in a change in angular momentum $d\vec{L}$ in a direction parallel to the torque vector. The gyroscope axle sweeps out an angle $d\phi$ in a time interval dt.

a   **b**   **c**

Figure 11.14 (a) A spinning gyroscope is placed on a pivot at the right end. (b) Diagram for the spinning gyroscope showing forces, torque, and angular momentum. (c) Overhead view (looking down the z axis) of the gyroscope's initial and final angular momentum vectors for an infinitesimal time interval dt.

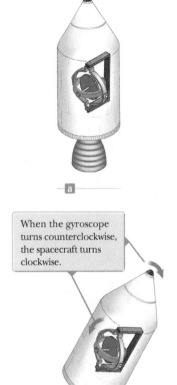

When the gyroscope turns counterclockwise, the spacecraft turns clockwise.

Figure 11.15 (a) A spacecraft carries a gyroscope that is not spinning. (b) The gyroscope is set into rotation.

The angular speed ω_p is called the **precessional frequency**. This result is valid only when $\omega_p \ll \omega$. Otherwise, a much more complicated motion is involved. As you can see from Equation 11.20, the condition $\omega_p \ll \omega$ is met when ω is large, that is, when the wheel spins rapidly. Furthermore, notice that the precessional frequency decreases as ω increases, that is, as the wheel spins faster about its axis of symmetry.

As an example of the usefulness of gyroscopes, suppose you are in a spacecraft in deep space and you need to alter your trajectory. To fire the engines in the correct direction, you need to turn the spacecraft. How, though, do you turn a spacecraft in empty space? One way is to have small rocket engines that fire perpendicularly out the side of the spacecraft, providing a torque around its center of mass. Such a setup is desirable, and many spacecraft have such rockets.

Let us consider another method, however, that does not require the consumption of rocket fuel. Suppose the spacecraft carries a gyroscope that is not rotating as in Figure 11.15a. In this case, the angular momentum of the spacecraft about its center of mass is zero. Suppose the gyroscope is set into rotation, giving the gyroscope a nonzero angular momentum. There is no external torque on the isolated system (spacecraft and gyroscope), so the angular momentum of this system must remain zero according to the isolated system (angular momentum) model. The zero value can be satisfied if the spacecraft rotates in the direction opposite that of the gyroscope so that the angular momentum vectors of the gyroscope and the spacecraft cancel, resulting in no angular momentum of the system. The result of rotating the gyroscope, as in Figure 11.15b, is that the spacecraft turns around! By including three gyroscopes with mutually perpendicular axles, any desired rotation in space can be achieved.

This effect created an undesirable situation with the *Voyager 2* spacecraft during its flight. The spacecraft carried a tape recorder whose reels rotated at high speeds. Each time the tape recorder was turned on, the reels acted as gyroscopes and the spacecraft started an undesirable rotation in the opposite direction. This rotation had to be counteracted by Mission Control by using the sideward-firing jets to *stop* the rotation!

Summary

Definitions

Given two vectors $\vec{A}$ and $\vec{B}$, the **vector product** $\vec{A} \times \vec{B}$ is a vector $\vec{C}$ having a magnitude

$$C = AB \sin \theta \qquad \text{(11.3)}$$

where θ is the angle between $\vec{A}$ and $\vec{B}$. The direction of the vector $\vec{C} = \vec{A} \times \vec{B}$ is perpendicular to the plane formed by $\vec{A}$ and $\vec{B}$, and this direction is determined by the right-hand rule.

The **torque** $\vec{\tau}$ on a particle due to a force $\vec{F}$ about an axis through the origin in an inertial frame is defined to be

$$\vec{\tau} \equiv \vec{r} \times \vec{F} \qquad \text{(11.1)}$$

The **angular momentum** $\vec{L}$ about an axis through the origin of a particle having linear momentum $\vec{p} = m\vec{v}$ is

$$\vec{L} \equiv \vec{r} \times \vec{p} \qquad \text{(11.10)}$$

where $\vec{r}$ is the vector position of the particle relative to the origin.

Concepts and Principles

The z component of angular momentum of a rigid object rotating about a fixed z axis is

$$L_z = I\omega \tag{11.14}$$

where I is the moment of inertia of the object about the axis of rotation and ω is its angular speed.

Analysis Models for Problem Solving

Nonisolated System (Angular Momentum). If a system interacts with its environment in the sense that there is an external torque on the system, the net external torque acting on a system is equal to the time rate of change of its angular momentum:

$$\sum \vec{\tau}_{ext} = \frac{d\vec{L}_{tot}}{dt} \tag{11.13}$$

Isolated System (Angular Momentum). If a system experiences no external torque from the environment, the total angular momentum of the system is conserved:

$$\Delta \vec{L}_{tot} = 0 \tag{11.18}$$

Applying this law of conservation of angular momentum to a system whose moment of inertia changes gives

$$I_i \omega_i = I_f \omega_f = \text{constant} \tag{11.19}$$

Objective Questions   **1.** denotes answer available in *Student Solutions Manual/Study Guide*

1. An ice skater starts a spin with her arms stretched out to the sides. She balances on the tip of one skate to turn without friction. She then pulls her arms in so that her moment of inertia decreases by a factor of 2. In the process of her doing so, what happens to her kinetic energy? (a) It increases by a factor of 4. (b) It increases by a factor of 2. (c) It remains constant. (d) It decreases by a factor of 2. (e) It decreases by a factor of 4.

2. A pet mouse sleeps near the eastern edge of a stationary, horizontal turntable that is supported by a frictionless, vertical axle through its center. The mouse wakes up and starts to walk north on the turntable. **(i)** As it takes its first steps, what is the direction of the mouse's displacement relative to the stationary ground below? (a) north (b) south (c) no displacement. **(ii)** In this process, the spot on the turntable where the mouse had been snoozing undergoes a displacement in what direction relative to the ground below? (a) north (b) south (c) no displacement. Answer yes or no for the

following questions. **(iii)** In this process, is the mechanical energy of the mouse–turntable system constant? **(iv)** Is the momentum of the system constant? **(v)** Is the angular momentum of the system constant?

3. Let us name three perpendicular directions as right, up, and toward you as you might name them when you are facing a television screen that lies in a vertical plane. Unit vectors for these directions are $\hat{r}$, $\hat{u}$, and $\hat{t}$, respectively. Consider the quantity $(-3\hat{u} \times 2\hat{t})$. **(i)** Is the magnitude of this vector (a) 6, (b) 3, (c) 2, or (d) 0? **(ii)** Is the direction of this vector (a) down, (b) toward you, (c) up, (d) away from you, or (e) left?

4. Let the four compass directions north, east, south, and west be represented by unit vectors $\hat{n}$, $\hat{e}$, $\hat{s}$, and $\hat{w}$, respectively. Vertically up and down are represented as $\hat{u}$ and $\hat{d}$. Let us also identify unit vectors that are halfway between these directions such as $\hat{ne}$ for northeast. Rank the magnitudes of the following cross products from largest to smallest. If any are equal in magnitude

or are equal to zero, show that in your ranking. (a) $\hat{\mathbf{n}} \times \hat{\mathbf{n}}$ (b) $\hat{\mathbf{w}} \times \hat{\mathbf{ne}}$ (c) $\hat{\mathbf{u}} \times \hat{\mathbf{ne}}$ (d) $\hat{\mathbf{n}} \times \hat{\mathbf{nw}}$ (e) $\hat{\mathbf{n}} \times \hat{\mathbf{e}}$

5. Answer yes or no to the following questions. (a) Is it possible to calculate the torque acting on a rigid object without specifying an axis of rotation? (b) Is the torque independent of the location of the axis of rotation?

6. Vector $\vec{\mathbf{A}}$ is in the negative y direction, and vector $\vec{\mathbf{B}}$ is in the negative x direction. (i) What is the direction of $\vec{\mathbf{A}} \times \vec{\mathbf{B}}$? (a) no direction because it is a scalar (b) x (c) $-y$ (d) z (e) $-z$ (ii) What is the direction of $\vec{\mathbf{B}} \times \vec{\mathbf{A}}$? Choose from the same possibilities (a) through (e).

7. Two ponies of equal mass are initially at diametrically opposite points on the rim of a large horizontal turntable that is turning freely on a frictionless, vertical axle through its center. The ponies simultaneously start walking toward each other across the turntable. (i) As they walk, what happens to the angular speed of the turntable? (a) It increases. (b) It decreases. (c) It stays constant. Consider the ponies–turntable system in this process and answer yes or no for the following questions. (ii) Is the mechanical energy of the system conserved? (iii) Is the momentum of the system conserved? (iv) Is the angular momentum of the system conserved?

8. Consider an isolated system moving through empty space. The system consists of objects that interact with each other and can change location with respect to one another. Which of the following quantities can change in time? (a) The angular momentum of the system. (b) The linear momentum of the system. (c) Both the angular momentum and linear momentum of the system. (d) Neither the angular momentum nor linear momentum of the system.

Conceptual Questions

1. denotes answer available in *Student Solutions Manual/Study Guide*

1. Stars originate as large bodies of slowly rotating gas. Because of gravity, these clumps of gas slowly decrease in size. What happens to the angular speed of a star as it shrinks? Explain.

2. A scientist arriving at a hotel asks a bellhop to carry a heavy suitcase. When the bellhop rounds a corner, the suitcase suddenly swings away from him for some unknown reason. The alarmed bellhop drops the suitcase and runs away. What might be in the suitcase?

3. Why does a long pole help a tightrope walker stay balanced?

4. Two children are playing with a roll of paper towels. One child holds the roll between the index fingers of her hands so that it is free to rotate, and the second child pulls at constant speed on the free end of the paper towels. As the child pulls the paper towels, the radius of the roll of remaining towels decreases. (a) How does the torque on the roll change with time? (b) How does the angular speed of the roll change in time? (c) If the child suddenly jerks the end paper towel with a large force, is the towel more likely to break from the others when it is being pulled from a nearly full roll or from a nearly empty roll?

5. Both torque and work are products of force and displacement. How are they different? Do they have the same units?

6. In some motorcycle races, the riders drive over small hills and the motorcycle becomes airborne for a short time interval. If the motorcycle racer keeps the throttle open while leaving the hill and going into the air, the motorcycle tends to nose upward. Why?

7. If the torque acting on a particle about an axis through a certain origin is zero, what can you say about its angular momentum about that axis?

8. A ball is thrown in such a way that it does not spin about its own axis. Does this statement imply that the angular momentum is zero about an arbitrary axis? Explain.

9. If global warming continues over the next one hundred years, it is likely that some polar ice will melt and the water will be distributed closer to the equator. (a) How would that change the moment of inertia of the Earth? (b) Would the duration of the day (one revolution) increase or decrease?

10. A cat usually lands on its feet regardless of the position from which it is dropped. A slow-motion film of a cat falling shows that the upper half of its body twists in one direction while the lower half twists in the opposite direction. (See Fig. CQ11.10.) Why does this type of rotation occur?

Figure CQ11.10

11. In Chapters 7 and 8, we made use of energy bar charts to analyze physical situations. Why have we not used bar charts for angular momentum in this chapter?

Problems

Section 11.1 The Vector Product and Torque

1. Given $\vec{\mathbf{M}} = 2\hat{\mathbf{i}} - 3\hat{\mathbf{j}} + \hat{\mathbf{k}}$ and $\vec{\mathbf{N}} = 4\hat{\mathbf{i}} + 5\hat{\mathbf{j}} - 2\hat{\mathbf{k}}$, calculate the vector product $\vec{\mathbf{M}} \times \vec{\mathbf{N}}$.

2. The displacement vectors 42.0 cm at 15.0° and 23.0 cm at 65.0° both start from the origin and form two sides of a parallelogram. Both angles are measured counterclockwise from the *x* axis. (a) Find the area of the parallelogram. (b) Find the length of its longer diagonal.

3. Two vectors are given by $\vec{\mathbf{A}} = \hat{\mathbf{i}} + 2\hat{\mathbf{j}}$ and $\vec{\mathbf{B}} = -2\hat{\mathbf{i}} + 3\hat{\mathbf{j}}$. Find (a) $\vec{\mathbf{A}} \times \vec{\mathbf{B}}$ and (b) the angle between $\vec{\mathbf{A}}$ and $\vec{\mathbf{B}}$.

4. Use the definition of the vector product and the definitions of the unit vectors $\hat{\mathbf{i}}$, $\hat{\mathbf{j}}$, and $\hat{\mathbf{k}}$ to prove Equations 11.7. You may assume the *x* axis points to the right, the *y* axis up, and the *z* axis horizontally toward you (not away from you). This choice is said to make the coordinate system a *right-handed system*.

5. Calculate the net torque (magnitude and direction) on the beam in Figure P11.5 about (a) an axis through *O* perpendicular to the page and (b) an axis through *C* perpendicular to the page.

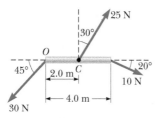

Figure P11.5

6. Two vectors are given by these expressions: $\vec{\mathbf{A}} = -3\hat{\mathbf{i}} + 7\hat{\mathbf{j}} - 4\hat{\mathbf{k}}$ and $\vec{\mathbf{B}} = 6\hat{\mathbf{i}} - 10\hat{\mathbf{j}} + 9\hat{\mathbf{k}}$. Evaluate the quantities (a) $\cos^{-1}[\vec{\mathbf{A}} \cdot \vec{\mathbf{B}}/AB]$ and (b) $\sin^{-1}[|\vec{\mathbf{A}} \times \vec{\mathbf{B}}|/AB]$. (c) Which give(s) the angle between the vectors?

7. If $|\vec{\mathbf{A}} \times \vec{\mathbf{B}}| = \vec{\mathbf{A}} \cdot \vec{\mathbf{B}}$, what is the angle between $\vec{\mathbf{A}}$ and $\vec{\mathbf{B}}$?

8. A particle is located at the vector position $\vec{\mathbf{r}} = (4.00\hat{\mathbf{i}} + 6.00\hat{\mathbf{j}})$ m, and a force exerted on it is given by $\vec{\mathbf{F}} = (3.00\hat{\mathbf{i}} + 2.00\hat{\mathbf{j}})$ N. (a) What is the torque acting on the particle about the origin? (b) Can there be another point about which the torque caused by this force on this particle will be in the opposite direction and half as large in magnitude? (c) Can there be more than one such point? (d) Can such a point lie on the *y* axis? (e) Can more than one such point lie on the *y* axis? (f) Determine the position vector of one such point.

9. Two forces $\vec{\mathbf{F}}_1$ and $\vec{\mathbf{F}}_2$ act along the two sides of an equilateral triangle as shown in Figure P11.9. Point *O* is the intersection of the altitudes of the triangle. (a) Find a third force $\vec{\mathbf{F}}_3$ to be applied at *B* and along *BC* that will make the total torque zero about the point *O*. (b) **What If?** Will the total torque change if $\vec{\mathbf{F}}_3$ is applied not at *B* but at any other point along *BC*?

Figure P11.9

10. A student claims that he has found a vector $\vec{\mathbf{A}}$ such that $(2\hat{\mathbf{i}} - 3\hat{\mathbf{j}} + 4\hat{\mathbf{k}}) \times \vec{\mathbf{A}} = (4\hat{\mathbf{i}} + 3\hat{\mathbf{j}} - \hat{\mathbf{k}})$. (a) Do you believe this claim? (b) Explain why or why not.

Section 11.2 Analysis Model: Nonisolated System (Angular Momentum)

11. A light, rigid rod of length $\ell = 1.00$ m joins two particles, with masses $m_1 = 4.00$ kg and $m_2 = 3.00$ kg, at its ends. The combination rotates in the *xy* plane about a pivot through the center of the rod (Fig. P11.11). Determine the angular momentum of the system about the origin when the speed of each particle is 5.00 m/s.

Figure P11.11

12. A 1.50-kg particle moves in the *xy* plane with a velocity of $\vec{\mathbf{v}} = (4.20\hat{\mathbf{i}} - 3.60\hat{\mathbf{j}})$ m/s. Determine the angular momentum of the particle about the origin when its position vector is $\vec{\mathbf{r}} = (1.50\hat{\mathbf{i}} + 2.20\hat{\mathbf{j}})$ m.

13. A particle of mass *m* moves in the *xy* plane with a velocity of $\vec{\mathbf{v}} = v_x\hat{\mathbf{i}} + v_y\hat{\mathbf{j}}$. Determine the angular momentum

of the particle about the origin when its position vector is $\vec{r} = x\hat{i} + y\hat{j}$.

14. Heading straight toward the summit of Pike's Peak, an airplane of mass 12 000 kg flies over the plains of Kansas at nearly constant altitude 4.30 km with constant velocity 175 m/s west. (a) What is the airplane's vector angular momentum relative to a wheat farmer on the ground directly below the airplane? (b) Does this value change as the airplane continues its motion along a straight line? (c) **What If?** What is its angular momentum relative to the summit of Pike's Peak?

15. **Review.** A projectile of mass m is launched with an initial velocity $\vec{v}_i$ making an angle θ with the horizontal as shown in Figure P11.15. The projectile moves in the gravitational field of the Earth. Find the angular momentum of the projectile about the origin (a) when the projectile is at the origin, (b) when it is at the highest point of its trajectory, and (c) just before it hits the ground. (d) What torque causes its angular momentum to change?

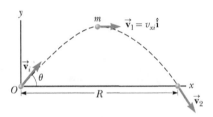

Figure P11.15

16. **Review.** A conical pendulum consists of a bob of mass m in motion in a circular path in a horizontal plane as shown in Figure P11.16. During the motion, the supporting wire of length ℓ maintains a constant angle θ with the vertical. Show that the magnitude of the angular momentum of the bob about the vertical dashed line is

Figure P11.16

$$L = \left(\frac{m^2 g \ell^3 \sin^4 \theta}{\cos \theta}\right)^{1/2}$$

17. A particle of mass m moves in a circle of radius R at a constant speed v as shown in Figure P11.17. The motion begins at point Q at time $t = 0$. Determine the angular momentum of the particle about the axis perpendicular to the page through point P as a function of time.

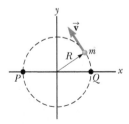

Figure P11.17 Problems 17 and 32.

18. A counterweight of mass $m = 4.00$ kg is attached to a light cord that is wound around a pulley as in Figure P11.18. The pulley is a thin hoop of radius $R = 8.00$ cm and mass $M = 2.00$ kg. The spokes have negligible mass. (a) What is the magnitude of the net torque on the system about the axle of the pulley? (b) When the counterweight has a speed v, the pulley has an angular speed $\omega = v/R$. Determine the magnitude of the total angular momentum of the system about the axle of the pulley. (c) Using your result from part (b) and $\vec{\tau} = d\vec{L}/dt$, calculate the acceleration of the counterweight.

Figure P11.18

19. The position vector of a particle of mass 2.00 kg as a function of time is given by $\vec{r} = (6.00\hat{i} + 5.00t\hat{j})$, where $\vec{r}$ is in meters and t is in seconds. Determine the angular momentum of the particle about the origin as a function of time.

20. A 5.00-kg particle starts from the origin at time zero. Its velocity as a function of time is given by

$$\vec{v} = 6t^2\hat{i} + 2t\hat{j}$$

where $\vec{v}$ is in meters per second and t is in seconds. (a) Find its position as a function of time. (b) Describe its motion qualitatively. Find (c) its acceleration as a function of time, (d) the net force exerted on the particle as a function of time, (e) the net torque about the origin exerted on the particle as a function of time, (f) the angular momentum of the particle as a function of time, (g) the kinetic energy of the particle as a function of time, and (h) the power injected into the system of the particle as a function of time.

21. A ball having mass m is fastened at the end of a flagpole that is connected to the side of a tall building at point P as shown in Figure P11.21. The length of the flagpole is ℓ, and it makes an angle θ with the x axis. The ball becomes loose and starts to fall with acceleration $-g\hat{j}$. (a) Determine the angular momentum of the ball about point P as a function of time. (b) For what physical reason does the angular momentum change? (c) What is the rate of change of the angular momentum of the ball about point P?

Figure P11.21

Section 11.3 Angular Momentum of a Rotating Rigid Object

22. A uniform solid sphere of radius $r = 0.500$ m and mass $m = 15.0$ kg turns counterclockwise about a vertical axis through its center. Find its vector angular momentum about this axis when its angular speed is 3.00 rad/s.

23. Big Ben (Fig. P10.49, page 328), the Parliament tower clock in London, has hour and minute hands with lengths of 2.70 m and 4.50 m and masses of 60.0 kg and 100 kg, respectively. Calculate the total angular momentum of these hands about the center point. (You may model the hands as long, thin rods rotating about one end. Assume the hour and minute hands are rotating at a constant rate of one revolution per 12 hours and 60 minutes, respectively.)

24. Show that the kinetic energy of an object rotating about a fixed axis with angular momentum $L = I\omega$ can be written as $K = L^2/2I$.

25. A uniform solid disk of mass $m = 3.00$ kg and radius ▣ $r = 0.200$ m rotates about a fixed axis perpendicular to its face with angular frequency 6.00 rad/s. Calculate the magnitude of the angular momentum of the disk when the axis of rotation (a) passes through its center of mass and (b) passes through a point midway between the center and the rim.

26. Model the Earth as a uniform sphere. (a) Calculate the angular momentum of the Earth due to its spinning motion about its axis. (b) Calculate the angular momentum of the Earth due to its orbital motion about the Sun. (c) Explain why the answer in part (b) is larger than that in part (a) even though it takes significantly longer for the Earth to go once around the Sun than to rotate once about its axis.

27. A particle of mass 0.400 kg is attached to the 100-cm ▣ mark of a meterstick of mass 0.100 kg. The meterstick rotates on the surface of a frictionless, horizontal table with an angular speed of 4.00 rad/s. Calculate the angular momentum of the system when the stick is pivoted about an axis (a) perpendicular to the table through the 50.0-cm mark and (b) perpendicular to the table through the 0-cm mark.

28. The distance between the centers of the wheels of a motorcycle is 155 cm. The center of mass of the motorcycle, including the rider, is 88.0 cm above the ground and halfway between the wheels. Assume the mass of each wheel is small compared with the body of the motorcycle. The engine drives the rear wheel only. What horizontal acceleration of the motorcycle will make the front wheel rise off the ground?

29. A space station is constructed in the shape of a hollow ▣ ring of mass 5.00×10^4 kg. Members of the crew walk on a deck formed by the inner surface of the outer cylindrical wall of the ring, with radius $r = 100$ m. At rest when constructed, the ring is set rotating about its axis so that the people inside experience an effective free-fall acceleration equal to g. (See Fig. P11.29.) The rotation is achieved by firing two small rockets attached tangentially to opposite points on the rim of

the ring. (a) What angular momentum does the space station acquire? (b) For what time interval must the rockets be fired if each exerts a thrust of 125 N?

Figure P11.29 Problems 29 and 40.

Section 11.4 Analysis Model: Isolated System (Angular Momentum)

30. A disk with moment of inertia I_1 rotates about a fric- ▣ tionless, vertical axle with angular speed ω_i. A second disk, this one having moment of inertia I_2 and initially not rotating, drops onto the first disk (Fig. P11.30). Because of friction between the surfaces, the two eventually reach the same angular speed ω_f. (a) Calculate ω_f. (b) Calculate the ratio of the final to the initial rotational energy.

Figure P11.30

31. A playground merry-go-round of radius $R = 2.00$ m ▣ has a moment of inertia $I = 250$ kg $\cdot$ m^2 and is rotating ▣ at 10.0 rev/min about a frictionless, vertical axle. Facing the axle, a 25.0-kg child hops onto the merry-go-round and manages to sit down on the edge. What is the new angular speed of the merry-go-round?

32. Figure P11.17 represents a small, flat puck with mass $m = 2.40$ kg sliding on a frictionless, horizontal surface. It is held in a circular orbit about a fixed axis by a rod with negligible mass and length $R = 1.50$ m, pivoted at one end. Initially, the puck has a speed of $v = 5.00$ m/s. A 1.30-kg ball of putty is dropped vertically onto the puck from a small distance above it and immediately sticks to the puck. (a) What is the new period of rotation? (b) Is the angular momentum of the puck–putty system about the axis of rotation constant in this process? (c) Is the momentum of the system constant in the process of the putty sticking to the puck? (d) Is the mechanical energy of the system constant in the process?

33. A 60.0-kg woman stands at the western rim of a horizontal turntable having a moment of inertia of 500 kg · m² and a radius of 2.00 m. The turntable is initially at rest and is free to rotate about a frictionless, vertical axle through its center. The woman then starts walking around the rim clockwise (as viewed from above the system) at a constant speed of 1.50 m/s relative to the Earth. Consider the woman–turntable system as motion begins. (a) Is the mechanical energy of the system constant? (b) Is the momentum of the system constant? (c) Is the angular momentum of the system constant? (d) In what direction and with what angular speed does the turntable rotate? (e) How much chemical energy does the woman's body convert into mechanical energy of the woman–turntable system as the woman sets herself and the turntable into motion?

34. A student sits on a freely rotating stool holding two dumbbells, each of mass 3.00 kg (Fig. P11.34). When his arms are extended horizontally (Fig. P11.34a), the dumbbells are 1.00 m from the axis of rotation and the student rotates with an angular speed of 0.750 rad/s. The moment of inertia of the student plus stool is 3.00 kg · m² and is assumed to be constant. The student pulls the dumbbells inward horizontally to a position 0.300 m from the rotation axis (Fig. P11.34b). (a) Find the new angular speed of the student. (b) Find the kinetic energy of the rotating system before and after he pulls the dumbbells inward.

Figure P11.34

35. A uniform cylindrical turntable of radius 1.90 m and mass 30.0 kg rotates counterclockwise in a horizontal plane with an initial angular speed of 4π rad/s. The fixed turntable bearing is frictionless. A lump of clay of mass 2.25 kg and negligible size is dropped onto the turntable from a small distance above it and immediately sticks to the turntable at a point 1.80 m to the east of the axis. (a) Find the final angular speed of the clay and turntable. (b) Is the mechanical energy of the turntable–clay system constant in this process? Explain and use numerical results to verify your answer. (c) Is the momentum of the system constant in this process? Explain your answer.

36. A puck of mass m_1 = 80.0 g and radius r_1 = 4.00 cm glides across an air table at a speed of $\vec{v}$ = 1.50 m/s as shown in Figure P11.36a. It makes a glancing collision with a second puck of radius r_2 = 6.00 cm and mass m_2 = 120 g (initially at rest) such that their rims just touch. Because their rims are coated with instant-acting glue, the pucks stick together and rotate after the collision (Fig. P11.36b). (a) What is the angular momentum of the system relative to the center of mass? (b) What is the angular speed about the center of mass?

Figure P11.36

37. A wooden block of mass M resting on a frictionless, horizontal surface is attached to a rigid rod of length ℓ and of negligible mass (Fig. P11.37). The rod is pivoted at the other end. A bullet of mass m traveling parallel to the horizontal surface and perpendicular to the rod with speed v hits the block and becomes embedded in it. (a) What is the angular momentum of the bullet–block system about a vertical axis through the pivot? (b) What fraction of the original kinetic energy of the bullet is converted into internal energy in the system during the collision?

Figure P11.37

38. **Review.** A thin, uniform, rectangular signboard hangs vertically above the door of a shop. The sign is hinged to a stationary horizontal rod along its top edge. The mass of the sign is 2.40 kg, and its vertical dimension is 50.0 cm. The sign is swinging without friction, so it is a tempting target for children armed with snowballs. The maximum angular displacement of the sign is 25.0° on both sides of the vertical. At a moment when the sign is vertical and moving to the left, a snowball of mass 400 g, traveling horizontally with a velocity of 160 cm/s to the right, strikes perpendicularly at the lower edge of the sign and sticks there. (a) Calculate the angular speed of the sign immediately before the impact. (b) Calculate its angular speed immediately after the impact. (c) The spattered sign will swing up through what maximum angle?

39. A wad of sticky clay with mass m and velocity $\vec{v}_i$ is fired at a solid cylinder of mass M and radius R (Fig. P11.39). The cylinder is initially at rest and is mounted on a fixed horizontal axle that runs through its center of mass. The line of motion of the projectile is perpendicular to the axle and at a distance $d < R$ from the center. (a) Find the angular speed of the system just after the clay strikes and sticks to the surface of the cylin-

der. (b) Is the mechanical energy of the clay–cylinder system constant in this process? Explain your answer. (c) Is the momentum of the clay–cylinder system constant in this process? Explain your answer.

Figure P11.39

40. *Why is the following situation impossible?* A space station shaped like a giant wheel has a radius of $r = 100$ m and a moment of inertia of 5.00×10^8 kg · m². A crew of 150 people of average mass 65.0 kg is living on the rim, and the station's rotation causes the crew to experience an apparent free-fall acceleration of g (Fig. P11.29). A research technician is assigned to perform an experiment in which a ball is dropped at the rim of the station every 15 minutes and the time interval for the ball to drop a given distance is measured as a test to make sure the apparent value of g is correctly maintained. One evening, 100 average people move to the center of the station for a union meeting. The research technician, who has already been performing his experiment for an hour before the meeting, is disappointed that he cannot attend the meeting, and his mood sours even further by his boring experiment in which every time interval for the dropped ball is identical for the entire evening.

41. A 0.005 00-kg bullet traveling horizontally with speed 1.00×10^3 m/s strikes an 18.0-kg door, embedding itself 10.0 cm from the side opposite the hinges as shown in Figure P11.41. The 1.00-m wide door is free to swing on its frictionless hinges. (a) Before it hits the door, does the bullet have angular momentum relative to the door's axis of rotation? (b) If so, evaluate this angular momentum. If not, explain why there is no angular momentum. (c) Is the mechanical energy of the bullet–door system constant during this collision? Answer without doing a calculation. (d) At what angular speed does the door swing open immediately after the collision? (e) Calculate the total energy of the bullet–door system and determine whether it is less than or equal to the kinetic energy of the bullet before the collision.

Hinge

18.0 kg

0.005 00 kg

Figure P11.41 An overhead view of a bullet striking a door.

Section 11.5 The Motion of Gyroscopes and Tops

42. A spacecraft is in empty space. It carries on board a gyroscope with a moment of inertia of $I_g = 20.0$ kg · m² about the axis of the gyroscope. The moment of inertia

of the spacecraft around the same axis is $I_s = 5.00 \times 10^5$ kg · m². Neither the spacecraft nor the gyroscope is originally rotating. The gyroscope can be powered up in a negligible period of time to an angular speed of 100 rad/s. If the orientation of the spacecraft is to be changed by 30.0°, for what time interval should the gyroscope be operated?

43. The angular momentum vector of a precessing gyroscope sweeps out a cone as shown in Figure P11.43. The angular speed of the tip of the angular momentum vector, called its precessional frequency, is given by $\omega_p = \tau/L$, where τ is the magnitude of the torque on the gyroscope and L is the magnitude of its angular momentum. In the motion called *precession of the equinoxes*, the Earth's axis of rotation precesses about the perpendicular to its orbital plane with a period of 2.58×10^4 yr. Model the Earth as a uniform sphere and calculate the torque on the Earth that is causing this precession.

Figure P11.43 A precessing angular momentum vector sweeps out a cone in space.

Additional Problems

44. A light rope passes over a light, frictionless pulley. One end is fastened to a bunch of bananas of mass M, and a monkey of mass M clings to the other end (Fig. P11.44). The monkey climbs the rope in an attempt to reach the bananas. (a) Treating the system as consisting of the monkey, bananas, rope, and pulley, find the net torque on the system about the pulley axis. (b) Using the result of part (a), determine the total angular momentum about the pulley axis and describe the motion of the system. (c) Will the monkey reach the bananas?

Figure P11.44

45. Comet Halley moves about the Sun in an elliptical orbit, with its closest approach to the Sun being about 0.590 AU and its greatest distance 35.0 AU (1 AU = the Earth–Sun distance). The angular momentum of the comet about the Sun is constant, and the gravitational force exerted by the Sun has zero moment arm. The comet's speed at closest approach is 54.0 km/s. What is its speed when it is farthest from the Sun?

46. **Review.** Two boys are sliding toward each other on a frictionless, ice-covered parking lot. Jacob, mass 45.0 kg, is gliding to the right at 8.00 m/s, and Ethan, mass 31.0 kg, is gliding to the left at 11.0 m/s along the same

line. When they meet, they grab each other and hang on. (a) What is their velocity immediately thereafter? (b) What fraction of their original kinetic energy is still mechanical energy after their collision? That was so much fun that the boys repeat the collision with the same original velocities, this time moving along parallel lines 1.20 m apart. At closest approach, they lock arms and start rotating about their common center of mass. Model the boys as particles and their arms as a cord that does not stretch. (c) Find the velocity of their center of mass. (d) Find their angular speed. (e) What fraction of their original kinetic energy is still mechanical energy after they link arms? (f) Why are the answers to parts (b) and (e) so different?

47. We have all complained that there aren't enough hours in a day. In an attempt to fix that, suppose all the people in the world line up at the equator and all start running east at 2.50 m/s relative to the surface of the Earth. By how much does the length of a day increase? Assume the world population to be 7.00×10^9 people with an average mass of 55.0 kg each and the Earth to be a solid homogeneous sphere. In addition, depending on the details of your solution, you may need to use the approximation $1/(1 - x) \approx 1 + x$ for small x.

48. A skateboarder with his board can be modeled as a particle of mass 76.0 kg, located at his center of mass, 0.500 m above the ground. As shown in Figure P11.48, the skateboarder starts from rest in a crouching position at one lip of a half-pipe (point Ⓐ). The half-pipe forms one half of a cylinder of radius 6.80 m with its axis horizontal. On his descent, the skateboarder moves without friction and maintains his crouch so that his center of mass moves through one quarter of a circle. (a) Find his speed at the bottom of the half-pipe (point Ⓑ). (b) Find his angular momentum about the center of curvature at this point. (c) Immediately after passing point Ⓑ, he stands up and raises his arms, lifting his center of gravity to 0.950 m above the concrete (point Ⓒ). Explain why his angular momentum is constant in this maneuver, whereas the kinetic energy of his body is not constant. (d) Find his speed immediately after he stands up. (e) How much chemical energy in the skateboarder's legs was converted into mechanical energy in the skateboarder–Earth system when he stood up?

Figure P11.48

49. A rigid, massless rod has three particles with equal masses attached to it as shown in Figure P11.49. The rod is free to rotate in a vertical plane about a frictionless axle perpendicular to the rod through the point P and is released from rest in the horizontal position at $t = 0$.

Assuming m and d are known, find (a) the moment of inertia of the system of three particles about the pivot, (b) the torque acting on the system at $t = 0$, (c) the angular acceleration of the system at $t = 0$, (d) the linear acceleration of the particle labeled 3 at $t = 0$, (e) the maximum kinetic energy of the system, (f) the maximum angular speed reached by the rod, (g) the maximum angular momentum of the system, and (h) the maximum speed reached by the particle labeled 2.

Figure P11.49

50. Two children are playing on stools at a restaurant counter. Their feet do not reach the footrests, and the tops of the stools are free to rotate without friction on pedestals fixed to the floor. One of the children catches a tossed ball, in a process described by the equation

$$(0.730 \text{ kg} \cdot \text{m}^2)(2.40\hat{\mathbf{j}} \text{ rad/s})$$

$$+ (0.120 \text{ kg})(0.350\hat{\mathbf{i}} \text{ m}) \times (4.30\hat{\mathbf{k}} \text{ m/s})$$

$$= [0.730 \text{ kg} \cdot \text{m}^2 + (0.120 \text{ kg})(0.350 \text{ m})^2]\vec{\boldsymbol{\omega}}$$

(a) Solve the equation for the unknown $\vec{\boldsymbol{\omega}}$. (b) Complete the statement of the problem to which this equation applies. Your statement must include the given numerical information and specification of the unknown to be determined. (c) Could the equation equally well describe the other child throwing the ball? Explain your answer.

51. A projectile of mass m moves to the right with a speed v_i **GP** (Fig. P11.51a). The projectile strikes and sticks to the end of a stationary rod of mass M, length d, pivoted about a frictionless axle perpendicular to the page through O (Fig. P11.51b). We wish to find the fractional change of kinetic energy in the system due to the collision. (a) What is the appropriate analysis model to describe the projectile and the rod? (b) What is the angular momentum of the system before the collision about an axis through O? (c) What is the moment of inertia of the system about an axis through O after the projectile sticks to the rod? (d) If the angular speed of the system after the collision is ω, what is the angular momentum of the system after the collision? (e) Find the angular speed ω after the collision in terms of the given quanti-

Figure P11.51

ties. (f) What is the kinetic energy of the system before the collision? (g) What is the kinetic energy of the system after the collision? (h) Determine the fractional change of kinetic energy due to the collision.

52. A puck of mass $m = 50.0$ g is attached to a taut cord passing through a small hole in a frictionless, horizontal surface (Fig. P11.52). The puck is initially orbiting with speed $v_i = 1.50$ m/s in a circle of radius $r_i = 0.300$ m. The cord is then slowly pulled from below, decreasing the radius of the circle to $r = 0.100$ m. (a) What is the puck's speed at the smaller radius? (b) Find the tension in the cord at the smaller radius. (c) How much work is done by the hand in pulling the cord so that the radius of the puck's motion changes from 0.300 m to 0.100 m?

Figure P11.52 Problems 52 and 53.

53. A puck of mass m is attached to a taut cord passing through a small hole in a frictionless, horizontal surface (Fig. P11.52). The puck is initially orbiting with speed v_i in a circle of radius r_i. The cord is then slowly pulled from below, decreasing the radius of the circle to r. (a) What is the puck's speed when the radius is r? (b) Find the tension in the cord as a function of r. (c) How much work is done by the hand in pulling the cord so that the radius of the puck's motion changes from r_i to r?

54. *Why is the following situation impossible?* A meteoroid strikes the Earth directly on the equator. At the time it lands, it is traveling exactly vertical and downward. Due to the impact, the time for the Earth to rotate once increases by 0.5 s, so the day is 0.5 s longer, undetectable to laypersons. After the impact, people on the Earth ignore the extra half-second each day and life goes on as normal. (Assume the density of the Earth is uniform.)

55. Two astronauts (Fig. P11.55), each having a mass of 75.0 kg, are connected by a 10.0-m rope of negligible mass. They are isolated in space, orbiting their center

Figure P11.55 Problems 55 and 56.

of mass at speeds of 5.00 m/s. Treating the astronauts as particles, calculate (a) the magnitude of the angular momentum of the two-astronaut system and (b) the rotational energy of the system. By pulling on the rope, one astronaut shortens the distance between them to 5.00 m. (c) What is the new angular momentum of the system? (d) What are the astronauts' new speeds? (e) What is the new rotational energy of the system? (f) How much chemical potential energy in the body of the astronaut was converted to mechanical energy in the system when he shortened the rope?

56. Two astronauts (Fig. P11.55), each having a mass M, are connected by a rope of length d having negligible mass. They are isolated in space, orbiting their center of mass at speeds v. Treating the astronauts as particles, calculate (a) the magnitude of the angular momentum of the two-astronaut system and (b) the rotational energy of the system. By pulling on the rope, one of the astronauts shortens the distance between them to $d/2$. (c) What is the new angular momentum of the system? (d) What are the astronauts' new speeds? (e) What is the new rotational energy of the system? (f) How much chemical potential energy in the body of the astronaut was converted to mechanical energy in the system when he shortened the rope?

57. Native people throughout North and South America used a *bola* to hunt for birds and animals. A bola can consist of three stones, each with mass m, at the ends of three light cords, each with length ℓ. The other ends of the cords are tied together to form a Y. The hunter holds one stone and swings the other two above his head (Figure P11.57a). Both these stones move together in a horizontal circle of radius 2ℓ with speed v_0. At a moment when the horizontal component of their velocity is directed toward the quarry, the hunter releases the stone in his hand. As the bola flies through the air, the cords quickly take a stable arrangement with constant 120-degree angles between them (Fig. P11.57b). In the vertical direction, the bola is in free fall. Gravitational forces exerted by the Earth make the junction of the cords move with the downward acceleration $\vec{g}$. You may ignore the vertical motion as you proceed to describe the horizontal motion of the bola. In terms of m, ℓ, and v_0, calculate (a) the magnitude of the momentum of the bola at the moment of release and, after release, (b) the horizontal speed of the center of mass of the bola and (c) the angular momentum of the bola about its center of mass. (d) Find the angular speed of the bola about its center of mass after it has settled into its Y shape. Calculate

Figure P11.57

the kinetic energy of the bola (e) at the instant of release and (f) in its stable Y shape. (g) Explain how the conservation laws apply to the bola as its configuration changes. Robert Beichner suggested the idea for this problem.

58. A uniform rod of mass 300 g and length 50.0 cm rotates in a horizontal plane about a fixed, frictionless, vertical pin through its center. Two small, dense beads, each of mass m, are mounted on the rod so that they can slide without friction along its length. Initially, the beads are held by catches at positions 10.0 cm on each side of the center and the system is rotating at an angular speed of 36.0 rad/s. The catches are released simultaneously, and the beads slide outward along the rod. (a) Find an expression for the angular speed ω_f of the system at the instant the beads slide off the ends of the rod as it depends on m. (b) What are the maximum and the minimum possible values for ω_f and the values of m to which they correspond?

59. Global warming is a cause for concern because even small changes in the Earth's temperature can have significant consequences. For example, if the Earth's polar ice caps were to melt entirely, the resulting additional water in the oceans would flood many coastal areas. Model the polar ice as having mass 2.30×10^{19} kg and forming two flat disks of radius 6.00×10^5 m. Assume the water spreads into an unbroken thin, spherical shell after it melts. Calculate the resulting change in the duration of one day both in seconds and as a percentage.

60. The puck in Figure P11.60 has a mass of 0.120 kg. The distance of the puck from the center of rotation is originally 40.0 cm, and the puck is sliding with a speed of 80.0 cm/s. The string is pulled downward 15.0 cm through the hole in the frictionless table. Determine the work done on the puck. (*Suggestion:* Consider the change of kinetic energy.)

Figure P11.60

Challenge Problems

61. A uniform solid disk of radius R is set into rotation with an angular speed ω_i about an axis through its center. While still rotating at this speed, the disk is placed into contact with a horizontal surface and immediately released as shown in Figure P11.61. (a) What is the angular speed of the disk once pure rolling takes place? (b) Find the fractional change in kinetic energy from the moment the disk is set down until pure

rolling occurs. (c) Assume the coefficient of friction between disk and surface is μ. What is the time interval after setting the disk down before pure rolling motion begins? (d) How far does the disk travel before pure rolling begins?

Figure P11.61

62. In Example 11.9, we investigated an elastic collision between a disk and a stick lying on a frictionless surface. Suppose everything is the same as in the example except that the collision is perfectly inelastic so that the disk adheres to the stick at the endpoint at which it strikes. Find (a) the speed of the center of mass of the system and (b) the angular speed of the system after the collision.

63. A solid cube of side $2a$ and mass M is sliding on a frictionless surface with uniform velocity $\vec{v}$ as shown in Figure P11.63a. It hits a small obstacle at the end of the table that causes the cube to tilt as shown in Figure P11.63b. Find the minimum value of the magnitude of $\vec{v}$ such that the cube tips over and falls off the table. *Note:* The cube undergoes an inelastic collision at the edge.

Figure P11.63

64. A solid cube of wood of side $2a$ and mass M is resting on a horizontal surface. The cube is constrained to rotate about a fixed axis AB (Fig. P11.64). A bullet of mass m and speed v is shot at the face opposite $ABCD$ at a height of $4a/3$. The bullet becomes embedded in the cube. Find the minimum value of v required to tip the cube so that it falls on face $ABCD$. Assume $m \ll M$.

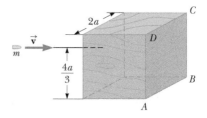

Figure P11.64

Static Equilibrium and Elasticity

In Chapters 10 and 11, we studied the dynamics of rigid objects. Part of this chapter addresses the conditions under which a rigid object is in equilibrium. The term *equilibrium* implies that the object moves with both constant velocity and constant angular velocity relative to an observer in an inertial reference frame. We deal here only with the special case in which both of these velocities are equal to zero. In this case, the object is in what is called *static equilibrium*. Static equilibrium represents a common situation in engineering practice, and the principles it involves are of special interest to civil engineers, architects, and mechanical engineers. If you are an engineering student, you will undoubtedly take an advanced course in statics in the near future.

The last section of this chapter deals with how objects deform under load conditions. An *elastic* object returns to its original shape when the deforming forces are removed. Several elastic constants are defined, each corresponding to a different type of deformation.

Balanced Rock in Arches National Park, Utah, is a 3 000 000-kg boulder that has been in stable equilibrium for several millennia. It had a smaller companion nearby, called "Chip Off the Old Block," that fell during the winter of 1975. Balanced Rock appeared in an early scene of the movie *Indiana Jones and the Last Crusade*. We will study the conditions under which an object is in equilibrium in this chapter. *(John W. Jewett, Jr.)*

12.1 Analysis Model: Rigid Object in Equilibrium

In Chapter 5, we discussed the particle in equilibrium model, in which a particle moves with constant velocity because the net force acting on it is zero. The situation with real (extended) objects is more complex because these objects often cannot be modeled as particles. For an extended object to be in equilibrium, a second condition must be satisfied. This second condition involves the rotational motion of the extended object.

363

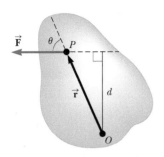

Figure 12.1 A single force $\vec{\mathbf{F}}$ acts on a rigid object at the point P.

Consider a single force $\vec{\mathbf{F}}$ acting on a rigid object as shown in Figure 12.1. Recall that the torque associated with the force $\vec{\mathbf{F}}$ about an axis through O is given by Equation 11.1:

$$\vec{\tau} = \vec{\mathbf{r}} \times \vec{\mathbf{F}}$$

The magnitude of $\vec{\tau}$ is Fd (see Equation 10.14), where d is the moment arm shown in Figure 12.1. According to Equation 10.18, the net torque on a rigid object causes it to undergo an angular acceleration.

In this discussion, we investigate those rotational situations in which the angular acceleration of a rigid object is zero. Such an object is in **rotational equilibrium.** Because $\Sigma \tau_{\text{ext}} = I\alpha$ for rotation about a fixed axis, the necessary condition for rotational equilibrium is that the net torque about any axis must be zero. We now have two necessary conditions for equilibrium of a rigid object:

1. The net external force on the object must equal zero:

$$\sum \vec{\mathbf{F}}_{\text{ext}} = 0 \qquad (12.1)$$

2. The net external torque on the object about *any* axis must be zero:

$$\sum \vec{\tau}_{\text{ext}} = 0 \qquad (12.2)$$

These conditions describe the **rigid object in equilibrium** analysis model. The first condition is a statement of translational equilibrium; it states that the translational acceleration of the object's center of mass must be zero when viewed from an inertial reference frame. The second condition is a statement of rotational equilibrium; it states that the angular acceleration about any axis must be zero. In the special case of **static equilibrium,** which is the main subject of this chapter, the object in equilibrium is at rest relative to the observer and so has no translational or angular speed (that is, $v_{\text{CM}} = 0$ and $\omega = 0$).

> **Pitfall Prevention 12.1**
> **Zero Torque** Zero net torque does not mean an absence of rotational motion. An object that is rotating at a constant angular speed can be under the influence of a net torque of zero. This possibility is analogous to the translational situation: zero net force does not mean an absence of translational motion.

Quick Quiz 12.1 Consider the object subject to the two forces of equal magnitude in Figure 12.2. Choose the correct statement with regard to this situation. **(a)** The object is in force equilibrium but not torque equilibrium. **(b)** The object is in torque equilibrium but not force equilibrium. **(c)** The object is in both force equilibrium and torque equilibrium. **(d)** The object is in neither force equilibrium nor torque equilibrium.

Quick Quiz 12.2 Consider the object subject to the three forces in Figure 12.3. Choose the correct statement with regard to this situation. **(a)** The object is in force equilibrium but not torque equilibrium. **(b)** The object is in torque equilibrium but not force equilibrium. **(c)** The object is in both force equilibrium and torque equilibrium. **(d)** The object is in neither force equilibrium nor torque equilibrium.

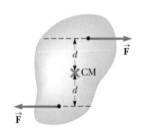

Figure 12.2 (Quick Quiz 12.1) Two forces of equal magnitude are applied at equal distances from the center of mass of a rigid object.

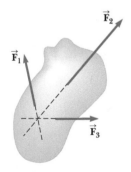

Figure 12.3 (Quick Quiz 12.2) Three forces act on an object. Notice that the lines of action of all three forces pass through a common point.

The two vector expressions given by Equations 12.1 and 12.2 are equivalent, in general, to six scalar equations: three from the first condition for equilibrium and three from the second (corresponding to x, y, and z components). Hence, in a complex system involving several forces acting in various directions, you could be faced with solving a set of equations with many unknowns. Here, we restrict our discussion to situations in which all the forces lie in the xy plane. (Forces whose vector representations are in the same plane are said to be *coplanar.*) With this restriction, we must deal with only three scalar equations. Two come from balancing the forces in the x and y directions. The third comes from the torque equation, namely that the net torque about a perpendicular axis through *any* point in the xy plane must be zero. This perpendicular axis will necessarily be parallel to the z axis, so the two conditions of the rigid object in equilibrium model provide the equations

$$\sum F_x = 0 \quad \sum F_y = 0 \quad \sum \tau_z = 0 \qquad \textbf{(12.3)}$$

where the location of the axis of the torque equation is arbitrary.

Analysis Model   **Rigid Object in Equilibrium**

Imagine an object that can rotate, but is exhibiting no translational acceleration a and no rotational acceleration α. Such an object is in both translational *and* rotational equilibrium, so the net force *and* the net torque about any axis are both equal to zero:

$$\sum \vec{F}_{ext} = 0 \qquad \textbf{(12.1)}$$

$$\sum \vec{\tau}_{ext} = 0 \qquad \textbf{(12.2)}$$

Examples:

- a balcony juts out from a building and must support the weight of several humans without collapsing
- a gymnast performs the difficult *iron cross* maneuver in an Olympic event
- a ship moves at constant speed through calm water and maintains a perfectly level orientation (Chapter 14)
- polarized molecules in a dielectric material in a constant electric field take on an average equilibrium orientation that remains fixed in time (Chapter 26)

12.2 More on the Center of Gravity

Whenever we deal with a rigid object, one of the forces we must consider is the gravitational force acting on it, and we must know the point of application of this force. As we learned in Section 9.5, associated with every object is a special point called its center of gravity. The combination of the various gravitational forces acting on all the various mass elements of the object is equivalent to a single gravitational force acting through this point. Therefore, to compute the torque due to the gravitational force on an object of mass M, we need only consider the force $M\vec{g}$ acting at the object's center of gravity.

How do we find this special point? As mentioned in Section 9.5, if we assume $\vec{g}$ is uniform over the object, the center of gravity of the object coincides with its center of mass. To see why, consider an object of arbitrary shape lying in the xy plane as illustrated in Figure 12.4. Suppose the object is divided into a large number of particles of masses $m_1, m_2, m_3, \ldots$ having coordinates $(x_1, y_1), (x_2, y_2), (x_3, y_3), \ldots$. In Equation 9.29, we defined the x coordinate of the center of mass of such an object to be

$$x_{CM} = \frac{m_1 x_1 + m_2 x_2 + m_3 x_3 + \cdots}{m_1 + m_2 + m_3 + \cdots} = \frac{\sum\limits_i m_i x_i}{\sum\limits_i m_i}$$

We use a similar equation to define the y coordinate of the center of mass, replacing each x with its y counterpart.

Let us now examine the situation from another point of view by considering the gravitational force exerted on each particle as shown in Figure 12.5. Each particle contributes a torque about an axis through the origin equal in magnitude to the particle's weight mg multiplied by its moment arm. For example, the magnitude of the torque due to the force $m_1\vec{g}_1$ is $m_1 g_1 x_1$, where g_1 is the value of the gravitational acceleration at the position of the particle of mass m_1. We wish to locate the center of gravity, the point at which application of the single gravitational force $M\vec{g}_{CG}$ (where $M = m_1 + m_2 + m_3 + \cdots$ is the total mass of the object and $\vec{g}_{CG}$ is the acceleration due to gravity at the location of the center of gravity) has the same effect on

Figure 12.4 An object can be divided into many small particles. These particles can be used to locate the center of mass.

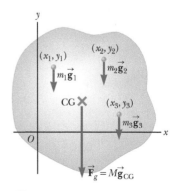

Figure 12.5 By dividing an object into many particles, we can find its center of gravity.

rotation as does the combined effect of all the individual gravitational forces $m_i \vec{\mathbf{g}}_i$. Equating the torque resulting from $M\vec{\mathbf{g}}_{CG}$ acting at the center of gravity to the sum of the torques acting on the individual particles gives

$$(m_1 + m_2 + m_3 + \cdots)g_{CG}\, x_{CG} = m_1 g_1 x_1 + m_2 g_2 x_2 + m_3 g_3 x_3 + \cdots$$

This expression accounts for the possibility that the value of g can in general vary over the object. If we assume uniform g over the object (as is usually the case), the g factors cancel and we obtain

$$x_{CG} = \frac{m_1 x_1 + m_2 x_2 + m_3 x_3 + \cdots}{m_1 + m_2 + m_3 + \cdots} \qquad \text{(12.4)}$$

Comparing this result with Equation 9.29 shows that the center of gravity is located at the center of mass as long as $\vec{\mathbf{g}}$ is uniform over the entire object. Several examples in the next section deal with homogeneous, symmetric objects. The center of gravity for any such object coincides with its geometric center.

Quick Quiz 12.3 A meterstick of uniform density is hung from a string tied at the 25-cm mark. A 0.50-kg object is hung from the zero end of the meterstick, and the meterstick is balanced horizontally. What is the mass of the meterstick? **(a)** 0.25 kg **(b)** 0.50 kg **(c)** 0.75 kg **(d)** 1.0 kg **(e)** 2.0 kg **(f)** impossible to determine

12.3 Examples of Rigid Objects in Static Equilibrium

The photograph of the one-bottle wine holder in Figure 12.6 shows one example of a balanced mechanical system that seems to defy gravity. For the system (wine holder plus bottle) to be in equilibrium, the net external force must be zero (see Eq. 12.1) and the net external torque must be zero (see Eq. 12.2). The second condition can be satisfied only when the center of gravity of the system is directly over the support point.

The center of gravity of the system (bottle plus holder) is directly over the support point.

© Cengage Learning/Charles D. Winters

Figure 12.6 This one-bottle wine holder is a surprising display of static equilibrium.

Problem-Solving Strategy   Rigid Object in Equilibrium

When analyzing a rigid object in equilibrium under the action of several external forces, use the following procedure.

1. Conceptualize. Think about the object that is in equilibrium and identify all the forces on it. Imagine what effect each force would have on the rotation of the object if it were the only force acting.

2. Categorize. Confirm that the object under consideration is indeed a rigid object in equilibrium. The object must have zero translational acceleration and zero angular acceleration.

3. Analyze. Draw a diagram and label all external forces acting on the object. Try to guess the correct direction for any forces that are not specified. When using the particle under a net force model, the object on which forces act can be represented in a free-body diagram with a dot because it does not matter where on the object the forces are applied. When using the rigid object in equilibrium model, however, we cannot use a dot to represent the object because the location where forces act is important in the calculation. Therefore, in a diagram showing the forces on an object, we must show the actual object or a simplified version of it.

Resolve all forces into rectangular components, choosing a convenient coordinate system. Then apply the first condition for equilibrium, Equation 12.1. Remember to keep track of the signs of the various force components.

▶ **Problem-Solving Strategy** c o n t i n u e d

Choose a convenient axis for calculating the net torque on the rigid object. Remember that the choice of the axis for the torque equation is arbitrary; therefore, choose an axis that simplifies your calculation as much as possible. Usually, the most convenient axis for calculating torques is one through a point through which the lines of action of several forces pass, so their torques around this axis are zero. If you don't know a force or don't need to know a force, it is often beneficial to choose an axis through the point at which this force acts. Apply the second condition for equilibrium, Equation 12.2.

Solve the simultaneous equations for the unknowns in terms of the known quantities.

4. Finalize. Make sure your results are consistent with your diagram. If you selected a direction that leads to a negative sign in your solution for a force, do not be alarmed; it merely means that the direction of the force is the opposite of what you guessed. Add up the vertical and horizontal forces on the object and confirm that each set of components adds to zero. Add up the torques on the object and confirm that the sum equals zero.

Example 12.1   **The Seesaw Revisited**   AM

A seesaw consisting of a uniform board of mass M and length ℓ supports at rest a father and daughter with masses m_f and m_d, respectively, as shown in Figure 12.7. The support (called the *fulcrum*) is under the center of gravity of the board, the father is a distance d from the center, and the daughter is a distance $\ell/2$ from the center.

(A) Determine the magnitude of the upward force $\vec{\mathbf{n}}$ exerted by the support on the board.

Figure 12.7 (Example 12.1) A balanced system.

SOLUTION

Conceptualize Let us focus our attention on the board and consider the gravitational forces on the father and daughter as forces applied directly to the board. The daughter would cause a clockwise rotation of the board around the support, whereas the father would cause a counterclockwise rotation.

Categorize Because the text of the problem states that the system is at rest, we model the board as a *rigid object in equilibrium*. Because we will only need the first condition of equilibrium to solve this part of the problem, however, we could also simply model the board as a *particle in equilibrium*.

⋯⋯

Analyze Define upward as the positive y direction and substitute the forces on the board into Equation 12.1:

$$n - m_f g - m_d g - M g = 0$$

Solve for the magnitude of the force $\vec{\mathbf{n}}$:

$$(1) \quad n = m_f g + m_d g + M g = \boxed{(m_f + m_d + M)g}$$

(B) Determine where the father should sit to balance the system at rest.

SOLUTION

Categorize This part of the problem requires the introduction of torque to find the position of the father, so we model the board as a *rigid object in equilibrium*.

⋯⋯

Analyze The board's center of gravity is at its geometric center because we are told that the board is uniform. If we choose a rotation axis perpendicular to the page through the center of gravity of the board, the torques produced by $\vec{\mathbf{n}}$ and the gravitational force on the board about this axis are zero.

continued

▶ **12.1** continued

Substitute expressions for the torques on the board due to the father and daughter into Equation 12.2:

$$(m_f g)(d) - (m_d g)\frac{\ell}{2} = 0$$

Solve for d:

$$d = \left(\frac{m_d}{m_f}\right)\frac{\ell}{2}$$

Finalize This result is the same one we obtained in Example 11.6 by evaluating the angular acceleration of the system and setting the angular acceleration equal to zero.

WHAT IF? Suppose we had chosen another point through which the rotation axis were to pass. For example, suppose the axis is perpendicular to the page and passes through the location of the father. Does that change the results to parts (A) and (B)?

Answer Part (A) is unaffected because the calculation of the net force does not involve a rotation axis. In part (B), we would conceptually expect there to be no change if a different rotation axis is chosen because the second condition of equilibrium claims that the torque is zero about *any* rotation axis.

Let's verify this answer mathematically. Recall that the sign of the torque associated with a force is positive if that force tends to rotate the system counterclockwise, whereas the sign of the torque is negative if the force tends to rotate the system clockwise. Let's choose a rotation axis perpendicular to the page and passing through the location of the father.

Substitute expressions for the torques on the board around this axis into Equation 12.2:

$$n(d) - (Mg)(d) - (m_d g)\left(d + \frac{\ell}{2}\right) = 0$$

Substitute from Equation (1) in part (A) and solve for d:

$$(m_f + m_d + M)g(d) - (Mg)(d) - (m_d g)\left(d + \frac{\ell}{2}\right) = 0$$

$$(m_f g)(d) - (m_d g)\left(\frac{\ell}{2}\right) = 0 \quad \rightarrow \quad d = \left(\frac{m_d}{m_f}\right)\frac{\ell}{2}$$

This result is in agreement with the one obtained in part (B).

Example 12.2   Standing on a Horizontal Beam   AM

A uniform horizontal beam with a length of $\ell = 8.00$ m and a weight of $W_b = 200$ N is attached to a wall by a pin connection. Its far end is supported by a cable that makes an angle of $\phi = 53.0°$ with the beam (Fig. 12.8a). A person of weight $W_p = 600$ N stands a distance $d = 2.00$ m from the wall. Find the tension in the cable as well as the magnitude and direction of the force exerted by the wall on the beam.

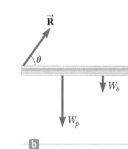

Figure 12.8 (Example 12.2)
(a) A uniform beam supported by a cable. A person walks outward on the beam.
(b) The force diagram for the beam. (c) The force diagram for the beam showing the components of $\vec{R}$ and $\vec{T}$.

SOLUTION

Conceptualize Imagine the person in Figure 12.8a moving outward on the beam. It seems reasonable that the farther he moves outward, the larger the torque he applies about the pivot and the larger the tension in the cable must be to balance this torque.

Categorize Because the system is at rest, we categorize the beam as a *rigid object in equilibrium*.

Analyze We identify all the external forces acting on the beam: the 200-N gravitational force, the force $\vec{T}$ exerted by the cable, the force $\vec{R}$ exerted by the wall at the pivot, and the 600-N force that the person exerts on the beam. These forces are all indicated in the force diagram for the beam shown in Figure 12.8b. When we assign directions for forces, it is sometimes helpful to imagine what would happen if a force were suddenly removed. For example, if the wall were to vanish suddenly, the left end of the beam would move to the left as it begins to fall. This scenario tells us that the wall is not only holding the beam up but is also pressing outward against it. Therefore, we draw the vector $\vec{R}$ in the direction shown in Figure 12.8b. Figure 12.8c shows the horizontal and vertical components of $\vec{T}$ and $\vec{R}$.

Applying the first condition of equilibrium, substitute expressions for the forces on the beam into component equations from Equation 12.1:

$$(1) \quad \sum F_x = R\cos\theta - T\cos\phi = 0$$

$$(2) \quad \sum F_y = R\sin\theta + T\sin\phi - W_p - W_b = 0$$

where we have chosen rightward and upward as our positive directions. Because R, T, and θ are all unknown, we cannot obtain a solution from these expressions alone. (To solve for the unknowns, the number of simultaneous equations must generally equal the number of unknowns.)

Now let's invoke the condition for rotational equilibrium. A convenient axis to choose for our torque equation is the one that passes through the pin connection. The feature that makes this axis so convenient is that the force $\vec{R}$ and the horizontal component of $\vec{T}$ both have a moment arm of zero; hence, these forces produce no torque about this axis.

Substitute expressions for the torques on the beam into Equation 12.2:

$$\sum \tau_z = (T\sin\phi)(\ell) - W_p d - W_b\left(\frac{\ell}{2}\right) = 0$$

This equation contains only T as an unknown because of our choice of rotation axis. Solve for T and substitute numerical values:

$$T = \frac{W_p d + W_b(\ell/2)}{\ell\sin\phi} = \frac{(600\text{ N})(2.00\text{ m}) + (200\text{ N})(4.00\text{ m})}{(8.00\text{ m})\sin 53.0°} = \boxed{313\text{ N}}$$

Rearrange Equations (1) and (2) and then divide:

$$\frac{R\sin\theta}{R\cos\theta} = \tan\theta = \frac{W_p + W_b - T\sin\phi}{T\cos\phi}$$

Solve for θ and substitute numerical values:

$$\theta = \tan^{-1}\left(\frac{W_p + W_b - T\sin\phi}{T\cos\phi}\right)$$

$$= \tan^{-1}\left[\frac{600\text{ N} + 200\text{ N} - (313\text{ N})\sin 53.0°}{(313\text{ N})\cos 53.0°}\right] = \boxed{71.1°}$$

Solve Equation (1) for R and substitute numerical values:

$$R = \frac{T\cos\phi}{\cos\theta} = \frac{(313\text{ N})\cos 53.0°}{\cos 71.1°} = \boxed{581\text{ N}}$$

Finalize The positive value for the angle θ indicates that our estimate of the direction of $\vec{R}$ was accurate.

Had we selected some other axis for the torque equation, the solution might differ in the details but the answers would be the same. For example, had we chosen an axis through the center of gravity of the beam, the torque equation would involve both T and R. This equation, coupled with Equations (1) and (2), however, could still be solved for the unknowns. Try it!

WHAT IF? What if the person walks farther out on the beam? Does T change? Does R change? Does θ change?

Answer T must increase because the gravitational force on the person exerts a larger torque about the pin connection, which must be countered by a larger torque in the opposite direction due to an increased value of T. If T increases, the vertical component of $\vec{R}$ decreases to maintain force equilibrium in the vertical direction. Force equilibrium in the horizontal direction, however, requires an increased horizontal component of $\vec{R}$ to balance the horizontal component of the increased $\vec{T}$. This fact suggests that θ becomes smaller, but it is hard to predict what happens to R. Problem 66 asks you to explore the behavior of R.

Example 12.3   The Leaning Ladder   AM

A uniform ladder of length ℓ rests against a smooth, vertical wall (Fig. 12.9a). The mass of the ladder is m, and the coefficient of static friction between the ladder and the ground is $\mu_s = 0.40$. Find the minimum angle θ_{min} at which the ladder does not slip.

SOLUTION

Conceptualize Think about any ladders you have climbed. Do you want a large friction force between the bottom of the ladder and the surface or a small one? If the friction force is zero, will the ladder stay up? Simulate a ladder with a ruler leaning against a vertical surface. Does the ruler slip at some angles and stay up at others?

Categorize We do not wish the ladder to slip, so we model it as a *rigid object in equilibrium*.

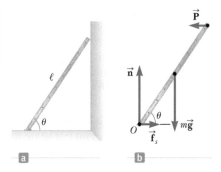

Figure 12.9 (Example 12.3) (a) A uniform ladder at rest, leaning against a smooth wall. The ground is rough. (b) The forces on the ladder.

Analyze A diagram showing all the external forces acting on the ladder is illustrated in Figure 12.9b. The force exerted by the ground on the ladder is the vector sum of a normal force $\vec{n}$ and the force of static friction $\vec{f}_s$. The wall exerts a normal force $\vec{P}$ on the top of the ladder, but there is no friction force here because the wall is smooth. So the net force on the top of the ladder is perpendicular to the wall and of magnitude P.

Apply the first condition for equilibrium to the ladder in both the x and the y directions:

$$(1) \quad \sum F_x = f_s - P = 0$$
$$(2) \quad \sum F_y = n - mg = 0$$

Solve Equation (1) for P:

$$(3) \quad P = f_s$$

Solve Equation (2) for n:

$$(4) \quad n = mg$$

When the ladder is on the verge of slipping, the force of static friction must have its maximum value, which is given by $f_{s,max} = \mu_s n$. Combine this equation with Equations (3) and (4):

$$(5) \quad P_{max} = f_{s,max} = \mu_s n = \mu_s mg$$

Apply the second condition for equilibrium to the ladder, evaluating torques about an axis perpendicular to the page through O:

$$\sum \tau_O = P\ell \sin\theta - mg\frac{\ell}{2}\cos\theta = 0$$

Solve for $\tan\theta$:

$$\frac{\sin\theta}{\cos\theta} = \tan\theta = \frac{mg}{2P} \;\rightarrow\; \theta = \tan^{-1}\left(\frac{mg}{2P}\right)$$

Under the conditions that the ladder is just ready to slip, θ becomes θ_{min} and P_{max} is given by Equation (5). Substitute:

$$\theta_{min} = \tan^{-1}\left(\frac{mg}{2P_{max}}\right) = \tan^{-1}\left(\frac{1}{2\mu_s}\right) = \tan^{-1}\left[\frac{1}{2(0.40)}\right] = \boxed{51°}$$

Finalize Notice that the angle depends only on the coefficient of friction, not on the mass or length of the ladder.

Example 12.4   Negotiating a Curb   AM

(A) Estimate the magnitude of the force $\vec{F}$ a person must apply to a wheelchair's main wheel to roll up over a sidewalk curb (Fig. 12.10a). This main wheel that comes in contact with the curb has a radius r, and the height of the curb is h.

▶ **12.4** continued

SOLUTION

Conceptualize Think about wheelchair access to buildings. Generally, there are ramps built for individuals in wheelchairs. Steplike structures such as curbs are serious barriers to a wheelchair.

Categorize Imagine the person exerts enough force so that the bottom of the main wheel just loses contact with the lower surface and hovers at rest. We model the wheel in this situation as a *rigid object in equilibrium.*
....

Analyze Usually, the person's hands supply the required force to a slightly smaller wheel that is concentric with the main wheel. For simplicity, let's assume the radius of this second wheel is the same as the radius of the main wheel. Let's estimate a combined gravitational force of magnitude $mg = 1\,400$ N for the person and the wheelchair, acting along a line of action passing through the axle of the main wheel, and choose a wheel radius of $r = 30$ cm. We also pick a curb height of $h = 10$ cm. Let's also assume the wheelchair and occupant are symmetric and each wheel supports a weight of 700 N. We then proceed to analyze only one of the main wheels. Figure 12.10b shows the geometry for a single wheel.

When the wheel is just about to be raised from the street, the normal force exerted by the ground on the wheel at point B goes to zero. Hence, at this time only three forces act on the wheel as shown in the force diagram in Figure 12.10c. The force $\vec{\mathbf{R}}$, which is the force exerted by the curb on the wheel, acts at point A, so if we choose to have our axis of rotation be perpendicular to the page and pass through point A, we do not need to include $\vec{\mathbf{R}}$ in our torque equation. The moment arm of $\vec{\mathbf{F}}$ relative to an axis through A is given by $2r - h$ (see Fig. 12.10c).

Figure 12.10 (Example 12.4) (a) A person in a wheelchair attempts to roll up over a curb. (b) Details of the wheel and curb. The person applies a force $\vec{\mathbf{F}}$ to the top of the wheel. (c) A force diagram for the wheel when it is just about to be raised. Three forces act on the wheel at this instant: $\vec{\mathbf{F}}$, which is exerted by the hand; $\vec{\mathbf{R}}$, which is exerted by the curb; and the gravitational force $m\vec{\mathbf{g}}$. (d) The vector sum of the three external forces acting on the wheel is zero.

Use the triangle OAC in Figure 12.10b to find the moment arm d of the gravitational force $m\vec{\mathbf{g}}$ acting on the wheel relative to an axis through point A:

$$(1) \quad d = \sqrt{r^2 - (r - h)^2} = \sqrt{2rh - h^2}$$

Apply the second condition for equilibrium to the wheel, taking torques about an axis through A:

$$(2) \quad \sum \tau_A = mgd - F(2r - h) = 0$$

Substitute for d from Equation (1):

$$mg\sqrt{2rh - h^2} - F(2r - h) = 0$$

Solve for F:

$$(3) \quad F = \frac{mg\sqrt{2rh - h^2}}{2r - h}$$

Simplify:

$$F = mg\frac{\sqrt{h}\sqrt{2r - h}}{2r - h} = mg\sqrt{\frac{h}{2r - h}}$$

Substitute the known values:

$$F = (700\ \text{N})\sqrt{\frac{0.1\ \text{m}}{2(0.3\ \text{m}) - 0.1\ \text{m}}}$$

$$= 3 \times 10^2\ \text{N}$$

continued

▶ **12.4** continued

(B) Determine the magnitude and direction of $\vec{\mathbf{R}}$.

SOLUTION

Apply the first condition for equilibrium to the x and y components of the forces on the wheel:

(4) $\sum F_x = F - R\cos\theta = 0$

(5) $\sum F_y = R\sin\theta - mg = 0$

Divide Equation (5) by Equation (4):

$$\frac{R\sin\theta}{R\cos\theta} = \tan\theta = \frac{mg}{F}$$

Solve for the angle θ:

$$\theta = \tan^{-1}\left(\frac{mg}{F}\right) = \tan^{-1}\left(\frac{700\text{ N}}{300\text{ N}}\right) = \boxed{70°}$$

Solve Equation (5) for R and substitute numerical values:

$$R = \frac{mg}{\sin\theta} = \frac{700\text{ N}}{\sin 70°} = \boxed{8 \times 10^2\text{ N}}$$

Finalize Notice that we have kept only one digit as significant. (We have written the angle as 70° because $7 \times 10^{1°}$ is awkward!) The results indicate that the force that must be applied to each wheel is substantial. You may want to estimate the force required to roll a wheelchair up a typical sidewalk accessibility ramp for comparison.

WHAT IF? Would it be easier to negotiate the curb if the person grabbed the wheel at point D in Figure 12.10c and pulled *upward*?

Answer If the force $\vec{\mathbf{F}}$ in Figure 12.10c is rotated counterclockwise by 90° and applied at D, its moment arm about an axis through A is $d + r$. Let's call the magnitude of this new force F'.

Modify Equation (2) for this situation:

$$\sum \tau_A = mgd - F'(d + r) = 0$$

Solve this equation for F' and substitute for d:

$$F' = \frac{mgd}{d + r} = \frac{mg\sqrt{2rh - h^2}}{\sqrt{2rh - h^2} + r}$$

Take the ratio of this force to the original force from Equation (3) and express the result in terms of h/r, the ratio of the curb height to the wheel radius:

$$\frac{F'}{F} = \frac{\dfrac{mg\sqrt{2rh - h^2}}{\sqrt{2rh - h^2} + r}}{\dfrac{mg\sqrt{2rh - h^2}}{2r - h}} = \frac{2r - h}{\sqrt{2rh - h^2} + r} = \frac{2 - \left(\dfrac{h}{r}\right)}{\sqrt{2\left(\dfrac{h}{r}\right) - \left(\dfrac{h}{r}\right)^2} + 1}$$

Substitute the ratio $h/r = 0.33$ from the given values:

$$\frac{F'}{F} = \frac{2 - 0.33}{\sqrt{2(0.33) - (0.33)^2} + 1} = 0.96$$

This result tells us that, *for these values*, it is slightly easier to pull upward at D than horizontally at the top of the wheel. For very high curbs, so that h/r is close to 1, the ratio F'/F drops to about 0.5 because point A is located near the right edge of the wheel in Figure 12.10b. The force at D is applied at a distance of about $2r$ from A, whereas the force at the top of the wheel has a moment arm of only about r. For high curbs, then, it is best to pull upward at D, although a large value of the force is required. For small curbs, it is best to apply the force at the top of the wheel. The ratio F'/F becomes larger than 1 at about $h/r = 0.3$ because point A is now close to the bottom of the wheel and the force applied at the top of the wheel has a larger moment arm than when applied at D.

Finally, let's comment on the validity of these mathematical results. Consider Figure 12.10d and imagine that the vector $\vec{\mathbf{F}}$ is upward instead of to the right. There is no way the three vectors can add to equal zero as required by the first equilibrium condition. Therefore, our results above may be qualitatively valid, but not exact quantitatively. To cancel the horizontal component of $\vec{\mathbf{R}}$, the force at D must be applied at an angle to the vertical rather than straight upward. This feature makes the calculation more complicated and requires both conditions of equilibrium.

12.4 Elastic Properties of Solids

Except for our discussion about springs in earlier chapters, we have assumed objects remain rigid when external forces act on them. In Section 9.8, we explored deformable systems. In reality, all objects are deformable to some extent. That is, it is possible to change the shape or the size (or both) of an object by applying external forces. As these changes take place, however, internal forces in the object resist the deformation.

We shall discuss the deformation of solids in terms of the concepts of *stress* and *strain*. **Stress** is a quantity that is proportional to the force causing a deformation; more specifically, stress is the external force acting on an object per unit cross-sectional area. The result of a stress is **strain,** which is a measure of the degree of deformation. It is found that, for sufficiently small stresses, stress is proportional to strain; the constant of proportionality depends on the material being deformed and on the nature of the deformation. We call this proportionality constant the **elastic modulus.** The elastic modulus is therefore defined as the ratio of the stress to the resulting strain:

$$\text{Elastic modulus} \equiv \frac{\text{stress}}{\text{strain}} \qquad \textbf{(12.5)}$$

The elastic modulus in general relates what is done to a solid object (a force is applied) to how that object responds (it deforms to some extent). It is similar to the spring constant k in Hooke's law (Eq. 7.9) that relates a force applied to a spring and the resultant deformation of the spring, measured by its extension or compression.

We consider three types of deformation and define an elastic modulus for each:

1. **Young's modulus** measures the resistance of a solid to a change in its length.
2. **Shear modulus** measures the resistance to motion of the planes within a solid parallel to each other.
3. **Bulk modulus** measures the resistance of solids or liquids to changes in their volume.

Young's Modulus: Elasticity in Length

Consider a long bar of cross-sectional area A and initial length L_i that is clamped at one end as in Figure 12.11. When an external force is applied perpendicular to the cross section, internal molecular forces in the bar resist distortion ("stretching"), but the bar reaches an equilibrium situation in which its final length L_f is greater than L_i and in which the external force is exactly balanced by the internal forces. In such a situation, the bar is said to be stressed. We define the **tensile stress** as the ratio of the magnitude of the external force F to the cross-sectional area A, where the cross section is perpendicular to the force vector. The **tensile strain** in this case is defined as the ratio of the change in length ΔL to the original length L_i. We define **Young's modulus** by a combination of these two ratios:

$$Y \equiv \frac{\text{tensile stress}}{\text{tensile strain}} = \frac{F/A}{\Delta L / L_i} \qquad \textbf{(12.6)} \qquad \blacktriangleleft \text{ Young's modulus}$$

Young's modulus is typically used to characterize a rod or wire stressed under either tension or compression. Because strain is a dimensionless quantity, Y has units of force per unit area. Typical values are given in Table 12.1 on page 374.

For relatively small stresses, the bar returns to its initial length when the force is removed. The **elastic limit** of a substance is defined as the maximum stress that can be applied to the substance before it becomes permanently deformed and does not return to its initial length. It is possible to exceed the elastic limit of a substance by

The amount by which the length of the bar changes due to the applied force is ΔL.

Figure 12.11 A force $\vec{\mathbf{F}}$ is applied to the free end of a bar clamped at the other end.

Table 12.1	Typical Values for Elastic Moduli		
Substance	Young's Modulus (N/m²)	Shear Modulus (N/m²)	Bulk Modulus (N/m²)
Tungsten	35×10^{10}	14×10^{10}	20×10^{10}
Steel	20×10^{10}	8.4×10^{10}	6×10^{10}
Copper	11×10^{10}	4.2×10^{10}	14×10^{10}
Brass	9.1×10^{10}	3.5×10^{10}	6.1×10^{10}
Aluminum	7.0×10^{10}	2.5×10^{10}	7.0×10^{10}
Glass	$6.5–7.8 \times 10^{10}$	$2.6–3.2 \times 10^{10}$	$5.0–5.5 \times 10^{10}$
Quartz	5.6×10^{10}	2.6×10^{10}	2.7×10^{10}
Water	—	—	0.21×10^{10}
Mercury	—	—	2.8×10^{10}

Figure 12.12 Stress-versus-strain curve for an elastic solid.

applying a sufficiently large stress as seen in Figure 12.12. Initially, a stress-versus-strain curve is a straight line. As the stress increases, however, the curve is no longer a straight line. When the stress exceeds the elastic limit, the object is permanently distorted and does not return to its original shape after the stress is removed. As the stress is increased even further, the material ultimately breaks.

Shear Modulus: Elasticity of Shape

Another type of deformation occurs when an object is subjected to a force parallel to one of its faces while the opposite face is held fixed by another force (Fig. 12.13a). The stress in this case is called a *shear stress*. If the object is originally a rectangular block, a shear stress results in a shape whose cross section is a parallelogram. A book pushed sideways as shown in Figure 12.13b is an example of an object subjected to a shear stress. To a first approximation (for small distortions), no change in volume occurs with this deformation.

We define the **shear stress** as F/A, the ratio of the tangential force to the area A of the face being sheared. The **shear strain** is defined as the ratio $\Delta x/h$, where Δx is the horizontal distance that the sheared face moves and h is the height of the object. In terms of these quantities, the **shear modulus** is

◄ Shear modulus

$$S \equiv \frac{\text{shear stress}}{\text{shear strain}} = \frac{F/A}{\Delta x/h} \qquad (12.7)$$

Values of the shear modulus for some representative materials are given in Table 12.1. Like Young's modulus, the unit of shear modulus is the ratio of that for force to that for area.

Bulk Modulus: Volume Elasticity

Bulk modulus characterizes the response of an object to changes in a force of uniform magnitude applied perpendicularly over the entire surface of the object as shown in Figure 12.14. (We assume here the object is made of a single substance.)

Figure 12.13 (a) A shear deformation in which a rectangular block is distorted by two forces of equal magnitude but opposite directions applied to two parallel faces. (b) A book is under shear stress when a hand placed on the cover applies a horizontal force away from the spine.

As we shall see in Chapter 14, such a uniform distribution of forces occurs when an object is immersed in a fluid. An object subject to this type of deformation undergoes a change in volume but no change in shape. The **volume stress** is defined as the ratio of the magnitude of the total force F exerted on a surface to the area A of the surface. The quantity $P = F/A$ is called **pressure,** which we shall study in more detail in Chapter 14. If the pressure on an object changes by an amount $\Delta P = \Delta F/A$, the object experiences a volume change ΔV. The **volume strain** is equal to the change in volume ΔV divided by the initial volume V_i. Therefore, from Equation 12.5, we can characterize a volume ("bulk") compression in terms of the **bulk modulus,** which is defined as

$$B \equiv \frac{\text{volume stress}}{\text{volume strain}} = -\frac{\Delta F/A}{\Delta V/V_i} = -\frac{\Delta P}{\Delta V/V_i} \qquad \text{(12.8)}$$

◀ **Bulk modulus**

A negative sign is inserted in this defining equation so that B is a positive number. This maneuver is necessary because an increase in pressure (positive ΔP) causes a decrease in volume (negative ΔV) and vice versa.

Table 12.1 lists bulk moduli for some materials. If you look up such values in a different source, you may find the reciprocal of the bulk modulus listed. The reciprocal of the bulk modulus is called the **compressibility** of the material.

Notice from Table 12.1 that both solids and liquids have a bulk modulus. No shear modulus and no Young's modulus are given for liquids, however, because a liquid does not sustain a shearing stress or a tensile stress. If a shearing force or a tensile force is applied to a liquid, the liquid simply flows in response.

Ⓠuick Quiz 12.4 For the three parts of this Quick Quiz, choose from the following choices the correct answer for the elastic modulus that describes the relationship between stress and strain for the system of interest, which is in italics: (a) Young's modulus (b) shear modulus (c) bulk modulus (d) none of those choices **(i)** A *block of iron* is sliding across a horizontal floor. The friction force between the sliding block and the floor causes the block to deform. **(ii)** A trapeze artist swings through a circular arc. At the bottom of the swing, the *wires* supporting the trapeze are longer than when the trapeze artist simply hangs from the trapeze due to the increased tension in them. **(iii)** A spacecraft carries a *steel sphere* to a planet on which atmospheric pressure is much higher than on the Earth. The higher pressure causes the radius of the sphere to decrease.

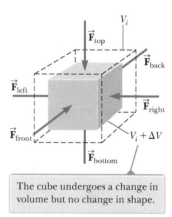

The cube undergoes a change in volume but no change in shape.

Figure 12.14 A cube is under uniform pressure and is therefore compressed on all sides by forces normal to its six faces. The arrowheads of force vectors on the sides of the cube that are not visible are hidden by the cube.

Prestressed Concrete

If the stress on a solid object exceeds a certain value, the object fractures. The maximum stress that can be applied before fracture occurs—called the *tensile strength, compressive strength,* or *shear strength*—depends on the nature of the material and on the type of applied stress. For example, concrete has a tensile strength of about 2×10^6 N/m², a compressive strength of 20×10^6 N/m², and a shear strength of 2×10^6 N/m². If the applied stress exceeds these values, the concrete fractures. It is common practice to use large safety factors to prevent failure in concrete structures.

Concrete is normally very brittle when it is cast in thin sections. Therefore, concrete slabs tend to sag and crack at unsupported areas as shown in Figure 12.15a. The slab can be strengthened by the use of steel rods to reinforce the concrete as illustrated in Figure 12.15b. Because concrete is much stronger under compression (squeezing) than under tension (stretching) or shear, vertical columns of concrete can support

Figure 12.15 (a) A concrete slab with no reinforcement tends to crack under a heavy load. (b) The strength of the concrete is increased by using steel reinforcement rods. (c) The concrete is further strengthened by prestressing it with steel rods under tension.

very heavy loads, whereas horizontal beams of concrete tend to sag and crack. A significant increase in shear strength is achieved, however, if the reinforced concrete is prestressed as shown in Figure 12.15c. As the concrete is being poured, the steel rods are held under tension by external forces. The external forces are released after the concrete cures; the result is a permanent tension in the steel and hence a compressive stress on the concrete. The concrete slab can now support a much heavier load.

Example 12.5 Stage Design

In Example 8.2, we analyzed a cable used to support an actor as he swings onto the stage. Now suppose the tension in the cable is 940 N as the actor reaches the lowest point. What diameter should a 10-m-long steel cable have if we do not want it to stretch more than 0.50 cm under these conditions?

SOLUTION

Conceptualize Look back at Example 8.2 to recall what is happening in this situation. We ignored any stretching of the cable there, but we wish to address this phenomenon in this example.

Categorize We perform a simple calculation involving Equation 12.6, so we categorize this example as a substitution problem.

Solve Equation 12.6 for the cross-sectional area of the cable:

$$A = \frac{FL_i}{Y\Delta L}$$

Assuming the cross section is circular, find the diameter of the cable from $d = 2r$ and $A = \pi r^2$:

$$d = 2r = 2\sqrt{\frac{A}{\pi}} = 2\sqrt{\frac{FL_i}{\pi Y\Delta L}}$$

Substitute numerical values:

$$d = 2\sqrt{\frac{(940 \text{ N})(10 \text{ m})}{\pi(20 \times 10^{10} \text{ N/m}^2)(0.005\,0 \text{ m})}} = 3.5 \times 10^{-3} \text{ m} = 3.5 \text{ mm}$$

To provide a large margin of safety, you would probably use a flexible cable made up of many smaller wires having a total cross-sectional area substantially greater than our calculated value.

Example 12.6 Squeezing a Brass Sphere

A solid brass sphere is initially surrounded by air, and the air pressure exerted on it is 1.0×10^5 N/m² (normal atmospheric pressure). The sphere is lowered into the ocean to a depth where the pressure is 2.0×10^7 N/m². The volume of the sphere in air is 0.50 m³. By how much does this volume change once the sphere is submerged?

SOLUTION

Conceptualize Think about movies or television shows you have seen in which divers go to great depths in the water in submersible vessels. These vessels must be very strong to withstand the large pressure under water. This pressure squeezes the vessel and reduces its volume.

Categorize We perform a simple calculation involving Equation 12.8, so we categorize this example as a substitution problem.

Solve Equation 12.8 for the volume change of the sphere:

$$\Delta V = -\frac{V_i \Delta P}{B}$$

Substitute numerical values:

$$\Delta V = -\frac{(0.50 \text{ m}^3)(2.0 \times 10^7 \text{ N/m}^2 - 1.0 \times 10^5 \text{ N/m}^2)}{6.1 \times 10^{10} \text{ N/m}^2}$$

$$= -1.6 \times 10^{-4} \text{ m}^3$$

The negative sign indicates that the volume of the sphere decreases.

Summary

Definitions

The gravitational force exerted on an object can be considered as acting at a single point called the **center of gravity.** An object's center of gravity coincides with its center of mass if the object is in a uniform gravitational field.

We can describe the elastic properties of a substance using the concepts of stress and strain. **Stress** is a quantity proportional to the force producing a deformation; **strain** is a measure of the degree of deformation. Stress is proportional to strain, and the constant of proportionality is the **elastic modulus:**

$$\text{Elastic modulus} \equiv \frac{\text{stress}}{\text{strain}} \qquad \text{(12.5)}$$

Concepts and Principles

Three common types of deformation are represented by (1) the resistance of a solid to elongation under a load, characterized by **Young's modulus** Y; (2) the resistance of a solid to the motion of internal planes sliding past each other, characterized by the **shear modulus** S; and (3) the resistance of a solid or fluid to a volume change, characterized by the **bulk modulus** B.

Analysis Model for Problem Solving

Rigid Object in Equilibrium A rigid object in equilibrium exhibits no translational or angular acceleration. The net external force acting on it is zero, and the net external torque on it is zero about any axis:

$$\sum \vec{\mathbf{F}}_{\text{ext}} = 0 \qquad \text{(12.1)}$$

$$\sum \vec{\tau}_{\text{ext}} = 0 \qquad \text{(12.2)}$$

The first condition is the condition for translational equilibrium, and the second is the condition for rotational equilibrium.

Objective Questions

1.] denotes answer available in *Student Solutions Manual/Study Guide*

1. The acceleration due to gravity becomes weaker by about three parts in ten million for each meter of increased elevation above the Earth's surface. Suppose a skyscraper is 100 stories tall, with the same floor plan for each story and with uniform average density. Compare the location of the building's center of mass and the location of its center of gravity. Choose one: (a) Its center of mass is higher by a distance of several meters. (b) Its center of mass is higher by a distance of several millimeters. (c) Its center of mass and its center of gravity are in the same location. (d) Its center of gravity is higher by a distance of several millimeters. (e) Its center of gravity is higher by a distance of several meters.

2. A rod 7.0 m long is pivoted at a point 2.0 m from the left end. A downward force of 50 N acts at the left end, and a downward force of 200 N acts at the right end. At what distance to the right of the pivot can a third force of 300 N acting upward be placed to produce rotational equilibrium? *Note:* Neglect the weight of the rod. (a) 1.0 m (b) 2.0 m (c) 3.0 m (d) 4.0 m (e) 3.5 m

3. Consider the object in Figure OQ12.3. A single force is exerted on the object. The line of action of the force does not pass through the object's center of mass. The acceleration of the object's center of mass due to this force (a) is the same as if the force were applied at the

center of mass, (b) is larger than the acceleration would be if the force were applied at the center of mass, (c) is smaller than the acceleration would be if the force were applied at the center of mass, or (d) is zero because the force causes only angular acceleration about the center of mass.

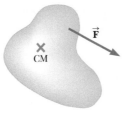

Figure OQ12.3

4. Two forces are acting on an object. Which of the following statements is correct? (a) The object is in equilibrium if the forces are equal in magnitude and opposite in direction. (b) The object is in equilibrium if the net torque on the object is zero. (c) The object is in equilibrium if the forces act at the same point on the object. (d) The object is in equilibrium if the net force and the net torque on the object are both zero. (e) The object cannot be in equilibrium because more than one force acts on it.

5. In the cabin of a ship, a soda can rests in a saucer-shaped indentation in a built-in counter. The can tilts as the ship slowly rolls. In which case is the can most stable against tipping over? (a) It is most stable when it is full. (b) It is most stable when it is half full. (c) It is most stable when it is empty. (d) It is most stable in two of these cases. (e) It is equally stable in all cases.

6. A 20.0-kg horizontal plank 4.00 m long rests on two supports, one at the left end and a second 1.00 m from the right end. What is the magnitude of the force exerted on the plank by the support near the right end? (a) 32.0 N (b) 45.2 N (c) 112 N (d) 131 N (e) 98.2 N

7. Assume a single 300-N force is exerted on a bicycle frame as shown in Figure OQ12.7. Consider the torque produced by this force about axes perpendicular to the plane of the paper and through each of the points

A through E, where E is the center of mass of the frame. Rank the torques τ_A, τ_B, τ_C, τ_D, and τ_E from largest to smallest, noting that zero is greater than a negative quantity. If two torques are equal, note their equality in your ranking.

Figure OQ12.7

8. In analyzing the equilibrium of a flat, rigid object, you are about to choose an axis about which you will calculate torques. Which of the following describes the choice you should make? (a) The axis should pass through the object's center of mass. (b) The axis should pass through one end of the object. (c) The axis should be either the x axis or the y axis. (d) The axis should pass through any point within the object. (e) Any axis within or outside the object can be chosen.

9. A certain wire, 3 m long, stretches by 1.2 mm when under tension 200 N. (i) Does an equally thick wire 6 m long, made of the same material and under the same tension, stretch by (a) 4.8 mm, (b) 2.4 mm, (c) 1.2 mm, (d) 0.6 mm, or (e) 0.3 mm? (ii) A wire with twice the diameter, 3 m long, made of the same material and under the same tension, stretches by what amount? Choose from the same possibilities (a) through (e).

10. The center of gravity of an ax is on the centerline of the handle, close to the head. Assume you saw across the handle through the center of gravity and weigh the two parts. What will you discover? (a) The handle side is heavier than the head side. (b) The head side is heavier than the handle side. (c) The two parts are equally heavy. (d) Their comparative weights cannot be predicted.

Conceptual Questions $\boxed{1.}$ denotes answer available in *Student Solutions Manual/Study Guide*

$\boxed{1.}$ A ladder stands on the ground, leaning against a wall. Would you feel safer climbing up the ladder if you were told that the ground is frictionless but the wall is rough or if you were told that the wall is frictionless but the ground is rough? Explain your answer.

2. The center of gravity of an object may be located outside the object. Give two examples for which that is the case.

3. (a) Give an example in which the net force acting on an object is zero and yet the net torque is nonzero. (b) Give an example in which the net torque acting on an object is zero and yet the net force is nonzero.

4. Stand with your back against a wall. Why can't you put your heels firmly against the wall and then bend forward without falling?

5. An arbitrarily shaped piece of plywood can be suspended from a string attached to the ceiling. Explain how you could use a plumb bob to find its center of gravity.

6. A girl has a large, docile dog she wishes to weigh on a small bathroom scale. She reasons that she can determine her dog's weight with the following method. First she puts the dog's two front feet on the scale and records the scale reading. Then she places only the dog's two back feet on the scale and records the reading. She thinks that the sum of the readings will be the dog's weight. Is she correct? Explain your answer.

7. Can an object be in equilibrium if it is in motion? Explain.

8. What kind of deformation does a cube of Jell-O exhibit when it jiggles?

Problems

Section 12.1 Analysis Model: Rigid Object in Equilibrium

1. What are the necessary conditions for equilibrium of the object shown in Figure P12.1? Calculate torques about an axis through point O.

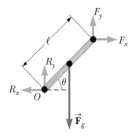

Figure P12.1

2. *Why is the following situation impossible?* A uniform beam of mass $m_b = 3.00$ kg and length $\ell = 1.00$ m supports blocks with masses $m_1 = 5.00$ kg and $m_2 = 15.0$ kg at two positions as shown in Figure P12.2. The beam rests on two triangular blocks, with point P a distance $d = 0.300$ m to the right of the center of gravity of the beam. The position of the object of mass m_2 is adjusted along the length of the beam until the normal force on the beam at O is zero.

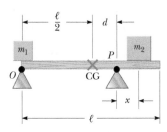

Figure P12.2

Section 12.2 More on the Center of Gravity

Problems 45, 48, 49, and 92 in Chapter 9 can also be assigned with this section.

3. A carpenter's square has the shape of an L as shown in Figure P12.3. Locate its center of gravity.

Figure P12.3

4. Consider the following distribution of objects: a 5.00-kg object with its center of gravity at (0, 0) m, a 3.00-kg object at (0, 4.00) m, and a 4.00-kg object at (3.00, 0) m. Where should a fourth object of mass 8.00 kg be placed so that the center of gravity of the four-object arrangement will be at (0, 0)?

5. Pat builds a track for his model car out of solid wood as shown in Figure P12.5. The track is 5.00 cm wide, 1.00 m high, and 3.00 m long. The runway is cut so that it forms a parabola with the equation $y = (x - 3)^2/9$. Locate the horizontal coordinate of the center of gravity of this track.

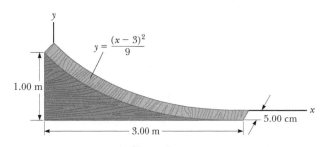

Figure P12.5

6. A circular pizza of radius R has a circular piece of radius $R/2$ removed from one side as shown in Figure P12.6. The center of gravity has moved from C to C' along the x axis. Show that the distance from C to C' is $R/6$. Assume the thickness and density of the pizza are uniform throughout.

Figure P12.6

7. Figure P12.7 on page 380 shows three uniform objects: a rod with $m_1 = 6.00$ kg, a right triangle with $m_2 = 3.00$ kg, and a square with $m_3 = 5.00$ kg. Their coordinates in meters are given. Determine the center of gravity for the three-object system.

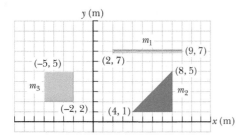

Figure P12.7

Section 12.3 Examples of Rigid Objects in Static Equilibrium

Problems 14, 26, 27, 28, 31, 33, 34, 60, 66, 85, 89, 97, and 100 in Chapter 5 can also be assigned with this section.

8. A 1 500-kg automobile has a wheel base (the distance between the axles) of 3.00 m. The automobile's center of mass is on the centerline at a point 1.20 m behind the front axle. Find the force exerted by the ground on each wheel.

9. Find the mass m of the counterweight needed to balance a truck with mass $M = 1\,500$ kg on an incline of $\theta = 45°$ (Fig. P12.9). Assume both pulleys are frictionless and massless.

Figure P12.9

10. A mobile is constructed of light rods, light strings, and beach souvenirs as shown in Figure P12.10. If $m_4 = 12.0$ g, find values for (a) m_1, (b) m_2, and (c) m_3.

Figure P12.10

11. A uniform beam of length 7.60 m and weight 4.50×10^2 N is carried by two workers, Sam and Joe, as shown in Figure P12.11. Determine the force that each person exerts on the beam.

Figure P12.11

12. A vaulter holds a 29.4-N pole in equilibrium by exerting an upward force $\vec{U}$ with her leading hand and a downward force $\vec{D}$ with her trailing hand as shown in Figure P12.12. Point C is the center of gravity of the pole. What are the magnitudes of (a) $\vec{U}$ and (b) $\vec{D}$?

Figure P12.12

13. A 15.0-m uniform ladder weighing 500 N rests against a frictionless wall. The ladder makes a 60.0° angle with the horizontal. (a) Find the horizontal and vertical forces the ground exerts on the base of the ladder when an 800-N firefighter has climbed 4.00 m along the ladder from the bottom. (b) If the ladder is just on the verge of slipping when the firefighter is 9.00 m from the bottom, what is the coefficient of static friction between ladder and ground?

14. A uniform ladder of length L and mass m_1 rests against a frictionless wall. The ladder makes an angle θ with the horizontal. (a) Find the horizontal and vertical forces the ground exerts on the base of the ladder when a firefighter of mass m_2 has climbed a distance x along the ladder from the bottom. (b) If the ladder is just on the verge of slipping when the firefighter is a distance d along the ladder from the bottom, what is the coefficient of static friction between ladder and ground?

15. A flexible chain weighing 40.0 N hangs between two hooks located at the same height (Fig. P12.15). At each hook, the tangent to the chain makes an angle $\theta = 42.0°$ with the horizontal. Find (a) the magnitude of the force each hook exerts on the chain and (b) the

Figure P12.15

tension in the chain at its midpoint. *Suggestion:* For part (b), make a force diagram for half of the chain.

16. A uniform beam of length L and mass m shown in Figure P12.16 is inclined at an angle θ to the horizontal. Its upper end is connected to a wall by a rope, and its lower end rests on a rough, horizontal surface. The coefficient of static friction between the beam and surface is μ_s. Assume the angle θ is such that the static friction force is at its *maximum* value. (a) Draw a force diagram for the beam. (b) Using the condition of rotational equilibrium, find an expression for the tension T in the rope in terms of m, g, and θ. (c) Using the condition of translational equilibrium, find a second expression for T in terms of μ_s, m, and g. (d) Using the results from parts (a) through (c), obtain an expression for μ_s involving only the angle θ. (e) What happens if the ladder is lifted upward and its base is placed back on the ground slightly to the left of its position in Figure P12.16? Explain.

Figure P12.16

17. Figure P12.17 shows a claw hammer being used to pull a nail out of a horizontal board. The mass of the hammer is 1.00 kg. A force of 150 N is exerted horizontally as shown, and the nail does not yet move relative to the board. Find (a) the force exerted by the hammer claws on the nail and (b) the force exerted by the surface on the point of contact with the hammer head. Assume the force the hammer exerts on the nail is parallel to the nail.

$\vec{F}$

30.0 cm

Single point of contact

30.0°

5.00 cm

Figure P12.17

18. A 20.0-kg floodlight in a park is supported at the end of a horizontal beam of negligible mass that is hinged to a pole as shown in Figure P12.18. A cable at an angle of $\theta = 30.0°$ with the beam helps support the light. (a) Draw a force diagram for the beam. By computing torques about an axis at the hinge at the left-hand end of the beam, find (b) the tension in the cable, (c) the horizontal component of the force exerted by the pole on the beam, and (d) the

θ

Figure P12.18

vertical component of this force. Now solve the same problem from the force diagram from part (a) by computing torques around the junction between the cable and the beam at the right-hand end of the beam. Find (e) the vertical component of the force exerted by the pole on the beam, (f) the tension in the cable, and (g) the horizontal component of the force exerted by the pole on the beam. (h) Compare the solution to parts (b) through (d) with the solution to parts (e) through (g). Is either solution more accurate?

19. Sir Lost-a-Lot dons his armor and sets out from the castle on his trusty steed (Fig. P12.19). Usually, the drawbridge is lowered to a horizontal position so that the end of the bridge rests on the stone ledge. Unfortunately, Lost-a-Lot's squire didn't lower the drawbridge far enough and stopped it at $\theta = 20.0°$ above the horizontal. The knight and his horse stop when their combined center of mass is $d = 1.00$ m from the end of the bridge. The uniform bridge is $\ell = 8.00$ m long and has mass 2 000 kg. The lift cable is attached to the bridge 5.00 m from the hinge at the castle end and to a point on the castle wall $h = 12.0$ m above the bridge. Lost-a-Lot's mass combined with his armor and steed is 1 000 kg. Determine (a) the tension in the cable and (b) the horizontal and (c) the vertical force components acting on the bridge at the hinge.

h

θ

d

ℓ

Figure P12.19 Problems 19 and 20.

20. **Review.** While Lost-a-Lot ponders his next move in the situation described in Problem 19 and illustrated in Figure P12.19, the enemy attacks! An incoming projectile breaks off the stone ledge so that the end of the drawbridge can be lowered past the wall where it usually rests. In addition, a fragment of the projectile bounces up and cuts the drawbridge cable! The hinge between the castle wall and the bridge is frictionless, and the bridge swings down freely until it is vertical and smacks into the vertical castle wall below the castle entrance. (a) How long does Lost-a-Lot stay in contact with the bridge while it swings downward? (b) Find the angular acceleration of the bridge just as it starts to move. (c) Find the angular speed of the bridge when it strikes the wall below the hinge. Find the force exerted by the hinge on the bridge (d) immediately after the cable breaks and (e) immediately before it strikes the castle wall.

21. John is pushing his daughter Rachel in a wheelbarrow when it is stopped by a brick 8.00 cm high (Fig. P12.21). The handles make an angle of $\theta = 15.0°$ with the ground. Due to the weight of Rachel and the wheelbarrow, a downward force of 400 N is exerted at the center of the wheel, which has a radius of 20.0 cm. (a) What force must John apply along the handles to just start the wheel over the brick? (b) What is the force (magnitude and direction) that the brick exerts on the wheel just as the wheel begins to lift over the brick? In both parts, assume the brick remains fixed and does not slide along the ground. Also assume the force applied by John is directed exactly toward the center of the wheel.

Figure P12.21 Problems 21 and 22.

22. John is pushing his daughter Rachel in a wheelbarrow when it is stopped by a brick of height h (Fig. P12.21). The handles make an angle of θ with the ground. Due to the weight of Rachel and the wheelbarrow, a downward force mg is exerted at the center of the wheel, which has a radius R. (a) What force F must John apply along the handles to just start the wheel over the brick? (b) What are the components of the force that the brick exerts on the wheel just as the wheel begins to lift over the brick? In both parts, assume the brick remains fixed and does not slide along the ground. Also assume the force applied by John is directed exactly toward the center of the wheel.

23. One end of a uniform 4.00-m-long rod of weight F_g is [W] supported by a cable at an angle of $\theta = 37°$ with the rod. The other end rests against the wall, where it is held by friction as shown in Figure P12.23. The coefficient of static friction between the wall and the rod is $\mu_s = 0.500$. Determine the minimum distance x from point A at which an additional object, also with the same weight F_g, can be hung without causing the rod to slip at point A.

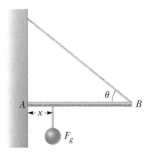

Figure P12.23

24. A 10.0-kg monkey climbs a uniform ladder with weight 1.20×10^2 N and length $L = 3.00$ m as shown in Figure P12.24. The ladder rests against the wall

and makes an angle of $\theta = 60.0°$ with the ground. The upper and lower ends of the ladder rest on frictionless surfaces. The lower end is connected to the wall by a horizontal rope that is frayed and can support a maximum tension of only 80.0 N. (a) Draw a force diagram for the ladder. (b) Find the normal force exerted on the bottom of the ladder. (c) Find the tension in the rope when the monkey is two-thirds of the way up the ladder. (d) Find the maximum distance d that the monkey can climb up the ladder before the rope breaks. (e) If the horizontal surface were rough and the rope were removed, how would your analysis of the problem change? What other information would you need to answer parts (c) and (d)?

Figure P12.24

25. A uniform plank of length 2.00 m and mass 30.0 kg is supported by three ropes as indicated by the blue vectors in Figure P12.25. Find the tension in each rope when a 700-N person is $d = 0.500$ m from the left end.

Figure P12.25

Section 12.4 Elastic Properties of Solids

26. A steel wire of diameter 1 mm can support a tension of 0.2 kN. A steel cable to support a tension of 20 kN should have diameter of what order of magnitude?

27. The deepest point in the ocean is in the Mariana Trench, about 11 km deep, in the Pacific. The pressure at this depth is huge, about 1.13×10^8 N/m². (a) Calculate the change in volume of 1.00 m³ of seawater carried from the surface to this deepest point. (b) The density of seawater at the surface is 1.03×10^3 kg/m³. Find its density at the bottom. (c) Explain whether or when it is a good approximation to think of water as incompressible.

28. Assume Young's modulus for bone is 1.50×10^{10} N/m². The bone breaks if stress greater than 1.50×10^8 N/m² is imposed on it. (a) What is the maximum force that can be exerted on the femur bone in the leg if it has a minimum effective diameter of 2.50 cm? (b) If this much force is applied compressively, by how much does the 25.0-cm-long bone shorten?

29. A child slides across a floor in a pair of rubber-soled shoes. The friction force acting on each foot is 20.0 N. The footprint area of each shoe sole is 14.0 cm², and the thickness of each sole is 5.00 mm. Find the horizontal distance by which the upper and lower surfaces of each sole are offset. The shear modulus of the rubber is 3.00 MN/m².

30. Evaluate Young's modulus for the material whose stress–strain curve is shown in Figure 12.12.

31. Assume if the shear stress in steel exceeds about 4.00×10^8 N/m², the steel ruptures. Determine the shearing force necessary to (a) shear a steel bolt 1.00 cm in diameter and (b) punch a 1.00-cm-diameter hole in a steel plate 0.500 cm thick.

32. When water freezes, it expands by about 9.00%. What pressure increase would occur inside your automobile engine block if the water in it froze? (The bulk modulus of ice is 2.00×10^9 N/m².)

33. A 200-kg load is hung on a wire of length 4.00 m, cross-sectional area 0.200×10^{-4} m², and Young's modulus 8.00×10^{10} N/m². What is its increase in length?

34. A walkway suspended across a hotel lobby is supported at numerous points along its edges by a vertical cable above each point and a vertical column underneath. The steel cable is 1.27 cm in diameter and is 5.75 m long before loading. The aluminum column is a hollow cylinder with an inside diameter of 16.14 cm, an outside diameter of 16.24 cm, and an unloaded length of 3.25 m. When the walkway exerts a load force of 8 500 N on one of the support points, how much does the point move down?

35. **Review.** A 2.00-m-long cylindrical steel wire with a cross-sectional diameter of 4.00 mm is placed over a light, frictionless pulley. An object of mass $m_1 = 5.00$ kg is hung from one end of the wire and an object of mass $m_2 = 3.00$ kg from the other end as shown in Figure P12.35. The objects are released and allowed to move freely. Compared with its length before the objects were attached, by how much has the wire stretched while the objects are in motion?

Figure P12.35

36. **Review.** A 30.0-kg hammer, moving with speed 20.0 m/s, strikes a steel spike 2.30 cm in diameter. The hammer rebounds with speed 10.0 m/s after 0.110 s. What is the average strain in the spike during the impact?

Additional Problems

37. A bridge of length 50.0 m and mass 8.00×10^4 kg is supported on a smooth pier at each end as shown in Figure P12.37. A truck of mass 3.00×10^4 kg is located 15.0 m from one end. What are the forces on the bridge at the points of support?

Figure P12.37

38. A uniform beam resting on two pivots has a length $L = 6.00$ m and mass $M = 90.0$ kg. The pivot under the left end exerts a normal force n_1 on the beam, and the second pivot located a distance $\ell = 4.00$ m from the left end exerts a normal force n_2. A woman of mass $m = 55.0$ kg steps onto the left end of the beam and begins walking to the right as in Figure P12.38. The goal is to find the woman's position when the beam begins to tip. (a) What is the appropriate analysis model for the beam before it begins to tip? (b) Sketch a force diagram for the beam, labeling the gravitational and normal forces acting on the beam and placing the woman a distance x to the right of the first pivot, which is the origin. (c) Where is the woman when the normal force n_1 is the greatest? (d) What is n_1 when the beam is about to tip? (e) Use Equation 12.1 to find the value of n_2 when the beam is about to tip. (f) Using the result of part (d) and Equation 12.2, with torques computed around the second pivot, find the woman's position x when the beam is about to tip. (g) Check the answer to part (e) by computing torques around the first pivot point.

Figure P12.38

39. In exercise physiology studies, it is sometimes important to determine the location of a person's center of mass. This determination can be done with the arrangement shown in Figure P12.39. A light plank rests on two scales, which read $F_{g1} = 380$ N and $F_{g2} = 320$ N. A distance of 1.65 m separates the scales. How far from the woman's feet is her center of mass?

Figure P12.39

40. The lintel of prestressed reinforced concrete in Figure P12.40 is 1.50 m long. The concrete encloses one steel reinforcing rod with cross-sectional area 1.50 cm². The rod joins two strong end plates. The cross-sectional area of the concrete perpendicular to the rod is 50.0 cm². Young's modulus for the concrete is 30.0×10^9 N/m². After the concrete cures and the original tension T_1 in the rod is released, the concrete is to be under compressive stress 8.00×10^6 N/m². (a) By what distance will the rod compress the concrete when the original tension in the rod is released? (b) What

Figure P12.40

is the new tension T_2 in the rod? (c) The rod will then be how much longer than its unstressed length? (d) When the concrete was poured, the rod should have been stretched by what extension distance from its unstressed length? (e) Find the required original tension T_1 in the rod.

41. The arm in Figure P12.41 weighs 41.5 N. The gravitational force on the arm acts through point A. Determine the magnitudes of the tension force $\vec{F}_t$ in the deltoid muscle and the force $\vec{F}_s$ exerted by the shoulder on the humerus (upper-arm bone) to hold the arm in the position shown.

Figure P12.41

42. When a person stands on tiptoe on one foot (a strenuous position), the position of the foot is as shown in Figure P12.42a. The total gravitational force $\vec{F}_g$ on the body is supported by the normal force $\vec{n}$ exerted by the floor on the toes of one foot. A mechanical model of the situation is shown in Figure P12.42b, where $\vec{T}$ is the force exerted on the foot by the Achilles tendon and $\vec{R}$ is the force exerted on the foot by the tibia. Find the values of T, R, and θ when $F_g = 700$ N.

Figure P12.42

43. A hungry bear weighing 700 N walks out on a beam
AMT in an attempt to retrieve a basket of goodies hanging at the end of the beam (Fig. P12.43). The beam is uni-

Figure P12.43

form, weighs 200 N, and is 6.00 m long, and it is supported by a wire at an angle of $\theta = 60.0°$. The basket weighs 80.0 N. (a) Draw a force diagram for the beam. (b) When the bear is at $x = 1.00$ m, find the tension in the wire supporting the beam and the components of the force exerted by the wall on the left end of the beam. (c) **What If?** If the wire can withstand a maximum tension of 900 N, what is the maximum distance the bear can walk before the wire breaks?

44. The following equations are obtained from a force diagram of a rectangular farm gate, supported by two hinges on the left-hand side. A bucket of grain is hanging from the latch.

$$-A + C = 0$$

$$+B - 392\text{ N} - 50.0\text{ N} = 0$$

$$A(0) + B(0) + C(1.80\text{ m}) - 392\text{ N}(1.50\text{ m})$$

$$- 50.0\text{ N}(3.00\text{ m}) = 0$$

(a) Draw the force diagram and complete the statement of the problem, specifying the unknowns. (b) Determine the values of the unknowns and state the physical meaning of each.

45. A uniform sign of weight F_g and width $2L$ hangs from a light, horizontal beam hinged at the wall and supported by a cable (Fig. P12.45). Determine (a) the tension in the cable and (b) the components of the reaction force exerted by the wall on the beam in terms of F_g, d, L, and θ.

Figure P12.45

46. A 1 200-N uniform boom at $\phi = 65°$ to the vertical is supported by a cable at an angle $\theta = 25.0°$ to the horizontal as shown in Figure P12.46. The boom is pivoted at the bottom, and an object of weight $m = 2\,000$ N hangs from its top. Find (a) the tension in the support cable and (b) the components of the reaction force exerted by the floor on the boom.

Figure P12.46

47. A crane of mass $m_1 = 3\,000$ kg supports a load of mass $m_2 = 10\,000$ kg as shown in Figure P12.47. The crane

is pivoted with a frictionless pin at A and rests against a smooth support at B. Find the reaction forces at (a) point A and (b) point B.

Figure P12.47

48. Assume a person bends forward to lift a load "with his back" as shown in Figure P12.48a. The spine pivots mainly at the fifth lumbar vertebra, with the principal supporting force provided by the erector spinalis muscle in the back. To see the magnitude of the forces involved, consider the model shown in Figure P12.48b for a person bending forward to lift a 200-N object. The spine and upper body are represented as a uniform horizontal rod of weight 350 N, pivoted at the base of the spine. The erector spinalis muscle, attached at a point two-thirds of the way up the spine, maintains the position of the back. The angle between the spine and this muscle is $\theta = 12.0°$. Find (a) the tension T in the back muscle and (b) the compressional force in the spine. (c) Is this method a good way to lift a load? Explain your answer, using the results of parts (a) and (b). (d) Can you suggest a better method to lift a load?

Figure P12.48

49. A 10 000-N shark is supported by a rope attached to a 4.00-m rod that can pivot at the base. (a) Calculate the tension in the cable between the rod and the wall, assuming the cable is holding the system in the position shown in Figure P12.49. Find (b) the horizontal force and (c) the vertical force exerted on the base of the rod. Ignore the weight of the rod.

Figure P12.49

50. *Why is the following situation impossible?* A worker in a factory pulls a cabinet across the floor using a rope as

shown in Figure P12.50a. The rope make an angle $\theta = 37.0°$ with the floor and is tied $h_1 = 10.0$ cm from the bottom of the cabinet. The uniform rectangular cabinet has height $\ell = 100$ cm and width $w = 60.0$ cm, and it weighs 400 N. The cabinet slides with constant speed when a force $F = 300$ N is applied through the rope. The worker tires of walking backward. He fastens the rope to a point on the cabinet $h_2 = 65.0$ cm off the floor and lays the rope over his shoulder so that he can walk forward and pull as shown in Figure P12.50b. In this way, the rope again makes an angle of $\theta = 37.0°$ with the horizontal and again has a tension of 300 N. Using this technique, the worker is able to slide the cabinet over a long distance on the floor without tiring.

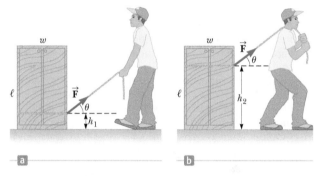

Figure P12.50 Problems 50 and 62.

51. A uniform beam of mass m is inclined at an angle θ to the horizontal. Its upper end (point P) produces a 90° bend in a very rough rope tied to a wall, and its lower end rests on a rough floor (Fig. P12.51). Let μ_s represent the coefficient of static friction between beam and floor. Assume μ_s is less than the cotangent of θ. (a) Find an expression for the maximum mass M that can be suspended from the top before the beam slips. Determine (b) the magnitude of the reaction force at the floor and (c) the magnitude of the force exerted by the beam on the rope at P in terms of m, M, and μ_s.

Figure P12.51

52. The large quadriceps muscle in the upper leg terminates at its lower end in a tendon attached to the upper end of the tibia (Fig. P12.52a, page 386). The forces on the lower leg when the leg is extended are modeled as in Figure P12.52b, where $\vec{T}$ is the force in the tendon, $\vec{F}_{g,\text{leg}}$ is the gravitational force acting on the lower leg, and $\vec{F}_{g,\text{foot}}$ is the gravitational force acting on the foot. Find T when the tendon is at an angle of $\phi = 25.0°$ with the tibia, assuming $F_{g,\text{leg}} = 30.0$ N, $F_{g,\text{foot}} = 12.5$ N, and the leg is extended at an angle $\theta = 40.0°$ with respect to the vertical. Also assume the center of gravity of the

tibia is at its geometric center and the tendon attaches to the lower leg at a position one-fifth of the way down the leg.

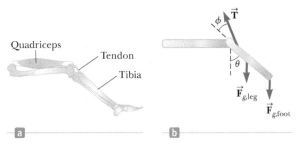

Figure P12.52

53. When a gymnast performing on the rings executes the *iron cross*, he maintains the position at rest shown in Figure P12.53a. In this maneuver, the gymnast's feet (not shown) are off the floor. The primary muscles involved in supporting this position are the latissimus dorsi ("lats") and the pectoralis major ("pecs"). One of the rings exerts an upward force $\vec{F}_h$ on a hand as shown in Figure P12.53b. The force $\vec{F}_s$ is exerted by the shoulder joint on the arm. The latissimus dorsi and pectoralis major muscles exert a total force $\vec{F}_m$ on the arm. (a) Using the information in the figure, find the magnitude of the force $\vec{F}_m$ for an athlete of weight 750 N. (b) Suppose an athlete in training cannot perform the iron cross but can hold a position similar to the figure in which the arms make a 45° angle with the horizontal rather than being horizontal. Why is this position easier for the athlete?

Figure P12.53

54. Figure P12.54 shows a light truss formed from three struts lying in a plane and joined by three smooth hinge pins at their ends. The truss supports a downward force of $\vec{F} = 1\,000$ N applied at the point *B*. The truss has negligible weight. The piers at *A* and *C*

are smooth. (a) Given $\theta_1 = 30.0°$ and $\theta_2 = 45.0°$, find n_A and n_C. (b) One can show that the force any strut exerts on a pin must be directed along the length of the strut as a force of tension or compression. Use that fact to identify the directions of the forces that the struts exert on the pins joining them. Find the force of tension or of compression in each of the three bars.

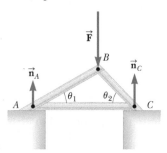

Figure P12.54

55. One side of a plant shelf is supported by a bracket mounted on a vertical wall by a single screw as shown in Figure P12.55. Ignore the weight of the bracket. (a) Find the horizontal component of the force that the screw exerts on the bracket when an 80.0 N vertical force is applied as shown. (b) As your grandfather waters his geraniums, the 80.0-N load force is increasing at the rate 0.150 N/s. At what rate is the force exerted by the screw changing? *Suggestion:* Imagine that the bracket is slightly loose.

Figure P12.55

56. A stepladder of negligible weight is constructed as shown in Figure P12.56, with $AC = BC = \ell = 4.00$ m. A painter of mass $m = 70.0$ kg stands on the ladder $d = 3.00$ m from the bottom. Assuming the floor is frictionless, find (a) the tension in the horizontal bar *DE* connecting the two halves of the ladder, (b) the normal forces at *A* and *B*, and (c) the components of the reaction force at the single hinge *C* that the left half of the ladder exerts on the right half. *Suggestion:* Treat the ladder as a single object, but also treat each half of the ladder separately.

Figure P12.56
Problems 56 and 57.

57. A stepladder of negligible weight is constructed as shown in Figure P12.56, with $AC = BC = \ell$. A painter of mass m stands on the ladder a distance d from the bottom. Assuming the floor is frictionless, find (a) the tension in the horizontal bar *DE* connecting the two

halves of the ladder, (b) the normal forces at A and B, and (c) the components of the reaction force at the single hinge C that the left half of the ladder exerts on the right half. *Suggestion:* Treat the ladder as a single object, but also treat each half of the ladder separately.

58. (a) Estimate the force with which a karate master strikes a board, assuming the hand's speed at the moment of impact is 10.0 m/s and decreases to 1.00 m/s during a 0.002 00-s time interval of contact between the hand and the board. The mass of his hand and arm is 1.00 kg. (b) Estimate the shear stress, assuming this force is exerted on a 1.00-cm-thick pine board that is 10.0 cm wide. (c) If the maximum shear stress a pine board can support before breaking is 3.60×10^6 N/m^2, will the board break?

59. Two racquetballs, each having a mass of 170 g, are placed in a glass jar as shown in Figure P12.59. Their centers lie on a straight line that makes a 45° angle with the horizontal. (a) Assume the walls are frictionless and determine P_1, P_2, and P_3. (b) Determine the magnitude of the force exerted by the left ball on the right ball.

Figure P12.59

60. **Review.** A wire of length L, Young's modulus Y, and cross-sectional area A is stretched elastically by an amount ΔL. By Hooke's law, the restoring force is $-k\,\Delta L$. (a) Show that $k = YA/L$. (b) Show that the work done in stretching the wire by an amount ΔL is $W = \frac{1}{2}YA(\Delta L)^2/L$.

61. **Review.** An aluminum wire is 0.850 m long and has a circular cross section of diameter 0.780 mm. Fixed at the top end, the wire supports a 1.20-kg object that swings in a horizontal circle. Determine the angular speed of the object required to produce a strain of 1.00×10^{-3}.

62. Consider the rectangular cabinet of Problem 50 shown in Figure P12.50, but with a force $\vec{F}$ applied horizontally at the upper edge. (a) What is the minimum force required to start to tip the cabinet? (b) What is the minimum coefficient of static friction required for the cabinet not to slide with the application of a force of this magnitude? (c) Find the magnitude and direction of the minimum force required to tip the cabinet if the point of application can be chosen *anywhere* on the cabinet.

63. A 500-N uniform rectangular sign 4.00 m wide and 3.00 m high is suspended from a horizontal, 6.00-m-long, uniform, 100-N rod as indicated in Figure P12.63. The left end of the rod is supported by a hinge, and the right end is supported by a thin cable making a 30.0° angle with the vertical. (a) Find the tension T in the cable. (b) Find the horizontal and vertical compo-

Figure P12.63

nents of force exerted on the left end of the rod by the hinge.

64. A steel cable 3.00 cm^2 in cross-sectional area has a mass of 2.40 kg per meter of length. If 500 m of the cable is hung over a vertical cliff, how much does the cable stretch under its own weight? Take $Y_{\text{steel}} = 2.00 \times 10^{11}$ N/m^2.

Challenge Problems

65. A uniform pole is propped between the floor and the ceiling of a room. The height of the room is 7.80 ft, and the coefficient of static friction between the pole and the ceiling is 0.576. The coefficient of static friction between the pole and the floor is greater than that between the pole and the ceiling. What is the length of the longest pole that can be propped between the floor and the ceiling?

66. In the What If? section of Example 12.2, let d represent the distance in meters between the person and the hinge at the left end of the beam. (a) Show that the cable tension is given by $T = 93.9d + 125$, with T in newtons. (b) Show that the direction angle θ of the hinge force is described by

$$\tan\theta = \left(\frac{32}{3d+4} - 1\right)\tan 53.0°$$

(c) Show that the magnitude of the hinge force is given by

$$R = \sqrt{8.82 \times 10^3 d^2 - 9.65 \times 10^4 d + 4.96 \times 10^5}$$

(d) Describe how the changes in T, θ, and R as d increases differ from one another.

67. Figure P12.67 shows a vertical force applied tangentially to a uniform cylinder of weight F_g. The coefficient of static friction between the cylinder and all surfaces is 0.500. The force $\vec{P}$ is increased in magnitude until the cylinder begins to rotate. In terms of F_g, find the maximum force magnitude P that can be applied without causing the cylinder to rotate. *Suggestion:* Show that both friction forces will be at their maximum values when the cylinder is on the verge of slipping.

Figure P12.67

68. A uniform rod of weight F_g and length L is supported at its ends by a frictionless trough as shown in Figure P12.68. (a) Show that the center of gravity of the rod must be vertically over point O when the rod is in equilibrium. (b) Determine the equilibrium value of the angle θ. (c) Is the equilibrium of the rod stable or unstable?

Figure P12.68

CHAPTER
13

Universal Gravitation

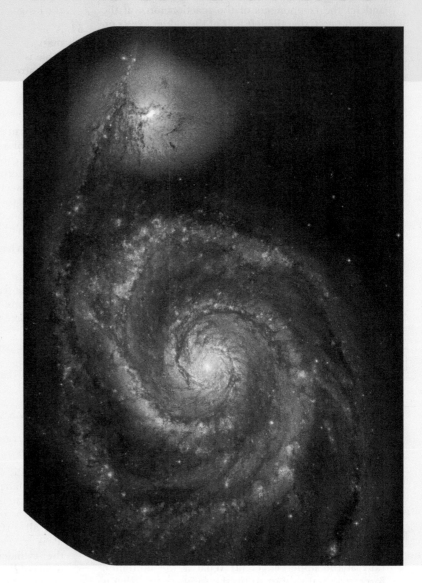

Hubble Space Telescope image of the Whirlpool Galaxy, M51, taken in 2005. The arms of this spiral galaxy compress hydrogen gas and create new clusters of stars. Some astronomers believe that the arms are prominent due to a close encounter with the small, yellow galaxy, NGC 5195, at the tip of one of its arms. *(NASA, Hubble Heritage Team, (STScI/AURA), ESA, S. Beckwith (STScI). Additional Processing: Robert Gendler)*

Before 1687, a large amount of data had been collected on the motions of the Moon and the planets, but a clear understanding of the forces related to these motions was not available. In that year, Isaac Newton provided the key that unlocked the secrets of the heavens. He knew, from his first law, that a net force had to be acting on the Moon because without such a force the Moon would move in a straight-line path rather than in its almost circular orbit. Newton reasoned that this force was the gravitational attraction exerted by the Earth on the Moon. He realized that the forces involved in the Earth–Moon attraction and in the Sun–planet attraction were not something special to those systems, but rather were particular cases of a general and universal attraction between objects. In other words, Newton saw that the same force of attraction that causes the Moon to follow its path around the Earth also causes an apple to fall from a tree. It was the first time that "earthly" and "heavenly" motions were unified.

In this chapter, we study the law of universal gravitation. We emphasize a description of planetary motion because astronomical data provide an important test of this law's validity. We then show that the laws of planetary motion developed by Johannes Kepler follow from

388

the law of universal gravitation and the principle of conservation of angular momentum for an isolated system. We conclude by deriving a general expression for the gravitational potential energy of a system and examining the energetics of planetary and satellite motion.

13.1 Newton's Law of Universal Gravitation

You may have heard the legend that, while napping under a tree, Newton was struck on the head by a falling apple. This alleged accident supposedly prompted him to imagine that perhaps all objects in the Universe were attracted to each other in the same way the apple was attracted to the Earth. Newton analyzed astronomical data on the motion of the Moon around the Earth. From that analysis, he made the bold assertion that the force law governing the motion of planets was the *same* as the force law that attracted a falling apple to the Earth.

In 1687, Newton published his work on the law of gravity in his treatise *Mathematical Principles of Natural Philosophy*. **Newton's law of universal gravitation** states that

> every particle in the Universe attracts every other particle with a force that is directly proportional to the product of their masses and inversely proportional to the square of the distance between them.

◀ **The law of universal gravitation**

If the particles have masses m_1 and m_2 and are separated by a distance r, the magnitude of this gravitational force is

$$F_g = G \frac{m_1 m_2}{r^2} \tag{13.1}$$

where G is a constant, called the *universal gravitational constant*. Its value in SI units is

$$G = 6.674 \times 10^{-11} \text{ N} \cdot \text{m}^2/\text{kg}^2 \tag{13.2}$$

The universal gravitational constant G was first evaluated in the late nineteenth century, based on results of an important experiment by Sir Henry Cavendish (1731–1810) in 1798. The law of universal gravitation was not expressed by Newton in the form of Equation 13.1, and Newton did not mention a constant such as G. In fact, even by the time of Cavendish, a unit of force had not yet been included in the existing system of units. Cavendish's goal was to measure the density of the Earth. His results were then used by other scientists 100 years later to generate a value for G.

Cavendish's apparatus consists of two small spheres, each of mass m, fixed to the ends of a light, horizontal rod suspended by a fine fiber or thin metal wire as illustrated in Figure 13.1. When two large spheres, each of mass M, are placed near the smaller ones, the attractive force between smaller and larger spheres causes the rod to rotate and twist the wire suspension to a new equilibrium orientation. The angle of rotation is measured by the deflection of a light beam reflected from a mirror attached to the vertical suspension.

The form of the force law given by Equation 13.1 is often referred to as an **inverse-square law** because the magnitude of the force varies as the inverse square of the separation of the particles.[1] We shall see other examples of this type of force law in subsequent chapters. We can express this force in vector form by defining a unit vector $\hat{\mathbf{r}}_{12}$ (Fig. 13.2). Because this unit vector is directed from particle 1 toward particle 2, the force exerted by particle 1 on particle 2 is

$$\vec{\mathbf{F}}_{12} = -G \frac{m_1 m_2}{r^2} \hat{\mathbf{r}}_{12} \tag{13.3}$$

The dashed line represents the original position of the rod.

Mirror   Light source

Figure 13.1 Cavendish apparatus for measuring gravitational forces.

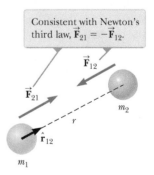

Consistent with Newton's third law, $\vec{\mathbf{F}}_{21} = -\vec{\mathbf{F}}_{12}$.

Figure 13.2 The gravitational force between two particles is attractive. The unit vector $\hat{\mathbf{r}}_{12}$ is directed from particle 1 toward particle 2.

---

[1]An *inverse* proportionality between two quantities x and y is one in which $y = k/x$, where k is a constant. A *direct* proportion between x and y exists when $y = kx$.

where the negative sign indicates that particle 2 is attracted to particle 1; hence, the force on particle 2 must be directed toward particle 1. By Newton's third law, the force exerted by particle 2 on particle 1, designated $\vec{\mathbf{F}}_{21}$, is equal in magnitude to $\vec{\mathbf{F}}_{12}$ and in the opposite direction. That is, these forces form an action–reaction pair, and $\vec{\mathbf{F}}_{21} = -\vec{\mathbf{F}}_{12}$.

Two features of Equation 13.3 deserve mention. First, the gravitational force is a field force that always exists between two particles, regardless of the medium that separates them. Second, because the force varies as the inverse square of the distance between the particles, it decreases rapidly with increasing separation.

Equation 13.3 can also be used to show that the gravitational force exerted by a finite-size, spherically symmetric mass distribution on a particle outside the distribution is the same as if the entire mass of the distribution were concentrated at the center. For example, the magnitude of the force exerted by the Earth on a particle of mass m near the Earth's surface is

$$F_g = G \frac{M_E m}{R_E^{\,2}} \tag{13.4}$$

where M_E is the Earth's mass and R_E its radius. This force is directed toward the center of the Earth.

Quick Quiz 13.1 A planet has two moons of equal mass. Moon 1 is in a circular orbit of radius r. Moon 2 is in a circular orbit of radius $2r$. What is the magnitude of the gravitational force exerted by the planet on Moon 2? **(a)** four times as large as that on Moon 1 **(b)** twice as large as that on Moon 1 **(c)** equal to that on Moon 1 **(d)** half as large as that on Moon 1 **(e)** one-fourth as large as that on Moon 1

Example 13.1   Billiards, Anyone?

Three 0.300-kg billiard balls are placed on a table at the corners of a right triangle as shown in Figure 13.3. The sides of the triangle are of lengths $a = 0.400$ m, $b = 0.300$ m, and $c = 0.500$ m. Calculate the gravitational force vector on the cue ball (designated m_1) resulting from the other two balls as well as the magnitude and direction of this force.

SOLUTION

Conceptualize Notice in Figure 13.3 that the cue ball is attracted to both other balls by the gravitational force. We can see graphically that the net force should point upward and toward the right. We locate our coordinate axes as shown in Figure 13.3, placing our origin at the position of the cue ball.

Figure 13.3 (Example 13.1) The resultant gravitational force acting on the cue ball is the vector sum $\vec{\mathbf{F}}_{21} + \vec{\mathbf{F}}_{31}$.

Categorize This problem involves evaluating the gravitational forces on the cue ball using Equation 13.3. Once these forces are evaluated, it becomes a vector addition problem to find the net force.

Analyze Find the force exerted by m_2 on the cue ball:

$$\vec{\mathbf{F}}_{21} = G \frac{m_2 m_1}{a^2} \hat{\mathbf{j}}$$

$$= (6.674 \times 10^{-11} \text{ N} \cdot \text{m}^2/\text{kg}^2) \frac{(0.300 \text{ kg})(0.300 \text{ kg})}{(0.400 \text{ m})^2} \hat{\mathbf{j}}$$

$$= 3.75 \times 10^{-11} \hat{\mathbf{j}} \text{ N}$$

Find the force exerted by m_3 on the cue ball:

$$\vec{\mathbf{F}}_{31} = G \frac{m_3 m_1}{b^2} \hat{\mathbf{i}}$$

$$= (6.674 \times 10^{-11} \text{ N} \cdot \text{m}^2/\text{kg}^2) \frac{(0.300 \text{ kg})(0.300 \text{ kg})}{(0.300 \text{ m})^2} \hat{\mathbf{i}}$$

$$= 6.67 \times 10^{-11} \hat{\mathbf{i}} \text{ N}$$

▶ **13.1** continued

Find the net gravitational force on the cue ball by adding these force vectors:

$$\vec{F} = \vec{F}_{31} + \vec{F}_{21} = \boxed{(6.67\,\hat{i} + 3.75\,\hat{j}) \times 10^{-11}\,\text{N}}$$

Find the magnitude of this force:

$$F = \sqrt{F_{31}^{2} + F_{21}^{2}} = \sqrt{(6.67)^{2} + (3.75)^{2}} \times 10^{-11}\,\text{N}$$

$$= \boxed{7.66 \times 10^{-11}\,\text{N}}$$

Find the tangent of the angle θ for the net force vector:

$$\tan\theta = \frac{F_y}{F_x} = \frac{F_{21}}{F_{31}} = \frac{3.75 \times 10^{-11}\,\text{N}}{6.67 \times 10^{-11}\,\text{N}} = 0.562$$

Evaluate the angle θ:

$$\theta = \tan^{-1}(0.562) = \boxed{29.4°}$$

Finalize The result for F shows that the gravitational forces between everyday objects have extremely small magnitudes.

13.2 Free-Fall Acceleration and the Gravitational Force

We have called the magnitude of the gravitational force on an object near the Earth's surface the *weight* of the object, where the weight is given by Equation 5.6. Equation 13.4 is another expression for this force. Therefore, we can set Equations 5.6 and 13.4 equal to each other to obtain

$$mg = G\frac{M_E m}{R_E^{2}}$$

$$g = G\frac{M_E}{R_E^{2}} \tag{13.5}$$

Equation 13.5 relates the free-fall acceleration g to physical parameters of the Earth—its mass and radius—and explains the origin of the value of 9.80 m/s² that we have used in earlier chapters. Now consider an object of mass m located a distance h above the Earth's surface or a distance r from the Earth's center, where $r = R_E + h$. The magnitude of the gravitational force acting on this object is

$$F_g = G\frac{M_E m}{r^{2}} = G\frac{M_E m}{(R_E + h)^{2}}$$

The magnitude of the gravitational force acting on the object at this position is also $F_g = mg$, where g is the value of the free-fall acceleration at the altitude h. Substituting this expression for F_g into the last equation shows that g is given by

$$g = \frac{GM_E}{r^{2}} = \frac{GM_E}{(R_E + h)^{2}} \tag{13.6}$$

◀ **Variation of *g* with altitude**

Therefore, it follows that g *decreases* with *increasing altitude*. Values of g for the Earth at various altitudes are listed in Table 13.1. Because an object's weight is mg, we see that as $r \to \infty$, the weight of the object approaches zero.

Table 13.1 **Free-Fall Acceleration *g* at Various Altitudes Above the Earth's Surface**

Altitude h (km)	g (m/s²)
1 000	7.33
2 000	5.68
3 000	4.53
4 000	3.70
5 000	3.08
6 000	2.60
7 000	2.23
8 000	1.93
9 000	1.69
10 000	1.49
50 000	0.13
∞	0

Q uick Quiz 13.2 Superman stands on top of a very tall mountain and throws a baseball horizontally with a speed such that the baseball goes into a circular orbit around the Earth. While the baseball is in orbit, what is the magnitude of the acceleration of the ball? **(a)** It depends on how fast the baseball is thrown. **(b)** It is zero because the ball does not fall to the ground. **(c)** It is slightly less than 9.80 m/s². **(d)** It is equal to 9.80 m/s².

Example 13.2   The Density of the Earth

Using the known radius of the Earth and that $g = 9.80$ m/s^2 at the Earth's surface, find the average density of the Earth.

SOLUTION

Conceptualize Assume the Earth is a perfect sphere. The density of material in the Earth varies, but let's adopt a simplified model in which we assume the density to be uniform throughout the Earth. The resulting density is the average density of the Earth.

Categorize This example is a relatively simple substitution problem.

Using Equation 13.5, solve for the mass of the Earth:

$$M_E = \frac{gR_E{}^2}{G}$$

Substitute this mass and the volume of a sphere into the definition of density (Eq. 1.1):

$$\rho_E = \frac{M_E}{V_E} = \frac{gR_E{}^2/G}{\frac{4}{3}\pi R_E{}^3} = \frac{3}{4}\frac{g}{\pi G R_E}$$

$$= \frac{3}{4}\frac{9.80 \text{ m/s}^2}{\pi (6.674 \times 10^{-11} \text{ N} \cdot \text{m}^2/\text{kg}^2)(6.37 \times 10^6 \text{ m})} = \boxed{5.50 \times 10^3 \text{ kg/m}^3}$$

WHAT IF? What if you were told that a typical density of granite at the Earth's surface is 2.75×10^3 kg/m^3? What would you conclude about the density of the material in the Earth's interior?

Answer Because this value is about half the density we calculated as an average for the entire Earth, we would conclude that the inner core of the Earth has a density much higher than the average value. It is most amazing that the Cavendish experiment—which can be used to determine G and can be done today on a tabletop—combined with simple free-fall measurements of g provides information about the core of the Earth!

13.3 Analysis Model: Particle in a Field (Gravitational)

When Newton published his theory of universal gravitation, it was considered a success because it satisfactorily explained the motion of the planets. It represented strong evidence that the same laws that describe phenomena on the Earth can be used on large objects like planets and throughout the Universe. Since 1687, Newton's theory has been used to account for the motions of comets, the deflection of a Cavendish balance, the orbits of binary stars, and the rotation of galaxies. Nevertheless, both Newton's contemporaries and his successors found it difficult to accept the concept of a force that acts at a distance. They asked how it was possible for two objects such as the Sun and the Earth to interact when they were not in contact with each other. Newton himself could not answer that question.

An approach to describing interactions between objects that are not in contact came well after Newton's death. This approach enables us to look at the gravitational interaction in a different way, using the concept of a **gravitational field** that exists at every point in space. When a particle is placed at a point where the gravitational field exists, the particle experiences a gravitational force. In other words, we imagine that the field exerts a force on the particle rather than consider a direct interaction between two particles. The gravitational field $\vec{\mathbf{g}}$ is defined as

Gravitational field ▶

$$\vec{\mathbf{g}} \equiv \frac{\vec{\mathbf{F}}_g}{m_0} \tag{13.7}$$

That is, the gravitational field at a point in space equals the gravitational force $\vec{\mathbf{F}}_g$ experienced by a *test particle* placed at that point divided by the mass m_0 of the test particle. We call the object creating the field the *source particle*. (Although the Earth is not a particle, it is possible to show that we can model the Earth as a particle for the purpose of finding the gravitational field that it creates.) Notice that the presence of the test particle is not necessary for the field to exist: the source particle creates the gravitational field. We can detect the presence of the field and measure its strength by placing a test particle in the field and noting the force exerted on it. In essence, we are describing the "effect" that any object (in this case, the Earth) has on the empty space around itself in terms of the force that *would* be present *if* a second object were somewhere in that space.[2]

The concept of a field is at the heart of the **particle in a field** analysis model. In the general version of this model, a particle resides in an area of space in which a field exists. Because of the existence of the field and a property of the particle, the particle experiences a force. In the gravitational version of the particle in a field model discussed here, the type of field is gravitational, and the property of the particle that results in the force is the particle's mass m. The mathematical representation of the gravitational version of the particle in a field model is Equation 5.5:

$$\vec{\mathbf{F}}_g = m\vec{\mathbf{g}} \tag{5.5}$$

In future chapters, we will see two other versions of the particle in a field model. In the electric version, the property of a particle that results in a force is *electric charge:* when a charged particle is placed in an *electric field*, it experiences a force. The magnitude of the force is the product of the electric charge and the field, in analogy with the gravitational force in Equation 5.5. In the magnetic version of the particle in a field model, a charged particle is placed in a *magnetic field*. One other property of this particle is required for the particle to experience a force: the particle must have a *velocity* at some nonzero angle to the magnetic field. The electric and magnetic versions of the particle in a field model are critical to the understanding of the principles of *electromagnetism*, which we will study in Chapters 23–34.

Because the gravitational force acting on the object has a magnitude $GM_E m/r^2$ (see Eq. 13.4), the gravitational field $\vec{\mathbf{g}}$ at a distance r from the center of the Earth is

$$\vec{\mathbf{g}} = \frac{\vec{\mathbf{F}}_g}{m} = -\frac{GM_E}{r^2}\hat{\mathbf{r}} \tag{13.8}$$

where $\hat{\mathbf{r}}$ is a unit vector pointing radially outward from the Earth and the negative sign indicates that the field points toward the center of the Earth as illustrated in Figure 13.4a. The field vectors at different points surrounding the Earth vary in both direction and magnitude. In a small region near the Earth's surface, the downward field $\vec{\mathbf{g}}$ is approximately constant and uniform as indicated in Figure 13.4b. Equation 13.8 is valid at all points *outside* the Earth's surface, assuming the Earth is spherical. At the Earth's surface, where $r = R_E$, $\vec{\mathbf{g}}$ has a magnitude of 9.80 N/kg. (The unit N/kg is the same as m/s[2].)

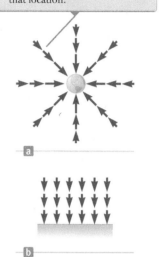

The field vectors point in the direction of the acceleration a particle would experience if it were placed in the field. The magnitude of the field vector at any location is the magnitude of the free-fall acceleration at that location.

Figure 13.4 (a) The gravitational field vectors in the vicinity of a uniform spherical mass such as the Earth vary in both direction and magnitude. (b) The gravitational field vectors in a small region near the Earth's surface are uniform in both direction and magnitude.

Analysis Model Particle in a Field (Gravitational)

Imagine an object with mass that we call a *source particle*. The source particle establishes a **gravitational field** $\vec{\mathbf{g}}$ throughout space. The gravitational field is evaluated by measuring the force on a test particle of mass m_0 and then using Equation 13.7. Now imagine a particle of mass m is placed in that field. The particle interacts with the gravitational field so that it experiences a gravitational force given by

$$\vec{\mathbf{F}}_g = m\vec{\mathbf{g}} \tag{5.5}$$

continued

[2]We shall return to this idea of mass affecting the space around it when we discuss Einstein's theory of gravitation in Chapter 39.

Analysis Model   Particle in a Field (Gravitational) *(continued)*

Examples:

- an object of mass m near the surface of the Earth has a *weight*, which is the result of the gravitational field established in space by the Earth
- a planet in the solar system is in orbit around the Sun, due to the gravitational force on the planet exerted by the gravitational field established by the Sun
- an object near a black hole is drawn into the black hole, never to escape, due to the tremendous gravitational field established by the black hole (Section 13.6)
- in the general theory of relativity, the gravitational field of a massive object is imagined to be described by a *curvature of space–time* (Chapter 39)
- the gravitational field of a massive object is imagined to be mediated by particles called *gravitons*, which have never been detected (Chapter 46)

Example 13.3   The Weight of the Space Station   **AM**

The International Space Station operates at an altitude of 350 km. Plans for the final construction show that material of weight 4.22×10^6 N, measured at the Earth's surface, will have been lifted off the surface by various spacecraft during the construction process. What is the weight of the space station when in orbit?

SOLUTION

Conceptualize The mass of the space station is fixed; it is independent of its location. Based on the discussions in this section and Section 13.2, we realize that the value of g will be reduced at the height of the space station's orbit. Therefore, the weight of the Space Station will be smaller than that at the surface of the Earth.

Categorize We model the Space Station as a *particle in a gravitational field*.

Analyze From the particle in a field model, find the mass of the space station from its weight at the surface of the Earth:

$$m = \frac{F_g}{g} = \frac{4.22 \times 10^6 \text{ N}}{9.80 \text{ m/s}^2} = 4.31 \times 10^5 \text{ kg}$$

Use Equation 13.6 with $h = 350$ km to find the magnitude of the gravitational field at the orbital location:

$$g = \frac{GM_E}{(R_E + h)^2}$$
$$= \frac{(6.674 \times 10^{-11} \text{ N} \cdot \text{m}^2/\text{kg}^2)(5.97 \times 10^{24} \text{ kg})}{(6.37 \times 10^6 \text{ m} + 0.350 \times 10^6 \text{ m})^2} = 8.82 \text{ m/s}^2$$

Use the particle in a field model again to find the space station's weight in orbit:

$$F_g = mg = (4.31 \times 10^5 \text{ kg})(8.82 \text{ m/s}^2) = \boxed{3.80 \times 10^6 \text{ N}}$$

Finalize Notice that the weight of the Space Station is less when it is in orbit, as we expected. It has about 10% less weight than it has when on the Earth's surface, representing a 10% decrease in the magnitude of the gravitational field.

13.4 Kepler's Laws and the Motion of Planets

Humans have observed the movements of the planets, stars, and other celestial objects for thousands of years. In early history, these observations led scientists to regard the Earth as the center of the Universe. This *geocentric model* was elaborated and formalized by the Greek astronomer Claudius Ptolemy (c. 100–c. 170) in the second century and was accepted for the next 1400 years. In 1543, Polish astronomer Nicolaus Copernicus (1473–1543) suggested that the Earth and the other planets revolved in circular orbits around the Sun (the *heliocentric model*).

Danish astronomer Tycho Brahe (1546–1601) wanted to determine how the heavens were constructed and pursued a project to determine the positions of both stars and planets. Those observations of the planets and 777 stars visible to the naked eye were carried out with only a large sextant and a compass. (The telescope had not yet been invented.)

German astronomer Johannes Kepler was Brahe's assistant for a short while before Brahe's death, whereupon he acquired his mentor's astronomical data and spent 16 years trying to deduce a mathematical model for the motion of the planets. Such data are difficult to sort out because the moving planets are observed from a moving Earth. After many laborious calculations, Kepler found that Brahe's data on the revolution of Mars around the Sun led to a successful model.

Kepler's complete analysis of planetary motion is summarized in three statements known as **Kepler's laws:**

◀ **Kepler's laws**

1. All planets move in elliptical orbits with the Sun at one focus.
2. The radius vector drawn from the Sun to a planet sweeps out equal areas in equal time intervals.
3. The square of the orbital period of any planet is proportional to the cube of the semimajor axis of the elliptical orbit.

Kepler's First Law

The geocentric and original heliocentric models of the solar system both suggested circular orbits for heavenly bodies. Kepler's first law indicates that the circular orbit is a very special case and elliptical orbits are the general situation. This notion was difficult for scientists of the time to accept because they believed that perfect circular orbits of the planets reflected the perfection of heaven.

Figure 13.5 shows the geometry of an ellipse, which serves as our model for the elliptical orbit of a planet. An ellipse is mathematically defined by choosing two points F_1 and F_2, each of which is a called a **focus,** and then drawing a curve through points for which the sum of the distances r_1 and r_2 from F_1 and F_2, respectively, is a constant. The longest distance through the center between points on the ellipse (and passing through each focus) is called the **major axis,** and this distance is $2a$. In Figure 13.5, the major axis is drawn along the x direction. The distance a is called the **semimajor axis.** Similarly, the shortest distance through the center between points on the ellipse is called the **minor axis** of length $2b$, where the distance b is the **semiminor axis.** Either focus of the ellipse is located at a distance c from the center of the ellipse, where $a^2 = b^2 + c^2$. In the elliptical orbit of a planet around the Sun, the Sun is at one focus of the ellipse. There is nothing at the other focus.

The **eccentricity** of an ellipse is defined as $e = c/a$, and it describes the general shape of the ellipse. For a circle, $c = 0$, and the eccentricity is therefore zero. The smaller b is compared with a, the shorter the ellipse is along the y direction compared with its extent in the x direction in Figure 13.5. As b decreases, c increases and the eccentricity e increases. Therefore, higher values of eccentricity correspond to longer and thinner ellipses. The range of values of the eccentricity for an ellipse is $0 < e < 1$.

Eccentricities for planetary orbits vary widely in the solar system. The eccentricity of the Earth's orbit is 0.017, which makes it nearly circular. On the other hand, the eccentricity of Mercury's orbit is 0.21, the highest of the eight planets. Figure 13.6a on page 396 shows an ellipse with an eccentricity equal to that of Mercury's orbit. Notice that even this highest-eccentricity orbit is difficult to distinguish from a circle, which is one reason Kepler's first law is an admirable accomplishment. The eccentricity of the orbit of Comet Halley is 0.97, describing an orbit whose major axis is much longer than its minor axis, as shown in Figure 13.6b. As a result, Comet Halley spends much of its 76-year period far from the Sun and invisible from the Earth. It is only visible to the naked eye during a small part of its orbit when it is near the Sun.

Now imagine a planet in an elliptical orbit such as that shown in Figure 13.5, with the Sun at focus F_2. When the planet is at the far left in the diagram, the distance

Johannes Kepler
German astronomer (1571–1630)
Kepler is best known for developing the laws of planetary motion based on the careful observations of Tycho Brahe.

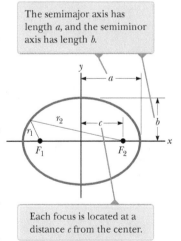

The semimajor axis has length a, and the semiminor axis has length b.

Each focus is located at a distance c from the center.

Figure 13.5 Plot of an ellipse.

Pitfall Prevention 13.2
Where Is the Sun? The Sun is located at one focus of the elliptical orbit of a planet. It is *not* located at the center of the ellipse.

Figure 13.6 (a) The shape of the orbit of Mercury, which has the highest eccentricity ($e = 0.21$) among the eight planets in the solar system. (b) The shape of the orbit of Comet Halley. The shape of the orbit is correct; the comet and the Sun are shown larger than in reality for clarity.

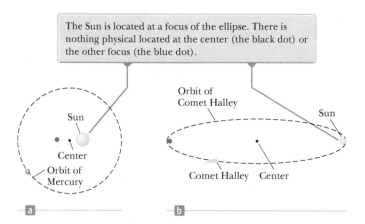

The Sun is located at a focus of the ellipse. There is nothing physical located at the center (the black dot) or the other focus (the blue dot).

Orbit of Comet Halley

Sun

Sun

Center

Orbit of Mercury

Comet Halley   Center

a

b

between the planet and the Sun is $a + c$. At this point, called the *aphelion*, the planet is at its maximum distance from the Sun. (For an object in orbit around the Earth, this point is called the *apogee*.) Conversely, when the planet is at the right end of the ellipse, the distance between the planet and the Sun is $a - c$. At this point, called the *perihelion* (for an Earth orbit, the *perigee*), the planet is at its minimum distance from the Sun.

Kepler's first law is a direct result of the inverse-square nature of the gravitational force. Circular and elliptical orbits correspond to objects that are *bound* to the gravitational force center. These objects include planets, asteroids, and comets that move repeatedly around the Sun as well as moons orbiting a planet. There are also *unbound* objects, such as a meteoroid from deep space that might pass by the Sun once and then never return. The gravitational force between the Sun and these objects also varies as the inverse square of the separation distance, and the allowed paths for these objects include parabolas ($e = 1$) and hyperbolas ($e > 1$).

Kepler's Second Law

Kepler's second law can be shown to be a result of the isolated system model for angular momentum. Consider a planet of mass M_p moving about the Sun in an elliptical orbit (Fig. 13.7a). Let's consider the planet as a system. We model the Sun to be so much more massive than the planet that the Sun does not move. The gravitational force exerted by the Sun on the planet is a central force, always along the radius vector, directed toward the Sun (Fig. 13.7a). The torque on the planet due to this central force about an axis through the Sun is zero because $\vec{\mathbf{F}}_g$ is parallel to $\vec{\mathbf{r}}$.

Therefore, because the external torque on the planet is zero, it is modeled as an isolated system for angular momentum, and the angular momentum $\vec{\mathbf{L}}$ of the planet is a constant of the motion:

$$\Delta \vec{\mathbf{L}} = 0 \quad \rightarrow \quad \vec{\mathbf{L}} = \text{constant}$$

Evaluating $\vec{\mathbf{L}}$ for the planet,

$$\vec{\mathbf{L}} = \vec{\mathbf{r}} \times \vec{\mathbf{p}} = M_p \vec{\mathbf{r}} \times \vec{\mathbf{v}} \quad \rightarrow \quad L = M_p |\vec{\mathbf{r}} \times \vec{\mathbf{v}}| \tag{13.9}$$

We can relate this result to the following geometric consideration. In a time interval dt, the radius vector $\vec{\mathbf{r}}$ in Figure 13.7b sweeps out the area dA, which equals half the area $|\vec{\mathbf{r}} \times d\vec{\mathbf{r}}|$ of the parallelogram formed by the vectors $\vec{\mathbf{r}}$ and $d\vec{\mathbf{r}}$. Because the displacement of the planet in the time interval dt is given by $d\vec{\mathbf{r}} = \vec{\mathbf{v}}\, dt$,

$$dA = \tfrac{1}{2}|\vec{\mathbf{r}} \times d\vec{\mathbf{r}}| = \tfrac{1}{2}|\vec{\mathbf{r}} \times \vec{\mathbf{v}}\, dt| = \tfrac{1}{2}|\vec{\mathbf{r}} \times \vec{\mathbf{v}}|\, dt$$

Substitute for the absolute value of the cross product from Equation 13.9:

$$dA = \tfrac{1}{2}\left(\frac{L}{M_p}\right) dt$$

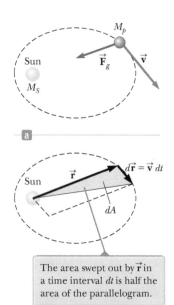

M_p

Sun

M_S

$\vec{\mathbf{F}}_g$

$\vec{\mathbf{v}}$

a

Sun

$\vec{\mathbf{r}}$

$d\vec{\mathbf{r}} = \vec{\mathbf{v}}\, dt$

dA

The area swept out by $\vec{\mathbf{r}}$ in a time interval dt is half the area of the parallelogram.

b

Figure 13.7 (a) The gravitational force acting on a planet is directed toward the Sun. (b) During a time interval dt, a parallelogram is formed by the vectors $\vec{\mathbf{r}}$ and $d\vec{\mathbf{r}} = \vec{\mathbf{v}}\, dt$.

Divide both sides by dt to obtain

$$\frac{dA}{dt} = \frac{L}{2M_p} \qquad \text{(13.10)}$$

where L and M_p are both constants. This result shows that that the derivative dA/dt is constant—the radius vector from the Sun to any planet sweeps out equal areas in equal time intervals as stated in Kepler's second law.

This conclusion is a result of the gravitational force being a central force, which in turn implies that angular momentum of the planet is constant. Therefore, the law applies to *any* situation that involves a central force, whether inverse square or not.

Kepler's Third Law

Kepler's third law can be predicted from the inverse-square law for circular orbits and our analysis models. Consider a planet of mass M_p that is assumed to be moving about the Sun (mass M_S) in a circular orbit as in Figure 13.8. Because the gravitational force provides the centripetal acceleration of the planet as it moves in a circle, we model the planet as a particle under a net force and as a particle in uniform circular motion and incorporate Newton's law of universal gravitation,

$$F_g = M_p a \quad \rightarrow \quad \frac{GM_S M_p}{r^2} = M_p \left(\frac{v^2}{r} \right)$$

The orbital speed of the planet is $2\pi r/T$, where T is the period; therefore, the preceding expression becomes

$$\frac{GM_S}{r^2} = \frac{(2\pi r/T)^2}{r}$$

$$T^2 = \left(\frac{4\pi^2}{GM_S} \right) r^3 = K_S r^3$$

where K_S is a constant given by

$$K_S = \frac{4\pi^2}{GM_S} = 2.97 \times 10^{-19} \text{ s}^2/\text{m}^3$$

This equation is also valid for elliptical orbits if we replace r with the length a of the semimajor axis (Fig. 13.5):

$$T^2 = \left(\frac{4\pi^2}{GM_S} \right) a^3 = K_S a^3 \qquad \text{(13.11)}$$ ◀ **Kepler's third law**

Equation 13.11 is Kepler's third law: the square of the period is proportional to the cube of the semimajor axis. Because the semimajor axis of a circular orbit is its radius, this equation is valid for both circular and elliptical orbits. Notice that the constant of proportionality K_S is independent of the mass of the planet.[3] Equation 13.11 is therefore valid for *any* planet. If we were to consider the orbit of a satellite such as the Moon about the Earth, the constant would have a different value, with the Sun's mass replaced by the Earth's mass; that is, $K_E = 4\pi^2/GM_E$.

Table 13.2 on page 398 is a collection of useful data for planets and other objects in the solar system. The far-right column verifies that the ratio T^2/r^3 is constant for all objects orbiting the Sun. The small variations in the values in this column are the result of uncertainties in the data measured for the periods and semimajor axes of the objects.

Recent astronomical work has revealed the existence of a large number of solar system objects beyond the orbit of Neptune. In general, these objects lie in the *Kuiper belt*, a region that extends from about 30 AU (the orbital radius of Neptune) to 50 AU. (An AU is an *astronomical unit*, equal to the radius of the Earth's orbit.) Current

Figure 13.8 A planet of mass M_p moving in a circular orbit around the Sun. The orbits of all planets except Mercury are nearly circular.

---

[3]Equation 13.11 is indeed a proportion because the ratio of the two quantities T^2 and a^3 is a constant. The variables in a proportion are not required to be limited to the first power only.

Table 13.2   Useful Planetary Data

Body	Mass (kg)	Mean Radius (m)	Period of Revolution (s)	Mean Distance from the Sun (m)	$\dfrac{T^2}{r^3}$ (s^2/m^3)
Mercury	3.30×10^{23}	2.44×10^{6}	7.60×10^{6}	5.79×10^{10}	2.98×10^{-19}
Venus	4.87×10^{24}	6.05×10^{6}	1.94×10^{7}	1.08×10^{11}	2.99×10^{-19}
Earth	5.97×10^{24}	6.37×10^{6}	3.156×10^{7}	1.496×10^{11}	2.97×10^{-19}
Mars	6.42×10^{23}	3.39×10^{6}	5.94×10^{7}	2.28×10^{11}	2.98×10^{-19}
Jupiter	1.90×10^{27}	6.99×10^{7}	3.74×10^{8}	7.78×10^{11}	2.97×10^{-19}
Saturn	5.68×10^{26}	5.82×10^{7}	9.29×10^{8}	1.43×10^{12}	2.95×10^{-19}
Uranus	8.68×10^{25}	2.54×10^{7}	2.65×10^{9}	2.87×10^{12}	2.97×10^{-19}
Neptune	1.02×10^{26}	2.46×10^{7}	5.18×10^{9}	4.50×10^{12}	2.94×10^{-19}
Pluto[a]	1.25×10^{22}	1.20×10^{6}	7.82×10^{9}	5.91×10^{12}	2.96×10^{-19}
Moon	7.35×10^{22}	1.74×10^{6}	—	—	—
Sun	1.989×10^{30}	6.96×10^{8}	—	—	—

[a]In August 2006, the International Astronomical Union adopted a definition of a planet that separates Pluto from the other eight planets. Pluto is now defined as a "dwarf planet" like the asteroid Ceres.

estimates identify at least 70 000 objects in this region with diameters larger than 100 km. The first Kuiper belt object (KBO) is Pluto, discovered in 1930 and formerly classified as a planet. Starting in 1992, many more have been detected. Several have diameters in the 1 000-km range, such as Varuna (discovered in 2000), Ixion (2001), Quaoar (2002), Sedna (2003), Haumea (2004), Orcus (2004), and Makemake (2005). One KBO, Eris, discovered in 2005, is believed to be significantly larger than Pluto. Other KBOs do not yet have names, but are currently indicated by their year of discovery and a code, such as 2009 YE7 and 2010 EK139.

A subset of about 1 400 KBOs are called "Plutinos" because, like Pluto, they exhibit a resonance phenomenon, orbiting the Sun two times in the same time interval as Neptune revolves three times. The contemporary application of Kepler's laws and such exotic proposals as planetary angular momentum exchange and migrating planets suggest the excitement of this active area of current research.

Ⓠuick Quiz 13.3 An asteroid is in a highly eccentric elliptical orbit around the Sun. The period of the asteroid's orbit is 90 days. Which of the following statements is true about the possibility of a collision between this asteroid and the Earth? **(a)** There is no possible danger of a collision. **(b)** There is a possibility of a collision. **(c)** There is not enough information to determine whether there is danger of a collision.

Example 13.4   The Mass of the Sun

Calculate the mass of the Sun, noting that the period of the Earth's orbit around the Sun is 3.156×10^7 s and its distance from the Sun is 1.496×10^{11} m.

SOLUTION

Conceptualize Based on the mathematical representation of Kepler's third law expressed in Equation 13.11, we realize that the mass of the central object in a gravitational system is related to the orbital size and period of objects in orbit around the central object.

Categorize This example is a relatively simple substitution problem.

Solve Equation 13.11 for the mass of the Sun:

$$M_S = \frac{4\pi^2 r^3}{GT^2}$$

Substitute the known values:

$$M_S = \frac{4\pi^2 (1.496 \times 10^{11}\ \text{m})^3}{(6.674 \times 10^{-11}\ \text{N} \cdot \text{m}^2/\text{kg}^2)(3.156 \times 10^7\ \text{s})^2} = \boxed{1.99 \times 10^{30}\ \text{kg}}$$

▶ **13.4** c o n t i n u e d

In Example 13.2, an understanding of gravitational forces enabled us to find out something about the density of the Earth's core, and now we have used this understanding to determine the mass of the Sun!

Example 13.5   A Geosynchronous Satellite   **AM**

Consider a satellite of mass m moving in a circular orbit around the Earth at a constant speed v and at an altitude h above the Earth's surface as illustrated in Figure 13.9.

(A) Determine the speed of satellite in terms of G, h, R_E (the radius of the Earth), and M_E (the mass of the Earth).

SOLUTION

Conceptualize Imagine the satellite moving around the Earth in a circular orbit under the influence of the gravitational force. This motion is similar to that of the International Space Station, the Hubble Space Telescope, and other objects in orbit around the Earth.

Categorize The satellite moves in a circular orbit at a constant speed. Therefore, we categorize the satellite as a *particle in uniform circular motion* as well as a *particle under a net force*.

Figure 13.9 (Example 13.5) A satellite of mass m moving around the Earth in a circular orbit of radius r with constant speed v. The only force acting on the satellite is the gravitational force $\vec{F}_g$. (Not drawn to scale.)

Analyze The only external force acting on the satellite is the gravitational force from the Earth, which acts toward the center of the Earth and keeps the satellite in its circular orbit.

Apply the particle under a net force and particle in uniform circular motion models to the satellite:

$$F_g = ma \quad \rightarrow \quad G\frac{M_E m}{r^2} = m\left(\frac{v^2}{r}\right)$$

Solve for v, noting that the distance r from the center of the Earth to the satellite is $r = R_E + h$:

$$(1) \quad v = \sqrt{\frac{GM_E}{r}} = \boxed{\sqrt{\frac{GM_E}{R_E + h}}}$$

(B) If the satellite is to be *geosynchronous* (that is, appearing to remain over a fixed position on the Earth), how fast is it moving through space?

SOLUTION

To appear to remain over a fixed position on the Earth, the period of the satellite must be 24 h = 86 400 s and the satellite must be in orbit directly over the equator.

Solve Kepler's third law (Equation 13.11, with $a = r$ and $M_S \rightarrow M_E$) for r:

$$r = \left(\frac{GM_E T^2}{4\pi^2}\right)^{1/3}$$

Substitute numerical values:

$$r = \left[\frac{(6.674 \times 10^{-11}\,\text{N} \cdot \text{m}^2/\text{kg}^2)(5.97 \times 10^{24}\,\text{kg})(86\,400\,\text{s})^2}{4\pi^2}\right]^{1/3}$$

$$= 4.22 \times 10^7\,\text{m}$$

Use Equation (1) to find the speed of the satellite:

$$v = \sqrt{\frac{(6.674 \times 10^{-11}\,\text{N} \cdot \text{m}^2/\text{kg}^2)(5.97 \times 10^{24}\,\text{kg})}{4.22 \times 10^7\,\text{m}}}$$

$$= \boxed{3.07 \times 10^3\,\text{m/s}}$$

Finalize The value of r calculated here translates to a height of the satellite above the surface of the Earth of almost 36 000 km. Therefore, geosynchronous satellites have the advantage of allowing an earthbound antenna to be aimed

continued

▶ **13.5** continued

in a fixed direction, but there is a disadvantage in that the signals between the Earth and the satellite must travel a long distance. It is difficult to use geosynchronous satellites for optical observation of the Earth's surface because of their high altitude.

WHAT IF? What if the satellite motion in part (A) were taking place at height h above the surface of another planet more massive than the Earth but of the same radius? Would the satellite be moving at a higher speed or a lower speed than it does around the Earth?

Answer If the planet exerts a larger gravitational force on the satellite due to its larger mass, the satellite must move with a higher speed to avoid moving toward the surface. This conclusion is consistent with the predictions of Equation (1), which shows that because the speed v is proportional to the square root of the mass of the planet, the speed increases as the mass of the planet increases.

13.5 Gravitational Potential Energy

In Chapter 8, we introduced the concept of gravitational potential energy, which is the energy associated with the configuration of a system of objects interacting via the gravitational force. We emphasized that the gravitational potential energy function $U = mgy$ for a particle–Earth system is valid only when the particle of mass m is near the Earth's surface, where the gravitational force is independent of y. This expression for the gravitational potential energy is also restricted to situations where a very massive object (such as the Earth) establishes a gravitational field of magnitude g and a particle of much smaller mass m resides in that field. Because the gravitational force between two particles varies as $1/r^2$, we expect that a more general potential energy function—one that is valid without the restrictions mentioned above—will be different from $U = mgy$.

Recall from Equation 7.27 that the change in the potential energy of a system associated with a given displacement of a member of the system is defined as the negative of the internal work done by the force on that member during the displacement:

$$\Delta U = U_f - U_i = -\int_{r_i}^{r_f} F(r)\, dr \tag{13.12}$$

We can use this result to evaluate the general gravitational potential energy function. Consider a particle of mass m moving between two points Ⓐ and Ⓑ above the Earth's surface (Fig. 13.10). The particle is subject to the gravitational force given by Equation 13.1. We can express this force as

$$F(r) = -\frac{GM_E m}{r^2}$$

where the negative sign indicates that the force is attractive. Substituting this expression for $F(r)$ into Equation 13.12, we can compute the change in the gravitational potential energy function for the particle–Earth system as the separation distance r changes:

$$U_f - U_i = GM_E m \int_{r_i}^{r_f} \frac{dr}{r^2} = GM_E m \left[-\frac{1}{r} \right]_{r_i}^{r_f}$$

$$U_f - U_i = -GM_E m \left(\frac{1}{r_f} - \frac{1}{r_i} \right) \tag{13.13}$$

As always, the choice of a reference configuration for the potential energy is completely arbitrary. It is customary to choose the reference configuration for zero

Figure 13.10 As a particle of mass m moves from Ⓐ to Ⓑ above the Earth's surface, the gravitational potential energy of the particle–Earth system changes according to Equation 13.12.

potential energy to be the same as that for which the force is zero. Taking $U_i = 0$ at $r_i = \infty$, we obtain the important result

$$U(r) = -\frac{GM_E m}{r} \qquad (13.14)$$

This expression applies when the particle is separated from the center of the Earth by a distance r, provided that $r \geq R_E$. The result is not valid for particles inside the Earth, where $r < R_E$. Because of our choice of U_i, the function U is always negative (Fig. 13.11).

Although Equation 13.14 was derived for the particle–Earth system, a similar form of the equation can be applied to any two particles. That is, the gravitational potential energy associated with any pair of particles of masses m_1 and m_2 separated by a distance r is

$$U = -\frac{Gm_1 m_2}{r} \qquad (13.15)$$

This expression shows that the gravitational potential energy for any pair of particles varies as $1/r$, whereas the force between them varies as $1/r^2$. Furthermore, the potential energy is negative because the force is attractive and we have chosen the potential energy as zero when the particle separation is infinite. Because the force between the particles is attractive, an external agent must do positive work to increase the separation between the particles. The work done by the external agent produces an increase in the potential energy as the two particles are separated. That is, U becomes less negative as r increases.

When two particles are at rest and separated by a distance r, an external agent has to supply an energy at least equal to $+Gm_1 m_2/r$ to separate the particles to an infinite distance. It is therefore convenient to think of the absolute value of the potential energy as the *binding energy* of the system. If the external agent supplies an energy greater than the binding energy, the excess energy of the system is in the form of kinetic energy of the particles when the particles are at an infinite separation.

We can extend this concept to three or more particles. In this case, the total potential energy of the system is the sum over all pairs of particles. Each pair contributes a term of the form given by Equation 13.15. For example, if the system contains three particles as in Figure 13.12,

$$U_{\text{total}} = U_{12} + U_{13} + U_{23} = -G\left(\frac{m_1 m_2}{r_{12}} + \frac{m_1 m_3}{r_{13}} + \frac{m_2 m_3}{r_{23}}\right)$$

The absolute value of U_{total} represents the work needed to separate the particles by an infinite distance.

◀ **Gravitational potential energy of the Earth–particle system**

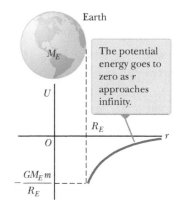

Figure 13.11 Graph of the gravitational potential energy U versus r for the system of an object above the Earth's surface.

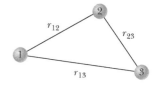

Figure 13.12 Three interacting particles.

| Example 13.6 | **The Change in Potential Energy** |

A particle of mass m is displaced through a small vertical distance Δy near the Earth's surface. Show that in this situation the general expression for the change in gravitational potential energy given by Equation 13.13 reduces to the familiar relationship $\Delta U = mg\,\Delta y$.

SOLUTION

Conceptualize Compare the two different situations for which we have developed expressions for gravitational potential energy: (1) a planet and an object that are far apart for which the energy expression is Equation 13.14 and (2) a small object at the surface of a planet for which the energy expression is Equation 7.19. We wish to show that these two expressions are equivalent.

continued

▶ **13.6** continued

Categorize This example is a substitution problem.

Combine the fractions in Equation 13.13:

$$(1) \quad \Delta U = -GM_E m \left(\frac{1}{r_f} - \frac{1}{r_i} \right) = GM_E m \left(\frac{r_f - r_i}{r_i r_f} \right)$$

Evaluate $r_f - r_i$ and $r_i r_f$ if both the initial and final positions of the particle are close to the Earth's surface:

$$r_f - r_i = \Delta y \quad r_i r_f \approx R_E^2$$

Substitute these expressions into Equation (1):

$$\Delta U \approx \frac{GM_E m}{R_E^2} \Delta y = mg \, \Delta y$$

where $g = GM_E/R_E^2$ (Eq. 13.5).

WHAT IF? Suppose you are performing upper-atmosphere studies and are asked by your supervisor to find the height in the Earth's atmosphere at which the "surface equation" $\Delta U = mg \, \Delta y$ gives a 1.0% error in the change in the potential energy. What is this height?

Answer Because the surface equation assumes a constant value for g, it will give a ΔU value that is larger than the value given by the general equation, Equation 13.13.

Set up a ratio reflecting a 1.0% error:

$$\frac{\Delta U_{\text{surface}}}{\Delta U_{\text{general}}} = 1.010$$

Substitute the expressions for each of these changes ΔU:

$$\frac{mg \, \Delta y}{GM_E m (\Delta y / r_i r_f)} = \frac{g r_i r_f}{GM_E} = 1.010$$

Substitute for r_i, r_f, and g from Equation 13.5:

$$\frac{(GM_E/R_E^2) R_E (R_E + \Delta y)}{GM_E} = \frac{R_E + \Delta y}{R_E} = 1 + \frac{\Delta y}{R_E} = 1.010$$

Solve for Δy:

$$\Delta y = 0.010 R_E = 0.010(6.37 \times 10^6 \text{ m}) = 6.37 \times 10^4 \text{ m} = 63.7 \text{ km}$$

13.6 Energy Considerations in Planetary and Satellite Motion

Given the general expression for gravitational potential energy developed in Section 13.5, we can now apply our energy analysis models to gravitational systems. Consider an object of mass m moving with a speed v in the vicinity of a massive object of mass M, where $M \gg m$. The system might be a planet moving around the Sun, a satellite in orbit around the Earth, or a comet making a one-time flyby of the Sun. If we assume the object of mass M is at rest in an inertial reference frame, the total mechanical energy E of the two-object system when the objects are separated by a distance r is the sum of the kinetic energy of the object of mass m and the potential energy of the system, given by Equation 13.15:

$$E = K + U$$

$$E = \tfrac{1}{2}mv^2 - \frac{GMm}{r} \tag{13.16}$$

If the system of objects of mass m and M is isolated, and there are no nonconservative forces acting within the system, the mechanical energy of the system given by Equation 13.16 is the total energy of the system and this energy is conserved:

$$\Delta E_{\text{system}} = 0 \quad \rightarrow \quad \Delta K + \Delta U_g = 0 \quad \rightarrow \quad E_i = E_f$$

Therefore, as the object of mass m moves from Ⓐ to Ⓑ in Figure 13.10, the total energy remains constant and Equation 13.16 gives

$$\tfrac{1}{2}mv_i^2 - \frac{GMm}{r_i} = \tfrac{1}{2}mv_f^2 - \frac{GMm}{r_f} \tag{13.17}$$

Combining this statement of energy conservation with our earlier discussion of conservation of angular momentum, we see that both the total energy and the total angular momentum of a gravitationally bound, two-object system are constants of the motion.

Equation 13.16 shows that E may be positive, negative, or zero, depending on the value of v. For a bound system such as the Earth–Sun system, however, E is necessarily *less than zero* because we have chosen the convention that $U \to 0$ as $r \to \infty$.

We can easily establish that $E < 0$ for the system consisting of an object of mass m moving in a circular orbit about an object of mass $M \gg m$ (Fig. 13.13). Modeling the object of mass m as a particle under a net force and a particle in uniform circular motion gives

$$F_g = ma \quad \rightarrow \quad \frac{GMm}{r^2} = \frac{mv^2}{r}$$

Multiplying both sides by r and dividing by 2 gives

$$\tfrac{1}{2}mv^2 = \frac{GMm}{2r} \tag{13.18}$$

Substituting this equation into Equation 13.16, we obtain

$$E = \frac{GMm}{2r} - \frac{GMm}{r}$$

$$\boxed{E = -\frac{GMm}{2r}} \quad \text{(circular orbits)} \tag{13.19}$$

◀ **Total energy for circular orbits of an object of mass m around an object of mass $M \gg m$**

This result shows that the total mechanical energy is negative in the case of circular orbits. Notice that the kinetic energy is positive and equal to half the absolute value of the potential energy. The absolute value of E is also equal to the binding energy of the system because this amount of energy must be provided to the system to move the two objects infinitely far apart.

The total mechanical energy is also negative in the case of elliptical orbits. The expression for E for elliptical orbits is the same as Equation 13.19 with r replaced by the semimajor axis length a:

$$\boxed{E = -\frac{GMm}{2a}} \quad \text{(elliptical orbits)} \tag{13.20}$$

◀ **Total energy for elliptical orbits of an object of mass m around an object of mass $M \gg m$**

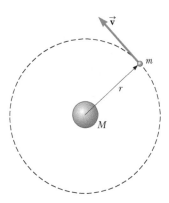

Figure 13.13 An object of mass m moving in a circular orbit about a much larger object of mass M.

Quick Quiz 13.4 A comet moves in an elliptical orbit around the Sun. Which point in its orbit (perihelion or aphelion) represents the highest value of (a) the speed of the comet, (b) the potential energy of the comet–Sun system, (c) the kinetic energy of the comet, and (d) the total energy of the comet–Sun system?

Example 13.7 | **Changing the Orbit of a Satellite**

A space transportation vehicle releases a 470-kg communications satellite while in an orbit 280 km above the surface of the Earth. A rocket engine on the satellite boosts it into a geosynchronous orbit. How much energy does the engine have to provide?

SOLUTION

Conceptualize Notice that the height of 280 km is much lower than that for a geosynchronous satellite, 36 000 km, as mentioned in Example 13.5. Therefore, energy must be expended to raise the satellite to this much higher position.

Categorize This example is a substitution problem.

Find the initial radius of the satellite's orbit when it is still in the vehicle's cargo bay:
$$r_i = R_E + 280 \text{ km} = 6.65 \times 10^6 \text{ m}$$

continued

▶ **13.7** continued

Use Equation 13.19 to find the difference in energies for the satellite–Earth system with the satellite at the initial and final radii:

$$\Delta E = E_f - E_i = -\frac{GM_E m}{2r_f} - \left(-\frac{GM_E m}{2r_i}\right) = -\frac{GM_E m}{2}\left(\frac{1}{r_f} - \frac{1}{r_i}\right)$$

Substitute numerical values, using $r_f = 4.22 \times 10^7$ m from Example 13.5:

$$\Delta E = -\frac{(6.674 \times 10^{-11}\ \text{N} \cdot \text{m}^2/\text{kg}^2)(5.97 \times 10^{24}\ \text{kg})(470\ \text{kg})}{2} \times$$
$$\left(\frac{1}{4.22 \times 10^7\ \text{m}} - \frac{1}{6.65 \times 10^6\ \text{m}}\right) = \boxed{1.19 \times 10^{10}\ \text{J}}$$

which is the energy equivalent of 89 gal of gasoline. NASA engineers must account for the changing mass of the spacecraft as it ejects burned fuel, something we have not done here. Would you expect the calculation that includes the effect of this changing mass to yield a greater or a lesser amount of energy required from the engine?

Escape Speed

Figure 13.14 An object of mass m projected upward from the Earth's surface with an initial speed v_i reaches a maximum altitude h.

Suppose an object of mass m is projected vertically upward from the Earth's surface with an initial speed v_i as illustrated in Figure 13.14. We can use energy considerations to find the value of the initial speed needed to allow the object to reach a certain distance away from the center of the Earth. Equation 13.16 gives the total energy of the system for any configuration. As the object is projected upward from the surface of the Earth, $v = v_i$ and $r = r_i = R_E$. When the object reaches its maximum altitude, $v = v_f = 0$ and $r = r_f = r_{\text{max}}$. Because the object–Earth system is isolated, we substitute these values into the isolated-system model expression given by Equation 13.17:

$$\tfrac{1}{2}mv_i^2 - \frac{GM_E m}{R_E} = -\frac{GM_E m}{r_{\text{max}}}$$

Solving for v_i^2 gives

$$v_i^2 = 2GM_E\left(\frac{1}{R_E} - \frac{1}{r_{\text{max}}}\right) \tag{13.21}$$

For a given maximum altitude $h = r_{\text{max}} - R_E$, we can use this equation to find the required initial speed.

We are now in a position to calculate the **escape speed**, which is the minimum speed the object must have at the Earth's surface to approach an infinite separation distance from the Earth. Traveling at this minimum speed, the object continues to move farther and farther away from the Earth as its speed asymptotically approaches zero. Letting $r_{\text{max}} \to \infty$ in Equation 13.21 and identifying v_i as v_{esc} gives

$$v_{\text{esc}} = \sqrt{\frac{2GM_E}{R_E}} \tag{13.22}$$

Pitfall Prevention 13.3
You Can't Really Escape Although Equation 13.22 provides the "escape speed" from the Earth, *complete* escape from the Earth's gravitational influence is impossible because the gravitational force is of infinite range.

This expression for v_{esc} is independent of the mass of the object. In other words, a spacecraft has the same escape speed as a molecule. Furthermore, the result is independent of the direction of the velocity and ignores air resistance.

If the object is given an initial speed equal to v_{esc}, the total energy of the system is equal to zero. Notice that when $r \to \infty$, the object's kinetic energy and the potential energy of the system are both zero. If v_i is greater than v_{esc}, however, the total energy of the system is greater than zero and the object has some residual kinetic energy as $r \to \infty$.

Example 13.8   **Escape Speed of a Rocket**

Calculate the escape speed from the Earth for a 5 000-kg spacecraft and determine the kinetic energy it must have at the Earth's surface to move infinitely far away from the Earth.

▶ **13.8** continued

SOLUTION

Conceptualize Imagine projecting the spacecraft from the Earth's surface so that it moves farther and farther away, traveling more and more slowly, with its speed approaching zero. Its speed will never reach zero, however, so the object will never turn around and come back.

Categorize This example is a substitution problem.

Use Equation 13.22 to find the escape speed:

$$v_{esc} = \sqrt{\frac{2GM_E}{R_E}} = \sqrt{\frac{2(6.674 \times 10^{-11}\,\text{N} \cdot \text{m}^2/\text{kg}^2)(5.97 \times 10^{24}\,\text{kg})}{6.37 \times 10^6\,\text{m}}}$$

$$= 1.12 \times 10^4\,\text{m/s}$$

Evaluate the kinetic energy of the spacecraft from Equation 7.16:

$$K = \tfrac{1}{2}mv_{esc}^2 = \tfrac{1}{2}(5.00 \times 10^3\,\text{kg})(1.12 \times 10^4\,\text{m/s})^2$$

$$= 3.13 \times 10^{11}\,\text{J}$$

The calculated escape speed corresponds to about 25 000 mi/h. The kinetic energy of the spacecraft is equivalent to the energy released by the combustion of about 2 300 gal of gasoline.

WHAT IF? What if you want to launch a 1 000-kg spacecraft at the escape speed? How much energy would that require?

Answer In Equation 13.22, the mass of the object moving with the escape speed does not appear. Therefore, the escape speed for the 1 000-kg spacecraft is the same as that for the 5 000-kg spacecraft. The only change in the kinetic energy is due to the mass, so the 1 000-kg spacecraft requires one-fifth of the energy of the 5 000-kg spacecraft:

$$K = \tfrac{1}{5}(3.13 \times 10^{11}\,\text{J}) = 6.25 \times 10^{10}\,\text{J}$$

Equations 13.21 and 13.22 can be applied to objects projected from any planet. That is, in general, the escape speed from the surface of any planet of mass M and radius R is

$$v_{esc} = \sqrt{\frac{2GM}{R}} \qquad (13.23)$$

◀ Escape speed from the surface of a planet of mass M and radius R

Escape speeds for the planets, the Moon, and the Sun are provided in Table 13.3. The values vary from 2.3 km/s for the Moon to about 618 km/s for the Sun. These results, together with some ideas from the kinetic theory of gases (see Chapter 21), explain why some planets have atmospheres and others do not. As we shall see later, at a given temperature the average kinetic energy of a gas molecule depends only on the mass of the molecule. Lighter molecules, such as hydrogen and helium, have a higher average speed than heavier molecules at the same temperature. When the average speed of the lighter molecules is not much less than the escape speed of a planet, a significant fraction of them have a chance to escape.

This mechanism also explains why the Earth does not retain hydrogen molecules and helium atoms in its atmosphere but does retain heavier molecules, such as oxygen and nitrogen. On the other hand, the very large escape speed for Jupiter enables that planet to retain hydrogen, the primary constituent of its atmosphere.

Black Holes

In Example 11.7, we briefly described a rare event called a supernova, the catastrophic explosion of a very massive star. The material that remains in the central core of such an object continues to collapse, and the core's ultimate fate depends on its mass. If the core has a mass less than 1.4 times the mass of our Sun, it gradually cools down and ends its life as a white dwarf star. If the core's mass is greater than this value, however, it may collapse further due to gravitational forces. What

Table 13.3 **Escape Speeds from the Surfaces of the Planets, Moon, and Sun**

Planet	v_{esc} (km/s)
Mercury	4.3
Venus	10.3
Earth	11.2
Mars	5.0
Jupiter	60
Saturn	36
Uranus	22
Neptune	24
Moon	2.3
Sun	618

remains is a neutron star, discussed in Example 11.7, in which the mass of a star is compressed to a radius of about 10 km. (On the Earth, a teaspoon of this material would weigh about 5 billion tons!)

An even more unusual star death may occur when the core has a mass greater than about three solar masses. The collapse may continue until the star becomes a very small object in space, commonly referred to as a **black hole.** In effect, black holes are remains of stars that have collapsed under their own gravitational force. If an object such as a spacecraft comes close to a black hole, the object experiences an extremely strong gravitational force and is trapped forever.

The escape speed for a black hole is very high because of the concentration of the star's mass into a sphere of very small radius (see Eq. 13.23). If the escape speed exceeds the speed of light c, radiation from the object (such as visible light) cannot escape and the object appears to be black (hence the origin of the terminology "black hole"). The critical radius R_S at which the escape speed is c is called the **Schwarzschild radius** (Fig. 13.15). The imaginary surface of a sphere of this radius surrounding the black hole is called the **event horizon,** which is the limit of how close you can approach the black hole and hope to escape.

There is evidence that supermassive black holes exist at the centers of galaxies, with masses very much larger than the Sun. (There is strong evidence of a super-massive black hole of mass 2–3 million solar masses at the center of our galaxy.)

Dark Matter

Equation (1) in Example 13.5 shows that the speed of an object in orbit around the Earth decreases as the object is moved farther away from the Earth:

$$v = \sqrt{\frac{GM_E}{r}} \tag{13.24}$$

Using data in Table 13.2 to find the speeds of planets in their orbits around the Sun, we find the same behavior for the planets. Figure 13.16 shows this behavior for the eight planets of our solar system. The theoretical prediction of the planet speed as a function of distance from the Sun is shown by the red-brown curve, using Equation 13.24 with the mass of the Earth replaced by the mass of the Sun. Data for the individual planets lie right on this curve. This behavior results from the vast majority of the mass of the solar system being concentrated in a small space, i.e., the Sun.

Extending this concept further, we might expect the same behavior in a galaxy. Much of the visible galactic mass, including that of a supermassive black hole, is near the central core of a galaxy. The opening photograph for this chapter shows the central core of the Whirlpool galaxy as a very bright area surrounded by the "arms" of the galaxy, which contain material in orbit around the central core. Based on this distribution of matter in the galaxy, the speed of an object in the outer part of the galaxy would be smaller than that for objects closer to the center, just like for the planets of the solar system.

That is *not* what is observed, however. Figure 13.17 shows the results of measurements of the speeds of objects in the Andromeda galaxy as a function of distance from the galaxy's center.[4] The red-brown curve shows the expected speeds for these objects if they were traveling in circular orbits around the mass concentrated in the central core. The data for the individual objects in the galaxy shown by the black dots are all well above the theoretical curve. These data, as well as an extensive amount of data taken over the past half century, show that for objects outside the central core of the galaxy, the curve of speed versus distance from the center of the galaxy is approximately flat rather than decreasing at larger distances. Therefore, these objects (including our own Solar System in the Milky Way) are rotating faster than can be accounted for by gravity due to the visible galaxy! This surprising

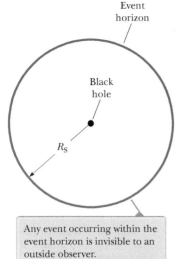

Event horizon

Black hole

R_S

Any event occurring within the event horizon is invisible to an outside observer.

Figure 13.15 A black hole. The distance R_S equals the Schwarzschild radius.

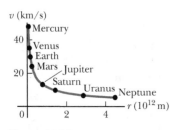

v (km/s)

Mercury

40 Venus
    Earth
    Mars Jupiter
20   Saturn
        Uranus Neptune
                  r (10^{12} m)
0   2   4

Figure 13.16 The orbital speed v as a function of distance r from the Sun for the eight planets of the solar system. The theoretical curve is in red-brown, and the data points for the planets are in black.

---

[4]V. C. Rubin and W. K. Ford, "Rotation of the Andromeda Nebula from a Spectroscopic Survey of Emission Regions," *Astrophysical Journal* **159:** 379–403 (1970).

result means that there must be additional mass in a more extended distribution, causing these objects to orbit so fast, and has led scientists to propose the existence of **dark matter.** This matter is proposed to exist in a large halo around each galaxy (with a radius up to 10 times as large as the visible galaxy's radius). Because it is not luminous (i.e., does not emit electromagnetic radiation) it must be either very cold or electrically neutral. Therefore, we cannot "see" dark matter, except through its gravitational effects.

The proposed existence of dark matter is also implied by earlier observations made on larger gravitationally bound structures known as galaxy clusters.[5] These observations show that the orbital speeds of galaxies in a cluster are, on average, too large to be explained by the luminous matter in the cluster alone. The speeds of the individual galaxies are so high, they suggest that there is 50 times as much dark matter in galaxy clusters as in the galaxies themselves!

Why doesn't dark matter affect the orbital speeds of planets like it does those of a galaxy? It seems that a solar system is too small a structure to contain enough dark matter to affect the behavior of orbital speeds. A galaxy or galaxy cluster, on the other hand, contains huge amounts of dark matter, resulting in the surprising behavior.

What, though, *is* dark matter? At this time, no one knows. One theory claims that dark matter is based on a particle called a weakly interacting massive particle, or WIMP. If this theory is correct, calculations show that about 200 WIMPs pass through a human body at any given time. The new Large Hadron Collider in Europe (see Chapter 46) is the first particle accelerator with enough energy to possibly generate and detect the existence of WIMPs, which has generated much current interest in dark matter. Keeping an eye on this research in the future should be exciting.

Figure 13.17 The orbital speed v of a galaxy object as a function of distance r from the center of the central core of the Andromeda galaxy. The theoretical curve is in red-brown, and the data points for the galaxy objects are in black. No data are provided on the left because the behavior inside the central core of the galaxy is more complicated.

Summary

Definitions

■ The **gravitational field** at a point in space is defined as the gravitational force $\vec{F}_g$ experienced by any test particle located at that point divided by the mass m_0 of the test particle:

$$\vec{g} \equiv \frac{\vec{F}_g}{m_0} \tag{13.7}$$

Concepts and Principles

■ **Newton's law of universal gravitation** states that the gravitational force of attraction between any two particles of masses m_1 and m_2 separated by a distance r has the magnitude

$$F_g = G\frac{m_1 m_2}{r^2} \tag{13.1}$$

where $G = 6.674 \times 10^{-11} \ \text{N} \cdot \text{m}^2/\text{kg}^2$ is the **universal gravitational constant.** This equation enables us to calculate the force of attraction between masses under many circumstances.

■ An object at a distance h above the Earth's surface experiences a gravitational force of magnitude mg, where g is the free-fall acceleration at that elevation:

$$g = \frac{GM_E}{r^2} = \frac{GM_E}{(R_E + h)^2} \tag{13.6}$$

In this expression, M_E is the mass of the Earth and R_E is its radius. Therefore, the weight of an object decreases as the object moves away from the Earth's surface.

[5]F. Zwicky, "On the Masses of Nebulae and of Clusters of Nebulae," *Astrophysical Journal* **86**: 217–246 (1937).

Kepler's laws of planetary motion state:

1. All planets move in elliptical orbits with the Sun at one focus.
2. The radius vector drawn from the Sun to a planet sweeps out equal areas in equal time intervals.
3. The square of the orbital period of any planet is proportional to the cube of the semimajor axis of the elliptical orbit.

Kepler's third law can be expressed as

$$T^2 = \left(\frac{4\pi^2}{GM_S}\right)a^3 \qquad \textbf{(13.11)}$$

where M_S is the mass of the Sun and a is the semimajor axis. For a circular orbit, a can be replaced in Equation 13.11 by the radius r. Most planets have nearly circular orbits around the Sun.

The **gravitational potential energy** associated with a system of two particles of mass m_1 and m_2 separated by a distance r is

$$U = -\frac{Gm_1m_2}{r} \qquad \textbf{(13.15)}$$

where U is taken to be zero as $r \rightarrow \infty$.

If an isolated system consists of an object of mass m moving with a speed v in the vicinity of a massive object of mass M, the total energy E of the system is the sum of the kinetic and potential energies:

$$E = \tfrac{1}{2}mv^2 - \frac{GMm}{r} \qquad \textbf{(13.16)}$$

The total energy of the system is a constant of the motion. If the object moves in an elliptical orbit of semimajor axis a around the massive object and $M \gg m$, the total energy of the system is

$$E = -\frac{GMm}{2a} \qquad \textbf{(13.20)}$$

For a circular orbit, this same equation applies with $a = r$.

The **escape speed** for an object projected from the surface of a planet of mass M and radius R is

$$v_{\text{esc}} = \sqrt{\frac{2GM}{R}} \qquad \textbf{(13.23)}$$

Analysis Model for Problem Solving

Particle in a Field (Gravitational) A source particle with some mass establishes a **gravitational field** $\vec{g}$ throughout space. When a particle of mass m is placed in that field, it experiences a gravitational force given by

$$\vec{F}_g = m\vec{g} \qquad \textbf{(5.5)}$$

Objective Questions   1. denotes answer available in *Student Solutions Manual/Study Guide*

1. A system consists of five particles. How many terms appear in the expression for the total gravitational potential energy of the system? (a) 4 (b) 5 (c) 10 (d) 20 (e) 25

2. Rank the following quantities of energy from largest to smallest. State if any are equal. (a) the absolute value of the average potential energy of the Sun–Earth system (b) the average kinetic energy of the Earth in its orbital motion relative to the Sun (c) the absolute value of the total energy of the Sun–Earth system

3. A satellite moves in a circular orbit at a constant speed around the Earth. Which of the following statements is true? (a) No force acts on the satellite. (b) The satellite moves at constant speed and hence doesn't accelerate. (c) The satellite has an acceleration directed away from the Earth. (d) The satellite has an acceleration directed toward the Earth. (e) Work is done on the satellite by the gravitational force.

4. Suppose the gravitational acceleration at the surface of a certain moon A of Jupiter is 2 m/s². Moon B has twice the mass and twice the radius of moon A. What is the gravitational acceleration at its surface? Neglect the gravitational acceleration due to Jupiter. (a) 8 m/s² (b) 4 m/s² (c) 2 m/s² (d) 1 m/s² (e) 0.5 m/s²

5. Imagine that nitrogen and other atmospheric gases were more soluble in water so that the atmosphere of the Earth is entirely absorbed by the oceans. Atmospheric pressure would then be zero, and outer space would start at the planet's surface. Would the Earth then have a gravitational field? (a) Yes, and at the surface it would be larger in magnitude than 9.8 N/kg. (b) Yes, and it would be essentially the same as the current value. (c) Yes, and it would be somewhat less than 9.8 N/kg. (d) Yes, and it would be much less than 9.8 N/kg. (e) No, it would not.

6. An object of mass m is located on the surface of a spherical planet of mass M and radius R. The escape speed from the planet does not depend on which of the following? (a) M (b) m (c) the density of the planet (d) R (e) the acceleration due to gravity on that planet

7. A satellite originally moves in a circular orbit of radius R around the Earth. Suppose it is moved into a circular orbit of radius $4R$. **(i)** What does the force exerted on the satellite then become? (a) eight times larger (b) four times larger (c) one-half as large (d) one-eighth as large (e) one-sixteenth as large **(ii)** What happens to the satellite's speed? Choose from the same possibilities (a) through (e). **(iii)** What happens to its period? Choose from the same possibilities (a) through (e).

8. The vernal equinox and the autumnal equinox are associated with two points 180° apart in the Earth's orbit. That is, the Earth is on precisely opposite sides of the Sun when it passes through these two points. From the vernal equinox, 185.4 days elapse before the autumnal equinox. Only 179.8 days elapse from the autumnal equinox until the next vernal equinox. Why is the interval from the March (vernal) to the September (autumnal) equinox (which contains the summer solstice) longer than the interval from the September to the March equinox rather than being equal to that interval? Choose one of the following reasons. (a) They are really the same, but the Earth spins faster during the "summer" interval, so the days are shorter. (b) Over the "summer" interval, the Earth moves slower because it is farther from the Sun. (c) Over the March-to-September interval, the Earth moves slower because it is closer to the Sun. (d) The Earth has less kinetic energy when it is warmer. (e) The Earth has less orbital angular momentum when it is warmer.

9. Rank the magnitudes of the following gravitational forces from largest to smallest. If two forces are equal, show their equality in your list. (a) the force exerted by a 2-kg object on a 3-kg object 1 m away (b) the force exerted by a 2-kg object on a 9-kg object 1 m away (c) the force exerted by a 2-kg object on a 9-kg object 2 m away (d) the force exerted by a 9-kg object on a 2-kg object 2 m away (e) the force exerted by a 4-kg object on another 4-kg object 2 m away

10. The gravitational force exerted on an astronaut on the Earth's surface is 650 N directed downward. When she is in the space station in orbit around the Earth, is the gravitational force on her (a) larger, (b) exactly the same, (c) smaller, (d) nearly but not exactly zero, or (e) exactly zero?

11. Halley's comet has a period of approximately 76 years, and it moves in an elliptical orbit in which its distance from the Sun at closest approach is a small fraction of its maximum distance. Estimate the comet's maximum distance from the Sun in astronomical units (AUs) (the distance from the Earth to the Sun). (a) 6 AU (b) 12 AU (c) 20 AU (d) 28 AU (e) 35 AU

Conceptual Questions   1. denotes answer available in *Student Solutions Manual/Study Guide*

1. Each *Voyager* spacecraft was accelerated toward escape speed from the Sun by the gravitational force exerted by Jupiter on the spacecraft. (a) Is the gravitational force a conservative or a nonconservative force? (b) Does the interaction of the spacecraft with Jupiter meet the definition of an elastic collision? (c) How could the spacecraft be moving faster after the collision?

2. In his 1798 experiment, Cavendish was said to have "weighed the Earth." Explain this statement.

3. Why don't we put a geosynchronous weather satellite in orbit around the 45th parallel? Wouldn't such a satellite be more useful in the United States than one in orbit around the equator?

4. (a) Explain why the force exerted on a particle by a uniform sphere must be directed toward the center of the sphere. (b) Would this statement be true if the mass distribution of the sphere were not spherically symmetric? Explain.

5. (a) At what position in its elliptical orbit is the speed of a planet a maximum? (b) At what position is the speed a minimum?

6. You are given the mass and radius of planet X. How would you calculate the free-fall acceleration on this planet's surface?

7. (a) If a hole could be dug to the center of the Earth, would the force on an object of mass m still obey Equation 13.1 there? (b) What do you think the force on m would be at the center of the Earth?

8. Explain why it takes more fuel for a spacecraft to travel from the Earth to the Moon than for the return trip. Estimate the difference.

9. A satellite in low-Earth orbit is not truly traveling through a vacuum. Rather, it moves through very thin air. Does the resulting air friction cause the satellite to slow down?

Section 13.1 Newton's Law of Universal Gravitation

Problem 12 in Chapter 1 can also be assigned with this section.

1. In introductory physics laboratories, a typical Cavendish balance for measuring the gravitational constant G uses lead spheres with masses of 1.50 kg and 15.0 g whose centers are separated by about 4.50 cm. Calculate the gravitational force between these spheres, treating each as a particle located at the sphere's center.

2. Determine the order of magnitude of the gravitational force that you exert on another person 2 m away. In your solution, state the quantities you measure or estimate and their values.

3. A 200-kg object and a 500-kg object are separated by 4.00 m. (a) Find the net gravitational force exerted by these objects on a 50.0-kg object placed midway between them. (b) At what position (other than an infinitely remote one) can the 50.0-kg object be placed so as to experience a net force of zero from the other two objects?

4. During a solar eclipse, the Moon, the Earth, and the Sun all lie on the same line, with the Moon between the Earth and the Sun. (a) What force is exerted by the Sun on the Moon? (b) What force is exerted by the Earth on the Moon? (c) What force is exerted by the Sun on the Earth? (d) Compare the answers to parts (a) and (b). Why doesn't the Sun capture the Moon away from the Earth?

5. Two ocean liners, each with a mass of 40 000 metric tons, are moving on parallel courses 100 m apart. What is the magnitude of the acceleration of one of the liners toward the other due to their mutual gravitational attraction? Model the ships as particles.

6. Three uniform spheres of masses $m_1 = 2.00$ kg, $m_2 = 4.00$ kg, and $m_3 = 6.00$ kg are placed at the corners of a right triangle as shown in Figure P13.6. Calculate the resultant gravitational force on the object of mass m_2, assuming the spheres are isolated from the rest of the Universe.

Figure P13.6

7. Two identical isolated particles, each of mass 2.00 kg, are separated by a distance of 30.0 cm. What is the magnitude of the gravitational force exerted by one particle on the other?

8. *Why is the following situation impossible?* The centers of two homogeneous spheres are 1.00 m apart. The spheres are each made of the same element from the periodic table. The gravitational force between the spheres is 1.00 N.

9. Two objects attract each other with a gravitational force of magnitude 1.00×10^{-8} N when separated by 20.0 cm. If the total mass of the two objects is 5.00 kg, what is the mass of each?

10. Review. A student proposes to study the gravitational force by suspending two 100.0-kg spherical objects at the lower ends of cables from the ceiling of a tall cathedral and measuring the deflection of the cables from the vertical. The 45.00-m-long cables are attached to the ceiling 1.000 m apart. The first object is suspended, and its position is carefully measured. The second object is suspended, and the two objects attract each other gravitationally. By what distance has the first object moved horizontally from its initial position due to the gravitational attraction to the other object? *Suggestion:* Keep in mind that this distance will be very small and make appropriate approximations.

Section 13.2 Free-Fall Acceleration and the Gravitational Force

11. When a falling meteoroid is at a distance above the Earth's surface of 3.00 times the Earth's radius, what is its acceleration due to the Earth's gravitation?

12. The free-fall acceleration on the surface of the Moon is about one-sixth that on the surface of the Earth. The radius of the Moon is about $0.250R_E$ (R_E = Earth's radius = 6.37×10^6 m). Find the ratio of their average densities, $\rho_{\text{Moon}}/\rho_{\text{Earth}}$.

13. Review. Miranda, a satellite of Uranus, is shown in Figure P13.13a. It can be modeled as a sphere of radius 242 km and mass 6.68×10^{19} kg. (a) Find the free-fall acceleration on its surface. (b) A cliff on Miranda is 5.00 km high. It appears on the limb at the 11 o'clock position in Figure P13.13a and is magnified in Figure P13.13b. If a devotee of extreme sports runs horizontally off the top of the cliff at 8.50 m/s, for what time interval is he in flight? (c) How far from the base of the vertical cliff does he strike the icy surface of Miranda? (d) What will be his vector impact velocity?

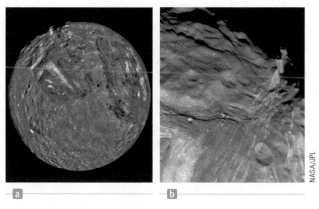

Figure P13.13

Section 13.3 Analysis Model: Particle in a Field (Gravitational)

14. (a) Compute the vector gravitational field at a point P on the perpendicular bisector of the line joining two objects of equal mass separated by a distance $2a$ as shown in Figure P13.14. (b) Explain physically why the field should approach zero as $r \to 0$. (c) Prove mathematically that the answer to part (a) behaves in this way. (d) Explain physically why the magnitude of the field should approach $2GM/r^2$ as $r \to \infty$. (e) Prove mathematically that the answer to part (a) behaves correctly in this limit.

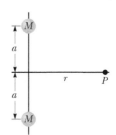

Figure P13.14

15. Three objects of equal mass are located at three corners of a square of edge length ℓ as shown in Figure P13.15. Find the magnitude and direction of the gravitational field at the fourth corner due to these objects.

Figure P13.15

16. A spacecraft in the shape of a long cylinder has a length
AMT of 100 m, and its mass with occupants is 1 000 kg.
W It has strayed too close to a black hole having a mass 100 times that of the Sun (Fig. P13.16). The nose of the spacecraft points toward the black hole, and the distance between the nose and the center of the black hole is 10.0 km. (a) Determine the total force on the spacecraft. (b) What is the difference in the gravita-

tional fields acting on the occupants in the nose of the ship and on those in the rear of the ship, farthest from the black hole? (This difference in accelerations grows rapidly as the ship approaches the black hole. It puts the body of the ship under extreme tension and eventually tears it apart.)

Figure P13.16

Section 13.4 Kepler's Laws and the Motion of Planets

17. An artificial satellite circles the Earth in a circular orbit at a location where the acceleration due to gravity is 9.00 m/s^2. Determine the orbital period of the satellite.

18. Io, a satellite of Jupiter, has an orbital period of 1.77 days and an orbital radius of 4.22×10^5 km. From these data, determine the mass of Jupiter.

19. A minimum-energy transfer orbit to an outer planet consists of putting a spacecraft on an elliptical trajectory with the departure planet corresponding to the perihelion of the ellipse, or the closest point to the Sun, and the arrival planet at the aphelion, or the farthest point from the Sun. (a) Use Kepler's third law to calculate how long it would take to go from Earth to Mars on such an orbit as shown in Figure P13.19. (b) Can such an orbit be undertaken at any time? Explain.

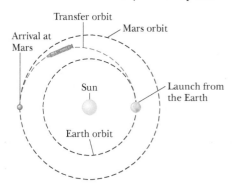

Figure P13.19

20. A particle of mass m moves along a straight line with constant velocity $\vec{\mathbf{v}}_0$ in the x direction, a distance b from the x axis (Fig. P13.20). (a) Does the particle possess any angular momentum about the origin? (b) Explain why the amount of its angular momentum should change or should stay constant. (c) Show that Kepler's second law is satisfied by showing that the two shaded triangles in the figure have the same area when $t_{\text{D}} - t_{\text{C}} = t_{\text{B}} - t_{\text{A}}$.

Figure P13.20

21. Plaskett's binary system consists of two stars that revolve in a circular orbit about a center of mass midway between them. This statement implies that the masses of the two stars are equal (Fig. P13.21). Assume the orbital speed of each star is $|\vec{v}| = 220$ km/s and the orbital period of each is 14.4 days. Find the mass M of each star. (For comparison, the mass of our Sun is 1.99×10^{30} kg.)

Figure P13.21

22. Two planets X and Y travel counterclockwise in circular orbits about a star as shown in Figure P13.22. The radii of their orbits are in the ratio 3:1. At one moment, they are aligned as shown in Figure P13.22a, making a straight line with the star. During the next five years, the angular displacement of planet X is 90.0° as shown in Figure P13.22b. What is the angular displacement of planet Y at this moment?

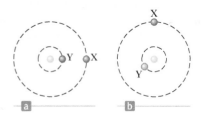

Figure P13.22

23. Comet Halley (Fig. P13.23) approaches the Sun to within 0.570 AU, and its orbital period is 75.6 yr. (AU is the symbol for astronomical unit, where 1 AU = 1.50×10^{11} m is the mean Earth–Sun distance.) How far from the Sun will Halley's comet travel before it starts its return journey?

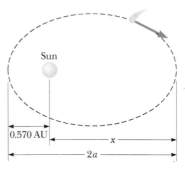

Figure P13.23 (Orbit is not drawn to scale.)

24. The *Explorer VIII* satellite, placed into orbit November 3, 1960, to investigate the ionosphere, had the following orbit parameters: perigee, 459 km; apogee, 2 289 km (both distances above the Earth's surface); period, 112.7 min. Find the ratio v_p/v_a of the speed at perigee to that at apogee.

25. Use Kepler's third law to determine how many days it takes a spacecraft to travel in an elliptical orbit from a point 6 670 km from the Earth's center to the Moon, 385 000 km from the Earth's center.

26. Neutron stars are extremely dense objects formed from the remnants of supernova explosions. Many rotate very rapidly. Suppose the mass of a certain spherical neutron star is twice the mass of the Sun and its radius is 10.0 km. Determine the greatest possible angular speed it can have so that the matter at the surface of the star on its equator is just held in orbit by the gravitational force.

27. A synchronous satellite, which always remains above the same point on a planet's equator, is put in orbit around Jupiter to study that planet's famous red spot. Jupiter rotates once every 9.84 h. Use the data of Table 13.2 to find the altitude of the satellite above the surface of the planet.

28. (a) Given that the period of the Moon's orbit about the Earth is 27.32 days and the nearly constant distance between the center of the Earth and the center of the Moon is 3.84×10^8 m, use Equation 13.11 to calculate the mass of the Earth. (b) Why is the value you calculate a bit too large?

29. Suppose the Sun's gravity were switched off. The planets would leave their orbits and fly away in straight lines as described by Newton's first law. (a) Would Mercury ever be farther from the Sun than Pluto? (b) If so, find how long it would take Mercury to achieve this passage. If not, give a convincing argument that Pluto is always farther from the Sun than is Mercury.

Section 13.5 Gravitational Potential Energy

Note: In Problems 30 through 50, assume $U = 0$ at $r = \infty$.

30. A satellite in Earth orbit has a mass of 100 kg and is at an altitude of 2.00×10^6 m. (a) What is the potential energy of the satellite–Earth system? (b) What is the magnitude of the gravitational force exerted by the Earth on the satellite? (c) **What If?** What force, if any, does the satellite exert on the Earth?

31. How much work is done by the Moon's gravitational field on a 1 000-kg meteor as it comes in from outer space and impacts on the Moon's surface?

32. How much energy is required to move a 1 000-kg object from the Earth's surface to an altitude twice the Earth's radius?

33. After the Sun exhausts its nuclear fuel, its ultimate fate will be to collapse to a *white dwarf* state. In this state, it would have approximately the same mass as it has now, but its radius would be equal to the radius of the Earth. Calculate (a) the average density of the white dwarf, (b) the surface free-fall acceleration, and (c) the gravitational potential energy associated with a 1.00-kg object at the surface of the white dwarf.

34. An object is released from rest at an altitude h above the surface of the Earth. (a) Show that its speed at a distance r from the Earth's center, where $R_E \le r \le R_E + h$, is

$$v = \sqrt{2GM_E\left(\frac{1}{r} - \frac{1}{R_E + h}\right)}$$

(b) Assume the release altitude is 500 km. Perform the integral

$$\Delta t = \int_i^f dt = -\int_i^f \frac{dr}{v}$$

to find the time of fall as the object moves from the release point to the Earth's surface. The negative sign appears because the object is moving opposite to the radial direction, so its speed is $v = -dr/dt$. Perform the integral numerically.

35. A system consists of three particles, each of mass 5.00 g, located at the corners of an equilateral triangle with sides of 30.0 cm. (a) Calculate the potential energy of the system. (b) Assume the particles are released simultaneously. Describe the subsequent motion of each. Will any collisions take place? Explain.

Section 13.6 Energy Considerations in Planetary and Satellite Motion

36. A space probe is fired as a projectile from the Earth's surface with an initial speed of 2.00×10^4 m/s. What will its speed be when it is very far from the Earth? Ignore atmospheric friction and the rotation of the Earth.

37. A 500-kg satellite is in a circular orbit at an altitude of 500 km above the Earth's surface. Because of air friction, the satellite eventually falls to the Earth's surface, where it hits the ground with a speed of 2.00 km/s. How much energy was transformed into internal energy by means of air friction?

38. A "treetop satellite" moves in a circular orbit just above the surface of a planet, assumed to offer no air resistance. Show that its orbital speed v and the escape speed from the planet are related by the expression $v_{esc} = \sqrt{2}v$.

39. A 1 000-kg satellite orbits the Earth at a constant altitude of 100 km. (a) How much energy must be added to the system to move the satellite into a circular orbit with altitude 200 km? What are the changes in the system's (b) kinetic energy and (c) potential energy?

40. A comet of mass 1.20×10^{10} kg moves in an elliptical orbit around the Sun. Its distance from the Sun ranges between 0.500 AU and 50.0 AU. (a) What is the eccentricity of its orbit? (b) What is its period? (c) At aphelion, what is the potential energy of the comet–Sun system? *Note:* 1 AU = one astronomical unit = the average distance from the Sun to the Earth = 1.496×10^{11} m.

41. An asteroid is on a collision course with Earth. An astronaut lands on the rock to bury explosive charges that will blow the asteroid apart. Most of the small fragments will miss the Earth, and those that fall into the atmo-

sphere will produce only a beautiful meteor shower. The astronaut finds that the density of the spherical asteroid is equal to the average density of the Earth. To ensure its pulverization, she incorporates into the explosives the rocket fuel and oxidizer intended for her return journey. What maximum radius can the asteroid have for her to be able to leave it entirely simply by jumping straight up? On Earth she can jump to a height of 0.500 m.

42. Derive an expression for the work required to move an Earth satellite of mass m from a circular orbit of radius $2R_E$ to one of radius $3R_E$.

43. (a) Determine the amount of work that must be done on a 100-kg payload to elevate it to a height of 1 000 km above the Earth's surface. (b) Determine the amount of additional work that is required to put the payload into circular orbit at this elevation.

44. (a) What is the minimum speed, relative to the Sun, necessary for a spacecraft to escape the solar system if it starts at the Earth's orbit? (b) *Voyager 1* achieved a maximum speed of 125 000 km/h on its way to photograph Jupiter. Beyond what distance from the Sun is this speed sufficient to escape the solar system?

45. A satellite of mass 200 kg is placed into Earth orbit at a height of 200 km above the surface. (a) Assuming a circular orbit, how long does the satellite take to complete one orbit? (b) What is the satellite's speed? (c) Starting from the satellite on the Earth's surface, what is the minimum energy input necessary to place this satellite in orbit? Ignore air resistance but include the effect of the planet's daily rotation.

46. A satellite of mass m, originally on the surface of the Earth, is placed into Earth orbit at an altitude h. (a) Assuming a circular orbit, how long does the satellite take to complete one orbit? (b) What is the satellite's speed? (c) What is the minimum energy input necessary to place this satellite in orbit? Ignore air resistance but include the effect of the planet's daily rotation. Represent the mass and radius of the Earth as M_E and R_E, respectively.

47. Ganymede is the largest of Jupiter's moons. Consider a rocket on the surface of Ganymede, at the point farthest from the planet (Fig. P13.47). Model the rocket as a particle. (a) Does the presence of Ganymede make Jupiter exert a larger, smaller, or same size force on the rocket compared with the force it would exert if Ganymede were not interposed? (b) Determine the escape speed for the rocket from the planet–satellite system. The radius of Ganymede is 2.64×10^6 m, and its mass

Jupiter

Ganymede

Figure P13.47

is 1.495×10^{23} kg. The distance between Jupiter and Ganymede is 1.071×10^9 m, and the mass of Jupiter is 1.90×10^{27} kg. Ignore the motion of Jupiter and Ganymede as they revolve about their center of mass.

48. A satellite moves around the Earth in a circular orbit of radius r. (a) What is the speed v_i of the satellite? (b) Suddenly, an explosion breaks the satellite into two pieces, with masses m and $4m$. Immediately after the explosion, the smaller piece of mass m is stationary with respect to the Earth and falls directly toward the Earth. What is the speed v of the larger piece immediately after the explosion? (c) Because of the increase in its speed, this larger piece now moves in a new elliptical orbit. Find its distance away from the center of the Earth when it reaches the other end of the ellipse.

49. At the Earth's surface, a projectile is launched straight up at a speed of 10.0 km/s. To what height will it rise? Ignore air resistance.

Additional Problems

50. A rocket is fired straight up through the atmosphere from the South Pole, burning out at an altitude of 250 km when traveling at 6.00 km/s. (a) What maximum distance from the Earth's surface does it travel before falling back to the Earth? (b) Would its maximum distance from the surface be larger if the same rocket were fired with the same fuel load from a launch site on the equator? Why or why not?

51. **Review.** A cylindrical habitat in space 6.00 km in diameter and 30.0 km long has been proposed (by G. K. O'Neill, 1974). Such a habitat would have cities, land, and lakes on the inside surface and air and clouds in the center. They would all be held in place by rotation of the cylinder about its long axis. How fast would the cylinder have to rotate to imitate the Earth's gravitational field at the walls of the cylinder?

52. *Voyager 1* and *Voyager 2* surveyed the surface of Jupiter's moon Io and photographed active volcanoes spewing liquid sulfur to heights of 70 km above the surface of this moon. Find the speed with which the liquid sulfur left the volcano. Io's mass is 8.9×10^{22} kg, and its radius is 1 820 km.

53. A satellite is in a circular orbit around the Earth at an altitude of 2.80×10^6 m. Find (a) the period of the orbit, (b) the speed of the satellite, and (c) the acceleration of the satellite.

54. *Why is the following situation impossible?* A spacecraft is launched into a circular orbit around the Earth and circles the Earth once an hour.

55. Let Δg_M represent the difference in the gravitational fields produced by the Moon at the points on the Earth's surface nearest to and farthest from the Moon. Find the fraction $\Delta g_M/g$, where g is the Earth's gravitational field. (This difference is responsible for the occurrence of the *lunar tides* on the Earth.)

56. A sleeping area for a long space voyage consists of two cabins each connected by a cable to a central hub as shown in Figure P13.56. The cabins are set spinning around the hub axis, which is connected to the rest of the spacecraft to generate artificial gravity in the cabins. A space traveler lies in a bed parallel to the outer wall as shown in Figure P13.56. (a) With $r = 10.0$ m, what would the angular speed of the 60.0-kg traveler need to be if he is to experience half his normal Earth weight? (b) If the astronaut stands up perpendicular to the bed, without holding on to anything with his hands, will his head be moving at a faster, a slower, or the same tangential speed as his feet? Why? (c) Why is the action in part (b) dangerous?

Figure P13.56

57. (a) A space vehicle is launched vertically upward from the Earth's surface with an initial speed of 8.76 km/s, which is less than the escape speed of 11.2 km/s. What maximum height does it attain? (b) A meteoroid falls toward the Earth. It is essentially at rest with respect to the Earth when it is at a height of 2.51×10^7 m above the Earth's surface. With what speed does the meteorite (a meteoroid that survives to impact the Earth's surface) strike the Earth?

58. (a) A space vehicle is launched vertically upward from the Earth's surface with an initial speed of v_i that is comparable to but less than the escape speed v_{esc}. What maximum height does it attain? (b) A meteoroid falls toward the Earth. It is essentially at rest with respect to the Earth when it is at a height h above the Earth's surface. With what speed does the meteorite (a meteoroid that survives to impact the Earth's surface) strike the Earth? (c) **What If?** Assume a baseball is tossed up with an initial speed that is very small compared to the escape speed. Show that the result from part (a) is consistent with Equation 4.12.

59. Assume you are agile enough to run across a horizontal surface at 8.50 m/s, independently of the value of the gravitational field. What would be (a) the radius and (b) the mass of an airless spherical asteroid of uniform density 1.10×10^3 kg/m^3 on which you could launch yourself into orbit by running? (c) What would be your period? (d) Would your running significantly affect the rotation of the asteroid? Explain.

60. Two spheres having masses M and $2M$ and radii R and $3R$, respectively, are simultaneously released from rest when the distance between their centers is $12R$. Assume the two spheres interact only with each other and we wish to find the speeds with which they collide. (a) What *two* isolated system models are appropriate for this system? (b) Write an equation from one of the models and solve it for $\vec{v}_1$, the velocity of the sphere of mass M at any time after release in terms of $\vec{v}_2$, the veloc-

ity of 2*M*. (c) Write an equation from the other model and solve it for speed v_1 in terms of speed v_2 when the spheres collide. (d) Combine the two equations to find the two speeds v_1 and v_2 when the spheres collide.

61. Two hypothetical planets of masses m_1 and m_2 and radii r_1 and r_2, respectively, are nearly at rest when they are an infinite distance apart. Because of their gravitational attraction, they head toward each other on a collision course. (a) When their center-to-center separation is d, find expressions for the speed of each planet and for their relative speed. (b) Find the kinetic energy of each planet just before they collide, taking $m_1 = 2.00 \times 10^{24}$ kg, $m_2 = 8.00 \times 10^{24}$ kg, $r_1 = 3.00 \times 10^6$ m, and $r_2 = 5.00 \times 10^6$ m. *Note:* Both the energy and momentum of the isolated two-planet system are constant.

62. (a) Show that the rate of change of the free-fall acceleration with vertical position near the Earth's surface is

$$\frac{dg}{dr} = -\frac{2GM_E}{R_E^{\,3}}$$

This rate of change with position is called a *gradient*. (b) Assuming h is small in comparison to the radius of the Earth, show that the difference in free-fall acceleration between two points separated by vertical distance h is

$$|\Delta g| = \frac{2GM_E h}{R_E^{\,3}}$$

(c) Evaluate this difference for $h = 6.00$ m, a typical height for a two-story building.

63. A ring of matter is a familiar structure in planetary and stellar astronomy. Examples include Saturn's rings and a ring nebula. Consider a uniform ring of mass 2.36×10^{20} kg and radius 1.00×10^8 m. An object of mass 1 000 kg is placed at a point A on the axis of the ring, 2.00×10^8 m from the center of the ring (Fig. P13.63). When the object is released, the attraction of the ring makes the object move along the axis toward the center of the ring (point B). (a) Calculate the gravitational

Figure P13.63

potential energy of the object–ring system when the object is at A. (b) Calculate the gravitational potential energy of the system when the object is at B. (c) Calculate the speed of the object as it passes through B.

64. A spacecraft of mass 1.00×10^4 kg is in a circular orbit at an altitude of 500 km above the Earth's surface. Mission Control wants to fire the engines in a direction tangent to the orbit so as to put the spacecraft in an elliptical orbit around the Earth with an apogee of 2.00×10^4 km, measured from the Earth's center. How much energy must be used from the fuel to achieve this orbit? (Assume that all the fuel energy goes into increasing the orbital energy. This model will give a lower limit to the required energy because some of the energy from the fuel will appear as internal energy in the hot exhaust gases and engine parts.)

65. **Review.** As an astronaut, you observe a small planet to be spherical. After landing on the planet, you set off, walking always straight ahead, and find yourself returning to your spacecraft from the opposite side after completing a lap of 25.0 km. You hold a hammer and a falcon feather at a height of 1.40 m, release them, and observe that they fall together to the surface in 29.2 s. Determine the mass of the planet.

66. A certain quaternary star system consists of three stars, each of mass m, moving in the same circular orbit of radius r about a central star of mass M. The stars orbit in the same sense and are positioned one-third of a revolution apart from one another. Show that the period of each of the three stars is given by

$$T = 2\pi \sqrt{\frac{r^3}{G(M + m/\sqrt{3})}}$$

67. Studies of the relationship of the Sun to our galaxy—the Milky Way—have revealed that the Sun is located near the outer edge of the galactic disc, about 30 000 ly (1 ly = 9.46×10^{15} m) from the center. The Sun has an orbital speed of approximately 250 km/s around the galactic center. (a) What is the period of the Sun's galactic motion? (b) What is the order of magnitude of the mass of the Milky Way galaxy? (c) Suppose the galaxy is made mostly of stars of which the Sun is typical. What is the order of magnitude of the number of stars in the Milky Way?

68. **Review.** Two identical hard spheres, each of mass m and radius r, are released from rest in otherwise empty space with their centers separated by the distance R. They are allowed to collide under the influence of their gravitational attraction. (a) Show that the magnitude of the impulse received by each sphere before they make contact is given by $[Gm^3(1/2r - 1/R)]^{1/2}$. (b) **What If?** Find the magnitude of the impulse each receives during their contact if they collide elastically.

69. The maximum distance from the Earth to the Sun (at aphelion) is 1.521×10^{11} m, and the distance of closest approach (at perihelion) is 1.471×10^{11} m. The Earth's orbital speed at perihelion is 3.027×10^4 m/s. Determine (a) the Earth's orbital speed at aphelion and the kinetic and potential energies of the Earth–Sun system

(b) at perihelion and (c) at aphelion. (d) Is the total energy of the system constant? Explain. Ignore the effect of the Moon and other planets.

70. Many people assume air resistance acting on a moving object will always make the object slow down. It can, however, actually be responsible for making the object speed up. Consider a 100-kg Earth satellite in a circular orbit at an altitude of 200 km. A small force of air resistance makes the satellite drop into a circular orbit with an altitude of 100 km. (a) Calculate the satellite's initial speed. (b) Calculate its final speed in this process. (c) Calculate the initial energy of the satellite–Earth system. (d) Calculate the final energy of the system. (e) Show that the system has lost mechanical energy and find the amount of the loss due to friction. (f) What force makes the satellite's speed increase? *Hint:* You will find a free-body diagram useful in explaining your answer.

71. X-ray pulses from Cygnus X-1, the first black hole to be identified and a celestial x-ray source, have been recorded during high-altitude rocket flights. The signals can be interpreted as originating when a blob of ionized matter orbits a black hole with a period of 5.0 ms. If the blob is in a circular orbit about a black hole whose mass is $20M_{Sun}$, what is the orbit radius?

72. Show that the minimum period for a satellite in orbit around a spherical planet of uniform density ρ is

$$T_{min} = \sqrt{\frac{3\pi}{G\rho}}$$

independent of the planet's radius.

73. Astronomers detect a distant meteoroid moving along a straight line that, if extended, would pass at a distance $3R_E$ from the center of the Earth, where R_E is the Earth's radius. What minimum speed must the meteoroid have if it is *not* to collide with the Earth?

74. Two stars of masses M and m, separated by a distance d, revolve in circular orbits about their center of mass (Fig. P13.74). Show that each star has a period given by

$$T^2 = \frac{4\pi^2 d^3}{G(M + m)}$$

Figure P13.74

75. Two identical particles, each of mass 1 000 kg, are coasting in free space along the same path, one in front of the other by 20.0 m. At the instant their separation distance has this value, each particle has precisely the same velocity of 800 $\hat{i}$ m/s. What are their precise velocities when they are 2.00 m apart?

76. Consider an object of mass m, not necessarily small compared with the mass of the Earth, released at a distance of 1.20×10^7 m from the center of the Earth. Assume the Earth and the object behave as a pair of

particles, isolated from the rest of the Universe. (a) Find the magnitude of the acceleration a_{rel} with which each starts to move relative to the other as a function of m. Evaluate the acceleration (b) for $m = 5.00$ kg, (c) for $m = 2\,000$ kg, and (d) for $m = 2.00 \times 10^{24}$ kg. (e) Describe the pattern of variation of a_{rel} with m.

77. As thermonuclear fusion proceeds in its core, the Sun loses mass at a rate of 3.64×10^9 kg/s. During the 5 000-yr period of recorded history, by how much has the length of the year changed due to the loss of mass from the Sun? *Suggestions:* Assume the Earth's orbit is circular. No external torque acts on the Earth–Sun system, so the angular momentum of the Earth is constant.

Challenge Problems

78. The Solar and Heliospheric Observatory (SOHO) spacecraft has a special orbit, located between the Earth and the Sun along the line joining them, and it is always close enough to the Earth to transmit data easily. Both objects exert gravitational forces on the observatory. It moves around the Sun in a near-circular orbit that is smaller than the Earth's circular orbit. Its period, however, is not less than 1 yr but just equal to 1 yr. Show that its distance from the Earth must be 1.48×10^9 m. In 1772, Joseph Louis Lagrange determined theoretically the special location allowing this orbit. *Suggestions:* Use data that are precise to four digits. The mass of the Earth is 5.974×10^{24} kg. You will not be able to easily solve the equation you generate; instead, use a computer to verify that 1.48×10^9 m is the correct value.

79. The oldest artificial satellite still in orbit is *Vanguard I*, launched March 3, 1958. Its mass is 1.60 kg. Neglecting atmospheric drag, the satellite would still be in its initial orbit, with a minimum distance from the center of the Earth of 7.02 Mm and a speed at this perigee point of 8.23 km/s. For this orbit, find (a) the total energy of the satellite–Earth system and (b) the magnitude of the angular momentum of the satellite. (c) At apogee, find the satellite's speed and its distance from the center of the Earth. (d) Find the semimajor axis of its orbit. (e) Determine its period.

80. A spacecraft is approaching Mars after a long trip from the Earth. Its velocity is such that it is traveling along a parabolic trajectory under the influence of the gravitational force from Mars. The distance of closest approach will be 300 km above the Martian surface. At this point of closest approach, the engines will be fired to slow down the spacecraft and place it in a circular orbit 300 km above the surface. (a) By what percentage must the speed of the spacecraft be reduced to achieve the desired orbit? (b) How would the answer to part (a) change if the distance of closest approach and the desired circular orbit altitude were 600 km instead of 300 km? (*Note:* The energy of the spacecraft–Mars system for a parabolic orbit is $E = 0$.)

Fluid Mechanics

Matter is normally classified as being in one of three states: solid, liquid, or gas. From everyday experience we know that a solid has a definite volume and shape, a liquid has a definite volume but no definite shape, and an unconfined gas has neither a definite volume nor a definite shape. These descriptions help us picture the states of matter, but they are somewhat artificial. For example, asphalt and plastics are normally considered solids, but over long time intervals they tend to flow like liquids. Likewise, most substances can be a solid, a liquid, or a gas (or a combination of any of these three), depending on the temperature and pressure. In general, the time interval required for a particular substance to change its shape in response to an external force determines whether we treat the substance as a solid, a liquid, or a gas.

A **fluid** is a collection of molecules that are randomly arranged and held together by weak cohesive forces and by forces exerted by the walls of a container. Both liquids and gases are fluids.

In our treatment of the mechanics of fluids, we'll be applying principles and analysis models that we have already discussed. First, we consider the mechanics of a fluid at rest, that is, *fluid statics*, and then study fluids in motion, that is, *fluid dynamics*.

Fish congregate around a reef in Hawaii searching for food. How do fish such as the lined butterflyfish (*Chaetodon lineolatus*) at the upper left control their movements up and down in the water? We'll find out in this chapter. (*Vlad61/Shutterstock.com*)

14.1 Pressure

Fluids do not sustain shearing stresses or tensile stresses such as those discussed in Chapter 12; therefore, the only stress that can be exerted on an object submerged in a static fluid is one that tends to compress the object from all sides. In other words, the force exerted by a static fluid on an object is always perpendicular to the surfaces of the object as shown in Figure 14.1. We discussed this situation in Section 12.4.

At any point on the surface of the object, the force exerted by the fluid is perpendicular to the surface of the object.

Figure 14.1 The forces exerted by a fluid on the surfaces of a submerged object.

417

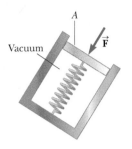

Figure 14.2 A simple device for measuring the pressure exerted by a fluid.

The pressure in a fluid can be measured with the device pictured in Figure 14.2. The device consists of an evacuated cylinder that encloses a light piston connected to a spring. As the device is submerged in a fluid, the fluid presses on the top of the piston and compresses the spring until the inward force exerted by the fluid is balanced by the outward force exerted by the spring. The fluid pressure can be measured directly if the spring is calibrated in advance. If F is the magnitude of the force exerted on the piston and A is the surface area of the piston, the **pressure** P of the fluid at the level to which the device has been submerged is defined as the ratio of the force to the area:

$$P \equiv \frac{F}{A} \tag{14.1}$$

Pressure is a scalar quantity because it is proportional to the magnitude of the force on the piston.

If the pressure varies over an area, the infinitesimal force dF on an infinitesimal surface element of area dA is

$$dF = P\, dA \tag{14.2}$$

where P is the pressure at the location of the area dA. To calculate the total force exerted on a surface of a container, we must integrate Equation 14.2 over the surface.

The units of pressure are newtons per square meter (N/m^2) in the SI system. Another name for the SI unit of pressure is the **pascal** (Pa):

$$1\ \text{Pa} \equiv 1\ N/m^2 \tag{14.3}$$

Pitfall Prevention 14.1

Force and Pressure Equations 14.1 and 14.2 make a clear distinction between force and pressure. Another important distinction is that *force is a vector* and *pressure is a scalar*. There is no direction associated with pressure, but the direction of the force associated with the pressure is perpendicular to the surface on which the pressure acts.

For a tactile demonstration of the definition of pressure, hold a tack between your thumb and forefinger, with the point of the tack on your thumb and the head of the tack on your forefinger. Now *gently* press your thumb and forefinger together. Your thumb will begin to feel pain immediately while your forefinger will not. The tack is exerting the same force on both your thumb and forefinger, but the pressure on your thumb is much larger because of the small area over which the force is applied.

Quick Quiz 14.1 Suppose you are standing directly behind someone who steps back and accidentally stomps on your foot with the heel of one shoe. Would you be better off if that person were **(a)** a large, male professional basketball player wearing sneakers or **(b)** a petite woman wearing spike-heeled shoes?

Example 14.1   The Water Bed

The mattress of a water bed is 2.00 m long by 2.00 m wide and 30.0 cm deep.

(A) Find the weight of the water in the mattress.

SOLUTION

Conceptualize Think about carrying a jug of water and how heavy it is. Now imagine a sample of water the size of a water bed. We expect the weight to be relatively large.

Categorize This example is a substitution problem.

Find the volume of the water filling the mattress: $V = (2.00\ \text{m})(2.00\ \text{m})(0.300\ \text{m}) = 1.20\ m^3$

Use Equation 1.1 and the density of fresh water (see Table 14.1) to find the mass of the water bed: $M = \rho V = (1\ 000\ kg/m^3)(1.20\ m^3) = 1.20 \times 10^3\ \text{kg}$

Find the weight of the bed: $Mg = (1.20 \times 10^3\ \text{kg})(9.80\ m/s^2) = 1.18 \times 10^4\ N$

which is approximately 2 650 lb. (A regular bed, including mattress, box spring, and metal frame, weighs approximately 300 lb.) Because this load is so great, it is best to place a water bed in the basement or on a sturdy, well-supported floor.

▶ **14.1** continued

(B) Find the pressure exerted by the water bed on the floor when the bed rests in its normal position. Assume the entire lower surface of the bed makes contact with the floor.

SOLUTION

When the water bed is in its normal position, the area in contact with the floor is 4.00 m². Use Equation 14.1 to find the pressure:

$$P = \frac{1.18 \times 10^4 \, \text{N}}{4.00 \, \text{m}^2} = 2.94 \times 10^3 \, \text{Pa}$$

WHAT IF? What if the water bed is replaced by a 300-lb regular bed that is supported by four legs? Each leg has a circular cross section of radius 2.00 cm. What pressure does this bed exert on the floor?

Answer The weight of the regular bed is distributed over four circular cross sections at the bottom of the legs. Therefore, the pressure is

$$P = \frac{F}{A} = \frac{mg}{4(\pi r^2)} = \frac{300 \, \text{lb}}{4\pi(0.020 \, 0 \, \text{m})^2} \left(\frac{1 \, \text{N}}{0.225 \, \text{lb}} \right)$$

$$= 2.65 \times 10^5 \, \text{Pa}$$

This result is almost 100 times larger than the pressure due to the water bed! The weight of the regular bed, even though it is much less than the weight of the water bed, is applied over the very small area of the four legs. The high pressure on the floor at the feet of a regular bed could cause dents in wood floors or permanently crush carpet pile.

14.2 Variation of Pressure with Depth

As divers well know, water pressure increases with depth. Likewise, atmospheric pressure decreases with increasing altitude; for this reason, aircraft flying at high altitudes must have pressurized cabins for the comfort of the passengers.

We now show how the pressure in a liquid increases with depth. As Equation 1.1 describes, the *density* of a substance is defined as its mass per unit volume; Table 14.1 lists the densities of various substances. These values vary slightly with temperature because the volume of a substance is dependent on temperature (as shown in Chapter 19). Under standard conditions (at 0°C and at atmospheric pressure), the densities of gases are about $\frac{1}{1\,000}$ the densities of solids and liquids. This difference in densities implies that the average molecular spacing in a gas under these conditions is about ten times greater than that in a solid or liquid.

Table 14.1 **Densities of Some Common Substances at Standard Temperature (0°C) and Pressure (Atmospheric)**

Substance	ρ (kg/m³)	Substance	ρ (kg/m³)
Air	1.29	Iron	7.86×10^3
Air (at 20°C and		Lead	11.3×10^3
atmospheric pressure)	1.20	Mercury	13.6×10^3
Aluminum	2.70×10^3	Nitrogen gas	1.25
Benzene	0.879×10^3	Oak	0.710×10^3
Brass	8.4×10^3	Osmium	22.6×10^3
Copper	8.92×10^3	Oxygen gas	1.43
Ethyl alcohol	0.806×10^3	Pine	0.373×10^3
Fresh water	1.00×10^3	Platinum	21.4×10^3
Glycerin	1.26×10^3	Seawater	1.03×10^3
Gold	19.3×10^3	Silver	10.5×10^3
Helium gas	1.79×10^{-1}	Tin	7.30×10^3
Hydrogen gas	8.99×10^{-2}	Uranium	19.1×10^3
Ice	0.917×10^3		

The parcel of fluid is in equilibrium, so the net force on it is zero.

$-P_0 A\hat{\mathbf{j}}$

d

$d + h$

$-Mg\hat{\mathbf{j}}$ $PA\hat{\mathbf{j}}$

Figure 14.3 A parcel of fluid in a larger volume of fluid is singled out.

Variation of pressure ▶ with depth

Now consider a liquid of density ρ at rest as shown in Figure 14.3. We assume ρ is uniform throughout the liquid, which means the liquid is incompressible. Let us select a parcel of the liquid contained within an imaginary block of cross-sectional area A extending from depth d to depth $d + h$. The liquid external to our parcel exerts forces at all points on the surface of the parcel, perpendicular to the surface. The pressure exerted by the liquid on the bottom face of the parcel is P, and the pressure on the top face is P_0. Therefore, the upward force exerted by the outside fluid on the bottom of the parcel has a magnitude PA, and the downward force exerted on the top has a magnitude $P_0 A$. The mass of liquid in the parcel is $M = \rho V = \rho Ah$; therefore, the weight of the liquid in the parcel is $Mg = \rho Ahg$. Because the parcel is at rest and remains at rest, it can be modeled as a particle in equilibrium, so that the net force acting on it must be zero. Choosing upward to be the positive y direction, we see that

$$\sum \vec{\mathbf{F}} = PA\hat{\mathbf{j}} - P_0 A\hat{\mathbf{j}} - Mg\hat{\mathbf{j}} = 0$$

or

$$PA - P_0 A - \rho Ahg = 0$$

$$P = P_0 + \rho gh \qquad (14.4)$$

That is, the pressure P at a depth h below a point in the liquid at which the pressure is P_0 is greater by an amount ρgh. If the liquid is open to the atmosphere and P_0 is the pressure at the surface of the liquid, then P_0 is **atmospheric pressure.** In our calculations and working of end-of-chapter problems, we usually take atmospheric pressure to be

$$P_0 = 1.00 \text{ atm} = 1.013 \times 10^5 \text{ Pa}$$

Equation 14.4 implies that the pressure is the same at all points having the same depth, independent of the shape of the container.

Because the pressure in a fluid depends on depth and on the value of P_0, any increase in pressure at the surface must be transmitted to every other point in the fluid. This concept was first recognized by French scientist Blaise Pascal (1623–

Pascal's law ▶

1662) and is called **Pascal's law: a change in the pressure applied to a fluid is transmitted undiminished to every point of the fluid and to the walls of the container.**

An important application of Pascal's law is the hydraulic press illustrated in Figure 14.4a. A force of magnitude F_1 is applied to a small piston of surface area A_1. The pressure is transmitted through an incompressible liquid to a larger piston of surface area A_2. Because the pressure must be the same on both sides, $P = F_1/A_1 = F_2/A_2$. Therefore, the force F_2 is greater than the force F_1 by a factor of A_2/A_1. By designing a hydraulic press with appropriate areas A_1 and A_2, a large out-

Because the increase in pressure is the same on the two sides, a small force $\vec{\mathbf{F}}_1$ at the left produces a much greater force $\vec{\mathbf{F}}_2$ at the right.

Δx_1 $\vec{\mathbf{F}}_1$ A_1 A_2 Δx_2

$\vec{\mathbf{F}}_2$

Sam Jordash/Digital Vision/Getty Images

Figure 14.4 (a) Diagram of a hydraulic press. (b) A vehicle undergoing repair is supported by a hydraulic lift in a garage.

a   b

put force can be applied by means of a small input force. Hydraulic brakes, car lifts, hydraulic jacks, and forklifts all make use of this principle (Fig. 14.4b).

Because liquid is neither added to nor removed from the system, the volume of liquid pushed down on the left in Figure 14.4a as the piston moves downward through a displacement Δx_1 equals the volume of liquid pushed up on the right as the right piston moves upward through a displacement Δx_2. That is, $A_1 \Delta x_1 = A_2 \Delta x_2$; therefore, $A_2/A_1 = \Delta x_1/\Delta x_2$. We have already shown that $A_2/A_1 = F_2/F_1$. Therefore, $F_2/F_1 = \Delta x_1/\Delta x_2$, so $F_1 \Delta x_1 = F_2 \Delta x_2$. Each side of this equation is the work done by the force on its respective piston. Therefore, the work done by $\vec{F}_1$ on the input piston equals the work done by $\vec{F}_2$ on the output piston, as it must to conserve energy. (The process can be modeled as a special case of the nonisolated system model: the *nonisolated system in steady state*. There is energy transfer into and out of the system, but these energy transfers balance, so that there is no net change in the energy of the system.)

Quick Quiz 14.2 The pressure at the bottom of a filled glass of water ($\rho = 1\,000$ kg/m^3) is P. The water is poured out, and the glass is filled with ethyl alcohol ($\rho = 806$ kg/m^3). What is the pressure at the bottom of the glass? **(a)** smaller than P **(b)** equal to P **(c)** larger than P **(d)** indeterminate

---

Example 14.2   The Car Lift

In a car lift used in a service station, compressed air exerts a force on a small piston that has a circular cross section of radius 5.00 cm. This pressure is transmitted by a liquid to a piston that has a radius of 15.0 cm.

(A) What force must the compressed air exert to lift a car weighing 13 300 N?

SOLUTION

Conceptualize Review the material just discussed about Pascal's law to understand the operation of a car lift.

Categorize This example is a substitution problem.

Solve $F_1/A_1 = F_2/A_2$ for F_1:

$$F_1 = \left(\frac{A_1}{A_2}\right)F_2 = \frac{\pi(5.00 \times 10^{-2}\ \text{m})^2}{\pi(15.0 \times 10^{-2}\ \text{m})^2}(1.33 \times 10^4\ \text{N})$$

$$= 1.48 \times 10^3\ \text{N}$$

(B) What air pressure produces this force?

SOLUTION

Use Equation 14.1 to find the air pressure that produces this force:

$$P = \frac{F_1}{A_1} = \frac{1.48 \times 10^3\ \text{N}}{\pi(5.00 \times 10^{-2}\ \text{m})^2}$$

$$= 1.88 \times 10^5\ \text{Pa}$$

This pressure is approximately twice atmospheric pressure.

Example 14.3   A Pain in Your Ear

Estimate the force exerted on your eardrum due to the water when you are swimming at the bottom of a pool that is 5.0 m deep.

SOLUTION

Conceptualize As you descend in the water, the pressure increases. You may have noticed this increased pressure in your ears while diving in a swimming pool, a lake, or the ocean. We can find the pressure difference exerted on the

continued

▶ **14.3** c o n t i n u e d

eardrum from the depth given in the problem; then, after estimating the ear drum's surface area, we can determine the net force the water exerts on it.

Categorize This example is a substitution problem.

The air inside the middle ear is normally at atmospheric pressure P_0. Therefore, to find the net force on the eardrum, we must consider the difference between the total pressure at the bottom of the pool and atmospheric pressure. Let's estimate the surface area of the eardrum to be approximately $1\ cm^2 = 1 \times 10^{-4}\ m^2$.

Use Equation 14.4 to find this pressure difference:

$$P_{bot} - P_0 = \rho g h$$

$$= (1.00 \times 10^3\ kg/m^3)(9.80\ m/s^2)(5.0\ m) = 4.9 \times 10^4\ Pa$$

Use Equation 14.1 to find the magnitude of the net force on the ear:

$$F = (P_{bot} - P_0)A = (4.9 \times 10^4\ Pa)(1 \times 10^{-4}\ m^2) \approx \boxed{5\ N}$$

Because a force of this magnitude on the eardrum is extremely uncomfortable, swimmers often "pop their ears" while under water, an action that pushes air from the lungs into the middle ear. Using this technique equalizes the pressure on the two sides of the eardrum and relieves the discomfort.

Example 14.4   **The Force on a Dam**

Water is filled to a height H behind a dam of width w (Fig. 14.5). Determine the resultant force exerted by the water on the dam.

S O L U T I O N

Conceptualize Because pressure varies with depth, we cannot calculate the force simply by multiplying the area by the pressure. As the pressure in the water increases with depth, the force on the adjacent portion of the dam also increases.

Categorize Because of the variation of pressure with depth, we must use integration to solve this example, so we categorize it as an analysis problem.

Figure 14.5 (Example 14.4) Water exerts a force on a dam.

Analyze Let's imagine a vertical y axis, with $y = 0$ at the bottom of the dam. We divide the face of the dam into narrow horizontal strips at a distance y above the bottom, such as the red strip in Figure 14.5. The pressure on each such strip is due only to the water; atmospheric pressure acts on both sides of the dam.

Use Equation 14.4 to calculate the pressure due to the water at the depth h:

$$P = \rho g h = \rho g(H - y)$$

Use Equation 14.2 to find the force exerted on the shaded strip of area $dA = w\ dy$:

$$dF = P\ dA = \rho g(H - y)w\ dy$$

Integrate to find the total force on the dam:

$$F = \int P\ dA = \int_0^H \rho g(H - y)w\ dy = \boxed{\tfrac{1}{2}\rho g w H^2}$$

Finalize Notice that the thickness of the dam shown in Figure 14.5 increases with depth. This design accounts for the greater force the water exerts on the dam at greater depths.

WHAT IF?   What if you were asked to find this force without using calculus? How could you determine its value?

Answer We know from Equation 14.4 that pressure varies linearly with depth. Therefore, the average pressure due to the water over the face of the dam is the average of the pressure at the top and the pressure at the bottom:

$$P_{avg} = \frac{P_{top} + P_{bottom}}{2} = \frac{0 + \rho g H}{2} = \tfrac{1}{2}\rho g H$$

▶ **14.4** continued

The total force on the dam is equal to the product of the average pressure and the area of the face of the dam:

$$F = P_{avg}A = (\tfrac{1}{2}\rho gH)(Hw) = \tfrac{1}{2}\rho gwH^2$$

which is the same result we obtained using calculus.

14.3 Pressure Measurements

During the weather report on a television news program, the *barometric pressure* is often provided. This reading is the current local pressure of the atmosphere, which varies over a small range from the standard value provided earlier. How is this pressure measured?

One instrument used to measure atmospheric pressure is the common barometer, invented by Evangelista Torricelli (1608–1647). A long tube closed at one end is filled with mercury and then inverted into a dish of mercury (Fig. 14.6a). The closed end of the tube is nearly a vacuum, so the pressure at the top of the mercury column can be taken as zero. In Figure 14.6a, the pressure at point A, due to the column of mercury, must equal the pressure at point B, due to the atmosphere. If that were not the case, there would be a net force that would move mercury from one point to the other until equilibrium is established. Therefore, $P_0 = \rho_{Hg}gh$, where ρ_{Hg} is the density of the mercury and h is the height of the mercury column. As atmospheric pressure varies, the height of the mercury column varies, so the height can be calibrated to measure atmospheric pressure. Let us determine the height of a mercury column for one atmosphere of pressure, $P_0 = 1$ atm $= 1.013 \times 10^5$ Pa:

$$P_0 = \rho_{Hg}gh \rightarrow h = \frac{P_0}{\rho_{Hg}g} = \frac{1.013 \times 10^5 \text{ Pa}}{(13.6 \times 10^3 \text{ kg/m}^3)(9.80 \text{ m/s}^2)} = 0.760 \text{ m}$$

Based on such a calculation, one atmosphere of pressure is defined to be the pressure equivalent of a column of mercury that is exactly 0.760 0 m in height at 0°C.

A device for measuring the pressure of a gas contained in a vessel is the open-tube manometer illustrated in Figure 14.6b. One end of a U-shaped tube containing a liquid is open to the atmosphere, and the other end is connected to a container of gas at pressure P. In an equilibrium situation, the pressures at points A and B must be the same (otherwise, the curved portion of the liquid would experience a net force and would accelerate), and the pressure at A is the unknown pressure of the gas. Therefore, equating the unknown pressure P to the pressure at point B, we see that $P = P_0 + \rho gh$. Again, we can calibrate the height h to the pressure P.

The difference in the pressures in each part of Figure 14.6 (that is, $P - P_0$) is equal to ρgh. The pressure P is called the **absolute pressure,** and the difference $P - P_0$ is called the **gauge pressure.** For example, the pressure you measure in your bicycle tire is gauge pressure.

Quick Quiz 14.3 Several common barometers are built, with a variety of fluids. For which of the following fluids will the column of fluid in the barometer be the highest? **(a)** mercury **(b)** water **(c)** ethyl alcohol **(d)** benzene

Figure 14.6 Two devices for measuring pressure: (a) a mercury barometer and (b) an open-tube manometer.

14.4 Buoyant Forces and Archimedes's Principle

Have you ever tried to push a beach ball down under water (Fig. 14.7a, p. 424)? It is extremely difficult to do because of the large upward force exerted by the water on the ball. The upward force exerted by a fluid on any immersed object is called

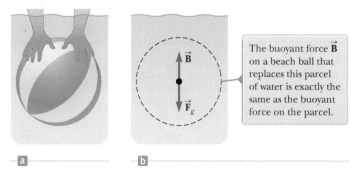

Figure 14.7 (a) A swimmer pushes a beach ball under water. (b) The forces on a beach ball–sized parcel of water.

Archimedes
Greek Mathematician, Physicist, and Engineer (c. 287–212 BC)
Archimedes was perhaps the greatest scientist of antiquity. He was the first to compute accurately the ratio of a circle's circumference to its diameter, and he also showed how to calculate the volume and surface area of spheres, cylinders, and other geometric shapes. He is well known for discovering the nature of the buoyant force and was also a gifted inventor. One of his practical inventions, still in use today, is Archimedes's screw, an inclined, rotating, coiled tube used originally to lift water from the holds of ships. He also invented the catapult and devised systems of levers, pulleys, and weights for raising heavy loads. Such inventions were successfully used to defend his native city, Syracuse, during a two-year siege by Romans.

The buoyant force on the cube is the resultant of the forces exerted on its top and bottom faces by the liquid.

Figure 14.8 The external forces acting on an immersed cube are the gravitational force $\vec{F}_g$ and the buoyant force $\vec{B}$.

a **buoyant force.** We can determine the magnitude of a buoyant force by applying some logic. Imagine a beach ball–sized parcel of water beneath the water surface as in Figure 14.7b. Because this parcel is in equilibrium, there must be an upward force that balances the downward gravitational force on the parcel. This upward force is the buoyant force, and its magnitude is equal to the weight of the water in the parcel. The buoyant force is the resultant force on the parcel due to all forces applied by the fluid surrounding the parcel.

Now imagine replacing the beach ball–sized parcel of water with a beach ball of the same size. The net force applied by the fluid surrounding the beach ball is the same, regardless of whether it is applied to a beach ball or to a parcel of water. Consequently, **the magnitude of the buoyant force on an object always equals the weight of the fluid displaced by the object.** This statement is known as **Archimedes's principle.**

With the beach ball under water, the buoyant force, equal to the weight of a beach ball–sized parcel of water, is much larger than the weight of the beach ball. Therefore, there is a large net upward force, which explains why it is so hard to hold the beach ball under the water. Note that Archimedes's principle does not refer to the makeup of the object experiencing the buoyant force. The object's composition is not a factor in the buoyant force because the buoyant force is exerted by the surrounding fluid.

To better understand the origin of the buoyant force, consider a cube of solid material immersed in a liquid as in Figure 14.8. According to Equation 14.4, the pressure P_{bot} at the bottom of the cube is greater than the pressure P_{top} at the top by an amount $\rho_{fluid}gh$, where h is the height of the cube and ρ_{fluid} is the density of the fluid. The pressure at the bottom of the cube causes an *upward* force equal to $P_{bot}A$, where A is the area of the bottom face. The pressure at the top of the cube causes a *downward* force equal to $P_{top}A$. The resultant of these two forces is the buoyant force $\vec{B}$ with magnitude

$$B = (P_{bot} - P_{top})A = (\rho_{fluid}gh)A$$

$$B = \rho_{fluid}gV_{disp} \tag{14.5}$$

where $V_{disp} = Ah$ is the volume of the fluid displaced by the cube. Because the product $\rho_{fluid}V_{disp}$ is equal to the mass of fluid displaced by the object,

$$B = Mg$$

where Mg is the weight of the fluid displaced by the cube. This result is consistent with our initial statement about Archimedes's principle above, based on the discussion of the beach ball.

Under normal conditions, the weight of a fish in the opening photograph for this chapter is slightly greater than the buoyant force on the fish. Hence, the fish would sink if it did not have some mechanism for adjusting the buoyant force. The fish accomplishes that by internally regulating the size of its air-filled swim bladder to increase its volume and the magnitude of the buoyant force acting on it, according to Equation 14.5. In this manner, fish are able to swim to various depths.

Before we proceed with a few examples, it is instructive to discuss two common situations: a totally submerged object and a floating (partly submerged) object.

Case 1: Totally Submerged Object When an object is totally submerged in a fluid of density ρ_{fluid}, the volume V_{disp} of the displaced fluid is equal to the volume V_{obj} of the object; so, from Equation 14.5, the magnitude of the upward buoyant force is $B = \rho_{\text{fluid}} g V_{\text{obj}}$. If the object has a mass M and density ρ_{obj}, its weight is equal to $F_g = Mg = \rho_{\text{obj}} g V_{\text{obj}}$, and the net force on the object is $B - F_g = (\rho_{\text{fluid}} - \rho_{\text{obj}}) g V_{\text{obj}}$. Hence, if the density of the object is less than the density of the fluid, the downward gravitational force is less than the buoyant force and the unsupported object accelerates upward (Fig. 14.9a). If the density of the object is greater than the density of the fluid, the upward buoyant force is less than the downward gravitational force and the unsupported object sinks (Fig. 14.9b). If the density of the submerged object equals the density of the fluid, the net force on the object is zero and the object remains in equilibrium. Therefore, the direction of motion of an object submerged in a fluid is determined *only* by the densities of the object and the fluid.

Case 2: Floating Object Now consider an object of volume V_{obj} and density $\rho_{\text{obj}} < \rho_{\text{fluid}}$ in static equilibrium floating on the surface of a fluid, that is, an object that is only *partially* submerged (Fig. 14.10). In this case, the upward buoyant force is balanced by the downward gravitational force acting on the object. If V_{disp} is the volume of the fluid displaced by the object (this volume is the same as the volume of that part of the object beneath the surface of the fluid), the buoyant force has a magnitude $B = \rho_{\text{fluid}} g V_{\text{disp}}$. Because the weight of the object is $F_g = Mg = \rho_{\text{obj}} g V_{\text{obj}}$ and because $F_g = B$, we see that $\rho_{\text{fluid}} g V_{\text{disp}} = \rho_{\text{obj}} g V_{\text{obj}}$, or

$$\frac{V_{\text{disp}}}{V_{\text{obj}}} = \frac{\rho_{\text{obj}}}{\rho_{\text{fluid}}} \qquad \text{(14.6)}$$

This equation shows that the fraction of the volume of a floating object that is below the fluid surface is equal to the ratio of the density of the object to that of the fluid.

> **Pitfall Prevention 14.2**
> **Buoyant Force Is Exerted by the Fluid** Remember that **the buoyant force is exerted by the fluid.** It is not determined by properties of the object except for the amount of fluid displaced by the object. Therefore, if several objects of different densities but the same volume are immersed in a fluid, they will all experience the same buoyant force. Whether they sink or float is determined by the relationship between the buoyant force and the gravitational force.

Quick Quiz 14.4 You are shipwrecked and floating in the middle of the ocean on a raft. Your cargo on the raft includes a treasure chest full of gold that you found before your ship sank, and the raft is just barely afloat. To keep you floating as high as possible in the water, should you **(a)** leave the treasure chest on top of the raft, **(b)** secure the treasure chest to the underside of the raft, or **(c)** hang the treasure chest in the water with a rope attached to the raft? (Assume throwing the treasure chest overboard is not an option you wish to consider.)

Figure 14.9 (a) A totally submerged object that is less dense than the fluid in which it is submerged experiences a net upward force and rises to the surface after it is released. (b) A totally submerged object that is denser than the fluid experiences a net downward force and sinks.

Figure 14.10 An object floating on the surface of a fluid experiences two forces, the gravitational force $\vec{F}_g$ and the buoyant force $\vec{B}$.

Example 14.5   Eureka!   AM

Archimedes supposedly was asked to determine whether a crown made for the king consisted of pure gold. According to legend, he solved this problem by weighing the crown first in air and then in water as shown in Figure 14.11. Suppose the scale read 7.84 N when the crown was in air and 6.84 N when it was in water. What should Archimedes have told the king?

SOLUTION

Conceptualize Figure 14.11 helps us imagine what is happening in this example. Because of the buoyant force, the scale reading is smaller in Figure 14.11b than in Figure 14.11a.

Categorize This problem is an example of Case 1 discussed earlier because the crown is completely submerged. The scale reading is a measure of one of the forces on the crown, and the crown is stationary. Therefore, we can categorize the crown as a *particle in equilibrium*.

Figure 14.11 (Example 14.5) (a) When the crown is suspended in air, the scale reads its true weight because $T_1 = F_g$ (the buoyancy of air is negligible). (b) When the crown is immersed in water, the buoyant force $\vec{\textbf{B}}$ changes the scale reading to a lower value $T_2 = F_g - B$.

Analyze When the crown is suspended in air, the scale reads the true weight $T_1 = F_g$ (neglecting the small buoyant force due to the surrounding air). When the crown is immersed in water, the buoyant force $\vec{\textbf{B}}$ reduces the scale reading to an *apparent* weight of $T_2 = F_g - B$.

Apply the particle in equilibrium model to the crown in water:

$$\sum F = B + T_2 - F_g = 0$$

Solve for B:

$$B = F_g - T_2$$

Because this buoyant force is equal in magnitude to the weight of the displaced water, $B = \rho_w g V_{disp}$, where V_{disp} is the volume of the displaced water and ρ_w is its density. Also, the volume of the crown V_c is equal to the volume of the displaced water because the crown is completely submerged, so $B = \rho_w g V_c$.

Find the density of the crown from Equation 1.1:

$$\rho_c = \frac{m_c}{V_c} = \frac{m_c g}{V_c g} = \frac{m_c g}{(B/\rho_w)} = \frac{m_c g \rho_w}{B} = \frac{m_c g \rho_w}{F_g - T_2}$$

Substitute numerical values:

$$\rho_c = \frac{(7.84\ \text{N})(1\ 000\ \text{kg/m}^3)}{7.84\ \text{N} - 6.84\ \text{N}} = 7.84 \times 10^3\ \text{kg/m}^3$$

Finalize From Table 14.1, we see that the density of gold is $19.3 \times 10^3\ \text{kg/m}^3$. Therefore, Archimedes should have reported that the king had been cheated. Either the crown was hollow, or it was not made of pure gold.

WHAT IF?   Suppose the crown has the same weight but is indeed pure gold and not hollow. What would the scale reading be when the crown is immersed in water?

Answer Find the buoyant force on the crown:

$$B = \rho_w g V_w = \rho_w g V_c = \rho_w g \left(\frac{m_c}{\rho_c}\right) = \rho_w \left(\frac{m_c g}{\rho_c}\right)$$

Substitute numerical values:

$$B = (1.00 \times 10^3\ \text{kg/m}^3)\frac{7.84\ \text{N}}{19.3 \times 10^3\ \text{kg/m}^3} = 0.406\ \text{N}$$

Find the tension in the string hanging from the scale:

$$T_2 = F_g - B = 7.84\ \text{N} - 0.406\ \text{N} = 7.43\ \text{N}$$

Example 14.6	A Titanic Surprise

An iceberg floating in seawater as shown in Figure 14.12a is extremely dangerous because most of the ice is below the surface. This hidden ice can damage a ship that is still a considerable distance from the visible ice. What fraction of the iceberg lies below the water level?

SOLUTION

Conceptualize You are likely familiar with the phrase, "That's only the tip of the iceberg." The origin of this popular saying is that most of the volume of a floating iceberg is beneath the surface of the water (Fig. 14.12b).

Figure 14.12 (Example 14.6) (a) Much of the volume of this iceberg is beneath the water. (b) A ship can be damaged even when it is not near the visible ice.

Categorize This example corresponds to Case 2 because only part of the iceberg is underneath the water. It is also a simple substitution problem involving Equation 14.6.

Evaluate Equation 14.6 using the densities of ice and seawater (Table 14.1):

$$ f = \frac{V_{disp}}{V_{ice}} = \frac{\rho_{ice}}{\rho_{seawater}} = \frac{917 \text{ kg/m}^3}{1\,030 \text{ kg/m}^3} = \boxed{0.890 \text{ or } 89.0\%} $$

Therefore, the visible fraction of ice above the water's surface is about 11%. It is the unseen 89% below the water that represents the danger to a passing ship.

14.5 Fluid Dynamics

Thus far, our study of fluids has been restricted to fluids at rest. We now turn our attention to fluids in motion. When fluid is in motion, its flow can be characterized as being one of two main types. The flow is said to be **steady,** or **laminar,** if each particle of the fluid follows a smooth path such that the paths of different particles never cross each other as shown in Figure 14.13. In steady flow, every fluid particle arriving at a given point in space has the same velocity.

Figure 14.13 Laminar flow around an automobile in a test wind tunnel.

Above a certain critical speed, fluid flow becomes **turbulent.** Turbulent flow is irregular flow characterized by small whirlpool-like regions as shown in Figure 14.14.

The term **viscosity** is commonly used in the description of fluid flow to characterize the degree of internal friction in the fluid. This internal friction, or *viscous force,* is associated with the resistance that two adjacent layers of fluid have to moving relative to each other. Viscosity causes part of the fluid's kinetic energy to be transformed to internal energy. This mechanism is similar to the one by which the kinetic energy of an object sliding over a rough, horizontal surface decreases as discussed in Sections 8.3 and 8.4.

Because the motion of real fluids is very complex and not fully understood, we make some simplifying assumptions in our approach. In our simplification model of **ideal fluid flow,** we make the following four assumptions:

1. **The fluid is nonviscous.** In a nonviscous fluid, internal friction is neglected. An object moving through the fluid experiences no viscous force.
2. **The flow is steady.** In steady (laminar) flow, all particles passing through a point have the same velocity.
3. **The fluid is incompressible.** The density of an incompressible fluid is constant.
4. **The flow is irrotational.** In irrotational flow, the fluid has no angular momentum about any point. If a small paddle wheel placed anywhere in the fluid does not rotate about the wheel's center of mass, the flow is irrotational.

Figure 14.14 Hot gases from a cigarette made visible by smoke particles. The smoke first moves in laminar flow at the bottom and then in turbulent flow above.

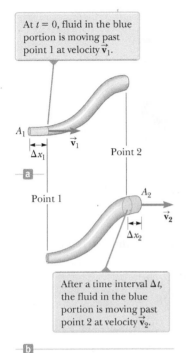

At $t = 0$, fluid in the blue portion is moving past point 1 at velocity $\vec{v}_1$.

A_1

Δx_1 $\vec{v}_1$

Point 2

a

Point 1 A_2

$\vec{v}_2$

Δx_2

After a time interval Δt, the fluid in the blue portion is moving past point 2 at velocity $\vec{v}_2$.

b

Figure 14.16 A fluid moving with steady flow through a pipe of varying cross-sectional area. (a) At $t = 0$, the small blue-colored portion of the fluid at the left is moving through area A_1. (b) After a time interval Δt, the blue-colored portion shown here is that fluid that has moved through area A_2.

▶ **Equation of Continuity for Fluids**

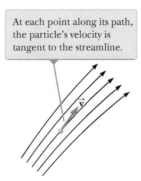

At each point along its path, the particle's velocity is tangent to the streamline.

$\vec{v}$

Figure 14.15 A particle in laminar flow follows a streamline.

The path taken by a fluid particle under steady flow is called a **streamline.** The velocity of the particle is always tangent to the streamline as shown in Figure 14.15. A set of streamlines like the ones shown in Figure 14.15 form a *tube of flow.* Fluid particles cannot flow into or out of the sides of this tube; if they could, the streamlines would cross one another.

Consider ideal fluid flow through a pipe of nonuniform size as illustrated in Figure 14.16. Let's focus our attention on a segment of fluid in the pipe. Figure 14.16a shows the segment at time $t = 0$ consisting of the gray portion between point 1 and point 2 and the short blue portion to the left of point 1. At this time, the fluid in the short blue portion is flowing through a cross section of area A_1 at speed v_1. During the time interval Δt, the small length Δx_1 of fluid in the blue portion moves past point 1. During the same time interval, fluid at the right end of the segment moves past point 2 in the pipe. Figure 14.16b shows the situation at the end of the time interval Δt. The blue portion at the right end represents the fluid that has moved past point 2 through an area A_2 at a speed v_2.

The mass of fluid contained in the blue portion in Figure 14.16a is given by $m_1 = \rho A_1\,\Delta x_1 = \rho A_1 v_1\,\Delta t$, where ρ is the (unchanging) density of the ideal fluid. Similarly, the fluid in the blue portion in Figure 14.16b has a mass $m_2 = \rho A_2\,\Delta x_2 = \rho A_2 v_2\,\Delta t$. Because the fluid is incompressible and the flow is steady, however, the mass of fluid that passes point 1 in a time interval Δt must equal the mass that passes point 2 in the same time interval. That is, $m_1 = m_2$ or $\rho A_1 v_1\,\Delta t = \rho A_2 v_2\,\Delta t$, which means that

$$A_1 v_1 = A_2 v_2 = \text{constant} \qquad (14.7)$$

This expression is called the **equation of continuity for fluids.** It states that the product of the area and the fluid speed at all points along a pipe is constant for an incompressible fluid. Equation 14.7 shows that the speed is high where the tube is constricted (small A) and low where the tube is wide (large A). The product Av, which has the dimensions of volume per unit time, is called either the *volume flux* or the *flow rate.* The condition $Av = \text{constant}$ is equivalent to the statement that the volume of fluid that enters one end of a tube in a given time interval equals the volume leaving the other end of the tube in the same time interval if no leaks are present.

You demonstrate the equation of continuity each time you water your garden with your thumb over the end of a garden hose as in Figure 14.17. By partially block-

© Cengage Learning/George Semple

Figure 14.17 The speed of water spraying from the end of a garden hose increases as the size of the opening is decreased with the thumb.

ing the opening with your thumb, you reduce the cross-sectional area through which the water passes. As a result, the speed of the water increases as it exits the hose, and the water can be sprayed over a long distance.

Example 14.7   Watering a Garden   AM

A gardener uses a water hose to fill a 30.0-L bucket. The gardener notes that it takes 1.00 min to fill the bucket. A nozzle with an opening of cross-sectional area 0.500 cm² is then attached to the hose. The nozzle is held so that water is projected horizontally from a point 1.00 m above the ground. Over what horizontal distance can the water be projected?

SOLUTION

Conceptualize Imagine any past experience you have with projecting water from a horizontal hose or a pipe using either your thumb or a nozzle, which can be attached to the end of the hose. The faster the water is traveling as it leaves the hose, the farther it will land on the ground from the end of the hose.

Categorize Once the water leaves the hose, it is in free fall. Therefore, we categorize a given element of the water as a projectile. The element is modeled as a *particle under constant acceleration* (due to gravity) in the vertical direction and a *particle under constant velocity* in the horizontal direction. The horizontal distance over which the element is projected depends on the speed with which it is projected. This example involves a change in area for the pipe, so we also categorize it as one in which we use the continuity equation for fluids.

Analyze

Express the volume flow rate R in terms of area and speed of the water in the hose:

$$R = A_1 v_1$$

Solve for the speed of the water in the hose:

$$v_1 = \frac{R}{A_1}$$

We have labeled this speed v_1 because we identify point 1 within the hose. We identify point 2 in the air just outside the nozzle. We must find the speed $v_2 = v_{xi}$ with which the water exits the nozzle. The subscript i anticipates that it will be the *initial* velocity component of the water projected from the hose, and the subscript x indicates that the initial velocity vector of the projected water is horizontal.

Solve the continuity equation for fluids for v_2:

$$(1) \quad v_2 = v_{xi} = \frac{A_1}{A_2} v_1 = \frac{A_1}{A_2}\left(\frac{R}{A_1}\right) = \frac{R}{A_2}$$

We now shift our thinking away from fluids and to projectile motion. In the vertical direction, an element of the water starts from rest and falls through a vertical distance of 1.00 m.

Write Equation 2.16 for the vertical position of an element of water, modeled as a particle under constant acceleration:

$$y_f = y_i + v_{yi}t - \tfrac{1}{2}gt^2$$

Call the initial position of the water $y_i = 0$ and recognize that the water begins with a vertical velocity component of zero. Solve for the time at which the water reaches the ground:

$$(2) \quad y_f = 0 + 0 - \tfrac{1}{2}gt^2 \rightarrow t = \sqrt{\frac{-2y_f}{g}}$$

Use Equation 2.7 to find the horizontal position of the element at this time, modeled as a particle under constant velocity:

$$x_f = x_i + v_{xi}t = 0 + v_2 t = v_2 t$$

Substitute from Equations (1) and (2):

$$x_f = \frac{R}{A_2}\sqrt{\frac{-2y_f}{g}}$$

Substitute numerical values:

$$x_f = \frac{30.0\ \text{L/min}}{0.500\ \text{cm}^2}\sqrt{\frac{-2(-1.00\ \text{m})}{9.80\ \text{m/s}^2}}\left(\frac{10^3\ \text{cm}^3}{1\ \text{L}}\right)\left(\frac{1\ \text{min}}{60\ \text{s}}\right) = 452\ \text{cm} = 4.52\ \text{m}$$

continued

▶ **14.7** continued

Finalize The time interval for the element of water to fall to the ground is unchanged if the projection speed is changed because the projection is horizontal. Increasing the projection speed results in the water hitting the ground farther from the end of the hose, but requires the same time interval to strike the ground.

Daniel Bernoulli
Swiss physicist (1700–1782)
Bernoulli made important discoveries in fluid dynamics. Bernoulli's most famous work, *Hydrodynamica*, was published in 1738; it is both a theoretical and a practical study of equilibrium, pressure, and speed in fluids. He showed that as the speed of a fluid increases, its pressure decreases. Referred to as "Bernoulli's principle," Bernoulli's work is used to produce a partial vacuum in chemical laboratories by connecting a vessel to a tube through which water is running rapidly.

© iStockphoto.com/ZU_09

14.6 Bernoulli's Equation

You have probably experienced driving on a highway and having a large truck pass you at high speed. In this situation, you may have had the frightening feeling that your car was being pulled in toward the truck as it passed. We will investigate the origin of this effect in this section.

As a fluid moves through a region where its speed or elevation above the Earth's surface changes, the pressure in the fluid varies with these changes. The relationship between fluid speed, pressure, and elevation was first derived in 1738 by Swiss physicist Daniel Bernoulli. Consider the flow of a segment of an ideal fluid through a nonuniform pipe in a time interval Δt as illustrated in Figure 14.18. This figure is very similar to Figure 14.16, which we used to develop the continuity equation. We have added two features: the forces on the outer ends of the blue portions of fluid and the heights of these portions above the reference position $y = 0$.

The force exerted on the segment by the fluid to the left of the blue portion in Figure 14.18a has a magnitude P_1A_1. The work done by this force on the segment in a time interval Δt is $W_1 = F_1 \Delta x_1 = P_1A_1 \Delta x_1 = P_1V$, where V is the volume of the blue portion of fluid passing point 1 in Figure 14.18a. In a similar manner, the work done on the segment by the fluid to the right of the segment in the same time interval Δt is $W_2 = -P_2A_2 \Delta x_2 = -P_2V$, where V is the volume of the blue portion of fluid passing point 2 in Figure 14.18b. (The volumes of the blue portions of fluid in Figures 14.18a and 14.18b are equal because the fluid is incompressible.) This work is negative because the force on the segment of fluid is to the left and the displacement of the point of application of the force is to the right. Therefore, the net work done on the segment by these forces in the time interval Δt is

$$W = (P_1 - P_2)V$$

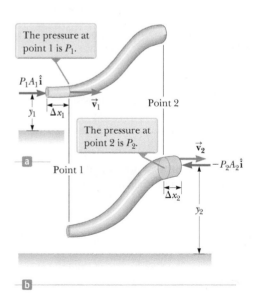

Figure 14.18 A fluid in laminar flow through a pipe. (a) A segment of the fluid at time $t = 0$. A small portion of the blue-colored fluid is at height y_1 above a reference position. (b) After a time interval Δt, the entire segment has moved to the right. The blue-colored portion of the fluid is that which has passed point 2 and is at height y_2.

Part of this work goes into changing the kinetic energy of the segment of fluid, and part goes into changing the gravitational potential energy of the segment–Earth system. Because we are assuming streamline flow, the kinetic energy K_{gray} of the gray portion of the segment is the same in both parts of Figure 14.18. Therefore, the change in the kinetic energy of the segment of fluid is

$$\Delta K = \left(\tfrac{1}{2}mv_2^2 + K_{gray}\right) - \left(\tfrac{1}{2}mv_1^2 + K_{gray}\right) = \tfrac{1}{2}mv_2^2 - \tfrac{1}{2}mv_1^2$$

where m is the mass of the blue portions of fluid in both parts of Figure 14.18. (Because the volumes of both portions are the same, they also have the same mass.)

Considering the gravitational potential energy of the segment–Earth system, once again there is no change during the time interval for the gravitational potential energy U_{gray} associated with the gray portion of the fluid. Consequently, the change in gravitational potential energy of the system is

$$\Delta U = \left(mgy_2 + U_{gray}\right) - \left(mgy_1 + U_{gray}\right) = mgy_2 - mgy_1$$

From Equation 8.2, the total work done on the system by the fluid outside the segment is equal to the change in mechanical energy of the system: $W = \Delta K + \Delta U$. Substituting for each of these terms gives

$$(P_1 - P_2)V = \tfrac{1}{2}mv_2^2 - \tfrac{1}{2}mv_1^2 + mgy_2 - mgy_1$$

If we divide each term by the portion volume V and recall that $\rho = m/V$, this expression reduces to

$$P_1 - P_2 = \tfrac{1}{2}\rho v_2^2 - \tfrac{1}{2}\rho v_1^2 + \rho gy_2 - \rho gy_1$$

Rearranging terms gives

$$P_1 + \tfrac{1}{2}\rho v_1^2 + \rho gy_1 = P_2 + \tfrac{1}{2}\rho v_2^2 + \rho gy_2 \qquad \textbf{(14.8)}$$

which is **Bernoulli's equation** as applied to an ideal fluid. This equation is often expressed as

$$P + \tfrac{1}{2}\rho v^2 + \rho gy = \text{constant} \qquad \textbf{(14.9)}$$ ◀ Bernoulli's equation

Bernoulli's equation shows that the pressure of a fluid decreases as the speed of the fluid increases. In addition, the pressure decreases as the elevation increases. This latter point explains why water pressure from faucets on the upper floors of a tall building is weak unless measures are taken to provide higher pressure for these upper floors.

When the fluid is at rest, $v_1 = v_2 = 0$ and Equation 14.8 becomes

$$P_1 - P_2 = \rho g(y_2 - y_1) = \rho gh$$

This result is in agreement with Equation 14.4.

Although Equation 14.9 was derived for an incompressible fluid, the general behavior of pressure with speed is true even for gases: as the speed increases, the pressure decreases. This *Bernoulli effect* explains the experience with the truck on the highway at the opening of this section. As air passes between you and the truck, it must pass through a relatively narrow channel. According to the continuity equation, the speed of the air is higher. According to the Bernoulli effect, this higher-speed air exerts less pressure on your car than the slower-moving air on the other side of your car. Therefore, there is a net force pushing you toward the truck!

Quick Quiz 14.5 You observe two helium balloons floating next to each other at the ends of strings secured to a table. The facing surfaces of the balloons are separated by 1–2 cm. You blow through the small space between the balloons. What happens to the balloons? **(a)** They move toward each other. **(b)** They move away from each other. **(c)** They are unaffected.

Example 14.8   The Venturi Tube

The horizontal constricted pipe illustrated in Figure 14.19, known as a *Venturi tube*, can be used to measure the flow speed of an incompressible fluid. Determine the flow speed at point 2 of Figure 14.19a if the pressure difference $P_1 - P_2$ is known.

SOLUTION

Conceptualize Bernoulli's equation shows how the pressure of an ideal fluid decreases as its speed increases. Therefore, we should be able to calibrate a device to give us the fluid speed if we can measure pressure.

Categorize Because the problem states that the fluid is incompressible, we can categorize it as one in which we can use the equation of continuity for fluids and Bernoulli's equation.

Figure 14.19 (Example 14.8) (a) Pressure P_1 is greater than pressure P_2 because $v_1 < v_2$. This device can be used to measure the speed of fluid flow. (b) A Venturi tube, located at the top of the photograph. The higher level of fluid in the middle column shows that the pressure at the top of the column, which is in the constricted region of the Venturi tube, is lower.

Analyze Apply Equation 14.8 to points 1 and 2, noting that $y_1 = y_2$ because the pipe is horizontal:

$$(1) \quad P_1 + \tfrac{1}{2}\rho v_1^2 = P_2 + \tfrac{1}{2}\rho v_2^2$$

Solve the equation of continuity for v_1:

$$v_1 = \frac{A_2}{A_1} v_2$$

Substitute this expression into Equation (1):

$$P_1 + \tfrac{1}{2}\rho\left(\frac{A_2}{A_1}\right)^2 v_2^2 = P_2 + \tfrac{1}{2}\rho v_2^2$$

Solve for v_2:

$$v_2 = A_1 \sqrt{\frac{2(P_1 - P_2)}{\rho(A_1^2 - A_2^2)}}$$

Finalize From the design of the tube (areas A_1 and A_2) and measurements of the pressure difference $P_1 - P_2$, we can calculate the speed of the fluid with this equation. To see the relationship between fluid speed and pressure difference, place two empty soda cans on their sides about 2 cm apart on a table. Gently blow a stream of air horizontally between the cans and watch them roll together slowly due to a modest pressure difference between the stagnant air on their outside edges and the moving air between them. Now blow more strongly and watch the increased pressure difference move the cans together more rapidly.

Example 14.9   Torricelli's Law   AM

An enclosed tank containing a liquid of density ρ has a hole in its side at a distance y_1 from the tank's bottom (Fig. 14.20). The hole is open to the atmosphere, and its diameter is much smaller than the diameter of the tank. The air above the liquid is maintained at a pressure P. Determine the speed of the liquid as it leaves the hole when the liquid's level is a distance h above the hole.

SOLUTION

Conceptualize Imagine that the tank is a fire extinguisher. When the hole is opened, liquid leaves the hole with a certain speed. If the pressure P at the top of the liquid is increased, the liquid leaves with a higher speed. If the pressure P falls too low, the liquid leaves with a low speed and the extinguisher must be replaced.

Figure 14.20 (Example 14.9) A liquid leaves a hole in a tank at speed v_1.

▶ **14.9** continued

Categorize Looking at Figure 14.20, we know the pressure at two points and the velocity at one of those points. We wish to find the velocity at the second point. Therefore, we can categorize this example as one in which we can apply Bernoulli's equation.

....

Analyze Because $A_2 \gg A_1$, the liquid is approximately at rest at the top of the tank, where the pressure is P. At the hole, P_1 is equal to atmospheric pressure P_0.

Apply Bernoulli's equation between points 1 and 2:

$$P_0 + \tfrac{1}{2}\rho v_1^{\,2} + \rho g y_1 = P + \rho g y_2$$

Solve for v_1, noting that $y_2 - y_1 = h$:

$$v_1 = \sqrt{\frac{2(P - P_0)}{\rho} + 2gh}$$

....

Finalize When P is much greater than P_0 (so that the term $2gh$ can be neglected), the exit speed of the water is mainly a function of P. If the tank is open to the atmosphere, then $P = P_0$ and $v_1 = \sqrt{2gh}$. In other words, for an open tank, the speed of the liquid leaving a hole a distance h below the surface is equal to that acquired by an object falling freely through a vertical distance h. This phenomenon is known as *Torricelli's law*.

WHAT IF?   What if the position of the hole in Figure 14.20 could be adjusted vertically? If the tank is open to the atmosphere and sitting on a table, what position of the hole would cause the water to land on the table at the farthest distance from the tank?

Answer Model a parcel of water exiting the hole as a projectile. From the *particle under constant acceleration* model, find the time at which the parcel strikes the table from a hole at an arbitrary position y_1:

$$y_f = y_i + v_{yi}t - \tfrac{1}{2}gt^2$$
$$0 = y_1 + 0 - \tfrac{1}{2}gt^2$$
$$t = \sqrt{\frac{2y_1}{g}}$$

From the *particle under constant velocity* model, find the horizontal position of the parcel at the time it strikes the table:

$$x_f = x_i + v_{xi}t = 0 + \sqrt{2g(y_2 - y_1)}\,\sqrt{\frac{2y_1}{g}}$$
$$= 2\sqrt{(y_2 y_1 - y_1^{\,2})}$$

Maximize the horizontal position by taking the derivative of x_f with respect to y_1 (because y_1, the height of the hole, is the variable that can be adjusted) and setting it equal to zero:

$$\frac{dx_f}{dy_1} = \tfrac{1}{2}(2)(y_2 y_1 - y_1^{\,2})^{-1/2}(y_2 - 2y_1) = 0$$

Solve for y_1:

$$y_1 = \tfrac{1}{2}y_2$$

Therefore, to maximize the horizontal distance, the hole should be halfway between the bottom of the tank and the upper surface of the water. Below this location, the water is projected at a higher speed but falls for a short time interval, reducing the horizontal range. Above this point, the water is in the air for a longer time interval but is projected with a smaller horizontal speed.

14.7 Other Applications of Fluid Dynamics

Consider the streamlines that flow around an airplane wing as shown in Figure 14.21 on page 434. Let's assume the airstream approaches the wing horizontally from the right with a velocity $\vec{v}_1$. The tilt of the wing causes the airstream to be deflected downward with a velocity $\vec{v}_2$. Because the airstream is deflected by the wing, the wing must exert a force on the airstream. According to Newton's third law, the airstream exerts a force $\vec{F}$ on the wing that is equal in magnitude and

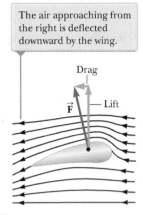

The air approaching from the right is deflected downward by the wing.

Figure 14.21 Streamline flow around a moving airplane wing. By Newton's third law, the air deflected by the wing results in an upward force on the wing from the air: *lift*. Because of air resistance, there is also a force opposite the velocity of the wing: *drag*.

opposite in direction. This force has a vertical component called **lift** (or aerodynamic lift) and a horizontal component called **drag.** The lift depends on several factors, such as the speed of the airplane, the area of the wing, the wing's curvature, and the angle between the wing and the horizontal. The curvature of the wing surfaces causes the pressure above the wing to be lower than that below the wing due to the Bernoulli effect. This pressure difference assists with the lift on the wing. As the angle between the wing and the horizontal increases, turbulent flow can set in above the wing to reduce the lift.

In general, an object moving through a fluid experiences lift as the result of any effect that causes the fluid to change its direction as it flows past the object. Some factors that influence lift are the shape of the object, its orientation with respect to the fluid flow, any spinning motion it might have, and the texture of its surface. For example, a golf ball struck with a club is given a rapid backspin due to the slant of the club. The dimples on the ball increase the friction force between the ball and the air so that air adheres to the ball's surface. Figure 14.22 shows air adhering to the ball and being deflected downward as a result. Because the ball pushes the air down, the air must push up on the ball. Without the dimples, the friction force is lower and the golf ball does not travel as far. It may seem counterintuitive to increase the range by increasing the friction force, but the lift gained by spinning the ball more than compensates for the loss of range due to the effect of friction on the translational motion of the ball. For the same reason, a baseball's cover helps the spinning ball "grab" the air rushing by and helps deflect it when a "curve ball" is thrown.

A number of devices operate by means of the pressure differentials that result from differences in a fluid's speed. For example, a stream of air passing over one end of an open tube, the other end of which is immersed in a liquid, reduces the pressure above the tube as illustrated in Figure 14.23. This reduction in pressure causes the liquid to rise into the airstream. The liquid is then dispersed into a fine spray of droplets. You might recognize that this *atomizer* is used in perfume bottles and paint sprayers.

Figure 14.22 Because of the deflection of air, a spinning golf ball experiences a lifting force that allows it to travel much farther than it would if it were not spinning.

Figure 14.23 A stream of air passing over a tube dipped into a liquid causes the liquid to rise in the tube.

Summary

Definitions

The **pressure** P in a fluid is the force per unit area exerted by the fluid on a surface:

$$P \equiv \frac{F}{A} \tag{14.1}$$

In the SI system, pressure has units of newtons per square meter (N/m^2), and $1 \ N/m^2 = 1$ **pascal** (Pa).

Concepts and Principles

The pressure in a fluid at rest varies with depth h in the fluid according to the expression

$$P = P_0 + \rho g h \qquad (14.4)$$

where P_0 is the pressure at $h = 0$ and ρ is the density of the fluid, assumed uniform.

Pascal's law states that when pressure is applied to an enclosed fluid, the pressure is transmitted undiminished to every point in the fluid and to every point on the walls of the container.

When an object is partially or fully submerged in a fluid, the fluid exerts on the object an upward force called the **buoyant force.** According to **Archimedes's principle,** the magnitude of the buoyant force is equal to the weight of the fluid displaced by the object:

$$B = \rho_{\text{fluid}} g V_{\text{disp}} \qquad (14.5)$$

The flow rate (volume flux) through a pipe that varies in cross-sectional area is constant; that is equivalent to stating that the product of the cross-sectional area A and the speed v at any point is a constant. This result is expressed in the **equation of continuity for fluids:**

$$A_1 v_1 = A_2 v_2 = \text{constant} \qquad (14.7)$$

The sum of the pressure, kinetic energy per unit volume, and gravitational potential energy per unit volume has the same value at all points along a streamline for an ideal fluid. This result is summarized in **Bernoulli's equation:**

$$P + \tfrac{1}{2}\rho v^2 + \rho g y = \text{constant} \qquad (14.9)$$

Objective Questions

1. denotes answer available in *Student Solutions Manual/Study Guide*

1. Figure OQ14.1 shows aerial views from directly above two dams. Both dams are equally wide (the vertical dimension in the diagram) and equally high (into the page in the diagram). The dam on the left holds back a very large lake, and the dam on the right holds back a narrow river. Which dam has to be built more strongly? (a) the dam on the left (b) the dam on the right (c) both the same (d) cannot be predicted

Dam   Dam

Figure OQ14.1

2. A beach ball filled with air is pushed about 1 m below the surface of a swimming pool and released from rest. Which of the following statements are valid, assuming the size of the ball remains the same? (Choose all correct statements.) (a) As the ball rises in the pool, the buoyant force on it increases. (b) When the ball is released, the buoyant force exceeds the gravitational force, and the ball accelerates upward. (c) The buoyant force on the ball decreases as the ball approaches the surface of the pool. (d) The buoyant force on the ball equals its weight and remains constant as the ball rises. (e) The buoyant force on the ball while it is submerged is approximately equal to the weight of a volume of water that could fill the ball.

3. A wooden block floats in water, and a steel object is attached to the bottom of the block by a string as in Figure OQ14.3. If the block remains floating, which

of the following statements are valid? (Choose all correct statements.) (a) The buoyant force on the steel object is equal to its weight. (b) The buoyant force on the block is equal to its weight. (c) The tension in the string is equal to the weight of the steel object. (d) The tension in the string is less than the weight of the steel object. (e) The buoyant force on the block is equal to the volume of water it displaces.

Figure OQ14.3

4. An apple is held completely submerged just below the surface of water in a container. The apple is then moved to a deeper point in the water. Compared with the force needed to hold the apple just below the surface, what is the force needed to hold it at the deeper point? (a) larger (b) the same (c) smaller (d) impossible to determine

5. A beach ball is made of thin plastic. It has been inflated with air, but the plastic is not stretched. By swimming with fins on, you manage to take the ball from the surface of a pool to the bottom. Once the ball is completely submerged, what happens to the buoyant force exerted on the beach ball as you take it deeper? (a) It increases. (b) It remains constant. (c) It decreases. (d) It is impossible to determine.

6. A solid iron sphere and a solid lead sphere of the same size are each suspended by strings and are submerged in a tank of water. (Note that the density of lead is greater than that of iron.) Which of the following statements are valid? (Choose all correct statements.) (a) The buoyant force on each is the same. (b) The buoyant force on the lead sphere is greater than the buoyant force on the iron sphere because lead has the greater density. (c) The tension in the string supporting the lead sphere is greater than the tension in the string supporting the iron sphere. (d) The buoyant force on the iron sphere is greater than the buoyant force on the lead sphere because lead displaces more water. (e) None of those statements is true.

7. Three vessels of different shapes are filled to the same level with water as in Figure OQ14.7. The area of the base is the same for all three vessels. Which of the following statements are valid? (Choose all correct statements.) (a) The pressure at the top surface of vessel A is greatest because it has the largest surface area. (b) The pressure at the bottom of vessel A is greatest because it contains the most water. (c) The pressure at the bottom of each vessel is the same. (d) The force on the bottom of each vessel is not the same. (e) At a given depth below the surface of each vessel, the pressure on the side of vessel A is greatest because of its slope.

Figure OQ14.7

8. One of the predicted problems due to global warming is that ice in the polar ice caps will melt and raise sea levels everywhere in the world. Is that more of a worry for ice (a) at the north pole, where most of the ice floats on water; (b) at the south pole, where most of the ice sits on land; (c) both at the north and south pole equally; or (d) at neither pole?

9. A boat develops a leak and, after its passengers are rescued, eventually sinks to the bottom of a lake. When the boat is at the bottom, what is the force of the lake bottom on the boat? (a) greater than the weight of the boat (b) equal to the weight of the boat (c) less than

the weight of the boat (d) equal to the weight of the displaced water (e) equal to the buoyant force on the boat

10. A small piece of steel is tied to a block of wood. When the wood is placed in a tub of water with the steel on top, half of the block is submerged. Now the block is inverted so that the steel is under water. **(i)** Does the amount of the block submerged (a) increase, (b) decrease, or (c) remain the same? **(ii)** What happens to the water level in the tub when the block is inverted? (a) It rises. (b) It falls. (c) It remains the same.

11. A piece of unpainted porous wood barely floats in an open container partly filled with water. The container is then sealed and pressurized above atmospheric pressure. What happens to the wood? (a) It rises in the water. (b) It sinks lower in the water. (c) It remains at the same level.

12. A person in a boat floating in a small pond throws an anchor overboard. What happens to the level of the pond? (a) It rises. (b) It falls. (c) It remains the same.

13. Rank the buoyant forces exerted on the following five objects of equal volume from the largest to the smallest. Assume the objects have been dropped into a swimming pool and allowed to come to mechanical equilibrium. If any buoyant forces are equal, state that in your ranking. (a) a block of solid oak (b) an aluminum block (c) a beach ball made of thin plastic and inflated with air (d) an iron block (e) a thin-walled, sealed bottle of water

14. A water supply maintains a constant rate of flow for water in a hose. You want to change the opening of the nozzle so that water leaving the nozzle will reach a height that is four times the current maximum height the water reaches with the nozzle vertical. To do so, should you (a) decrease the area of the opening by a factor of 16, (b) decrease the area by a factor of 8, (c) decrease the area by a factor of 4, (d) decrease the area by a factor of 2, or (e) give up because it cannot be done?

15. A glass of water contains floating ice cubes. When the ice melts, does the water level in the glass (a) go up, (b) go down, or (c) remain the same?

16. An ideal fluid flows through a horizontal pipe whose diameter varies along its length. Measurements would indicate that the sum of the kinetic energy per unit volume and pressure at different sections of the pipe would (a) decrease as the pipe diameter increases, (b) increase as the pipe diameter increases, (c) increase as the pipe diameter decreases, (d) decrease as the pipe diameter decreases, or (e) remain the same as the pipe diameter changes.

Conceptual Questions **1.** denotes answer available in *Student Solutions Manual/Study Guide*

1. When an object is immersed in a liquid at rest, why is the net force on the object in the horizontal direction equal to zero?

2. Two thin-walled drinking glasses having equal base areas but different shapes, with very different cross-sectional areas above the base, are filled to the same

level with water. According to the expression $P = P_0 + \rho g h$, the pressure is the same at the bottom of both glasses. In view of this equality, why does one weigh more than the other?

3. Because atmospheric pressure is about 10^5 N/m² and the area of a person's chest is about 0.13 m², the force of the atmosphere on one's chest is around 13 000 N. In view of this enormous force, why don't our bodies collapse?

4. A fish rests on the bottom of a bucket of water while the bucket is being weighed on a scale. When the fish begins to swim around, does the scale reading change? Explain your answer.

5. You are a passenger on a spacecraft. For your survival and comfort, the interior contains air just like that at the surface of the Earth. The craft is coasting through a very empty region of space. That is, a nearly perfect vacuum exists just outside the wall. Suddenly, a meteoroid pokes a hole, about the size of a large coin, right through the wall next to your seat. (a) What happens? (b) Is there anything you can or should do about it?

6. If the airstream from a hair dryer is directed over a table-tennis ball, the ball can be levitated. Explain.

7. A water tower is a common sight in many communities. Figure CQ14.7 shows a collection of colorful water towers in Kuwait City, Kuwait. Notice that the large weight of the water results in the center of mass of the system being high above the ground. Why is it desirable for a water tower to have this highly unstable shape rather than being shaped as a tall cylinder?

Figure CQ14.7

8. If you release a ball while inside a freely falling elevator, the ball remains in front of you rather than falling to the floor because the ball, the elevator, and you all experience the same downward gravitational acceleration. What happens if you repeat this experiment with a helium-filled balloon?

9. (a) Is the buoyant force a conservative force? (b) Is a potential energy associated with the buoyant force? (c) Explain your answers to parts (a) and (b).

10. An empty metal soap dish barely floats in water. A bar of Ivory soap floats in water. When the soap is stuck in the soap dish, the combination sinks. Explain why.

11. How would you determine the density of an irregularly shaped rock?

12. Place two cans of soft drinks, one regular and one diet, in a container of water. You will find that the diet drink floats while the regular one sinks. Use Archimedes's principle to devise an explanation.

13. The water supply for a city is often provided from reservoirs built on high ground. Water flows from the reservoir, through pipes, and into your home when you turn the tap on your faucet. Why does water flow more rapidly out of a faucet on the first floor of a building than in an apartment on a higher floor?

14. Does a ship float higher in the water of an inland lake or in the ocean? Why?

15. When ski jumpers are airborne (Fig. CQ14.15), they bend their bodies forward and keep their hands at their sides. Why?

Figure CQ14.15

16. Why do airplane pilots prefer to take off with the airplane facing into the wind?

17. Prairie dogs ventilate their burrows by building a mound around one entrance, which is open to a stream of air when wind blows from any direction. A second entrance at ground level is open to almost stagnant air. How does this construction create an airflow through the burrow?

18. In Figure CQ14.18, an airstream moves from right to left through a tube that is constricted at the middle. Three table-tennis balls are levitated in equilibrium above the vertical columns through which the air escapes. (a) Why is the ball at the right higher than the one in the middle? (b) Why is the ball at the left lower than the ball at the right even though the horizontal tube has the same dimensions at these two points?

Figure CQ14.18

19. A typical silo on a farm has many metal bands wrapped around its perimeter for support as shown in Figure CQ14.19. Why is the spacing between successive bands smaller for the lower portions of the silo on the left, and why are double bands used at lower portions of the silo on the right?

Figure CQ14.19

Problems

Problems

WebAssign The problems found in this chapter may be assigned online in Enhanced WebAssign

1. straightforward; **2.** intermediate; **3.** challenging

1. full solution available in the *Student Solutions Manual/Study Guide*

AMT Analysis Model tutorial available in Enhanced WebAssign

GP Guided Problem

M Master It tutorial available in Enhanced WebAssign

W Watch It video solution available in Enhanced WebAssign

Note: In all problems, assume the density of air is the 20°C value from Table 14.1, 1.20 kg/m³, unless noted otherwise.

Section 14.1 Pressure

1. A large man sits on a four-legged chair with his feet off the floor. The combined mass of the man and chair is 95.0 kg. If the chair legs are circular and have a radius of 0.500 cm at the bottom, what pressure does each leg exert on the floor?

2. The nucleus of an atom can be modeled as several protons and neutrons closely packed together. Each particle has a mass of 1.67×10^{-27} kg and radius on the order of 10^{-15} m. (a) Use this model and the data provided to estimate the density of the nucleus of an atom. (b) Compare your result with the density of a material such as iron. What do your result and comparison suggest concerning the structure of matter?

3. A 50.0-kg woman wearing high-heeled shoes is invited into a home in which the kitchen has vinyl floor covering. The heel on each shoe is circular and has a radius of 0.500 cm. (a) If the woman balances on one heel, what pressure does she exert on the floor? (b) Should the homeowner be concerned? Explain your answer.

4. Estimate the total mass of the Earth's atmosphere. (The radius of the Earth is 6.37×10^6 m, and atmospheric pressure at the surface is 1.013×10^5 Pa.)

5. Calculate the mass of a solid gold rectangular bar that has dimensions of 4.50 cm × 11.0 cm × 26.0 cm.

Section 14.2 Variation of Pressure with Depth

6. (a) A very powerful vacuum cleaner has a hose 2.86 cm in diameter. With the end of the hose placed perpendicularly on the flat face of a brick, what is the weight of the heaviest brick that the cleaner can lift? (b) **What If?** An octopus uses one sucker of diameter 2.86 cm on each of the two shells of a clam in an attempt to pull the shells apart. Find the greatest force the octopus can exert on a clamshell in salt water 32.3 m deep.

7. The spring of the pressure gauge shown in Figure P14.7 has a force constant of 1 250 N/m, and the piston has a diameter of 1.20 cm. As the gauge is lowered into water in a lake, what change in depth causes the piston to move in by 0.750 cm?

Figure P14.7

8. The small piston of a hydraulic lift (Fig. P14.8) has a cross-sectional area of 3.00 cm², and its large piston has a cross-sectional area of 200 cm². What downward force of magnitude F_1 must be applied to the small piston for the lift to raise a load whose weight is $F_g = 15.0$ kN?

Figure P14.8

9. What must be the contact area between a suction cup (completely evacuated) and a ceiling if the cup is to support the weight of an 80.0-kg student?

10. A swimming pool has dimensions 30.0 m × 10.0 m and a flat bottom. When the pool is filled to a depth of 2.00 m with fresh water, what is the force exerted by the water on (a) the bottom? (b) On each end? (c) On each side?

11. (a) Calculate the absolute pressure at the bottom of a freshwater lake at a point whose depth is 27.5 m. Assume the density of the water is 1.00×10^3 kg/m³ and that the air above is at a pressure of 101.3 kPa. (b) What force is exerted by the water on the window of an underwater vehicle at this depth if the window is circular and has a diameter of 35.0 cm?

12. *Why is the following situation impossible?* Figure P14.12 shows Superman attempting to drink cold water through a straw of length $\ell = 12.0$ m. The walls of the tubular straw are very strong and do not collapse. With his great strength, he achieves maximum possible suction and enjoys drinking the cold water.

Figure P14.12

13. For the cellar of a new house, a hole is dug in the ground, with vertical sides going down 2.40 m. A concrete foundation wall is built all the way across the 9.60-m width of the excavation. This foundation wall is 0.183 m away from the front of the cellar hole. During a rainstorm, drainage from the street fills up the space in front of the concrete wall, but not the cellar behind the wall. The water does not soak into the clay soil. Find the force the water causes on the foundation wall. For comparison, the weight of the water is given by 2.40 m $\times$ 9.60 m $\times$ 0.183 m $\times$ 1 000 kg/m^3 $\times$ 9.80 m/s^2 = 41.3 kN.

14. A container is filled to a depth of 20.0 cm with water. On top of the water floats a 30.0-cm-thick layer of oil with specific gravity 0.700. What is the absolute pressure at the bottom of the container?

15. **Review.** The tank in Figure P14.15 is filled with water of depth $d = 2.00$ m. At the bottom of one sidewall is a rectangular hatch of height $h = 1.00$ m and width $w = 2.00$ m that is hinged at the top of the hatch. (a) Determine the magnitude of the force the water exerts on the hatch. (b) Find the magnitude of the torque exerted by the water about the hinges.

Figure P14.15
Problems 15 and 16.

16. **Review.** The tank in Figure P14.15 is filled with water of depth d. At the bottom of one sidewall is a rectangular hatch of height h and width w that is hinged at the top of the hatch. (a) Determine the magnitude of the force the water exerts on the hatch. (b) Find the magnitude of the torque exerted by the water about the hinges.

17. **Review.** Piston ① in Figure P14.17 has a diameter of 0.250 in. Piston ② has a diameter of 1.50 in. Determine the magnitude F of the force necessary to support the 500-lb load in the absence of friction.

Figure P14.17

18. **Review.** A solid sphere of brass (bulk modulus of 14.0×10^{10} N/m^2) with a diameter of 3.00 m is thrown into the ocean. By how much does the diameter of the sphere decrease as it sinks to a depth of 1.00 km?

Section 14.3 Pressure Measurements

19. Normal atmospheric pressure is 1.013×10^5 Pa. The approach of a storm causes the height of a mercury barometer to drop by 20.0 mm from the normal height. What is the atmospheric pressure?

20. The human brain and spinal cord are immersed in the cerebrospinal fluid. The fluid is normally continuous between the cranial and spinal cavities and exerts a pressure of 100 to 200 mm of H$_2$O above the prevailing atmospheric pressure. In medical work, pressures are often measured in units of millimeters of H$_2$O because body fluids, including the cerebrospinal fluid, typically have the same density as water. The pressure of the cerebrospinal fluid can be measured by means of a *spinal tap* as illustrated in Figure P14.20. A hollow tube is inserted into the spinal column, and the height to which the fluid rises is observed. If the fluid rises to a height of 160 mm, we write its gauge pressure as 160 mm H$_2$O. (a) Express this pressure in pascals, in atmospheres, and in millimeters of mercury. (b) Some conditions that block or inhibit the flow of cerebrospinal fluid can be investigated by means of *Queckenstedt's test*. In this procedure, the veins in the patient's neck are compressed to make the blood pressure rise in the brain, which in turn should be transmitted to the cerebrospinal fluid. Explain how the level of fluid in the spinal tap can be used as a diagnostic tool for the condition of the patient's spine.

Figure P14.20

21. Blaise Pascal duplicated Torricelli's barometer using a red Bordeaux wine, of density 984 kg/m^3, as the working liquid (Fig. P14.21). (a) What was the height h of the wine column for normal atmospheric pressure? (b) Would you expect the vacuum above the column to be as good as for mercury?

Figure P14.21

22. Mercury is poured into a U-tube as shown in Figure P14.22a. The left arm of the tube has cross-sectional area A_1 of 10.0 cm^2, and the right arm has a cross-sectional area A_2 of 5.00 cm^2. One hundred grams of water are then poured into the right arm as shown in Figure P14.22b. (a) Determine the length of the water column in the right arm of the U-tube. (b) Given that the density of mercury is 13.6 g/cm^3, what distance h does the mercury rise in the left arm?

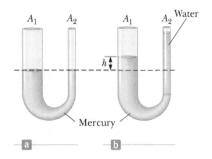

Figure P14.22

23. A backyard swimming pool with a circular base of diameter 6.00 m is filled to depth 1.50 m. (a) Find the absolute pressure at the bottom of the pool. (b) Two persons with combined mass 150 kg enter the pool and float quietly there. No water overflows. Find the pressure increase at the bottom of the pool after they enter the pool and float.

24. A tank with a flat bottom of area A and vertical sides is filled to a depth h with water. The pressure is P_0 at the top surface. (a) What is the absolute pressure at the bottom of the tank? (b) Suppose an object of mass M and density less than the density of water is placed into the tank and floats. No water overflows. What is the resulting increase in pressure at the bottom of the tank?

Section 14.4 Buoyant Forces and Archimedes's Principle

25. A table-tennis ball has a diameter of 3.80 cm and average density of 0.084 0 g/cm^3. What force is required to hold it completely submerged under water?

26. The gravitational force exerted on a solid object is 5.00 N. When the object is suspended from a spring scale and submerged in water, the scale reads 3.50 N (Fig. P14.26). Find the density of the object.

Figure P14.26 Problems 26 and 27.

27. A 10.0-kg block of metal measuring 12.0 cm by 10.0 cm by 10.0 cm is suspended from a scale and immersed in water as shown in Figure P14.26b. The 12.0-cm dimension is vertical, and the top of the block is 5.00 cm below the surface of the water. (a) What are the magnitudes of the forces acting on the top and on the bottom of the block due to the surrounding water? (b) What is the reading of the spring scale? (c) Show that the buoyant force equals the difference between the forces at the top and bottom of the block.

28. A light balloon is filled with 400 m^3 of helium at atmospheric pressure. (a) At 0°C, the balloon can lift a payload of what mass? (b) **What If?** In Table 14.1, observe that the density of hydrogen is nearly half the density of helium. What load can the balloon lift if filled with hydrogen?

29. A cube of wood having an edge dimension of 20.0 cm and a density of 650 kg/m^3 floats on water. (a) What is the distance from the horizontal top surface of the cube to the water level? (b) What mass of lead should be placed on the cube so that the top of the cube will be just level with the water surface?

30. The United States possesses the ten largest warships in the world, aircraft carriers of the *Nimitz* class. Suppose one of the ships bobs up to float 11.0 cm higher in the ocean water when 50 fighters take off from it in a time interval of 25 min, at a location where the free-fall acceleration is 9.78 m/s^2. The planes have an average laden mass of 29 000 kg. Find the horizontal area enclosed by the waterline of the ship.

31. A plastic sphere floats in water with 50.0% of its volume submerged. This same sphere floats in glycerin with 40.0% of its volume submerged. Determine the densities of (a) the glycerin and (b) the sphere.

32. A spherical vessel used for deep-sea exploration has a radius of 1.50 m and a mass of 1.20×10^4 kg. To dive, the vessel takes on mass in the form of seawater. Determine the mass the vessel must take on if it is to descend at a constant speed of 1.20 m/s, when the resistive force on it is 1 100 N in the upward direction. The density of seawater is equal to 1.03×10^3 kg/m^3.

33. A wooden block of volume 5.24×10^{-4} m^3 floats in water, and a small steel object of mass m is placed on top of the block. When $m = 0.310$ kg, the system is in equilibrium and the top of the wooden block is at the level of the water. (a) What is the density of the wood? (b) What happens to the block when the steel object is replaced by an object whose mass is less than 0.310 kg? (c) What happens to the block when the steel object is replaced by an object whose mass is greater than 0.310 kg?

34. The weight of a rectangular block of low-density material is 15.0 N. With a thin string, the center of the horizontal bottom face of the block is tied to the bottom of a beaker partly filled with water. When 25.0% of the block's volume is submerged, the tension in the string is 10.0 N. (a) Find the buoyant force on the block. (b) Oil of density 800 kg/m^3 is now steadily added to the beaker, forming a layer above the water and surrounding the block. The oil exerts forces on each of the four sidewalls of the block that the oil touches. What are the directions of these forces? (c) What happens to the string tension as the oil is added? Explain how the oil has this effect on the string tension. (d) The string breaks when its tension reaches 60.0 N. At this moment, 25.0% of the block's volume is still below the water line. What additional fraction of the block's volume is below the top surface of the oil?

35. A large weather balloon whose mass is 226 kg is filled with helium gas until its volume is 325 m^3. Assume the density of air is 1.20 kg/m^3 and the density of helium is 0.179 kg/m^3. (a) Calculate the buoyant force acting on the balloon. (b) Find the net force on the balloon and determine whether the balloon will rise or fall after it is released. (c) What additional mass can the balloon support in equilibrium?

36. A *hydrometer* is an instrument used to determine liquid density. A simple one is sketched in Figure P14.36. The bulb of a syringe is squeezed and released to let the atmosphere lift a sample of the liquid of interest into a tube containing a calibrated rod of known density. The rod, of length L and average density ρ_0, floats partially immersed in the liquid of density ρ. A length h of the rod protrudes above the surface of the liquid. Show that the density of the liquid is given by

$$\rho = \frac{\rho_0 L}{L - h}$$

Figure P14.36 Problems 36 and 37.

37. Refer to Problem 36 and Figure P14.36. A hydrometer is to be constructed with a cylindrical floating rod. Nine

fiduciary marks are to be placed along the rod to indicate densities of 0.98 g/cm^3, 1.00 g/cm^3, 1.02 g/cm^3, 1.04 g/cm^3, ..., 1.14 g/cm^3. The row of marks is to start 0.200 cm from the top end of the rod and end 1.80 cm from the top end. (a) What is the required length of the rod? (b) What must be its average density? (c) Should the marks be equally spaced? Explain your answer.

38. On October 21, 2001, Ian Ashpole of the United Kingdom achieved a record altitude of 3.35 km (11 000 ft) powered by 600 toy balloons filled with helium. Each filled balloon had a radius of about 0.50 m and an estimated mass of 0.30 kg. (a) Estimate the total buoyant force on the 600 balloons. (b) Estimate the net upward force on all 600 balloons. (c) Ashpole parachuted to the Earth after the balloons began to burst at the high altitude and the buoyant force decreased. Why did the balloons burst?

39. How many cubic meters of helium are required to lift a light balloon with a 400-kg payload to a height of 8 000 m? Take $\rho_{\text{He}} = 0.179$ kg/m^3. Assume the balloon maintains a constant volume and the density of air decreases with the altitude z according to the expression $\rho_{\text{air}} = \rho_0 e^{-z/8\,000}$, where z is in meters and $\rho_0 = 1.20$ kg/m^3 is the density of air at sea level.

Section 14.5 **Fluid Dynamics**
Section 14.6 **Bernoulli's Equation**

40. Water flowing through a garden hose of diameter 2.74 cm fills a 25-L bucket in 1.50 min. (a) What is the speed of the water leaving the end of the hose? (b) A nozzle is now attached to the end of the hose. If the nozzle diameter is one-third the diameter of the hose, what is the speed of the water leaving the nozzle?

41. A large storage tank, open at the top and filled with water, develops a small hole in its side at a point 16.0 m below the water level. The rate of flow from the leak is found to be 2.50×10^{-3} m^3/min. Determine (a) the speed at which the water leaves the hole and (b) the diameter of the hole.

42. Water moves through a constricted pipe in steady, ideal flow. At the lower point shown in Figure P14.42, the pressure is $P_1 = 1.75 \times 10^4$ Pa and the pipe diameter is 6.00 cm. At another point $y = 0.250$ m higher, the pressure is $P_2 = 1.20 \times 10^4$ Pa and the pipe diameter is 3.00 cm. Find the speed of flow (a) in the lower section and (b) in the upper section. (c) Find the volume flow rate through the pipe.

Figure P14.42

43. Figure P14.43 on page 442 shows a stream of water in steady flow from a kitchen faucet. At the faucet, the

diameter of the stream is 0.960 cm. The stream fills a 125-cm³ container in 16.3 s. Find the diameter of the stream 13.0 cm below the opening of the faucet.

© Cengage Learning/George Semple

Figure P14.43

44. A village maintains a large tank with an open top, containing water for emergencies. The water can drain from the tank through a hose of diameter 6.60 cm. The hose ends with a nozzle of diameter 2.20 cm. A rubber stopper is inserted into the nozzle. The water level in the tank is kept 7.50 m above the nozzle. (a) Calculate the friction force exerted on the stopper by the nozzle. (b) The stopper is removed. What mass of water flows from the nozzle in 2.00 h? (c) Calculate the gauge pressure of the flowing water in the hose just behind the nozzle.

45. A legendary Dutch boy saved Holland by plugging a hole of diameter 1.20 cm in a dike with his finger. If the hole was 2.00 m below the surface of the North Sea (density 1 030 kg/m³), (a) what was the force on his finger? (b) If he pulled his finger out of the hole, during what time interval would the released water fill 1 acre of land to a depth of 1 ft? Assume the hole remained constant in size.

46. Water falls over a dam of height h with a mass flow rate of R, in units of kilograms per second. (a) Show that the power available from the water is

$$P = Rgh$$

where g is the free-fall acceleration. (b) Each hydroelectric unit at the Grand Coulee Dam takes in water at a rate of 8.50×10^5 kg/s from a height of 87.0 m. The power developed by the falling water is converted to electric power with an efficiency of 85.0%. How much electric power does each hydroelectric unit produce?

47. Water is pumped up from the Colorado River to supply Grand Canyon Village, located on the rim of the canyon. The river is at an elevation of 564 m, and the village is at an elevation of 2 096 m. Imagine that the water is pumped through a single long pipe 15.0 cm in diameter, driven by a single pump at the bottom end. (a) What is the minimum pressure at which the

water must be pumped if it is to arrive at the village? (b) If 4 500 m³ of water is pumped per day, what is the speed of the water in the pipe? *Note:* Assume the free-fall acceleration and the density of air are constant over this range of elevations. The pressures you calculate are too high for an ordinary pipe. The water is actually lifted in stages by several pumps through shorter pipes.

48. In ideal flow, a liquid of density 850 kg/m³ moves from a horizontal tube of radius 1.00 cm into a second horizontal tube of radius 0.500 cm at the same elevation as the first tube. The pressure differs by ΔP between the liquid in one tube and the liquid in the second tube. (a) Find the volume flow rate as a function of ΔP. Evaluate the volume flow rate for (b) $\Delta P = 6.00$ kPa and (c) $\Delta P = 12.0$ kPa.

49. The Venturi tube discussed in Example 14.8 and shown in Figure P14.49 may be used as a fluid flowmeter. Suppose the device is used at a service station to measure the flow rate of gasoline ($\rho = 7.00 \times 10^2$ kg/m³) through a hose having an outlet radius of 1.20 cm. If the difference in pressure is measured to be $P_1 - P_2 = 1.20$ kPa and the radius of the inlet tube to the meter is 2.40 cm, find (a) the speed of the gasoline as it leaves the hose and (b) the fluid flow rate in cubic meters per second.

P_1 P_2

Figure P14.49

50. **Review.** Old Faithful Geyser in Yellowstone National Park erupts at approximately one-hour intervals, and the height of the water column reaches 40.0 m (Fig. P14.50). (a) Model the rising stream as a series of separate droplets. Analyze the free-fall motion of

Videowokart/Shutterstock.com

Figure P14.50

one of the droplets to determine the speed at which the water leaves the ground. (b) **What If?** Model the rising stream as an ideal fluid in streamline flow. Use Bernoulli's equation to determine the speed of the water as it leaves ground level. (c) How does the answer from part (a) compare with the answer from part (b)? (d) What is the pressure (above atmospheric) in the heated underground chamber if its depth is 175 m? Assume the chamber is large compared with the geyser's vent.

Section 14.7 Other Applications of Fluid Dynamics

51. An airplane is cruising at altitude 10 km. The pressure outside the craft is 0.287 atm; within the passenger compartment, the pressure is 1.00 atm and the temperature is 20°C. A small leak occurs in one of the window seals in the passenger compartment. Model the air as an ideal fluid to estimate the speed of the airstream flowing through the leak.

52. An airplane has a mass of 1.60×10^4 kg, and each wing has an area of 40.0 m². During level flight, the pressure on the lower wing surface is 7.00×10^4 Pa. (a) Suppose the lift on the airplane were due to a pressure difference alone. Determine the pressure on the upper wing surface. (b) More realistically, a significant part of the lift is due to deflection of air downward by the wing. Does the inclusion of this force mean that the pressure in part (a) is higher or lower? Explain.

53. A siphon is used to drain water from a tank as illustrated in Figure P14.53. Assume steady flow without friction. (a) If $h = 1.00$ m, find the speed of outflow at the end of the siphon. (b) **What If?** What is the limitation on the height of the top of the siphon above the end of the siphon? *Note:* For the flow of the liquid to be continuous, its pressure must not drop below its vapor pressure. Assume the water is at 20.0°C, at which the vapor pressure is 2.3 kPa.

Figure P14.53

54. The Bernoulli effect can have important consequences for the design of buildings. For example, wind can blow around a skyscraper at remarkably high speed, creating low pressure. The higher atmospheric pressure in the still air inside the buildings can cause windows to pop out. As originally constructed, the John Hancock Building in Boston popped windowpanes that fell many stories to the sidewalk below. (a) Suppose a horizontal wind blows with a speed of 11.2 m/s outside a large pane of plate glass with dimensions 4.00 m × 1.50 m. Assume the density of the air to be constant at 1.20 kg/m³. The air inside the building is at atmospheric pressure. What is the total force exerted by air on the windowpane? (b) **What If?** If a second skyscraper is built nearby, the airspeed can be especially high where wind passes through the narrow separation between the buildings. Solve part (a) again with a wind speed of 22.4 m/s, twice as high.

55. A hypodermic syringe contains a medicine with the density of water (Fig. P14.55). The barrel of the syringe has a cross-sectional area $A = 2.50 \times 10^{-5}$ m², and the needle has a cross-sectional area $a = 1.00 \times 10^{-8}$ m². In the absence of a force on the plunger, the pressure everywhere is 1.00 atm. A force $\vec{F}$ of magnitude 2.00 N acts on the plunger, making medicine squirt horizontally from the needle. Determine the speed of the medicine as it leaves the needle's tip.

Figure P14.55

Additional Problems

56. Decades ago, it was thought that huge herbivorous dinosaurs such as *Apatosaurus* and *Brachiosaurus* habitually walked on the bottom of lakes, extending their long necks up to the surface to breathe. *Brachiosaurus* had its nostrils on the top of its head. In 1977, Knut Schmidt-Nielsen pointed out that breathing would be too much work for such a creature. For a simple model, consider a sample consisting of 10.0 L of air at absolute pressure 2.00 atm, with density 2.40 kg/m³, located at the surface of a freshwater lake. Find the work required to transport it to a depth of 10.3 m, with its temperature, volume, and pressure remaining constant. This energy investment is greater than the energy that can be obtained by metabolism of food with the oxygen in that quantity of air.

57. (a) Calculate the absolute pressure at an ocean depth of 1 000 m. Assume the density of seawater is 1 030 kg/m³ and the air above exerts a pressure of 101.3 kPa. (b) At this depth, what is the buoyant force on a spherical submarine having a diameter of 5.00 m?

58. In about 1657, Otto von Guericke, inventor of the air pump, evacuated a sphere made of two brass hemispheres (Fig. P14.58). Two teams of eight horses each could pull the hemispheres apart only on some trials and then "with greatest difficulty," with the resulting

Figure P14.58

sound likened to a cannon firing. Find the force F required to pull the thin-walled evacuated hemispheres apart in terms of R, the radius of the hemispheres; P, the pressure inside the hemispheres; and atmospheric pressure P_0.

59. A spherical aluminum ball of mass 1.26 kg contains an empty spherical cavity that is concentric with the ball. The ball barely floats in water. Calculate (a) the outer radius of the ball and (b) the radius of the cavity.

60. **GP** A helium-filled balloon (whose envelope has a mass of $m_b = 0.250$ kg) is tied to a uniform string of length $\ell = 2.00$ m and mass $m = 0.050\ 0$ kg. The balloon is spherical with a radius of $r = 0.400$ m. When released in air of temperature 20°C and density $\rho_{air} = 1.20$ kg/m³, it lifts a length h of string and then remains stationary as shown in Figure P14.60. We wish to find the length of string lifted by the balloon. (a) When the balloon remains stationary, what is the appropriate analysis model to describe it? (b) Write a force equation for the balloon from this model in terms of the buoyant force B, the weight F_b of the balloon, the weight F_{He} of the helium, and the weight F_s of the segment of string of length h. (c) Make an appropriate substitution for each of these forces and solve symbolically for the mass m_s of the segment of string of length h in terms of m_b, r, ρ_{air}, and the density of helium ρ_{He}. (d) Find the numerical value of the mass m_s. (e) Find the length h numerically.

Figure P14.60

61. **Review.** Figure P14.61 shows a valve separating a reservoir from a water tank. If this valve is opened, what is the maximum height above point B attained by the water stream coming out of the right side of the tank? Assume $h = 10.0$ m, $L = 2.00$ m, and $\theta = 30.0°$, and assume the cross-sectional area at A is very large compared with that at B.

Figure P14.61

62. The true weight of an object can be measured in a vacuum, where buoyant forces are absent. A measurement in air, however, is disturbed by buoyant forces. An object of volume V is weighed in air on an equal-arm balance with the use of counterweights of density ρ. Representing the density of air as ρ_{air} and the balance reading as F_g', show that the true weight F_g is

$$F_g = F_g' + \left(V - \frac{F_g'}{\rho g}\right)\rho_{air}g$$

63. Water is forced out of a fire extinguisher by air pressure as shown in Figure P14.63. How much gauge air pressure in the tank is required for the water jet to have a speed of 30.0 m/s when the water level is 0.500 m below the nozzle?

Figure P14.63

64. **Review.** Assume a certain liquid, with density 1 230 kg/m³, exerts no friction force on spherical objects. A ball of mass 2.10 kg and radius 9.00 cm is dropped from rest into a deep tank of this liquid from a height of 3.30 m above the surface. (a) Find the speed at which the ball enters the liquid. (b) Evaluate the magnitudes of the two forces that are exerted on the ball as it moves through the liquid. (c) Explain why the ball moves down only a limited distance into the liquid and calculate this distance. (d) With what speed will the ball pop up out of the liquid? (e) How does the time interval Δt_{down}, during which the ball moves from the surface down to its lowest point, compare with the time interval Δt_{up} for the return trip between the same two points? (f) **What If?** Now modify the model to suppose the liquid exerts a small friction force on the ball, opposite in direction to its motion. In this case, how do the time intervals Δt_{down} and Δt_{up} compare? Explain your answer with a conceptual argument rather than a numerical calculation.

65. **Review.** A light spring of constant $k = 90.0$ N/m is attached vertically to a table (Fig. P14.65a). A 2.00-g balloon is filled with helium (density = 0.179 kg/m³)

Figure P14.65

to a volume of 5.00 m³ and is then connected with a light cord to the spring, causing the spring to stretch as shown in Figure P14.65b. Determine the extension distance *L* when the balloon is in equilibrium.

66. To an order of magnitude, how many helium-filled toy balloons would be required to lift you? Because helium is an irreplaceable resource, develop a theoretical answer rather than an experimental answer. In your solution, state what physical quantities you take as data and the values you measure or estimate for them.

67. A 42.0-kg boy uses a solid block of Styrofoam as a raft while fishing on a pond. The Styrofoam has an area of 1.00 m² and is 0.050 0 m thick. While sitting on the surface of the raft, the boy finds that the raft just supports him so that the top of the raft is at the level of the pond. Determine the density of the Styrofoam.

68. A common parameter that can be used to predict turbulence in fluid flow is called the *Reynolds number*. The Reynolds number for fluid flow in a pipe is a dimensionless quantity defined as

$$\mathrm{Re} = \frac{\rho v d}{\mu}$$

where ρ is the density of the fluid, v is its speed, d is the inner diameter of the pipe, and μ is the viscosity of the fluid. Viscosity is a measure of the internal resistance of a liquid to flow and has units of Pa · s. The criteria for the type of flow are as follows:

• If Re < 2 300, the flow is laminar.
• If 2 300 < Re < 4 000, the flow is in a transition region between laminar and turbulent.
• If Re > 4 000, the flow is turbulent.

(a) Let's model blood of density 1.06×10^3 kg/m³ and viscosity 3.00×10^{-3} Pa · s as a pure liquid, that is, ignore the fact that it contains red blood cells. Suppose it is flowing in a large artery of radius 1.50 cm with a speed of 0.067 0 m/s. Show that the flow is laminar. (b) Imagine that the artery ends in a *single* capillary so that the radius of the artery reduces to a much smaller value. What is the radius of the capillary that would cause the flow to become turbulent? (c) Actual capillaries have radii of about 5–10 micrometers, much smaller than the value in part (b). Why doesn't the flow in actual capillaries become turbulent?

69. Evangelista Torricelli was the first person to realize that we live at the bottom of an ocean of air. He correctly surmised that the pressure of our atmosphere is attributable to the weight of the air. The density of air at 0°C at the Earth's surface is 1.29 kg/m³. The density decreases with increasing altitude (as the atmosphere thins). On the other hand, if we assume the density is constant at 1.29 kg/m³ up to some altitude *h* and is zero above that altitude, then *h* would represent the depth of the ocean of air. (a) Use this model to determine the value of *h* that gives a pressure of 1.00 atm at the surface of the Earth. (b) Would the peak of Mount Everest rise above the surface of such an atmosphere?

70. **Review.** With reference to the dam studied in Example 14.4 and shown in Figure 14.5, (a) show that the total torque exerted by the water behind the dam about a horizontal axis through *O* is $\frac{1}{6}\rho g w H^3$. (b) Show that the effective line of action of the total force exerted by the water is at a distance $\frac{1}{3}H$ above *O*.

71. A 1.00-kg beaker containing 2.00 kg of oil (density = 916.0 kg/m³) rests on a scale. A 2.00-kg block of iron suspended from a spring scale is completely submerged in the oil as shown in Figure P14.71. Determine the equilibrium readings of both scales.

Figure P14.71 Problems 71 and 72.

72. A beaker of mass m_b containing oil of mass m_o and density ρ_o rests on a scale. A block of iron of mass m_{Fe} suspended from a spring scale is completely submerged in the oil as shown in Figure P14.71. Determine the equilibrium readings of both scales.

73. In 1983, the United States began coining the one-cent piece out of copper-clad zinc rather than pure copper. The mass of the old copper penny is 3.083 g and that of the new cent is 2.517 g. The density of copper is 8.920 g/cm³ and that of zinc is 7.133 g/cm³. The new and old coins have the same volume. Calculate the percent of zinc (by volume) in the new cent.

74. **Review.** A long, cylindrical rod of radius *r* is weighted on one end so that it floats upright in a fluid having a density ρ. It is pushed down a distance *x* from its equilibrium position and released. Show that the rod will execute simple harmonic motion if the resistive effects of the fluid are negligible, and determine the period of the oscillations.

75. **Review.** Figure P14.75 shows the essential parts of a hydraulic brake system. The area of the piston in the master cylinder is 1.8 cm² and that of the piston

Figure P14.75

in the brake cylinder is 6.4 cm². The coefficient of friction between shoe and wheel drum is 0.50. If the wheel has a radius of 34 cm, determine the frictional torque about the axle when a force of 44 N is exerted on the brake pedal.

76. The *spirit-in-glass thermometer*, invented in Florence, Italy, around 1654, consists of a tube of liquid (the spirit) containing a number of submerged glass spheres with slightly different masses (Fig. P14.76). At sufficiently low temperatures, all the spheres float, but as the temperature rises, the spheres sink one after another. The device is a crude but interesting tool for measuring temperature. Suppose the tube is filled with ethyl alcohol, whose density is 0.789 45 g/cm³ at 20.0°C and decreases to 0.780 97 g/cm³ at 30.0°C. (a) Assuming that one of the spheres has a radius of 1.000 cm and is in equilibrium halfway up the tube at 20.0°C, determine its mass. (b) When the temperature increases to 30.0°C, what mass must a second sphere of the same radius have to be in equilibrium at the halfway point? (c) At 30.0°C, the first sphere has fallen to the bottom of the tube. What upward force does the bottom of the tube exert on this sphere?

Figure P14.76

77. **Review.** A uniform disk of mass 10.0 kg and radius 0.250 m spins at 300 rev/min on a low-friction axle. It must be brought to a stop in 1.00 min by a brake pad that makes contact with the disk at an average distance 0.220 m from the axis. The coefficient of friction between pad and disk is 0.500. A piston in a cylinder of diameter 5.00 cm presses the brake pad against the disk. Find the pressure required for the brake fluid in the cylinder.

78. **Review.** In a water pistol, a piston drives water through a large tube of area A_1 into a smaller tube of area A_2 as shown in Figure P14.78. The radius of the large tube is 1.00 cm and that of the small tube is 1.00 mm. The smaller tube is 3.00 cm above the larger tube. (a) If the pistol is fired horizontally at a height of 1.50 m, determine the time interval required for the water to

travel from the nozzle to the ground. Neglect air resistance and assume atmospheric pressure is 1.00 atm. (b) If the desired range of the stream is 8.00 m, with what speed v_2 must the stream leave the nozzle? (c) At what speed v_1 must the plunger be moved to achieve the desired range? (d) What is the pressure at the nozzle? (e) Find the pressure needed in the larger tube. (f) Calculate the force that must be exerted on the trigger to achieve the desired range. (The force that must be exerted is due to pressure over and above atmospheric pressure.)

Figure P14.78

79. An incompressible, nonviscous fluid is initially at rest in the vertical portion of the pipe shown in Figure P14.79a, where $L = 2.00$ m. When the valve is opened, the fluid flows into the horizontal section of the pipe. What is the fluid's speed when all the fluid is in the horizontal section as shown in Figure P14.79b? Assume the cross-sectional area of the entire pipe is constant.

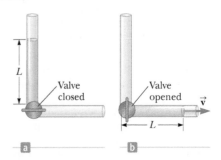

Figure P14.79

80. The water supply of a building is fed through a main pipe 6.00 cm in diameter. A 2.00-cm-diameter faucet tap, located 2.00 m above the main pipe, is observed to fill a 25.0-L container in 30.0 s. (a) What is the speed at which the water leaves the faucet? (b) What is the gauge pressure in the 6-cm main pipe? Assume the faucet is the only "leak" in the building.

81. A U-tube open at both ends is partially filled with water (Fig. P14.81a). Oil having a density 750 kg/m³ is then poured into the right arm and forms a column $L = 5.00$ cm high (Fig. P14.81b). (a) Determine the difference h in the heights of the two liquid surfaces. (b) The right arm is then shielded from any air motion while air is blown across the top of the left arm until the surfaces of the two liquids are at the same height (Fig. P14.81c). Determine the speed of the air being

blown across the left arm. Take the density of air as constant at 1.20 kg/m^3.

Figure P14.81

82. A woman is draining her fish tank by siphoning the water into an outdoor drain as shown in Figure P14.82. The rectangular tank has footprint area A and depth h. The drain is located a distance d below the surface of the water in the tank, where $d \gg h$. The cross-sectional area of the siphon tube is A'. Model the water as flowing without friction. Show that the time interval required to empty the tank is given by

$$\Delta t = \frac{Ah}{A'\sqrt{2gd}}$$

Figure P14.82

83. The hull of an experimental boat is to be lifted above the water by a hydrofoil mounted below its keel as shown in Figure P14.83. The hydrofoil has a shape like that of an airplane wing. Its area projected onto a horizontal surface is A. When the boat is towed at sufficiently high speed, water of density ρ moves in streamline flow so that its average speed at the top of the hydrofoil is n times larger than its speed v_b below the hydrofoil. (a) Ignoring the buoyant force, show that the upward lift force exerted by the water on the hydrofoil has a magnitude

$$F \approx \tfrac{1}{2}(n^2 - 1)\rho v_b^2 A$$

Figure P14.83

(b) The boat has mass M. Show that the liftoff speed is given by

$$v \approx \sqrt{\frac{2Mg}{(n^2 - 1)A\rho}}$$

84. A jet of water squirts out horizontally from a hole near the bottom of the tank shown in Figure P14.84. If the hole has a diameter of 3.50 mm, what is the height h of the water level in the tank?

Figure P14.84

Challenge Problems

85. An ice cube whose edges measure 20.0 mm is floating in a glass of ice-cold water, and one of the ice cube's faces is parallel to the water's surface. (a) How far below the water surface is the bottom face of the block? (b) Ice-cold ethyl alcohol is gently poured onto the water surface to form a layer 5.00 mm thick above the water. The alcohol does not mix with the water. When the ice cube again attains hydrostatic equilibrium, what is the distance from the top of the water to the bottom face of the block? (c) Additional cold ethyl alcohol is poured onto the water's surface until the top surface of the alcohol coincides with the top surface of the ice cube (in hydrostatic equilibrium). How thick is the required layer of ethyl alcohol?

86. *Why is the following situation impossible?* A barge is carrying a load of small pieces of iron along a river. The iron pile is in the shape of a cone for which the radius r of the base of the cone is equal to the central height h of the cone. The barge is square in shape, with vertical sides of length $2r$, so that the pile of iron comes just up to the edges of the barge. The barge approaches a low bridge, and the captain realizes that the top of the pile of iron is not going to make it under the bridge. The captain orders the crew to shovel iron pieces from the pile into the water to reduce the height of the pile. As iron is shoveled from the pile, the pile always has the shape of a cone whose diameter is equal to the side length of the barge. After a certain volume of iron is removed from the barge, it makes it under the bridge without the top of the pile striking the bridge.

87. Show that the variation of atmospheric pressure with altitude is given by $P = P_0 e^{-\alpha y}$, where $\alpha = \rho_0 g/P_0$, P_0

is atmospheric pressure at some reference level $y = 0$, and ρ_0 is the atmospheric density at this level. Assume the decrease in atmospheric pressure over an infinitesimal change in altitude (so that the density is approximately uniform over the infinitesimal change) can be expressed from Equation 14.4 as $dP = -\rho g \, dy$. Also assume the density of air is proportional to the pressure, which, as we will see in Chapter 20, is equivalent to assuming the temperature of the air is the same at all altitudes.

Tables

Table A.1   Conversion Factors

Length

	m	cm	km	in.	ft	mi
1 meter	1	10^2	10^{-3}	39.37	3.281	6.214×10^{-4}
1 centimeter	10^{-2}	1	10^{-5}	0.393 7	3.281×10^{-2}	6.214×10^{-6}
1 kilometer	10^3	10^5	1	3.937×10^4	3.281×10^3	0.621 4
1 inch	2.540×10^{-2}	2.540	2.540×10^{-5}	1	8.333×10^{-2}	1.578×10^{-5}
1 foot	0.304 8	30.48	3.048×10^{-4}	12	1	1.894×10^{-4}
1 mile	1 609	1.609×10^5	1.609	6.336×10^4	5 280	1

Mass

	kg	g	slug	u
1 kilogram	1	10^3	6.852×10^{-2}	6.024×10^{26}
1 gram	10^{-3}	1	6.852×10^{-5}	6.024×10^{23}
1 slug	14.59	1.459×10^4	1	8.789×10^{27}
1 atomic mass unit	1.660×10^{-27}	1.660×10^{-24}	1.137×10^{-28}	1

Note: 1 metric ton = 1 000 kg.

Time

	s	min	h	day	yr
1 second	1	1.667×10^{-2}	2.778×10^{-4}	1.157×10^{-5}	3.169×10^{-8}
1 minute	60	1	1.667×10^{-2}	6.994×10^{-4}	1.901×10^{-6}
1 hour	3 600	60	1	4.167×10^{-2}	1.141×10^{-4}
1 day	8.640×10^4	1 440	24	1	2.738×10^{-5}
1 year	3.156×10^7	5.259×10^5	8.766×10^3	365.2	1

Speed

	m/s	cm/s	ft/s	mi/h
1 meter per second	1	10^2	3.281	2.237
1 centimeter per second	10^{-2}	1	3.281×10^{-2}	2.237×10^{-2}
1 foot per second	0.304 8	30.48	1	0.681 8
1 mile per hour	0.447 0	44.70	1.467	1

Note: 1 mi/min = 60 mi/h = 88 ft/s.

Force

	N	lb
1 newton	1	0.224 8
1 pound	4.448	1

(Continued)

Table A.1   Conversion Factors *(continued)*

Energy, Energy Transfer

	J	ft · lb	eV
1 joule	1	0.737 6	6.242×10^{18}
1 foot-pound	1.356	1	8.464×10^{18}
1 electron volt	1.602×10^{-19}	1.182×10^{-19}	1
1 calorie	4.186	3.087	2.613×10^{19}
1 British thermal unit	1.055×10^3	7.779×10^2	6.585×10^{21}
1 kilowatt-hour	3.600×10^6	2.655×10^6	2.247×10^{25}

	cal	Btu	kWh
1 joule	0.238 9	9.481×10^{-4}	2.778×10^{-7}
1 foot-pound	0.323 9	1.285×10^{-3}	3.766×10^{-7}
1 electron volt	3.827×10^{-20}	1.519×10^{-22}	4.450×10^{-26}
1 calorie	1	3.968×10^{-3}	1.163×10^{-6}
1 British thermal unit	2.520×10^2	1	2.930×10^{-4}
1 kilowatt-hour	8.601×10^5	3.413×10^2	1

Pressure

	Pa	atm
1 pascal	1	9.869×10^{-6}
1 atmosphere	1.013×10^5	1
1 centimeter mercury[a]	1.333×10^3	1.316×10^{-2}
1 pound per square inch	6.895×10^3	6.805×10^{-2}
1 pound per square foot	47.88	4.725×10^{-4}

	cm Hg	lb/in.²	lb/ft²
1 pascal	7.501×10^{-4}	1.450×10^{-4}	2.089×10^{-2}
1 atmosphere	76	14.70	2.116×10^3
1 centimeter mercury[a]	1	0.194 3	27.85
1 pound per square inch	5.171	1	144
1 pound per square foot	3.591×10^{-2}	6.944×10^{-3}	1

[a]At 0°C and at a location where the free-fall acceleration has its "standard" value, 9.806 65 m/s².

Table A.2   Symbols, Dimensions, and Units of Physical Quantities

Quantity	Common Symbol	Unit[a]	Dimensions[b]	Unit in Terms of Base SI Units
Acceleration	$\vec{a}$	m/s²	L/T^2	m/s²
Amount of substance	n	MOLE		mol
Angle	θ, ϕ	radian (rad)	1	
Angular acceleration	$\vec{\alpha}$	rad/s²	T^{-2}	s⁻²
Angular frequency	ω	rad/s	T^{-1}	s⁻¹
Angular momentum	$\vec{L}$	kg · m²/s	ML^2/T	kg · m²/s
Angular velocity	$\vec{\omega}$	rad/s	T^{-1}	s⁻¹
Area	A	m²	L^2	m²
Atomic number	Z			
Capacitance	C	farad (F)	Q^2T^2/ML^2	$A^2 \cdot s^4/kg \cdot m^2$
Charge	q, Q, e	coulomb (C)	Q	A · s

(Continued)

Table A.2   Symbols, Dimensions, and Units of Physical Quantities (continued)

Quantity	Common Symbol	Unit[a]	Dimensions[b]	Unit in Terms of Base SI Units
Charge density				
Line	λ	C/m	Q/L	$A \cdot s/m$
Surface	σ	C/m^2	Q/L^2	$A \cdot s/m^2$
Volume	ρ	C/m^3	Q/L^3	$A \cdot s/m^3$
Conductivity	σ	$1/\Omega \cdot m$	$Q^2 T/ML^3$	$A^2 \cdot s^3/kg \cdot m^3$
Current	I	AMPERE	Q/T	A
Current density	J	A/m^2	Q/TL^2	A/m^2
Density	ρ	kg/m^3	M/L^3	kg/m^3
Dielectric constant	κ			
Electric dipole moment	$\vec{\mathbf{p}}$	$C \cdot m$	QL	$A \cdot s \cdot m$
Electric field	$\vec{\mathbf{E}}$	V/m	ML/QT^2	$kg \cdot m/A \cdot s^3$
Electric flux	Φ_E	$V \cdot m$	ML^3/QT^2	$kg \cdot m^3/A \cdot s^3$
Electromotive force	$\mathcal{E}$	volt (V)	ML^2/QT^2	$kg \cdot m^2/A \cdot s^3$
Energy	E, U, K	joule (J)	ML^2/T^2	$kg \cdot m^2/s^2$
Entropy	S	J/K	$ML^2/T^2 K$	$kg \cdot m^2/s^2 \cdot K$
Force	$\vec{\mathbf{F}}$	newton (N)	ML/T^2	$kg \cdot m/s^2$
Frequency	f	hertz (Hz)	T^{-1}	s^{-1}
Heat	Q	joule (J)	ML^2/T^2	$kg \cdot m^2/s^2$
Inductance	L	henry (H)	ML^2/Q^2	$kg \cdot m^2/A^2 \cdot s^2$
Length	ℓ, L	METER	L	m
Displacement	$\Delta x, \Delta \vec{\mathbf{r}}$			
Distance	d, h			
Position	$x, y, z, \vec{\mathbf{r}}$			
Magnetic dipole moment	$\vec{\boldsymbol{\mu}}$	$N \cdot m/T$	QL^2/T	$A \cdot m^2$
Magnetic field	$\vec{\mathbf{B}}$	tesla (T) $(= Wb/m^2)$	M/QT	$kg/A \cdot s^2$
Magnetic flux	Φ_B	weber (Wb)	ML^2/QT	$kg \cdot m^2/A \cdot s^2$
Mass	m, M	KILOGRAM	M	kg
Molar specific heat	C	$J/mol \cdot K$		$kg \cdot m^2/s^2 \cdot mol \cdot K$
Moment of inertia	I	$kg \cdot m^2$	ML^2	$kg \cdot m^2$
Momentum	$\vec{\mathbf{p}}$	$kg \cdot m/s$	ML/T	$kg \cdot m/s$
Period	T	s	T	s
Permeability of free space	μ_0	$N/A^2 (= H/m)$	ML/Q^2	$kg \cdot m/A^2 \cdot s^2$
Permittivity of free space	ϵ_0	$C^2/N \cdot m^2 (= F/m)$	$Q^2 T^2/ML^3$	$A^2 \cdot s^4/kg \cdot m^3$
Potential	V	volt (V)$(= J/C)$	ML^2/QT^2	$kg \cdot m^2/A \cdot s^3$
Power	P	watt (W)$(= J/s)$	ML^2/T^3	$kg \cdot m^2/s^3$
Pressure	P	pascal (Pa)$(= N/m^2)$	M/LT^2	$kg/m \cdot s^2$
Resistance	R	ohm $(\Omega)(= V/A)$	$ML^2/Q^2 T$	$kg \cdot m^2/A^2 \cdot s^3$
Specific heat	c	$J/kg \cdot K$	$L^2/T^2 K$	$m^2/s^2 \cdot K$
Speed	v	m/s	L/T	m/s
Temperature	T	KELVIN	K	K
Time	t	SECOND	T	s
Torque	$\vec{\boldsymbol{\tau}}$	$N \cdot m$	ML^2/T^2	$kg \cdot m^2/s^2$
Velocity	$\vec{\mathbf{v}}$	m/s	L/T	m/s
Volume	V	m^3	L^3	m^3
Wavelength	λ	m	L	m
Work	W	joule (J)$(= N \cdot m)$	ML^2/T^2	$kg \cdot m^2/s^2$

[a]The base SI units are given in uppercase letters.

[b]The symbols M, L, T, K, and Q denote mass, length, time, temperature, and charge, respectively.

Mathematics Review

This appendix in mathematics is intended as a brief review of operations and methods. Early in this course, you should be totally familiar with basic algebraic techniques, analytic geometry, and trigonometry. The sections on differential and integral calculus are more detailed and are intended for students who have difficulty applying calculus concepts to physical situations.

B.1 Scientific Notation

Many quantities used by scientists often have very large or very small values. The speed of light, for example, is about 300 000 000 m/s, and the ink required to make the dot over an i in this textbook has a mass of about 0.000 000 001 kg. Obviously, it is very cumbersome to read, write, and keep track of such numbers. We avoid this problem by using a method incorporating powers of the number 10:

$$10^0 = 1$$
$$10^1 = 10$$
$$10^2 = 10 \times 10 = 100$$
$$10^3 = 10 \times 10 \times 10 = 1\ 000$$
$$10^4 = 10 \times 10 \times 10 \times 10 = 10\ 000$$
$$10^5 = 10 \times 10 \times 10 \times 10 \times 10 = 100\ 000$$

and so on. The number of zeros corresponds to the power to which ten is raised, called the **exponent** of ten. For example, the speed of light, 300 000 000 m/s, can be expressed as 3.00×10^8 m/s.

In this method, some representative numbers smaller than unity are the following:

$$10^{-1} = \frac{1}{10} = 0.1$$

$$10^{-2} = \frac{1}{10 \times 10} = 0.01$$

$$10^{-3} = \frac{1}{10 \times 10 \times 10} = 0.001$$

$$10^{-4} = \frac{1}{10 \times 10 \times 10 \times 10} = 0.000\ 1$$

$$10^{-5} = \frac{1}{10 \times 10 \times 10 \times 10 \times 10} = 0.000\ 01$$

In these cases, the number of places the decimal point is to the left of the digit 1 equals the value of the (negative) exponent. Numbers expressed as some power of ten multiplied by another number between one and ten are said to be in **scientific notation.** For example, the scientific notation for 5 943 000 000 is 5.943×10^9 and that for 0.000 083 2 is 8.32×10^{-5}.

When numbers expressed in scientific notation are being multiplied, the following general rule is very useful:

$$10^n \times 10^m = 10^{n+m} \qquad \text{(B.1)}$$

where n and m can be *any* numbers (not necessarily integers). For example, $10^2 \times 10^5 = 10^7$. The rule also applies if one of the exponents is negative: $10^3 \times 10^{-8} = 10^{-5}$.

When dividing numbers expressed in scientific notation, note that

$$\frac{10^n}{10^m} = 10^n \times 10^{-m} = 10^{n-m} \qquad \text{(B.2)}$$

Exercises

With help from the preceding rules, verify the answers to the following equations:

1. $86\,400 = 8.64 \times 10^4$
2. $9\,816\,762.5 = 9.816\,762\,5 \times 10^6$
3. $0.000\,000\,039\,8 = 3.98 \times 10^{-8}$
4. $(4.0 \times 10^8)(9.0 \times 10^9) = 3.6 \times 10^{18}$
5. $(3.0 \times 10^7)(6.0 \times 10^{-12}) = 1.8 \times 10^{-4}$
6. $\dfrac{75 \times 10^{-11}}{5.0 \times 10^{-3}} = 1.5 \times 10^{-7}$
7. $\dfrac{(3 \times 10^6)(8 \times 10^{-2})}{(2 \times 10^{17})(6 \times 10^5)} = 2 \times 10^{-18}$

B.2 Algebra

Some Basic Rules

When algebraic operations are performed, the laws of arithmetic apply. Symbols such as x, y, and z are usually used to represent unspecified quantities, called the **unknowns.**

First, consider the equation

$$8x = 32$$

If we wish to solve for x, we can divide (or multiply) each side of the equation by the same factor without destroying the equality. In this case, if we divide both sides by 8, we have

$$\frac{8x}{8} = \frac{32}{8}$$

$$x = 4$$

Next consider the equation

$$x + 2 = 8$$

In this type of expression, we can add or subtract the same quantity from each side. If we subtract 2 from each side, we have

$$x + 2 - 2 = 8 - 2$$

$$x = 6$$

In general, if $x + a = b$, then $x = b - a$.

Now consider the equation

$$\frac{x}{5} = 9$$

If we multiply each side by 5, we are left with x on the left by itself and 45 on the right:

$$\left(\frac{x}{5}\right)(5) = 9 \times 5$$
$$x = 45$$

In all cases, *whatever operation is performed on the left side of the equality must also be performed on the right side.*

The following rules for multiplying, dividing, adding, and subtracting fractions should be recalled, where a, b, c, and d are four numbers:

	Rule	Example
Multiplying	$\left(\dfrac{a}{b}\right)\left(\dfrac{c}{d}\right) = \dfrac{ac}{bd}$	$\left(\dfrac{2}{3}\right)\left(\dfrac{4}{5}\right) = \dfrac{8}{15}$
Dividing	$\dfrac{(a/b)}{(c/d)} = \dfrac{ad}{bc}$	$\dfrac{2/3}{4/5} = \dfrac{(2)(5)}{(4)(3)} = \dfrac{10}{12}$
Adding	$\dfrac{a}{b} \pm \dfrac{c}{d} = \dfrac{ad \pm bc}{bd}$	$\dfrac{2}{3} - \dfrac{4}{5} = \dfrac{(2)(5) - (4)(3)}{(3)(5)} = -\dfrac{2}{15}$

Exercises

In the following exercises, solve for x.

Answers

1. $a = \dfrac{1}{1 + x}$ $\qquad x = \dfrac{1 - a}{a}$

2. $3x - 5 = 13$ $\qquad x = 6$

3. $ax - 5 = bx + 2$ $\qquad x = \dfrac{7}{a - b}$

4. $\dfrac{5}{2x + 6} = \dfrac{3}{4x + 8}$ $\qquad x = -\dfrac{11}{7}$

Powers

When powers of a given quantity x are multiplied, the following rule applies:

$$x^n x^m = x^{n+m} \tag{B.3}$$

For example, $x^2 x^4 = x^{2+4} = x^6$.

When dividing the powers of a given quantity, the rule is

$$\frac{x^n}{x^m} = x^{n-m} \tag{B.4}$$

For example, $x^8/x^2 = x^{8-2} = x^6$.

A power that is a fraction, such as $\frac{1}{3}$, corresponds to a root as follows:

$$x^{1/n} = \sqrt[n]{x} \tag{B.5}$$

For example, $4^{1/3} = \sqrt[3]{4} = 1.587\,4$. (A scientific calculator is useful for such calculations.)

Finally, any quantity x^n raised to the mth power is

$$(x^n)^m = x^{nm} \tag{B.6}$$

Table B.1 summarizes the rules of exponents.

Table B.1	**Rules of Exponents**
	$x^0 = 1$
	$x^1 = x$
	$x^n x^m = x^{n+m}$
	$x^n/x^m = x^{n-m}$
	$x^{1/n} = \sqrt[n]{x}$
	$(x^n)^m = x^{nm}$

Exercises

Verify the following equations:

1. $3^2 \times 3^3 = 243$
2. $x^5 x^{-8} = x^{-3}$

3. $x^{10}/x^{-5} = x^{15}$

4. $5^{1/3} = 1.709\,976$ (Use your calculator.)

5. $60^{1/4} = 2.783\,158$ (Use your calculator.)

6. $(x^4)^3 = x^{12}$

Factoring

Some useful formulas for factoring an equation are the following:

$ax + ay + az = a(x + y + z)$ common factor

$a^2 + 2ab + b^2 = (a + b)^2$ perfect square

$a^2 - b^2 = (a + b)(a - b)$ differences of squares

Quadratic Equations

The general form of a quadratic equation is

$$ax^2 + bx + c = 0 \qquad \text{(B.7)}$$

where x is the unknown quantity and a, b, and c are numerical factors referred to as **coefficients** of the equation. This equation has two roots, given by

$$x = \frac{-b \pm \sqrt{b^2 - 4ac}}{2a} \qquad \text{(B.8)}$$

If $b^2 \geq 4ac$, the roots are real.

Example B.1

The equation $x^2 + 5x + 4 = 0$ has the following roots corresponding to the two signs of the square-root term:

$$x = \frac{-5 \pm \sqrt{5^2 - (4)(1)(4)}}{2(1)} = \frac{-5 \pm \sqrt{9}}{2} = \frac{-5 \pm 3}{2}$$

$$x_+ = \frac{-5 + 3}{2} = \boxed{-1} \quad x_- = \frac{-5 - 3}{2} = \boxed{-4}$$

where x_+ refers to the root corresponding to the positive sign and x_- refers to the root corresponding to the negative sign.

Exercises

Solve the following quadratic equations:

Answers

1. $x^2 + 2x - 3 = 0$	$x_+ = 1$	$x_- = -3$
2. $2x^2 - 5x + 2 = 0$	$x_+ = 2$	$x_- = \frac{1}{2}$
3. $2x^2 - 4x - 9 = 0$	$x_+ = 1 + \sqrt{22}/2$	$x_- = 1 - \sqrt{22}/2$

Linear Equations

A linear equation has the general form

$$y = mx + b \qquad \text{(B.9)}$$

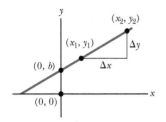

Figure B.1 A straight line graphed on an *xy* coordinate system. The slope of the line is the ratio of Δy to Δx.

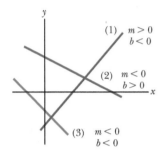

Figure B.2 The brown line has a positive slope and a negative *y*-intercept. The blue line has a negative slope and a positive *y*-intercept. The green line has a negative slope and a negative *y*-intercept.

where *m* and *b* are constants. This equation is referred to as linear because the graph of *y* versus *x* is a straight line as shown in Figure B.1. The constant *b*, called the **y-intercept,** represents the value of *y* at which the straight line intersects the *y* axis. The constant *m* is equal to the **slope** of the straight line. If any two points on the straight line are specified by the coordinates (x_1, y_1) and (x_2, y_2) as in Figure B.1, the slope of the straight line can be expressed as

$$\text{Slope} = \frac{y_2 - y_1}{x_2 - x_1} = \frac{\Delta y}{\Delta x} \qquad \text{(B.10)}$$

Note that *m* and *b* can have either positive or negative values. If $m > 0$, the straight line has a *positive* slope as in Figure B.1. If $m < 0$, the straight line has a *negative* slope. In Figure B.1, both *m* and *b* are positive. Three other possible situations are shown in Figure B.2.

Exercises

1. Draw graphs of the following straight lines: (a) $y = 5x + 3$ (b) $y = -2x + 4$ (c) $y = -3x - 6$
2. Find the slopes of the straight lines described in Exercise 1.

Answers   (a) 5   (b) -2 (c) -3

3. Find the slopes of the straight lines that pass through the following sets of points: (a) $(0, -4)$ and $(4, 2)$ (b) $(0, 0)$ and $(2, -5)$ (c) $(-5, 2)$ and $(4, -2)$

Answers   (a) $\frac{3}{2}$ (b) $-\frac{5}{2}$ (c) $-\frac{4}{9}$

Solving Simultaneous Linear Equations

Consider the equation $3x + 5y = 15$, which has two unknowns, *x* and *y*. Such an equation does not have a unique solution. For example, $(x = 0, y = 3)$, $(x = 5, y = 0)$, and $(x = 2, y = \frac{9}{5})$ are all solutions to this equation.

If a problem has two unknowns, a unique solution is possible only if we have *two* pieces of information. In most common cases, those two pieces of information are equations. In general, if a problem has *n* unknowns, its solution requires *n* equations. To solve two simultaneous equations involving two unknowns, *x* and *y*, we solve one of the equations for *x* in terms of *y* and substitute this expression into the other equation.

In some cases, the two pieces of information may be (1) one equation and (2) a condition on the solutions. For example, suppose we have the equation $m = 3n$ and the condition that *m* and *n* must be the smallest positive nonzero integers possible. Then, the single equation does not allow a unique solution, but the addition of the condition gives us that $n = 1$ and $m = 3$.

Example B.2

Solve the two simultaneous equations

$$(1) \quad 5x + y = -8$$

$$(2) \quad 2x - 2y = 4$$

SOLUTION

From Equation (2), $x = y + 2$. Substitution of this equation into Equation (1) gives

$$5(y + 2) + y = -8$$

$$6y = -18$$

▶ **B.2** continued

$$y = -3$$
$$x = y + 2 = -1$$

Alternative Solution Multiply each term in Equation (1) by the factor 2 and add the result to Equation (2):

$$10x + 2y = -16$$
$$\underline{2x - 2y = 4}$$
$$12x \qquad = -12$$
$$x = -1$$
$$y = x - 2 = -3$$

Two linear equations containing two unknowns can also be solved by a graphical method. If the straight lines corresponding to the two equations are plotted in a conventional coordinate system, the intersection of the two lines represents the solution. For example, consider the two equations

$$x - y = 2$$
$$x - 2y = -1$$

These equations are plotted in Figure B.3. The intersection of the two lines has the coordinates $x = 5$ and $y = 3$, which represents the solution to the equations. You should check this solution by the analytical technique discussed earlier.

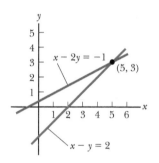

Figure B.3 A graphical solution for two linear equations.

Exercises

Solve the following pairs of simultaneous equations involving two unknowns:

	Answers
1. $x + y = 8$	$x = 5, y = 3$
$x - y = 2$	
2. $98 - T = 10a$	$T = 65, a = 3.27$
$T - 49 = 5a$	
3. $6x + 2y = 6$	$x = 2, y = -3$
$8x - 4y = 28$	

Logarithms

Suppose a quantity x is expressed as a power of some quantity a:

$$x = a^y \qquad \text{(B.11)}$$

The number a is called the **base** number. The **logarithm** of x with respect to the base a is equal to the exponent to which the base must be raised to satisfy the expression $x = a^y$:

$$y = \log_a x \qquad \text{(B.12)}$$

Conversely, the **antilogarithm** of y is the number x:

$$x = \text{antilog}_a y \qquad \text{(B.13)}$$

In practice, the two bases most often used are base 10, called the *common* logarithm base, and base $e = 2.718\ 282$, called Euler's constant or the *natural* logarithm base. When common logarithms are used,

$$y = \log_{10} x \quad (\text{or } x = 10^y) \qquad \text{(B.14)}$$

When natural logarithms are used,

$$y = \ln x \quad (\text{or } x = e^y) \tag{B.15}$$

For example, $\log_{10} 52 = 1.716$, so antilog$_{10}$ $1.716 = 10^{1.716} = 52$. Likewise, $\ln 52 = 3.951$, so antiln $3.951 = e^{3.951} = 52$.

In general, note you can convert between base 10 and base e with the equality

$$\ln x = (2.302\ 585) \log_{10} x \tag{B.16}$$

Finally, some useful properties of logarithms are the following:

$$\left.\begin{array}{l} \log(ab) = \log a + \log b \\[4pt] \log(a/b) = \log a - \log b \\[4pt] \log(a^n) = n \log a \end{array}\right\} \text{any base}$$

$$\ln e = 1$$

$$\ln e^a = a$$

$$\ln\left(\frac{1}{a}\right) = -\ln a$$

B.3 Geometry

The **distance** d between two points having coordinates (x_1, y_1) and (x_2, y_2) is

$$d = \sqrt{(x_2 - x_1)^2 + (y_2 - y_1)^2} \tag{B.17}$$

Two angles are equal if their sides are perpendicular, right side to right side and left side to left side. For example, the two angles marked θ in Figure B.4 are the same because of the perpendicularity of the sides of the angles. To distinguish the left and right sides of an angle, imagine standing at the angle's apex and facing into the angle.

Radian measure: The arc length s of a circular arc (Fig. B.5) is proportional to the radius r for a fixed value of θ (in radians):

$$s = r\theta$$
$$\theta = \frac{s}{r} \tag{B.18}$$

Table B.2 gives the **areas** and **volumes** for several geometric shapes used throughout this text.

The equation of a **straight line** (Fig. B.6) is

$$y = mx + b \tag{B.19}$$

where b is the y-intercept and m is the slope of the line.

The equation of a **circle** of radius R centered at the origin is

$$x^2 + y^2 = R^2 \tag{B.20}$$

The equation of an **ellipse** having the origin at its center (Fig. B.7) is

$$\frac{x^2}{a^2} + \frac{y^2}{b^2} = 1 \tag{B.21}$$

where a is the length of the semimajor axis (the longer one) and b is the length of the semiminor axis (the shorter one).

The equation of a **parabola** the vertex of which is at $y = b$ (Fig. B.8) is

$$y = ax^2 + b \tag{B.22}$$

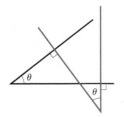

Figure B.4 The angles are equal because their sides are perpendicular.

Figure B.5 The angle θ in radians is the ratio of the arc length s to the radius r of the circle.

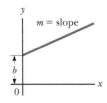

Figure B.6 A straight line with a slope of m and a y-intercept of b.

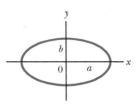

Figure B.7 An ellipse with semimajor axis a and semiminor axis b.

| Table B.2 | **Useful Information for Geometry** |

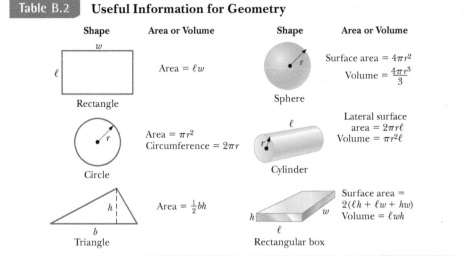

Shape	Area or Volume	Shape	Area or Volume
Rectangle	Area $= \ell w$	Sphere	Surface area $= 4\pi r^2$ Volume $= \frac{4\pi r^3}{3}$
Circle	Area $= \pi r^2$ Circumference $= 2\pi r$	Cylinder	Lateral surface area $= 2\pi r \ell$ Volume $= \pi r^2 \ell$
Triangle	Area $= \frac{1}{2}bh$	Rectangular box	Surface area $= 2(\ell h + \ell w + hw)$ Volume $= \ell wh$

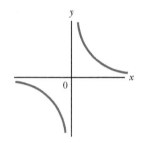

Figure B.8 A parabola with its vertex at $y = b$.

The equation of a **rectangular hyperbola** (Fig. B.9) is

$$xy = \text{constant} \qquad \text{(B.23)}$$

Figure B.9 A hyperbola.

B.4 Trigonometry

That portion of mathematics based on the special properties of the right triangle is called trigonometry. By definition, a right triangle is a triangle containing a 90° angle. Consider the right triangle shown in Figure B.10, where side a is opposite the angle θ, side b is adjacent to the angle θ, and side c is the hypotenuse of the triangle. The three basic trigonometric functions defined by such a triangle are the sine (sin), cosine (cos), and tangent (tan). In terms of the angle θ, these functions are defined as follows:

a = opposite side
b = adjacent side
c = hypotenuse

Figure B.10 A right triangle, used to define the basic functions of trigonometry.

$$\sin \theta = \frac{\text{side opposite } \theta}{\text{hypotenuse}} = \frac{a}{c} \qquad \text{(B.24)}$$

$$\cos \theta = \frac{\text{side adjacent to } \theta}{\text{hypotenuse}} = \frac{b}{c} \qquad \text{(B.25)}$$

$$\tan \theta = \frac{\text{side opposite } \theta}{\text{side adjacent to } \theta} = \frac{a}{b} \qquad \text{(B.26)}$$

The Pythagorean theorem provides the following relationship among the sides of a right triangle:

$$c^2 = a^2 + b^2 \qquad \text{(B.27)}$$

From the preceding definitions and the Pythagorean theorem, it follows that

$$\sin^2 \theta + \cos^2 \theta = 1$$

$$\tan \theta = \frac{\sin \theta}{\cos \theta}$$

The cosecant, secant, and cotangent functions are defined by

$$\csc \theta = \frac{1}{\sin \theta} \quad \sec \theta = \frac{1}{\cos \theta} \quad \cot \theta = \frac{1}{\tan \theta}$$

Table B.3	Some Trigonometric Identities

$$\sin^2 \theta + \cos^2 \theta = 1 \qquad\qquad \csc^2 \theta = 1 + \cot^2 \theta$$

$$\sec^2 \theta = 1 + \tan^2 \theta \qquad\qquad \sin^2 \frac{\theta}{2} = \tfrac{1}{2}(1 - \cos \theta)$$

$$\sin 2\theta = 2 \sin \theta \cos \theta \qquad\qquad \cos^2 \frac{\theta}{2} = \tfrac{1}{2}(1 + \cos \theta)$$

$$\cos 2\theta = \cos^2 \theta - \sin^2 \theta \qquad\qquad 1 - \cos \theta = 2 \sin^2 \frac{\theta}{2}$$

$$\tan 2\theta = \frac{2 \tan \theta}{1 - \tan^2 \theta} \qquad\qquad \tan \frac{\theta}{2} = \sqrt{\frac{1 - \cos \theta}{1 + \cos \theta}}$$

$$\sin (A \pm B) = \sin A \cos B \pm \cos A \sin B$$

$$\cos (A \pm B) = \cos A \cos B \mp \sin A \sin B$$

$$\sin A \pm \sin B = 2 \sin \left[\tfrac{1}{2}(A \pm B)\right] \cos \left[\tfrac{1}{2}(A \mp B)\right]$$

$$\cos A + \cos B = 2 \cos \left[\tfrac{1}{2}(A + B)\right] \cos \left[\tfrac{1}{2}(A - B)\right]$$

$$\cos A - \cos B = 2 \sin \left[\tfrac{1}{2}(A + B)\right] \sin \left[\tfrac{1}{2}(B - A)\right]$$

The following relationships are derived directly from the right triangle shown in Figure B.10:

$$\sin \theta = \cos (90° - \theta)$$

$$\cos \theta = \sin (90° - \theta)$$

$$\cot \theta = \tan (90° - \theta)$$

Some properties of trigonometric functions are the following:

$$\sin (-\theta) = -\sin \theta$$

$$\cos (-\theta) = \cos \theta$$

$$\tan (-\theta) = -\tan \theta$$

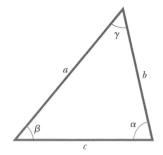

Figure B.11 An arbitrary, non-right triangle.

The following relationships apply to *any* triangle as shown in Figure B.11:

$$\alpha + \beta + \gamma = 180°$$

$$\text{Law of cosines} \begin{cases} a^2 = b^2 + c^2 - 2bc \cos \alpha \\ b^2 = a^2 + c^2 - 2ac \cos \beta \\ c^2 = a^2 + b^2 - 2ab \cos \gamma \end{cases}$$

$$\text{Law of sines} \qquad \frac{a}{\sin \alpha} = \frac{b}{\sin \beta} = \frac{c}{\sin \gamma}$$

Table B.3 lists a number of useful trigonometric identities.

Example B.3

Consider the right triangle in Figure B.12 in which $a = 2.00$, $b = 5.00$, and c is unknown. From the Pythagorean theorem, we have

$$c^2 = a^2 + b^2 = 2.00^2 + 5.00^2 = 4.00 + 25.0 = 29.0$$

$$c = \sqrt{29.0} = 5.39$$

To find the angle θ, note that

$$\tan \theta = \frac{a}{b} = \frac{2.00}{5.00} = 0.400$$

Figure B.12 (Example B.3)

▶ **B.3** continued

Using a calculator, we find that

$$\theta = \tan^{-1}(0.400) = \boxed{21.8°}$$

where $\tan^{-1}(0.400)$ is the notation for "angle whose tangent is 0.400," sometimes written as arctan (0.400).

Exercises

1. In Figure B.13, identify (a) the side opposite θ (b) the side adjacent to ϕ and then find (c) $\cos\theta$, (d) $\sin\phi$, and (e) $\tan\phi$.

Answers   (a) 3   (b) 3   (c) $\frac{4}{5}$ (d) $\frac{4}{5}$ (e) $\frac{4}{3}$

2. In a certain right triangle, the two sides that are perpendicular to each other are 5.00 m and 7.00 m long. What is the length of the third side?

Answer   8.60 m

3. A right triangle has a hypotenuse of length 3.0 m, and one of its angles is 30°. (a) What is the length of the side opposite the 30° angle? (b) What is the side adjacent to the 30° angle?

Answers   (a) 1.5 m   (b) 2.6 m

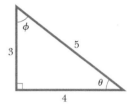

Figure B.13 (Exercise 1)

B.5 Series Expansions

$$(a + b)^n = a^n + \frac{n}{1!} a^{n-1} b + \frac{n(n-1)}{2!} a^{n-2} b^2 + \cdots$$

$$(1 + x)^n = 1 + nx + \frac{n(n-1)}{2!} x^2 + \cdots$$

$$e^x = 1 + x + \frac{x^2}{2!} + \frac{x^3}{3!} + \cdots$$

$$\ln(1 \pm x) = \pm x - \tfrac{1}{2} x^2 \pm \tfrac{1}{3} x^3 - \cdots$$

$$\left.\begin{array}{l} \sin x = x - \dfrac{x^3}{3!} + \dfrac{x^5}{5!} - \cdots \\[2mm] \cos x = 1 - \dfrac{x^2}{2!} + \dfrac{x^4}{4!} - \cdots \\[2mm] \tan x = x + \dfrac{x^3}{3} + \dfrac{2x^5}{15} + \cdots \quad |x| < \dfrac{\pi}{2} \end{array}\right\} x \text{ in radians}$$

For $x \ll 1$, the following approximations can be used:[1]

$$(1 + x)^n \approx 1 + nx \qquad \sin x \approx x$$
$$e^x \approx 1 + x \qquad \cos x \approx 1$$
$$\ln(1 \pm x) \approx \pm x \qquad \tan x \approx x$$

B.6 Differential Calculus

In various branches of science, it is sometimes necessary to use the basic tools of calculus, invented by Newton, to describe physical phenomena. The use of calculus is fundamental in the treatment of various problems in Newtonian mechanics, electricity, and magnetism. In this section, we simply state some basic properties and "rules of thumb" that should be a useful review to the student.

---

[1]The approximations for the functions $\sin x$, $\cos x$, and $\tan x$ are for $x \le 0.1$ rad.

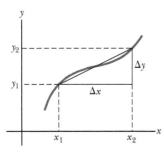

Figure B.14 The lengths Δx and Δy are used to define the derivative of this function at a point.

First, a **function** must be specified that relates one variable to another (e.g., a coordinate as a function of time). Suppose one of the variables is called y (the dependent variable), and the other x (the independent variable). We might have a function relationship such as

$$y(x) = ax^3 + bx^2 + cx + d$$

If a, b, c, and d are specified constants, y can be calculated for any value of x. We usually deal with continuous functions, that is, those for which y varies "smoothly" with x.

The **derivative** of y with respect to x is defined as the limit as Δx approaches zero of the slopes of chords drawn between two points on the y versus x curve. Mathematically, we write this definition as

$$\frac{dy}{dx} = \lim_{\Delta x \to 0} \frac{\Delta y}{\Delta x} = \lim_{\Delta x \to 0} \frac{y(x + \Delta x) - y(x)}{\Delta x} \tag{B.28}$$

where Δy and Δx are defined as $\Delta x = x_2 - x_1$ and $\Delta y = y_2 - y_1$ (Fig. B.14). Note that dy/dx *does not* mean dy divided by dx, but rather is simply a notation of the limiting process of the derivative as defined by Equation B.28.

A useful expression to remember when $y(x) = ax^n$, where a is a *constant* and n is *any* positive or negative number (integer or fraction), is

$$\frac{dy}{dx} = nax^{n-1} \tag{B.29}$$

If $y(x)$ is a polynomial or algebraic function of x, we apply Equation B.29 to *each* term in the polynomial and take $d[\text{constant}]/dx = 0$. In Examples B.4 through B.7, we evaluate the derivatives of several functions.

Special Properties of the Derivative

A. Derivative of the product of two functions If a function $f(x)$ is given by the product of two functions—say, $g(x)$ and $h(x)$—the derivative of $f(x)$ is defined as

$$\frac{d}{dx} f(x) = \frac{d}{dx} [g(x)h(x)] = g\frac{dh}{dx} + h\frac{dg}{dx} \tag{B.30}$$

B. Derivative of the sum of two functions If a function $f(x)$ is equal to the sum of two functions, the derivative of the sum is equal to the sum of the derivatives:

$$\frac{d}{dx} f(x) = \frac{d}{dx} [g(x) + h(x)] = \frac{dg}{dx} + \frac{dh}{dx} \tag{B.31}$$

C. Chain rule of differential calculus If $y = f(x)$ and $x = g(z)$, then dy/dz can be written as the product of two derivatives:

$$\frac{dy}{dz} = \frac{dy}{dx}\frac{dx}{dz} \tag{B.32}$$

D. The second derivative The second derivative of y with respect to x is defined as the derivative of the function dy/dx (the derivative of the derivative). It is usually written as

$$\frac{d^2y}{dx^2} = \frac{d}{dx}\left(\frac{dy}{dx}\right) \tag{B.33}$$

Some of the more commonly used derivatives of functions are listed in Table B.4.

Table B.4 **Derivative for Several Functions**

$$\frac{d}{dx}(a) = 0$$

$$\frac{d}{dx}(ax^n) = nax^{n-1}$$

$$\frac{d}{dx}(e^{ax}) = ae^{ax}$$

$$\frac{d}{dx}(\sin ax) = a\cos ax$$

$$\frac{d}{dx}(\cos ax) = -a\sin ax$$

$$\frac{d}{dx}(\tan ax) = a\sec^2 ax$$

$$\frac{d}{dx}(\cot ax) = -a\csc^2 ax$$

$$\frac{d}{dx}(\sec x) = \tan x \sec x$$

$$\frac{d}{dx}(\csc x) = -\cot x \csc x$$

$$\frac{d}{dx}(\ln ax) = \frac{1}{x}$$

$$\frac{d}{dx}(\sin^{-1} ax) = \frac{a}{\sqrt{1 - a^2x^2}}$$

$$\frac{d}{dx}(\cos^{-1} ax) = \frac{-a}{\sqrt{1 - a^2x^2}}$$

$$\frac{d}{dx}(\tan^{-1} ax) = \frac{a}{1 + a^2x^2}$$

Note: The symbols a and n represent constants.

Example B.4

Suppose $y(x)$ (that is, y as a function of x) is given by

$$y(x) = ax^3 + bx + c$$

where a and b are constants. It follows that

$$y(x + \Delta x) = a(x + \Delta x)^3 + b(x + \Delta x) + c$$

$$= a(x^3 + 3x^2\,\Delta x + 3x\,\Delta x^2 + \Delta x^3) + b(x + \Delta x) + c$$

so

$$\Delta y = y(x + \Delta x) - y(x) = a(3x^2\,\Delta x + 3x\,\Delta x^2 + \Delta x^3) + b\Delta x$$

Substituting this into Equation B.28 gives

$$\frac{dy}{dx} = \lim_{\Delta x \to 0}\frac{\Delta y}{\Delta x} = \lim_{\Delta x \to 0}\left[3ax^2 + 3ax\,\Delta x + a\,\Delta x^2\right] + b$$

$$\frac{dy}{dx} = \boxed{3ax^2 + b}$$

Example B.5

Find the derivative of

$$y(x) = 8x^5 + 4x^3 + 2x + 7$$

SOLUTION

Applying Equation B.29 to each term independently and remembering that d/dx (constant) $= 0$, we have

$$\frac{dy}{dx} = 8(5)x^4 + 4(3)x^2 + 2(1)x^0 + 0$$

$$\frac{dy}{dx} = \boxed{40x^4 + 12x^2 + 2}$$

Example B.6

Find the derivative of $y(x) = x^3/(x + 1)^2$ with respect to x.

SOLUTION

We can rewrite this function as $y(x) = x^3(x + 1)^{-2}$ and apply Equation B.30:

$$\frac{dy}{dx} = (x + 1)^{-2}\frac{d}{dx}(x^3) + x^3\frac{d}{dx}(x + 1)^{-2}$$

$$= (x + 1)^{-2}\,3x^2 + x^3\,(-2)(x + 1)^{-3}$$

$$\frac{dy}{dx} = \boxed{\frac{3x^2}{(x + 1)^2} - \frac{2x^3}{(x + 1)^3}} = \boxed{\frac{x^2(x + 3)}{(x + 1)^3}}$$

Example B.7

A useful formula that follows from Equation B.30 is the derivative of the quotient of two functions. Show that

$$\frac{d}{dx}\left[\frac{g(x)}{h(x)}\right] = \frac{h\dfrac{dg}{dx} - g\dfrac{dh}{dx}}{h^2}$$

SOLUTION

We can write the quotient as gh^{-1} and then apply Equations B.29 and B.30:

$$\frac{d}{dx}\left(\frac{g}{h}\right) = \frac{d}{dx}(gh^{-1}) = g\frac{d}{dx}(h^{-1}) + h^{-1}\frac{d}{dx}(g)$$

$$= -gh^{-2}\frac{dh}{dx} + h^{-1}\frac{dg}{dx}$$

$$= \frac{h\dfrac{dg}{dx} - g\dfrac{dh}{dx}}{h^2}$$

B.7 Integral Calculus

We think of integration as the inverse of differentiation. As an example, consider the expression

$$f(x) = \frac{dy}{dx} = 3ax^2 + b \tag{B.34}$$

which was the result of differentiating the function

$$y(x) = ax^3 + bx + c$$

in Example B.4. We can write Equation B.34 as $dy = f(x)\,dx = (3ax^2 + b)\,dx$ and obtain $y(x)$ by "summing" over all values of x. Mathematically, we write this inverse operation as

$$y(x) = \int f(x)\,dx$$

For the function $f(x)$ given by Equation B.34, we have

$$y(x) = \int (3ax^2 + b)\,dx = ax^3 + bx + c$$

where c is a constant of the integration. This type of integral is called an *indefinite integral* because its value depends on the choice of c.

A general **indefinite integral** $I(x)$ is defined as

$$I(x) = \int f(x)\,dx \tag{B.35}$$

where $f(x)$ is called the *integrand* and $f(x) = dI(x)/dx$.

For a *general continuous* function $f(x)$, the integral can be interpreted geometrically as the area under the curve bounded by $f(x)$ and the x axis, between two specified values of x, say, x_1 and x_2, as in Figure B.15.

The area of the blue element in Figure B.15 is approximately $f(x_i)\,\Delta x_i$. If we sum all these area elements between x_1 and x_2 and take the limit of this sum as $\Delta x_i \to 0$,

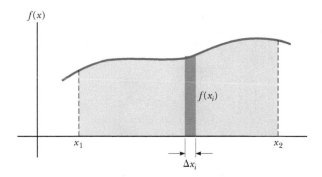

Figure B.15 The definite integral of a function is the area under the curve of the function between the limits x_1 and x_2.

we obtain the *true* area under the curve bounded by $f(x)$ and the x axis, between the limits x_1 and x_2:

$$\text{Area} = \lim_{\Delta x_i \to 0} \sum_i f(x_i)\Delta x_i = \int_{x_1}^{x_2} f(x)\, dx \qquad \text{(B.36)}$$

Integrals of the type defined by Equation B.36 are called **definite integrals.**
One common integral that arises in practical situations has the form

$$\int x^n\, dx = \frac{x^{n+1}}{n+1} + c \quad (n \neq -1) \qquad \text{(B.37)}$$

This result is obvious, being that differentiation of the right-hand side with respect to x gives $f(x) = x^n$ directly. If the limits of the integration are known, this integral becomes a *definite integral* and is written

$$\int_{x_1}^{x_2} x^n\, dx = \frac{x^{n+1}}{n+1}\bigg|_{x_1}^{x_2} = \frac{x_2^{\,n+1} - x_1^{\,n+1}}{n+1} \quad (n \neq -1) \qquad \text{(B.38)}$$

Examples

1. $\displaystyle\int_0^a x^2\, dx = \frac{x^3}{3}\bigg]_0^a = \frac{a^3}{3}$

2. $\displaystyle\int_0^b x^{3/2}\, dx = \frac{x^{5/2}}{5/2}\bigg]_0^b = \tfrac{2}{5} b^{5/2}$

3. $\displaystyle\int_3^5 x\, dx = \frac{x^2}{2}\bigg]_3^5 = \frac{5^2 - 3^2}{2} = 8$

Partial Integration

Sometimes it is useful to apply the method of *partial integration* (also called "integrating by parts") to evaluate certain integrals. This method uses the property

$$\int u\, dv = uv - \int v\, du \qquad \text{(B.39)}$$

where u and v are *carefully* chosen so as to reduce a complex integral to a simpler one. In many cases, several reductions have to be made. Consider the function

$$I(x) = \int x^2 e^x\, dx$$

which can be evaluated by integrating by parts twice. First, if we choose $u = x^2$, $v = e^x$, we obtain

$$\int x^2 e^x\, dx = \int x^2\, d(e^x) = x^2 e^x - 2\int e^x x\, dx + c_1$$

Now, in the second term, choose $u = x$, $v = e^x$, which gives

$$\int x^2 e^x \, dx = x^2 e^x - 2x e^x + 2 \int e^x \, dx + c_1$$

or

$$\int x^2 e^x \, dx = x^2 e^x - 2xe^x + 2e^x + c_2$$

The Perfect Differential

Another useful method to remember is that of the *perfect differential*, in which we look for a change of variable such that the differential of the function is the differential of the independent variable appearing in the integrand. For example, consider the integral

$$I(x) = \int \cos^2 x \sin x \, dx$$

This integral becomes easy to evaluate if we rewrite the differential as $d(\cos x) = -\sin x \, dx$. The integral then becomes

$$\int \cos^2 x \sin x \, dx = -\int \cos^2 x \, d(\cos x)$$

If we now change variables, letting $y = \cos x$, we obtain

$$\int \cos^2 x \sin x \, dx = -\int y^2 \, dy = -\frac{y^3}{3} + c = -\frac{\cos^3 x}{3} + c$$

Table B.5 lists some useful indefinite integrals. Table B.6 gives Gauss's probability integral and other definite integrals. A more complete list can be found in various handbooks, such as *The Handbook of Chemistry and Physics* (Boca Raton, FL: CRC Press, published annually).

Table B.5 **Some Indefinite Integrals (An arbitrary constant should be added to each of these integrals.)**

$$\int x^n \, dx = \frac{x^{n+1}}{n+1} \text{ (provided } n \neq 1\text{)}$$

$$\int \frac{dx}{x} = \int x^{-1} \, dx = \ln x$$

$$\int \frac{dx}{a + bx} = \frac{1}{b} \ln (a + bx)$$

$$\int \frac{x \, dx}{a + bx} = \frac{x}{b} - \frac{a}{b^2} \ln (a + bx)$$

$$\int \frac{dx}{x(x + a)} = -\frac{1}{a} \ln \frac{x + a}{x}$$

$$\int \frac{dx}{(a + bx)^2} = -\frac{1}{b(a + bx)}$$

$$\int \frac{dx}{a^2 + x^2} = \frac{1}{a} \tan^{-1} \frac{x}{a}$$

$$\int \frac{dx}{a^2 - x^2} = \frac{1}{2a} \ln \frac{a + x}{a - x} \, (a^2 - x^2 > 0)$$

$$\int \frac{dx}{x^2 - a^2} = \frac{1}{2a} \ln \frac{x - a}{x + a} \, (x^2 - a^2 > 0)$$

$$\int \ln ax \, dx = (x \ln ax) - x$$

$$\int xe^{ax} \, dx = \frac{e^{ax}}{a^2} (ax - 1)$$

$$\int \frac{dx}{a + be^{cx}} = \frac{x}{a} - \frac{1}{ac} \ln (a + be^{cx})$$

$$\int \sin ax \, dx = -\frac{1}{a} \cos ax$$

$$\int \cos ax \, dx = \frac{1}{a} \sin ax$$

$$\int \tan ax \, dx = -\frac{1}{a} \ln (\cos ax) = \frac{1}{a} \ln (\sec ax)$$

$$\int \cot ax \, dx = \frac{1}{a} \ln (\sin ax)$$

$$\int \sec ax \, dx = \frac{1}{a} \ln (\sec ax + \tan ax) = \frac{1}{a} \ln \left[\tan \left(\frac{ax}{2} + \frac{\pi}{4} \right) \right]$$

$$\int \csc ax \, dx = \frac{1}{a} \ln (\csc ax - \cot ax) = \frac{1}{a} \ln \left(\tan \frac{ax}{2} \right)$$

(Continued)

Table B.5 Some Indefinite Integrals *(continued)*

$$\int \frac{x\,dx}{a^2 \pm x^2} = \pm\frac{1}{2}\ln\left(a^2 \pm x^2\right)$$

$$\int \frac{dx}{\sqrt{a^2 - x^2}} = \sin^{-1}\frac{x}{a} = -\cos^{-1}\frac{x}{a}\ (a^2 - x^2 > 0)$$

$$\int \frac{dx}{\sqrt{x^2 \pm a^2}} = \ln\left(x + \sqrt{x^2 \pm a^2}\right)$$

$$\int \frac{x\,dx}{\sqrt{a^2 - x^2}} = -\sqrt{a^2 - x^2}$$

$$\int \frac{x\,dx}{\sqrt{x^2 \pm a^2}} = \sqrt{x^2 \pm a^2}$$

$$\int \sqrt{a^2 - x^2}\,dx = \frac{1}{2}\left(x\sqrt{a^2 - x^2} + a^2\sin^{-1}\frac{x}{|a|}\right)$$

$$\int x\sqrt{a^2 - x^2}\,dx = -\frac{1}{3}\left(a^2 - x^2\right)^{3/2}$$

$$\int \sqrt{x^2 \pm a^2}\,dx = \frac{1}{2}x\sqrt{x^2 \pm a^2} \pm a^2\ln\left(x + \sqrt{x^2 \pm a^2}\right)$$

$$\int x\left(\sqrt{x^2 \pm a^2}\right)dx = \frac{1}{3}\left(x^2 \pm a^2\right)^{3/2}$$

$$\int e^{ax}\,dx = \frac{1}{a}e^{ax}$$

$$\int \sin^2 ax\,dx = \frac{x}{2} - \frac{\sin 2ax}{4a}$$

$$\int \cos^2 ax\,dx = \frac{x}{2} + \frac{\sin 2ax}{4a}$$

$$\int \frac{dx}{\sin^2 ax} = -\frac{1}{a}\cot ax$$

$$\int \frac{dx}{\cos^2 ax} = \frac{1}{a}\tan ax$$

$$\int \tan^2 ax\,dx = \frac{1}{a}\left(\tan ax\right) - x$$

$$\int \cot^2 ax\,dx = -\frac{1}{a}\left(\cot ax\right) - x$$

$$\int \sin^{-1} ax\,dx = x(\sin^{-1} ax) + \frac{\sqrt{1 - a^2 x^2}}{a}$$

$$\int \cos^{-1} ax\,dx = x(\cos^{-1} ax) - \frac{\sqrt{1 - a^2 x^2}}{a}$$

$$\int \frac{dx}{\left(x^2 + a^2\right)^{3/2}} = \frac{x}{a^2\sqrt{x^2 + a^2}}$$

$$\int \frac{x\,dx}{\left(x^2 + a^2\right)^{3/2}} = -\frac{1}{\sqrt{x^2 + a^2}}$$

Table B.6 Gauss's Probability Integral and Other Definite Integrals

$$\int_0^\infty x^n\,e^{-ax}\,dx = \frac{n!}{a^{n+1}}$$

$$I_0 = \int_0^\infty e^{-ax^2}\,dx = \frac{1}{2}\sqrt{\frac{\pi}{a}}\quad\text{(Gauss's probability integral)}$$

$$I_1 = \int_0^\infty xe^{-ax^2}\,dx = \frac{1}{2a}$$

$$I_2 = \int_0^\infty x^2\,e^{-ax^2}\,dx = -\frac{dI_0}{da} = \frac{1}{4}\sqrt{\frac{\pi}{a^3}}$$

$$I_3 = \int_0^\infty x^3\,e^{-ax^2}\,dx = -\frac{dI_1}{da} = \frac{1}{2a^2}$$

$$I_4 = \int_0^\infty x^4\,e^{-ax^2}\,dx = \frac{d^2 I_0}{da^2} = \frac{3}{8}\sqrt{\frac{\pi}{a^5}}$$

$$I_5 = \int_0^\infty x^5\,e^{-ax^2}\,dx = \frac{d^2 I_1}{da^2} = \frac{1}{a^3}$$

$$\vdots$$

$$I_{2n} = (-1)^n\frac{d^n}{da^n}I_0$$

$$I_{2n+1} = (-1)^n\frac{d^n}{da^n}I_1$$

B.8 Propagation of Uncertainty

In laboratory experiments, a common activity is to take measurements that act as raw data. These measurements are of several types—length, time interval, temperature, voltage, and so on—and are taken by a variety of instruments. Regardless of the measurement and the quality of the instrumentation, **there is always uncertainty associated with a physical measurement.** This uncertainty is a combination of that associated with the instrument and that related to the system being measured. An example of the former is the inability to exactly determine the position of a length measurement between the lines on a meterstick. An example of uncertainty related to the system being measured is the variation of temperature within a sample of water so that a single temperature for the sample is difficult to determine.

Uncertainties can be expressed in two ways. **Absolute uncertainty** refers to an uncertainty expressed in the same units as the measurement. Therefore, the length of a computer disk label might be expressed as (5.5 ± 0.1) cm. The uncertainty of ± 0.1 cm by itself is not descriptive enough for some purposes, however. This uncertainty is large if the measurement is 1.0 cm, but it is small if the measurement is 100 m. To give a more descriptive account of the uncertainty, **fractional uncertainty** or **percent uncertainty** is used. In this type of description, the uncertainty is divided by the actual measurement. Therefore, the length of the computer disk label could be expressed as

$$\ell = 5.5 \text{ cm} \pm \frac{0.1 \text{ cm}}{5.5 \text{ cm}} = 5.5 \text{ cm} \pm 0.018 \quad \text{(fractional uncertainty)}$$

or as

$$\ell = 5.5 \text{ cm} \pm 1.8\% \quad \text{(percent uncertainty)}$$

When combining measurements in a calculation, the percent uncertainty in the final result is generally larger than the uncertainty in the individual measurements. This is called **propagation of uncertainty** and is one of the challenges of experimental physics.

Some simple rules can provide a reasonable estimate of the uncertainty in a calculated result:

Multiplication and division: When measurements with uncertainties are multiplied or divided, add the *percent uncertainties* to obtain the percent uncertainty in the result.

Example: The Area of a Rectangular Plate

$$A = \ell w = (5.5 \text{ cm} \pm 1.8\%) \times (6.4 \text{ cm} \pm 1.6\%) = 35 \text{ cm}^2 \pm 3.4\%$$

$$= (35 \pm 1) \text{ cm}^2$$

Addition and subtraction: When measurements with uncertainties are added or subtracted, add the *absolute uncertainties* to obtain the absolute uncertainty in the result.

Example: A Change in Temperature

$$\Delta T = T_2 - T_1 = (99.2 \pm 1.5)°\text{C} - (27.6 \pm 1.5)°\text{C} = (71.6 \pm 3.0)°\text{C}$$

$$= 71.6°\text{C} \pm 4.2\%$$

Powers: If a measurement is taken to a power, the percent uncertainty is multiplied by that power to obtain the percent uncertainty in the result.

Example: The Volume of a Sphere

$$V = \tfrac{4}{3}\pi r^3 = \tfrac{4}{3}\pi (6.20 \text{ cm} \pm 2.0\%)^3 = 998 \text{ cm}^3 \pm 6.0\%$$

$$= (998 \pm 60) \text{ cm}^3$$

For complicated calculations, many uncertainties are added together, which can cause the uncertainty in the final result to be undesirably large. Experiments should be designed such that calculations are as simple as possible.

Notice that uncertainties in a calculation always add. As a result, an experiment involving a subtraction should be avoided if possible, especially if the measurements being subtracted are close together. The result of such a calculation is a small difference in the measurements and uncertainties that add together. It is possible that the uncertainty in the result could be larger than the result itself!

Periodic Table of the Elements

Group I Group II Transition elements

H 1 1.007 9 1*s*	

Symbol — **Ca** 20 — Atomic number
Atomic mass† — 40.078
4*s*² — Electron configuration

Li 3 6.941 2*s*¹	**Be** 4 9.0122 2*s*²
Na 11 22.990 3*s*¹	**Mg** 12 24.305 3*s*²

K 19 39.098 4*s*¹	**Ca** 20 40.078 4*s*²	**Sc** 21 44.956 3*d*¹4*s*²	**Ti** 22 47.867 3*d*²4*s*²	**V** 23 50.942 3*d*³4*s*²	**Cr** 24 51.996 3*d*⁵4*s*¹	**Mn** 25 54.938 3*d*⁵4*s*²	**Fe** 26 55.845 3*d*⁶4*s*²	**Co** 27 58.933 3*d*⁷4*s*²
Rb 37 85.468 5*s*¹	**Sr** 38 87.62 5*s*²	**Y** 39 88.906 4*d*¹5*s*²	**Zr** 40 91.224 4*d*²5*s*²	**Nb** 41 92.906 4*d*⁴5*s*¹	**Mo** 42 95.94 4*d*⁵5*s*¹	**Tc** 43 (98) 4*d*⁵5*s*²	**Ru** 44 101.07 4*d*⁷5*s*¹	**Rh** 45 102.91 4*d*⁸5*s*¹
Cs 55 132.91 6*s*¹	**Ba** 56 137.33 6*s*²	57–71*	**Hf** 72 178.49 5*d*²6*s*²	**Ta** 73 180.95 5*d*³6*s*²	**W** 74 183.84 5*d*⁴6*s*²	**Re** 75 186.21 5*d*⁵6*s*²	**Os** 76 190.23 5*d*⁶6*s*²	**Ir** 77 192.2 5*d*⁷6*s*²
Fr 87 (223) 7*s*¹	**Ra** 88 (226) 7*s*²	89–103**	**Rf** 104 (261) 6*d*²7*s*²	**Db** 105 (262) 6*d*³7*s*²	**Sg** 106 (266)	**Bh** 107 (264)	**Hs** 108 (277)	**Mt** 109 (268)

*Lanthanide series

La 57 138.91 5*d*¹6*s*²	**Ce** 58 140.12 5*d*¹4*f*¹6*s*²	**Pr** 59 140.91 4*f*³6*s*²	**Nd** 60 144.24 4*f*⁴6*s*²	**Pm** 61 (145) 4*f*⁵6*s*²	**Sm** 62 150.36 4*f*⁶6*s*²

**Actinide series

Ac 89 (227) 6*d*¹7*s*²	**Th** 90 232.04 6*d*²7*s*²	**Pa** 91 231.04 5*f*²6*d*¹7*s*²	**U** 92 238.03 5*f*³6*d*¹7*s*²	**Np** 93 (237) 5*f*⁴6*d*¹7*s*²	**Pu** 94 (244) 5*f*⁶7*s*²

Note: Atomic mass values given are averaged over isotopes in the percentages in which they exist in nature.
†For an unstable element, mass number of the most stable known isotope is given in parentheses.

			Group III	Group IV	Group V	Group VI	Group VII	Group 0
							H 1 1.007 9 $1s^1$	**He** 2 4.002 6 $1s^2$
			B 5 10.811 $2p^1$	**C** 6 12.011 $2p^2$	**N** 7 14.007 $2p^3$	**O** 8 15.999 $2p^4$	**F** 9 18.998 $2p^5$	**Ne** 10 20.180 $2p^6$
			Al 13 26.982 $3p^1$	**Si** 14 28.086 $3p^2$	**P** 15 30.974 $3p^3$	**S** 16 32.066 $3p^4$	**Cl** 17 35.453 $3p^5$	**Ar** 18 39.948 $3p^6$
Ni 28 58.693 $3d^84s^2$	**Cu** 29 63.546 $3d^{10}4s^1$	**Zn** 30 65.41 $3d^{10}4s^2$	**Ga** 31 69.723 $4p^1$	**Ge** 32 72.64 $4p^2$	**As** 33 74.922 $4p^3$	**Se** 34 78.96 $4p^4$	**Br** 35 79.904 $4p^5$	**Kr** 36 83.80 $4p^6$
Pd 46 106.42 $4d^{10}$	**Ag** 47 107.87 $4d^{10}5s^1$	**Cd** 48 112.41 $4d^{10}5s^2$	**In** 49 114.82 $5p^1$	**Sn** 50 118.71 $5p^2$	**Sb** 51 121.76 $5p^3$	**Te** 52 127.60 $5p^4$	**I** 53 126.90 $5p^5$	**Xe** 54 131.29 $5p^6$
Pt 78 195.08 $5d^96s^1$	**Au** 79 196.97 $5d^{10}6s^1$	**Hg** 80 200.59 $5d^{10}6s^2$	**Tl** 81 204.38 $6p^1$	**Pb** 82 207.2 $6p^2$	**Bi** 83 208.98 $6p^3$	**Po** 84 (209) $6p^4$	**At** 85 (210) $6p^5$	**Rn** 86 (222) $6p^6$
Ds 110 (271)	**Rg** 111 (272)	**Cn** 112 (285)	113[††] (284)	**Fl** 114 (289)	115[††] (288)	**Lv** 116 (293)	117[††] (294)	118[††] (294)

Eu 63 151.96 $4f^76s^2$	**Gd** 64 157.25 $4f^75d^16s^2$	**Tb** 65 158.93 $4f^85d^16s^2$	**Dy** 66 162.50 $4f^{10}6s^2$	**Ho** 67 164.93 $4f^{11}6s^2$	**Er** 68 167.26 $4f^{12}6s^2$	**Tm** 69 168.93 $4f^{13}6s^2$	**Yb** 70 173.04 $4f^{14}6s^2$	**Lu** 71 174.97 $4f^{14}5d^16s^2$
Am 95 (243) $5f^77s^2$	**Cm** 96 (247) $5f^76d^17s^2$	**Bk** 97 (247) $5f^86d^17s^2$	**Cf** 98 (251) $5f^{10}7s^2$	**Es** 99 (252) $5f^{11}7s^2$	**Fm** 100 (257) $5f^{12}7s^2$	**Md** 101 (258) $5f^{13}7s^2$	**No** 102 (259) $5f^{14}7s^2$	**Lr** 103 (262) $5f^{14}6d^17s^2$

[††]Elements 113, 115, 117, and 118 have not yet been officially named. Only small numbers of atoms of these elements have been observed.
Note: For a description of the atomic data, visit *physics.nist.gov/PhysRefData/Elements/per_text.html*.

D

SI Units

Table D.1 — SI Units

Base Quantity	SI Base Unit	
	Name	Symbol
Length	meter	m
Mass	kilogram	kg
Time	second	s
Electric current	ampere	A
Temperature	kelvin	K
Amount of substance	mole	mol
Luminous intensity	candela	cd

Table D.2 — Some Derived SI Units

Other Quantity	Name	Symbol	Expression in Terms of Base Units	Expression in Terms of SI Units
Plane angle	radian	rad	m/m	
Frequency	hertz	Hz	s^{-1}	
Force	newton	N	$kg \cdot m/s^2$	J/m
Pressure	pascal	Pa	$kg/m \cdot s^2$	N/m^2
Energy	joule	J	$kg \cdot m^2/s^2$	$N \cdot m$
Power	watt	W	$kg \cdot m^2/s^3$	J/s
Electric charge	coulomb	C	$A \cdot s$	
Electric potential	volt	V	$kg \cdot m^2/A \cdot s^3$	W/A
Capacitance	farad	F	$A^2 \cdot s^4/kg \cdot m^2$	C/V
Electric resistance	ohm	Ω	$kg \cdot m^2/A^2 \cdot s^3$	V/A
Magnetic flux	weber	Wb	$kg \cdot m^2/A \cdot s^2$	$V \cdot s$
Magnetic field	tesla	T	$kg/A \cdot s^2$	
Inductance	henry	H	$kg \cdot m^2/A^2 \cdot s^2$	$T \cdot m^2/A$

Answers to Quick Quizzes and Odd-Numbered Problems

Chapter 1

Answers to Quick Quizzes

1. (a)
2. False
3. (b)

Answers to Odd-Numbered Problems

1. (a) 5.52×10^3 kg/m^3 (b) It is between the density of aluminum and that of iron and is greater than the densities of typical surface rocks.
3. 23.0 kg
5. 7.69 cm
7. 0.141 nm
9. (b) only
11. (a) kg · m/s (b) N · s
13. No.
15. 11.4×10^3 kg/m^3
17. 871 m^2
19. By measuring the pages, we find that each page has area 0.277 m × 0.217 m = 0.060 m^2. The room has wall area 37 m^2, requiring 616 sheets that would be counted as 1 232 pages. Volume 1 of this textbook contains only 784 pages.
21. 1.00×10^{10} lb
23. 4.05×10^3 m^2
25. 2.86 cm
27. 151 μm
29. (a) 507 years (b) 2.48×10^9 bills
31. ~ 10^6 balls in a room 4 m by 4 m by 3 m
33. ~ 10^2 piano tuners
35. (209 ± 4) cm^2
37. 31 556 926.0 s
39. 19
41. 8.80%
43. 63
45. (a) 6.71 m (b) 0.894 (c) 0.745
47. 48.6 kg
49. ±3.46
51. Answers may vary somewhat due to variation in reading precise numbers off the graph. (a) 0.015 g (b) 8% (c) 5.2 g/m^2 (d) For shapes cut from this copy paper, the mass of the cutout is proportional to its area. The proportionality constant is 5.2 g/m^2 ± 8%, where the uncertainty is estimated. (e) This result is to be expected if the paper has thickness and density that are uniform within

the experimental uncertainty. (f) The slope is the areal density of the paper, its mass per unit area.
53. 5.2 m^3, 3%
55. 316 m
57. 5.0 m
59. 3.41 m
61. (a) aluminum, 2.75 g/cm^3; copper, 9.36 g/cm^3; brass, 8.91 g/cm^3; tin, 7.68 g/cm^3; iron, 7.88 g/cm^3 (b) The tabulated values are smaller by 2% for aluminum, by 5% for copper, by 6% for brass, by 5% for tin, and by 0.3% for iron.
63. 1×10^{10} gal/yr
65. Answers may vary. (a) ~ 10^{29} prokaryotes (b) ~ 10^{14} kg
67. (a) $B = 2.70$ g/cm^3, $C = 1.19$ g/cm^4 (b) 1.39 kg
69. $V = 0.579t + (1.19 \times 10^{-9})t^2$, where V is in cubic feet and t is in seconds
71. (a) 0.529 cm/s (b) 11.5 cm/s
73. (a) 12.1 m (b) 135° (c) 25.2° (d) 135°

Chapter 2

Answers to Quick Quizzes

1. (c)
2. (b)
3. False. Your graph should look something like the one shown below. This v_x–t graph shows that the maximum speed is about 5.0 m/s, which is 18 km/h (= 11 mi/h), so the driver was not speeding.

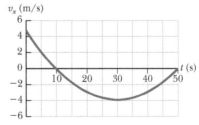

4. (b)
5. (c)
6. (a)–(e), (b)–(d), (c)–(f)
7. (i) (e) (ii) (d)

Answers to Odd-Numbered Problems

1. (a) 5 m/s (b) 1.2 m/s (c) −2.5 m/s (d) −3.3 m/s (e) 0
3. (a) 3.75 m/s (b) 0

A-25

5. (a) 2.30 m/s   (b) 16.1 m/s   (c) 11.5 m/s
7. (a) -2.4 m/s   (b) -3.8 m/s   (c) 4.0 s
9. (a) 5.0 m/s   (b) -2.5 m/s   (c) 0   (d) $+5.0$ m/s
11. (a) 5.00 m   (b) 4.88×10^3 s
13. (a) 2.80 h   (b) 218 km
15. (a)

(b) 1.60 m/s²   (c) 0.800 m/s²
17. (a) 1.3 m/s²   (b) $t = 3$ s, $a = 2$ m/s²   (c) $t = 6$ s, $t > 10$ s
   (d) $a = -1.5$ m/s², $t = 8$ s
19. (a) 20 m/s, 5 m/s   (b) 263 m
21. (a) 2.00 m   (b) -3.00 m/s   (c) -2.00 m/s²
23.

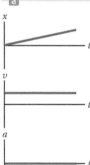

25. (a) 4.98×10^{-9} s   (b) 1.20×10^{15} m/s²
27. (a) 9.00 m/s   (b) -3.00 m/s   (c) 17.0 m/s   (d) The graph
   of velocity versus time is a straight line passing through
   13 m/s at 10:05 a.m. and sloping downward, decreasing
   by 4 m/s for each second thereafter.   (e) If and only if we
   know the object's velocity at one instant of time, knowing
   its acceleration tells us its velocity at every other moment
   as long as the acceleration is constant.
29. -16.0 cm/s²
31. (a) -202 m/s²   (b) 198 m
33. (a) 35.0 s   (b) 15.7 m/s
35. 3.10 m/s
37. (a)

(b) Particle under constant acceleration
(c) $v_f^2 = v_i^2 + 2a(x_f - x_i)$ (Equation 2.17)
(d) $a = \dfrac{v_f^2 - v_i^2}{2(x_f - x_i)}$ (e) 1.25 m/s²   (f) 8.00 s

39. (a) The idea is false unless the acceleration is zero. We
   define constant acceleration to mean that the velocity
   is changing steadily in time. So, the velocity cannot be
   changing steadily in space.
   (b) This idea is true. Because the velocity is changing
   steadily in time, the velocity halfway through an interval
   is equal to the average of its initial and final values.
41. (a) 13.5 m   (b) 13.5 m   (c) 13.5 m   (d) 22.5 m
43. (a) 1.88 km   (b) 1.46 km

(c)

(d) 0a: $x = 1.67t^2$; ab: $x = 50t - 375$; bc: $x = 250t - 2.5t^2 - 4\,375$ (In all three expressions, x is in meters and t is in
   seconds.)   (e) 37.5 m/s
45. (a) 0.231 m   (b) 0.364 m   (c) 0.399 m   (d) 0.175 m
47. David will be unsuccessful. The average human reaction
   time is about 0.2 s (research on the Internet) and a dollar bill is about 15.5 cm long, so David's fingers are about
   8 cm from the end of the bill before it is dropped. The
   bill will fall about 20 cm before he can close his fingers.
49. (a) 510 m   (b) 20.4 s
51. 1.79 s
53. (a) 10.0 m/s up   (b) 4.68 m/s down
55. (a) 7.82 m   (b) 0.782 s
57. (a) $a_x(t) = a_{xi} + Jt$; $v_x(t) = v_{xi} + a_{xi}t + \frac{1}{2}Jt^2$;
   $x(t) = x_i + v_{xi}t + \frac{1}{2}a_{xi}t^2 + \frac{1}{6}Jt^3$
59. (a) $a = -(10.0 \times 10^7)t + 3.00 \times 10^5$; $x = -(1.67 \times 10^7)t^3 + (1.50 \times 10^5)t^2$ (In these expressions, a is in m/s², x is in
   meters, and t is in seconds.)   (b) 3.00 ms   (c) 450 m/s
   (d) 0.900 m
61. (a) 4.00 m/s   (b) 1.00 ms   (c) 0.816 m
63. (a) 3.00 s   (b) -15.3 m/s
   (c) 31.4 m/s down and 34.8 m/s down
65. (a) 3.00 m/s   (b) 6.00 s   (c) -0.300 m/s²   (d) 2.05 m/s
67. (a) 2.83 s   (b) It is exactly the same situation as in Example
   2.8 except that this problem is in the vertical direction. The
   descending elevator plays the role of the speeding car, and
   the falling bolt plays the role of the accelerating trooper.
   Turn Figure 2.13 through 90° clockwise to visualize the
   elevator–bolt problem!   (c) If each floor is 3 m high, the
   highest floor that can be reached is the 13th floor.
69. (a) From the graph, we see that the Acela is cruising at a
   constant positive velocity in the positive x direction from
   about -50 s to 50 s. From 50 s to 200 s, the Acela accelerates in the positive x direction reaching a top speed
   of about 170 mi/h. Around 200 s, the engineer applies
   the brakes, and the train, still traveling in the positive x
   direction, slows down and then stops at 350 s. Just after

350 s, the train reverses direction (v becomes negative) and steadily gains speed in the negative x direction. (b) approximately 2.2 mi/h/s (c) approximately 6.7 mi

71. (a) Here, v_1 must be greater than v_2 and the distance between the leading athlete and the finish line must be great enough so that the trailing athlete has time to catch up.

(b) $t = \dfrac{d_1}{v_1 - v_2}$ (c) $d_2 = \dfrac{v_2 d_1}{v_1 - v_2}$

73. (a) 5.46 s   (b) 73.0 m
(c) $v_{\text{Stan}} = 22.6$ m/s, $v_{\text{Kathy}} = 26.7$ m/s

75. (a) $v_B = (1/\tan\theta)v$ (b) The velocity v_B starts off larger than v for small values of θ and then decreases, approaching zero as θ approaches 90°.

77. (a) 15.0 s   (b) 30.0 m/s   (c) 225 m

79. 1.60 m/s^2

81. (a) 35.9 m   (b) 4.04 s   (c) 45.8 m   (d) 22.6 m/s

83. (a) 5.32 m/s^2 for Laura and 3.75 m/s^2 for Healan
(b) 10.6 m/s for Laura and 11.2 m/s for Healan
(c) Laura, by 2.63 m   (d) 4.47 m at $t = 2.84$ s

85. (a) 26.4 m   (b) 6.8%

Chapter 3

Answers to Quick Quizzes

1. vectors: (b), (c); scalars: (a), (d), (e)
2. (c)
3. (b) and (c)
4. (b)
5. (c)

Answers to Odd-Numbered Problems

1. $(-2.75, -4.76)$ m
3. (a) 8.60 m   (b) 4.47 m, $-63.4°$; 4.24 m, 135°
5. (a) $(-3.56$ cm, -2.40 cm)   (b) $(r = 4.30$ cm, $\theta = 326°)$
(c) $(r = 8.60$ cm, $\theta = 34.0°)$ (d) $(r = 12.9$ cm, $\theta = 146°)$
7. 70.0 m
9. This situation can *never* be true because the distance is the length of an arc of a circle between two points, whereas the magnitude of the displacement vector is a straight-line chord of the circle between the same points.
11. (a) 5.2 m at 60°   (b) 3.0 m at 330°   (c) 3.0 m at 150°
(d) 5.2 m at 300°
13. approximately 420 ft at $-3°$
15. 47.2 units at 122°
17. (a) yes   (b) The speed of the camper should be 28.3 m/s or more to satisfy this requirement.
19. (a) $(-11.1\hat{\mathbf{i}} + 6.40\hat{\mathbf{j}})$ m   (b) $(1.65\hat{\mathbf{i}} + 2.86\hat{\mathbf{j}})$ cm
(c) $(-18.0\hat{\mathbf{i}} - 12.6\hat{\mathbf{j}})$ in.
21. 358 m at 2.00° S of E
23. (a) $2.00\hat{\mathbf{i}} - 6.00\hat{\mathbf{j}}$ (b) $4.00\hat{\mathbf{i}} + 2.00\hat{\mathbf{j}}$ (c) 6.32   (d) 4.47
(e) 288°; 26.6°
25. 9.48 m at 166°
27. 4.64 m at 78.6° N of E
29. (a) 185 N at 77.8° from the positive x axis
(b) $(-39.3\hat{\mathbf{i}} - 181\hat{\mathbf{j}})$ N
31. (a) 2.83 m at $\theta = 315°$ (b) 13.4 m at $\theta = 117°$
33. (a) $8.00\hat{\mathbf{i}} + 12.0\hat{\mathbf{j}} - 4.00\hat{\mathbf{k}}$ (b) $2.00\hat{\mathbf{i}} + 3.00\hat{\mathbf{j}} - 1.00\hat{\mathbf{k}}$
(c) $-24.0\hat{\mathbf{i}} - 36.0\hat{\mathbf{j}} + 12.0\hat{\mathbf{k}}$

35. (a) $-3.00\hat{\mathbf{i}} + 2.00\hat{\mathbf{j}}$ (b) 3.61 at 146°   (c) $3.00\hat{\mathbf{i}} - 6.00\hat{\mathbf{j}}$
37. (a) $a = 5.00$ and $b = 7.00$ (b) For vectors to be equal, all their components must be equal. A vector equation contains more information than a scalar equation.
39. 196 cm at 345°
41. (a) $\vec{\mathbf{E}} = (15.1\hat{\mathbf{i}} + 7.72\hat{\mathbf{j}})$ cm   (b) $\vec{\mathbf{F}} = (-7.72\hat{\mathbf{i}} + 15.1\hat{\mathbf{j}})$ cm
(c) $\vec{\mathbf{G}} = (-7.72\hat{\mathbf{i}} - 15.1\hat{\mathbf{j}})$ cm
43. (a) $(-20.5\hat{\mathbf{i}} + 35.5\hat{\mathbf{j}})$ m/s   (b) $25.0\hat{\mathbf{j}}$ m/s
(c) $(-61.5\hat{\mathbf{i}} + 107\hat{\mathbf{j}})$ m   (d) $37.5\hat{\mathbf{j}}$ m   (e) 157 km
45. 1.43×10^4 m at 32.2° above the horizontal
47. (a) 10.4 cm   (b) $\theta = 35.5°$
49. (a)

(b) 18.3 b   (c) 12.4 b at 233° counterclockwise from east
51. 240 m at 237°
53. (a) 25.4 s   (b) 15.0 km/h
55. (a) 0.079 8 N   (b) 57.9°   (c) 32.1°
57. (a) The x, y, and z components are, respectively, 2.00, 1.00, and 3.00.   (b) 3.74   (c) $\theta_x = 57.7°$, $\theta_y = 74.5°$, $\theta_z = 36.7°$
59. 1.15°
61. (a) $(10\,000 - 9\,600\sin\theta)^{1/2}$ cm   (b) 270°; 140 cm   (c) 90°; 20.0 cm   (d) They do make sense. The maximum value is attained when $\vec{\mathbf{A}}$ and $\vec{\mathbf{B}}$ are in the same direction, and it is 60 cm + 80 cm. The minimum value is attained when $\vec{\mathbf{A}}$ and $\vec{\mathbf{B}}$ are in opposite directions, and it is 80 cm − 60 cm.
63. (a) $-2.00\hat{\mathbf{k}}$ m/s   (b) its velocity vector
65. (a) $\vec{\mathbf{R}}_1 = a\hat{\mathbf{i}} + b\hat{\mathbf{j}}$ (b) $R_1 = (a^2 + b^2)^{1/2}$
(c) $\vec{\mathbf{R}}_2 = a\hat{\mathbf{i}} + b\hat{\mathbf{j}} + c\hat{\mathbf{k}}$
67. (a) (10.0 m, 16.0 m)   (b) This center of mass of the tree distribution is the same location whatever order we take the trees in. (We will study center of mass in Chapter 9.)

Chapter 4

Answers to Quick Quizzes

1. (a)
2. (i) (b)   (ii) (a)
3. 15°, 30°, 45°, 60°, 75°
4. (i) (d)   (ii) (b)
5. (i) (b)   (ii) (d)

Answers to Odd-Numbered Problems

1. (a) 4.87 km at 209° from east   (b) 23.3 m/s
(c) 13.5 m/s at 209°
3. (a) $(1.00\hat{\mathbf{i}} + 0.750\hat{\mathbf{j}})$ m/s   (b) $(1.00\hat{\mathbf{i}} + 0.500\hat{\mathbf{j}})$ m/s, 1.12 m/s
5. (a) $\vec{\mathbf{r}} = 18.0t\hat{\mathbf{i}} + (4.00t - 4.90t^2)\hat{\mathbf{j}}$, where $\vec{\mathbf{r}}$ is in meters and t is in seconds
(b) $\vec{\mathbf{v}} = 18.0\hat{\mathbf{i}} + (4.00 - 9.80t)\hat{\mathbf{j}}$, where $\vec{\mathbf{v}}$ is in meters per second and t is in seconds
(c) $\vec{\mathbf{a}} = -9.80\hat{\mathbf{j}}$ m/s^2
(d) $\vec{\mathbf{r}} = (54.0\hat{\mathbf{i}} - 32.1\hat{\mathbf{j}})$ m; $\vec{\mathbf{v}} = (18.0\hat{\mathbf{i}} - 25.4\hat{\mathbf{j}})$ m/s;
$\vec{\mathbf{a}} = -9.80\hat{\mathbf{j}}$ m/s^2

7. (a) $\vec{v} = -12.0t\hat{j}$, where $\vec{v}$ is in meters per second and t is in seconds   (b) $\vec{a} = -12.0\hat{j}$ m/s^2 (c) $\vec{r} = (3.00\hat{i} - 6.00\hat{j})$ m; $\vec{v} = -12.0\hat{j}$ m/s

9. (a) $(0.800\hat{i} - 0.300\hat{j})$ m/s^2 (b) 339°
   (c) $(360\hat{i} - 72.7\hat{j})$ m, −15.2°

11. 12.0 m/s

13. (a) 2.81 m/s horizontal   (b) 60.2° below the horizontal

15. 53.1°

17. (a) 3.96 m/s horizontally forward   (b) 9.6%

19. 67.8°

21. $d\tan\theta_i - \dfrac{gd^2}{2v_i^2\cos^2\theta_i}$

23. (a) The ball clears by 0.89 m.   (b) while descending

25. (a) 18.1 m/s   (b) 1.13 m   (c) 2.79 m

27. 9.91 m/s

29. (a) (0, 50.0 m)   (b) $v_{xi} = 18.0$ m/s; $v_{yi} = 0$ (c) Particle under constant acceleration   (d) Particle under constant velocity   (e) $v_{xf} = v_{xi}$; $v_{yf} = -gt$ (f) $x_f = v_{xi}t$; $y_f = y_i - \frac{1}{2}gt^2$ (g) 3.19 s   (h) 36.1 m/s, −60.1°

31. 1.92 s

33. 377 m/s^2

35. 2.06×10^3 rev/min

37. 0.749 rev/s

39. 7.58×10^3 m/s, 5.80×10^3 s

41. 1.48 m/s^2 inward and 29.9° backward

43. (a) Yes. The particle can be either speeding up or slowing down, with a tangential component of acceleration of magnitude $\sqrt{6^2 - 4.5^2} = 3.97$ m/s^2.   (b) No. The magnitude of the acceleration cannot be less than $v^2/r = 4.5$ m/s^2.

45. (a) 1.26 h   (b) 1.13 h   (c) 1.19 h

47. (a) 15.0 km/h east   (b) 15.0 km/h west   (c) 0.016 7 h = 60.0 s

49. (a) 9.80 m/s^2 down and 2.50 m/s^2 south   (b) 9.80 m/s^2 down   (c) The bolt moves on a parabola with its axis downward and tilting to the south. It lands south of the point directly below its starting point.   (d) The bolt moves on a parabola with a vertical axis.

51. (a) $\dfrac{2d/c}{1 - v^2/c^2}$ (b) $\dfrac{2d}{c}$

   (c) The trip in flowing water takes a longer time interval. The swimmer travels at the low upstream speed for a longer time interval, so his average speed is reduced below c. Mathematically, $1/(1 - v^2/c^2)$ is always greater than 1. In the extreme, as $v \to c$, the time interval becomes infinite. In that case, the student can never return to the starting point because he cannot swim fast enough to overcome the river current.

53. 15.3 m

55. 54.4 m/s^2

57. The relationship between the height h and the walking speed is $h = (4.16 \times 10^{-3})v_x^2$, where h is in meters and v_x is in meters per second. At a typical walking speed of 4 to 5 km/h, the ball would have to be dropped from a height of about 1 cm, clearly much too low for a person's hand. Even at Olympic-record speed for the 100-m run (confirm on the Internet), this situation would only occur if the ball is dropped from about 0.4 m, which is also below the hand of a normally proportioned person.

59. (a) 101 m/s   (b) 3.27×10^4 ft   (c) 20.6 s

61. (a) 26.9 m/s   (b) 67.3 m   (c) $(2.00\hat{i} - 5.00\hat{j})$ m/s^2

63. (a) $(7.62\hat{i} - 6.48\hat{j})$ cm   (b) $(10.0\hat{i} - 7.05\hat{j})$ cm

65. (a) 1.52 km   (b) 36.1 s   (c) 4.05 km

67. The initial height of the ball when struck is 3.94 m, which is too high for the batter to hit the ball.

69. (a) 1.69 km/s   (b) 1.80 h

71. (a) 46.5 m/s   (b) −77.6°   (c) 6.34 s

73. (a) $x = v_i(0.164\ 3 + 0.002\ 299v_i^2)^{1/2} + 0.047\ 94v_i^2$, where x is in meters and v_i is in meters per second   (b) 0.041 0 m   (c) 961 m   (d) $x \approx 0.405v_i$ (e) $x \approx 0.095\ 9v_i^2$ (f) The graph of x versus v_i starts from the origin as a straight line with slope 0.405 s. Then it curves upward above this tangent line, becoming closer and closer to the parabola $x = 0.095\ 9v_i^2$, where x is in meters and v_i is in meters per second.

75. (a) 6.80 km   (b) 3.00 km vertically above the impact point   (c) 66.2°

77. (a) 20.0 m/s   (b) 5.00 s   (c) $(16.0\hat{i} - 27.1\hat{j})$ m/s   (d) 6.53 s   (e) $24.5\hat{i}$ m

79. (a) 4.00 km/h   (b) 4.00 km/h

81. (a) 43.2 m   (b) $(9.66\hat{i} - 25.6\hat{j})$ m/s   (c) Air resistance would ordinarily make the jump distance smaller and the final horizontal and vertical velocity components both somewhat smaller. If a skilled jumper shapes her body into an airfoil, however, she can deflect downward the air through which she passes so that it deflects her upward, giving her more time in the air and a longer jump.

83. (a) swim perpendicular to the banks   (b) 133 m   (c) 53.1°   (d) 107 m

85. 33.5° below the horizontal

87. $\tan^{-1}\left(\dfrac{\sqrt{2gh}}{v}\right)$

89. Safe distances are less than 270 m or greater than 3.48×10^3 m from the western shore.

Chapter 5

Answers to Quick Quizzes

1. (d)
2. (a)
3. (d)
4. (b)
5. (i) (c)   (ii) (a)
6. (b)
7. (b) Pulling up on the rope decreases the normal force, which, in turn, decreases the force of kinetic friction.

Answers to Odd-Numbered Problems

1. (a) 534 N   (b) 54.5 kg
3. (a) $(6.00\hat{i} + 15.0\hat{j})$ N   (b) 16.2 N
5. (a) $(2.50\hat{i} + 5.00\hat{j})$ N   (b) 5.59 N
7. 2.58 N
9. (a) 1.53 m   (b) 24.0 N forward and upward at 5.29° with the horizontal
11. (a) 3.64×10^{-18} N   (b) 8.93×10^{-30} N is 408 billion times smaller
13. (a) force exerted by spring on hand, to the left; force exerted by spring on wall, to the right   (b) force exerted

by wagon on handle, downward to the left; force exerted by wagon on planet, upward; force exerted by wagon on ground, downward   (c) force exerted by football on player, downward to the right; force exerted by football on planet, upward   (d) force exerted by small-mass object on large-mass object, to the left   (e) force exerted by negative charge on positive charge, to the left   (f) force exerted by iron on magnet, to the left

15. (a) $(-45.0\hat{\mathbf{i}} + 15.0\hat{\mathbf{j}})$ m/s   (b) $162°$ from the $+x$ axis
(c) $(-225\hat{\mathbf{i}} + 75.0\hat{\mathbf{j}})$ m   (d) $(-227\hat{\mathbf{i}} + 79.0\hat{\mathbf{j}})$ m

17. (a) $t = \sqrt{\dfrac{2h}{g}}$ (b) $a_x = \dfrac{F}{m}$ (c) $x = \dfrac{Fh}{mg}$
(d) $a = \sqrt{(F/m)^2 + g^2}$

19. (a) 5.00 m/s² at $36.9°$ (b) 6.08 m/s² at $25.3°$

21. (a) 15.0 lb up   (b) 5.00 lb up   (c) 0

23. (a) 2.15×10^3 N forward   (b) 645 N forward   (c) 645 N toward the rear   (d) 1.02×10^4 N at $74.1°$ below the horizontal and rearward

25. (a) 3.43 kN   (b) 0.967 m/s horizontally forward

27. (a) $P\cos 40° - n = 0$ and $P\sin 40° - 220$ N $= 0$; $P = 342$ N and $n = 262$ N   (b) $P - n\cos 40° - (220$ N$)\sin 40° = 0$ and $n\sin 40 - (220$ N$)\cos 40° = 0$; $n = 262$ N and $P = 342$ N   (c) The results agree. The methods are of the same level of difficulty. Each involves one equation in one unknown and one equation in two unknowns. If we are interested in finding n without finding P, method (b) is simpler.

29. (a) 7.0 m/s² horizontal and to the right   (b) 21 N
(c) 14 N horizontal and to the right

31. (a)

(b) 613 N

33. $T_1 = 253$ N, $T_2 = 165$ N, $T_3 = 325$ N

35. 100 N and 204 N

37. 8.66 N east

39. (a) $a = g\tan\theta$ (b) 4.16 m/s²

41. (a) 646 N up   (b) 646 N up   (c) 627 N up   (d) 589 N up

43. (a) $T_1 = 79.8$ N, $T_2 = 39.9$ N   (b) 2.34 m/s²

45. (a) $F_x > 19.6$ N   (b) $F_x \leq -78.4$ N
(c)

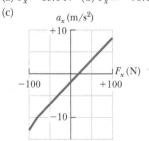

47. 3.73 m

49. (a) 2.20 m/s²   (b) 27.4 N

51. (a) 706 N   (b) 814 N   (c) 706 N   (d) 648 N

53. 1.76 kN to the left

55. (a) 0.306 (b) 0.245

57. $\mu_s = 0.727$, $\mu_k = 0.577$

59. (a) 1.11 s   (b) 0.875 s

61. (a) 1.78 m/s²   (b) 0.368 (c) 9.37 N   (d) 2.67 m/s

63. 37.8 N

65. (a)

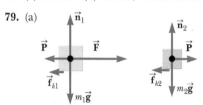

(b) 1.29 m/s² to the right   (c) 27.2 N

67. 6.84 m

69. $0.060\,0$ m

71. (a) $0.087\,1$ (b) 27.4 N

73. (a) Removing mass   (b) 13.7 mi/h · s

75. (a) $a = -\mu_k g$ (b) $d = \dfrac{v_i^2}{2\mu_k g}$

77. (a) 2.22 m   (b) 8.74 m/s down the incline

79. (a)

(b) F (c) $F - P$ (d) P (e) m_1: $F - P = m_1 a$; m_2: $P = m_2 a$
(f) $a = \dfrac{F - \mu_1 m_1 g - \mu_2 m_2 g}{m_1 + m_2}$
(g) $P = \dfrac{m_2}{m_1 + m_2}[F + m_1(\mu_2 - \mu_1)g]$

81. (a)

(b) 0.408 m/s²   (c) 83.3 N

83. (a)

(b) 2.00 m/s² to the right   (c) 4.00 N on m_1, 6.00 N right on m_2, 8.00 N right on m_3 (d) 14.0 N between m_1 and m_2, 8.00 N between m_2 and m_3 (e) The m_2 block models the heavy block of wood. The contact force on your back is modeled by the force between the m_2 and the m_3 blocks, which is much less than the force F. The difference between F and this contact force is the net force causing the acceleration of the 5-kg pair of objects. The acceleration is real and nonzero, but it lasts for so short a time that it is never associated with a large velocity. The frame of the building and your legs exert forces, small in magnitude relative to the hammer blow, to bring the partition, block, and you to rest again over a time interval large relative to the hammer blow.

85. (a) *Upper pulley:*   *Lower pulley:*

(b) $Mg/2$, $Mg/2$, $Mg/2$, $3Mg/2$, Mg (c) $Mg/2$

87. $\mu_k = 0.287$

89. (b) If θ is greater than $\tan^{-1}(1/\mu_s)$, motion is impossible.

91. (a) The net force on the cushion is in a fixed direction, downward and forward making angle $\tan^{-1}(F/mg)$ with the vertical. Starting from rest, it will move along this line with (b) increasing speed. Its velocity changes in magnitude.   (c) 1.63 m   (d) It will move along a parabola. The axis of the parabola is parallel to the line described in part (a). If the cushion is thrown in a direction above this line, its path will be concave downward, making its velocity become more and more nearly parallel to the line over time. If the cushion is thrown down more steeply, its path will be concave upward, again making its velocity turn toward the fixed direction of its acceleration.

93. $(M + m_1 + m_2)(m_1 g/m_2)$

95. (a) 30.7°   (b) 0.843 N

97. 72.0 N

99. (a) 0.931 m/s²   (b) From a value of 0.625 m/s² for large x, the acceleration gradually increases, passes through a maximum, and then drops more rapidly, becoming negative and reaching −2.10 m/s² at $x = 0$.
(c) 0.976 m/s² at $x = 25.0$ cm   (d) 6.10 cm

101. (a) 4.90 m/s²   (b) 3.13 m/s at 30.0° below the horizontal
(c) 1.35 m   (d) 1.14 s
(e) The mass of the block makes no difference.

103. (a) 2.13 s   (b) 1.66 m

Chapter 6

Answers to Quick Quizzes

1. (i) (a)   **(ii)** (b)

2. (i) Because the speed is constant, the only direction the force can have is that of the centripetal acceleration. The force is larger at Ⓒ than at Ⓐ because the radius at Ⓒ is smaller. There is no force at Ⓑ because the wire is straight.   **(ii)** In addition to the forces in the centripetal direction in part (a), there are now tangential forces to provide the tangential acceleration. The tangential force is the same at all three points because the tangential acceleration is constant.

3. (c)

4. (a)

Answers to Odd-Numbered Problems

1. any speed up to 8.08 m/s

3. (a) 8.33×10^{-8} N toward the nucleus
(b) 9.15×10^{22} m/s² inward

5. 6.22×10^{-12} N

7. 2.14 rev/min

9. (a) static friction   (b) 0.085 0

11. $v \le 14.3$ m/s

13. (a) 1.33 m/s²   (b) 1.79 m/s² at 48.0° inward from the direction of the velocity

15. (a) $v = \sqrt{R\left(\dfrac{2T}{m} - g\right)}$ (b) $2T$ up

17. (a) 8.62 m   (b) Mg, downward   (c) 8.45 m/s²   (d) Calculation of the normal force shows it to be negative, which is impossible. We interpret it to mean that the normal force goes to zero at some point and the passengers will fall out of their seats near the top of the ride if they are not restrained in some way. We could arrive at this same result without calculating the normal force by noting that the acceleration in part (c) is smaller than that due to gravity. The teardrop shape has the advantage of a larger acceleration of the riders at the top of the arc for a path having the same height as the circular path, so the passengers stay in the cars.

19. No. The archeologist needs a vine of tensile strength equal to or greater than 1.38 kN to make it across.

21. (a) 17.0°   (b) 5.12 N

23. (a) 491 N   (b) 50.1 kg   (c) 2.00 m/s²

25. 0.527°

27. 0.212 m/s², opposite the velocity vector

29. 3.01 N up

31. (a) 1.47 N · s/m   (b) 2.04×10^{-3} s   (c) 2.94×10^{-2} N

35. (a) $0.034\,7$ s⁻¹   (b) 2.50 m/s   (c) $a = -cv$

37. (a) At Ⓐ, the velocity is eastward and the acceleration is southward.   (b) At Ⓑ, the velocity is southward and the acceleration is westward.

39. 781 N

41. (a) $mg - \dfrac{mv^2}{R}$ (b) $\sqrt{gR}$

43. (a) $v = v_i\, e^{-bt/m}$ (b)

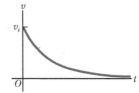

(c) In this model, the object keeps moving forever.   (d) It travels a finite distance in an infinite time interval.

45. (a) the downward gravitational force and the tension force in the string, always directed toward the center of the path

(b)

(c) 6.05 N   (d) 7.82 m/s

47. (a) 106 N up the incline   (b) 0.396
49. (a) 0.016 2 kg/m   (b) $\frac{1}{2}D\rho A$ (c) 0.778   (d) 1.5%   (e) For nested coffee filters falling in air at terminal speed, the graph of air resistance force as a function of the square of speed demonstrates that the force is proportional to the speed squared, within the experimental uncertainty estimated as 2%. This proportionality agrees with the theoretical model of air resistance at high speeds. The drag coefficient of a coffee filter is $D = 0.78 \pm 2\%$.
51. $g(\cos \phi \tan \theta - \sin \phi)$
53. (a) The only horizontal force on the car is the force of friction, with a maximum value determined by the surface roughness (described by the coefficient of static friction) and the normal force (here equal to the gravitational force on the car). (b) 34.3 m   (c) 68.6 m   (d) Braking is better. You should not turn the wheel. If you used any of the available friction force to change the direction of the car, it would be unavailable to slow the car and the stopping distance would be greater.   (e) The conclusion is true in general. The radius of the curve you can barely make is twice your minimum stopping distance.
55. (a) 735 N   (b) 732 N   (c) The gravitational force is larger. The normal force is smaller, just like it is when going over the top of a Ferris wheel.
57. (a) 5.19 m/s   (b)

(c) 555 N

59. (b) The gravitational and friction forces remain constant, the normal force increases, and the person remains in motion with the wall.   (c) The gravitational force remains constant, the normal and friction forces decrease, and the person slides relative to the wall and downward into the pit.
61. (a) $v_{\min} = \sqrt{\dfrac{Rg(\tan \theta - \mu_s)}{1 + \mu_s \tan \theta}}$, $v_{\max} = \sqrt{\dfrac{Rg(\tan \theta + \mu_s)}{1 - \mu_s \tan \theta}}$
(b) $\mu_s = \tan \theta$
63. 12.8 N
65. (a) 78.3 m/s   (b) 11.1 s   (c) 121 m
67. (a) 8.04 s   (b) 379 m/s   (c) 1.19×10^{-2} m/s   (d) 9.55 cm
69. (a) 0.013 2 m/s   (b) 1.03 m/s   (c) 6.87 m/s

Chapter 7

Answers to Quick Quizzes

1. (a)
2. (c), (a), (d), (b)
3. (d)
4. (a)
5. (b)
6. (c)
7. (i) (c)   (ii) (a)
8. (d)

Answers to Odd-Numbered Problems

1. (a) 1.59×10^3 J   (b) smaller   (c) the same
3. (a) 472 J   (b) 2.76 kN
5. (a) 31.9 J   (b) 0   (c) 0   (d) 31.9 J
9. 16.0
11. (a) 16.0 J   (b) 36.9°
13. $\vec{\mathbf{A}}$ = 7.05 m at 28.4°
15. (a) 7.50 J   (b) 15.0 J   (c) 7.50 J   (d) 30.0 J
17. (a) 0.938 cm   (b) 1.25 J
19. (a) 575 N/m   (b) 46.0 J
21. (a) $x = mg\left(\dfrac{1}{k_1} + \dfrac{1}{k_2}\right)$ (b) $k = \left(\dfrac{1}{k_1} + \dfrac{1}{k_2}\right)^{-1}$
23. (a) Design the spring constant so that the weight of one tray removed from the pile causes an extension of the springs equal to the thickness of one tray.   (b) 316 N/m   (c) We do not need to know the length and width of the tray.
25. (b) mgR
27. (a)

(b) The slope of the line is 116 N/m.   (c) We use all the points listed and also the origin. There is no visible evidence for a bend in the graph or nonlinearity near either end.   (d) 116 N/m   (e) 12.7 N
29. 50.0 J
31. (a) 60.0 J   (b) 60.0 J
33. (a) 1.20 J   (b) 5.00 m/s   (c) 6.30 J
35. 878 kN up
37. (a) 4.56 kJ   (b) 4.56 kJ   (c) 6.34 kN   (d) 422 km/s²
(e) 6.34 kN   (f) The two theories agree.
39. (a) 97.8 J   (b) $(-4.31\hat{\mathbf{i}} + 31.6\hat{\mathbf{j}})$ N   (c) 8.73 m/s
41. (a) 2.5 J   (b) −9.8 J   (c) −12 J
43. (a) −196 J   (b) −196 J   (c) −196 J
(d) The gravitational force is conservative.
45. (a) 125 J   (b) 50.0 J   (c) 66.7 J   (d) nonconservative
(e) The work done on the particle depends on the path followed by the particle.
47. A/r^2 away from the other particle
49. $(7 - 9x^2y)\hat{\mathbf{i}} - (3x^3)\hat{\mathbf{j}}$
51. (a) 40.0 J   (b) −40.0 J   (c) 62.5 J

53.

Stable   Unstable   Neutral

55. 90.0 J

57. (a) 8×10^7 N/m   (b) It lasts for a time interval. If the interaction occupied no time interval, the force exerted by each ball on the other would be infinite, and that cannot happen.   (c) 0.8 J   (d) 0.15 mm   (e) 10^{-4} s

59. 0.299 m/s

61. (a) $\vec{\mathbf{F}}_1 = (20.5\hat{\mathbf{i}} + 14.3\hat{\mathbf{j}})$ N, $\vec{\mathbf{F}}_2 = (-36.4\hat{\mathbf{i}} + 21.0\hat{\mathbf{j}})$ N
(b) $\sum \vec{\mathbf{F}} = (-15.9\hat{\mathbf{i}} + 35.3\hat{\mathbf{j}})$ N
(c) $\vec{\mathbf{a}} = (-3.18\hat{\mathbf{i}} + 7.07\hat{\mathbf{j}})$ m/s^2
(d) $\vec{\mathbf{v}} = (-5.54\hat{\mathbf{i}} + 23.7\hat{\mathbf{j}})$ m/s
(e) $\vec{\mathbf{r}} = (-2.30\hat{\mathbf{i}} + 39.3\hat{\mathbf{j}})$ m   (f) 1.48 kJ   (g) 1.48 kJ
(h) The work–kinetic energy theorem is consistent with Newton's second law.

63. 0.131 m

65. (a) $U(x) = 1 + 4e^{-2x}$ (b) The force must be conservative because the work the force does on the particle on which it acts depends only on the original and final positions of the particle, not on the path between them.

67. (a) $x = 3.62m/(4.30 - 23.4m)$, where x is in meters and m is in kilograms   (b) 0.095 1 m   (c) 0.492 m (d) 6.85 m
(e) The situation is impossible.   (f) The extension is directly proportional to m when m is only a few grams. Then it grows faster and faster, diverging to infinity for $m = 0.184$ kg.

Chapter 8

Answers to Quick Quizzes

1. (a) For the television set, energy enters by electrical transmission (through the power cord). Energy leaves by heat (from hot surfaces into the air), mechanical waves (sound from the speaker), and electromagnetic radiation (from the screen).   (b) For the gasoline-powered lawn mower, energy enters by matter transfer (gasoline). Energy leaves by work (on the blades of grass), mechanical waves (sound), and heat (from hot surfaces into the air).   (c) For the hand-cranked pencil sharpener, energy enters by work (from your hand turning the crank). Energy leaves by work (done on the pencil), mechanical waves (sound), and heat due to the temperature increase from friction.

2. (i) (b)   **(ii)** (b)   **(iii)** (a)

3. (a)

4. $v_1 = v_2 = v_3$

5. (c)

Answers to Odd-Numbered Problems

1. (a) $\Delta E_{int} = Q + T_{ET} + T_{ER}$
(b) $\Delta K + \Delta U + \Delta E_{int} = W + Q + T_{MW} + T_{MT}$
(c) $\Delta U = Q + T_{MT}$ (d) $0 = Q + T_{MT} + T_{ET} + T_{ER}$

3. 10.2 m

5. (a) $v = (3gR)^{1/2}$ (b) 0.098 0 N down

7. (a) 4.43 m/s   (b) 5.00 m

9. 5.49 m/s

11. $\sqrt{\dfrac{8gh}{15}}$

13. $\dfrac{v^2}{2\mu_k g}$

15. (a) 0.791 m/s   (b) 0.531 m/s

17. (a) 5.60 J   (b) 2.29 rev

19. (a) 168 J

21. (a) 1.40 m/s   (b) 4.60 cm after release   (c) 1.79 m/s

23. (a) -160 J   (b) 73.5 J   (c) 28.8 N   (d) 0.679

25. (a) 4.12 m   (b) 3.35 m

27. (a) Isolated. The only external influence on the system is the normal force from the slide, but this force is always perpendicular to its displacement so it performs no work on the system.   (b) No, the slide is frictionless.
(c) $E_{system} = mgh$ (d) $E_{system} = \frac{1}{5}mgh + \frac{1}{2}mv_i^2$
(e) $E_{system} = mgy_{max} + \frac{1}{2}mv_{xi}^2$
(f) $v_i = \sqrt{\dfrac{8gh}{5}}$ (g) $y_{max} = h(1 - \frac{4}{5}\cos^2\theta)$ (h) If friction is present, mechanical energy of the system would *not* be conserved, so the child's kinetic energy at all points after leaving the top of the waterslide would be reduced when compared with the frictionless case. Consequently, her launch speed and maximum height would be reduced as well.

29. 1.23 kW

31. 4.5×10^3 N

33. \$145

35. $\sim 10^4$ W

37. (a) 423 mi/gal   (b) 776 mi/gal

39. 236 s or 3.93 min

41. (a) 10.2 kW   (b) 10.6 kW   (c) 5.82 MJ

43. (a) 0.588 J   (b) 0.588 J   (c) 2.42 m/s
(d) $K = 0.196$ J, $U = 0.392$ J

45. $H = h + \dfrac{d^2}{4h}$

47. (a) $K = 2 + 24t^2 + 72t^4$, where t is in seconds and K is in joules   (b) $a = 12t$ and $F = 48t$, where t is in seconds, a is in m/s^2, and F is in newtons   (c) $P = 48t + 288t^3$, where t is in seconds and P is in watts   (d) 1.25×10^3 J

49. (a) 11.1 m/s   (b) 1.00×10^3 J   (c) 1.35 m

51. (a) -6.08×10^3 J   (b) -4.59×10^3 J   (c) 4.59×10^3 J

53. (a) $x = -4.0$ mm   (b) -1.0 cm

55. (a) 2.17 kW   (b) 58.6 kW

57. (a) 1.38×10^4 J   (b) 5.51×10^3 W
(c) The value in part (b) represents only energy that leaves the engine and is transformed to kinetic energy of the car. Additional energy leaves the energy by sound and heat. More energy leaves the engine to do work against friction forces and air resistance.

59. (a) 1.53 J at $x = 6.00$ cm, 0 J at $x = 0$ (b) 1.75 m/s
(c) 1.51 m/s   (d) The answer to part (c) is not half the answer to part (b) because the equation for the speed of an oscillator is not linear in position

61. (a) 100 J   (b) 0.410 m   (c) 2.84 m/s   (d) -9.80 mm
(e) 2.85 m/s

63. 0.328

65. (a) 0.400 m   (b) 4.10 m/s   (c) The block stays on the track.

67. 33.4 kW

69. $2m$

71. 2.92 m/s
75. (b) 0.342
77. (a) 14.1 m/s (b) 800 N (c) 771 N (d) 1.57 kN up
79. (a) $-\mu_k gx/L$ (b) $(\mu_k gL)^{1/2}$
81. (a) 6.15 m/s (b) 9.87 m/s
83. less dangerous
85. (a) 25.8 m (b) 27.1 m/s^2

Chapter 9

Answers to Quick Quizzes

1. (d)
2. (b), (c), (a)
3. (i) (c), (e) (ii) (b), (d)
4. (a) All three are the same. (b) dashboard, seat belt, air bag
5. (a)
6. (b)
7. (b)
8. (i) (a) (ii) (b)

Answers to Odd-Numbered Problems

1. (b) $p = \sqrt{2mK}$
3. 7.00 N
5. $\vec{\mathbf{F}}_{on\,bat} = (+3.26\hat{\mathbf{i}} - 3.99\hat{\mathbf{j}})$ kN
7. (a) $\vec{\mathbf{v}}_{pi} = -\left(\dfrac{m_g}{m_g + m_p}\right)v_{gb}\hat{\mathbf{i}}$ (b) $\vec{\mathbf{v}}_{gi} = \left(\dfrac{m_p}{m_g + m_p}\right)v_{gb}\hat{\mathbf{i}}$
9. 40.5 g
11. (a) $-6.00\hat{\mathbf{i}}$ m/s (b) 8.40 J (c) The original energy is in the spring. (d) A force had to be exerted over a displacement to compress the spring, transferring energy into it by work. The cord exerts force, but over no displacement. (e) System momentum is conserved with the value zero. (f) The forces on the two blocks are internal forces, which cannot change the momentum of the system; the system is isolated. (g) Even though there is motion afterward, the final momenta are of equal magnitude in opposite directions, so the final momentum of the system is still zero.
13. (a) 13.5 N · s (b) 9.00 kN
15. (c) no difference
17. (a) 9.60×10^{-2} s (b) 3.65×10^5 N (c) 26.6g
19. (a) $12.0\hat{\mathbf{i}}$ N · s (b) $4.80\hat{\mathbf{i}}$ m/s (c) $2.80\hat{\mathbf{i}}$ m/s (d) $2.40\hat{\mathbf{i}}$ N
21. 16.5 N
23. 301 m/s
25. (a) 2.50 m/s (b) 37.5 kJ
27. (a) 0.284 (b) 1.15×10^{-13} J and 4.54×10^{-14} J
29. (a) 4.85 m/s (b) 8.41 m
31. 91.2 m/s
33. 0.556 m
35. (a) 1.07 m/s at $-29.7°$ (b) $\dfrac{\Delta K}{K_i} = -0.318$
37. $(3.00\hat{\mathbf{i}} - 1.20\hat{\mathbf{j}})$ m/s
39. $v_O = v_i \cos\theta,\ v_Y = v_i \sin\theta$
41. 2.50 m/s at $-60.0°$
43. (a) $(-9.33\hat{\mathbf{i}} - 8.33\hat{\mathbf{j}})$ Mm/s (b) 439 fJ
45. $\vec{\mathbf{r}}_{CM} = (0\hat{\mathbf{i}} + 1.00\hat{\mathbf{j}})$ m
47. 3.57×10^8 J
49. (a) 15.9 g (b) 0.153 m
51. (a) $(1.40\hat{\mathbf{i}} + 2.40\hat{\mathbf{j}})$ m/s (b) $(7.00\hat{\mathbf{i}} + 12.0\hat{\mathbf{j}})$ kg · m/s
53. 0.700 m

55. (a) $\vec{\mathbf{v}}_{1f} = -0.780\hat{\mathbf{i}}$ m/s, $\vec{\mathbf{v}}_{2f} = 1.12\hat{\mathbf{i}}$ m/s
(b) $\vec{\mathbf{v}}_{CM} = 0.360\hat{\mathbf{i}}$ m/s before and after the collision
57. (b) The bumper continues to exert a force to the left until the particle has swung down to its lowest point.
59. (a) $\sqrt{\dfrac{F(2d - \ell)}{2m}}$ (b) $\dfrac{F\ell}{2}$
61. 15.0 N in the direction of the initial velocity of the exiting water stream.
63. (a) 442 metric tons (b) 19.2 metric tons (c) It is much less than the suggested value of 442/2.50. Mathematically, the logarithm in the rocket propulsion equation is not a linear function. Physically, a higher exhaust speed has an extra-large cumulative effect on the rocket body's final speed by counting again and again in the speed the body attains second after second during its burn.
65. (a) zero (b) $\dfrac{mv_i}{\sqrt{2}}$ upward
67. 260 N normal to the wall
69. (a) $1.33\hat{\mathbf{i}}$ m/s (b) $-235\hat{\mathbf{i}}$ N (c) 0.680 s (d) $-160\hat{\mathbf{i}}$ N · s and $+160\hat{\mathbf{i}}$ N · s (e) 1.81 m (f) 0.454 m (g) -427 J (h) $+107$ J (i) The change in kinetic energy of one member of the system, according to Equation 8.2, will be equal to the negative of the change in internal energy for that member: $\Delta K = -\Delta E_{int}$. The change in internal energy, in turn, is the product of the friction force and the distance through which the member moves. Equal friction forces act on the person and the cart, but the forces move through different distances, as we see in parts (e) and (f). Therefore, there are different changes in internal energy for the person and the cart and, in turn, different changes in kinetic energy. The total change in kinetic energy of the system, -320 J, becomes $+320$ J of extra internal energy in the entire system in this perfectly inelastic collision.
71. (a) Momentum of the bullet–block system is conserved in the collision, so you can relate the speed of the block and bullet immediately after the collision to the initial speed of the bullet. Then, you can use conservation of mechanical energy for the bullet–block–Earth system to relate the speed after the collision to the maximum height. (b) 521 m/s upward
73. $2v_i$ for the particle with mass m and 0 for the particle with mass $3m$.
75. (a) $\dfrac{m_1 v_1 + m_2 v_2}{m_1 + m_2}$ (b) $(v_1 - v_2)\sqrt{\dfrac{m_1 m_2}{k(m_1 + m_2)}}$
(c) $v_{1f} = \dfrac{(m_1 - m_2)v_1 + 2m_2 v_2}{m_1 + m_2}$,
$v_{2f} = \dfrac{2m_1 v_1 + (m_2 - m_1)v_2}{m_1 + m_2}$
77. m_1: 13.9 m m_2: 0.556 m
79. 0.960 m
81. 143 m/s
83. (a) 0; inelastic (b) $(-0.250\hat{\mathbf{i}} + 0.75\hat{\mathbf{j}} - 2.00\hat{\mathbf{k}})$ m/s; perfectly inelastic (c) either $a = -6.74$ with $\vec{\mathbf{v}} = -0.419\hat{\mathbf{k}}$ m/s or $a = 2.74$ with $\vec{\mathbf{v}} = -3.58\hat{\mathbf{k}}$ m/s
85. 0.403
87. (a) $-0.256\hat{\mathbf{i}}$ m/s and $0.128\hat{\mathbf{i}}$ m/s (b) $-0.064\,2\hat{\mathbf{i}}$ m/s and 0 (c) 0 and 0
89. (a) 100 m/s (b) 374 J

91. (a) 2.67 m/s (incident particle), 10.7 m/s (target particle)   (b) −5.33 m/s (incident particle), 2.67 m/s (target particle)   (c) 7.11×10^{-3} J in case (a) and 2.84×10^{-2} J in case (b). The incident particle loses more kinetic energy in case (a), in which the target mass is 1.00 g.

93. (a) particle of mass m: $\sqrt{2}v_i$; particle of mass $3m$: $\sqrt{\frac{2}{3}}v_i$ (b) 35.3°

95. (a) $v_{\text{CM}} = \sqrt{\dfrac{F}{2m}(x_1 + x_2)}$

(b) $\theta = \cos^{-1}\left[1 - \dfrac{F}{2mgL}(x_1 - x_2)\right]$

Chapter 10

Answers to Quick Quizzes

1. (i) (c)   (ii) (b)
2. (b)
3. (i) (b)   (ii) (a)
4. (i) (b)   (ii) (a)
5. (b)
6. (a)
7. (b)

Answers to Odd-Numbered Problems

1. (a) 7.27×10^{-5} rad/s   (b) Because of its angular speed, the Earth bulges at the equator.
3. (a) 5.00 rad, 10.0 rad/s, 4.00 rad/s²
(b) 53.0 rad, 22.0 rad/s, 4.00 rad/s²
5. (a) 4.00 rad/s²   (b) 18.0 rad
7. (a) 5.24 s   (b) 27.4 rad
9. (a) 8.21×10^2 rad/s²   (b) 4.21×10^3 rad
11. 13.7 rad/s²
13. 3.10 rad/s
15. (a) 0.180 rad/s   (b) 8.10 m/s² radially inward
17. (a) 25.0 rad/s   (b) 39.8 rad/s²   (c) 0.628 s
19. (a) 8.00 rad/s   (b) 8.00 m/s   (c) 64.1 m/s² at an angle 3.58° from the radial line to point P (d) 9.00 rad
21. (a) 126 rad/s   (b) 3.77 m/s   (c) 1.26 km/s²   (d) 20.1 m
23. 0.572
25. (a) 3.47 rad/s   (b) 1.74 m/s   (c) 2.78 s   (d) 1.02 rotations
27. −3.55 N · m
29. 21.5 N
31. 177 N
33. (a) 24.0 N · m   (b) 0.035 6 rad/s²   (c) 1.07 m/s²
35. (a) 21.6 kg · m²   (b) 3.60 N · m   (c) 52.5 rev
37. 0.312
39. (a) 5.80 kg · m²
(b) Yes, knowing the height of the door is unnecessary.
41. 1.28 kg · m²
43. $\frac{11}{12}mL^2$
45. (a) 143 kg · m²   (b) 2.57 kJ
47. (a) 24.5 m/s   (b) no   (c) no   (d) no   (e) no   (f) yes
49. 1.03×10^{-3} J
51. 149 rad/s
53. (a) 1.59 m/s   (b) 53.1 rad/s
55. (a) 11.4 N   (b) 7.57 m/s²   (c) 9.53 m/s   (d) 9.53 m/s
57. (a) $2(Rg/3)^{1/2}$ (b) $4(Rg/3)^{1/2}$ (c) $(Rg)^{1/2}$
59. (a) 500 J   (b) 250 J   (c) 750 J
61. (a) $\frac{2}{3}g\sin\theta$ (b) The acceleration of $\frac{1}{2}g\sin\theta$ for the hoop is smaller than that for the disk.   (c) $\frac{1}{3}\tan\theta$

63. (a) The disk   (b) disk: $\sqrt{\frac{4}{3}gh}$; hoop: $\sqrt{gh}$
65. (a) 1.21×10^{-4} kg · m²   (b) Knowing the height of the can is unnecessary.   (c) The mass is not uniformly distributed; the density of the metal can is larger than that of the soup.
67. (a) 4.00 J   (b) 1.60 s   (c) 0.80 m
69. (a) 12.5 rad/s   (b) 128 rad
71. (a) 0.496 W   (b) 413 W
73. (a) $(3g/L)^{1/2}$ (b) $3g/2L$ (c) $-\frac{3}{2}g\hat{\mathbf{i}} - \frac{3}{4}g\hat{\mathbf{j}}$
(d) $-\frac{3}{2}Mg\hat{\mathbf{i}} + \frac{1}{4}Mg\hat{\mathbf{j}}$
75. $\dfrac{g(h_2 - h_1)}{2\pi R^2}$
77. (a) Particle under a net force   (b) Rigid object under a net torque   (c) 118 N   (d) 156 N   (e) $\dfrac{r^2}{a}(T_2 - T_1)$ (f) 1.17 kg · m²
79. $\omega = \sqrt{\dfrac{2mgd\sin\theta + kd^2}{I + mR^2}}$
81. $\sqrt{\dfrac{10}{7}\left[\dfrac{g(R - r)(1 - \cos\theta)}{r^2}\right]}$
83. (a) 2.70R   (b) $F_x = -20mg/7$, $F_y = -mg$
85. (a) $\sqrt{\frac{3}{4}gh}$ (b) $\sqrt{\frac{3}{4}gh}$
87. (a) 0.800 m/s²   (b) 0.400 m/s²
(c) 0.600 N, 0.200 N forward
89. (a) $\sigma = 0.060\,2$ s⁻¹, $\omega_0 = 3.50$ rad/s   (b) $\alpha = -0.176$ rad/s²
(c) 1.29 rev   (d) 9.26 rev
91. (b) to the left
93. (a) 2.88 s   (b) 12.8 s

Chapter 11

Answers to Quick Quizzes

1. (d)
2. (i) (a)   (ii) (c)
3. (b)
4. (a)

Answers to Odd-Numbered Problems

1. $\hat{\mathbf{i}} + 8.00\hat{\mathbf{j}} + 22.0\hat{\mathbf{k}}$
3. (a) $7.00\hat{\mathbf{k}}$ (b) 60.3°
5. (a) 30 N · m (counterclockwise)
(b) 36 N · m (counterclockwise)
7. 45.0°
9. (a) $F_3 = F_1 + F_2$ (b) no
11. $17.5\hat{\mathbf{k}}$ kg · m²/s
13. $m(xv_y - yv_x)\hat{\mathbf{k}}$
15. (a) zero   (b) $(-mv_i^3 \sin^2\theta\cos\theta/2g)\hat{\mathbf{k}}$
(c) $(-2mv_i^3 \sin^2\theta\cos\theta/g)\hat{\mathbf{k}}$
(d) The downward gravitational force exerts a torque on the projectile in the negative z direction.
17. $mvR[\cos(vt/R) + 1]\hat{\mathbf{k}}$
19. $60.0\hat{\mathbf{k}}$ kg · m²/s
21. (a) $-m\ell gt\cos\theta\,\hat{\mathbf{k}}$ (b) The Earth exerts a gravitational torque on the ball.   (c) $-mg\ell\cos\theta\,\hat{\mathbf{k}}$
23. 1.20 kg · m²/s
25. (a) 0.360 kg · m²/s   (b) 0.540 kg · m²/s
27. (a) 0.433 kg · m²/s   (b) 1.73 kg · m²/s
29. (a) 1.57×10^8 kg · m²/s   (b) 6.26×10^3 s = 1.74 h
31. 7.14 rev/min

33. (a) The mechanical energy of the system is not constant. Some chemical energy is converted into mechanical energy. (b) The momentum of the system is not constant. The turntable bearing exerts an external northward force on the axle. (c) The angular momentum of the system is constant. (d) 0.360 rad/s counterclockwise (e) 99.9 J

35. (a) 11.1 rad/s counterclockwise (b) No; 507 J is transformed into internal energy. (c) No; the turntable bearing promptly imparts impulse 44.9 kg · m/s north into the turntable–clay system and thereafter keeps changing the system momentum.

37. (a) $mv\ell$ down (b) $M/(M + m)$

39. (a) $\omega = 2mv_i d/[M + 2m]R^2$ (b) No; some mechanical energy of the system changes into internal energy. (c) The momentum of the system is not constant. The axle exerts a backward force on the cylinder when the clay strikes.

41. (a) yes (b) 4.50 kg · m²/s (c) No. In the perfectly inelastic collision, kinetic energy is transformed to internal energy. (d) 0.749 rad/s (e) The total energy of the system *must* be the same before and after the collision, assuming we ignore the energy leaving by mechanical waves (sound) and heat (from the newly-warmer door to the cooler air). The kinetic energies are as follows: $K_i = 2.50 \times 10^3$ J; $K_f = 1.69$ J. Most of the initial kinetic energy is transformed to internal energy in the collision.

43. 5.46×10^{22} N · m

45. 0.910 km/s

47. 7.50×10^{-11} s

49. (a) $7md^2/3$ (b) $mgd\hat{\mathbf{k}}$ (c) $3g/7d$ counterclockwise (d) $2g/7$ upward (e) mgd (f) $\sqrt{6g/7d}$ (g) $m\sqrt{14gd^3/3}$ (h) $\sqrt{2gd/21}$

51. (a) isolated system (angular momentum) (b) $mv_i d/2$ (c) $(\frac{1}{12}M + \frac{1}{4}m)d^2$ (d) $(\frac{1}{12}M + \frac{1}{4}m)d^2\omega$ (e) $\dfrac{6mv_i}{(M + 3m)d}$ (f) $\frac{1}{2}mv_i^2$ (g) $\dfrac{3m^2v_i^2}{2(M + 3m)}$ (h) $-\dfrac{M}{M + 3m}$

53. (a) $v_i r_i/r$ (b) $(mv_i^2 r_i^2)r^{-3}$ (c) $\frac{1}{2}mv_i^2(r_i^2/r^2 - 1)$

55. (a) 3 750 kg · m²/s (b) 1.88 kJ (c) 3 750 kg · m²/s (d) 10.0 m/s (e) 7.50 kJ (f) 5.62 kJ

57. (a) $2mv_0$ (b) $2v_0/3$ (c) $4m\ell v_0/3$ (d) $4v_0/9\ell$ (e) mv_0^2 (f) $26mv_0^2/27$ (g) No horizontal forces act on the bola from outside after release, so the horizontal momentum stays constant. Its center of mass moves steadily with the horizontal velocity it had at release. No torques about its axis of rotation act on the bola, so the angular momentum stays constant. Internal forces cannot affect momentum conservation and angular momentum conservation, but they can affect mechanical energy.

59. an increase of 6.368×10^{-4} % or 0.550 s, which is not significant

61. (a) $\frac{1}{3}\omega_i$ (b) $-\frac{2}{3}$ (c) $\dfrac{R\omega_i}{3\mu g}$ (d) $\dfrac{R^2\omega_i^2}{18\mu g}$

63. $4\sqrt{\frac{1}{3}ga}(\sqrt{2} - 1)$

Chapter 12

Answers to Quick Quizzes

1. (a)
2. (b)
3. (b)
4. (i) (b) (ii) (a) (iii) (c)

Answers to Odd-Numbered Problems

1. $F_x - R_x = 0, F_y + R_y - F_g = 0,$
$F_y \cos \theta - F_x \sin \theta - 0.5F_g \cos \theta = 0$

3. (3.85 cm, 6.85 cm)

5. 0.750 m

7. (2.54 m, 4.75 m)

9. 177 kg

11. Sam exerts an upward force of 176 N, and Joe exerts an upward force of 274 N.

13. (a) $f_s = 268$ N, $n = 1\ 300$ N (b) 0.324

15. (a) 29.9 N (b) 22.2 N

17. (a) 1.04 kN at 60.0° upward and to the right (b) $(370\hat{\mathbf{i}} + 910\hat{\mathbf{j}})$ N

19. (a) 27.7 kN (b) 11.5 kN (c) 4.19 kN

21. (a) 859 N (b) 1.04 kN at 36.9° to the left and upward

23. 2.81 m

25. $T_1 = 501$ N, $T_2 = 672$ N, $T_3 = 384$ N

27. (a) $-0.053\ 8$ m³ (b) 1.09×10^3 kg/m³ (c) With only a 5% change in volume in this extreme case, liquid water is indeed nearly incompressible in biological and student laboratory situations.

29. 23.8 μm

31. (a) 3.14×10^4 N (b) 6.28×10^4 N

33. 4.90 mm

35. 0.029 2 mm

37. $n_A = 5.98 \times 10^5$ N, $n_B = 4.80 \times 10^5$ N

39. 0.896 m

41. $F_t = 724$ N, $F_s = 716$ N

43. (a)

(b) $T = 343$ N, $R_x = 171$ N to the right, $R_y = 683$ N up (c) 5.14 m

45. (a) $T = F_g(L + d)/[\sin \theta\ (2L + d)]$ (b) $R_x = F_g(L + d) \cot \theta/(2L + d); R_y = F_g L/(2L + d)$

47. $\vec{\mathbf{F}}_A = (-6.47 \times 10^5 \hat{\mathbf{i}} + 1.27 \times 10^5\ \hat{\mathbf{j}})$ N, $\vec{\mathbf{F}}_B = 6.47 \times 10^5 \hat{\mathbf{i}}$ N

49. (a) 5.08 kN (b) 4.77 kN (c) 8.26 kN

51. (a) $\frac{1}{2}m\left(\dfrac{2\mu_s \sin \theta - \cos \theta}{\cos \theta - \mu_s \sin \theta}\right)$ (b) $(m + M)g\sqrt{1 + \mu_s^2}$ (c) $g\sqrt{M^2 + \mu_s^2(m + M)^2}$

53. (a) 9.28 kN (b) The moment arm of the force $\vec{\mathbf{F}}_h$ is no longer 70 cm from the shoulder joint but only 49.5 cm, therefore reducing $\vec{\mathbf{F}}_m$ to 6.56 kN.

55. (a) 66.7 N (b) increasing at 0.125 N/s

57. (a) $\dfrac{1}{\sqrt{15}}\dfrac{mgd}{\ell}$ (b) $n_A = mg\left(\dfrac{2\ell - d}{2\ell}\right), n_B = \dfrac{mgd}{2\ell}$ (c) $R_x = \dfrac{1}{\sqrt{15}}\dfrac{mgd}{\ell}, R_y = \dfrac{mgd}{2\ell}$ (to the right and downward on the right half of the ladder)

59. (a) $P_1 = P_3 = 1.67$ N, $P_2 = 3.33$ N (b) 2.36 N

61. 5.73 rad/s
63. (a) 443 N   (b) 221 N (to the right), 217 N (upward)
65. 9.00 ft
67. $3F_g/8$

Chapter 13

Answers to Quick Quizzes

1. (e)
2. (c)
3. (a)
4. (a) Perihelion (b) Aphelion (c) Perihelion (d) All points

Answers to Odd-Numbered Problems

1. 7.41×10^{-10} N
3. (a) 2.50×10^{-7} N toward the 500-kg object   (b) between the objects and 2.45 m from the 500-kg object
5. 2.67×10^{-7} m/s^2
7. 2.97 nN
9. 2.00 kg and 3.00 kg
11. 0.614 m/s^2, toward Earth
13. (a) 7.61 cm/s^2 (b) 363 s (c) 3.08 km
   (d) 28.9 m/s at 72.9° below the horizontal
15. $\dfrac{GM}{\ell^2}(\frac{1}{2} + \sqrt{2})$ at 45° to the positive x axis
17. 1.50 h or 90.0 min
19. (a) 0.71 yr   (b) The departure must be timed so that the spacecraft arrives at the aphelion when the target planet is there.
21. 1.26×10^{32} kg
23. 35.1 AU
25. 4.99 days
27. 8.92×10^7 m
29. (a) yes   (b) 3.93 yr
31. 2.82×10^9 J
33. (a) 1.84×10^9 kg/m^3 (b) 3.27×10^6 m/s^2
   (c) -2.08×10^{13} J
35. (a) -1.67×10^{-14} J (b) The particles collide at the center of the triangle.
37. 1.58×10^{10} J
39. (a) 4.69×10^8 J (b) -4.69×10^8 J (c) 9.38×10^8 J
41. 1.78×10^3 m
43. (a) 850 MJ (b) 2.71×10^9 J
45. (a) 5.30×10^3 s (b) 7.79 km/s (c) 6.43×10^9 J
47. (a) same size force (b) 15.6 km/s
49. 2.52×10^7 m
51. $\omega = 0.057\ 2$ rad/s or 1 rev in 110 s
53. (a) 2.43 h (b) 6.59 km/s (c) 4.74 m/s^2 toward the Earth
55. 2.25×10^{-7}
57. (a) 1.00×10^7 m (b) 1.00×10^4 m/s
59. (a) 15.3 km   (b) 1.66×10^{16} kg (c) 1.13×10^4 s (d) No; its mass is so large compared with yours that you would have a negligible effect on its rotation.
61. (a) $v_1 = m_2\sqrt{\dfrac{2G}{d(m_1 + m_2)}}$, $v_2 = m_1\sqrt{\dfrac{2G}{d(m_1 + m_2)}}$,

   $v_{\text{rel}} = \sqrt{\dfrac{2G(m_1 + m_2)}{d}}$ (b) 1.07×10^{32} J and 2.67×10^{31} J
63. (a) -7.04×10^4 J (b) -1.57×10^5 J (c) 13.2 m/s
65. 7.79×10^{14} kg

67. (a) 2×10^8 yr (b) $\sim 10^{41}$ kg (c) 10^{11}
69. (a) 2.93×10^4 m/s (b) $K = 2.74 \times 10^{33}$ J,
   $U = -5.39 \times 10^{33}$ J (c) $K = 2.56 \times 10^{33}$ J,
   $U = -5.21 \times 10^{33}$ J (d) Yes; $E = -2.65 \times 10^{33}$ J at both aphelion and perihelion.
71. 119 km
73. $\sqrt{\dfrac{GM}{4R_E}}$
75. $(800 + 1.73 \times 10^{-4})\hat{\mathbf{i}}$ m/s and $(800 - 1.73 \times 10^{-4})\hat{\mathbf{i}}$ m/s
77. 18.2 ms
79. (a) -3.67×10^7 J (b) 9.24×10^{10} kg · m^2/s
   (c) $v = 5.58$ km/s, $r = 1.04 \times 10^7$ m   (d) 8.69×10^6 m
   (e) 134 min

Chapter 14

Answers to Quick Quizzes

1. (a)
2. (a)
3. (c)
4. (b) or (c)
5. (a)

Answers to Odd-Numbered Problems

1. 2.96×10^6 Pa
3. (a) 6.24 MPa (b) Yes; this pressure could puncture the vinyl flooring.
5. 24.8 kg
7. 8.46 m
9. 7.74×10^{-3} m^2
11. (a) 3.71×10^5 Pa (b) 3.57×10^4 N
13. 2.71×10^5 N
15. (a) 2.94×10^4 N (b) 1.63×10^4 N · m
17. 2.31 lb
19. 98.6 kPa
21. (a) 10.5 m   (b) No. The vacuum is not as good because some alcohol and water will evaporate. The equilibrium vapor pressures of alcohol and water are higher than the vapor pressure of mercury.
23. (a) 116 kPa (b) 52.0 Pa
25. 0.258 N down
27. (a) 4.9 N down, 16.7 N up (b) 86.2 N   (c) By either method of evaluation, the buoyant force is 11.8 N up.
29. (a) 7.00 cm (b) 2.80 kg
31. (a) 1 250 kg/m^3 (b) 500 kg/m^3
33. (a) 408 kg/m^3 (b) When m is less than 0.310 kg, the wooden block will be only partially submerged in the water. (c) When m is greater than 0.310 kg, the wooden block and steel object will sink.
35. (a) 3.82×10^3 N   (b) 1.04×10^3 N; the balloon rises because the net force is positive: the upward buoyant force is greater than the downward gravitational force. (c) 106 kg
37. (a) 11.6 cm (b) 0.963 g/cm^3
   (c) No; the density ρ is not linear in h.
39. 1.52×10^3 m^3
41. (a) 17.7 m/s (b) 1.73 mm
43. 0.247 cm
45. (a) 2.28 N toward Holland   (b) 1.74×10^6 s
47. (a) 15.1 MPa (b) 2.95 m/s

49. (a) 1.91 m/s   (b) 8.65×10^{-4} m^3/s
51. 347 m/s
53. (a) 4.43 m/s   (b) 10.1 m
55. 12.6 m/s
57. (a) 1.02×10^7 Pa   (b) 6.61×10^5 N
59. (a) 6.70 cm   (b) 5.74 cm
61. 2.25 m
63. 455 kPa
65. 0.556 m
67. 160 kg/m^3
69. (a) 8.01 km   (b) yes
71. upper scale: 17.3 N; lower scale: 31.7 N
73. 91.64%
75. 27 N · m
77. 758 Pa
79. 4.43 m/s
81. (a) 1.25 cm   (b) 14.3 m/s
85. (a) 18.3 mm   (b) 14.3 mm   (c) 8.56 mm

Index

$m = 10.0 \, kg$

$h_i = 3.00 \, m$

$E_{pi} = mgh$

$\qquad = (10)(9.8)(3)$

$\qquad = 294 \, m$

$E_{pi} - \mu_k \cdot F_N = \frac{1}{2}k x^2$

$294 - \mu_k \cdot (10 \times 9.8) = \frac{1}{2}(2350)(6)^2$

$294 - \mu_k(98) = \cancel{42300}$

$\qquad \mu_k =$

$294 = 42300 + \mu_k 98$

$42006 = \mu_k \cancel{588}$

$\qquad \mu_k = 428.6$

71.4

$(10 \, kg \times 9.8 \times 3) = \frac{1}{2}(2350)(0.25)^2 + \mu_k \cdot (10 \cdot 9.8) \cdot (6)$

$294 \qquad = 73.4 + \mu_k (10 \times 9.8) \times (6)$

$294 \qquad = 73.4 + 588 \, \mu_k$